Compendium of Biophysics

Scrivener Publishing
100 Cummings Center, Suite 541J
Beverly, MA 01915-6106

Publishers at Scrivener
Martin Scrivener (martin@scrivenerpublishing.com)
Phillip Carmical (pcarmical@scrivenerpublishing.com)

Compendium of Biophysics

Andrey B.Rubin

Scrivener Publishing

WILEY

This edition first published 2017 by John Wiley & Sons, Inc., 111 River Street, Hoboken, NJ 07030, USA and Scrivener Publishing LLC, 100 Cummings Center, Suite 541J, Beverly, MA 01915, USA

For more information about Scrivener publications please visit www.scrivenerpublishing.com.

Wiley Global Headquarters
111 River Street, Hoboken, NJ 07030, USA

For details of our global editorial offices, customer services, and more information about Wiley products visit us at www.wiley.com.

Library of Congress Cataloging-in-Publication Data
ISBN 978-1-119-16025-0

Cover images: ibreakstock | Dreamstime.com
Cover design by Kris Hackerott

Set in size of 10pt and Minion Pro by Exeter Premedia Services Private Ltd., Chennai, India

10 9 8 7 6 5 4 3 2 1

Contents

Andrey B. Rubin is a professor of biophysics at Lomonosov Moscow State University in the Department of Biophysics. Born in Russia, he is chair of the National Committee for Biophysics in the Russian Academy of Science. He has been head of the Department of Biophysics at MSU, Governor of the Task Force on Education in Biophysics, and a member of the RAS Council on Space Biology and Biological Membranes since 2005. He has received many awards for his contributions to the science of biophysics, and he holds many patents and inventions, as well as having been the author of numerous papers. He is also on the editorial board of the journal, Biophysics, in the Russian language.

The book does not forgive you for being lazy, And like a hoover, will refresh your brain. Its aim is not to make you go crazy But save your gyri from chondrosis pain.

Introduction

Biophysics is a science about physical and physicochemical interactions which lie in the basis of biological processes. Modern theoretical constructions and biophysical models are based on physical notions of energy, force, types of interactions, on general principles of physical and formal kinetics, thermodynamics and information theory. These notions reflect the nature of fundamental interactions and laws of motion of matter that is the subject of physics as a basic natural science. As a biological science, biophysics has biological processes and phenomena in the focal point of its interests. The key challenge for up-to-date biophysics is the insight into the deepest elemental levels, which comprise the molecular basis of the structural organization of living organisms.

The present stage of biophysics development is characterized by principal advances, associated first of all with the great progress in biophysics of complex systems and molecular biophysics. It is namely in these fields, studying the laws of dynamic behavior of biological systems and mechanisms of molecular interaction in biological structures, that general results were obtained and then used to form the general theoretical basis of biophysics. Main ideas developed in such parts of biophysics as kinetics, thermodynamics, the theory of regulation of biological systems, structures of biopolymers and their electronic and conformational properties, provide a deep insight intomechanisms of important biological processes.

At the same time, the specificity of biological systems is also displayed in the uniqueness of the physical mechanisms of their molecular processes. A principal distinction is that specific parameters of elementary interactions can vary depending on the conditions in organisms where they proceed. For example, parameters of individual elementary acts of electron transfer in photosynthetic reaction centers not only change specifically in a life cycle, but vary also in different types of plants distinguished by physiological and biochemical parameters and fertility. This means that molecular interaction mechanisms do depend on the local environment in biological systems and are themselves exposed to the direct physiological and biochemical regulation. This forms an indissoluble connection between molecular interactions and characteristics of biological phenomena that develop on their basis. That is why studies of deep biophysicalmechanisms, associatedwith physiological and biochemical peculiarities of biological objects, are a base for practical application of the results of biophysical research. Suffice it to mention the development of different methods of early diagnostics of the state of biological systems, based on the data of molecular mechanisms of biological processes, which are widely used in diverse ranges of medicine and agriculture.

In this book, the main ideas of modern biophysics are presented in the form accessible to wide circles of readers. Biophysics (biological physics) is a science about physical and physico-chemical mechanisms of interactions which lie in the basis of biological processes. Physical properties of biopolymers and kinetics of cell metabolic reactions are responsible for molecular characteristics of biological processes. A biomacromolecule as the main structure element in a cell is considered in biophysics as a peculiar molecular machine where energy is transformed and conversed from one type of energy into another. It is pertinent to recall what Bruce Alberts, a well-known American biologist, said about a cell. Hewrote that "the entire cell can be viewed as a factory that contains an elaborate network of interlocking assembly lines, each of which is composed of a set of large protein machines" (Cell, 1998, vol. 92, pp. 291–294).

The real understanding of how these protein machines operate demands the knowledge of not only atomic equilibrium structure, but also our understanding of kinetic and energy characteristics of intermediate transformations. In the postgenome sequencing era, the first priority is given to the mechanisms of intramolecular mobility of macromolecular complexes as the base of their activity. Such an approach corresponds to the biophysical concept of directed electron-conformational interactions when energy transformation and reaction product generation become a result of internal interaction between separate parts within the whole macromolecular complex. In other words, this is the concept of a "physicalmachine" put forward in the 1970–1980s by D. S.Chernavsky, L.A. Blumenfeld, and M.V.Volkelshtein.

In theoretical biophysics, generalized kinetic and physical models of interactions allow us to describe different biological phenomena. However, the analysis of such models clearly demonstrates that different biological processes can very often be similar with respect to their molecular mechanisms. For example, mechanisms of primary photobiological processes (photosynthesis, visual reception), enzyme catalysis in the enzyme active center, and ion transfer through membrane channels are governed by similar physical principles. It follows that educational programs for biology at universities should necessarily include ideas of physics, mathematics and physical chemistry, thus illustrating their efficiency in solving biological problems. Biophysics bears the main responsibility of showing the important role of the regular application of ideas from exact sciences in studying biological processes.

PART I

BIOPHYSICS OF COMPLEX SYSTEMS

I
Kinetics of Biological Processes

1

Qualitative Methods
for Studying Dynamic Models
of Biological Processes

2

Types of Dynamic Behavior
of Biological Systems

3

Kinetics of Enzyme Processes

4

Self-organization Processes
in Distributed Biological Systems

1

Qualitative Methods for Studying Dynamic Models of Biological Processes

The functioning of the integrated biological system is a result of interactions of its components in time and space. Elucidation of the principles of regulation of such a system is a problem that can be solved only with the use of correctly chosen mathematical methods.

The kinetics of biological processes includes the time-dependent behavior of various processes proceeding at different levels of life organization: biochemical conversions, generation of electric potentials on biological membranes, cell cycles, accumulation of biomass or species reproduction, interactions of living populations in biocommunities.

1.1 General Principles of Description of Kinetic Behavior of Biological Systems

The kinetics of a system is characterized by a totality of variables and parameters expressed via measurable quantities, which at each instant of time have definite numerical values. □

In different biological systems, different measurable values can play the role of variables: those are concentrations of intermediate substances in biochemistry, the number of microorganisms or their overall biomass in microbiology, the species population number in ecology, membrane potentials in biophysics of membrane processes, etc. Parameters may be temperature, humidity, pH, electric conductance of membranes, etc.

This is sufficient to construct a general mathematical model representing a system of n differential equations:

$$\begin{aligned} dc_1/dt &= f_1(c_1, \dots, c_n); \\ &\dots\dots\dots\dots\dots\dots \\ dc_n/dt &= f_n(c_1, \dots, c_n), \end{aligned} \tag{1.1}$$

where $c_1(t), \dots, c_n(t)$ are unknown functions of time describing the system variables (for example, substance concentrations); dc_i/dt are rates of changes of these variables; f_i are functions dependent on external and internal parameters of the system. A comprehensive model of type (1.1) may contain a large number of equations, including nonlinear ones.

Many essential questions concerning the qualitative character of the system behavior, in particular, stability of stationary states and transition between them, oscillation modes and others, can be solved using methods of the qualitative theory of differential equations. These methods permit revealing important general properties of the model without determining explicitly the unknown functions $c_1(t), \dots, c_n(t)$. Such an approach gives good results when analyzing the models that consist of a small number of equations and reflect the most important dynamic features of the system.

The key approach in the qualitative theory of differential equations is to characterize the state of the system as a whole by variables $c_1, c_2, \dots, c_n$, which they aquire at each instant of time upon changing in accord with (1.1). If the values of variables $c_1, c_2, \dots, c_n$ are put on rectangular coordinate axes in the n-dimensional space, the system state will be described by some point M in this space with coordinates $M(c_1, c_2, \dots, c_n)$. The point M is called a representation point.

The change in the system state is comparable to the displacement of the point M in the n-dimensional space. The space with coordinates $c_1, c_2, \dots, c_n$ is a phase state; the curve, described in it by the point M, is a phase trajectory.

1.2 Qualitative Analysis of Elementary Models of Biological Processes

Let us consider qualitative methods of studying such systems represented as a system of two independent differential equations (the right-hand parts do not depend explicitly on time), that can be written as:

$$dx/dt = P(x, y), \quad dy/dt = Q(x, y). \quad \square \tag{1.2}$$

Here $P(x, y)$ and $Q(x, y)$ are continuous functions, determined in some range G of the Euclidean plane (x and y are Cartesian coordinates) and having continuous derivatives not lower than the first order.

The range may be both unlimited and limited. When variables x and y have a certain biological meaning (substance concentrations, species population number), some restrictions are usually superimposed on them. First of all, biological variables cannot be negative.

Accept the coordinates of the representation point M_0 to be (x_0, y_0) at $t = t_0$. At every next instant of time t, the representation point will move in compliance with the system of equations (1.2) and have the position $M(x, y)$, corresponding to $x(t)$, $y(t)$. The set of points on the phase plane x, y is a phase trajectory. □

The character of phase trajectories reflects general qualitative features of the system behavior in time. The phase plane, divided in trajectories, represents an easily visible "portrait" of the system. It allows grasping at once the whole set of possible motions (changes in variables x, y) corresponding to the initial conditions. The phase trajectory has tangents, the slopes of which in every point $M(x, y)$ equals the derivative value in this point dy/dx. Accordingly, to trace a phase trajectory through point $M_1(x_1, y_1)$ of the phase plane, it is enough to know the direction of the tangent in this point of the plane or the value of the derivative

$$\left.\frac{dy}{dx}\right|_{\substack{x=x_1 \\ y=y_1}}.$$

To this end, it is required to have an equation with variables x, y and without time t in an explicit form. For that, let us divide the second equation in system (1.2) by the first one. The following differential equation is obtained

$$\frac{dy}{dx} = \frac{Q(x, y)}{P(x, y)}, \tag{1.3}$$

which is frequently much more simple than the initial system (1.2). Solution of equation (1.3) $y = y(x, c)$ or in an explicit form $F(x, y) = C$, where C is the constant of integration, yields a family of integral curves — phase trajectories of system (1.2) on the plane x, y.

But generally, equation (1.3) may have no analytical solution, and then integral plotting should be done using qualitative methods.

Method of Isoclinic Lines. The method of isoclinic lines is typically used for qualitative plotting of a phase portrait of a system. In this case, lines, which intersect the integral lines at a certain angle, are plotted on the phase plane. The analysis of a number of isoclinic lines can show the probable course of the integral lines. □

The equation of isoclinic lines can be obtained from equation (1.3). Suppose $dy/dx = A$, where A is a definite constant value. The value of A is a slope of the tangent to the phase trajectory and, consequently, can have values from $-\infty$ to $+\infty$. Substituting the A value instead of dy/dx in (1.3), we get the equation of isoclinic lines:

$$A = \frac{Q(x, y)}{P(x, y)}. \tag{1.4}$$

By giving different definite numeric values to A, we obtain a family of curves. In any point of each of these curves, the tangent slope to the phase trajectory, passing through this point, is the same value, namely the value of A, which characterizes the given isoclinic line.

Note that in the case of linear systems, i.e. systems of the type

$$dx/dt = ax + by, \quad dy/dt = cx + dy, \tag{1.5}$$

isoclinic lines represent a bundle of straight lines, passing through the origin of coordinates:

$$\frac{cx+dy}{ax+by} = A \quad \text{or} \quad y = \frac{(Aa-c)x}{d-Ab}.$$

Singular Points. Equation (1.3) determines directly the singular tangent to the corresponding integral curve in each point of the plane. Exclusion is the point of intersection of all isoclinic lines $(\overline{x}, \overline{y})$, at which the tangent direction is indefinite, because in this case the value of the derivative is ambiguous:

$$\left.\frac{dy}{dx}\right|_{\substack{x=\overline{x}\\ y=\overline{y}}} = \frac{Q(\overline{x},\overline{y})}{P(\overline{x},\overline{y})} = \frac{0}{0}.$$

The points, in which time derivatives of variables x and y turn concurrently to zero

$$\left.\frac{dx}{dt}\right|_{\overline{x},\overline{y}} = P(\overline{x},\overline{y}) = 0, \quad \left.\frac{dy}{dt}\right|_{\overline{x},\overline{y}} = Q(\overline{x},\overline{y}) = 0 \tag{1.6}$$

and in which the direction of tangents to integral curves is indefinite, are singular points. The singular point in the equation of phase trajectories (1.3) complies with the stationary state of system (1.2), because the rates of changes of variables in this point are equal to zero, and its coordinates are stationary values of variables $\overline{x}, \overline{y}$.

For a qualitative study of a system, it is often possible not to go beyond plotting only some isoclinic lines on the phase plane. Of special interest are the so-called basic isoclinic lines: $dy/dx = 0$ is the isoclinic line of horizontal tangents to phase trajectories, the equation of which is $Q(x,y) = 0$, and the isoclinic of vertical tangents $dy/dx = \infty$, which is in line with equation $P(x,y) = 0$. □

The plotting of the basic isoclinic lines and the determination of their intersection point, the coordinates of which satisfy the following conditions

$$P(\overline{x},\overline{y}) = 0, \quad Q(\overline{x},\overline{y}) = 0, \tag{1.7}$$

gives the intersection point of all isoclinic lines on the phase plane. As mentioned above, this point is a singular point and corresponds to the stationary state of the system (Fig. 1.1).

Figure 1.1 demonstrates the case of one stationary point of intersection of basic isoclinic lines of the system. The figure shows directions of the tangents dy/dx to the trajectories on the phase plane.

The number of stationary states in system of equations (1.2) is equal to the number of intersection points of basic isoclinic lines on the phase plane.

Stability of Stationary States. Assume the considered system to be in the equilibrium state. Then the representation point on the phase plane is stationary in one of the singular points of the equation of integral curves (1.3), because, by definition, in these points $dx/dt = 0, dy/dt = 0$.

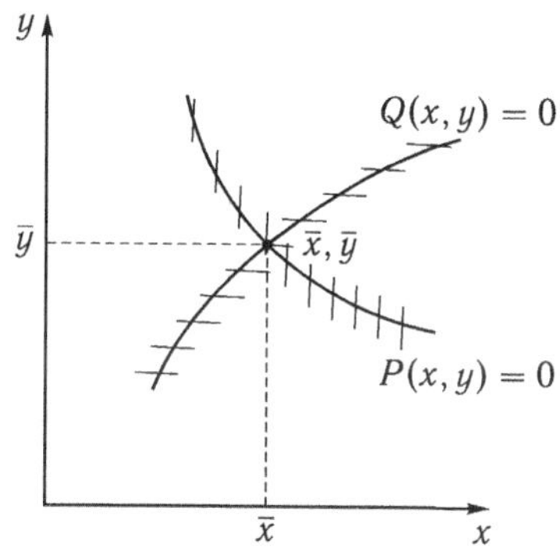

Figure 1.1. The stationary state is determined by the point of intersection of the basic isoclinic lines.

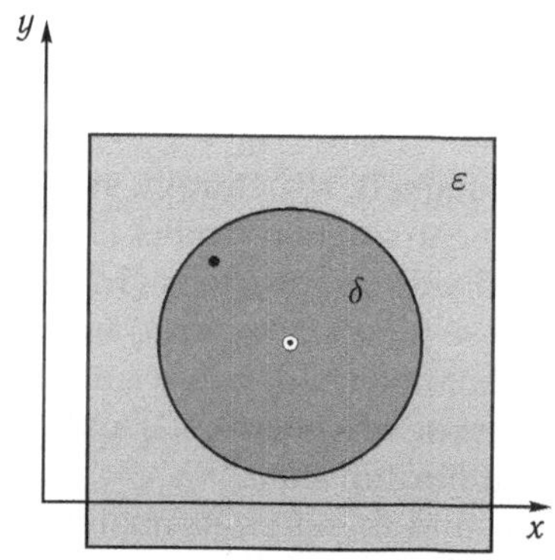

Figure 1.2. Illustration of determination of stability.

The state of equilibrium is stable (according to the Lyapunov theory) if for any given region of permissible deviations from the equilibrium state (region ε), region $\delta(\varepsilon)$, surrounding the equilibrium state and having such a property that neither of the representation point movements, beginning in δ, will never reach the boundary of region ε. On the contrary, the equilibrium state is unstable, if it is possible to indicate the region of deviations from the equilibrium state ε, for which there is no region δ surrounding the equilibrium state and having the property that neither of the motions, beginning inside region δ, will never reach the boundary of region ε.

Now if the system is displaced from the equilibrium state, the representation point will be displaced from the singular point and will move along the phase plane in compliance with equations of its motion (1.2). The question, if the analyzed point is stable, is determined correspondingly by whether the representation point is displaced from a given region, surrounding the singular point (this region can be larger of smaller depending on the statement of the problem) (Fig. 1.2).

Studies of stability of the equilibrium state (the point of intersection of basic isoclinic lines $P(x, y) = 0$, $Q(x, y) = 0$) are connected with the analysis of the character of displacements of the representation point upon deviation from the equilibrium state. To facilitate calculations, let us instead of variables x, y introduce new variables ξ, η determining them as displacements relative to the equilibrium position on the phase plane:

$$x = \bar{x} + \xi, \quad y = \bar{y} + \eta. \tag{1.8}$$

Substituting these expressions in (1.2), we get

$$d\overline{x}/dt + d\xi/dt = P(\overline{x}+\xi, \overline{y}+\eta),$$
$$d\overline{y}/dt + d\eta/dt = Q(\overline{x}+\xi, \overline{y}+\eta), \quad (1.9)$$

$d\overline{x}/dt = d\overline{y}/dt = 0$, because $\overline{x}, \overline{y}$ are the coordinates of the singular point.

Let us factorize the right-hand side of the above equations in Taylor series by variables ξ, η and cast out nonlinear members. The following system of linear equations will be obtained:

$$d\xi/dt = a\xi + b\eta, \quad d\eta/dt = c\xi + d\eta, \quad (1.10)$$

where coefficients a, b, c, and d are values of quotient derivatives in point $(\overline{x}, \overline{y})$:

$$a = P'_x(\overline{x}, \overline{y}), \quad b = P'_y(\overline{x}, \overline{y}), \quad c = Q'_x(\overline{x}, \overline{y}), \quad d = Q'_y(\overline{x}, \overline{y}).$$

System (1.10) is called a linearized system or the system of the first approximation. □

For a large class of systems, namely structurally stable, or "rough" systems, the character of phase trajectories near singular points is preserved at any sufficiently small changes in the right-hand side of equations (1.2) — functions P and Q, if the changes in the derivatives of these functions are also small. For such systems, studies of equations of the first approximation (1.10) give a correct answer to the question on the stability of the equilibrium state of system (1.2) and on the topological structure of the phase plane near this equilibrium state.

System (1.10) is a linear one, and therefore its analytical solution is possible. The general solution of the system is found as follows:

$$\xi = Ae^{\lambda t}, \quad \eta = Be^{\lambda t}. \quad (1.11)$$

By substitution of these expressions in (1.10) and reduction of the obtained expressions by $e^{\lambda t}$, the following expression is obtained:

$$\lambda A = aA + bB, \quad \lambda B = cA + dB. \quad (1.12)$$

Algebraic system of equations (1.12) with unknown members A and B has, as known, a nonzero solution only if its determinant, consisting of coefficients at the unknown members, is zero:

$$\begin{vmatrix} a-\lambda & b \\ c & d-\lambda \end{vmatrix} = 0.$$

Having uncovered this determinant, we get the so-called characteristic equation of the system:

$$\lambda^2 - (a+d)\lambda + (ad - bc) = 0. \quad \square \quad (1.13)$$

The solution of this equation yields indices $\lambda_{1,2}$ at which nonzero solutions for A and B of system (1.12) are possible:

$$\lambda_{1,2} = \frac{a+d}{2} \pm \sqrt{\frac{(a+d)^2}{4} + bc - ad}. \quad (1.14)$$

If the radicand is negative, $\lambda_{1,2}$ are complex conjugate values. Let us assume that both roots of equation (1.13) have real numbers varying from zero, and there are no multiple roots. Then the general solution of system (1.10) written as (1.11) may be represented as a linear combination of exponents with indices λ_1 and λ_2:

$$\xi = C_{11}e^{\lambda_1 t} + C_{12}e^{\lambda_2 t}, \quad \eta = C_{21}e^{\lambda_1 t} + C_{22}e^{\lambda_2 t}. \tag{1.15}$$

The behavior of variables ξ, η, in compliance with (1.15) and, consequently, the behavior of variables x and y near the singular point $(\overline{x}, \overline{y})$ depend on the type of indices of the exponents λ_1 and λ_2. When the indices λ_1 and λ_2 are real and have the same sign, the singular point is called a node (Fig. 1.3).

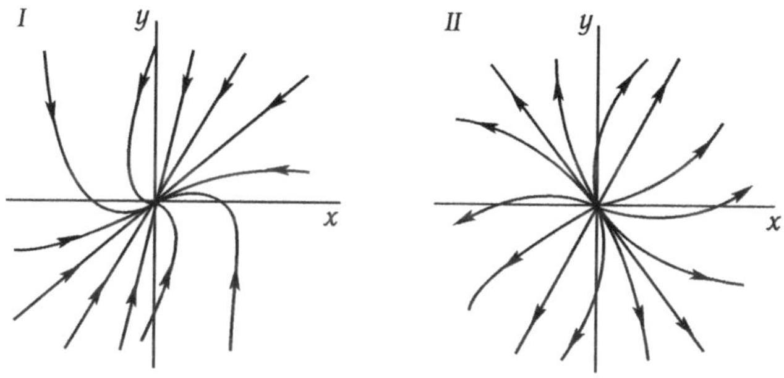

Figure 1.3. Stable (*I*) and unstable (*II*) nodes on phase plane.

If $\lambda_{1,2} < 0$, the values of variables ξ, η (deviations from the equilibrium position) decrease with time. In this case, singular point $(\overline{x}, \overline{y})$ is a stable node (*I*). If $\lambda_{1,2} > 0$, values ξ, η increase with time and the singular point is an unstable node (*II*).

Many biological systems are characterized by a "non-oscillatory" transition from an arbitrary initial state to the stationary one, which corresponds to a stationary solution of the stable node type in the model.

When roots of $\lambda_{1,2}$ are real, but have opposite signs, the behavior of variables is represented by hyperbolic-type curves on the phase plane (Fig. 1.4). Such a singular point is unstable and is called a singular point of the "saddle" type. It can be seen that independent of the position of the representation point at the initial time (with the exception of the singular point and the separatrix), in the long run it will always move away from the equilibrium.

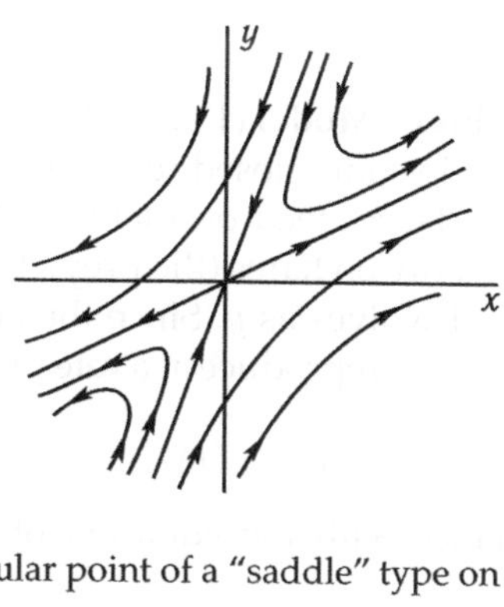

Figure 1.4. Singular point of a "saddle" type on phase plane (xy).

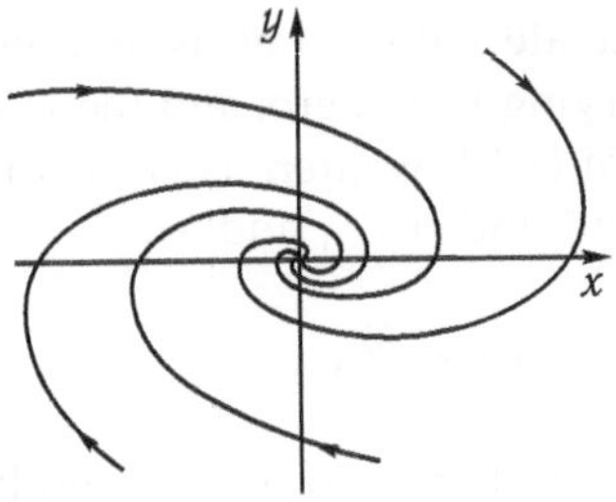

Figure 1.5. Singular point of a "focus" type on phase plane (xy).

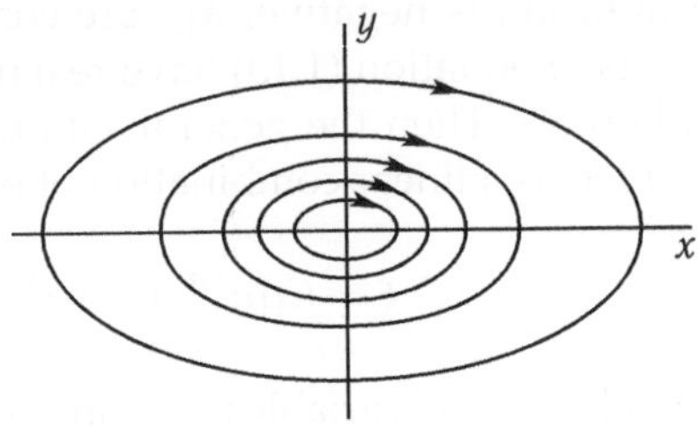

Figure 1.6. Singular point of a "center" type on phase plane (xy).

Singular points of the "saddle" type play an important role in the so-called "trigger" biological systems (see in detail in Section 1 of Chapter 2).

If λ_1 and λ_2 are complex conjugate, changes of variables x and y in time have an oscillation character, and the phase trajectories look like helices (Fig. 1.5). In this case, the singular point is called a focus. At the same time, if real numbers $\lambda_{1,2}$ are negative ($\mathrm{Re}\,\lambda_{1,2} < 0$), oscillations decay and the position of equilibrium is a stable focus. But if $\mathrm{Re}\,\lambda_{1,2} > 0$, the oscillation amplitude increases with time, and the singular point is an unstable focus.

When $\mathrm{Re}\,\lambda = 0$, phase trajectories near the singular point have the shape of ellipsoids (Fig. 1.6). In this case, no integrated curve passes through the singular point. Such an isolated singular point, near which integrated curves have the shape of closed curves, in particular ellipsoids, "mutually enclosed in each other" and including the singular point, is called a center.

Let us formulate the above classification of singular points of a linear system (1.10). If degeneration is absent ($ad - bc \neq 0$), six types of equilibrium states can exist depending on the character of the roots of characteristic equation (1.13) which are also called Lyapunov indices:

1) Stable node (λ_1 and λ_2 are real and negative);
2) Unstable node (λ_1 and λ_2 are real and positive);
3) Saddle (λ_1 and λ_2 are real and have opposite signs);
4) Stable focus (λ_1 and λ_2 are complex and $\mathrm{Re}\,\lambda < 0$);
5) Unstable focus (λ_1 and λ_2 are complex and $\mathrm{Re}\,\lambda > 0$);
6) Center (λ_1 and λ_2 are imaginary). □

Equilibrium states (1–5) are rough: their character does not change at rather small changes in the right-hand sides of equations (1.2) and their derivatives of the first order.

Analysis of the "Predator–Prey" Model (1.17). Now let us consider the ecological Volterra model. Assume that in some closed region there live prey and predators, for example, hares and wolves. Hares feed on plant food that is always abundant. Wolves (the predators) can feed only on hares (the prey). Let us designate the number of hares as x and the number of wolves as y. Since the amount of food for hares is unlimited, we can suggest that hares reproduce at a rate proportional to their amount:

$$\dot{x}_{\mathrm{dimens}} = \varepsilon_1 x. \tag{1.16}$$

(Equation (1.16) is in compliance with the equation of an autocatalytic chemical reaction of the first order.)

Accept the loss in the number of hares to be proportional to the probability of their encounter with wolves, i.e. proportional to the product $x \times y$. The number of wolves also increases the faster, the more frequent their encounters with hares, i.e. proportional to $x \times y$. In chemical kinetics, this corresponds to a bimolecular reaction, when the probability of appearance of a new molecule is proportional to the probability of encounter of two molecules, i.e. the product of their concentrations. In addition, natural death of wolves takes place, the rate of decrease in the number of species being proportional to their number. This is in compliance with the process of a chemical outflow from the reaction sphere. As a result, the following system of equations is obtained for changes in the number of hares x and wolves y:

$$dx/dt = x(\varepsilon_1 - \gamma_1 y), \quad dy/dt = -y(\varepsilon_2 - \gamma_2 x). \tag{1.17}$$

Let us study the singular point in the Volterra predator-prey model (1.17). Its coordinates are found promptly if the right-hand sides of equations in system (1.17) are equal to zero. This yields stationary non-zero values: $\overline{x} = \varepsilon_2/\gamma_2$, $\overline{y} = \varepsilon_1/\gamma_1$. As parameters $\varepsilon_1, \varepsilon_2, \gamma_1, \gamma_2$ are positive, point $(\overline{x}, \overline{y})$ lies in the positive quadrant of the phase plane. Linearization of this point yields

$$\frac{d\xi}{dt} = -\gamma_1 \overline{x}\eta = -\frac{\gamma_1\varepsilon_2}{\gamma_2}\eta; \quad \frac{d\eta}{dt} = -\gamma_2\overline{y}\xi = -\frac{\gamma_2\varepsilon_1}{\gamma_1}\xi.$$

Here $\xi(t)$, $\eta(t)$ are deviations from the singular point on the phase plane:

$$\xi(t) = x(t) - \overline{x}, \quad \eta(t) = y(t) - \overline{y}.$$

The characteristic equation of the system is as follows:

$$\begin{vmatrix} -\lambda & \dfrac{-\gamma_1\varepsilon_2}{\gamma_2} \\ \dfrac{\gamma_2\varepsilon_1}{\gamma_1} & -\lambda \end{vmatrix} = 0, \qquad \lambda^2 + \varepsilon_1\varepsilon_2 = 0.$$

The roots of this equation are purely imaginary: $\lambda_{1,2} = \pm i\sqrt{\varepsilon_1\varepsilon_2}$.

In this case, phase trajectories near the singular point look like concentric ellipsoids, and the singular point itself is the center (Fig. 1.7). Far from the singular point, phase trajectories are closed, though their shape varies from the ellipsoid one.

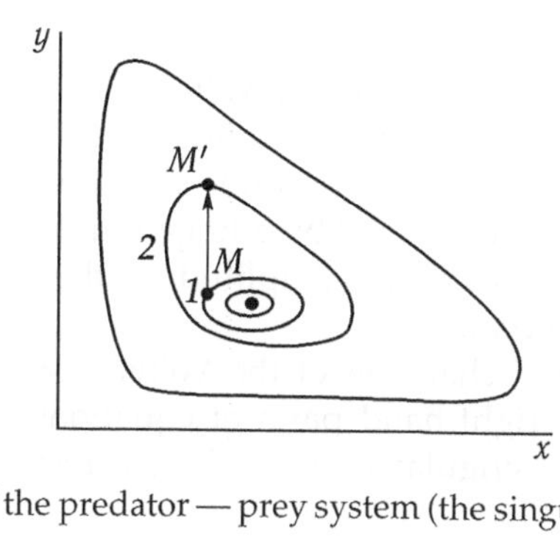

Figure 1.7. Phase portrait of the predator — prey system (the singular point of a "center" type).

On the whole, the singular point of the "center" type is unstable. Let oscillations $x(t)$ and $y(t)$ proceed so that the representation point moves along the phase trajectory *1* (Fig. 1.7). At the instant of time when the point is in position M, such a number of species Δy is added to the system from the outside that the representation point jumps from point M to point M'. After that, if the system is again left on its own, oscillations $x(t)$ and $y(t)$ will occur with larger amplitudes than previously, and the representation point will move along trajectory 2. So, upon external action the oscillations change their characteristics forever.

Figure 1.8 shows plots of functions $x(t)$ and $y(t)$. It is seen that $x(t)$ and $y(t)$ are periodic functions of time, the maximum of the prey number surpassing the maximum of the number of predators.

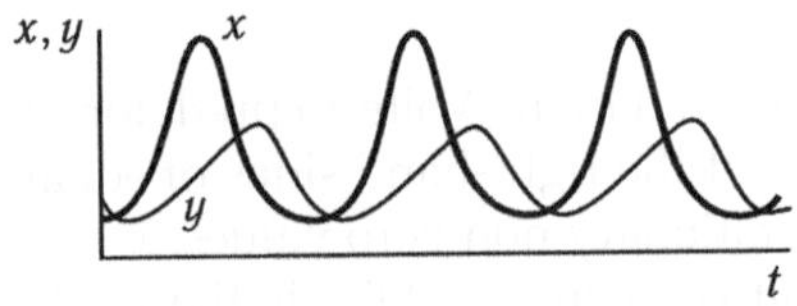

Figure 1.8. Dependence of the number of predators y and prey x on time.

Figure 1.9 shows curves of the number of North American hares and lynxes in Canada, plotted using the data on the number of harvested skins. The shape of real curves is much less correct than that of theoretical ones. But the model ensures the coincidence of the most essential characteristics of these curves — the values of amplitudes and the lagging of oscillations in the numbers of predators and prey.

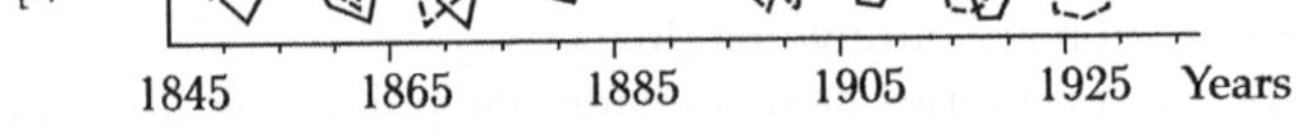

Figure 1.9. Curves of hare and lynx numbers in Canada (Villee, Dethier, 1971).

Periods of hare (prey) and bobcat (predators) population waves are approximately the same and make 9–10 years, the maximum of hare numbers surpasses that of boncts by a year.

A much more serious disadvantage of the Volterra model is instability of solutions of the system of equations, when any random change in the number of a species leads to a change in the oscillation amplitude of both types. Needless to say, in natural conditions, animals are affected by a huge number of such random actions. But as seen from Fig. 1.9, the oscillation amplitude of the number of species changes insignificantly from year to year.

Because of the "unrough" character of the Volterra system, an arbitrarily small change in the form of the right-hand parts of equations in system (1.17) leads to changes in the type of the singular point and, as a result, the character of phase trajectories of the system.

To eliminate this disadvantage, different modifications of system (1.17) were proposed. Let us analyze the model that takes into account self-restraints in the growth of both populations. It shows how the character of solutions can be changed upon alterations in the parameters of the system:

$$\begin{aligned} dx/dt &= x(\varepsilon_1 - \gamma_{12}y - \gamma_{11}x), \\ dy/dt &= y(-\varepsilon_2 + \gamma_{21}x - \gamma_{22}y). \end{aligned} \tag{1.18}$$

System (1.18) differs from the earlier analyzed system (1.17) by that the right-hand sides of equations contain members $-\gamma_{11}x$, $-\gamma_{22}y$, which reflect the fact that the number of the prey population cannot grow unlimitedly even in the absence of predators due to limited nature of food resources. The same restrictions are imposed also on the population of predators (Fig. 1.10).

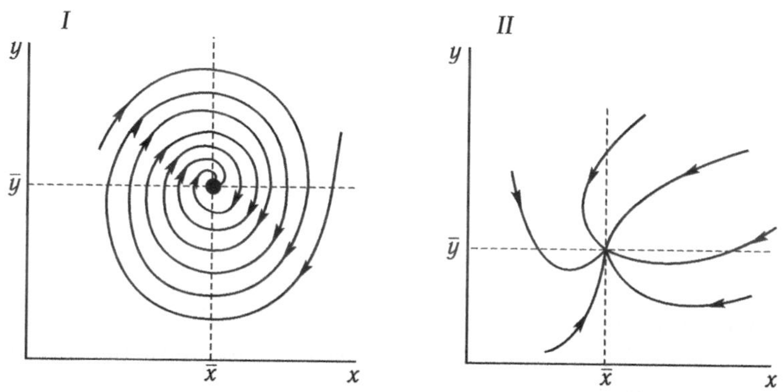

Figure 1.10. Phase portrait of system (1.18); see details in the text.

To determine stationary numbers of types $\overline{x}$ and $\overline{y}$, let us make the right-hand sides of equations in system (1.18) be equal to zero. Solutions with zero values of the numbers of prey and predators will be of no interest for us. Let us analyze the following system of algebraic equations:

$$\gamma_{11}\overline{x} + \gamma_{12}\overline{y} = \varepsilon_1, \quad \gamma_{12}\overline{x} - \gamma_{22}\overline{y} = \varepsilon_2.$$

The coordinates of the singular point are found using the expressions

$$\overline{x} = \frac{\varepsilon_1\gamma_{22} + \varepsilon_2\gamma_{12}}{\gamma_{11}\gamma_{22} + \gamma_{12}^2}, \quad \overline{y} = \frac{\varepsilon_1\gamma_{12} - \varepsilon_2\gamma_{11}}{\gamma_{11}\gamma_{22} + \gamma_{12}^2}. \tag{1.19}$$

The roots of the characteristic equation of system (1.18) linearized near the singular point (1.19) are as follows:

$$\begin{aligned} \lambda_{1,2} = \frac{1}{2}\Big\{ &-[\varepsilon_1\gamma_{22}(\gamma_{11} - \gamma_{22}) + \varepsilon_2\gamma_{11}(\gamma_{12} + \gamma_{22})] \pm \\ &\pm [\varepsilon_1\gamma_{22}(\gamma_{11} - \gamma_{22}) + \varepsilon_2\gamma_{11}(\gamma_{12} + \gamma_{22})]^2 - \\ &- 4\gamma_{12}\gamma_{21}[(\varepsilon_1\gamma_{22} + \varepsilon_2\gamma_{12})(\varepsilon_1\gamma_{21} - \varepsilon_2\gamma_{11})]^{1/2}\Big\}. \end{aligned}$$

From the expression for characteristic numbers, it is seen that if the below condition is fulfilled,

$$[\varepsilon_1\gamma_{22}(\gamma_{11} - \gamma_{22}) + \varepsilon_2\gamma_{11}(\gamma_{12} + \gamma_{22})]^2 \leqslant 4\gamma_{12}\gamma_{21}(\varepsilon_1\gamma_{22} + \varepsilon_2\gamma_{12})(\varepsilon_1\gamma_{21} - \varepsilon_2\gamma_{11}), \quad (1.20)$$

then the number of predators and prey performs damped oscillations in time, the system has a non-zero singular point — a stable focus. The phase portrait of the system is shown in Fig. 1.10, *I*.

Suppose that parameters of the system are changed so that condition (1.20) is equality. In this case, the singular point will lie on the boundary of stable focuses and nodes. When the sign of the inequality (1.20) changes to the opposite one, bifurcation takes place in the system — the singular point becomes a stable node (Fig. 1.10, *II*).

At $\gamma_{ii} = 0$, system (1.18) is reduced to unrough system (1.17) with a singular point of a "center" type. Thus, the appearance of even small negative nonlinear members in the right-hand sides of equations causes a qualitative change in the phase portrait and conversion of the unrough singular point of a "center" type to a rough singular point of the "stable focus" or "node" type depending on the correlation of the system parameters.

It is evident that specific values of parameters and the character of matching nonlinear members at the initial Lotka–Volterra model should correctly reflect real features of the ecological system. In such a case, mathematical modeling demonstrates that namely this defines the type of dynamic behavior of the system. Parameters γ_{ii} may be regarded as a kind of controlling parameters, causing a qualitative deformation of the phase portrait of the system with a change in the type of its stability. In Section 3 of Chapter 4, other cases of parametric dependence of the dynamics of species interactions are analyzed. □

2

Types of Dynamic Behavior of Biological Systems

2.1 Biological Triggers

An important distinctive feature of biological systems is their capacity to switch from one mode of functioning to another, which corresponds to several stable stationary states of the system. On the phase plane, such a system has two (and more) stable singular points. Regions affected by stable singular points are separated by separatrixes, which generally pass through an unstable singular point of the "saddle" type (Fig. 2.1).

A system with two or more stable stationary states, between which transitions may occur, is a trigger system. Unlike the systems considered in Chapter 1, which have a single stationary state, in trigger systems, stationary values of variables depend on initial conditions. If a system functions in a stable mode, it cannot be released from this mode by small deviations. However, in real biological systems, it is possible to switch from one stable stationary state to another. Suppose that the system functions in the stable mode a, and it should be switched to another stable mode c. This can be done in two ways.

External effects can so change the values of variables x and y, for example, increasing drastically x, that this will displace the system to some point c' (Fig. 2.1), that is on the right of the saddle separatrix in the region of attraction of the stable node c. After that, the system itself will move to point c along the phase trajectory and will be found in the target mode. This is the so-called forced way of switching the trigger; it is also known as a specific way. In the case of a chemical reaction,

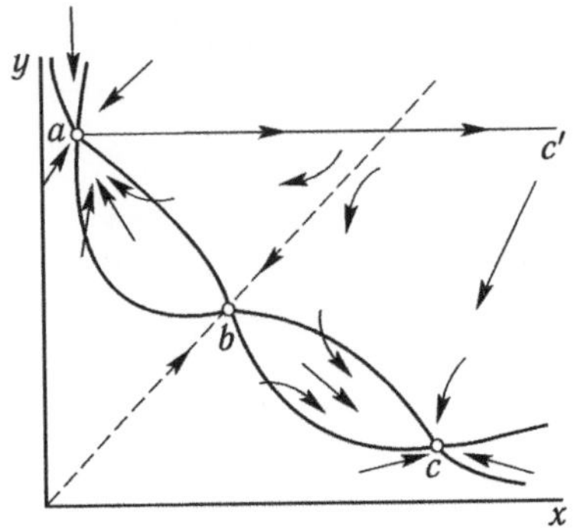

Figure 2.1. Phase portrait of a trigger system with two stable singular points.

The number of stationary states in the system is determined by the number of intersection points of basic isoclinic lines of vertical and horizontal tangents (thick lines). Intersection point of isoclinic lines b is a saddle, and the intersection points of basic isoclininc lines a and c on both sides of the saddle separatrix (the dashed line) are stable nodes. If the initial position of the representation point is on the left of the saddle separatrix, the system is in the region, affected by the singular point a, and approaches this stable stationary state. From the initial points on the right of the separatrix, the system will move to the stable singular point c.

for such switching it is necessary to add some amount of a definite chemical agent (here substance x) to the system.

Another, more delicate way consists in parametric nonspecific switching. Here not variables but system parameters are subjected to direct action, which may be achieved by different methods, for example, changes in temperature, pH or rate of substrate input. The main point of parametric switching is in the use of a characteristic dependence of the phase portrait on some modifier of the system (Fig. 2.2).

Upon a change in the phase portrait, the coordinates of the singular point c, undoubtedly, also change slightly because they depend on the system parameters. When the modifier returns to its previous values, the initial phase portrait of the system will be restored, but the system will already function in the target mode c.

The capacity of the trigger system to switch was a prerequisite for its use upon modeling the processes, leading to differentiation of tissues. From this point of view, every cell has a set of possible stable stationary regimes, but as a matter of fact, it functions only in one of them at the given instant of time. It is in the process of differentiation that the cell switches from one stationary regime of functioning to another.

A model of a genetic trigger, based on the biochemical scheme of protein synthesis regulation in prokaryotes, is shown in Fig. 2.3.

The mathematical model of the process in a non-dimensional form can be written as follows:

$$\frac{dx_1}{dt} = \frac{A_1}{1+x_2^n} - x_1, \quad \frac{dx_2}{dt} = \frac{A_2}{1+x_1^n} - x_2. \tag{2.1}$$

This model has been obtained by reduction of the entire system, taking into account processes occurring in compliance with the scheme in Fig. 2.3. The meaning of variables in system (2.1) is as follows: x_1 and x_2 are dimensionless concentrations of specific metabolites — co-repressors of products P_1 and P_2. Unit of time

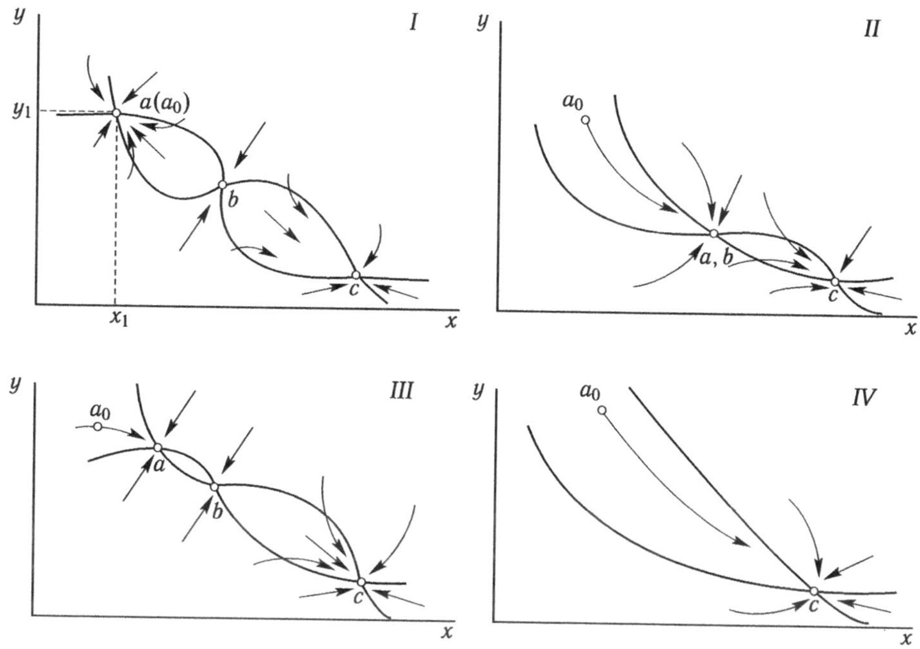

Figure 2.2. Process of parametric switching of a trigger system of a phase plane.

Upon changing the modifier, the system that is at the beginning of the process in point $a(a_0)$ with corresponding coordinates x and y on the phase plane (*I*) will be, due to changes in the phase portrait, in the region of attraction of stable node c (*IV*), to where it will move spontaneously (through stages shown in Fig. *II* and *III*).

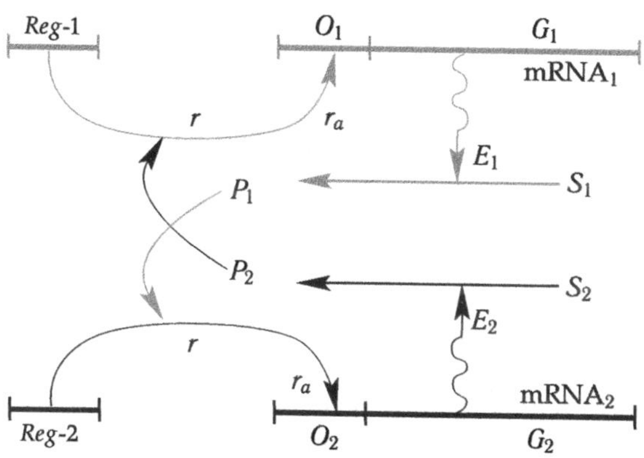

Figure 2.3. Scheme of mutual adaptation of two systems of enzymatic synthesis (Jacob and Monod model) (Romanovsky et al.,1975).

The regulator gene (*Reg*) in each system synthesizes the inactive repressor (r), which forms the active complex (r_a) by associating with the system product (P). The active complex, reversibly reacting with the site of the structural gene (G), called an operon (O), inhibits the synthesis of mRNA. Conversion of substrate S to product P occurs with participation of enzyme E. So, product P_2 of system II is a co-repressor of system I, and P_1 is a co-repressor of system II. The process of co-repression may involve one, two or more product molecules. It is clear that upon intensive operation of system I, this character of interaction promotes the blocking of system II, and vice versa.

is the characteristic time of enzyme reactions, which is as large as about some minutes. Parameter n reflects the order of the repression reactions. Parameters A_1 and A_2 are dependent on substrates S_1 and S_2, the activity and composition of enzymes of basic metabolism. In the case when both systems of synthesis are identical (they uptake the same amount of energy) and concentrations of substrates S_1 and S_2 are the same, parameters A_1 and A_2 are equal, and model (2.1) is symmetric.

Let us analyze the behavior of the system at different n values. At $n = 1$, the system has one symmetric stationary solution, determined as a positive root of the equation for stationary concentrations: $\overline{x}^2 + \overline{x} - A = 0$. A phase portrait of the system is shown in Fig. 2.4. It has one stable singular point of the node type.

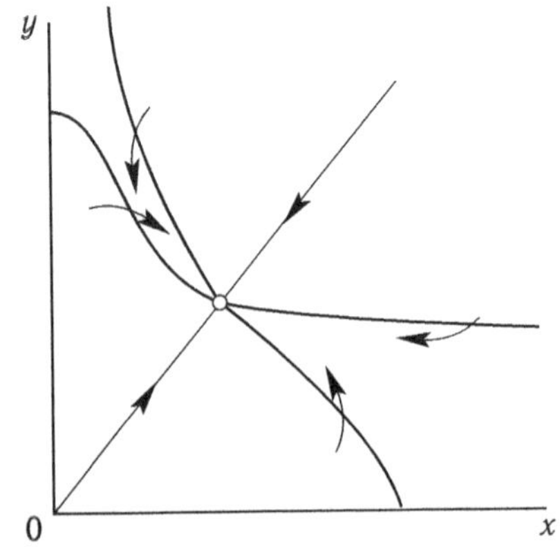

Figure 2.4. Phase portrait of model (2.1) at $n = 1$ (according to Yu. M. Romanovsky, N. V. Stepanova, D. S. Chernavsky, 1975).

At $n = 2$, the number of stationary states is equal to the number of positive real roots of the following equation

$$A\left[1 + A^2/(1 + x^2)^2\right]^{-1} - x = 0.$$

At $A < 2$, only one solution is possible $x = \overline{x} < 1$. It is stable, and the phase portrait of the system is the same as that in Fig. 2.4. At $A \geqslant 2$, three stationary states appear (as in Fig. 2.1), and the system becomes a trigger one. The value $A = 2$ can be considered as a bifurcation parameter, at which the stable node turns to a saddle, and two stables nodes are formed near it.

Thus, the triggering regime in the system appears, when two (or more) product molecules are involved in co-repression ($n \geqslant 2$) and when the level of basic metabolism is rather high ($A \geqslant 2$).

In the asymmetric model, the qualitative pattern is preserved, but the character of bifurcation changes slightly. In this case, there are two parameters (A_1 and A_2). The triggering regime takes place only when each of them is higher than 2, the phase portrait becoming asymmetric.

In conclusion, it should be noted that trigger systems describe adequately one of the key features of biological systems — their capacity to switch from one regime to another; namely because of this, trigger models have become widely used along with oscillation models. Some of them will be analyzed in detail in Section 2 of Chapter 3 during the description of kinetic models of enzyme catalysis.

2.2 Oscillatory Processes in Biology. Limit Cycles

At present a rather large number of oscillatory systems in biology have been studied experimentally: periodic biochemical reactions, oscillations in glycolysis, periodic processes in photosynthesis, oscillations in species populations etc. In all these processes, some values characterizing the system change periodically due to the features of the system itself without any periodic action from the outside. Such systems are called autooscillatory systems.

Autooscillatory systems are systems in which non-damped oscillations are established and preserved due to the forces, depending on the state of the system itself, the amplitude of these oscillations being determined by the system properties rather than by initial conditions. □

The analysis of equations describing autooscillatory systems demonstrates that on a phase plane a stationary solution of such a system is the so-called limit cycle (Fig. 2.5).

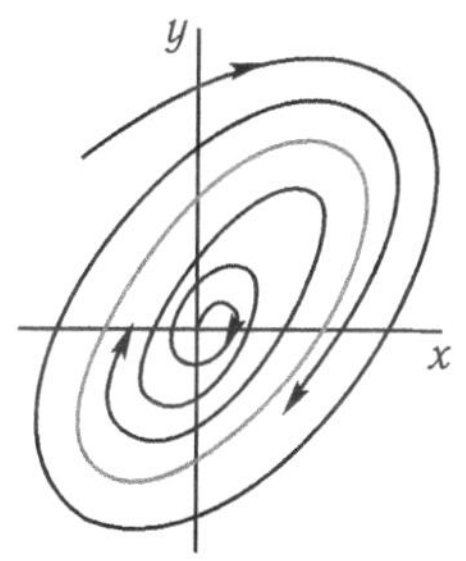

Figure 2.5. Stable limit cycle on phase plane xy.

A limit cycle is a closed curve on the phase plane, to which all integral curves tend within the range $t \to \infty$. Such a cycle represents a stationary regime of definite amplitude independent of the initial conditions, but determined by the form of equations of the system. The existence of a limit cycle on the phase plane is the main feature of an autooscillatory system. It is evident that during an autooscillating process the oscillation phase may vary.

A limit cycle is an isolated closed trajectory to the effect that all phase trajectories which are near the limit cycle but do not coincide with it, are not closed and represent helices. They "wound" on the limit cycle or "unwound" from it. This is the cardinal difference of a limit cycle from the infinite number of closed phase trajectories surrounding a singular point of the "center" type, which, as shown in Chapter 1, Fig. 1.6, is in a certain sense unstable. □

For motions, represented by a stable limit cycle, the period and amplitude (exactly, the whole range of amplitudes obtained upon expansion of periodic motion in a Fourier series) are independent of the initial conditions. All the neighboring motions (corresponding to a whole range of initial values) asymptotically approach the periodic motion by the limit cycle, which has a certain period and definite amplitude.

Model of Glycolysis. Today there are several dozens of autooscillation models of biological processes of different levels — from the cellular to the population one. Their description can be found in special literature. A classic example of an oscillatory biochemical system is the glycolytic chain.

It should be reminded that glucose and other sugars are decomposed in the process of glycolysis, during which the compounds containing six carbon atoms are converted to tricarboxylic acids including three carbon atoms. In this case, due to the free energy excess two ATP molecules are formed per one molecule of six-carbon sugar.

A decisive role in generation of oscillations in concentrations of fructoso-6-phosphate (F6P), fructoso-1,6-diphosphate (FDP) and reduction of NAD belongs to the key enzyme of the glycolytic pathway — phosphofructokinase (PFK). Construction of a mathematical model of the glycolytic chain including more than 20 stages is facilitated by the existence of several weak points which determine the kinetics of the process.

Let us analyze the elementary Higgins model, in which it is assumed that the key factor determining the rate of the process is linear activation of PFK with fructose diphosphate. Then the scheme of the process can be represented in a simplified form

$$\begin{array}{c}\text{Activation}\\ \downarrow\\ [\mathrm{Gl}] \underset{1}{\rightarrow} \mathrm{F6P} \underset{2}{\xrightarrow{E_1}} \mathrm{FDP} \underset{3}{\xrightarrow{E_2}}\ .\end{array} \tag{2.2}$$

Here [Gl] is glucose; F6P is the substrate of the key reaction, catalyzed by enzyme E_1 (PFK); FDP is the product of this reaction which is the substrate at the next stage, catalyzed by enzyme E_2. The arrow shows the effect of intermediate product FDP on the activity of the key enzyme E_1.

Let us introduce the following designations: v_1 is the rate of input of substrate F6P to the reaction analyzed; v_2 is the rate of conversion of F6P to FDP; v_3 is the rate of expenditure of FDP at the next stage. For simplicity, all reactions are accepted to be irreversible. With the above designations, the equations for slow variables (concentrations of substrate F6P and product FDP) look like these

$$dx/dt = v_1 - v_2 = P(x, y), \quad dy/dt = v_2 - v_3 = Q(x, y), \tag{2.3}$$

where v_2 is the quasistationary rate of the key enzyme reaction; x and y are concentrations of substrate F6P and product FDP, respectively.

Accept that substrate F6P enters the reaction sphere at a constant rate

$$v_1 = k; \tag{2.4}$$

v_2 is the quasistationary rate of the reaction determined by the following expression

$$v_2 = \kappa \frac{x}{K_{m_x} + x} \cdot \frac{y}{K_{m_y} + y}, \tag{2.5}$$

where κ is the maximal rate of the analyzed reaction at complete saturation with the substrate; K_{m_x} is the Michaelis constant; K_{m_y} characterizes the product activation in the key reaction.

The rate of product utilization is determined by a slow and irreversible reaction, catalyzed by pyruvate kinase (PK). The rate of decrease of y can be written as

$$v_3 = q \frac{y}{K'_{m_y} + y}, \tag{2.6}$$

where K'_{m_y} is the Michelis constant for the reaction of decrease of y.

Subject to (2.4)–(2.6), system (2.3) can be represented as

$$\begin{aligned}\frac{dx}{dt} &= k - \kappa\frac{x}{K_{m_x} + x}\cdot\frac{y}{K_{m_y} + y},\\ \frac{dy}{dt} &= \kappa\frac{x}{K_{m_x} + x}\cdot\frac{y}{K_{m_y} + y} - q\frac{y}{K'_{m_y} + y}.\end{aligned} \tag{2.7}$$

System (2.7) can be simplified if we assume that $K_{m_x} \gg x$, $K_{m_y} \gg y$ and make the following substitution of the variables:

$$t' = \frac{t'\kappa k K'_{m_y}}{K_{m_x}K_{m_y}(q-k)}, \quad x' = \frac{x\kappa K'_{m_y}}{K_{m_x}K_{m_y}(q-k)}, \quad y' = y\frac{q-k}{kK'_{m_y}}. \tag{2.8}$$

In this case, system (2.7) is transformed to

$$dx'/dt' = 1 - x'y', \quad dy'/dt' = \alpha y'\left(x' - \frac{1+r}{1+ry'}\right), \tag{2.9}$$

where

$$\alpha = \frac{(q-k)^2 K_{m_x}K_{m_y}}{K'^2 k\kappa}, \quad r = \frac{k}{q+k}.$$

Figure 2.6 shows the phase portrait of system (2.9).

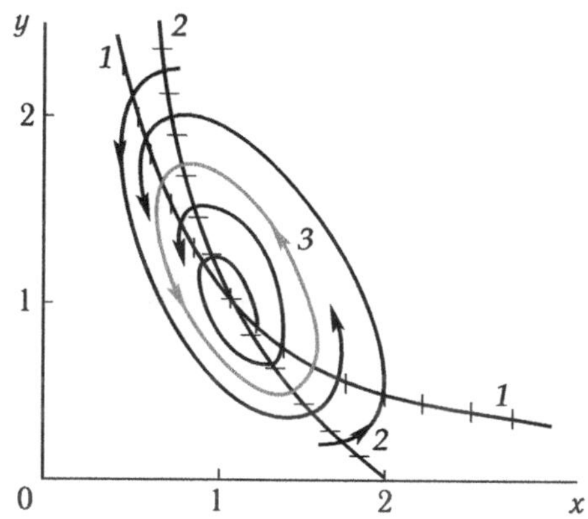

Figure 2.6. Phase portrait of oscillations in the glycolytic system (Romanovsky et al., 1971). The isoclinic line of vertical tangents (curve *1*) is hyperbola $y' = 1/x'$. Isoclinic lines of horizontal tangents is hyperbola $x' = (1+r)/(1+ry')$ (curve *2*) and axis $y' = 0$. Closed curve *3* is the limit cycle.

The character of stability of the singular point is determined by the expression for roots of the characteristic equation of the following linearized system

$$\lambda_{1,2} = -\frac{1}{2}\left(1 - \frac{\alpha r}{1+r}\right) \pm \frac{1}{2}\sqrt{\left(1 + \frac{\alpha r}{1+r}\right)^2 - \frac{4\alpha}{1+r}}. \tag{2.10}$$

From expression (2.10) it is seen that at $\frac{4\alpha}{1+r} > \left(1 + \frac{\alpha r}{1+r}\right)^2$ there is a singular point of the "focus" type. Given $\alpha r/(1+r) < 1$, the focus is stable, but if $\alpha r/(1+r) > 1$, the focus is unstable. In this case, the limit cycle may be in the vicinity of the singular point. The value of parameter $\alpha r/(1+r)$ is a bifurcation one. It separates the range of the system parameter values, at which the system can have only damped oscillations, from the range where autooscillations can occur.

Figure 2.7 shows the kinetics of changes in time of concentrations of variables x and y and phase portraits of system (2.7) at different values of the system parameters. When the parameters change so that transition proceeds through the bifurcation point ($\alpha r/(1+r) = 1$), a limit cycle appears in the system instead of the stable singular point (Fig. 2.7, *a* and *b*), i.e. a stable autooscillation regime is established (Fig. 2.7, *c* and *d*), to which the stability is now passed.

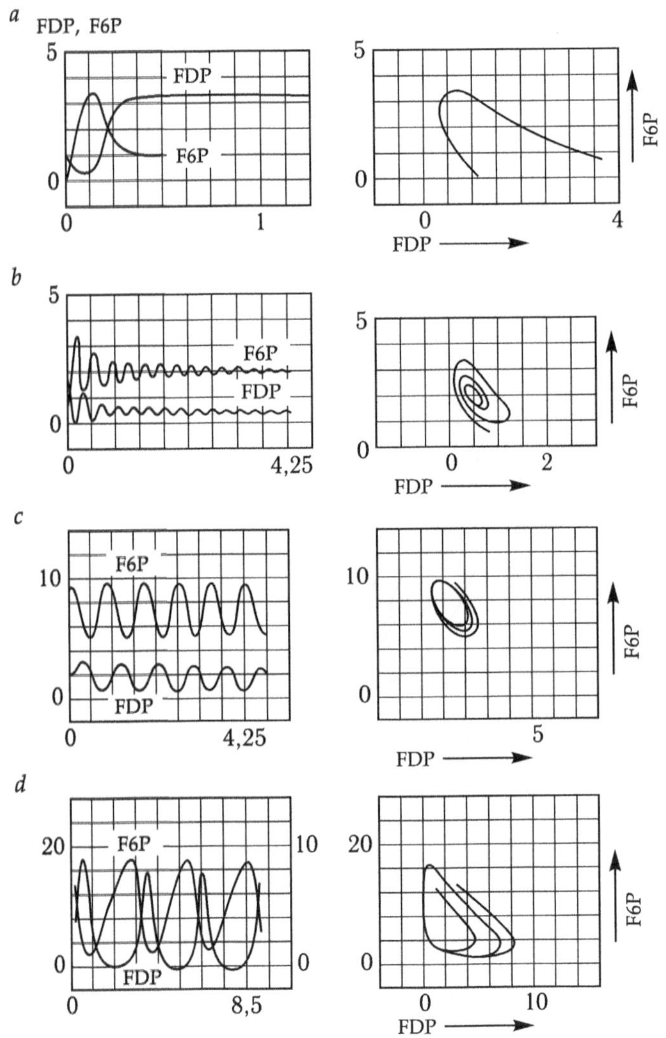

Figure 2.7. Kinetics of changes in concentrations of FDP (y) and F6P (x) calculated using a computer and phase portrait of the glycolytic model at different values of rate constants (reproduced from J. J. Higging, 1967).

a, Oscillation-free kinetics; *b*, damped oscillations; *c*, almost sinusoid oscillations; *d*, nonlinear oscillations.

As seen, system (2.7) does describe the generation of autooscillations at definite parameter values. Under experimental conditions, the rate of the substrate input to the cells (i.e. parameter k) can be readily changed. With a decrease of k, the α value in Eq. (2.9) will grow, which must result in generation of autooscillations (at $\alpha r/(1 + r) > 1$). As a matter of fact, substitutions of glucose by another sugar, for which value k and the rate of the substrate input decrease, lead to generation of autooscillations. The other conclusion, that values x' (F6P) and y' (FDP) fluctuate almost opposite in phase, is corroborated by the experimental data. The autocatalytic character of F6P conversion issues from the mathematical condition of generation of autooscillations — the presence of nonlinear members of the xy type. □

It was demonstrated experimentally that activation of enzyme PFK does occur, though not with its direct products. This enzyme is activated by ADP and AMP and inhibited by ATP, i.e. under conditions when the energy reserves are small and AMP and ADP are accumulated. Conversion of F6P is conjugated with phosphorylation of this compound due to the phosphate group of ATP

$$\mathrm{ATP} + \mathrm{F6P} \rightarrow \mathrm{ADP} + \mathrm{FDP},$$

as a result of which product A (ADP activator) is formed.

E. E. Selkov constructed a model in which AMP, formed in the following reaction, acts as an activator

$$2\,\mathrm{ADP} \rightleftarrows \mathrm{ATP} + \mathrm{AMP}.$$

This model produces oscillations, the nature of which has a relaxation character.

2.3 Time Hierarchy in Biological Systems

One of the main problems in mathematical modeling is the choice of variables essential for description of the object of variables, which are necessary and sufficient for construction of a correct mathematical model. Namely, in this case, it is possible to reproduce basic types of dynamic behavior of a complex object and to understand the principles of its self-regulation and control.

To achieve this, it is required to construct a model of the phenomenon, that would contain a possibly smaller number of variables and parameters and at the same time would reflect the basic properties of the phenomenon (for example, stability of stationary states, oscillations, triggers, quasistochastic features etc.). The problem of reduction of the initial number of variables (simplification of the model) turns out to be easily solved when time hierarchy exists in the system: concurrently running processes differ greatly in their characteristic times (see Section 1 in Chapter 1). □

An important feature of processes, occurring with participation of active intermediate particles, is the establishment of a regime for a short time period (during which the relative change in concentrations of initial substances is not great), at which the difference in the rates of formation of v_o and expenditure of v_p of intermediate compounds becomes small as compared to these rates. This means that the concentration of intermediate substances does not change. Such regime is called quasistationary, and the concentrations of active intermediate particles corresponding to it are called quasistationary concentrations. □

Thus, for a group of rapidly changing variables no differential equations may be written because, as compared to other slower variables, they practically instantaneously reach their stationary values. Then for fast variables, instead of differential equations describing their behavior in time, it is possible to write algebraic equations, determining their stationary values, which, in their turn, can be placed as parameters in differential equations for slow variables. In this way reduction is performed, i.e., decrease in the number of differential equations of a complete system which will include now only slow time-dependent variables. □

Let us consider, for example, some process described by the system of two differential equations

$$dx/dt = \phi(x, y), \quad dy/dt = G(x, y), \tag{2.11}$$

where y is a slow variable and x is a fast variable. This would mean that the incremental ratio of Δy and Δx for a short period of time Δt is much smaller than unity: $\Delta y/\Delta x \ll 1$.

Let us write system (2.11) in a way more convenient for analysis. Take advantage of the fact that the rate of changes in variable x exceeds greatly the rate of changes in variable y. This allows presenting the $\phi(x, y)$ function as the product of a high value $A \gg 1$ by the $F(x, y)$ function, by the order of magnitude corresponding to function $G(x, y)$.

So, the first equation of system (2.11) is transformed to $dx/dt = AF(x, y)$.

By dividing the left- and right-hand sides of this equation by A and designating $\epsilon = 1/A$, we obtain a complete system of equations identical with system (2.11):

$$dy/dt = G(x, y), \quad \epsilon\, dx/dt = F(x, y), \tag{2.12}$$

where $\epsilon \ll 1$. Complete system (2.11) can be simplified only if the character of solutions of this system does not change when the small parameter ϵ is approaching zero. In this case, the limit cycle $\epsilon \to 0$ can be performed, and an algebraic equation can be obtained from the second differential equation of system (2.11). A simplified (degenerated) system looks like this

$$dy/dt = G(x, y), \quad F(x, y) = 0. \tag{2.13}$$

Let us examine the phase portrait of complete system (2.11) in Fig. 2.8. An important feature of the phase portrait of the system is the presence of areas on plane xy differing greatly by the rates of changes of variables. Indeed, phase trajectories at any point of the phase plane, excluding the ϵ-vicinity of the curve $F(x, y) = 0$, have an incline, the tangent of its angle is determined by the following equation:

$$\frac{dy}{dx} = \epsilon \frac{G(x, y)}{F(x, y)} \simeq \epsilon \ll 1,$$

i. e., phase trajectories are positioned almost horizontally. These are the so-called areas of fast motions, in which along the phase trajectory the variable y is constant and variable x changes quickly. Having reached the ϵ-vicinity of the curve $F(x, y) = 0$ on one of such horizontals, the representation point will begin moving along this curve. The rate of movement along horizontal regions of the trajectory is $dx/dt \simeq$

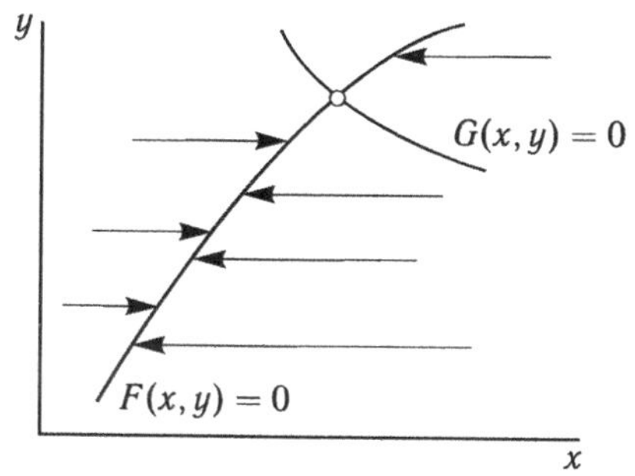

Figure 2.8. Phase portrait of complete system (2.11).

The character of phase trajectories of the system is determined by the position of basic isoclinic lines described by equations $G(x, y) = 0$ (the isoclinic line of horizontal tangents) and $F(x, y) = 0$ (the isoclinic line of vertical tangents). Their intersection point is a singular point of the complete system, and its coordinates are stationary values of variables x and y.

$1/\epsilon = A$, i.e. is very high as compared to the movement in the nearest vicinity of the curve $F(x, y) = 0$. That is why the total time of reaching the stationary state x, y on the curve $F(x, y) = 0$ is determined by the character of motion along this curve, i.e. depends on the initial values of slow variable y and does not depend on the initial value of fast variable x.

As has been shown above, if a system of equations has several equations with a small parameter before the derivative, all the equations describing changes in the fast variable can be substituted by the algebraic ones. If there are parameters of different levels of smallness, the reduction should be done sequentially.

In equations of chemical and biological kinetics, time constants of fast processes of different orders of magnitude frequently play the role of small parameters. In other cases, the ratio of small to high concentrations is a small parameter. This is also compatible with different rates of changes in variables, because the rate of changes in the high concentration is, as a rule, less than the rate of changes in the low concentration. Such a situation often takes place upon analysis of enzyme processes. □

3

Kinetics of Enzyme Processes

3.1 Elementary Enzyme Reactions

The majority of important biological processes occur with the involvement of enzymes, the chemical properties of which are considered in courses in biochemistry. Enzymes play a key role in cell metabolism, determining not only the pathways of substance conversion, but also the rates of reaction product conversion. The character of enzymatic processes makes allowance for a phenomenological description of their kinetics using systems of differential equations, in which variables are concentrations of interacting substances, substrates, products, enzymes. In this case it is sufficient to make use of general biochemical conceptions on the sequence of events in an enzyme reaction without going into physical details of mechanisms, i.e. to take into consideration that a necessary stage of enzyme catalysis is the formation of an enzyme-substrate complex (the Michaelis complex), as well as to employ notions on adaptation of enzymatic processes by inhibitors and activators.

Michaelis–Menten Equation. The simplest enzyme reaction with involvement of one substrate and formation of one product looks like this:

$$E + S \underset{k_{-1}}{\overset{k_1}{\rightleftharpoons}} ES \xrightarrow{k_2} E + P. \tag{3.1}$$

Here S is the substrate, P is the product, E is the enzyme, ES is the enzyme-substrate complex, k_1 and k_{-1} are constants of direct and reversible reactions of the formation of an enzyme-substrate complex, and k_2 is the rate constant of the product formation. At low concentrations of the product, this reaction is, as a rule, irreversible. The constant of effective disintegration of the enzyme-substrate complex k_2 shows how many catalytic acts per unit of time the enzyme can perform, and that is why it is called the number of enzyme turn-overs. Let us write the system of differential equations corresponding to the reaction scheme (3.1):

$$\begin{aligned} d[S]/dt &= -k_1[S][E] + k_{-1}[ES], \\ d[E]/dt &= -k_1[S][E] + k_{-1}[ES] + k_2[ES], \\ d[ES]/dt &= k_1[S][E] - k_{-1}[ES] - k_2[ES], \\ d[P]/dt &= k_2[ES]. \end{aligned} \tag{3.2}$$

By summing up the second and third equations of the above system, we get the condition for conservation of the total amount of the enzyme in the system:

$$\frac{d}{dt}([E] + [ES]) = 0, \quad \text{or} \quad [E] + [ES] = [E_0] = \text{const}. \tag{3.3}$$

In a closed system, the sum of the substrate and product mass remains also constant: $[S] + [P] = \text{const}$. As seen from Eqs. (3.2), $d[E]/dt = -d[ES]/dt$. Express $E = E_0 - (ES)$ and note that the last equation in system (3.2) for the product change is determined by the variable ES. Then, instead of the four equations in (3.2) the system of two differential equations for variables $[S]$ and $[ES]$ may be solved:

$$\begin{aligned} d[S]/dt &= -k_1[S][E_0 - (ES)] + k_{-1}[ES], \\ d[ES]/dt &= k_1[S][E_0 - (ES)] - k_{-1}[ES] - k_2[ES]. \end{aligned} \tag{3.4}$$

The characteristic time of changes in the substrate is evidently equal to $\tau_s = [S_0]/v_p$, where v_p is the rate of the enzyme reaction, S_0 is the total amount of the substrate. The maximal v_p value is $v_0 = k_2[E_0]$. Consequently, the minimal τ_s is $\tau_s^{\min} = [S_0]/k_2[E_0]$. Let us see how much these variables differ in the rates of their changes. The characteristic time of the enzyme revolution is determined primarily by the reaction of decomposition of $[ES]$ and constant k_2 ($k_2 \gg k_1$). It is $\tau_E = 1/k_2$.

In real biochemical processes, the substrate concentration is many-fold higher than the concentration of the unbound enzyme (usually $[E_0] = 10^{-6}$ M, and $[S_0] = 10^{-2}$ M). So, $[E_0]/[S_0] = 10^{-4}$. □

Hence, $\tau_S \gg \tau_E$, i.e. the variable S changes much slower than $[ES]$, and this means that the rate of changes in the substrate is low as compared to the rate of changes in the enzyme-substrate complex. Accordingly, upon analysis of the system behavior at rather large periods, the concentration of the enzyme-substrate complex $[ES]$ in the second equation of system (3.2) may be assumed to be quasistationary, and the second equation itself may be substituted for the algebraic one. As a result, the quasi-stationary value of the concentration of the enzyme-substrate complex $[ES]$ will be $[ES] = E_0S/(K_m + S)$.

The rate of reaction (the rate of product conversion equal to the rate of the substrate decrease) is expressed from the fourth equation of system (3.2):

$$v = -\frac{dS}{dt} = \frac{dP}{dt} = \frac{k_2E_0S}{K_m + S} = \frac{v_0S}{K_m + S}. \tag{3.5}$$

Equation (3.5) is called the Michaelis equation. It is seen from the equation that an increase in the substrate S concentration from 0 to ∞, the reaction rate (the slope of the initial regions in kinetic curves $S(t)$) increases from zero to its maximal value $v_0 = k_2 E_0$. Therefore enzymatic processes are processes with saturation. Figure 3.1 shows the dependence of the reaction rate on the substrate concentration (the Michaelis hyperbola). □

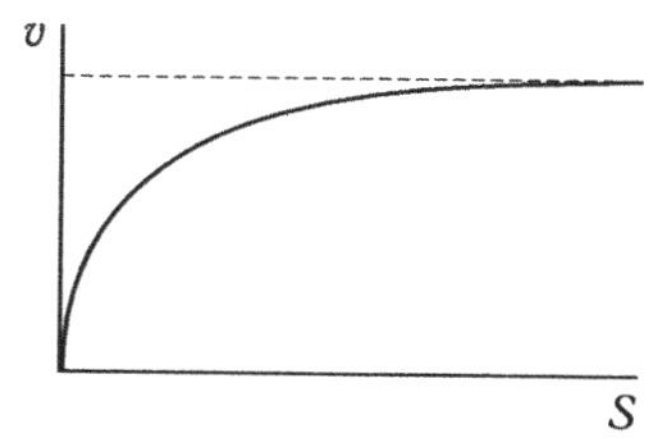

Figure 3.1. Stationary rate of the elementary enzyme reaction as a function of substrate concentration.

It is also seen from equation (3.5) that at $K_m = S$ the reaction rate is $v/2$. So, by its physical meaning and numerical value, the Michaelis constant is equal to the substrate concentration, at which the stationary reaction rate reaches its maximum, or, in other words, when half of enzyme molecules are in the state of a complex with the substrate.

In living systems, Michaelis constants of appropriate enzymatic processes and reagent concentrations are usually of an order of magnitude. The K_m value varies greatly (from 1 to 10^{-8} M). For example, $K_m = 3.5 \cdot 10^{-5}$ M for lactate dehydrogenase of pyruvic acid, $2.8 \cdot 10^{-2}$ M for invertase of sucrose, and $2.1 \cdot 10^{-1}$ M for maltase-maltose.

It should be noted that the Michaelis constant value corresponds to high activity of the enzyme. It is seen from expression $K_m = ((k_{-1} + k_2)/k_1)$ that at $k_{-1} \ll k_2$ and accepting that for most enzymes $k_2 \sim 10^2$–10^3 s^{-1}, the difference in K_m values is determined mainly by constant k_1. In other words, differences in enzyme activity are dependent on their variance in affinity to the substrate, which grows with an increase in k_1 and accordingly a decrease in K_m.

The Role of Inhibitors. In the presence of some substances, called inhibitors (I), the enzyme reaction may slow down. Thus, if the inhibitor can couple with the enzyme (E) in its active center by the scheme $E + I \rightleftarrows [EI]$, we have a reaction in the presence of a competing inhibitor substituting the substrate. In this case, formula (3.5) for the reaction rate becomes more complex:

$$\mu = \frac{\mu_0 S}{K_m + S + k_1 I}.$$

If the enzyme can accept both the inhibitor molecule and the substrate molecule with the formation of a complex, we have allosteric (noncompeting) inhibition. In this case, the rate of the product formation will be written as follows:

$$\mu = \frac{\mu_0' S}{(K_m' + S)(1 + k_1' I)}. \quad \square$$

Note that if the substrate is in excess, when the reaction rate is no longer dependent on its concentration, both formulas lead to integral qualitative dependence of the rate of product formation on the inhibitor concentration:

$$\mu = \frac{\mu_i}{1 + k_i I}.$$

Further analysis of allosteric features of enzymes allowed concluding that they may have a number of catalytic centers much greater than one. This means that n substrate molecules ($n > 1$) may couple to the enzyme molecule. Having changed the stoichiometry of the enzyme and substrate reaction, the following chain is obtained,

$$E + nS \rightleftarrows [ES^n] \rightarrow E + nP.$$

The rate of the product generation in this reaction will be $\mu = \mu_n S^n/(1 + k_n S^n)$. It is essential that for allosteric enzymes with the stoichiometric coefficient n varying from unity ($n > 1$), the character of the dependence $\mu(S)$ changes. The curve has a sigmoidal shape with a characteristic inflection (Fig. 3.2, curve *2*).

A similar peculiarity is observed upon taking into account the inhibiting action of the substrate excess. In this case, the rate is determined by the formula,

$$\mu = \frac{\mu_{ni}}{1 + k_{ni} I^n}.$$

In addition to inhibitors, there are substances that increase the intensity of the enzyme work; they are called activators (A). By forming a triple complex with the substrate and enzyme, they raise the rate of product formation:

$$\mu = \frac{\mu_A S}{K_m + S} \frac{A}{k_A + A}.$$

It can be seen that qualitatively the influence of activators on the rate of the enzyme reaction is described similar to the influence of the substrate concentration.

Up to the present, it was accepted that at high substrate concentrations, the rate of the enzyme reaction did not depend on the concentration. But there are enzyme reactions having a characteristic dependence of the stationary rate on the substrate concentration as a curve with its maximum. Such dependence is explained by the so-called substrate inhibition, which is a consequence of formation

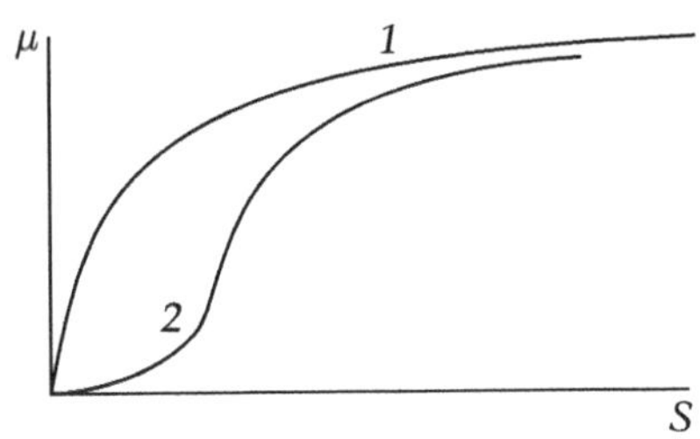

Figure 3.2. Two variants of dependence of the rate of enzyme reaction on the substrate concentration.

1, At $n = 1$; *2*, at $n > 1$.

of an inactive complex (along with the active one) with the enzyme. The ratio of probabilities of formation of active and inactive complexes changes with alterations of the substrate concentration. At high substrate concentrations, predominant is the probability of formation of inactive complexes ES^2, which include two substrate molecules simultaneously. As will be shown below, just the substrate inhibition of enzymes is the most typical reason for nonlinearity of biochemical systems. The existence of such a type of nonlinearity gives rise to important viewpoints on control mechanisms of the properties of enzymatic systems: multiplicity of stationary states and oscillation character of changes in variables. □

The stationary reaction rate in the system in which, in addition to an active complex ES, an inactive complex ES^2 is formed,

$$\begin{gathered} E + S \underset{k_{-2}}{\overset{k_2}{\rightleftarrows}} ES \xrightarrow{k_3} E + P, \\ ES + S \underset{k_{-4}}{\overset{k_4}{\rightleftarrows}} ES^2, \end{gathered} \tag{3.6}$$

is expressed like this

$$v = \frac{k_2[E_0][S]}{K_m + [S] + [S^2]/K_S}, \tag{3.7}$$

where $K_S = k_4/k_{-4}$.

Let us consider an open enzymatic system with substrate suppression and constant rate of substrate inflow to the reaction volume. If in addition outflow of the product from the reaction volume takes place, scheme (3.6) is extended by two other reactions:

$$S_0 \xrightarrow{k_1} S, \quad P \xrightarrow{k_5}. \tag{3.8}$$

The system of kinetic equations corresponding to schemes (3.6) and (3.8) are as follows:

$$\begin{aligned} &d[S]/dt = k_1[S_0] - k_2[S][E] + k_{-2}[SE] - k_4[SE][S] + k_{-4}[S^2E], \\ &d[ES]/dt = k_2[S][E] - k_{-2}[SE] - k_3[SE] - k_4[SE][S] + k_{-4}[S^2E], \\ &d[E]/dt = -k_2[S][E] + k_{-2}[SE] + k_3[SE], \\ &d[S^2E]/dt = k_4[SE][S] - k_{-4}[S^2E], \\ &d[P]/dt = k_3[SE] - k_5[P], \quad [E] + [SE] + [S^2E] = E_0. \end{aligned} \tag{3.9}$$

System (3.9) can be simplified with account of the fact that the enzyme concentration E_0 is much lower than that of the substrate, similar to the above case for an elementary enzymatic reaction.

The replacement of differential equations for the rate of changes in the concentrations of enzyme-substrate complexes by the algebraic equation and substitution of corresponding variables into the equation for changes in the substrate concentrations result in the following equation of enzymatic reaction with the substrate inhibition:

$$\frac{d\sigma}{d\tau} = \alpha - \frac{(c+1)a\sigma}{1 + a\sigma + \beta(a\sigma)^2} = f(\alpha, \sigma). \tag{3.10}$$

Here, $\sigma = [S]/K_m$ is the non-dimensional substrate concentration; $\tau = k_3[E_0]t/K_m$ is the non-dimensional time; $\alpha = k_1[S_0]/k_3[E_0]$, $c = k_{-2}/k_3$, $a = k_2[S_0]/k_3$, and $\beta = (k_3/k_2)(k_4/k_{-4})$.

Equation (3.10) differs from equation (3.7) by the free member α, which characterizes the rate of the substrate inflow to the reaction range. Stationary points of equation (3.10) are determined from the condition $d\sigma/d\tau = 0$ or

$$\frac{(c+1)a\sigma}{1 + a\sigma + \beta(a\sigma)^2} = \alpha. \tag{3.11}$$

To determine the number and character of singular points of this equation, it is convenient to use the graphic representation of the dependence on the σ value of the substrate inflow rate α and its uptake in the reaction v. Intersection points of the plot of the function

$$v(\sigma) = \frac{(c+1)a\sigma}{1 + a\sigma + \beta(a\sigma)^2} \tag{3.12}$$

with the straight line of the constant source α will correspond to the solutions of equation (3.11).

As has been shown earlier, the $v(\sigma)$ function is plotted as a curve with its maximum (Fig. 3.3).

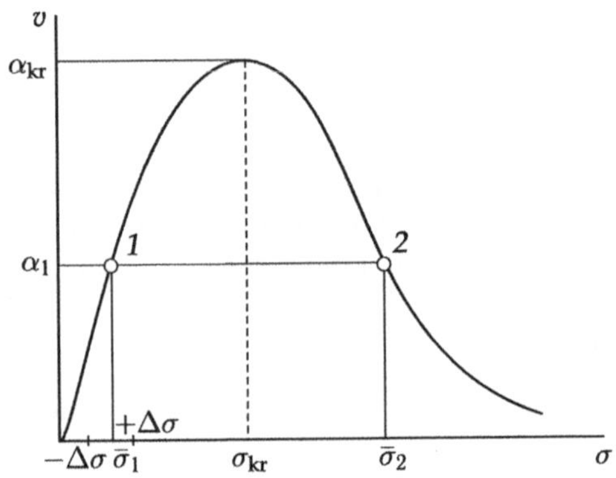

Figure 3.3. Dependence of the rate of enzyme reaction on the substrate concentration.

A family of straight lines parallel to the abscissa axis corresponds to different values of the rate of substrate inflow T. The plot of function $T(\sigma)$ may have one or two intersection points with the straight line α or have none. This corresponds to the existence of two or one stationary states in the system or to their absence. When the system has two stationary states, singular point $\bar{\sigma}_1$ is stable, and $\bar{\sigma}_2$ is unstable. This can be easily substantiated by the following considerations. Let as a result of some deviation $\Delta\sigma < 0$ from stationary point $\bar{\sigma}_1$, value σ has become smaller than the stationary value. At $\sigma < \bar{\sigma}_1$, the rate of substrate inflow is higher than the rate of its outflow ($\alpha > v$), and accordingly variable σ will increase approaching $\bar{\sigma}_1$. But if deviation from stationary point exceeds zero ($\Delta\sigma > 0$), the expenditure rate of substrate is higher than its inflow, and σ will decrease, again approaching the stationary value $\bar{\sigma}_1$. Thus, at any deviation from stationary state $\bar{\sigma}_1$, the system will return to it and, as a result, state $\bar{\sigma}_1$ is stable. Similar reasoning concerning stationary point $\bar{\sigma}_2$ leads to conclusion that it is unstable. This conclusion is easily made if the sign of the derivative by σ in the right-hand side of function $f(\alpha, \sigma)$ of Eq. (3.10) is determined; it is negative for $\sigma < \sigma_{kr}$ and positive for $\sigma > \sigma_{kr}$ (see the analysis of stationary state stability for one equation in Section 2 of Chapter 1).

The rate of the substrate inflow to the reaction range can be considered as the parameter most suitable for controlling the enzymatic system; in Eq. (3.10) value α corresponds to it. The dependence of the stationary value σ on the rate of the substrate inflow α to the system is presented by a bifurcation diagram (Fig. 3.4). □

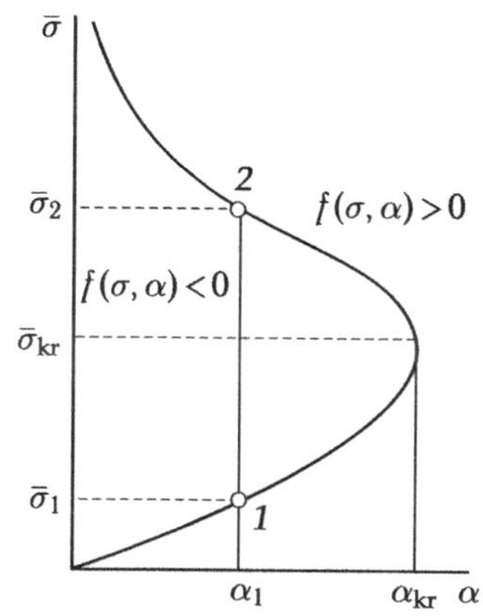

Figure 3.4. Bifurcation diagram of a system with substrate suppression.

The parametric curve of enzymatic system $\bar{\sigma}(\alpha)$ consists of two branches. Points on the lower branch of the curve are stable and those on the upper branch are unstable. Parameter α_{cr} is a bifurcation value, and the only stationary state at the junction of the upper and lower branches corresponds to it. At $\alpha > \alpha_{cr}$, its stationary state is unattainable, the unlimited amount of the substrate being accumulated in the system.

3.2 Multiplicity of Stationary States in Enzyme Systems

When the function of the substrate inflow to the reaction range is not constant and has a more complex character, the types of behavior of the enzyme system may be more complex. For example, in an open enzyme system with substrate suppression and reversible reaction of substrate inflow

$$S_0 \underset{k_{-1}}{\overset{k_1}{\rightleftarrows}} S \tag{3.13}$$

two stable stationary states may happen, i.e. such a system is a trigger one (see Section 1 in Chapter 2). □

Unlike the analyzed system (3.10), the non-dimensional rate of changes in the substrate concentration in Eq. (3.13) is as follows:

$$\frac{d\sigma}{d\tau} = \alpha - \beta\sigma - \frac{\sigma}{1+\sigma+\gamma\sigma^2} = f(\alpha,\sigma). \tag{3.14}$$

Here $\sigma = [S]/K_m$, $K_m = (k_{-2} + k_3)/k_2$, $\alpha = k_1[S_0]/(k_3E_0)$, $\beta = k_{-1}K_m/(k_3E_0)$, $\gamma = K_m k_4/k_{-4}$.

Let us use the graphic representation once again and find stationary solutions of equations (3.14) from the following equation

$$\alpha - \beta\sigma - \frac{\sigma}{1+\sigma+\gamma\sigma^2} = 0 \tag{3.15}$$

in the form of intersection points of the plot of the $v(\sigma)$ function of the substrate expenditure in the reaction and the straight line of the source $\bar{v}_i = \alpha - \beta\sigma$ (Fig. 3.5).

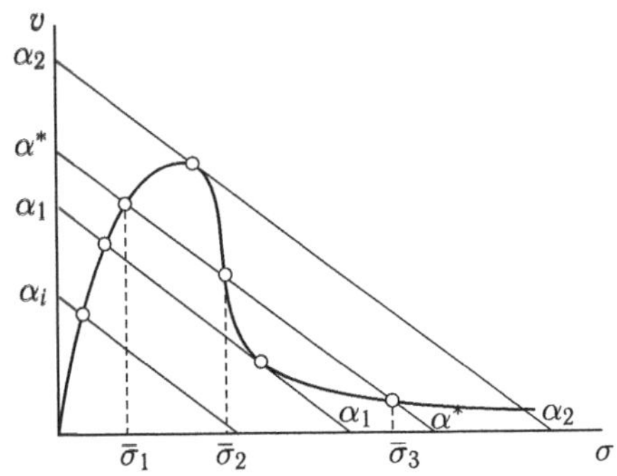

Figure 3.5. Graphic determination of a number of stationary states in a system with substrate suppression and reversible reaction of substrate inflow.

The family of parallel straight lines (α_i) corresponds to different α values. At some values of the rate of substrate inflow (α^*), Eq. (3.15) has three roots, which corresponds to three different states of the analyzed system.

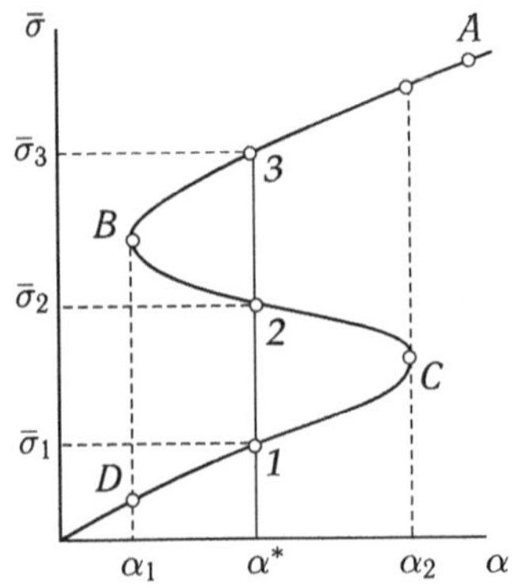

Figure 3.6. Curve of stationary states of a system with substrate suppression and reversible reaction of substrate inflow.

The system is characterized by an *S*-shaped curve of stationary states. When the controlling parameter changes in the range $\alpha_1 \leqslant \alpha \leqslant \alpha_2$, three different stationary states of the system correspond to each value. In accord with the sign of the derivative of function $f'_\sigma(\alpha, \sigma)$ in the right-hand side of Eq. (3.14), all points on the lower *DC* and upper *BA* branches refer to stable ($f'_\sigma < 0$) stationary states, while points on the intermediate region of curve *BC* refer to unstable stationary states ($f'_\sigma > 0$).

The same points can be determined by plotting the dependence $\bar{\sigma}$ of stationary solutions of Eq. (3.15) on parameter α, which characterizes the rate of the substrate inflow to the system (Fig. 3.6).

At all values of the controlling parameter in the range of $\alpha_1 \leqslant \alpha \leqslant \alpha_2$, where α_1 and α_2 correspond to bifurcation values of the parameter, the analyzed system can function in one of the two stable stationary states, i.e. has trigger properties. This means that upon changes in the controlling parameter, the system can switch from one stable regime to the other. Let us explain how this occurs with the help of the plot (Fig. 3.6).

Assume that stationary point *A* on the upper branch of curve $\bar{\sigma}(\alpha)$ corresponds to the initial state of the system. The rate of the substrate inflow (α) will be decreased, and the system will "move" to the left along the upper stable branch of stationary states. When parameter α_1 reaches the bifurcation value, the system will leave the unstable point B and, having performed the spasmodic transition $B \rightarrow D$, get to the lower branch of stable stationary states. By increasing further the value of the controlling parameter from α_1 to α_2, it is possible to move the system along the stable branch DC to the bifurcation point C, after reaching which the system will spontaneously return to the initial state A. At reversible changes in controlling parameters α (a decrease and then an increase to the previous values), a closed cycle of states of the analyzed system will take place.

Which of the two possible stable states is realized in the system and what are the directions of the spasmodic transitions depend on whether the parameter increases or decreases. This property of the system — to perform transition from one state to the other in different ways depending on the direction of changes in the parameter values — is called hysteresis (Fig. 3.7). □

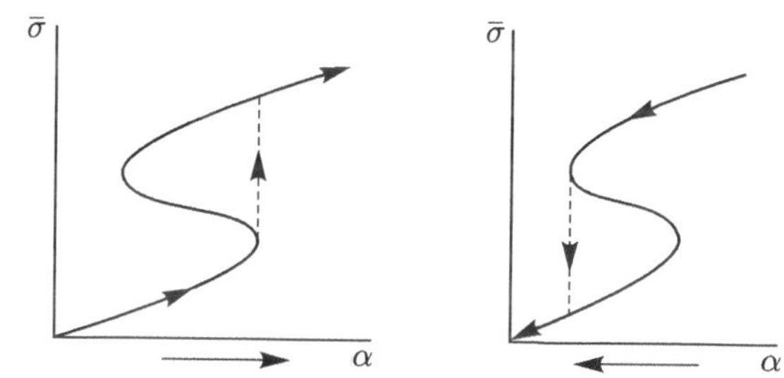

Figure 3.7. Hysteresis phenomenon in a system with substrate suppression and reversible reaction of substrate inflow.

The multiplicity of stationary states and concomitant hysteresis and trigger phenomena play a significant role as means of control of biological systems. Indeed, in the presence of hysteresis properties, the behavior of the system in response to changes of its parameters, which reflects the changes in the external conditions, will be dependent on its prehistory. So, hysteresis inherent in enzymatic systems may be the basis of the dynamic memory, peculiar to all biological objects.

Trigger properties of enzymatic systems play a critical role in controlling intracellular metabolism as well as in cell differentiation, when due to division daughter cells are generated differing qualitatively from precursor cells. At present trigger properties of enzymatic systems, accomplishing the transport function, are well-investigated. In particular, such phenomena were discovered upon studying the transfer of solutions across porous membranes. The system of membrane transfer, conjugated with chemical reactions involving the transported compound, has trigger properties. □

3.3 Oscillations in Enzyme Systems

Nonlinearity of the kinetics of most important biochemical processes is responsible for the existence of continuous periodic regimes (autooscillations) along with trigger regimes in biological systems. Qualitative methods of analysis of dynamic systems may be used for studying their properties.

The discovery of the "biological clock" with an endogenous rhythm in unicellular organisms was the convincing evidence that the source of biological rhythms is the autooscillation system of biochemical reactions, localized inside the cell.

The development of thermodynamics of irreversible processes demonstrated that oscillations in chemical reactions may occur only far from the thermodynamic equilibrium (see Chapter 4). It is known also that to provide a non-equilibrium stationary state in chemical reactions, it is necessary that the change in concentrations of two different groups of reagents would take place at two time scales greatly varying from each other. In enzymatic systems, this condition is always fulfilled, because as a rule the concentration of the substrate is much larger than that of the enzyme, which in its turn provides a possibility for existence of two strongly varying time scales for the substrate (τ_S) and enzyme-containing complexes ($\tau_S \gg \tau_E$) (see Section 1 in Chapter 3). The time hierarchy of agents and conditions of approximate stationary state for some of them (for example, for the enzyme-substrate complex) can be realized not only in open but also in closed enzymatic systems. But in closed systems, these conditions are realized only during a limited time interval, when the substrate concentration is rather high. That is why oscillations, occurring in closed enzymatic systems, are always damped, which is determined by the evolution of the closed system to the equilibrium state. □

Continuous oscillation changes in agent concentrations may occur only in open enzymatic systems, where the existence of two greatly varying time scales can be maintained for an unlimitedly long period due to the substrate inflow from the external source and the product outflow to the external environment. In a system of coupled polyenzyme reactions, in which the product of one reaction is the substrate for the following one, the time scale associated with the first substrate and the last product, can exceed greatly the time scale characteristic of intermediate agents. Therefore in such systems, it is most likely that undamped oscillations of concentrations would appear.

So, necessary conditions for generation of concentration oscillations in systems are nonlinear mechanism of reactions, the "openness" of enzymatic systems and significant differences in characteristic times of separate stages of the process. □

A widespread reason for nonlinearity of biochemical systems is the presence of the feedback in the chain of enzyme reactions, the product of which controls the enzyme activity subject to the general scheme

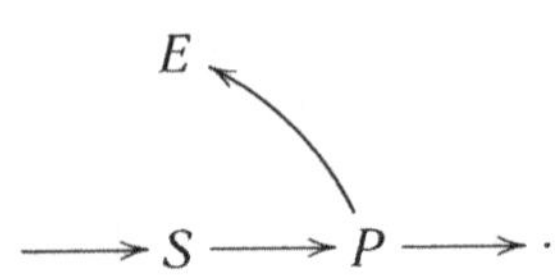

The character of such effect may vary: the most prevalent is the product inhibition, but there may be cases, when the product activates the enzyme at all values of its own concentrations or inhibits at some concentrations and activates at other ones. Another common mechanism of controlling biochemical systems is substrate inhibition of enzymes.

The variety of stationary states, hysteresis and autooscillations as the basic properties of the kinetics of the studied reaction are a direct consequence of strong nonlinearity caused by mechanisms of substrate and product inhibition.

As combined inhibition of enzymes by substrates and products is a quite common phenomenon, the analyzed model can be used for explaining periodic oscillations in different biochemical processes. □

Some papers are devoted to studying the generation of oscillations in polyenzymatic systems, where the final product affects the rate of the initial stage. This case corresponds to the cascade of sequential enzymatic reactions with the feedback:

$$\xrightarrow{v_0} S_0 \xrightarrow[E_1]{v_1} S_1 \xrightarrow[E_2]{v_2} S_2 \xrightarrow[E_3]{v_3} \ldots \rightarrow S_n \xrightarrow{v_{n+1}} .$$

According to this scheme, the substrate S_0 is supplied to the system at a rate v_0 and with the help of enzymes $E_1, E_2, \ldots, E_n$ is sequentially converted to the final product S_n, used in further reactions at a rate v_{n+1}.

The activity of the key enzyme E_1 is dependent on the final product S_n. The problem of autooscillations in such a system is solved by determining the ratio of its parameters, at which complex conjugate roots appear in characteristic equations of the linearized system (Chapter 1). The analysis made by E. E. Selkov (1967) demonstrated that oscillations depend largely on the values of parameters κ_1 including the effect of S_n on activity of E_1, the dependence of the rate of utilization of product S_n on its concentration as well as on the number of links n in the system. It was found that at $\kappa_1 < 0$ (product inhibition of the key reaction of E_1) and number of links $n = 2$, the system is stable (the stable node or focus). However, under the same conditions ($\kappa_1 < 0$), an increase in the number of intermediate stages $n = 3, 4, \ldots$ results in generation of oscillation instability (the unstable focus). □

3.4 Mathematical Modeling of Metabolic Pathways

General Approaches. Recent years have been marked by great advance in the understanding of processes of cell metabolism, genetic control, and signal transduction in cells.

Information on the sequence of genes was compiled in different databases allowing for the reconstruction of the whole array of metabolic pathways for an individual organism. For example, for the *E. coli* bacterium metabolic maps were plotted, which were, in their turn, compiled in databases, representing descriptions of all metabolic pathways and their control. Special mathematical models were required for unification and perception of the accumulated experimental data. As a result, a novel field appeared in mathematical modeling — metabolic modeling.

These models are successfully used for analyzing the phenotype of organisms at different gene deletions, for studying gene expression that leads to a certain structure of the metabolic network, and for determining regulatory genes.

The critical distinguishing feature of the above considered system of chemical reactions and the metabolic network is the presence of regulatory links and genetic control of the activity of metabolites. In biological systems, all levels of metabolic functioning (transcription, translation and catalytic activity) are closely interconnected, thereby providing a possibility of adaptation to varying external conditions.

If detailed information on the mechanisms of genetic and post-translation regulation is available, general kinetic models describing regulation of enzyme activity may be designed based on the mass action law.

Analysis of the Balance of Metabolic Fluxes (Problem of Linear Programming). For a mathematical description and analysis of metabolic fluxes, the approach is used based on the assumption on the balance of metabolites and the use of the linear programming method, which is called the Flux Balance Analysis (FBA). The metabolic flux is the rate of turnover of molecules through a metabolic pathway. Let us analyze the scheme of metabolic reactions, $v_1, \ldots, v_6$, resulting in the initial metabolite S_1 conversion into metabolite S_4 (the final product).

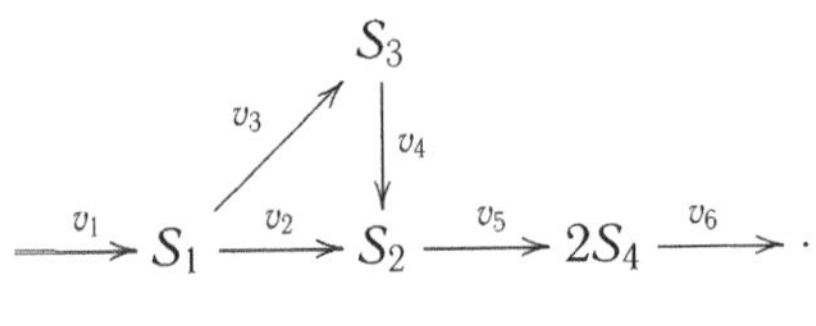

Here S_i is metabolites, v_i is the reaction rates. Accept that the rates directed to the formation of metabolites, have a positive sign, and the reaction rates in which the metabolites are expended have a negative sign. Let us write in general terms the system of differential equations describing time changes in all metabolites:

$$\begin{aligned} \frac{dS_1}{dt} &= v_1 - v_2 - v_3, \\ \frac{dS_2}{dt} &= v_2 + v_4 - v_5, \\ \frac{dS_3}{dt} &= v_3 - v_4, \\ \frac{dS_4}{dt} &= 2v_5 - v_6. \end{aligned} \tag{3.16}$$

The flux balance analysis (FBA) allows considering a stationary situation, when the system is in the equilibrium state. Every metabolite has metabolic flux balance, i.e. all fluxes, directed to the formation of a certain metabolite are balanced by fluxes, decreasing the concentration of this metabolite. The sum of all fluxes involved in changing the concentration of the given metabolite is equal to zero. For a system of differential equations this means that the sum of all rates of reactions in which this metabolite is involved is equal to zero. Then all right-hand sides of the equations are equal to zero, and the system of differential equations degenerates to a system of algebraic equations. This approach permits evaluating the distribution of metabolic fluxes (or stationary rates) and their relative contribution to the formation of the given metabolite.

The first step consists in plotting a stoichiometric matrix based on the scheme of metabolic fluxes. It is suggested that concentrations of all intermediates are at pseudostationary equilibrium, and the changes associated with growth processes are regarded as insignificant. A stoichiometric model yields a set of algebraic equations reflecting the existence of concentration balance for all metabolites studied:

$$\begin{aligned} x_{11} \cdot \nu_{11} + x_{12} \cdot \nu_{12} + \ldots + x_{1m} \cdot \nu_{1m} &= 0, \\ x_{21} \cdot \nu_{21} + x_{22} \cdot \nu_{22} + \ldots + x_{2m} \cdot \nu_{2m} &= 0, \\ &\ldots \\ x_{n1} \cdot \nu_{n1} + x_{n2} \cdot \nu_{n2} + \ldots + x_{nm} \cdot \nu_{nm} &= 0, \end{aligned}$$

where x_{ij} is the number of molecules entering the j-th reaction, v_0 is the rate of the j-th reaction, involving the i-th metabolite. The resultant system of equations can be written like this:

$$\mathbf{Nv} = 0.$$

Matrix **N** contains stoichiometric coefficients, and vector **v** represents values of fluxes of metabolic reactions. In general terms the dimension of **N** may be written as $m \times n$, where m is the number of metabolites involved in reactions, and n is the number of reactions.

As the number of reactions is usually larger than the number of intracellular metabolites ($n > m$), the system of equations obtained on the basis of the stoichiometric matrix, is underdetermined, and that is why there may be an infinite number of solutions, i.e. an infinite number of possible distributions of fluxes. In numerical experiments, a definite distribution of fluxes can be found by formulating a correct linear programming problem or an optimization problem, which is often called the flux balance analysis. To formulate a linear programming problem it is necessary to define the target function z, which, for example, can describe the maximal product yield with regard to all reactions v_i in which it is generated. For example, it is necessary to find the maximum function

$$z = \sum_i v_i \rightarrow \max \tag{3.17}$$

under the following conditions:

$$\mathbf{Nv} = 0, \tag{3.18}$$

$$\alpha_i \leqslant \nu_i \leqslant \beta_i. \tag{3.19}$$

Linear inequalities $\alpha_i \leqslant \nu_i \leqslant \beta_i$ are used for obtaining additional restrictions for separate fluxes. The restrictions determine the upper and lower boundaries in changes of the flux value for every reaction. α_i is the lower boundary of the flux, and β_i is its upper boundary. As an example, let us take the redox status of the cell, and namely the state associated with the presence of NAD(P)H.

The first step in studying the state of NAD(P)H in the cell is plotting a scheme of the central metabolism of a plant cell, namely glycolysis, and the pentozophosphate pathway. To analyze the state of NAD(P)H in the cell, let us analyze two possible distributions of metabolic fluxes. In the first case, the metabolic flux through the pentozophopaphate pathway makes 10 % of the flux through the glycolytic pathway. In the second case, let us study the possibility to realize such distribution of fluxes when the yield of NAD(P)H is maximal.

Variable v_i is compared to each i-th rate of reaction (Fig. 3.8). In this case, part of sequential stages from *GAP* to *Pyr* can be reduced to one, because the result of optimization here does not depend on the number of sequential stages.

Then, proceeding from the main principle of the flux balance analysis, it is assumed that all fluxes are in the quasistationary state. Taking into account that the magnitude of any flux is determined by a non-negative number, a system of equations and inequalities can be written.

After that, the target function is determined for the final formulation of the optimization problem. As can be seen from Fig. 3.9, the main stages of NAD(P)H

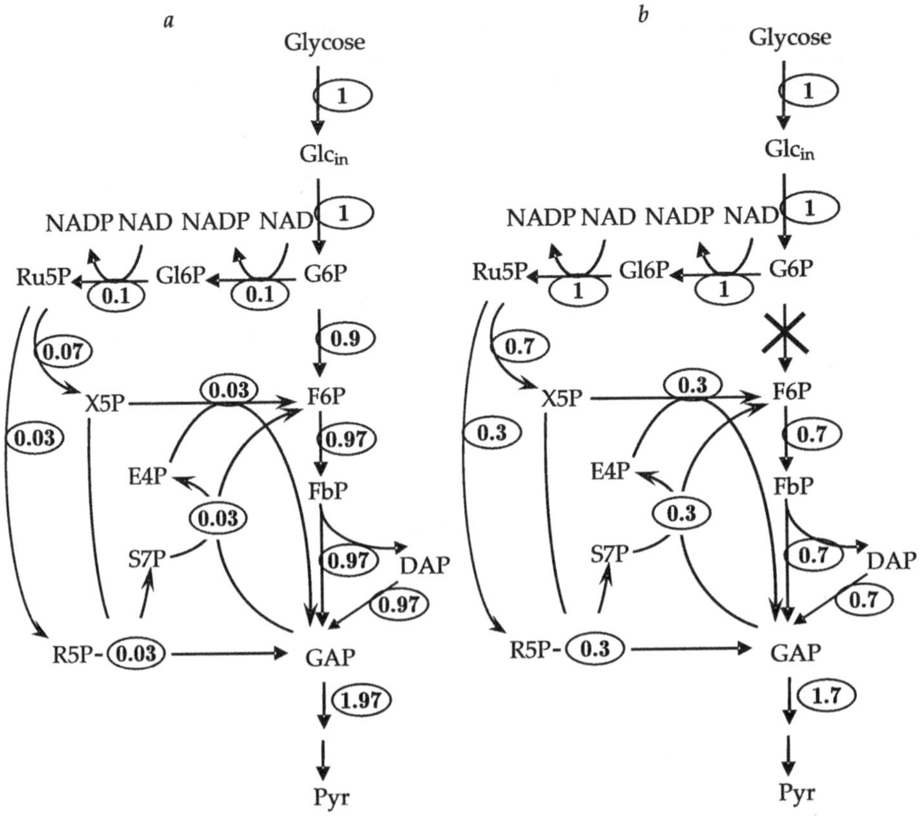

Figure 3.8. (*a*) Initial distribution of fluxes. (*b*) Distribution of fluxes in the case of maximal yield of NAD(P)H. Relative values of fluxes are encircled.

generation are reactions v_7, v_8 and v_9. Therefore, for analysis of the maximally probable yield of NAD(P)H we choose the following target functions: $z = v_7 + v_8 + v_9$.

As the initial distribution of fluxes, let us consider the case when 90 % of the metabolic flux proceeds through glycolysis and only 10 % through the pentozophosphate pathway. All the fluxes of the metabolic pathway are normalized by the value of the input flux to obtain relative units. The result of the initial distribution is shown in Fig. 3.8,*a*.

As can be seen, if the yield of NAD(P)H is maximal, the whole metabolic flux passes through the pentozophosphate shunt. In this case, the glycolytic reaction catalyzed by phophoglucoisomerase, terminates "working" which is in keeping with the silencing of the *pgi* gene. Using the analysis of flux balance in the central metabolic pathway, we have tried to evaluate the state of NAD(P)H in different situations. It can be suggested that when it is required to raise the level of NAD(P)H, the *pgi* gene is inactivated, which results in inhibiting the reaction in glycolysis and redistribution of metabolic fluxes in such a way that the maximal flux proceeds through the pentozophopaphate shunt.

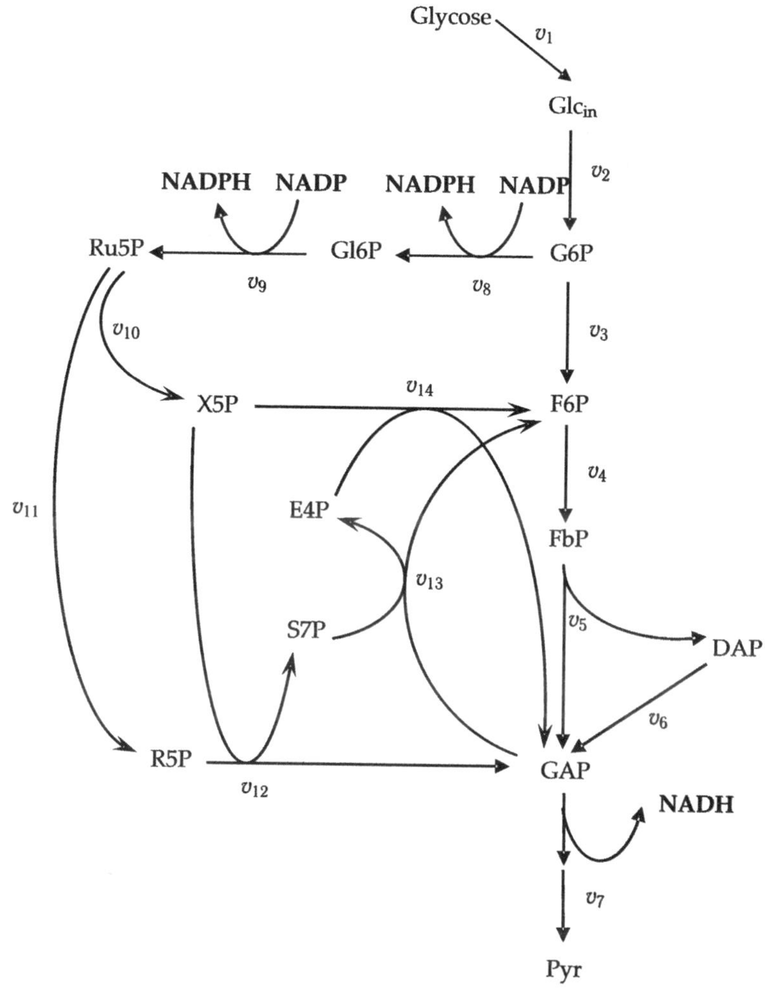

Figure 3.9. Reduced metabolic pathway for solving the optimization problem. Every i-th reaction has a corresponding flux v_i.

4

Self-Organization Processes in Distributed Biological Systems

4.1 General Characteristics of Autowave Processes

Thus far we have considered the so-called "point" systems, in which changes of variables occur equally in all points of the space occupied by the system. Such an approach is largely reasonable for homogeneous objects, their parameters (substance density, temperature) being similar in all points. These requirements are in line, for example, with microorganism culture in a continuous cultivator for overall agitation. However, in biological and chemical systems, it is necessary to take into account their spatial heterogeneity, which should be reflected in corresponding mathematical models.

Now let us study systems in which the state is changed in different points, and between separate points there are "diffusion bonds" or substance fluxes. They are called systems distributed in space. In such systems, time changes in substance concentrations proceed due to two processes: chemical conversions described by "point" kinetic models and substance diffusion from high-concentration regions to regions of lower concentrations. All biological systems (active membranes, cells and tissues, and communities of organisms) belong to the class of active distributed systems. Substance conversion and energy transformation in them occur in individual elementary volumes, interconnected by diffusion substance transfer. □

Thus, every elementary volume is, in its turn, a closed elementary system far from thermodynamic equilibrium (see Chapter 5 and 6). This system is characterized by distribution over its space of substances rich in energy, or distribution of the energy source, responsible for generation of different processes in elementary volumes and associated with diffusion of active substances. Among peculiar dynamic phenomena which take place in such systems of great importance is propagation of self-sustained excitation waves. They retain their characteristics (amplitude, period, shape, and rate) due to the energy source distributed in the environment. It is just because of this property that these systems have been called active distributed systems.

In biological systems, definite biochemical substances (excitation waves in nervous tissues and the cardiac muscle) play the role of a distributed energy source, maintaining wave propagation. It was found that a number of fundamental dynamic properties of biological systems are inherent to them as active distributed systems. Dynamic phenomena in active distributed systems are called autowave processes (AWP) including a whole class of spatial effects.

The description of AWP properties based on common physical and mathematical principles is a subject of the theory of active distributed systems.

It was found that chemical systems contain a class of processes resulting in almost all basic types of AWP. The most important of them is the reaction of Belousov–Zhabotinsky. It is a redox reaction with involvement of brommalonic acid, where cesium and manganese ions are catalysts. Changes in the valence of these ions result in local changes in the color of solution, which is a visual method for observing spatial phenomena in a reaction vessel. For example, it was demonstrated that for an active homogeneous medium this system has two-dimensional wave processes — reverberators and formation of dissipative structures.

In living systems, the most studied are AWP in excited membranes of nervous fibers. It was shown that in neuron networks excitation waves are propagated which is caused by changes in the intercellular concentration of the K^+ and Ca^{2+} ions. Waves of electric potential travel in the form of travelling pulses (TP) along the cardiac muscle fibers. Here pathological states such as arrhythmia and fibrillation (unsynchronized activity of muscles) are determined by generation of autonomous sources of reverberators.

Other types of AWP are clearly displayed in processes of morphogenesis upon tissue differentiation. In this case, the material basis is genetic systems of protein biosynthesis and active transport of substances across cell membranes. In communities of organisms in some cases interaction of cells occurs by means of generated attractant substances (cyclic AMP). The mutual movement of cells to the signal source and their aggregation has a wave character. In embryonic structures, this mechanism determines the motion of cells during tissue formation. In addition, AWP underlie motions in blood vessel walls and mechanical movement of cells along a plane surface.

It is known that cell differentiation plays a key role in development and is manifested as transitions of separate cells and whole tissues from one specialization type to another. Such transitions are a consequence of involvement of one type of genes and silencing of other types of genes. At early stages of embryogenesis, cells exchange with each other different metabolites, including morphogens as regulators of gene activity. In non-differentiated tissues, morphogens are distributed homogeneously. When the distribution of morphogens over the ensemble of cells becomes not uniform, morphological structures phenotypically different are formed and observed.

A critical role in the formation of spatially non-uniform distribution of morphogens belongs to instability which destroys the initial uniform state (A. M. Turing, 1952). This conclusion was done based on the analysis of the model describing formation and decomposition of morphogens. Turing demonstrated that uniform stationary distribution of agents $\overline{x}$ and $\overline{y}$ at definite values of parameters, characterizing the generation and decomposition kinetics of morphogens, may become unstable with respect to space-and-time perturbations.

These excitations, called unstable modes, having once appeared due to random fluctuations, will develop increasing their amplitude. But the amplitude will not increase unlimitedly — it will be stabilized by nonlinear effects. As a result, the system will generate stationary spatially non-uniform distribution of reagents (the dissipative structure), which determines the process of positional (depending on the spatial position) differentiation of tissues. □

As a result of accurate experimental studies on a simple organism — hydra, two types of morphogens were discovered: an activator and an inhibitor. The experiments showed that after decapitation, when during growth the size of the hydra exceeds some minimum value, a new head forms in the place occupied by the removed one. So, an increase in the hydra size contributed to the positional differentiation.

For the interpretation of this phenomenon, a model was proposed, the analysis of which demonstrated that an increase in the system size may lead to instability, destroying the uniform distribution of substances (A. Gierer and H. Meinhardt, 1972). In particular, this refers to the distribution over the hydra body of concentrations of two morphogens: a short-range activator and a long-range inhibitor of the head formation processes.

A new interdisciplinary field of sciences called synergetics (H. Haken, 1980) and its methods is involved in solving the problems of what mechanisms and conditions provide the appearance of instability and transition to a new ordered state. □

4.2 Mathematical Models of Self-Organized Structures

We will consider mathematical models reflecting key dynamic properties of AWP. Their principal distinction is that variables change concurrently due to chemical processes, having different character in different points of space. In a general form, this can be written as the dependence of the rate of concentration changes of i-th substance:

$$\dot{c}_i = f_i(c_1, c_2, \ldots, c_n, t).$$

This rate depends on the concentrations of all reaction components in point r as well as directly on r and t, because external parameters, for example, temperature, can vary in different points r and change with time.

Diffusion Equation. Let us study the role of diffusion processes in changing concentrations c_i at the instant of time t in point r, i.e. $c_i(t, r)$. For the sake of simplicity, take a reaction vessel in the form of a long narrow tube with the cross-section S (Fig. 4.1). Distinguish in it the elementary volume ΔV with coordinates r and $r + \Delta r$.

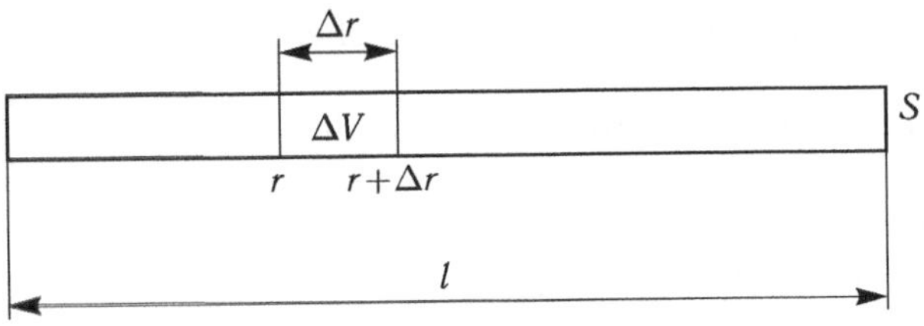

Figure 4.1. To the derivation of the diffusion equation.

Then $\Delta V = S\Delta r$. The mass ΔM_r of the substance running through the tube cross-section in point r is proportional to the concentration gradient $\Delta c_i/\Delta r$ of substance c_i towards r and to the time period from t to $t + \Delta t$, during which diffusion occurs:

$$\Delta M_r = -D_i \frac{\Delta c_i(r,t)}{\Delta r} S\Delta t.$$

Here D_i is the diffusion coefficient determined by properties of a dissolved substance and solvent. The following mass will diffuse through the other boundary with the coordinate $r + \Delta r$ during the same time period:

$$\Delta M_{r+\Delta r} = D_i \frac{\Delta c_i(r+\Delta r,t)}{\Delta r} S\Delta t.$$

Thus, due to diffusion the overall change in the substance mass in elementary volume ΔV will be as follows:

$$\Delta M = \Delta M_{r+\Delta r} + \Delta M_r = \frac{DS\Delta t}{\Delta r}[-\Delta c_i(r,t) + \Delta c_i(r+\Delta r,t)],$$

and the change in concentration c_i will be

$$\Delta c_i = \frac{\Delta M}{\Delta V} = \frac{\Delta M}{S\Delta r} = \frac{1}{\Delta r}\left[\frac{\Delta c_i(r+\Delta r,t)}{\Delta r} - \frac{\Delta c_i(r,t)}{\Delta r}\right] D_i\Delta t.$$

Passing to limit at $\Delta r \to 0$, we get

$$\Delta c_i = D_i \Delta t \frac{\partial^2 c_i(r,t)}{\partial r^2}.$$

Hence, it is easy to determine the rate of changes of concentration Δc_i in volume ΔV due to diffusion processes. At $\Delta t \to 0$, we have

$$\frac{\partial c_i}{\partial t} = D_i \frac{\partial^2 c_i(r,t)}{\partial r^2}.$$

So, if we assume that the rates of chemical reactions do not depend explicitly on the coordinate and time and the system has no "hydrodynamic flows" — systematic nonrandom motions of reacting substances — the distributed reaction–diffusion system in the one-dimensional case is described by differential equations with partial derivatives

$$\frac{\partial c_i}{\partial t} = f_i(c_1, c_2, \ldots, c_n) + D_i \frac{\partial^2 c_i(r,t)}{\partial r^2} \qquad (i = 1, 2, \ldots, n), \tag{4.1}$$

where $f_i(c_1, \ldots, c_n)$ corresponds to the point model and describes the concentration change c_i in every space point due to chemical conversions, and $D_i \dfrac{\partial^2 c_i(r,t)}{\partial r^2}$ corresponds to transfer processes between points neighboring in space.

Distributed system (4.1) may be reduced to the point system

$$dc_i/dt = f_i(c_1, c_2, \ldots, c_n) \qquad (i = 1, \ldots, n)$$

in the following cases. First, when all diffusion coefficients $D_i = 0$. In this case, molecules will not collide at all, and no chemical reaction is possible. Second, if diffusion coefficients are very high ($D_i \to \infty$), diffusion rates will be large as compared to the rates of chemical reactions. In this case, after some induction period the solution of system (4.1) will be close to the solution of the point system

$$\bar{c}_i = \bar{f}_i(c_1, c_2, \ldots, c_n) \qquad (i = 1, \ldots, n),$$

where $\bar{f}_i$ is volume-averaged rates. Finally, if external conditions and initial concentrations are uniform through the whole volume, $\varphi_i(r) = \varphi_i = \text{const}$, it is sufficient to analyze the point system.

Solution of Linear Diffusion Equations. At present, only with the aid of computers it is possible to find solutions of boundary-value problems for the system of nonlinear equations (4.1). Let us consider the basic properties of analytical solutions developed for linear equations

$$\frac{\partial c_i(r,t)}{\partial t} = f_i(c_1, c_2, \ldots, c_n) + D_i \frac{\partial^2 c_i(r)}{\partial r^2}, \tag{4.2}$$

containing no nonlinear functions relative to concentrations c_i.

Equation (4.1) is heterogeneous because its right-hand side contains "point", chemical $f_i(c_1, c_2, \ldots, c_n)$ and diffusion $D_i \dfrac{\partial^2 c_i(r)}{\partial r^2}$ members. The boundary-value problem of Eq. (4.1) is solved in several stages. We will restrict ourselves to one variable $c(r,t)$. First the solution of the homogeneous equation is found

$$\frac{\partial c(t,r)}{\partial t} = D \frac{\partial^2 c(r)}{\partial r^2} \tag{4.3}$$

without point members. This equation has only a diffusion part with zero boundary conditions (substance absorbance at the boundary)

$$c(0,t) = 0, \quad c(l,t) = 0$$

and initial conditions at $t = 0$

$$c(r,0) = \varphi(r).$$

Without discussing mathematical computations, let us note that here the method of separation of variables r and t is used, when the solution $c_i(r,t)$ is as follows

$$c(r,t) = R(r)T(t).$$

Here $R(r)$ is the function of only spatial variable r, and $T(t)$ is the function of only time variable t. It was found that the $R(r)$ function reflecting the character of spatial "irregularity" of the concentration distribution $c_i(r)$ has the shape of a sinusoid dependent on the reactor sizes l

$$R_n(r) = B_n \sin \frac{n\pi}{l} r,$$

where n is any integer, B_n is the coefficient, which will be accepted to be equal to unity (Fig. 4.2). The time character of the solution is assigned by the $T(t)$ function, also dependent on l and n:

$$T_n(t) = A_n \exp\left[-D\left(\frac{n\pi}{l}\right)^2 t\right].$$

Value $\sqrt{\lambda} = n\pi/l$ is usually called the wave number and designated by letter K:

$$K_n = \sqrt{\lambda_n} = n\pi/l.$$

Thus,

$$c_n(r,t) = R_n(r)T_n(t) = A_n \exp(-D\lambda_n t)\sin\sqrt{\lambda_n} r,$$

or

$$c_n(r,t) = A_n \exp(-DK_n^2 t)\sin K_n r, \tag{4.4}$$

are partial solutions of homogeneous equation (4.3) which satisfy zero boundary conditions.

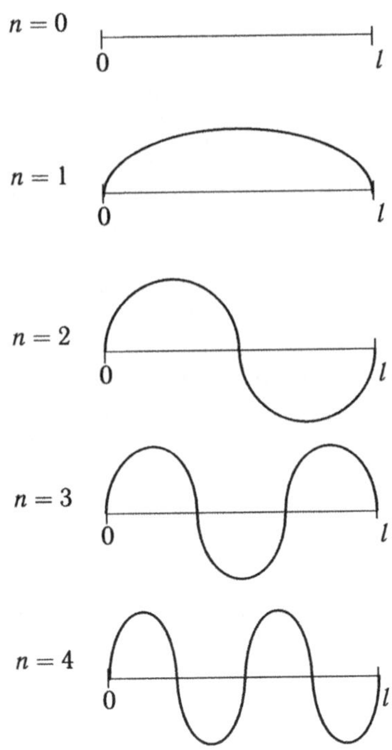

Figure 4.2. Types of function $\sin \frac{n\pi r}{l}$ at different n values.

It is seen that solutions of Eq. (4.4) represent dying with time sinusoid spatial distributions of concentration c — a set of n harmonics.

Obviously, the expression with the sign 'sin' is the product of the wave number $K_n = \sqrt{\lambda_n} = n\pi/l$ and the coordinate r. This means that the wave number K_n determines the frequency of fluctuations of variable c in space, or, which is the same, value $\Lambda_n = 2\pi/K_n$ is the "period" of fluctuations of c in space r. The larger the number of harmonic n taken from the common set of n harmonics — solutions, the smaller the period of sinusoid $\Lambda_n = 2\pi/K_n = 2l/n$ in space, but the larger also the attenuation rate of this sinusoid in time due to the $\exp[-(n\pi/l)^2 t]$ factor. □

So, the general solution is a superposition of partial solutions: (1) rapidly damping with time and spatially "high-frequency" ones, and (2) slowly dying with time and spatially "low-frequency" sinusoids. It is clear that at $t \to \infty$ the sinusoids in Fig. 4.2 will "die" and the distribution in space will become homogeneous. Under boundary conditions of impermeability of the termini of a one-dimensional reactor

$$\frac{\partial c}{\partial r}(0, t) = \frac{\partial c}{\partial r}(l, t) = 0$$

(absorbtion at the boundary), the irregularity in space is determined by the cosine

$$R_n = B_n \cos\left(\frac{n\pi}{l}\right) r.$$

In nonhomogeneous system (4.2), distribution of substance sources is determined by point chemical members $f(c_1, c_2, \ldots, c_n)$, which function continuously, and not only at the start time. Namely this means that here we have an active distributed system. Subject to the type of "point" functions $f(c_1, c_2, \ldots, c_n)$ different AWP are observed in active kinetic environments. But only in some cases important for biology it is possible to perform a qualitative study under most general assumptions. Thus, the problem of propagation of a concentration wave, occurring, for example, in ecology upon spread of species, was analyzed. The same phenomenon is observed in processes of information transfer in biological systems by means of concentration wave motion.

Stability of Stationary Regimes. The same as in the case of models in common differential equations, of special interest is the analysis of stationary regimes formed in the system with time. Stationary solutions are found from conditions when time derivatives are equal to zero:

$$D_i \frac{\partial^2 c_i(r)}{\partial r^2} + f_i(c_1, c_2, \ldots, c_n) = 0. \tag{4.5}$$

As a rule, investigations of stability of a stationary solution can be performed based on the analysis of the system of equations, linearized near the singular point. Let us consider the problem of stability of solutions homogeneous in space, limiting ourselves to an equation with one variable and taking, for short, that $D = 1$:

$$\frac{\partial c(t, r)}{\partial t} = f(c) + \frac{\partial^2 c(t, r)}{\partial r^2}. \tag{4.6}$$

Substituting the "excited" solution $c(t, r) = \bar{c}(r) + \xi(t, r)$ in the above equation and taking into account that $\bar{c}(r)$ satisfies the ratio $f(\bar{c}) + \partial^2 \bar{c}/\partial r^2 = 0$, the following equation for perturbations is obtained:

$$\frac{\partial \xi(t,r)}{\partial t} = Q\xi(t,r) + \frac{\partial^2 \xi(t,r)}{\partial r^2}, \tag{4.7}$$

where $Q = f_c'(\bar{c})$.

Solution of $\xi_n(t,r)$ has a simple form (cf. 4.4):

$$\xi_n(t,r) = a_n(t)\cos[(n\pi/l)r]. \tag{4.8}$$

Here we have to determine coefficients $a_n(t)$, as it was done in Eq. (4.4). By substituting Eq. (4.8) in Eq. (4.7), we will find that the coefficient

$$a_n(t) = \exp\left\{[-(n\pi/l)^2 + Q]t\right\} \tag{4.9}$$

determines the time character of changes in the initial perturbation $\xi(0,r)$.

As a result, we can write the expression for solution of Eq. (4.7) as

$$\xi(t,r) = \sum_{n=0}^{\infty} a_n l \exp\left\{[Q - (n\pi/l)^2]t\right\} \cos(n\pi/l)r. \tag{4.10}$$

Here values $\exp\{[Q - (n\pi/l)^2]t\}$ determine the time character of damping or growth of initial perturbation $\xi(0,r)$, and multipliers $\cos[(n\pi/l)r]$ determine the distribution of the initial deviation along coordinate r.

If in Eqs. (4.8)–(4.10) $Q = f_c'(c) < 0$, at any $n = 0, 1, 2, \ldots$ the perturbation will be damping, i.e. function $\xi(t,r) \to \infty$ at $t \to \infty$, whatever the initial spatial deviation of $\xi(0,r)$ from the stationary homogeneous distribution of concentration $\bar{c}(r)$ is. In this case, any small perturbation of spatially homogeneous stationary solution damps with time. If $Q = 0$, the exponential index is negative at all n values except for $n = 0$. In such systems all harmonics $\cos(n\pi r/l)$ of the initial perturbation for $n = 0, 1, 2, 3, \ldots$ (for the zero harmonic $n = 0$, the linear approximation does not yield any answer) will damp. If $Q > 0$, there is a finite number of harmonics of the $\cos(n\pi r/l)$ type, which lead to development of perturbation of a stationary solution initially homogeneous in space. It can be seen that these are only the harmonics in which exponential indices (4.9) are positive, i.e.

$$(n\pi/l)^2 < Q.$$

If the initial perturbation $\xi(0,r)$ has no such harmonics, it will disappear with time. During development, continuous harmonics can take the system away from the initially homogeneous stationary distribution of the substance and switch it to another regime, which includes generation of autowave processes or spatially nonhomogeneous dissipative structures.

Studies of stability of spatially nonhomogeneous solutions are of a more complex nature.

Basic Models. The main results, describing evolution of distributed systems in time and space, were obtained on the so-called basic models, where interactions of two variables x and y in one-dimensional space (spatial coordinate r) are considered. Such a system has the following general form

$$\frac{\partial x}{\partial t} = P(x,y) + D_x \frac{\partial^2 x}{\partial r^2}, \qquad \frac{\partial y}{\partial t} = Q(x,y) + D_y \frac{\partial^2 y}{\partial r^2}. \tag{4.11}$$

It was found that with a certain type of functions P and Q, a comparatively simple model (4.11) can describe qualitatively spontaneous generation of waves and structures as self-organization processes in living systems.

Self-organization processes are related to phenomena where, due to instabilities the systems, lose their initial organization in space and time and instead of it acquire a new one. For example, in a chemical reaction system, under definite conditions the stationary spatially homogeneous distribution of reagents may become unstable and will be substituted by autooscillations, i.e. stable self-sustained periodic oscillations or waves of chemical activity. □

The Brusselator. The behavior of distributed systems (4.11) of two equations may be extremely diverse and can demonstrate basic dynamic properties, such as propagating excitation in the form of travelling pulses, standing waves, quasistochastic waves, and dissipative structures. Studies of basic models are aimed at clarifying bifurcation values and ratios between model parameters, determining one or another type of its behavior. Like in the case of point systems (Chapter 1), here special attention is focused on the analysis of the stationary state stability. We will omit its details and indicate only key directions of the analysis and its probable results.

Let in a closed vessel — a tube of length L — a chain of chemical conversions of substances A, B, X, Y, C, and R takes place at a constant temperature following the hypothetical scheme:

$$A \to X, \quad 2X + Y \to 3X, \quad B + X \to Y + C, \quad X \to R. \tag{4.12}$$

Suppose that initial substances A and B are uniformly distributed in the tube and their reserve is large, so that they are not expended significantly. Substances R and C are precipitated. Substances X (x) and Y (y) may diffuse along the tube, their concentration changes during the chemical conversions, and they represent variable systems. The concentrations of A and B and the diffusion coefficients D_x and D_y are constant parameters of the system. The butt ends of the tube, the same as its walls, are supposed to be impermeable for the substances involved in the reaction. Given these assumptions, the model of the system looks like a system of the following equations in partial derivatives:

$$\frac{\partial x}{\partial t} = A + x^2 y - (B+1)x + D_x \frac{\partial^2 x}{\partial r^2}, \quad \frac{\partial y}{\partial t} = Bx - x^2 y + D_y \frac{\partial^2 y}{\partial r^2}. \tag{4.13}$$

The first step in examining the above system is determination of its solution stationary in time and homogeneous in space. To find this solution, let us consider the point system, which has no diffusion members, and equate the time derivatives to zero. The only stationary solution will be obtained:

$$\overline{x} = A, \quad \overline{y} = B/A. \tag{4.14}$$

Phase portraits of the point system are given in Fig. 4.3.

At different values of wave number $k = \pi n/\Lambda$, instability in the system will take place at different values of B.

In the definite range of negative values of the wave number, there may be spatially periodic and time-independent solutions corresponding to the emergence of dissipative structures. The emergence of these structures takes place when the coefficients D_x

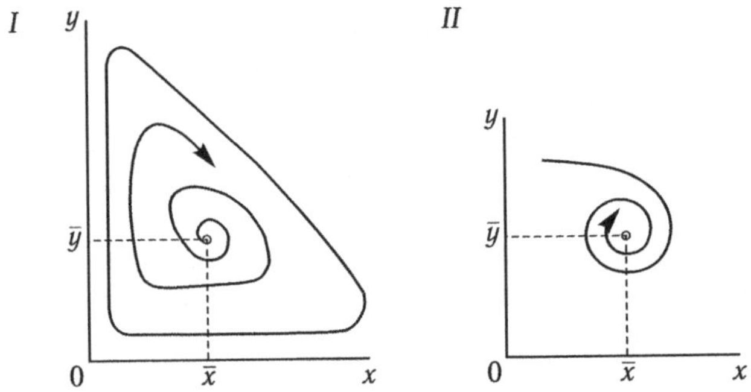

Figure 4.3. Phase portraits of the "brusselator" point system. (*I*) $B > (1 + A^2)$; (*II*) $B < (1 + A^2)$.
Studies of the type of stability of the singular point (4.14) showed that it is the stable focus at $B < 1 + A^2$ (*II*) and the unstable focus at $B > 1 + A^2$ (*I*). In this case the point system has a stable limit cycle, i.e. the point system is an autooscillation one.

and D_y differ greatly, and the parameters A and B are not too far from their bifurcation values. If diffusion rates are very high, instability arises at high $\lambda_c^2 = (1/A)(D_x D_y)^{1/2}$ values, and hence the system remains practically homogeneous. If in the brusselator point system, perturbations grow in a vibrational manner and the increase in the amplitude of these vibrations is restricted by the limit cycle, the distributed system has the vibrational type instability. In this case, the increase of perturbations in the distributed system (4.13) may also lead to stable in time and spatially nonhomogeneous distribution of x and y concentrations. As in the case of the saddle type instability, the perturation growth is restricted by dissipative processes in the system, described by nonlinear members in equations of chemical reactions (hence the name "dissipative structures"). In addition, autowave processes like standing and travelling waves can emerge in such systems.

Different autowave processes include sources of spontaneous waves ("driving centers"), spiral waves (reverberators), quasistochastic waves and other phenomena.

Investigation of the self-organization principles (synergetic mechanisms on elementary models) is of fundamental and applied importance. The development of the theory of autowave processes is crucial for understanding the dynamics and regulation of such cardinal phenomena in living organisms as processes of tissue differentiation and morphogenesis, and also nerve conduction of the cardiac muscle.

Wave processes can be observed clearly in the process of the Belousov–Zhabotinsky reaction in two-dimensional reactors — small flat-bottom vessels. This is a redox reaction with the involvement of brommalonic acid, cerium and manganese ions being the catalysts. Changes in the valence of these ions result in local changes in the color of the reactor solution. The mechanisms of emergence and development of autowave reactions are described by equations similar to Eq. (4.13). To this effect, the brusselator is the "basic" model explaining the main qualitative peculiarities of processes occurring in the system, the same as the Volterra model is basic in mathematical ecology.

The Belousov–Zhabotinsky Reaction. In 1959, B. P. Belousov observed periodic (the frequency of about 10^{-2} Hz) changes in the color of solution in oxidation reaction of citric acid with bromate; cerium ions were the catalyst. Changes in the color were determined by periodic turnovers $Ce^{3+} \rightleftarrows Ce^{4+}$. In 1964–1974, A. M. Zhabotinsky and coworkers carried out detailed experimental and theoretical studies of this and similar phenomena.

The reaction, called the Belousov–Zhabotinsky reaction, is a redox one. As an example let us describe the reaction in which bromate (BrO_3^- ion) is the oxidizer, malonic acid (MK, $H_2C(COOH)_2$) is the reducing agent, and cerium ions are catalysts. Changes in the concentration of Ce^{4+} define relaxation oscillations in the solution color (Fig. 4.4). Their period τ is divided in two phases: τ_1 and τ_2 denoting the rise and decrease of $[Ce^{4+}]$. A simplified scheme of the reaction has two corresponding stages:

$$Ce^{3+} \xrightarrow{BrO_3^-,\ H^+} Ce^{4+} \qquad \text{(I, oxidation)},$$

$$Ce^{4+} \xrightarrow{MK} Ce^{3+} \qquad \text{(II, reduction)}.$$

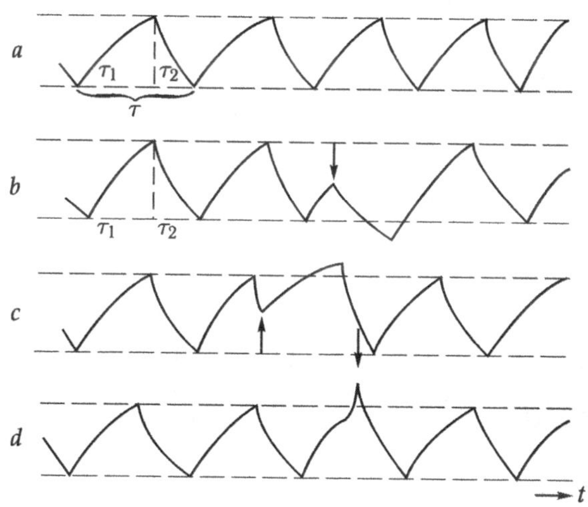

Figure 4.4. Autooscillation concentrations of Ce^{4+} (*a*). Phase jumps during oscillations in Ce^{4+} caused by single additions of Br^- (*b*), Ag^+ (*c*), and Ce^{4+} (*d*). Arrows indicate the moments at which additions were made.

The products of bromate reduction, formed at stage I, brominate malonic acid. As a result, brommalonic acid (BMA) is obtained, which is then destroyed with the release of Br^-. The bromide ion strongly inhibits the reaction.

The vibration process occurs as follows. The system has some initial amount of Ce^{4+}. At stage II, Br^- is generated which inhibits reaction I and at some rate disappears from the system. The concentration of Ce^{4+} decreases as a result of both the reduction of reaction II and slowing-down of reaction I with bromide. When the concentration of Ce^{4+}, expended during reaction II, decreases to the bottom threshold value, the concentration of Br^- also decreases. Reaction I starts anew, and the concentration of Ce^{4+} grows. With a decrease in the top threshold value of Ce^{4+}, Br^- increases drastically, which results in slowing down of reaction I. Then the cycle is repeated. □

A. M. Zhabotinsky and A. N. Zaikin observed and studied autowave processes in the described distributed chemical system in the absence of convection but in the presence of diffusion. This is realized in thin tubes (a one-dimensional system) or in thin layers (a two-dimensional system). As a catalyst a ferroin complex was used instead of Ce^{4+}. Excitation of the system is done by touching the solution surface with a needle, wetted in solution of $AgNO_3$ (Ag^+ binds Br^-) or a heated piece of wire. Waves are generated traveling at a rate of about 0.01 cm/s. The experiments allowed discovering point sources of autogeneration — driving centers (DC) emerging in the homogeneous environment as a result of local fluctuations in concentrations, i.e. special initial conditions. Concentric waves propagate from DC at a constant rate. Figure 4.5 shows DC photographed in the above system in every 30 seconds.

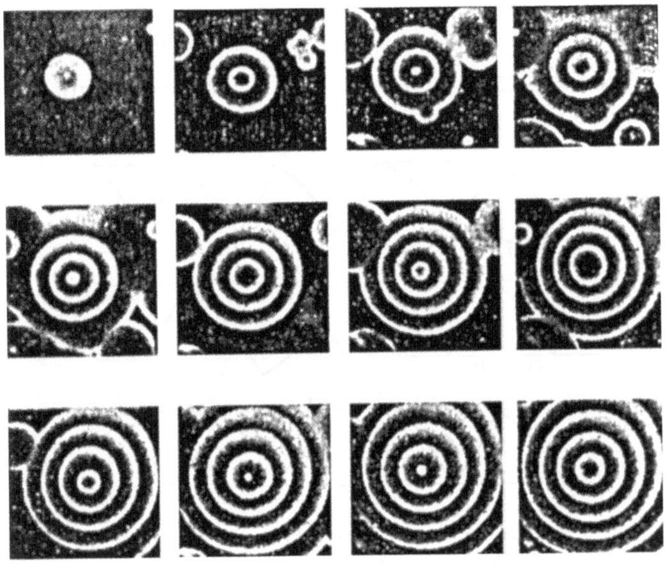

Figure 4.5. Driving centers.

Upon rupture of the wave front, reverberator (a spiral wave) may be generated. In particular, reverberator is formed upon wave movement in the two-dimensional environment around an opening — a spiral is the evolvent of an opening.

Non-Turing Systems. The main idea of A. M. Turing is that dissipative stationary structures, formed by diffused chemical substances interacting with each other, underlie the processes of self-organization of biological processes. Turing dissipative structures exist under conditions of continuous substance and energy fluxes, and the structure itself is formed at a bifurcation parameter value as a result of the loss of stability in the initial spatial homogeneous distribution of chemical agents.

At the same time in many biological processes, new structures are formed. They retain their stability for a rather long time even after termination of the substance flux upon exchange with the environment.

For example, the emergence of structures in bacterial communities in tissues upon ontogenesis can be attributed to such most clearly expressed shaping processes in biology.

These systems are called non-Turing systems. In Turing models, self-organization emerged at definite bifurcational parameter values of order in the initial homogeneous system. In non-Turing systems, peculiarities of the transition process, leading to self-organization, are explained by definite elements of organization of the initial state. For example, the formation of ring structures by mobile bacteria is conditioned by a delay in the response to the change in the concentration of the product released during metabolic processes. The ring structure is formed not at any rate of movement of the cell division front, but when it is rather low and is lower than the specific rate of product decomposition. □

There are cases when bacterial structures are formed by cells which release attractants to the environment. Cyclic AMP can play the role of an attractant. A component, directed along the concentration gradient of the attractant, appears in the chaotic motion of bacteria, the local decrease in this attractant being specified by the effect of chemotaxis — consumption of the attractant by bacteria.

Population waves of *E. coli* cells are an example of formation of stable structures. After point inoculation in a nutrient medium, bacterial cells of *E. coli* start dividing violently and thus form expanding wave fronts. At fluctuation of the density of bacterial cells in the direct vicinity of the expanding population front, these fluctuations can give rise to microsurges surpassing the basic front. At high rates of movement of the basic front (> 4 mm/h), such surges are absorbed by this front, and the propagating wave preserves the concentric form (Fig. 4.6, *a*). But at low rates of movement of the bacterial front, the generated surges can form a secondary population wave (protuberance) running ahead of the basic wave (Fig. 4.6, *b*, *c*). If new microsurges are in their turn formed on this protuberance, the wave structure becomes branched (Fig. 4.6, *d*, *e*). Given multiple sequential surges are formed, fractal structures are generated (Fig. 4.6, *f*).

Models of Processes in Cell Membrane of *Chara coralline* Alga. Nonlinear phenomena on different spatial-time scales play a considerable regulatory role

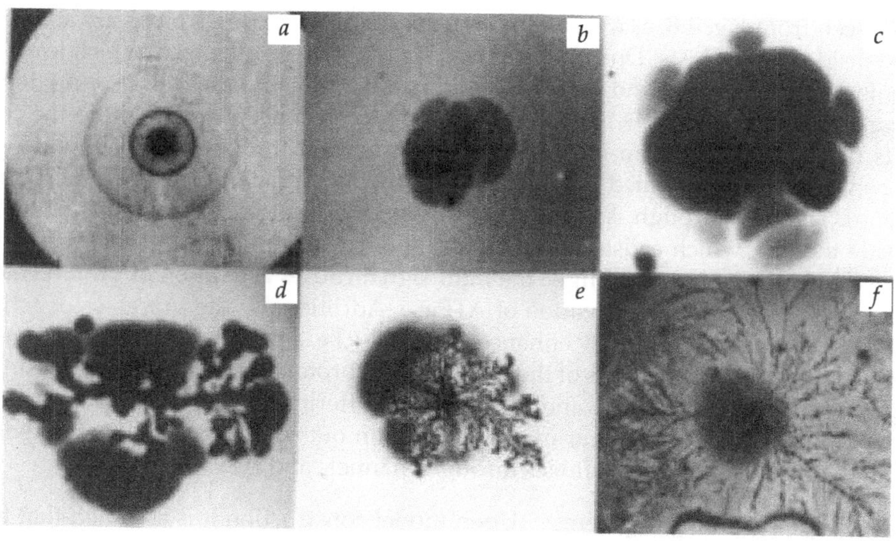

Figure 4.6. Typical structures formed upon growth of *E. coli* bacteria populations in a nutrient medium.

a, Symmetric concentric waves; *b* and *c*, symmetry breaking in form of multiple surges; *d* and *e*, fractal-like structures; *f*, fractal structure (reproduced from M. A. Tsyganov et al., 1999).

in processes of metabolism and living cell growth. Cells of *Nitella* and *Chara corallina* alga are unique model objects for studying nonlinear phenomena in biological systems. These cells have large dimensions (0.6–1.0 mm in diameter and 40–80 mm in length) and their structure is much simpler as compared to that of cells of higher plants or animals. During 10–40 min in the light, along the cell cytoplasmic membrane there emerge stable areas with different values of pH and electric potential, with currents of about 1 μA circulating between them (Fig. 4.7). Under illumination the pH value in the "alkaline" area grows to 9.5–10.0 from the initial pH = 7.0–8.0 in the dark, whereas in the "acidic" area pH decreases to 6.0–6.5. As a result, the difference in pH values of the formed structures may be 2.5–3.0.

Figure 4.7. Distribution of pH ranges along the cytoplasmic membrane of *Chara corallina* cells.

Local changes of transmembrane potential and pH values near the membrane may have a vibrational character; both spontaneous continuous oscillations of the transmembrane potential and damping oscillations can be generated under illumination of the cell. Oscillations can be associated with the areas, but can also appear independent of their formation.

Plyusnina et al. (2005) proposed the following scheme of processes leading to generation of structures of the potential and pH along the cytoplasmic membrane of the *Chara* alga (Fig. 4.8).

As seen from Fig. 4.8, as a result of photosynthetic processes protons are accumulated inside chloroplasts. Due to this the cytoplasm may be alkalized, which leads to opening of the channels and increase of the passive flux into the cell, causing depolarization of the membrane and growth of the proton concentration within the cell. In its turn, this leads to an increase in the proton flux through ATPase, which on the contrary is accompanied by a decrease of the membrane potential (hyperpolarization). The flux through ATPase leads to the growth of the proton concentration outside the cell, which causes its inhibition and activation of proton channels, and the cycle is repeated. Accordingly, the light signal causes activation of proton channels, which then leads to activation of ATPase. Additionally the processes of active and passive transport mutually enhance each other, i.e. a positive feedback is formed between them, and oscillations of the potential and protons are realized in the system. Local regions of acidification and alkalization, differing in the value of the transmembrane potential in the near-membrane region outside the cell, can emerge due to differently oriented proton fluxes through channels and ATPase.

Oscillations in Local Systems. Upon model construction it is assumed that the key process, determining the nonlinear character of changes in the membrane potential, is transfer of ions, particularly of protons, across the membrane. To describe the transmembrane transfer of protons in a model, data on the catalytic cycle of ATPase functioning and ion-conducting channels are used. The distribution of the potential along the membrane is described based on the cable properties of the cell membrane.

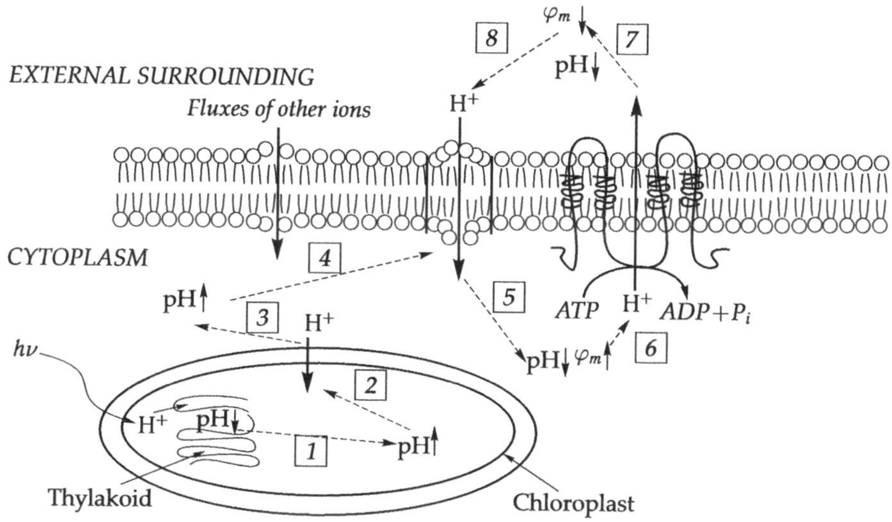

Figure 4.8. Scheme of processes near the cell membrane of *Chara coralline* alga. Dotted arrows show pathways of regulation of pH and φ_m (the membrane potential).

(1) Illumination initiates processes of photosynthesis, pH of the thylakoid decreases, pH of the chloroplast increases. (2) Increase of pH inside chloroplasts induces the proton flux from the cytoplasm to chloroplasts. (3) The proton flux from the cytoplasm to chloroplasts increases pH of the cytoplasm (pH $\uparrow$). (4) Activation of proton channels in the cytoplasmic membrane. (5) Increase of the proton flux through channels leads to decrease of pH (pH $\downarrow$) of the cytoplasm and depolarization of the membrane potential (φ_m $\uparrow$). (6) Activation of proton ATPase. (7) Increase of the protein flux through ATPase leads to decrease of pH (pH $\downarrow$) outside the cell and hyperpolarization of the membrane potential (φ_m $\downarrow$). (8) Activation of proton channels. The cycle is repeated again (from stage 5).

The model permits estimating the contribution of the transmembrane potential and ionic fluxes to the observed nonlinear regimes. Fisah et al. (1989) observed damping and continuous oscillations of the membrane potential and pH near the surface of the *Nitella* cells under illumination conditions. Damping oscillations of the membrane potential with a period of 2 min were observed upon substitution of darkness conditions for illumination ones, and continuous oscillations with a period of 15 min were generated spontaneously under illumination. A linear analysis of the local model allowed finding parameter ranges in which damping and continuous oscillating regimes can exist. Depending on initial conditions y at the same set of parameters, there can exist damping oscillations and the limit cycle of the large amplitude, which corresponds to continuous oscillations of the potential and proton concentration.

Parameters at which Turing instability emerges in the system were determined. It was demonstrated that at some critical value of parameters dissipative structures emerge in the system, which correspond to ranges with different pH in a real system (Fig. 4.9, *a* and *b*). Upon the decrease of parameters corresponding to growth of pH in the cell at stage 3 (Fig. 4.8) with increased illumination intensity, the amplitude of the structures grows. This result agrees with the experimental data on the growth of the amplitude of the structures along the cytoplasmic membrane at increased illumination intensity (Bulychev et al., 2001). Thus, the model describes both oscillating regimes and emergence of spatial inhomogeneity of pH discovered in the experiments.

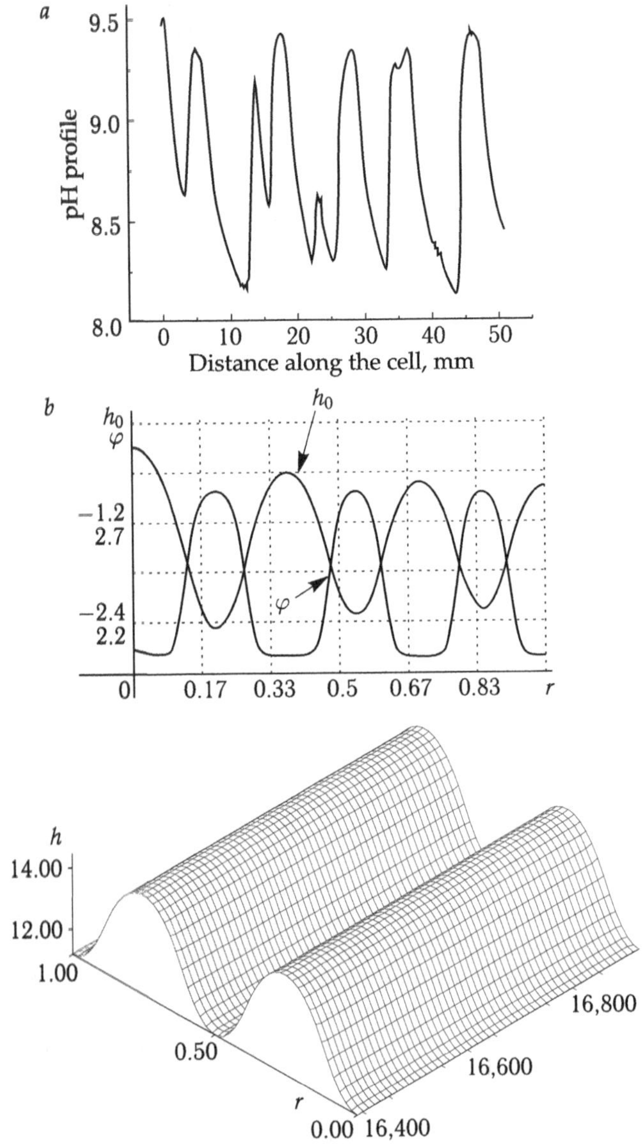

Figure 4.9. Acidic and alkaline areas along the cell membrane of the *Chara coralline* alga. *a*, Experiment (reproduced from A. A. Bulychev et al., 2002); *b*, model.

4.3 Chaotic Processes in Determined Systems

In the systems considered above, the determined character of their behavior is that small changes of initial conditions lead to small changes in the final result. Generally speaking, this is also valid for the behavior of systems in bifurcational points. Insignificant perturbations "push" the system to one of the *a priori* possible regimes. In such cases (trigger switching, stable cycle near the unstable focus), corresponding phase trajectories have a quite definite character, and only the "push" itself of the system to one of the regimes is accidental.

However, lately it has become clear that a rather common feature of many nonlinear determined systems is their chaotic behavior in time. It was discovered that under definite critical values of internal parameters chaos emerges in many biological, meteorological and economic nonlinear models with a number of degrees of freedom exceeding two. This means that during rather large time periods the behavior of these systems becomes unpredictable, and the systems themselves turn out to be in the state of chaos and irregularity. It should be specially underlined that in this case chaos emerges as a result of internal dynamics of the system and not due to some external sources or chaotic effects of the environment. At present it becomes clear that chaotic behavior is inherent to a wide class of determined systems and plays an important role in self-organization processes in nature. □

Models of Chaotic Systems. Let us give a brief description of a determined system model with chaotic behavior and discuss their possible significance in biology. As far back as at the end of the nineteenth century, French scientist Jules Henri Poincaré discovered that some mechanical systems governed by equations of classical mechanics demonstrate chaotic behavior. For example, a mechanical pendulum periodically forced by the external force $F_0 \cos \omega t$ has chaotic dependence of angle θ on time (Fig. 4.10), if the amplitude of the driving force exceeds some critical value F_c.

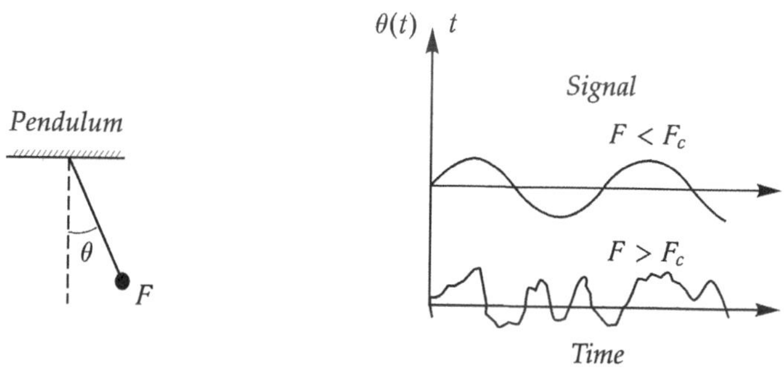

Figure 4.10. A periodically forced pendulum. Dependence of the driving force F and angle θ versus time (reproduced from H. G. Schuster, 1984).

In other experiments, the subject of studies was the motion in layers of a liquid placed in a covered vessel heated from the bottom. When the difference of temperatures ΔT between the top cold and bottom hot layers was large, the stationary convective motion disappears and transition to chaotic motion is observed (Fig. 4.11) (Bénard instabilities). In the Belousov–Zhabotinsky reaction, the stationary spatial distribution of colored agents (cerium ions) is destroyed at definite rates of the flux of the reactive mixture through the reactor, and a chaotic regime is established in the system. All these processes are described by systems of autonomous nonlinear differential equations of the first order. Analytical studies allowed determining qualitative characteristics of chaotic motion (the amplitude of the driving force F_0, the difference of temperatures ΔT) which appears with changes in the external driving parameter. Here a number of the following questions arise. Are there any general regularities in transition of determined systems to chaotic states? Is it possible to predict the probability of chaos by the form of differential equations of a determined model? What is the role of chaos in behavior and evolution of determined systems?

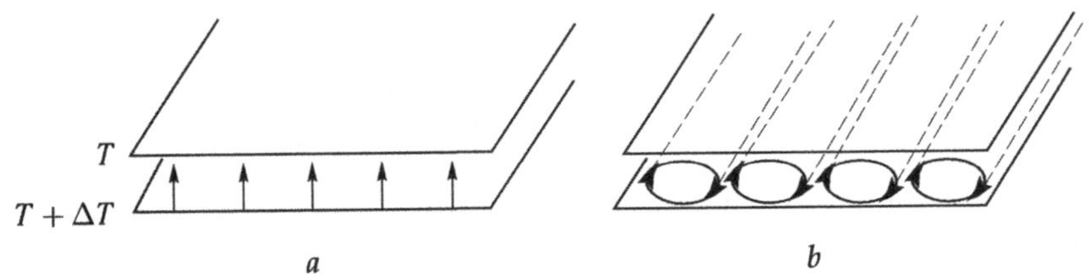

Figure 4.11. Bénard instabilities (reproduced from H. G. Schuster, 1984).
a, Heat fluxes; *b*, convection shafts formed in liquids when temperature gradient *T* exceeds a certain critical value.

The first strict determined model with chaotic behavior was evidently the system of equations proposed by Edward Lorenz (1963) in meteorology for weather forecasts. The model is based on conceptions of the relationship of air flows in the atmosphere and the difference of temperatures in its separate layers. The same approach can be used to describe the behavior of the liquid heated from the bottom in Bénard experiments (Fig. 4.11). □

The Lorentz model looks like this

$$\dot{x} = -\sigma x + \sigma y, \quad \dot{y} = rx - y - xz, \quad \dot{z} = xy - bz, \tag{4.15}$$

where σ and b are dimensionless constants, r is the controlling parameter proportional to the temperature difference. Variable x is proportional to the rate of the circulating liquid, y corresponds to the difference of temperatures between ascending and descending currents, z is proportional to the deviation of the vertical temperature profile from the equilibrium value.

In this model, the variables can have chaotic behavior upon an increase in the difference of temperatures ΔT, when the controlling parameter exceeds the critical r_c value ($r > r_c$). Figure 4.12 demonstrates time-dependent behavior of the variables. At $r < r_c$ the dependence $y(t)$ represents damping periodic motion. But given the critical boundary value $r > r_c$ is exceeded, irregular chaotic bursts emerge in the oscillations. As r grows, they become more frequent until the motion is randomized completely (Fig. 4.13). In the three-dimensional Lorenz model (4.15), the trajectory in the phase space can be calculated with the aid of computers. Figure 4.14 shows an example of such a trajectory computed at $r = 2$, $\sigma = 10$, and $b = 8/3$. As seen, the trajectory is attracted to the restricted range in the phase space. The system motion is roving, i.e. the trajectory makes irregular turns now to the right, now to the left. Small changes in the initial conditions shortly cause the new solution to deviate from the previous one and change the number of turns. Such behavior of the system is called a "strange attractor". The attractor area in the phase space is limited but may have a complex structure. The attractor *per se* is formed from the motion of one trajectory, which should pass through every point in the attractor area. But in this case, points initially arbitrarily close on the attractor, remove from each other to the final distance in a rather long period of time after the beginning of the motion. Apparently, Lorenz models can play the role of a basic model for explaining chaotization of the determined behavior of the system upon changes in controlling parameters. The possibility of emergence of chaos in biological systems can be illustrated by a known example of chaotic heartbeats at a definite frequency of stimulant pulses.

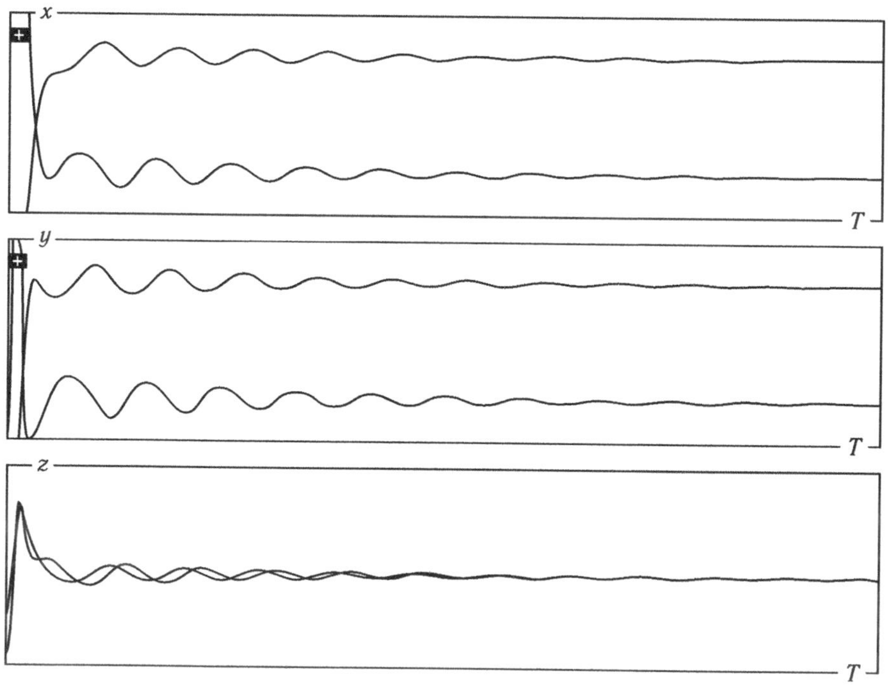

Figure 4.12. Dynamics of variables in the Lorentz model at the r value corresponding to the presence of two stationary states.

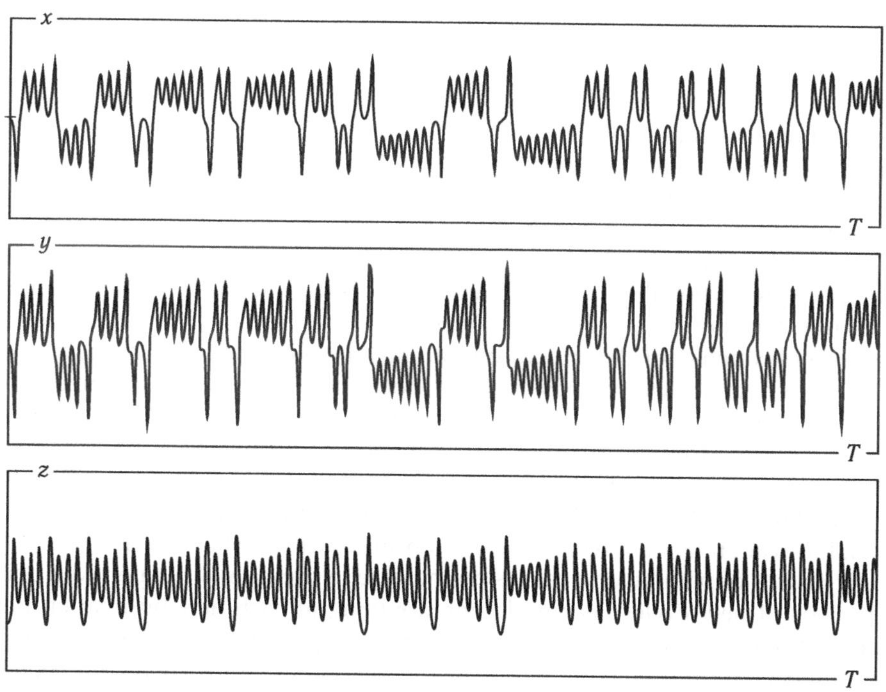

Figure 4.13. Dynamics of variables in the Lorentz model at the r value corresponding to a quasi-stochastic solution.

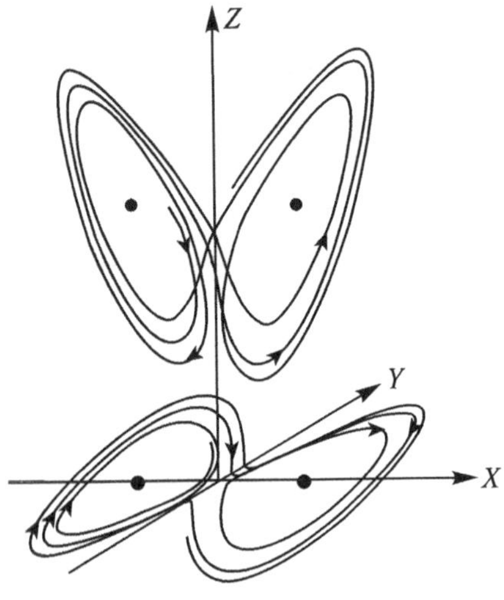

Figure 4.14. Phase trajectories in the Lorenz model. Top, trajectories projected onto the X–Z plane. Bottom, trajectories projected onto the X–Y plane. Points correspond to stationary solutions (reproduced from E. N. Lorenz, 1964).

Model of Population Dynamics. In closed environments, the model can also have chaotic properties. If the population number is not high and at the given instant of time depends on its number in the preceding instants of times, the population dynamics can be described using a discrete method with a logistic equation. In the simplest case, the population number after n sequential generations changes obeying the law

$$x_{n+1} = f(x_n) = rx_n(1 - x_n). \tag{4.16}$$

This equation is a differential equation, describing the change in the population number at discrete instants of time $0, 1, \ldots, t, t+1, t+2, \ldots$ in the form of corresponding numbers of the sequence $x_0, x_1, \ldots, x_n, x_{n+1}, x_{n+2}, \ldots$, where every x_n member is dependent on the preceding member (x_{n-1}). Expression $(1 - x_n)$ in the right-hand side of Eq. (4.16) self-restricts the population number due to the limited nature of the living space proportional to $(1 - x_n)$; parameter r is dependent on living conditions and fertility. Function $x_{n+1} = f(x_n)$, obtained at iteration of $x_1, x_2, \ldots, x_n, \ldots$, demonstrates complex behavior dependent on parameter r. Generally speaking, for $f(x_n)$ most various regimes can be observed: monotonic and oscillatory approximation to the equilibrium state, moving away from it, stable oscillations, quasistochastic behavior (chaos). In Eq. (4.16) parameter r determines the slope of the plot of the function in the right-hand side. At $r < 3$ the population number approaches the stable equilibrium state (Fig. 4.15). Given r grows and the plot of $f(x_n)$ becomes steeper, bifurcation takes place: stable equilibrium turns to stable cycles (Fig. 4.15). Further growth of parameter r causes an increase in the number of periodic points and periods of new cycles become larger, doubling at each new bifurcation point (Fig. 4.15). When r grows, bifurcations occur more frequently and the values of population numbers are repeated every $2, 4, 8, \ldots$ generations. And when r exceeds the critical value $r > r_c = 3.5699\ldots \simeq 3.570$, the solution is randomized and oscillations become completely chaotic. The value $r_c = 3.570$ is a constant characterizing the threshold of system chaotization. At the same time, even the chaotic regime has

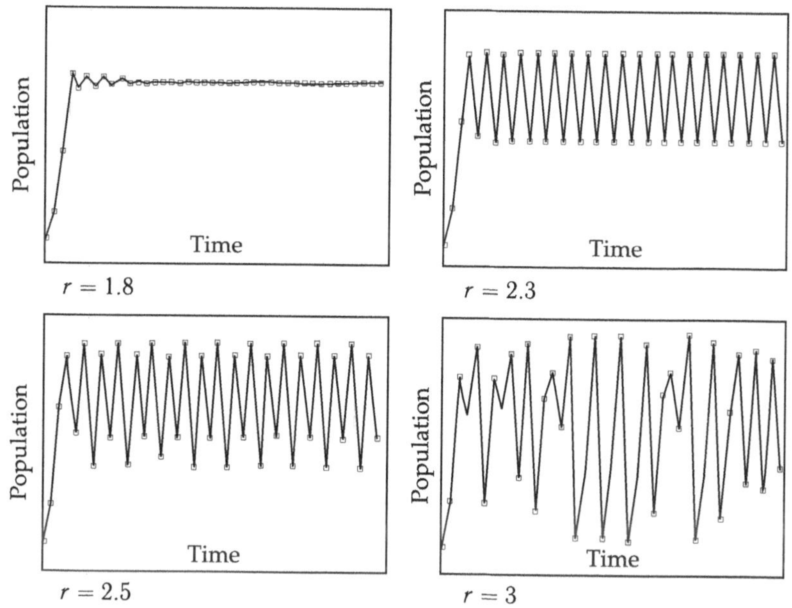

Figure 4.15. Regimes of behavior of variable x in Eq. (4.16).

"regularity windows", when upon growth of $r > r_c$ stable cycles with periods 3 or 7 unexpectedly come back. Then they once again undergo bifurcation doubling of periods $3, 6, 12, \ldots$ or $7, 14, 28, \ldots$, and chaotic regime recommences once more. Of interest is not only the chaos itself, but also the pathway or the scenario of its emergence in the determined system. As seen from Fig. 4.15, chaotic regime emerges through bifurcation with doubling of the cycles in the system. It is possible to determine ranges of the values for parameter r during its growth ($r \to r_c$), within the limits of which each period becomes stable prior to the following increase. It was found that sequential values of r, at which the number of stable periodic points is doubled and becomes equal to 2^n, change obeying the law

$$r_n = r_c - \text{const}\delta^{-n} \qquad \text{at } n \gg 1,$$

where $\delta = 4.669 \ldots \simeq 4.670$ is the Feigenbaum constant. It equals the ratio of sequential ranges in the values of parameters r within the limits of which each cycle retains its stability:

$$\delta_n = \frac{r_n - r_{n-1}}{r_{n+1} - r_n}. \tag{4.17}$$

At $n \to \infty$ $\delta_n \to \delta = 4.670$. The r value corresponds to the n-th bifurcation, where period 2^n loses stability and period 2^{n+1} acquires it. This relationship shows that the lengths of the ranges decrease in the row of points $r_{n-1} \to r_n \to r_{n+1}$, i.e. doublings occur more frequent at growth of $r \to r_c$. It was found that the constant δ has a universal character inherent to many other natural systems, where the cycle period is doubled prior to the emergence of chaos.

General Properties of Deterministic Chaos. In addition to the above described systems, there are a variety of other systems displaying chaotic behavior. The dynamical chaos is understood as irregular chaotic motion in nonlinear systems, for which the general system evolution in time nevertheless obeys the dynamic laws. The chaotic behavior observed with time emerges not as a result of external noise sources, infinite number of degrees of freedom (in model (4.15) they are only three), or quantum mechanical uncertainty (see (8.4)). It is determined by the capability of nonlinear systems with time to "diverge from each other" exponentially rapidly the trajectories which were initially close to each other within the limited volume of the phase space. Since in the reality the initial conditions may be set only with the limited precision, the initial error grows exponentially in time so that the real trajectory becomes unpredictable notwithstanding that definite dynamic laws (kinetics equations) control the system behavior. Hence we have the term "deterministic chaos".

Moreover, it has been demonstrated that even in classical mechanics, stable regular motion is rather an exception. At specific parameter values, mechanical motion in the phase space is easily randomized, each time the definite shape of the trajectory being dependent on the starting conditions. A mechanical system of this type is Sinai billiard where the inside table walls are convex so that at low initial deviations of the incident angles of seemingly the same starting trajectories, the consecutive bouncing of the ball from the walls leads to exponential "scattering" of the trajectories and, as a result, to their unpredictability at large time periods.

Let us consider the initial element of the phase volume in the form of a sphere of radius ε in which trajectories originating from it are placed. It can be accepted that these trajectories deviate from each other at the starting instant of time within the sphere's radius ε.

We have seen above that stability of the stationary state of kinetic systems is determined by the sign of Lyapunov index (Chapter 1). For any motion that is stable according to Lyapunov, a small starting deviation from the initial conditions does not increase but decrease with time. In other words, in systems stable by Lyapunov's definition, these starting deviations will damp with time, and the sphere's radius ε initially containing all starting points of the trajectories, will decrease and in the limit will decrease to zero upon transfer to the limit stable multitude at $t \to \infty$. All trajectories in stable systems are drawn to the common point (node, focus) or "are wound" on a line, i.e. the trajectory of the limit cycle.

We will examine how the starting phase volume of the autooscillatory system evolves with time upon the system transfer to stable limit multitude — the limit cycle trajectory.

Figure 4.16 shows that with time compression of the starting volume and changing in the position on the phase plane corresponding to this volume upon its decrease occur concurrently. In the limit the points of all trajectories assemble in one line (the stable limit cycle) and then at any deviation "wind up" on it. In other words, when all trajectories of the system in the limit lie on the limit cycle, the initial element of the phase volume, that contained starting trajectories, is shrinking to zero.

In systems unstable according to Lyapunov, positive values $\lambda > 0$ are present in the range of characteristic indices. This means that the system has instability at least in some trajectories, upon the motion along which the starting small perturbation $D(0)$ of the regime exponentially increases in time

$$D(t) = D(0)\exp(\lambda t), \tag{4.18}$$

and the trajectories move away from each other. In unstable nonlinear systems, the starting phase volume may increase infinitely, generally speaking, occupy in the limit the total phase space.

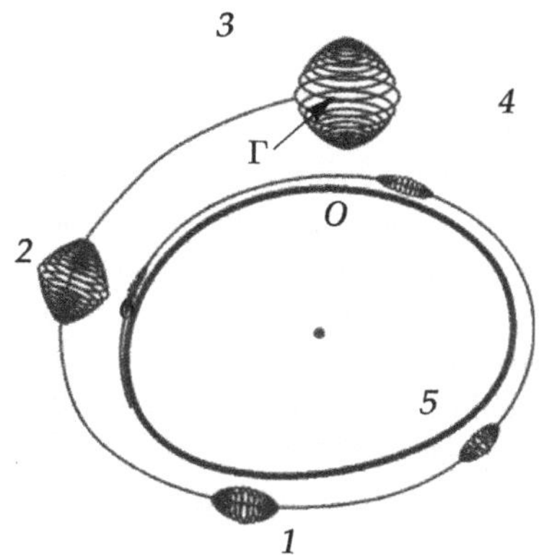

Figure 4.16. Compression of the phase space element of radius ϵ upon "winding" the trajectory on the stable limit cycle — trajectory Γ (reproduced from V. S. Anischenko et al., 1999).

However, in a nonlinear dissipative system, the change in the starting small volume will be quite complex. It has been found that the elementary phase volume in nonlinear dissipative systems should stretch in some directions, while in other directions is should shrink due to dissipation (loss) of part of the energy. In stable systems, the elementary volume of the element of the limit set that contained points of starting trajectories, with time tends in the limit to zero. But in unstable nonlinear dissipative systems, the small phase volume of radius r surrounding the starting point, in the limit turns to a structure, occupying the whole region where the trajectories of the system perform tangled nonperiodical motions without overlapping each other and closing on themselves (the absence of periodicity).

Figure 4.17 demonstrates deterministic chaos and behavior of a phase trajectory in the nonlinear model of a radio-frequency oscillator analyzed by V. S. Anischenko and coworkers:

$$\begin{aligned} \dot{x} &= mx + y - xz, \\ \dot{y} &= -x, \qquad\qquad I = \{1,\ \text{if } x > 0;\ 0,\ \text{if } x \leqslant 0\}. \\ \dot{z} &= -gz + gI(x)x^2, \end{aligned} \tag{4.19}$$

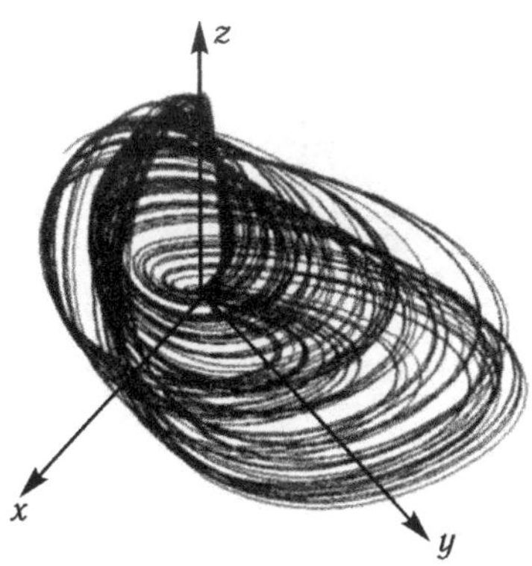

Figure 4.17. Deterministic chaos in the three-dimensional system (4.19). Phase trajectories for parameter values $m = 1.5$ and $g = 0.2$ (reproduced from V. S. Anischenko et al., 1999).

This pattern corresponds to chaotic behavior of the models considered above (Fig. 4.14 and 4.15). The trajectory is unwound in the three-dimensional space, moving away from the starting point in a spiraling way and returning to the range of the starting state. Then due to instability, the process will be repeated non-periodically, when at $t \to \infty$ the trajectory does not close on itself. If it were possible to once again set the starting conditions with absolute precision, the same trajectory would be obtained because of determinancy. However due to general instability and unavoidable small deviations from starting conditions, the new trajectory will differ from the previous one and will fill the limit nonzero volume.

Figure 4.18 shows how the phase volume element evolves in this system with time in the regime of deterministic chaos, stretching in some directions and shrinking in others. In a period of time, the point of the starting phase volume element can be found in any region of the phase space which is occupied by the attractor. Thus the points of the starting state are randomized through the whole region occupied by the attractor while the system moves along the trajectory.

In this case, an analog of the starting element of the phase volume is a drop of ink placed in a vessel filled with water. Upon stirring, ink particles could be found in any part of the water volume. Each of the elementary particles would move

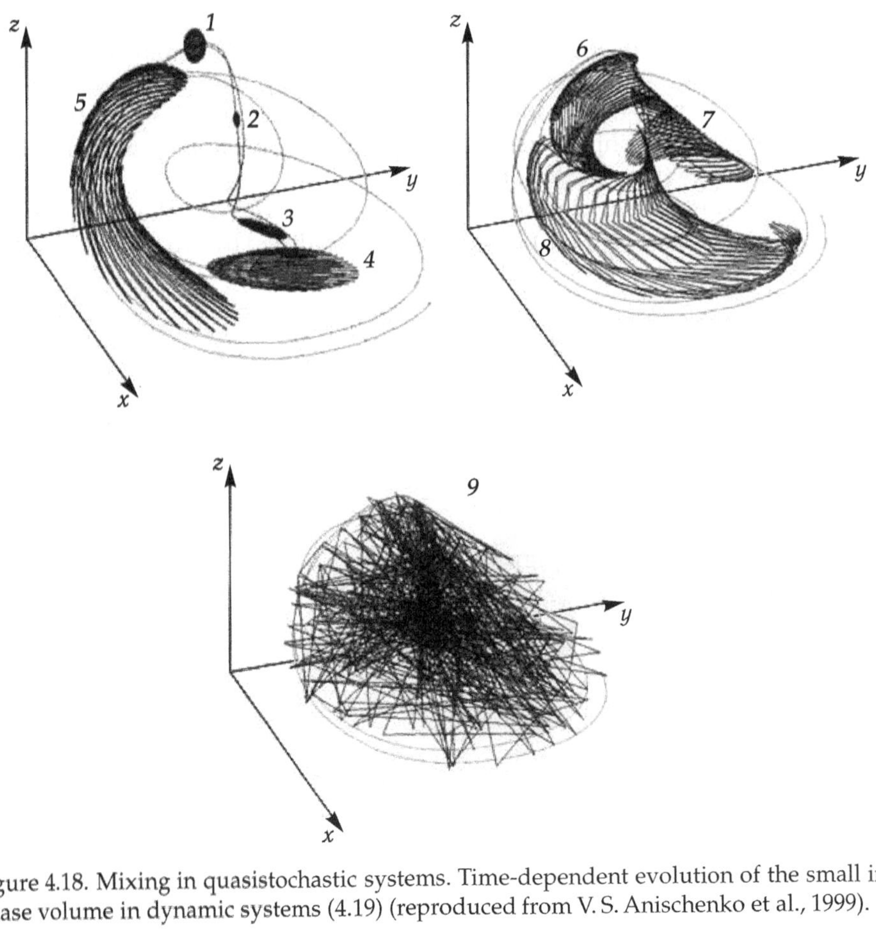

Figure 4.18. Mixing in quasistochastic systems. Time-dependent evolution of the small initial phase volume in dynamic systems (4.19) (reproduced from V. S. Anischenko et al., 1999).

along a complex determined mechanical trajectory affected by water fluxes. When this experiment is performed once more, we will again have randomization — the overall volume will be filled (the strange attractor) with ink particles (points of the starting element of the phase volume), though, naturally, no accurate reproduction of trajectories of each particle will be achieved. Note that the ambiguity in setting the starting state does always take place to some extent. But deterministic chaos is displayed in nonlinear systems since their behavior is sensitive to small changes in their initial conditions. As a result, the system controlled by deterministic laws behaves in a chaotic manner, becomes unpredictable with respect to a certain type of a trajectory, but rather definite relative to properties of its strange attractor as a region occupied by its unstable trajectories.

Complex geometric structure of a strange attractor is characterized by its dimensions. Parameters of the attractor dimensions depend both on its metric properties and statistic properties of dynamic fluxes leading to their formation. Different geometric structures of strange attractors can be arbitrary compared measuring their general volumes by the number of cubes required to cover all elements of these multitudes. The fractal dimension D_F of the arbitrarily limit multitude G of the strange attractor points in the N-meric phase space will be determined by the minimal number $M(\varepsilon)$ of N-meric cubes with side ε, required to cover all elements of the multitude G.

According to Kolmogorov – Hausdorf, the fractal dimension D_F is determined as follows:

$$D_F = \lim_{\varepsilon \to 0} \left[\frac{\ln M(\varepsilon)}{\ln(1/\varepsilon)} \right]. \tag{4.20}$$

According to Eq. (4.20), the dimensions of geometric shapes of the point, line and surface will have their usual values of 0, 1 and 2. However for nontrivial geometric complex multitudes, including dynamic complex attractors, the dimension D_F may be fractional. Apparently there is some probability P_i that the phase trajectory can pass through the i-th cube of those covering the attractor.

Let us introduce informational dimension D_I of the attractor:

$$D_I = \lim_{\varepsilon \to 0} \frac{I(\varepsilon)}{\ln(1/\varepsilon)}, \quad I(\varepsilon) = -\sum_{i=1}^{M(\varepsilon)} P_i \ln P_i, \tag{4.21}$$

where $I(\varepsilon)$ is the Shannon information required for determining the state of the system consisting of $M(\varepsilon)$ cubes with the side size ε, the value of which determines the accuracy of setting the boundaries of the system element. At small ε values and correspondingly an increase in the accuracy, the equality $I(\varepsilon) \approx D_I \ln(1/\varepsilon)$ is valid. It follows that D_I characterizes the growth of information contained in the system with a decrease in ε.

It was found, that there is correlation between the fractal dimension D_F (4.20) and the Lyapunov index (or Lyapunov dimension). As a matter of fact, in the presence of at least one positive Lyapunov index and upon stirring, classical properties of deterministic chaos reflecting the existence of aperiodic exponentially unstable solutions lead to geometric structures of the attractor with a fractional dimension.

At present only for relatively simple dynamic model systems (two- or three-dimensional) it is possible to determine quantitative correlation of the fractal dimension D_F of the chaotic multitude of points of the strange attractor with the Lyapunov index or Lyapunov dimension D_L which can be computer estimated.

Previously we have considered some models of biological systems, including both point systems and systems with distributed parameters. Let us see under what conditions their regular behavior can be randomized with the emergence of deterministic chaos in the phase space. It should be noted that chaos emerges in systems of autonomous differential equations of the third order (containing three equations, for example (4.15)) or in discrete systems of the first order (for example, discrete Verhulst functions (4.16)).

In Section 2 of Chapter 2 we have described an autooscillatory system of glycolysis with experimentally detected alternations in intermediate states (NADH, PPK, ADP, F6P, and FDP) in cells of yeast extracts. The model of autooscillations in glycolysis has two equations (2.7) responsible for the existence of a stable limit cycle as a regular dynamic attractor of this point system. Yet it may be expected that in such a complex system as glycolysis, even more complex dynamic behavior, including chaotic behavior, may be observed under certain conditions consistent with the addition to model (2.7) of a third equation. As a matter of fact, direct experiments on cells from yeast extract at continuous regular addition of glucose permitted recording aperiodic stochastic alterations in NADH concentrations detected by the changes in the fluorescence of this compound (Hess, 1995).

The model consistent with these experimental conditions looks like this

$$\begin{aligned} \frac{d(\mathrm{F6P})}{dt} &= v_{in} + A\sin(\omega_0 t) - v_{\mathrm{PFK}}, \\ \frac{d(\mathrm{ADP})}{dt} &= -\frac{d(\mathrm{ATP})}{dt} = v_{\mathrm{PFK}} - v_{\mathrm{PK}}, \end{aligned} \tag{4.22}$$

where $v_{in} + A\sin(\omega_0 t)$ corresponds to constant (v_{in}) and ($A\sin(\omega_0 t)$) to the sinusoidal "feeding" of the substrate. Accordingly, the concentration of F6P grows due to constant (v_{in}) or periodic influx of ($A\sin(\omega_0 t)$) and decreases in the reaction catalyzed by PPK at the rate v_{PFK}.

The dependence of ATP on the rate of substrate F6P consumption in glycolysis has been taken into account in the model. The ATP molecule is formed at the last stage of glycolysis with the involvement of phosphoenol piruvate (PEP) and the enzyme pyruvate kinase (PK) at the rate v_{PK}. The balance of ADP and ATP is determined by the difference in the rates of reactions v_{PFK} and v_{PK} in the second equation.

In the absence of periodic additions of the substrate at $\omega_0 = 0$, the limit cycle similar to (2.7) is generated in the model with a frequency of recurrent changes in ADP and NADH.

In a typical case, at $\nu_{\max(\mathrm{PK})} = 1.4$ mM the model yields oscillation period values $T_0 = 2\pi/\omega_0 = 11$ min, which is compatible with the experimental data ($T_0 = 12$ min). However upon periodic influx of F6P, the pattern of oscillations changes and their character becomes dependent on the relation of the frequencies of the limit cycle of the system and the frequency of F6P (ω_e) injections into the system. Now the period T of the observed oscillations of NADH and ADP becomes dependent on the period of $T_e = 2\pi/\omega_e$ injections, as $T = mT_e$, where m is the integer, and T_e is the period of injections.

Calculations show that when the relation ω_e/ω_0 increases from the starting value $\omega_e/\omega_0 = 0.43$ to $\omega_e/\omega_0 = 1.4$, the number m changes drastically from $m = 1$ to $m = 2$, i.e. the period of ADP oscillations increases two-fold. This was also corroborated in experiments where bifurcation of the oscillation period from $m = 1$ to $m = 2$ was observed at close values of the relation of frequencies $\omega_e/\omega_0 \approx 1.5$.

Further increase in the frequency of substrate injections results in randomization of the oscillations. The scenario of chaos formation in a real system is quite complex. In this case, chaos establishes with the doubling of the number of periodic points, the doubling of the cycling period in bifurcation points, and the emergence of "windows of regularity" at chaotic regimes upon variations in the parameter ϖ_e/ϖ_0. When this parameter reaches the value of 2.08, the system is randomized which is nonetheless associated with the appearance of separate "regularity windows" with a period of oscillations corresponding to the numbers $m = 5, 6, 7, 8$ ($T_0 = mT_e$). Then transition from chaos to regular oscillations ($m = 3$) is again observed at $\varpi_e/\varpi_0 = 2.58$, which is close to experimental values $\varpi_e/\varpi_0 \approx 3$.

Figure 4.19 shows experimentally observed aperiodic chaotic changes in the fluorescence of NADH in cells of yeast extracts with sinusoidal glucose influx at $\varpi_e/\varpi_0 = 2.76$.

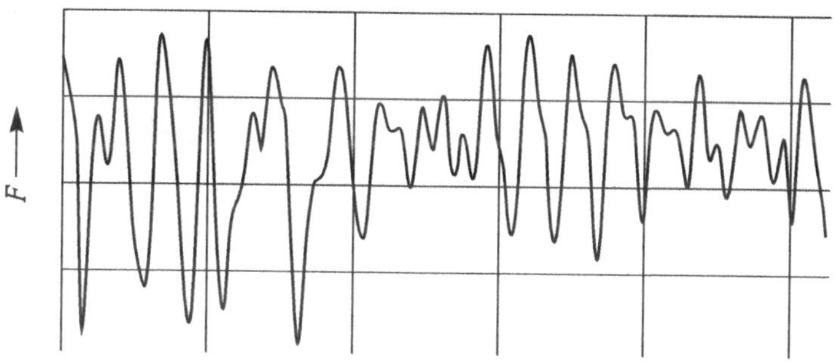

Figure 4.19. Measured fluorescence of NADH from yeast extract at a sinusoidal influx of glucose (reproduced from M. Marcus, D. Kuschmitz, B. Hess (1985) Properties of strange attractors in yeast glycolysis. Biophysical Chemistry, 22, 95–105).

The processing of these data allowed finding the power spectrum (F) of NADH oscillations which was quite even in the chaotic regime. It did not contain sharp peaks at certain frequencies as opposed to the regime of sinusoidal oscillations, where peaks were observed in the power spectrum at ϖ_0 and $2\varpi_0$.

The structure of the strange chaotic attractor of this system can be represented on the phase plane in the coordinates dF/dt and F (Fig. 4.20) where it is seen how the oscillation period of the system T ($T = mT_0$) changes consistent with the chaotization parameter ϖ_e/ϖ_0.

Chaotic oscillations were also observed at the stationary regime in a bioreactor where the influx of both the substrate and cells of yeast extracts and the outflow of the reacting medium from the bioreactor were maintained.

On the whole, the chaotic regime in glycolysis associated with other biochemical cycles through intermittent compounds of ATP and NADH can reflect fine regulatory effects of mutual influence of separate parts of cell metabolism. The presence of chaotic regimes may be associated with alterations in metabolism and may also characterize the intermediate state of metabolic processes established upon the change of dynamic regimes in response to external effects.

Mandelbrot Sets. More generalized regularities of transition from order to chaos were discovered in the so-called Mandelbrot sets (1980) and can be represented graphically.

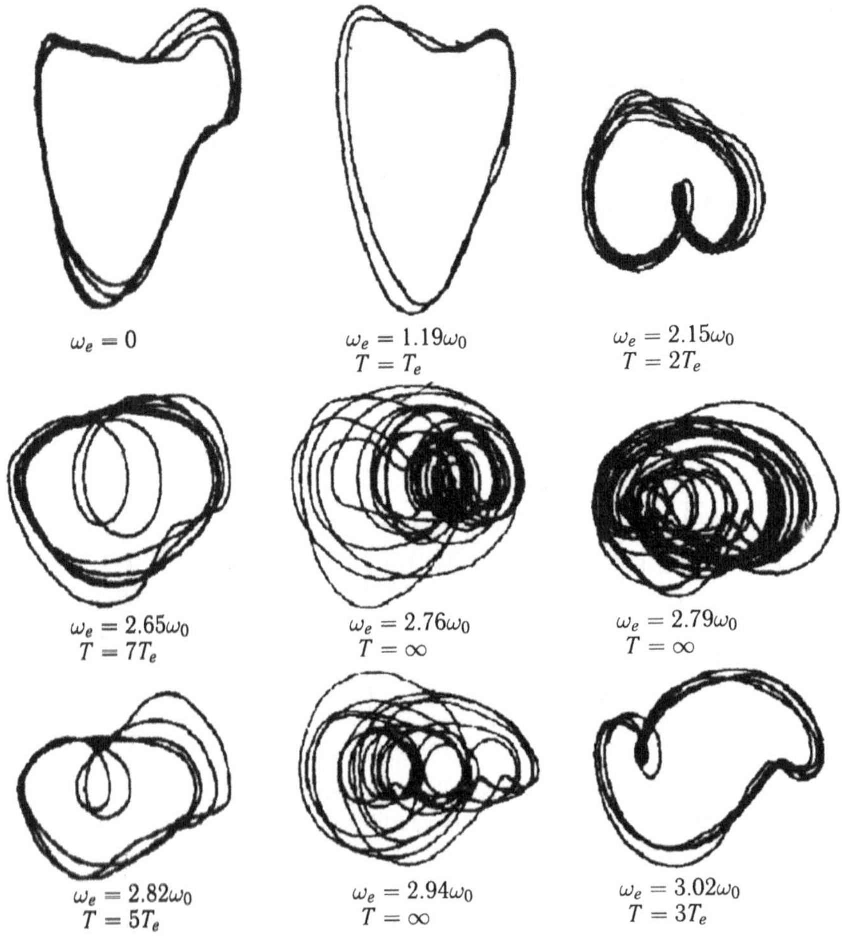

Figure 4.20. Schemes of dependence of dF/dt on F (measured fluorescence of NADH in relative units) for different values of the frequency of the input flux ϖ_e. ϖ_0 is the frequency of the system at a constant flux ($\omega_e = 0$), T and T_e are response and entry periods respectively (reproduced from M. Marcus, D. Kuschmitz, B. Hess (1985) Properties of strange attractors in yeast glycolysis. Biophysical Chemistry, 22, 95–105).

Let us analyze a prime sequence of complex numbers

$$z_{n+1} = f_c(z_n) = z_n^2 + c, \tag{4.23}$$

where z_n are numbers in a complex plane for complex values of parameter c. Equation (4.23) is reduced to logistic equation (4.16) with new variables

$$x = 1/2 - z/r, \quad c = (2r - r^4)/4.$$

Suppose that constant $c = 0$. Then at each iteration in Eq. (4.23), beginning from z_0 we get the square of the corresponding complex number z_n^2 according to the law

$$z_0 \to z_0^2 \to z_0^4 \to z_0^8 \to \ldots.$$

It is evident that numbers z_n^2 in the limit will either decrease tending to zero (at $|z_0| < 1$) or on the contrary increase to infinity (at $|z_0| > 1$). In the first case the attractor for the process $z \rightarrow z^2$ will be zero, and in the other case it will be infinity. The boundary between the two regions of these attractors is obviously a circumference with the radius 1 around point 0.

Accordingly, the overall plane is divided in two areas of influence with the boundary between them in the form of a smooth circumference. However, the situation changes drastically if the constant c differs from zero. For example, at $c = -0.12375 + 0.56508i$ the internal attractor is already not zero, and the boundary is not smooth (Fig. 4.21). The irregular, or as they say, fractal character of this boundary resembles the coastal line, the length of which is the larger the smaller is the scale with account of all "chaotic and disordered" imperfections that we choose upon measuring the distances.

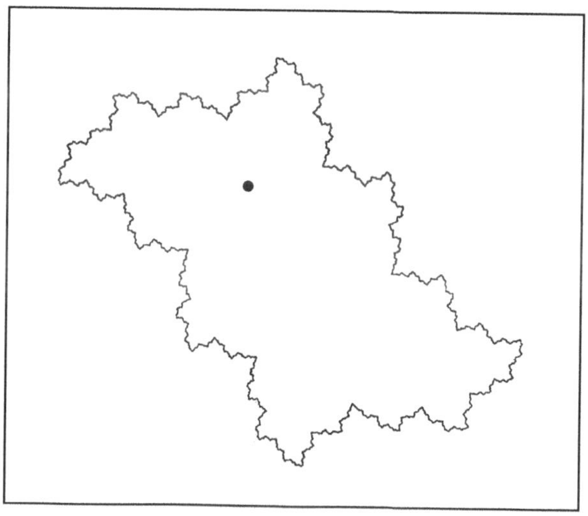

Figure 4.21. Domain of attraction of a fixed point (reproduced from H. O. Peitgen and P. H. Rechter, 1986).

Depending on the values of constant c, it is possible to make a computer-aided plot of fractal boundaries of the domains of attraction. There may be several such domains and their shape resembles complex intricate patterns (Fig. 4.22, *a–d*). The stochastic character of points hitting in one or another fractal region is determined by the natural small error in setting the initial conditions by computer for iteration in the sequence

$$z_0 \rightarrow z_0^2 + c_i \rightarrow z_0^4 + c_i \rightarrow \ldots .$$

The pattern becomes even more striking and demonstrable if different points on the plane are stained with various colors depending on how quickly they get to the domains of the attraction that depends on the number of iterations. The boundaries of domains of attraction are called Julia sets. These boundaries may be connected (Fig. 4.22) at certain values of constant c.

Figure 4.23 shows graphical representations of sets of constant c values in the complex area for which Julia sets are constrained. The image in Fig. 4.23 is called the Mandelbrot set (1980) or set M after the name of the scientist who was the first

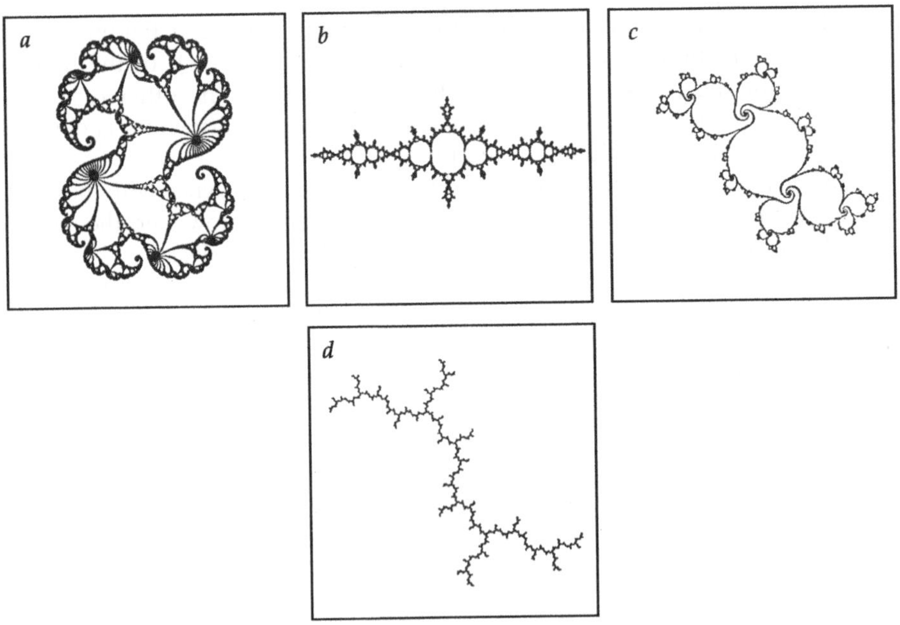

Figure 4.22. Examples of typical Julia sets for the process $x_{n+1} \rightarrow x_n^2 + c$ (reproduced from H. O. Peitgen and P. H. Rechter, 1986).

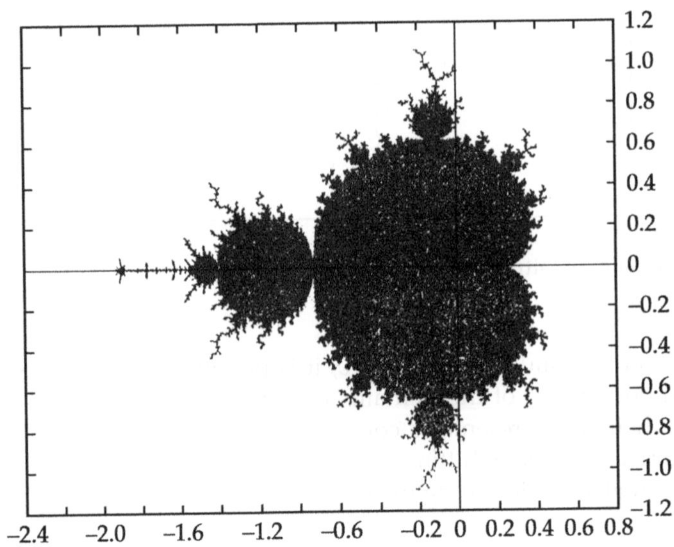

Figure 4.23. Mandelbrot sets for the process $x_{n+1} \rightarrow x_n^2 + c$ (reproduced from H. O. Peitgen and P. H. Rechter, 1986).

Part of the complex c plane — $2.25 < \mathrm{Re}\, c < 0.75$, $-1.5 < \mathrm{Im}\, c < 1.5$ is shown. The pattern reflects conformity with different c parameter values for different types of the boundary.

to publish it. For all c values of the black image, the corresponding boundaries of attraction or Julia sets are connected. If c lies outside the black image, the Julia sets are divided in an infinite number of separate fragments (Fig. 4.24).

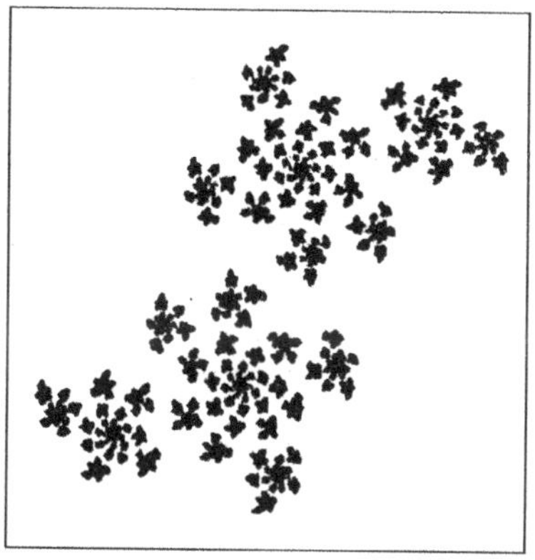

Figure 4.24. The unconstrained Julia set. The Fatou dust (reproduced from H. O. Peitgen and P. H. Rechter, 1986).

The exclusive richness of complex images "buried" in an elementary analytical formula

$$z \to z^2 + c$$

is striking. At present, studies in this field are being intensely developed.

We have already mentioned some examples of systems which demonstrate chaotic behavior under definite conditions: Bénard instability in liquid fluxes, chemical processes in distributed systems. A thorough study may reveal chaotic behavior in a number of systems far from each other, in which the chaotic state emerges in real conditions and plays a significant role in the dynamics. Irregular bursts in population numbers and unpredictable peaks of diseases among the population during epidemics which do not correlate with the external conditions and methods of treatment, heart arrhythmia and cardiac muscle fibrillation, chaotic movement of eye muscles at schizophrenia, chaotic small-scale intramolecular motions of proteins. From this point of view, considerable attention should be focused on the problem of shape formation in nature and particularly in biological systems. For example, in the process of morphogenesis the genetic program copied in DNA determines only the sequence of biosynthesis of macromolecules. Then they interact in the cell, obeying the dynamic regularities and as a result leading to the emergence of dissipative structures, including the fractal character. □

There are known examples (air vesicles, sectile leaves) of branched structures of different organs and tissues, which consist of chaotically compiled small details, though preserve specific shapes inherent to the entire formation on the whole.

It was found that to imitate on a computer a complex fractal shape, there are rather simple rules which allow us to reproduce its formation according to the laws of chaos. As a matter of fact, chaotic behavior reflects deep regularities in the dynamic organization of complex systems. The available models of deterministic chaos may represent elementary rules of chaotization and in this respect can be regarded as primary basic models of chaos. It is clear that in this case we are just at the beginning of studying the role of chaos in nature and in self-organization of biological systems.

At the moment, the basic generalized result is that the behavior of determined systems, including biological ones which are always considered as predictable, in fact displays chaotic properties under definite conditions.

II
Thermodynamics of Biological Processes

5

Thermodynamics of Irreversible Processes
in Biological Systems
Near Equilibrium (Linear Thermodynamics)

6

Thermodynamics of Systems
Far From Equilibrium (Nonlinear Thermodynamics)

The capacity to absorb and transform energy in different forms and use it in metabolic processes for providing growth, development and reproduction is one of the most important properties of living systems. General regularities of processes of energy exchange accompanying biochemical conversions are studied with the methods of classical chemical thermodynamics, which are used to analyze energy transformation in chemical processes upon performing useful work. Important results were obtained on estimation of changes (increase or decrease) in free energy in different biological processes, based on which the thermodynamic possibility of their coupling could be judged.

Metabolic processes (including redox reactions, synthesis and hydrolysis of high-energy compounds, transport of substances and ions through membranes, motor activity, and consumption of light energy in photosynthesis — all of them being associated with energy transformation) follow the law of energy conservation or the first law of thermodynamics. However, the direct consideration of this law does not involve the time factor, characterizing the process of transformation, because estimation of energy effects of any transformations is obtained by comparing the parameters of the initial and final states of the system.

In accordance with the second law of thermodynamics, the concept of entropy is introduced, which in an isolated system always increases upon reaching equilibrium prior to its maximal value. The law of entropy increase in isolated systems is a criterion of evolution on the pathway of achieving the final equilibrium state. Though in an open system in the equilibrium state no directed processes take place except for random fluctuations near the equilibrium position, which is equal to termination of the biological system existence.

In open systems a stationary state can emerge, the criteria of achieving which are of the greatest interest. This means that it is necessary to take into account the role of the time factor, because the stationary state will be a result of directed changes in variables (substance concentrations) characterizing chemical processes. Therefore energy conversions in open systems should be characterized not only from the point of view of their general statistic balance but also estimate the rate of energy conversion per unit of time. □

To this end, it is required to make an analysis of the interrelation of different kinetic processes associated with energy consumption and release. Estimations of the rates of changes in the free energy are based on the kinetic scheme which takes into account the character of the changes in corresponding concentrations of agents in chemical processes. Of special significance are the problems of stability of stationary states, the criteria for reaching them far from equilibrium and also the thermodynamics of autooscillatory processes, self-organization processes of great biological importance. Organic merging of the kinetic and thermodynamic approaches is a characteristic feature of the contemporary stage in the development of thermodynamics of irreversible processes.

In the systems close to the equilibrium, the results obtained using Onsager relations in the area of energy coupling. In systems far from the equilibrium, thermodynamics faces the challenge of searching for criteria of evolution and stability of stationary states. In this area thermodynamics is completely based on original mathematical models and its results can serve only as an additional illustration for understanding peculiarities of the dynamic behavior of open systems. This in full measure refers to autooscillating processes, trigger switching of the system from one regime to another, and finally to self-organization processes. All these problems are included in sections devoted to problems of nonlinear thermodynamics.

5

Thermodynamics of Irreversible Processes in Biological Systems Near Equilibrium (Linear Thermodynamics)

In general, classical thermodynamics considers equilibrium states of a system wherein system parameters and characteristics do not change with time. The actual transitions between different states are not among the subjects of classical thermodynamics, and its methods permit evaluating only typical energy effects of chemical transformations by comparing parameters of the initial and final states of the system. However in open systems, reactions and appropriate energy transformations occur regularly, therefore it is required to know the rates of energy transformation processes at every time moment. This means that in calculating energy effects it is necessary to take into account the time factor as well, thereby to somehow combine thermodynamic and kinetic approaches in describing the properties of an open system. We will analyze thermodynamic criteria of stability of stationary states and criteria of attaining them close to and far from equilibrium and will become acquainted with thermodynamic characteristics of autooscillatory and trigger regimes.

The First and Second Laws of Thermodynamics. Let us remember the basic laws of classical thermodynamics and results of their application in biology. According to the first law, the amount of the heat δQ absorbed by the system from the external environment is spent for increasing its internal energy dU and performing the overall work δA that includes both work against the external pressure P to change the volume dV of the system and the effective work $\delta A'_{max}$ accompanying chemical transformations:

$$\delta Q = dU + \delta A$$

where the work is expressed as

$$\delta A = p\,dV + \delta A'_{max}$$

or

$$\delta Q = dU + p\,dV + A'_{max}. \tag{5.1}$$

The experimental checking of the first law was performed in special calorimeters used to measure the heat generated by an organism in metabolism processes, upon evaporation and also in secreted products. The nutrients received by organisms are decomposed with the release of their free energy used for life activity. Experiments were carried out on the set of reactions of basal metabolism at a relatively short period of time when biomass accumulation does not take place as a result of growth and no noticeable work is performed. It was found that the heat released by the organism totally complies with the energy absorbed via nutrient consumption. The validity of the first law signifies that the organisms themselves are not an independent source of any new form of energy.

The second law of thermodynamics confers the criterion of the direction of spontaneous irreversible processes. Any alteration in the state of the system can be described by the corresponding change in the special function of the entropy state S which is determined by the total value of normalized heats Q/T absorbed by the system.

For equilibrium processes the small change of the entropy dS is equal to the elementary normalized heat $\delta Q/T$ absorbed by the system and is larger for non-equilibrium processes

$$dS \geqslant \delta Q/T. \tag{5.2}$$

In isolated systems $\delta Q = 0$ and hence

$$dS \geqslant 0. \tag{5.3}$$

This is the evolutionary criterion of direction of irreversible changes in isolated systems which in all cases proceed with an increase of entropy to its maximal values upon termination of the process and formation of thermodynamic equilibrium. The entropy increase testifies to a decrease in the level of ordering and organization of the system, i.e. its chaotization.

We will be interested in the relation between the thermodynamic parameters and the value of the maximally effective work $\delta A'_{max}$ which characterizes internal irreversible chemical transformations in the system. It is impossible to determine univalently the $\delta A'_{max}$ value using expression (5.1) for the first law upon transition from one state to the other because the δQ value depends on the transition pathway. But by combining expressions (5.1) and (5.2) one can find special characteristic functions, the change of which equals $\delta A'_{max}$ in definite conditions. In biochemical processes, the most important are the free energy F

$$F = U - TS$$

and the total thermodynamic potential G or Gibbs energy

$$G = U + pV - TS.$$

If the processes occur at constant T and V, then

$$\delta A'_{\text{max}} \leqslant -d(U - TS) = T\,dS - dU = -(dF)_{T,V},$$

and if T and p are constant, we have

$$\delta A'_{\text{max}} \leqslant -d(U + pV - TS) = T\,dS - dU - P\,dV = -(dG)_{T,P}, \tag{5.4}$$

where the inequality sign complies with irreversible processes.

So, performance of the effective work $\delta A'_{\text{max}}$ entails a decrease in the free energy and Gibbs energy in irreversible processes. Note that typically in biochemical transformations the change in the system volume can be neglected; $dV \approx 0$ and therefore the dF and dG values are compatible. Classical thermodynamics allows one to calculate energy effects and accordingly determine the direction and possibility of coupling different biological processes.

There are various methods for estimation of the ΔG and ΔF values that are described in detail in physical chemistry. So, given the equilibrium constant (K) of a chemical reaction identified, it is possible to estimate the ΔG_0 value, that would comply with the decrease in ΔG upon transition from the initial non-equilibrium state of the mixture, in which concentrations of its components are equal to unity, to the final equilibrium state

$$\Delta G_0 = -RT \ln K,$$

where R is the gas constant (1.987 cal/K/mol or 8.314 J/K/mol). For example, in the reaction

$$\text{Glucoso-1-phosphate} \rightleftarrows \text{Glucoso-6-phosphate}$$

the equilibrium constant $K = 17$, hence it follows that $\Delta G_0 = 17\,000$ cal/mol < 0. The negative value $\Delta G < 0$ shows that in standard conditions this reaction is spontaneous and is essentially shifted to the right, which is supported also by the high value $K = 17 \gg 1$.

Similarly, hydrolysis of ATP that proceeds with splitting of the residue of phosphorous acid and its transfer to water

$$\text{Adenosine} - \text{Ph} \sim \text{Ph} + H_2O \rightleftarrows \text{Adenosine Ph} \sim \text{Ph} + H_3PO_4,$$

is characterized by the negative value $\Delta G_0 = -7$ kcal/mol. It is the release of a relatively large portion of energy during hydrolysis of ATP that makes the Ph $\sim$ Ph bond in ATP a macro-energy bond as compared to other reactions of group transfer.

By comparing ΔG_0 values of different processes, one can determine whether their coupling is possible when one (coupled) process occurs with of the increase in ΔG at the expense of the ΔG reduction in the other (coupling) process. So, oxidation of a glucose molecule during respiration is accompanied by a decrease of $\Delta G_0 = -678$ kcal/mol. It is equal to the increase in ΔG_0 during photosynthesis upon the formation of a glucose molecule from water and CO_2. Thereby from the thermodynamic point of view, coupling of photosynthesis and respiration is possible. The same approach is used when the aim is to find the possibility of coupling other simpler processes as well. The formation of acetyl phosphate in the phosphorylation reaction

$$\text{Acetyl CoA} + \text{Phosphate} \rightleftarrows \text{Acetyl phosphate} + \text{CoA}$$

proceeds with an increase in $\Delta G_0 = 3$ kcal/mol > 0 and can be readily realized upon coupling to ATP hydrolysis or the reaction

$$\text{Glucoso-1-phosphate} + H_2O \rightleftarrows \text{Glucose} + \text{Phosphate}$$

where $\Delta G_0 = -4.8$ kcal/mol.

It should be mentioned that in all the above cases, one can bear in mind only the thermodynamic probability of the process rather than its actual occurrence by the given molecular mechanism. Moreover, the direct character of the temporal change of the free energy in the course of the reaction or upon coupling of different processes is not taken into consideration with such an approach.

The driving force of a chemical process is associated with the difference in chemical potentials of initial and final products. If this difference vanishes, the process is terminated and equilibrium is formed in the system when the rates of the direct and reverse reactions are equal. For example, in the simple case of a monomolecular reaction where the stoichiometric coefficients are equal to unity

$$C_1 \rightleftarrows C_2,$$

the driving force A merely equals the difference of chemical potentials μ_{C_1} and μ_{C_2}:

$$A = \mu_{C_1} - \mu_{C_2}.$$

Here $\mu_{C_i} = \mu_{o,C_i} + RT \ln C_i$ is the chemical potential of substance C_i. The A value is also called the chemical affinity of the reaction. It may be represented as

$$A = -\sum v_i \mu_i, \tag{5.5}$$

where v_i values are stoichiometric coefficients in the equation for the reaction

$$v_1 C_1 + v_2 C_2 + \ldots \rightleftarrows v_1' C_1' + v_2' C_2'.$$

They are included in equation (5.5) with the negative sign for substances from the left-hand side and with the positive sign for substances from the right-hand side of the reaction equation. At equilibrium when all $C_i = C_i^{\text{equi}}$ the A value vanishes.

Entropy Change in Open Systems. It may seem at first glance that the application of the classical expression of the second law of thermodynamics to biological systems leads to a paradoxical conclusion that the processes of life activity violate the thermodynamic principles. Indeed, complication and enhancement of the rate of ordering of organisms during their growth proceed spontaneously. But as follows from the second law, such spontaneous processes should be accompanied by a decrease rather than an increase in entropy. It is evident that only in isolated systems the entropy increase in irreversible spontaneous processes should take place, whereas biological systems are open. Therefore the problem is to perceive in what way the entropy change is connected with the parameters of the process in an open system and to clarify whether it is possible to predict the general direction of irreversible processes in such a system by evaluating the change of its entropy. The main difficulty in solving this problem is that one should take into account the alteration of all thermodynamic values in time just when the processes occur in the system. It is postulated that in an open system the overall entropy change dS may happen independently, or due to metabolism with the external environment d_eS, or because of internal irreversible processes (d_iS).

$$dS = d_eS + d_iS. \tag{5.6}$$

In all cases $d_iS > 0$, and only if the internal processes occur reversibly and at equilibrium, then $d_iS = 0$. For isolated systems $d_eS = 0$, and so we have the classical expression of the second law:

$$dS = d_iS \geqslant 0.$$

Cell metabolism always has two groups of such processes, for example, the glucose intake from outside, release of products of its oxidation outwards (d_eS) and oxidation of glucose in respiration processes (d_iS).

In photosynthesis, the inflow of free energy of light causes both the formation of a complex glucose molecule from simple compounds H_2O and CO_2 and reduction of the cell entropy $d_eS < 0$, while decomposition of glucose increases its entropy $d_iS > 0$. In compliance with the proportion of rates of changes in d_eS and d_iS the overall entropy dS in an open system may either increase or decrease with time. Let us differentiate expression (5.6).

$$\frac{dS}{dt} = \frac{d_eS}{dt} + \frac{d_iS}{dt}. \tag{5.7}$$

If $\frac{d_eS}{dt} < 0$ and $\left|\frac{d_eS}{dt}\right| > \frac{d_iS}{dt}$, the entropy of an open system will decrease $\frac{dS}{dt} < 0$. The negative value $\frac{d_eS}{dt} < 0$ or the inflow of negative entropy corresponds to the outflow of positive entropy into the external environment and the inflow of nutrients from the outside accompanied by the release of the free energy from the nutrients in the organism. But if $\left|\frac{d_eS}{dt}\right| < \frac{d_iS}{dt}$, then $\frac{dS}{dt} > 0$, which corresponds to the general degradation and decomposition of the system. In a stationary state $\frac{d_eS}{dt} < 0$, $\left|\frac{d_eS}{dt}\right| = \frac{d_iS}{dt}$ and $\frac{dS}{dt} = 0$.

Relationship between d_iS and Open System Parameters. Let us accept that the exchange processes of an open system with the environment occur at equilibrium, and the only reason for the irreversibility and entropy increase in the system are its internal processes. In this case

$$\frac{d_iS}{dt} = -\frac{1}{T}\left(\frac{dG}{dt}\right)_{T,p}. \tag{5.8}$$

It can be shown that the rate of positive entropy formation within an open system depends on the chemical affinity A and the reaction rate v

$$\frac{d_iS}{dt} = -\frac{1}{T}A\,v > 0. \tag{5.9}$$

Given $A = 0$ and $v = 0$, equilibrium is formed in the system and we have

$$\frac{d_i S}{dt} = 0. \tag{5.10}$$

Expression (5.9) has a simple meaning. It shows that the rate of the positive entropy formation in the system in irreversible chemical processes is directly proportional to its driving force A and rate v. It is obvious that the $\frac{d_i S}{dt}$ value is a variable because in the course of the chemical reaction the unstable concentrations of the reactants are always changing and hence values A and v dependent on them change accordingly. Below we will analyze how the value $\frac{d_i S}{dt}$ behaves at a stationary state where all concentrations become constant. When several chemical reactions proceed simultaneously in the system ($j = 1, 2, \ldots, n$), the overall rate of positive entropy generation is dependent on their driving forces and rates,

$$T\frac{d_i S}{dt} = A_1 v_1 + A_1 v_2 + \ldots + A_n v_n = \sum_j A_j v_j > 0. \tag{5.11}$$

Coupling of Processes. Let us have two irreversible reactions ($A_1 v_1$ and $A_2 v_2$). Then

$$T\frac{d_i S}{dt} = A_1 v_1 + A_2 v_2 > 0. \tag{5.12}$$

The positive $T\frac{d_i S}{dt}$ value in (5.12) is provided when

$$A_1 v_1 > 0, \quad A_2 v_2 > 0 \tag{5.13}$$

or when

$$A_1 v_1 < 0, \quad A_2 v_2 > 0, \quad A_2 v_2 > |A_1 v_1|. \tag{5.14}$$

In this case the both reactions may be coupled. The first (coupled) reaction proceeds contrary to the difference of chemical potentials of its reagents (A_1 and v_1 have opposite signs) at the expense of the second (coupling) reaction. The energy released in the coupling reaction is not dissipated into heat, but is spent to maintain a coupled process proceeding with the increase in the chemical potentials of its initial products, i.e. with the increase in the free energy. Conditions (5.13) and (5.14) permit us to find the upper limit of the coupled reaction rate

$$v_1 \leqslant \frac{A_2 v_2}{A_1}, \tag{5.15}$$

that is to relate the thermodynamic value of chemical affinity to the kinetic value of the reaction rate.

Onsager Relations. The driving forces and rates (inflows) should obviously be interrelated when the enhancement (reduction) of the driving force causes a corresponding increase (decrease) in the process rate. This concerns not only chemical reactions, but other irreversible processes as well. For example, the processes of heat transfer and substance diffusion across the membrane from one phase to the other involve driving forces (temperature and concentration gradients), and the fluxes denote the heat or substance transfer between the two phases. In all such cases the entropy increase can be written like this

$$T\frac{d_iS}{dt} = XI > 0, \tag{5.16}$$

where X is the driving force, and I is the flux value.

If the system is near equilibrium then the driving forces and fluxes which are very small, are directly proportional to each other

$$I = LX \tag{5.17}$$

where L is the constant linear coefficient.

Take a chemical reversible reaction

$$C_1 \underset{\overleftarrow{v}}{\overset{\overrightarrow{v}}{\leftrightarrows}} C_2$$

where the value of the total rate (or flux) equals the difference between the rates of direct $\overrightarrow{v} = \overrightarrow{k}\,C_1$ and reverse $\overleftarrow{v} = \overleftarrow{k}\,C_2$ reactions:

$$I = \overrightarrow{v} - \overleftarrow{v} = \overrightarrow{k}\,C_1 - \overleftarrow{k}\,C_2.$$

Obviously at equilibrium $\overrightarrow{v} = \overleftarrow{v}$ and $I = 0$, and near equilibrium $I \cong 0$. The value of the chemical affinity A near equilibrium is also very low

$$A = \mu_{C_1} - \mu_{C_2} \ll RT.$$

In this case I and A are proportional as well. If in an open system near equilibrium several processes occur at the same time, their thermodynamic proportions reflect their mutual effects. For two processes (I_1, X_1) and (I_2, X_2) these proportions are as follows

$$\begin{aligned} I_1 &= L_{11}X_1 + L_{12}X_2, \\ I_2 &= L_{21}X_1 + L_{22}X_2 \end{aligned} \tag{5.18}$$

where constant coefficients L_{11} and L_{22} show the dependence of a flux on its force, and coefficients L_{12} and L_{21} correspond to the mutual effect of the force of one process on the flux of the other process. They are called Onsager reciprocity coefficients and equations (5.18) are defined as Onsager linear relations. Near equilibrium we have

$$L_{12} = L_{21}.$$

Now we can determine the quantitative relationship between processes occurring concurrently in the cell without knowing their molecular mechanisms.

Let us consider the substance active transfer across the membrane that occurs at the expense of the energy of the coupled metabolic process and therefore can move against the concentration gradient of the transferred substance. Then

$$\begin{aligned} I_1 &= L_{11}X_1 + L_{12}X_2, \quad \text{and} \\ I_2 &= L_{21}X_1 + L_{22}X_2, \quad L_{12} = L_{21}, \end{aligned}$$

where the process (I_1, X_1) of coupled transfer moves against the force gradient X_1 $(I_1, X_1 < 0)$ due to the energy of the coupling process $(I_2, X_2 < 0)$. If no coupling takes place, then $L_{12} = L_{21} = 0$ and the processes occur independent of each other under the action of only "their own" driving forces

$$I_1 = L_{11}X_1, \quad I_2 = L_{22}X_2.$$

The following value is introduced as a coupling measure

$$q = \frac{L_{12}}{\sqrt{L_{11}L_{22}}}.$$

The q value is 0 when the coupling is absent $(L_{12} = 0)$, and $q = 1$ at completely coupled processes. At the initial moment of "triggering" the system, the high rate of the coupling process I_2 drops to minimal values and at the same time the X_1 value rises. On account of this, a stationary state is formed when the resultant coupled flux vanishes $(I_1 = 0)$. If the system is entirely coupled, stationary state $I_2 = 0$ is formed for the coupling flux too. In this case, the system has no noticeable changes and the total energy of the coupling flux is spent on maintaining the force X_1. Let us imagine a turbine wheel immersed in a water stream. Depending on the depth of immersion, the number of the turbine wheel blades in water (X_1), velocities of the turbine rotation (I_1) and water flux (I_2) would change. These examples are valid not only for an active transfer, but also for other cases. So, in the system of respiration control in mitochondria, the substrate oxidation rate (I_2) depends on the ADP/ATP ratio, i.e. on the driving force X_1. In the state of mitochondria when the ADP concentration equals zero and no discernible formation of ATP occurs $(I_1 = 0)$, the total energy is spent on maintaining the maximal phosphate potential (X_1^{max}). An addition of uncouplers results in a decrease in X_1, but then $I_1 \neq 0$ which causes acceleration of the coupling flux.

The coefficient of energy transformation in coupling processes is I_1X_1/I_2X_2 and in mitochondria its value can make 80–90 %. The use of Onsager equations permits obtaining characteristics of macromolecular complexes, i.e. biological energy transformers, without a detailed analysis of molecular mechanisms of their functioning.

Prigogine Theorem. It has already been shown that at a stationary state of an open system

$$\frac{dS}{dt} = \frac{d_eS}{dt} + \frac{d_iS}{dt} = 0,$$

$\frac{d_eS}{dt}$ and $\frac{d_iS}{dt}$ members differing from zero. The question arises whether it is possible to predict the formation of a stationary state in an open system by analyzing the change in $\frac{d_iS}{dt}$ with time. Let us have two processes occurring concurrently in an open system near equilibrium, for which Onsager relations (5.18) and (5.19) are valid. When one of the processes (I_1X_1) is in a stationary regime, then $I_1 = 0$ for it. Determine the following value

$$T\frac{d_iS}{dt} = I_1X_1 + I_2X_2 = L_{11}X_1^2 + 2L_{11}X_1X_2 + L_{22}X_2^2, \tag{5.19}$$

which has a positive quadratic form. When approaching the stationary state, the driving force X_1 and flux I_1 change in a certain way so that eventually $\bar{I} = 0$. Let us see how $T\frac{d_I S}{dt}$ depends on the change of X_I. To this end we take partial derivative $\frac{\partial\left(T\frac{d_I S}{dt}\right)}{dX_I}$ at constant X_2 and T. From (5.20) we get that

$$\left[\frac{\partial\left(T\frac{d_I S}{dt}\right)}{dX_I}\right]_{X_2=\text{const}} = 2(L_{11}X_I + L_{12}X_2) = 2I_I.$$

But at a stationary state, $I_I = \bar{I}_I = 0$. Consequently, the equations

$$\left[\frac{\partial\left(T\frac{d_I S}{dt}\right)}{dX_I}\right]_{X_2=\text{const}} = 0 \quad \text{and} \quad I_I = \bar{I}_I = 0$$

are equivalent. The vanishing of the partial derivative of $T\left(\frac{d_I S}{dt}\right)$ with respect to X_I in a stationary point $I_I = \bar{I}_I = 0$ shows that at the stationary state the positive function $\frac{d_i S}{dt}$ has the extremum and accordingly is positive. So, while reaching the stationary state, the rate of entropy formation within the open system is decreasing monotonously, gradually coming close to its maximal constant positive value. This is the criterion of the direction of irreversible processes in open systems that proceed near equilibrium where Onsager relations are effective. Provided the system has a stationary state where the $T\frac{d_i S}{dt}$ value is minimal, any deviation from the stationary point caused by disturbances would lead to its increase. But then in view of the Prigogine theorem, the values of forces and fluxes in the system should change so that the rate of the entropy formation $T\frac{d_i S}{dt}$ would again change, and the system would return to the stationary point. This result illustrates the stability of the stationary state in question.

As follows from the monotonic character of $T\frac{d_i S}{dt}$ changes near equilibrium the stationary state cannot represent an autooscillation regime. Indeed, in this case variable concentrations in the system and as a result, values I and X change regularly which is incompatible with the unidirectional monotonic change of $T\frac{d_i S}{dt}$ and its constancy in the stationary point. Experimental measuring of the rate of entropy formation in the system can be performed using calorimeters by studying heat fluxes accompanying the entropy formation upon compulsory changes in the system. It was demonstrated in experiments on biological objects that, for example, the rate of heat generation and respiration decrease continuously during embryo development beginning from the initial stages of organism development and attain constant values at the stationary phase of growth. However it should be noted that the level of thermogenesis can alter in the course of organism development not only due to changes in the values of driving forces and fluxes. The heat generation in organisms depends also on the state of membrane structures and the level of coupling of the oxidative phosphorylation processes. At length, essential is the fact that biological systems are far from equilibrium where the proportionality of I and X (5.18) or reciprocal relations (5.19) are broken. This is of special importance for biochemical processes where the most common transitions are accompanied by changes in ΔG by 1–2 kcal/mol, while the Onsager relations are effective at $\Delta G \leqslant$ kcal/mol. Given such conditions in stationary states far from equilibrium, the Prigogine theorem is not valid (an autooscillation regime).

Thermodynamics of Active Transport. Let us analyze the thermodynamics of active transport of one (sodium) ion not associated with the transfer of other substances. For the sake of simplicity of argumentation, accept that it is possible to indicate a metabolic process triggering the active transport.

Let us denote the rate of the active cation transport as I_+^a and the rate of metabolism as I_r then

$$\begin{aligned} I_+^a &= L_+^a X_+ + L_{+r}^a A, \\ I_r &= L_{+r}^a X_+ + L_r^a A, \end{aligned} \tag{5.20}$$

where X_+ is the negative difference of electrochemical cation potentials (moving against "its own" driving force gradient) and A is the affinity of the metabolic reaction necessary for the transport. In the case of one metabolic reaction (ATP hydrolysis), the rates of intake and generation of all metabolites are associated stoichiometrically. Therefore to estimate the metabolic rate one may take, e.g. the rate of O_2 consumption upon respiration. Then the affinity A may be expressed as a negative change in the entire thermodynamic potential of the metabolic reaction (ATP hydrolysis) per mole of O_2 consumed.

It should be reminded that the electrochemical potential of ion $\overline{\mu}$ is summed up of its chemical potential μ and electric potential φ on the membrane

$$\overline{\mu} = \mu_0 + RT \ln C_+ + ZF\varphi,$$

where F is the Faraday number (96,500 coulomb/mol), Z is the ionic valence, and C_+ is the ion concentration. Phenomenological coefficients L_r^a link the active transport (I_+^a) and metabolism (I_r) because of the coupling and interference of the fluxes and forces of these processes. So, for a monovalent ion, the driving force X_+ is determined as

$$X_+ \Delta\overline{\mu} = RT \ln \frac{C_+^i}{C_+^0} + F\Delta I. \tag{5.21}$$

By placing the same solutions ($\Delta C = C_+^i = C_+^o = 0$) on both sides of the membrane and by varying the $\Delta\varphi$ value, one can determine phenomenological coefficients from (5.20)

$$L_+^a = \frac{\partial I_+^a}{\partial X^+} = -\frac{\partial I_+^a}{\partial (F\Delta\varphi)} = -\frac{\Delta I_+^a}{\Delta (F\Delta\varphi)} \quad (\Delta C = 0,\ A = \text{const}),$$

$$L_{+r}^a = \frac{\partial I_r}{\partial X^+} = -\frac{\partial I_r}{\partial (F\Delta\varphi)} = -\frac{\Delta I_r}{\partial (F\Delta\varphi)} = -\frac{\Delta I}{\partial (F\Delta\varphi)}$$

$$(\Delta C = 0,\ A = \text{const}).$$

Coefficients L_+^a and L_{+r}^a are found from the slope of corresponding straight lines $\Delta I_+^a = -L_+^a \Delta (F\Delta\varphi)$ and $\Delta I_r = -L_{+r}^a \Delta (F\Delta\varphi)$. The constant A values at transitory changes in X_+ facilitate a successful thermodynamic analysis.

Thus, for a sodium flux in frog skin tissues, the equations are as follows

$$
\begin{aligned}
I^a_{\mathrm{Na}} &= L_{\mathrm{Na}} X_{\mathrm{Na}} + L_{\mathrm{Na},r} A, \\
I^{\mathrm{sb}}_r &= L_{\mathrm{Na},r} X_{\mathrm{Na}} + L_r A.
\end{aligned} \tag{5.22}
$$

The I^a_{Na} value is accepted to be positive when the flux moves from the external (mucous membrane) to the internal (serous membrane) surfaces of the tissue. The I^{sb}_r value is the rate that portion of the total respiration process (oxygen consumption) which is directly connected with the transport and exceeds the level of basal metabolism of substances. It denotes the suprabasal flux of the metabolic reaction. Obviously, it is critical to preserve the system parameters and be able to purposefully vary X_{Na} at constant A in experimental conditions in order to determine the coefficients and maintain linear dependencies in equations (5.22). The experiments were carried out on frog skin with varying the X_{Na} values caused by $\Delta\varphi$ changing. At the same time the composition of the washing liquid and sodium concentration remained unchanged. In such conditions $X_+ \cong -F\Delta\varphi$ and the equations for transport look like this

$$
\begin{aligned}
I^a_{\mathrm{Na}} &= L_{\mathrm{Na}}(-F\Delta\varphi) + L_{\mathrm{Na},r} A, \\
I^{\mathrm{sb}}_r &= L_{\mathrm{Na},r}(-F\Delta\varphi) + L_r A,
\end{aligned}
$$

where $\Delta\varphi = \varphi^i - \varphi^0$.

It was proved in direct experiments that the rate of active transport on frog skin I^a_{Na} depends linearly on the $\Delta\varphi$ value, $\Delta\varphi$ changing symmetrically in the range from 0 to ± 80 mV. The I_2 value can be determined from the oxygen consumption using oxygen electrodes. It was found that at symmetrical disturbances of the potential the relation of I_r and $\Delta\varphi$ was linear in the range from 0 to ± 70 mV. The dependency of I^a_{Na} and I^{sb}_r versus the external concentration of sodium in conditions of its constant internal concentration at constant zero difference of the electric potentials ($\Delta\varphi = 0$) was studied as well. Under such conditions, it was also observed that the rate of active transport I_{Na} and suprabasal oxygen consumption I^{sb}_r depend linearly on the difference of chemical potentials $\Delta\bar{\mu}_{\mathrm{Na}}$ on the membrane. But if X_{Na} is changed by varying the internal concentration of sodium, no linearity is observed any longer. This is the result of changes in the microstructure and composition of the membrane *per se*.

Such experiments were successfully performed for studying the active transport of protons using equations of non-equilibrium thermodynamics for two fluxes. In all cases, the varying of X_+ permits estimating phenomenological coefficients and affinity A of the driving metabolic reaction. Analogous formalism has been effectively applied to describe the processes of phosphorylation in mitochondria and chloroplasts. It is generally accepted that these objects have tight coupling of the three main processes underlying the bioenergetics of cell membranes: electron transport with substrate oxidation (I_0, A_0), phosphorylation of ADP with ATP formation (I_{P}, A_{P}), and translocation of protons across the coupling membrane ($I_{\mathrm{H}}\Delta\bar{\mu}_{\mathrm{H}}$). Of the key importance is the transmembrane circulation of protons, which is induced by electron transfer and, in its turn, "triggers" the ATP synthesis. The phenomenological description of the system includes three corresponding equations

$$
\begin{aligned}
I_p &= L_p A_p + L_{p\mathrm{H}}\Delta\bar{\mu}_{\mathrm{H}} + L_{p0} A_{p0}, \\
I_{\mathrm{H}} &= L_{p\mathrm{H}} A_p + L_{\mathrm{H}}\Delta\bar{\mu}_{\mathrm{H}} + L_{0\mathrm{H}} A_0, \\
I_0 &= L_{p0} A_p + L_{0\mathrm{H}}\Delta\bar{\mu}_{\mathrm{H}} + L_0 A_0,
\end{aligned} \tag{5.23}
$$

where the meaning of coefficients L is obvious. Equations (5.23) can be used when the affinity values A_0 and A_P may be varied within a wide range and the $\Delta\bar{\mu}_H$ value can be estimated from the difference of pH and electric potentials on the coupling membranes. It was determined that in this case there is also a linear dependence between the forces and fluxes, which should facilitate determining coefficients L experimentally. To achieve this, the experimental conditions can be simplified by maintaining $A_0 = \text{const}$ without changing the $\Delta\bar{\mu}_H$ value and by achieving a stationary state when $I_H = 0$. The $\Delta\bar{\mu}_H$ value may be kept equal to zero. In both cases, equations (5.23) will be simplified so that they can be used for the estimation of required data. This approach is used as a starting point for discussing various hypotheses of energetic coupling. Particularly, it is presumed in the chemosmotic hypothesis that only proton translocation is directly associated with the ATP formation rather than with the electron transfer. In extreme case, coefficient l_{p0} should equal zero. Undoubtedly it should be clearly perceived that the thermodynamic analysis can be helpful in estimating the energetic efficiency and the level of coupling of processes, but it does not suggest anything about their molecular mechanisms.

6

Thermodynamics of Systems Far From Equilibrium (Nonlinear Thermodynamics)

Methods of classical thermodynamics can be used for concluding whether an isolated system can perform a spontaneous transition from one state to the other. An open system has stationary states that can be either near or far from thermodynamic equilibrium. The problem of whether an open system is able to make transition from some initial state to the final stationary state can be solved by comparing the velocities of entropy formation of these states, provided they both are within the range of linear thermodynamics, that is near thermodynamic equilibrium. However far from equilibrium, it is impossible to make univalent conclusions on how the velocity of entropy formation changes. The evolution of such non-equilibrium dynamic systems is determined primarily by both the kinetics of interactions of constituents and the movement of the system along phase trajectories rather than by statistical ordering of its initial states. Such systems have a limited number of final states and behave like "chemical machines". That is why propagation of thermodynamic ideas over non-equilibrium system behavior can provide only an additional characteristic of stationary states far from equilibrium, while the position and modes of attaining such states depend mainly on their kinetic equations.

Stability of Stationary Points. Let us briefly discuss the thermodynamic features of stability of stationary points, the properties of which have been analyzed in Chapters 1–4. Assume that the stationary state of the system was disturbed bringing about deviations in the values of forces and fluxes from their stationary values

($\overline{X}$ and I). It appears that if the initial stationary state is stable, the product of the "disturbance" values δI and δX must be positive

$$\delta I \cdot \delta X > 0.$$

This is the criterion of stability of stationary states far from equilibrium. However, attempts to find general thermodynamic criteria for the motion to the stationary state far from equilibrium are not successful. The reason is the deterministic behavior of kinetic systems where contrary to equilibrium systems, the idea of entropy is not decisive in predicting the direction of transition processes.

Compare the type of stationary point stability to its distance from thermodynamic equilibrium. Only stable stationary states of the "node" type are allowed near equilibrium. As the distance from the equilibrium increases the values X and I will also increase (Fig. 6.1) and the system may leave the range of linear thermodynamics without losing its overall stability. The "stable focus" point corresponds to this event. Though it is possible that upon moving away from equilibrium, the system will undergo a bifurcation change and instability will take place. Unstable stationary points that are inherent to the system far from equilibrium are "saddles" or "unstable focuses". At the bifurcation point, where stability is lost, the product $\delta I \cdot \delta X$ becomes negative ($\delta I \cdot \delta X < 0$) which agrees with the thermodynamic threshold of instability origination in the system. In this case it is said that thermodynamic fluctuation appears leading the system away from the unstable point which may be a reason for the system collapse. However, at definite parameter values, this fluctuation seems to stimulate the system transfer to a new state to which the stability is passed over. For example, this may be a limit cycle near the unstable focus. The formation of dissipative structures in distributed systems is also preceded by the disturbance of thermodynamic stability far from equilibrium. Finally, trigger transitions between stable stationary states (see Fig. 3.4) occur at the boundary of stability on the curve of stationary states, when thermodynamic stability is broken, and the system makes an abrupt transition from one stable state to another.

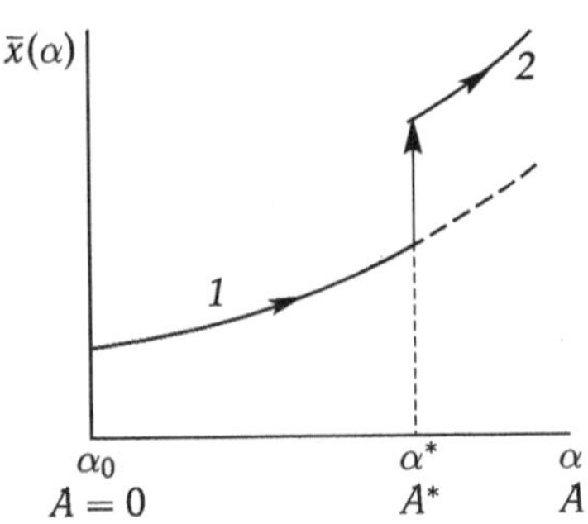

Figure 6.1. Dependence of the stationary concentration of a component element versus parameter α that is a measure of deviations.

So, thermodynamic features of stability of stationary states are compatible with corresponding mathematical features and can serve as their additional characteristic. But far from equilibrium, no general thermodynamic criteria exist for the direction of movement of the open system because its behavior is conditioned by its dynamic properties and kinetic regulatory mechanisms rather than by general statistical rules.

This specificity underlies the complexity of entropy conceptions and information when describing the general properties of biological systems.

Entropy and Information. In accordance with the Boltzmann formula, entropy is determined as the logarithm of the number of microstates possible in the given macroscopic system

$$S = k_B \ln W \tag{6.1}$$

where $k_B = 1.38 \cdot 10^{-16}$ erg $\cdot$ degr^{-1} or $3.31 \cdot 10^{-24}$ entropy units (1 e.u. = 1 cal $\cdot$ degr^{-1} = 4.1 J/K), or $1.36 \cdot 10^{-23}$ J/K is the Boltzmann constant, and W is the number of microstates (for example, the number of possible ways to distribute gas molecules in a vessel). Namely to this effect entropy is the measure of the system disorder and chaotization. Real systems possess stable and unstable degrees of freedom, for instance, related to solid walls of the vessel and the gas molecules in it. The notion of entropy is associated just with unsteady degrees of freedom when chaotization of a system may take place and the number of allowable microstates far exceeds unity. In totally stable systems only a single solution is fulfilled, i.e. the number of ways to realize this particular macrostate of the system equals unity ($W = 1$) and the entropy equals zero. The notion of entropy as well as thermodynamic conceptions can be used in biology with respect to certain metabolic processes rather than for describing at large the behavior and general biological properties of organisms. The relationship between entropy and information in the information theory was recognized for statistical degrees of freedom. Assume that we obtained information on what is a specific way of all the possible ones to realize the specified macrostate of the system. Apparently, the amount of the information obtained is larger the higher is the initial uncertainty or entropy of the system.

In keeping with the information theory, in this case the amount of information concerning a single actual state of the system will be as follows

$$I = \log_2 W. \tag{6.2}$$

A unity of amount of information (bit) is the information contained in a reliable message when the number of initial possible states is $W = 2$:

$$I = \log_2 2 = 1 \text{ bit}. \tag{6.3}$$

For example, the message on what side the coin fell down when thrown in the air contains 1 bit of information. A comparison of formulas (6.1) and (6.2) allows finding relationship of the entropy (in entropy units) and information (in bits)

$$S(\text{e.u.}) = 2.3 \cdot 10^{-24} = 1 \text{ bit}. \tag{6.4}$$

Now let us try to formally estimate the amount of information contained in the human body consisting of 10^{13} cells. Using formula (6.3) we get

$$I = \log_2 10^{13} \sim 10^{13} \log_2 10^{13} \sim 4 \cdot 10^{14} \text{ bits}.$$

This amount of information would be required initially to realize the only possible accurate location of cells in the organism. It is equivalent to a quite insignificant decrease in the entropy of the system

$$\Delta S = 2.3 \cdot 10^{-24} \cdot 4 \cdot 10^{+14} \sim 10^{-9} \text{ e.u.} \sim 4 \cdot 10^{-9} \text{ J/K}.$$

If it is accepted that the human organism has also a unique character of the sequence of amino acid residues in proteins and nucleotide residues in the DNA, the overall amount of information contained in the human body will be

$$I \sim 1.3 \cdot 10^{26} \text{ bits},$$

which is equivalent to a small reduction of entropy by $\Delta S \sim 300$ e.u. $= 1200$ J/K. In metabolic processes, this decrease of entropy is readily compensated with its increase upon oxidation of 900 glucose molecules. Thus, a comparison of formulas (6.1) and (6.2) demonstrates that biological systems do not possess formally any increased information capacity as compared to other non-living systems consisting of the same number of structural elements. At first glance, this conclusion conflicts with the significance of the role of informational processes in biology.

But the relation of I and S values in (6.4) is valid only with respect to the information on which of the total number of microstates W is realized at the moment. This microinformation associated with the arrangement of all atoms in the system cannot be recorded and stored because any of such microstates would rapidly transfer to another one due to thermal fluctuations. And the value of biological information is determined not by its amount, but first of all by the possibility to record, store, process, and further transmit the information to be used in the organism life activity.

The basic condition of perception and storage of information is the ability of the receptor system due to the obtained information to switch to one of the stable states a priori inherent to the system because of its organization. That is why informational processes in organized systems are associated only with definite degrees of freedom. The storage of information must proceed with a concurrent energy loss in the receptor system so that it could be stored for a sufficient time and not lost due to thermal fluctuations. It is here that microinformation, which the system could not store, turns into macroinformation, which the system stores and then can pass it to other acceptor systems. Entropy is recognized to be the measure of the diversity of microstates that cannot be memorized by the system, while macroinformation is the measure of the diversity of their states, which should be remembered by the system.

For instance, the information capacity of DNA is determined only by the amount of specific nucleotides and not by the total number of microstates including vibrations of all atoms in the DNA chain. The process of information storage in DNA consists in fixation of specific positions of nucleotides that are stable because of the chemical links formed in the chain. Further transmission of the genetic information occurs as a result of biochemical processes in which energy dissipation and formation of consistent steady chemical structures provides for the efficiency of biological processing of the information. On the whole, information processes are very frequent in biology. On the molecular level, they happen not only upon storage and processing of the genetic information, but also upon mutual recognition of macromolecules, provide for the specificity and direction of enzyme reactions, and are of importance for interaction of cellular membranes and surfaces. Physiological receptor processes, playing an independent information role in the organism life activity, are also based on interactions of macromolecules. In all cases, macroinformation emerges initially as conformational changes upon dissipation of a portion of energy on definite degrees of freedom in the interacting macromolecules. As a consequence, macroinformation is recorded as a set of energetically rather deep conformational substates, which permit storing this information for a time period required for its further processing. The biological importance of this macroinformation is realized with respect to the specificities of the biological system organization and actual cell structures, in which further processes take place causing corresponding physiological biochemical effects.

It can be stated that living systems regulate directionally biochemical reactions at the level of single macromolecules, the totality of which defines macroscopic properties of biological systems.

Such properties are not inherent even to the latest state-of-the-art devices such as submicron processors in which electron fluxes are monitored with unavoidable energy losses. Below it will be shown that in biomembranes, electron fluxes are regulated with respect to the transfer of every individual electron along the chain of macromolecular carriers.

Moreover, it will be demonstrated that energy transformation in biological processes occurs in macromolecular energy-transforming nanosized "machines".

Small dimensions specify low values of energy gradients and, as a result, approximate the operation of such machines to the conditions of thermodynamic reversibility. As known, this increases the energetic efficiency (coefficient of efficiency) of energy transformation. Such nanosized molecular machines optimally combine the maximal energy yield and the low level of energy dissipation corresponding to the low rate of entropy generation in the system.

Small differences in the values of redox potentials of separate carriers in the chain of photosynthesis and respiration illustrate the above statement providing for conditions close to reversibility of individual processes in electron transport.

The analysis of the operation of some molecular motors coupled to energy transformation challenges the development of thermodynamics of small-scale systems, in which the values of energy differences at elementary stages of working cycles are comparable to those of thermal fluctuations. Indeed, the average value for the overall energy of a macrosystem (ideal gas) consisting of N particles and distributed over them in accord with the Gauss law is $3/2Nk_BT$. The size of random fluctuations of this value is about $1/\sqrt{N}$ and is insignificant in respect to the average value for a system consisting of a large number of particles. However at low N, the size of fluctuations approaches the average energy value for such a small system that can be only of several k_BT units.

For example, a kinesin molecule smaller than 100 nm moves along microtubules carrying cell organelles and making 8-nm steps every 10–15 milliseconds due to the energy of ATP hydrolysis ($20k_BT$). At every step the "kinesin motor" makes work of $12k_BT$ with the coefficient of efficiency 60%. In this sense, kinesin is one of various molecular machines making use of the energy of phosphate bond hydrolysis in different processes including replication, transcription, repairing and the like. The small size of such machines can be helpful in absorbing the energy of large thermal fluctuations from the environmental space. Undoubtedly, on average when the molecular motor moves along its dynamic trajectory, the performance of work results in thermal energy release. But at separate stages of the working cycle, the accidentally absorbed energy of thermal fluctuations combined with the "directed" energy of hydrolysis of phosphate bonds may contribute to the relation of the free energy change and the work performed. In this case, thermal fluctuations can cause already evident deviations from the averaged dynamic trajectories. Consequently, such small systems cannot be adequately described in terms of classical thermodynamics. At present, these problems are being actively worked on together with the progress in nanotechnologies associated with the development of nanosized molecular machines.

It should be noted once again that biochemical processes of energy transformation, in which useful chemical work is performed, are per se only suppliers of elements for self-organization of biological structures and thus for generation of information in biological systems.

It is precisely to biological reactions that the basic principles of chemical thermodynamics and, in particular, the fundamental concept of chemical potential, as a measure of the dependence of the number of possible microstates on the number of particles in the system, are applicable.

A chemical reaction is regarded as a result of redistribution of the number of moles or the number of particles (molecules) of reagents and products in the reaction provided their total number is unchanged. These redistributions are associated with the splitting and formation of chemical bonds and thereby are coupled to thermal effects. It is in the range of linear thermodynamics that their general direction obeys the Prigogine theorem. Figuratively speaking, the biochemical reaction creates original elements and delivers them to the site of self-assembly of stable "informational" macromolecular complexes, i.e. carriers of information. The self-assembly itself occurs spontaneously and naturally proceeds with the general free energy decrease: $\Delta F = \Delta U - T\Delta S < 0$. Actually, upon formation of a stable ordered structure, the absolute value of the energy of formed structural bonding ($-\Delta U$) should be larger than the decrease in the entropy member ($-T\Delta S$) in the expression for the free energy $|\Delta U| > |T\Delta S|$, so that $\Delta F < 0$.

Let us remind that during prebiotic evolution, stable structural "bricks" of living matter (amino acids, nucleotides, sugars) were formed self-spontaneously, abiogenically from inorganic simple compounds, without involvement of living systems, at the expense of external sources of energy (light and electric charges) required for overcoming the activation barriers of reactions of synthesis.

Generally, the direct emergence of biological information on the macromolecular level leads to an appropriate decrease in the structural entropy (generation of negative entropy). This entropy reduction is compensated by the formation of steady associations in the information structure. At the same time, the balance of "thermodynamic" entropy in an open system is determined by the relation of the driving forces and fluxes in a group of chemical processes that create conditions for synthesis of information structures.

Apparently, the calculation of the total balance between structural and thermodynamic entropy in a living system has a purely arithmetic meaning. It is conditioned by two interrelated groups of processes, however differing in their nature, with no direct compensation for the entropy change between them.

PART II

MOLECULAR BIOPHYSICS

III
Three-dimensional Organization of Biopolymers

7

Three-dimensional Configurations of Polymer Molecules

7.1 Statistical Character of Polymer Structure

Macromolecules and most important biopolymers — proteins and nucleic acids — make the basis of cell structures. In contrast to small molecules, the specifics of polymer molecules are determined by a large number of single-type links (monomers) joined in a linear chain. The heat motion of atoms and groups of atoms contained in a polymer chain, their turns and rotations around single bonds are a condition for a large number of internal degrees of freedom in a macromolecule. This makes us regard macromolecules as a macroscopic system, whose statistical behavior is manifested by mean values of such parameters as dimensions, shape, and degree of folding of the macromolecule. At the same time, chemical bonds and short-range and long-range interactions between atoms superimpose certain restrictions on the number of possible conformations of macromolecules. Changes in the conformations of biopolymers occurring in the processes of cell metabolism and energy transformation also have a quite definite character and reflect the intramolecular dynamic organization of biopolymers. So, the particularity of a biological macromolecule as a physical object consists in a close combination of statistical and deterministic (mechanical) characteristics of its behavior: on the one hand, a large number of interacting atoms and intramolecular degrees of freedom and, as a consequence, the possibility to realize a vast number of different conformations, and on the other hand, a specific chemical character and directed conformational changes upon functioning of biopolymers.

Complex processes of energy transformation including migration of the energy of electron excitation energy and electron transport occur in macromolecules. The functioning of macromolecules is based on electron-conformational interactions (ECI) revealed in different processes involving biologically active macromolecular complexes.

The main problem is to reveal the nature of interactions of atomic groups that determine conformational peculiarities and internal dynamics of biological macromolecules, mechanisms of electron and conformational transitions, and their coupling and proceeding from this to understand the mechanisms of functioning of biopolymers in living systems. This is the subject of molecular biophysics. □

Every macrostate of a polymer is characterized by definite values of molecular parameters and can be realized by a large number of microstates or conformations. Heat motion and rotation around single bonds compiling a polymer chain of atoms should lead to an essential extent of chain folding and the formation of a coil. This coil has a large number of conformations which the chain can have in the process of microbrownian motions of its parts. Here the term "conformation" has a "microsense". It means a definite, with an accuracy to the values of amplitudes of valence oscillations, location of atoms in a polymer. It is clear that the interdependency of such motion of the links will be most noticeably displayed in neighboring links and will decay with an increase in the distance between them.

7.2 Volumetric Interactions and Globule–Coil Transitions in Polymer Macromolecules

Upon formation of different macromolecular structures and transitions between them, all interactions between atoms independent of their real physical nature can be separated into two types: short-range interactions between atoms of neighboring links and long-range interactions, or volumetric effects, between atoms which though are remote from each other, but have randomly drawn close together in space as a result of the chain bending. □

Coil and Globule. Let us analyze a homogeneous sequence of identical links in a single chain or a homopolymer (Lifshitz, 1968). Without defining concretely the physical nature of the links and their interactions, we will set geometrical dimensions of the homopolymer by the mean-square distance $\overline{h}^2$ between its ends, and its internal three-dimensional structure by spatial distribution of the density of the links $n(x)$. Due to the volumetric interactions, the density of the number of the links in the space occupied by the molecule can change from point to point. Because of the interconnection of links or the linear memory in a polymer filament, the change in the density in one point of the space is associated with the change in the density in another point, i.e. three-dimensional correlation of the density subsists. If a macromolecule has no volumetric interactions at all, it has no reliable three-dimensional structure. In this state, the fluctuations in the density are of an order of magnitude of the density itself. Such a state is called a coil. Here the correlation radius ξ is of the same order as the dimensions of the macromolecule $[R = (\overline{h}^2)^{1/2}]$: $\xi \sim R$, where $R \sim lN^{1/2}$.

The subsistence of volumetric interactions can lead to realization of such a state, in which density fluctuations are small as compared to the density itself. Such a dense formation is called a globule. The radius of correlation of density fluctuations is much smaller than the dimensions of the macromolecule: $\xi \ll R$, in this case $R \sim N^{1/3}$. In contrast to a coil, a globule has a definite three-dimensional structure. The core of a large globule is spatially homogeneous with a constant concentration of links n_0.

Conditions for Existence of Coil and Globule. Under what conditions are a coil and a globule formed? Due to volumetric interactions the regions drawn close together can either approach each other or repulse. At a temperature increase repulsion of monomers enhances while a temperature decrease promotes their mutual attraction. At a definite temperature the repulsion of monomers is completely compensated by their mutual attraction. This temperature point is called a Θ-point or Θ-temperature. In the Θ-point there are no volumetric interactions, and the macromolecule is a coil of $R \sim lN^{1/2}$ dimensions which is evidently preserved even when the temperature rises $T > \Theta$. But in the range $T > \Theta$, the dimensions of the coil increase due to the increased repulsive forces: $R > lN^{1/2}$. Thus,

$$\overline{R}^2 = \alpha(\overline{R}^2)_0, \tag{7.1}$$

where $(\overline{R}^2)_0$ is estimated without taking into account long-range volumetric interactions; α is the swelling coefficient of the macromolecule; $\alpha > 1$ at $T > \Theta$ and $\alpha = 1$ at $T = \Theta$.

In good-quality solvents, the attraction between the chain atoms and the solvent is higher than between the chain atoms, which is equal to an increase in their mutual repulsion in such solvent (at $T > \Theta$; here $\alpha > 1$). On the contrary, in low-quality solvents the mutual attraction of the polymer links is higher than their attraction to the solvent molecules (at $T < \Theta$; here $\alpha < 1$). In the range $T < \Theta$, attractive forces prevail in the volumetric attraction and can cause condensation of the polymer coil to a tight weakly fluctuating globule. This globule is stabilized by self-coordinated compressing field caused by attractive forces between monomers.

It was found that a globule has a discrete range of free energy values. A temperature above the critical one leads to the disappearance of the discrete range and the globular state, which accordingly can be formed by the external field only at rather low temperatures. In such a globule, the ends of the chain are statistically independent, i.e. the correlation radius is far smaller than the macromolecule dimensions and that correlates with the globular state of the chain. In real macromolecules, in the absence of an external field the volumetric interactions generate a self-coordinated field which invokes the formation of a globule. The character of density distribution in a globule is shown in Fig. 7.1.

Energy of Interaction of Links. A change of temperature results in changed dimensions of the macromolecule that is expressed in the corresponding change of the average density of the number of monomer links and, as a consequence, in the change in the energy of their interaction. Hence it is evident that the contribution of attractive and repulsive forces to the free energy of interaction F of links is dependent on the number n. At small n the free energy of interaction can be expanded in series according to degrees of freedom of n:

$$F = Vk_BT(Bn^2 + Cn^3 + \ldots), \tag{7.2}$$

where V is the volume of the system; B and C are expansion coefficients called virial coefficients (B is the second virial coefficient, C is the third virial coefficient, etc.). Obviously, the first term in the right-side part of equation (7.2) proportional to $\sim n^2$

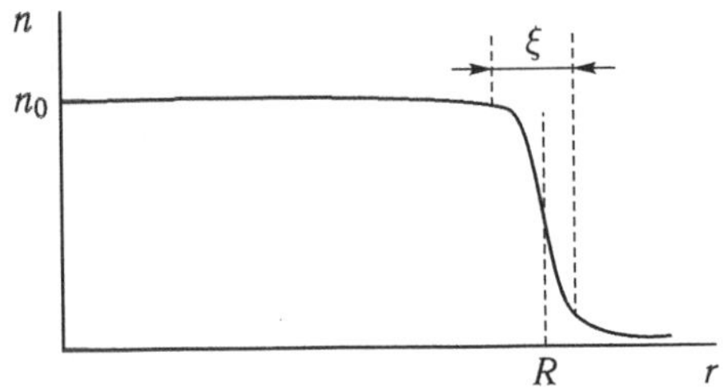

Figure 7.1. Distribution of the density of links in a globule.
$R \sim N^{1/3}$ is the radius of the globule, $R \ll lN^{1/2}$; N is the number of links, l is the length of a link, ξ is the radius of correlation of links in the globule ($\xi \ll R$). The density profile for a large globule is a "fuzzy step" with its core preserving some constant concentration of links n_0.

describes the contribution of paired collisions of links; the second term describes the contribution of triple collisions, etc. If upon collisions the attraction energy is high as compared to $k_B T$, the attractive forces of the colliding particles play the key role, making a negative contribution to the total energy of the system ($B < 0$). As a result, the macromolecule shrinks relative to the dimensions of an ideal coil. An inverse pattern is observed when the attraction energy is less than $k_B T$ and the repulsive forces make a positive contribution to the free energy and cause swelling of the coil ($B > 0$). In the absence of volumetric interactions at Θ-temperature

$$B(\Theta) = 0, \tag{7.3}$$

condition $B < 0$ corresponds to the case of a low-quality solvent, and $B > 0$ to the case of a good-quality solvent with repulsive links of the polymer. Figure 7.2 shows a typical view of the free energy dependence (of a globule) on n (curve *1*).

Evidently, the above plot agrees with the negative value of the second virial coefficient ($B < 0$) and the positive value of the third virial coefficient ($C > 0$). So, in the elemental case when $B < 0$, the contribution of repulsion ($C > 0$) dominates already upon collision of three particles. When the temperature and other conditions in solution change, virial coefficients change and correspondingly the value F changes also.

Coil–Globule Transitions. When F has no minimum any longer at n values, varying from zero (Fig. 7.2, curve *2*), the globule disintegrates and the globule–coil transition occurs. The condition $F = 0$ corresponds to the transition point.

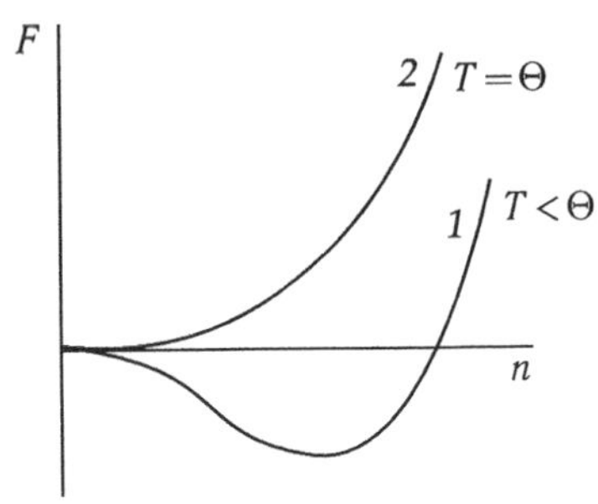

Figure 7.2. Dependence of the free energy of a globule (*1*) and a coil (*2*) on the density of links.

Transitions between the states of a coil and a globule are transitions between different phases accompanied by a change in the aggregate state of the macromolecule. As known, first order phase transitions take place between two states, each of which is stable on one side of the transition point and metastable on the other side. Upon a change of the external parameter (temperature) the transition between them entails a heat effect (for example, melting of ice and formation of a liquid aqueous phase). □

Second order phase transitions proceed without heat effects, with only one minimum of the free energy existing in the transition range. Upon first order phase transitions, the entropy and internal energy change very rapidly due to the expenditure of the final heat of the transition and the final change in the specific volume of the system. In the transition point of second order, the heat capacity changes unevenly, the entropy and the internal energy change continuously, while the specific volume of the system does not experience a sudden change.

On the whole it was shown (Lifshitz et al., 1979) that in the case of a rigid chain, the coil – globule transition occurs as a phase transition of the first order, though its transition heat is low. The final sudden change in the density is observed somewhat lower the Θ-point. In the case of flexible chains, the transition occurs as a gradual second order transition stretched over the entire Θ region (Fig. 7.3). □

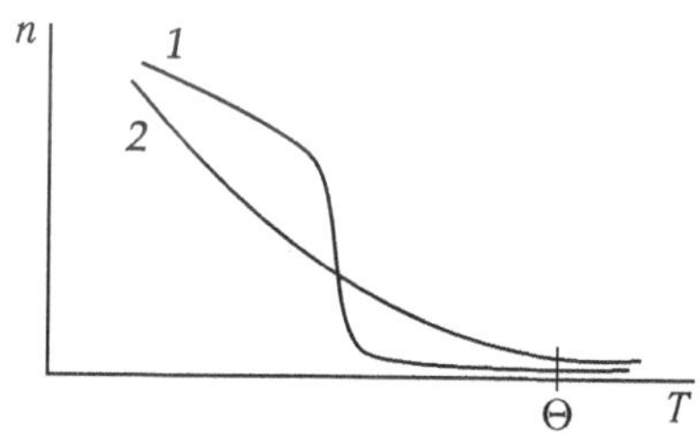

Figure 7.3. Temperature dependence of average density within a macromolecule in the range of the coil – globule transition.)

1, Rigid chain; *2*, flexible chain.

7.3 Phase Transitions in Proteins

Temperature Denaturation. Protein globules undergo transitions of the order – disorder type in a relatively narrow range of the disturbing variable (temperature) and in this regard resemble first order phase transitions. A straightforward thermodynamic interpretation of temperature transitions of biopolymers is that denaturation is a typical monomolecular reaction of transition from the native form A to the denatured form B: $A \rightleftarrows B$. Direct calorimetric measurements of heat effects allow determining the enthalpy change accompanying such transitions. The simplest thermodynamic interpretation of calorimetric data is based on the Kirchhoff equation of the dependence of the enthalpy ΔH of the process on temperature:

$$\Delta H = \int_{T_1}^{T_2} C_p(T)\, dT, \tag{7.4}$$

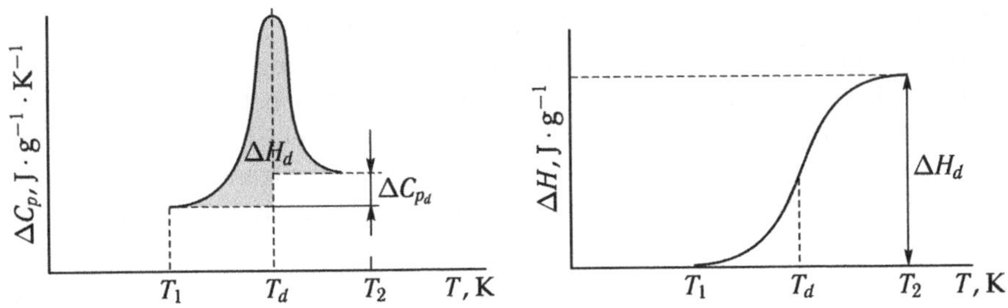

Figure 7.4. Changes of heat capacity ΔC_p and enthalpy ΔH in the range of heat denaturation of biopolymers.

T_1 is the temperature of the beginning of the melting process, T_d is the temperature of the heat absorption maximum ("melting temperature"), T_2 is the temperature of the termination of the melting process, ΔH_d is the melting enthalpy, ΔC_{p_d} is the possible change of heat capacity after heat denaturation.

where $C_p(T)$ is the temperature-dependent heat capacity at constant pressure; T_1 and T_2 are the limits of temperature changes. In standard measurements, heat is supplied to the sample at a constant rate $dQ/dt = v_Q$ and the rate of the temperature rise in the sample is recorded: $dT/dt = v_T$. The heat capacity is estimated using the formula

$$C_p = v_Q/v_T. \tag{7.5}$$

The sensitivity of up-to-date microcalorimeters permits recording changes in the heat flow up to 10^{-7} W. Figure 7.4 shows typical curves of changes in enthalpy ΔH and heat capacity in the range of heat denaturation of biopolymers. The whole temperature range (T_1, T_2) is separated in two regions: the denaturation (T_d, T_2) and non-denaturation (T_1, T_d) ones. Within each of these regions, the correlation of forms A and B changes in every point of transition depending on the temperature of the sample. During heat denaturation the heat capacity of a biopolymer changes upon its transition from the native (helical, globular) state to the denatured (coil-like) state. The change in enthalpy upon this transition is estimated by the formula

$$\Delta H_{\text{den}} = \int_{T_1}^{T_2} C_p\,dT - \int_{T_1}^{T_d} C_{p(\text{nat})}\,dT - \int_{T_d}^{T_2} C_{p(\text{den})}\,dT = \Delta H_d - \int_{T_1}^{T_d} \Delta C_{p,d}\,dT, \tag{7.6}$$

where ΔH_d is the molar enthalpy of transition; T_d is the temperature of transition (maximum of the melting peak); $C_{p\text{ nat}}$, $C_{p\text{ den}}$, $\Delta C_{p,d}$ are heat capacities of the biopolymer in the native and denatured states and the difference of these values at the temperature of transition T_d, respectively. In this process, the change of entropy makes

$$\Delta S_{\text{den}} = \frac{\Delta H_d}{T_d} - \int_{T_1}^{T_d} \frac{\Delta C_{p,d}}{T}\,dT, \tag{7.7}$$

and the Gibbs free energy ΔG changes as follows

$$\Delta G_{\text{den}} = \Delta H_{\text{den}} - T_1 \Delta S_{\text{den}},$$

$$\Delta G_{\text{den}} = \Delta H_d \frac{T_d - T_1}{T_d} - \int_{T_1}^{T_d} \frac{T - T_1}{T} \Delta C_{p,d}\,dT. \tag{7.8}$$

It was shown that a protein molecule can undergo reversible conformational transitions in a partially disorganized state, not only upon heating, but also upon cooling of the solution (P. L. Privalov, 1974). In this state (cold denaturation), the tertiary structure is disturbed because of the reduction of hydrophobic interactions upon cooling (see Section 5 in Chapter 9). On the contrary, upon heating the increase of dissipative forces proceeds faster than the hydrophobic ones, which induces denaturation. In both cases, denaturation occurs as a cooperative process between two states — native and denatured — by the "all-or-none" principle. But upon heat denaturation, the protein enthalpy and entropy grow, while upon cold denaturation they decrease.

Coordination with Structural Changes. Under conditions of strong hydrophobic interactions, the disturbance of secondary structure elements, immersed in the hydrophobic core, is energetically disadvantageous. That is why the change in the density of the hydrophobic core can take place only upon motions of the secondary structure regions (α-helix and β-sheets) as a whole. It is because of this that upon denaturation the change in the density of the hydrophobic core is not local, but in effect involves the entire hydrophobic core with a displacement of the "structural backbone" elements. The expansion of a molecule near the native state proceeds with preserving irregular features of the three-dimensional structures. In this case it appears that on the pathway of such an affinity expansion, a state is formed when the attractive forces are largely weakened and the corresponding increase in the mobility of side groups has not yet occurred (E. I. Shakhnovich, 1984). This leads to a spasmodic jump in the volume of the denaturation transition, i.e. results in intramolecular phase transition of the first order associated with disturbance of the tight packing of the core.

In the denatured state water penetrates within the protein destabilizing it and increasing its volume. On the contrary, when the protein is in the native state, water plays a stabilizing role due to hydrophobic interactions. In a molten globule independent rotations of side chains may occur, but water molecules have no access to it. Compactness of the molten globule is provided by residual hydrophobic interactions. When these interactions are weak, the solvent (water) interacts with proteins penetrating in pores and widening them, as a result of which the globule begins swelling. The stronger the interaction of water with the protein chain, i.e. the energetically more expedient these contacts as compared to contacts of the chain links with each other, the clearer the swelling of the globule. On this pathway a one-time transition of the globule to a "molten" globule may take place. The molten globule state can play a definite functional role providing protein interaction with ligands (V. E. Bychkova and O. B. Ptitsyn, 1993).

The thermodynamic description of temperature transitions in proteins is associated with critical temperatures. We have seen (Section 2) that the temperature T_θ of the coil – globule transition depends on the character of interactions in the chain. In a heteropolymer chain, where energies of interactions between individual links vary, a further transition from the unorganized globule with a great number of equally possible states to the "frozen" phase state occurs at the critical temperature $T_c < T_\theta$. Here a molten globule is already formed with transition to the native structure possessing a unique and single conformation. So, the temperature of chain folding T_c into the native conformation differs from temperature T_θ of coil condensation to a globule.

The extent of thermal stability of a protein and its behavior upon denaturation are dependent on the properties of domains which are connected with the regions of increased density in globular proteins. The distribution of charges and dipole moments in the space of globules affects the dimensions of cooperative regions — energy domains. Redistribution of the charges under the action of environmental conditions leads to changes in the domain dimensions and protein properties. By using point mutations, it is possible to substitute separate amino acids in proteins and in this way control the distribution of charges in critical points responsible for thermal stability. For example, the substitution of a histidine residue in tetracytochrome in the ligand of the iron atom for a methionine residue has resulted in directed modification and increase of the protein thermal stability due to association of iron with the sulfur of methionine.

Thus, the statistical pattern of phase transitions becomes more complicated due to superposition of structural rearrangements which depend on the physical nature of interactions of monomer links and are not necessarily averaged over the total space volume occupied by the macromolecule. From the phenomenological point of view, it can be accepted that curves of temperature transition of biopolymers correspond to first order phase transitions, though in contrast to melting of usual crystals there are no abrupt disruption of thermodynamic functions H and S because of heterogeneity and small dimensions of the protein molecule. At the same time, a drastic change in thermal capacity upon transition rather resembles second order transition. □

8

Different Types of Interactions in Macromolecules

The primary structure or the basic sequence of a polymer chain (amino acids in the protein chain, nucleotides in the chain of nucleic acids) is determined by chemical or valence interactions. Moreover, weak non-valence forces exist between molecules which result in attraction at large distances and repulsion at small ones. Figure 8.1 demonstrates a typical dependence of potential interaction energy $U(r)$

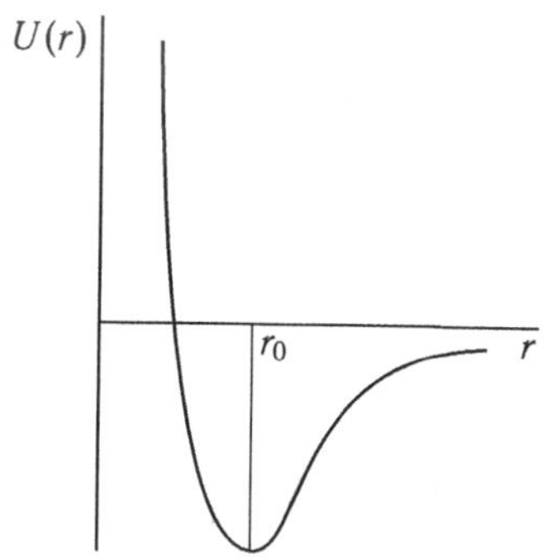

Figure 8.1. Typical view of the dependence of the interaction energy on the distance (see explanation in the text).

of two particles of molecular nature on the distance (r) between them. The interaction energy $U(r)$, or the interaction potential, is related to the interaction force $F(r)$:

$$F(r) = -dU(r)/dr.$$

At small distances due to repulsion of molecules upon mutual penetration of their electron shells, repulsive forces predominate, $dU_{\text{rep}}(r)/dr < 0$, while attraction predominates at large distances: $dU_{\text{att}}(r)/dr > 0$. The total interaction potential $U(R)$ is an algebraic sum:

$$U(r) = U_{\text{rep}}(r) - U_{\text{att}}(r).$$

At $r = r_0$ the minimum on curve $U(r)$ corresponds to the equilibrium position in which the attractive forces are equalized by repulsive forces.

The secondary structure of macromolecules is dependent on different types of interactions. Further, we will analyze the nature of these interactions and their contribution to stabilization of the biomacromolecule structure.

8.1 Van-der-Waals Interactions

Van-der-Waals forces play a great role in formation of condensed liquid and solid states as well as in interactions on the phase interface. They determine interactions in gases and the emerging deviations from the ideal gas laws. In the first approximation, these deviations obey the known van-der-Waals equation for gases

$$(p + a/v^2)(v - b) = nRT, \quad \square \qquad (8.1)$$

where a and b are constants determining the mutual attraction and repulsion of molecules. Subject to the correlation between pressure p and volume v different aggregate states of the liquid and gas and transitions between them can be realized.

As a rule, the value of the characteristic energy of van-der-Waals interactions is in the range from 4–8 kJ/mol and higher, whereas at room temperature ($T = 300$ K) the heat energy of the molecule is $RT \simeq 48$ J/mol · degr, and the energy of covalent bonds is 170–630 kJ/mol · degr. Van-der-Waals forces have an electromagnetic nature and are determined by interaction of electric dipoles in adjacent molecules. Depending on whether the interacting molecules have constant electric dipole moments or the latter appear due to polarizability of electron shells, different types of van-der-Waals forces may exist. □

Despersion Interactions

The most prevalent types of non-valence interactions are despersion, or polarization, interactions observed between molecules having no constant dipole moments. The major part of such an interaction is explained by the most weakly associated external electrons. The nature of depression interactions has a quantum-mechanical character and is a result of emergence of "instantaneous" dipoles, which are formed upon electron motion in molecules having no constant dipole moments. □

In the harmonic approximation, the electron motion in the molecule can be regarded as a linear harmonic oscillator, the energy of which, as known from quantum mechanics, equals

$$E_n = (n + 1/2)\hbar\omega_0 \qquad (n = 0, 1, 2, \ldots), \tag{8.2}$$

where ω_0 is the oscillation frequency of such an electron oscillator, estimated subject to the rules of classical mechanics, n is the quantum number, its magnitude determines the energetic level of the electron. It is seen that in the lowest quantum state at $n = 0$, the energy of the oscillator does not turn to zero and equals

$$E_0 = \hbar\omega_0/2. \tag{8.3}$$

The $E_0 = 1/2\hbar\omega_0$ value is called zero energy. Its existence is a consequence of the relationship of uncertainties according to which the electron as a quantum particle cannot have a definite coordinate x and pulse p at the same time. Pulse p of the particle is equal to the product of its mass m and rate v: $p = mv$, or $E = p^2/2m$. Uncertainties in coordinate magnitudes Δx and pulse Δp of electrons are related by the Heisenberg relationship

$$\Delta x \Delta p \simeq \hbar, \tag{8.4}$$

which is also called the uncertainty relationship. Subject to (8.4) the coordinate and pulse of the electron cannot have concurrently strictly definite values, i.e. $\Delta x \neq 0$, $\Delta p \neq 0$, and this is possible only provided the electron has some minimal energy ($E_0 = 1/2\hbar\omega_0$). □

In other words, at $n = 0$ the electron has zero oscillations with energy E_0 which are associated with fast motions and displacements of the charge from the equilibrium positions, and consequently with the emergence of "instantaneous" dipole moments in the molecule at rest. The emergence of a dipole moment in one molecule induces a field in the environmental space and thus a dipole moment in another molecule. The interaction of these fast changing dipole moments leads to the situation when a change in the dipole moment of one oscillator causes additional polarization in the other, and vice versa. Thus, the two oscillators become interrelated. A strict analysis shows that if each of them oscillates with frequency ω_0 in the system of connected oscillators, two frequencies ω_+ and ω_-, corresponding to symmetric and asymmetric oscillations, emerge (Fig. 8.2). The energies of connected oscillators have discrete quantum values

$$E_{n+} = \hbar\omega_+(n_+ + 1/2), \quad E_{n-} = \hbar\omega_-(n_- + 1/2) \tag{8.5}$$

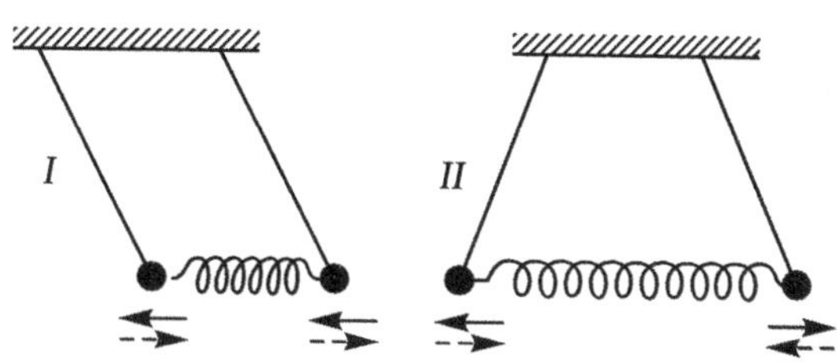

Figure 8.2. Symmetrical (*I*) and asymmetrical (*II*) oscillations of connected pendulums.

or in the zero state $n_+ = n_- = 0$

$$E_{n+} = \hbar\omega_+/2, \quad E_{n-} = \hbar\omega_-/2, \tag{8.6}$$

where

$$\omega_+ = \omega_0\left(1 - \frac{e^2}{R^3 f} - \frac{e^4}{2R^6 f^2}\right), \qquad f = m\omega_0^2.$$
$$\omega_- = \omega_0\left(1 + \frac{e^2}{R^3 f} - \frac{e^4}{2R^6 f^2}\right),$$

The overall energy of a system of connected oscillators will be

$$E = E_{n+} + E_{n-} = \hbar\omega_0\left(1 - \frac{e^4}{2R^6 f^2}\right). \tag{8.7}$$

A comparison of this expression with the sum of the values of zero energy of unconnected oscillators (8.3)

$$E = \hbar\omega_0/2 + \hbar\omega_0/2 = \hbar\omega_0$$

shows that the total energy of the system is smaller than this sum by the value $e^4\hbar\omega_0/2R^6 f^2$ which is the binding energy, or the energy of dispersion interaction U_{disp}:

$$U_{\text{disp}} = -\frac{e^4\hbar\omega_0}{2f^2}\frac{1}{R^6}. \tag{8.8}$$

So, the change in the level of energies associated with the emergence of instantaneous asymmetry in the distribution of charges is a physical reason for intermolecular interactions between molecules, which are in the ground state. A more strict estimation of the dispersion interaction yields the expression

$$U_{\text{disp}} = -\frac{3I_1 I_2}{2(I_1 + I_2)}\frac{\alpha_1\alpha_2}{R^6},$$

where I_1 and I_2 are ionization potentials, α_1 and α_2 are polarizabilities of the molecules. □

The relative role of different types of van-der-Waals forces for individual substances is given in Table 8.1.

Table 8.1. Values of polarizability, dipole moment, ionization potential and energy of different types of weak interactions between some atoms and molecules

Substance	α, cm$^3 \cdot 10^{24}$	P, e. u. $\times 10^{18}$	I, eW	U_{or}	U_{ind}	U_{disp}
					($\times 10^5$ J)	
H	0.667	0	13.6	0	0	6.1
O_2	1.57	0	13.6	0	0	39.8
N_2	1.74	0	15.8	0	0	57.2
H_2O	1.48	1.84	18.0	190	10	47
NH_3	2.24	1.5	11.7	84	10	70

Interaction Potentials. As seen from Fig. 8.1, these potentials are formed from attractive and repulsive energy; as a rule, they are not calculated strictly theoretically but are set as Buckingham interatomic potentials

$$U(R_{ik}) = -A/R_{ik}^6 + B\exp(-\mu R_{ik}) \tag{8.9}$$

or Lennard–Jones potentials

$$U(R_{ik}) = -A/R_{ik}^6 + B/R_{ik}^{12}, \tag{8.10}$$

where A, B, and μ are constants, R_{ik} is the distance between interacting atoms i and k. The initial members in Equations (8.9) and (8.10) correspond to attraction and the second members to repulsion. For example, the values of empirical parameters A and B of the Lennard–Jones potential for $O \cdots O$-interactions are $A = 1547.7$ kJ $\cdot$ nm^6/mol, $B = 895.23 \cdot 10^3$ kJ $\cdot$ nm^{12}/mol, and for $H \cdots C$-interactions they decrease accordingly to 536.7 kJ $\cdot$ nm^6/mol and 238.1 $\cdot$ 10^3 kJ $\cdot$ nm^{12}/mol. At the given values of constants, formulas (8.9) and (8.10) determine the minimal distance R_0 at which the system of two interacting atoms is stable. If only van-der-Waals forces are taken into account and neglect other interactions, the force field around the atom has the center of symmetry. Then it is possible to determine the effective volume around the atom, or the effective radius of interaction, within which the two atoms cannot stay simultaneously.

Ramachandran (1963) determined minimal distances for pairs of different atoms and small atomic groups (CH_3, CH_2). For example, for a pair of atoms $C \cdots C$ not bound covalently the typical distance is 0.32 nm, and the minimally possible is 0.3 nm; for $C \cdots O$ it is accordingly 0.28 and 0.27 nm, and for $H \cdots H$ it is 0.2 and 0.19 nm.

Van-der-Waals forces are the basis for the formation of a number of biostructures and, in particular, coiled-coil polynucleotides. The packing of planar molecules of nitrogen bases in vertical packages is provided by "vertical" (stacking) interactions, to which the dispersion member U_{disp} (8.8) makes an essential contribution. Accordingly, for stacking interactions of cytosine–guanine pairs $U_{\text{disp}} = -39.1$ kJ/mol and $U_{\text{rep}} = +18.14$ kJ/mol. Taking into consideration hydrogen bonds and electrostatic repulsion, in this case the total value of the interaction energy is -27.9 kJ/mol. "Horizontal" interactions are responsible for the packing of nitrogen bases in one plane and determine the formation of complementary pairs adenine–thymine (A–T) and guanine–cytosine (G–C). In this case, dispersion interactions play a less important role and the main contribution is made by the electrostatic attraction. Thus, for the A–T pair $U_{\text{disp}} = -7.18$ kJ/mol, $U_{\text{el-stat}} = 19.32$ kJ/mol and $U_{\text{rep}} = +16.00$ kJ/mol. In Chapter 15 we analyze van-der-Waals interactions between the two planes.

8.2 Hydrogen Bond. Charge–Dipole Interactions

Together with usual van-der-Waals forces, hydrogen bonds and electrostatic interactions between charged and polar groups play a great role in stabilization of biostructures. Particularly, it is known that hydrogen bonds stabilize the secondary structure of polypeptide chains. □

Hydrogen Bond. The hydrogen bond is formed between electro-negative atoms (O, N, F, Cl) with involvement of hydrogen which forms a covalent bond with one of them. The nature of the hydrogen bond is complex and is not reduced only to electrostatic attraction, though it makes a considerable contribution to the hydrogen bond energy. The data of the structural analysis of crystals show that the distance between electro-negative atoms, connected by a hydrogen bond, is typically smaller than van-der-Waals radii of corresponding atoms by about 0.02–0.09 nm. For example, the sum of van-der-Waals radii for the pair of atoms O $\cdots$ O is 0.32–0.36 nm, and upon formation of the hydrogen bond O–H $\cdots$ O this distance decreases to 0.25–0.3 nm. So, in the dimer of formic acid

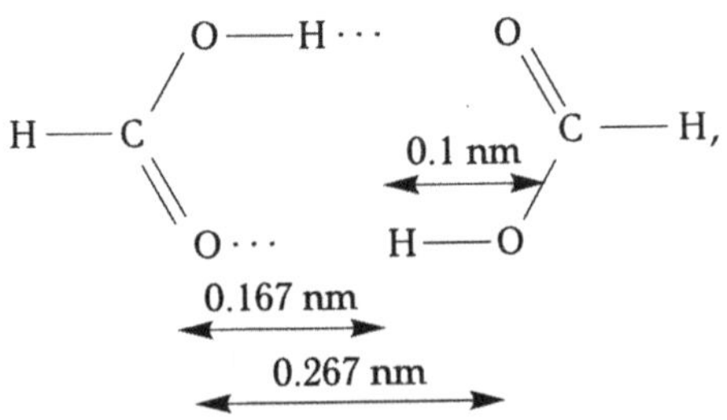

where the O–H distance in the hydroxyl is 0.1 nm, it seems the length of the hydrogen bond is 0.167 nm. The heat of dimerization of formic acid is 59.22 kJ/mol; consequently, 29.6 kJ/mol fall to the share of the energy of one H-bond.

In addition to electrostatic interaction, van-der-Waals attraction and repulsion, the energy of delocalization U_{deloc} of two electrons of the A–H bond and the lone-electron pair of the other electro-negative atom (B) also contributes to the energy of hydrogen bond U_{hydro}, which decreases its length as well. For example, for the O–H $\cdots$ O bond

$$U_{\text{hydro}} = U_{\text{el-stat}} + U_{\text{disp}} + U_{\text{deloc}} + U_{\text{rep}},$$

which gives accordingly $U_{\text{hydro}} = |-25.2 - 12.6 - 33.6 + 35.28| = |-36.12|$ kJ/mol. □

Electrostatic Interactions. This type of interactions not only makes a contribution to the energy of hydrogen bonds, but also plays an important independent role in stabilization of biostructures. Different atoms in peptide chains differ greatly by the character of distribution of charges in them. The interaction of partially charged atoms is characterized by an electrostatic potential the type of which is determined by the equation

$$U_{\text{el-stat}} = \sum \frac{q_i q_j}{\epsilon R_{ij}}, \qquad \square \qquad (8.11)$$

where q_i and q_j are values of partial charges in atoms (i and j), R_{ij} is the distance between atoms, ϵ is the dielectric constant dependent on polarizability of atoms and their mutual localization, its value for proteins is usually taken to be equal to ~ 3.5.

8.3 Internal Rotation and Rotational Isomerism

Let us consider the dependence of conformation of a linear polymer chain on the nature of atom group interactions discussed in the previous section.

Energy of Internal Rotation. Under normal conditions, the lengths of bonds and valence angles (between adjacent bonds) remain practically constant, because at standard temperatures they experience only insignificant heat fluctuations (not exceeding 3% of the length of the bonds (± 0.005 nm) and $\pm(3 \div 5)^\circ$ for valence angles). The remaining variables, which determine the conformation of a polymer molecule, are rotation angles around the bonds, or internal rotation angles φ_i. They are dihedral angles between the two planes formed by pairs of bonds $i-1, i$ and $i, i+1$.

As an example, Figure 8.3 shows a polymethylene chain in the trans-conformation where all φ_i angles of internal rotation of hydrogen atoms in CH-groups around the single i-th bond are zero. Here the sign of φ_I is taken positive for the clockwise rotation and is measured relative to the transform. Since all lengths of the bonds and valence angles θ_i remain constant, the configuration of the main chain is determined by the set $n-2$ of values $\varphi_2, \varphi_3, \ldots, \varphi_{n-1}$, or $\{\varphi\}$. Therefore, different conformations are formed because of the rotation of atomic groups around separate single bonds. The principal circumstance is that internal rotation of bonds is not independent. Hence it follows that the energy of conformation $E\{\varphi\}$ is dependent on the interaction energy of adjacent links, or on neighboring rotation angles,

$$E\{\varphi\} = \sum_{i=2}^{n-1} E_i(\varphi_{i-1}, \varphi_i), \tag{8.12}$$

where each of the members of the sum E_i is regarded as a contribution to the total conformation energy $E\{\varphi\}$, caused by the state of bonds i and $i-1$. The interaction of bonds i and $i+1$ is described by the following member of the sum $E_{i+1}(\varphi_i, \varphi_{i+1})$, depending on the corresponding pair of sequential angles of internal rotation.

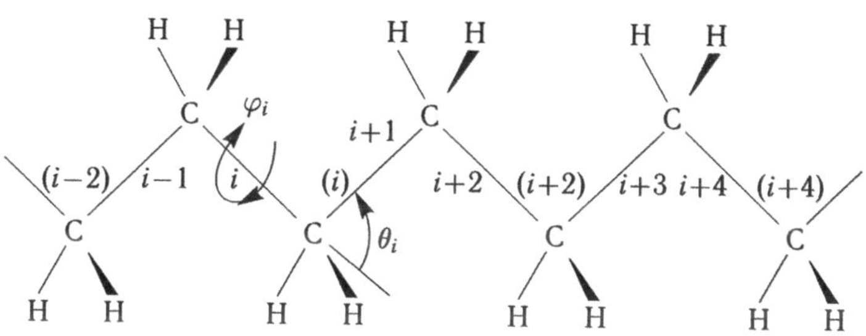

Figure 8.3. Region of a polymethylene chain.
θ_i is the valence angle, φ_i is the internal rotation angle.

Such a statistic system, where the energy of constituent elements depends on their interaction with each other, is called a cooperative system.

Internal Rotation Potential. The energy of short-range interactions of atomic groups revealed upon their rotation around singular bonds determines the total conformation energy of the polymer chain.

The nature of these interactions is associated mainly with repulsion of atoms, with overlapping electron shells that results in retarding energy potential of internal rotation, preventing rotation of atomic groups. Depending on the distance between rotating atomic groups upon changes in the angle of rotation around the singular bond, the magnitude of the inhibitory potential will differ, and consequently different rotational isomers make different contributions to the conformation energy of the whole molecule. □

As seen from Figures 8.4 and 8.5, the ethane molecule has the minimum of conformation energy in the trans-conformation and the maximum in the cys-conformation. During transition of one trans-conformation to another through a cys-form upon rotation around the C–C-bond by angle $\varphi = 120°$, the energy barrier is ~ 12.5 kJ/mol. Based only on van-der-Waals and electrostatic interactions, this value should be just ~ 2.1 kJ/mol. The minimum of energy is realized if the substitutes of corresponding groups are in cross (trans-) conformations, and the maxima correspond to (cys-) conformations. In this case, the dependence of the internal rotation potential on the rotation angle can be approximated by the expression

$$E(\varphi) = (E_0/2)(1 - \cos 3\varphi), \tag{8.13}$$

where E_0 is the barrier height.

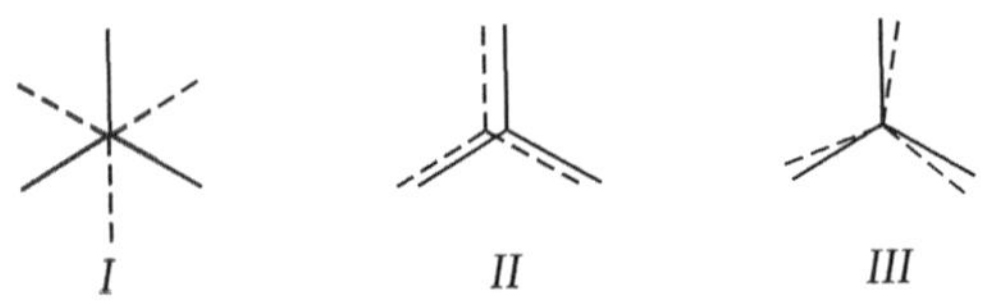

Figure 8.4. Localization of CH-bonds of ethane in trans- (*I*), cys- (*II*) and intermediate (*III*) conformations (projection to the plane perpendicular to C–C bonds).

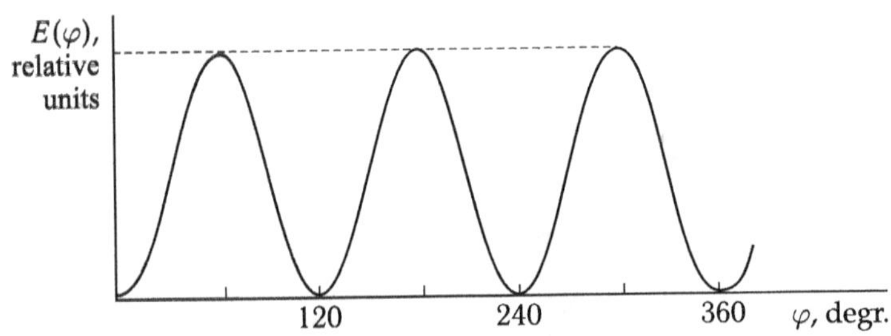

Figure 8.5. Dependence of potential energy of internal rotation in ethane on the rotation angle. Angles of 0, 120° etc. correspond to case *I* in Fig. 8.4, and angles of 60°, 180° etc. correspond to case *II* in Fig. 8.4.

9

Conformational Energy and Three-dimensional Structure of Biopolymers

9.1 Conformational Energy of Polypeptide Chains

Polypeptides form the structural basis of proteins and demonstrate unique structural and molecular characteristics.

Nature of the Peptide Bond. The general character of the structure of a polypeptide chain, consisting of sequentially connected amino acid residues, is shown in Fig. 9.1. The basis of the chain is the –CO–NH– bond, formed by the –COOH-group of one amino acid and the $-NH_2$-group of another amino acid

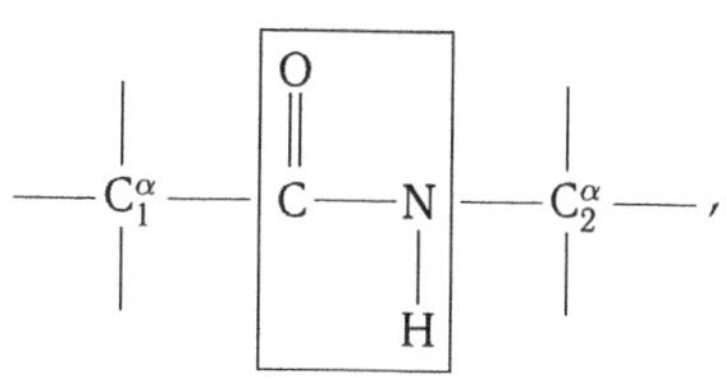

where C_1^α, C_2^α are α-carbon atoms belonging to these atoms.

The most important peculiarity of a peptide bond is its partially double character. The lone pair of $2s^2$-electrons of the nitrogen atom is generalized between carbons and nitrogen, thus yielding an admixture with a double bond between them. In this

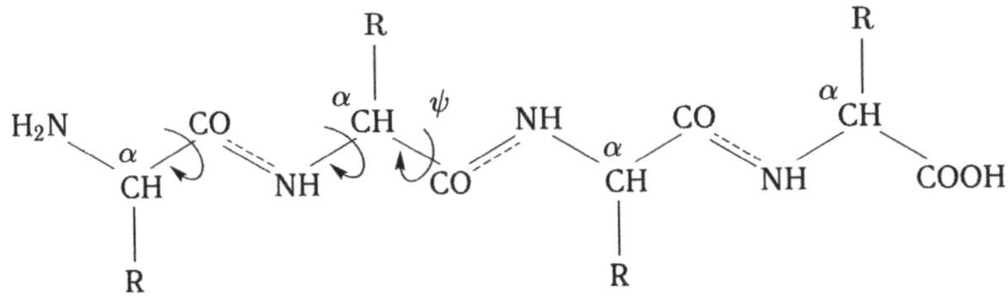

Figure 9.1. A fragment of a polypeptide chain.

case, the carbon electron is "pushed out" from the double π-bond C = O and is localized on the oxygen atom, partially converting the C = O bond into a single one and increasing its length from 0.121 to 0.124 nm:

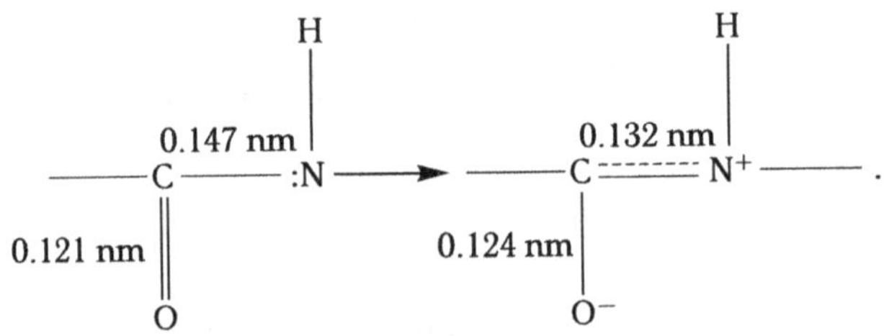

So, between atoms N, C and O delocalization of electrons takes place accompanied by a shift of the electron density in the directions N → C → O:

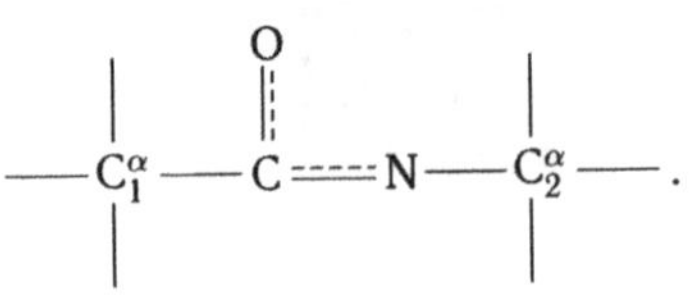

As long as all three atoms N, C, and O are involved in formation of a double bond, their s- and p-orbitals are hydrolyzed with the formation of sp-hybrid orbitals. Single σ-bonds are formed between atoms due to orbitals $O(sp^2)$ and $C(sp^2)$ as well as to orbitals $C(sp^2)$ and $N(sp^2)$. Three p_x-orbitals of atoms N, C, and O perpendicular to the plane of σ-bonds overlap and form a delocalized system of π-electrons (Fig. 9.2). The energy of delocalization of electrons, stabilizing the structure of the peptide group, depends on the extent of overlapping of electron orbitals. The stabilization is maximal when the atoms of the peptide group, including α-carbon atoms of adjacent amino acids, are located on the same plane. In accord with the isomeric theory, the spatial structure of a polypeptide chain is determined by rotation around singular bonds between neighboring atoms (see Fig. 9.1). Because of the double character of the peptide bond C = N the rotation around it (angle ω) is impeded around bonds N–C$^\alpha$ (angle φ) and C$^\alpha$–C (angle φ) (Fig. 9.3).

Energy of Internal Rotation of a Polypeptide Chain. The values of internal rotation barriers E_0 for the angles φ and ψ are set to be 0.84–2.52 kJ/mol, which is much smaller than $E_0 \sim 84$ kJ/mol for the peptide bond C–N. The peptide group CO–NH rotates around these bonds as a single whole, without disturbances in the positions of atoms in the single plane. In other words, pairwise cooperative interaction takes place upon rotation around single bonds belonging to the same α-carbon atom. However

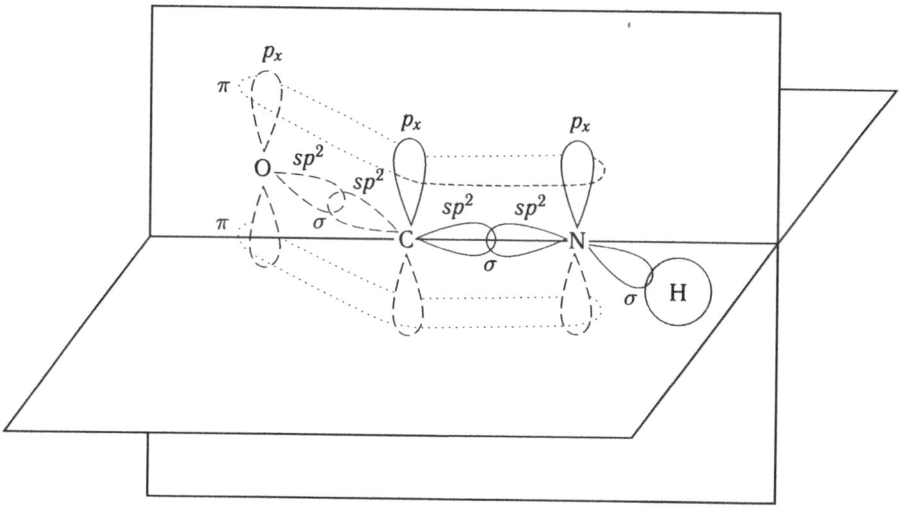

Figure 9.2. Electron configuration of a polypeptide chain.

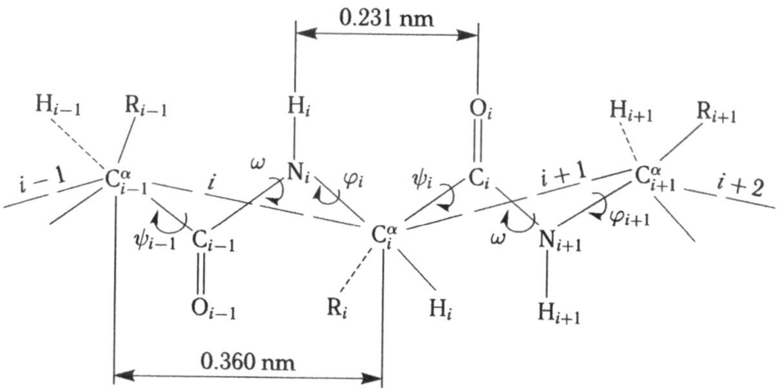

Figure 9.3. Spatial configuration of a polypeptide chain.

interrelated rotations within one pair of angles (φ_i and ψ_i) is utterly independent of the values of the adjacent pair of angles (φ_{i-1} and ψ_{i-1}) or (φ_{i+1} and ψ_{i+1}).

Accordingly, the polypeptide chain is a chain with "isolated" pairs of internal rotation angles, where each pair of angles can be considered independently.

Calculation of Conformational Energy. Calculation of conformational energy of large regions of protein structures is rather complex, it is associated with taking into account all interactions and will be discussed in the following sections. Here we will consider only separate portions of a polypeptide chain.

Subject to van-der-Waals interaction (8.10) between the internal rotation potential (8.13) and electrostatic effects (8.11), the conformational energy looks as follows:

$$E(\varphi,\psi) = \sum_{k,l} E_{kl}(\varphi,\psi) + \{E^0_\varphi(1-\cos 3\varphi)/2 + E^0_\psi(1-\cos 3\psi)/2 + E_{\text{el-stat}}\}, \quad (9.1)$$

where the members $E_{k,l}(\varphi,\psi)$ are determined by the Buckingham potential (8.9) with a distance between the atoms $r_{k,l}$ which depend on φ and ψ.

The conformational energy of a molecule in Eq. (9.1) is the sum of separate types of interactions in the pairwise additive approximation, when each pair of atoms interacts independently of their intramolecular environment. The potential energy optimum and the conformational state corresponding to it are estimated by minimizing the function $E(\varphi, \psi)$ with varying the values of dihedral angles of rotation around singular bonds. Naturally, taking into account the vast amount of calculations, the solution of this task can be only computer-aided. The determination of conformational states of separate amino acid residues using formula (9.1) is a necessary and important stage of plotting conformational maps of the three-dimensional protein structure. The conformational state of a separate amino acid residue is determined by the interaction of atoms of the side chain of the given residue with adjacent peptide groups. Calculations for amino acids with long side chains are associated with the requirement to take into consideration both rotation around their singular bonds (the bonds C_α–C_β and C_β–C_γ) and interaction of neighboring side chains. Rotation barriers in side chains (angles γ) are dependent on the nature of atomic groups. For aliphatic groups $E_0 \sim 12.6$ kJ/mol, and for the phenyl group in the side chain $E_0 \sim 2.52$ kJ/mol (H. A. Scheraga, 1968).

Convenient objects for studying such interactions are N-acetyl-α-amino acid methyl amides (CH_3–CO–NH–CHR–CO–NH–CH_3), where R is the radical corresponding to the side chain of the amino acid residue. Figure 9.4 shows a molecule of N-acetyl-L-alanine methyl amide ($R = CH_3$) in the planar conformation with zero values of the dihedral angles $\varphi(N-C^\alpha)$, $\chi(C^\alpha-C^\beta)$, $\psi(C^\alpha-C^\beta)$, and the angle $\omega = 180°$ (C′ ···· N).

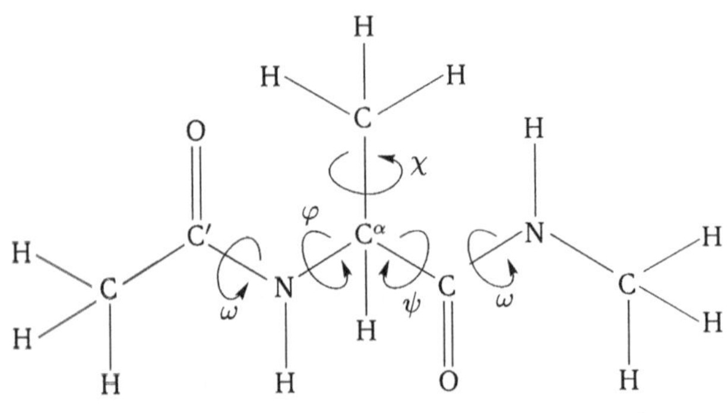

Figure 9.4. Molecule of methyl amide-N-acetyl-L-alanine in the planar conformation (see explanation in the text).

Figure 9.5 shows a diagram of the surface of conformational energy in the function of the angles φ and ψ (G. Ramachandran, 1963, 1968). The diagram exhibits the presence of four regions of low energy, the minima of which are in the regions bordered by contours of 4–8 kJ/mol. A similar pattern was obtained also for other molecules with one amino acid residue at optimal states of side chains. The theoretically calculated conformational states of individual N-acetyl-α-amino acid methyl amides were verified experimentally. This proves the validity of the above methods (see Eq. (9.1)) for taking into account the interactions of atoms within one amino acid residue. Thus, at constant lengths of bonds and valence angles the conformation is fully determined by the values of the dihedral angles (φ, ψ, χ, ω).

Such calculations were made also for more complicated systems, in particular, polypeptides.

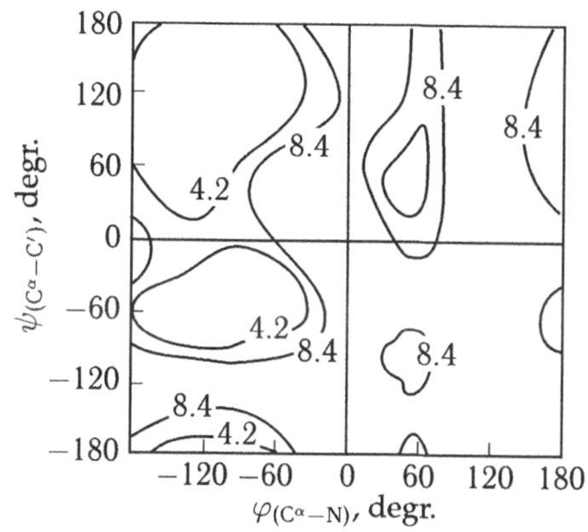

Figure 9.5. Diagram of potential surface of a molecule of N-acetyl-L-alanine methyl amide (numbers show the conformation energy values in kJ/mol).

On the whole it can be concluded that the available calculation methods allow taking into account the most essential interatomic non-valence interactions with a sufficient extent of accuracy, and thus determine the energetically optimal conformations of polypeptide chain regions comparatively small in size. However, as will be shown below, the situation is far more complicated as concerns the spatial structure of large protein molecules.

9.2 Numerical Methods for Estimation of Conformational Energy of Biopolymers

Numerical Methods for Estimation of Conformation Energy of Biopolymers. In a general case, any molecular system, consisting of a set of atomic nuclei and electrons, can be most accurately described using the mathematical apparatus of quantum mechanics. But quantum-mechanical calculations require vast computer capacities, the intensity of computations increasing nonlinearly depending on the number of electrons in the molecule. Recent advances in the development of algorithms for quantum chemistry by means of some approximations and simplifications have resulted in reducing the requirements to linear or almost linear scaling. This has permitted expanding the borders of using the numerical methods of quantum chemistry from simplest molecules containing tens of atoms to complex biomolecular systems. At present, relatively more accurate quantum-mechanical methods *ab initio* are used for systems with dimensions to 10^3 atoms, whereas rougher semi-empiric methods allow constructing models of complex systems with dimensions to 10^4 atoms. Nowadays, the classical model in which a molecule is considered as a mechanical system of interacting elements (atoms or groups of atoms) subject to Newton laws are widely used in molecular modeling. Classical models are much "less exigent" to computer capacities as compared to quantum methods, which makes it possible to study molecular systems consisting of about 10^6 atoms. In many cases, the problem can be solved using compound quantum-classical models: a small part of the system (for example, a functionally significant protein region) is described in terms of quantum mechanics and the remaining part in terms of classical mechanics.

Methods of Empirical Force Fields. To model a spatial structure and dynamic behavior of individual molecules and complex biomolecular systems in classical interpretation, it is necessary to have a possibility to calculate their potential energy E_{mol}. This permits searching for the states of a molecular system which correspond to its free energy. In a general case, E_{mol} of a molecule or a group of molecules can be represented as the sum of independent potentials (terms):

$$E_{mol} = E_{valence} + E_{nonbond}, \tag{9.2}$$

to be precise, potentials for valence ($E_{valence}$) and non-valence ($E_{nonbond}$) interactions between atoms of the system are introduced individually, which in their turn are detailed further depending on the used force field:

$$E_{valence} = E_{bond} + E_{angle} + E_{torsion}, \tag{9.3}$$

$$E_{nonbond} = E_{vdW} + E_{Coulomb}. \tag{9.4}$$

Each potential is accompanied by a detailed description based on the set of physical approximations. The constants used in equations for potentials terms depend on the initially chosen force field.

A rather large number of force fields (*AMBER, CHARM, ECEPP, GROMOS* and others) developed by different research teams are available. The principal difference between them is the analytical type of the expression for energy and different parametrization of energy terms, though on the whole the structure of the force field remains unchanged. The ideology of the force field energy allows making computations with the use of different methods, the most important of the latter being molecular mechanics (methods for energy minimization), molecular and stochastic dynamics, Monte Carlo methods.

Methods for Energy Minimization. The main task of this group of methods is the finding of points corresponding to stable states of the system (local minima) on the surface of potential energy. For multi-atom molecular systems where the number of local minima is immense, of the greatest interest is the task of determining the global energy minimum and states close to it, because it is supposed that namely these states are realized under native conditions. Unfortunately, so far there have been no efficient approaches to the general solution of this problem. According to the definition, minimization methods permit bringing the system only to one of the local minima near the start point. They do not allow overcoming the energy barriers, i.e. "traveling" over the surface of potential energy and getting to its various regions.

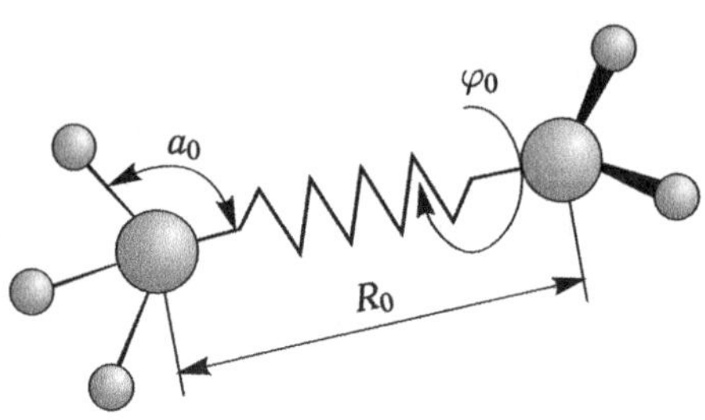

Figure 9.6. Key molecular structural parameters used in a force field. Indicated are the covalent bond (R_0) and valence (α_0) and dihedral (φ_0) angles.

Molecular dynamics and Monte Carlo methods are used to solve this important problem (see below).

Examples of problems of physicochemical biology where methods of energy minimization are used:

1) Obtaining optimal structures of protein – ligand complexes due to minimization of energy of intermolecular interactions.
2) Bringing the system to the state under the given set of conditions, such as, obtaining protein conformations satisfying a number of restrictions imposed on angles and distances as measured in the experiment.

Method of Molecular Dynamics (MD). This method allows following the evolution of a molecular system in time, and considering both systems in the state of thermodynamic equilibrium and non-equilibrium systems. This section gives a general description of the method, while a more detailed description will be given in Chapter 11. Some examples of using MD in solving the problems of physicochemical biology are the following:

1) Analysis of thermal oscillations in biological molecules and supramolecular systems, determination of conformationally labile and rigid regions of the structure and domain motions (for example, in proteins);
2) Modeling the processes of biopolymer folding, studies of denaturation of molecular systems in given conditions of the external environment (temperature, pressure, characteristics of the solvent, etc.);
3) Analysis of structure-dynamical aspects of intermolecular interactions, including solvation/desolvation effects;
4) Studies of fast processes in complex molecular systems, for example, transfer of ions and/or molecules across the membrane, fluctuations of the electrostatic potential in active centers of enzymes or on the surface of the lipid bilayer etc.

The aim of MD is the following: on the grounds of coordinate ($\vec{r}_i$) and rate ($\vec{v}_i = d\vec{r}_i/dt$) values of atoms in the system at the given instant of time, to determine these parameters (connected with the potential and kinetic energy of the system) at any subsequent instant of time. The method of MD is based on the numerical solution of Newton equations for a system consisting of moving classical particles of the mass m_i which interact with each other according to a definite law (in conformity with the chosen empirical force field (EFF)):

$$m_i \frac{d^2\vec{r}_i}{dt^2} = \vec{F}_i = -\frac{dE_{mol}}{d\vec{r}_i}. \tag{9.5}$$

The system of equations of this type is solved numerically, using small intervals (steps) in time (usually $\sim 10^{-15}$ s), a priori smaller than the fastest motion in molecules — proton oscillations. It is obvious that to study the system behavior during only 1 nanosecond (10^{-9} s) it is required to make $\sim 10^6$ iterations. For such complex systems as a protein molecule in water, a hydrated lipid bilayer etc., this requires enormous computing resources and calculation time. For example, it is possible to study the behavior of proteins in the medium only on the time scale limited to $\sim 10^2$ ns. Consequently, at present many most essential "slow" biological problems (for example, self-organization and global structural rearrangements in macro- and supramolecular systems) cannot yet be analyzed properly using the method of MD.

Conformational Search Using the Monte Carlo (MC) Method. It is believed that the author of the stochastic MK method is N. Metropolis, who together with E. Teller and M. Rosenblatt worked on the problem of observing the phase transition in a system of solid spheres at Los-Alamos National Laboratory (N. Metropolis et al., 1953). The idea of the MK method as applied to biopolymers consists in scanning their conformational space by "chance" changes in the geometry of the system (for example, coordinates of atoms or dihedral angles of the molecule). At such an approach it is necessary to control the new states to construct the given type of the ensemble. The control is realized as follows: if the energy of the new state is smaller than the energy of the initial state, the new state is added to the ensemble; if vice versa, the new state also has a chance to get into the ensemble with the probability $P = \exp(-\Delta E/k_{\mathrm{B}}T)$, where ΔE is the difference in the energies of the initial and new states. As a result, at every step one state is added to the ensemble: it may be either the new state or the previous one, if the new state was rejected. The Metropolis criterion for the probability to accept the state sets a canonical ensemble ($P = \rho_{i+1}/\rho_i$, where ρ_{i+1} and ρ_i are probabilities of consecutive states).

In contrast to the MD method, the MK method does not yield a dynamic pattern of the system behavior because it has no kinetic characteristics of the system. The MK method permits a more uniform filling of the phase space as compared to the MD method. But unlike MD, the method of transition between states has no clear physical interpretation. Therefore the MK method allows studying only equilibrium distributions and cannot yield information on dynamic processes in the system.

9.3 Predictions of Three-dimensional Structure of Proteins

The main peculiarity of the three-dimensional structure of proteins is the ability of a polypeptide chain to form a determined structure having dynamic properties essential for performing biological functions. From the macroscopic point of view, the protein molecules have a quite high average density of packing (1.33–1.42 g/cm^2) which is somewhat larger than the density of a solid dry protein and in effect is close to the density of crystals of low-molecular organic compounds. The values of the specific volume of protein molecules in solution approach those of a dry protein and are a little lower than in the case of synthetic polymers.

Molecular biophysics faces an important problem in studying the three-dimensional structure of proteins: to what extent is the protein (secondary and tertiary) structure dependent on its main sequence? This problem includes the following questions.

1) Does the given main sequence unambiguously determine the structure of a protein molecule upon chain folding, i.e. to what extent is the protein chain packing to the native form determined?
2) Is the packing of the random sequence of residues to a structure similar to the native protein structure possible?
3) How were the existing in nature primary sequences of amino acid residues formed and fixed?

The significance of this problem can be illustrated by a simple calculation. As known, the number of different proteins in nature is $\sim 10^5$–10^6 (given the proteins from each organism are unique, this number would be larger). If we accept that protein contains about 200 amino acid residues on average, the number of different sequences will be 20^{200}, which is immeasurably larger than the number of proteins selected by nature ($20^{200} \approx 10^{260} \gg 10^6$). As long ago as in the middle of the last century it was shown that the character of the amino acid sequence and the native protein conformation having the minimal free energy are interrelated (R. Lumry and H. Eyring, 1954). It is known from the works of Nobel Prize winner C. Anfinsen, who studied the reversibility of ribonuclease transition from the denatured state to the native one, that *the needed (for protein folding) information is contained in the linear amino acid sequence of the peptide chain and that no additional genetic information, larger than that contained in the DNA, is required.*

The existing approaches to the theoretical prediction of the three-dimensional structure of proteins include empirical, "physical" and bioinformation methods of modeling.

The empirical method is based on revealing the correlation between the secondary structure of protein and its amino acid composition and sequence. In terms of this correlation it is possible to formulate heuristic principles of the structure of protein molecules and empirical rules of polypeptide chain folding with the formation of secondary and tertiary structures.

We will consider the methods of "physical" modeling of protein structure based on calculations of conformation energy using the approach of empirical force fields (see Section 2 in this chapter).

And finally, methods of comparative modeling are based on the evolutionary similarity between proteins to predict their three-dimensional structure. This approach is also called the homology-based modeling.

Empirical Prediction of Protein Structure. The results of X-ray analysis of globular proteins in the crystalline state provide the initial data for the empirical rules to relate certain amino acid residues to a definite type of secondary structure.

Figure 9.7 shows a diagram of distribution of amino acid residues in the α- and β-regions enriched with hydrophobic groups, while the bends and loops which connect them contain predominantly short polar groups, charged groups, glycine and proline residues (O. B. Ptitsyn and A. V. Finkelstein, 1972). The internal branches of the middle part of a helix contain mainly hydrophobic residues (*Ala* and *Leu*). Residues with negative and positive charged side chains are localized at the N- and C-termini respectively.

Different types of protein structures including structural segments make the structural hierarchy shown in Fig. 9.8. Apparently, this hierarchy of protein structures reflects also the sequence of stages of the block mechanism of protein folding. It means that prediction of a segment of secondary structure based on the main sequence is equal to decoding the initial stage of folding of the protein globule of the polypeptide chain.

In their pioneering work, in the empirical statistical method for prediction of the secondary structure P. Y. Chou and G. D. Fasman (1974) used parameters deduced from known 3D-structures. The statistical analysis permits establishing that some residues are more frequently included in the α-helix than the others: as a rule, alanine, glutamine, leucine and methionine are found in α-helices, whereas proline, glycine, tyrosine and serine are not. Proline cannot occur within the α-helix due to steric hindrances and the impossibility to form a hydrogen bond with the nitrogen atom

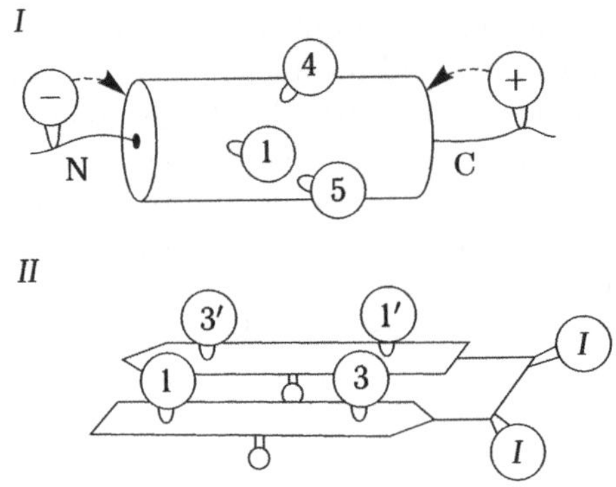

Figure 9.7. Typical distribution of amino acid residues in the α-helix (*I*) and β-hairpin (*II*) (reproduced from O. B. Ptitsyn and A. V. Finkelstein, 1972).

Circles with numbers designate hydrophobic groups, the numbers showing their mutual positions in the chain. Circles with "+" and "−" designate charged groups. At the ends of the helix the arrows indicate hydrogen bonds, which can be formed between polar side groups and open NH-groups of the N-terminus of the helix and CO-groups of its C-terminus.

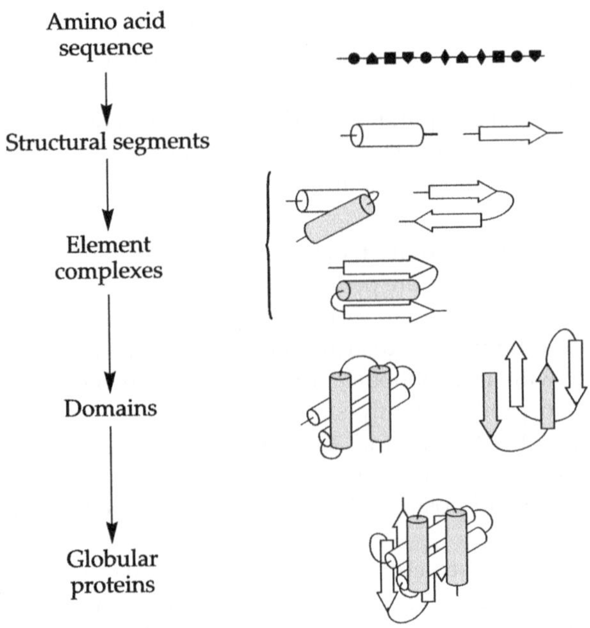

Figure 9.8. Hierarchy of protein structure and sequence of protein folding (reproduced from O. B. Ptitsyn and A. V. Finkelstein, 1979).

of the proline residue involved in the pyrrolidine ring, at the same time proline is frequently found at the N-terminus of the α-helix. The predictions also use the statistics on the lengths of secondary structure regions. So, the length of α-helices varies from 4 to 40 residues at an average length of 10 residues (which corresponds to three turns of a helix), and β-layers usually consist of β-strands the length of which varies from 4 to 10 residues.

Preferences (frequencies of occurrence) of 20 amino acid residues for three structural states (alpha-helix, beta-structure and coil) permit predicting secondary structure segments when the cluster of preferences over the sequence has an average preference value larger than some threshold value. This is the principle of the empirical statistical method developed by Chou and Fasman (1974).

From the historical point of view, the early methods of secondary structure prediction by the amino acid sequence of proteins were based on the determination of the frequency of occurrence of each residue in α-helices, β-structures and disordered structures (without taking into account mutual influence of residues on each other). They enabled researchers to make correct predictions approximately in half cases. Later models took into account the fact that it is not the residue itself but a certain motif in the sequence of residues that determines the probability of acquiring a certain type of secondary structure. Taking into account the influence of the environment, the increase of the size of the "training set" (the number of accessible protein structures) as well as the use of advanced mathematical methods, such as latent markovian models, the method of reference vectors and neutron networks, allow raising this accuracy to $\sim$ 80–90 %.

Method of Physical Modeling of protein structure is based on the step-by-step estimation of atoms not connected with each other and the solvent by valence bonds. It is proposed that the native conformation of proteins corresponds to the minimum of free energy and is characterized by consistency of all types of intramolecular interactions. The latter prevail over intermolecular interactions. It is expedient to make energy estimations of all types of interactions in macromolecules using methods of empirical force fields (see Section 2 in this chapter). However, a direct prediction of the most stable conformation of a protein molecule by minimization of the total energy of all types of interactions is impossible due to insurmountable mathematical difficulties.

So, if it is accepted that on average each amino acid residue has 10 low-energy forms, the total number of structural variants is 10^n (where n is the number of residues in the protein chain). An exhaustive search for all possible conformations, even computer-aided, can be performed only for about a 10-membered peptide fragment. It is clear that the only correct pathway in this case is a step-by-step consideration of separate types of interactions, determining the conformation of each residue with the nearest, neighboring and remote residues of the chain. On the whole, such an algorithm resembles the "block" model of folding, when separate protein structural elements formed at early stages of folding, do not disintegrate further (see below). For a number of globular proteins it was established experimentally that at early stages of folding the so-called folding nuclei are formed (typically stable hydrophobic fragments of secondary structure) — structural blocks (A. R. Fersht, 1990).

Short-range interactions with the nearest peptide groups were taken into account on molecules of acetyl amino acid methyl amides (see Fig. 9.4). Short-range interactions determine the set of low-energy conformations of the unbound monopeptide, and the middle-range and long-range interactions determine the choice of conformations from this set, which are included in the native protein structure. Due to the consistent character of interactions in proteins, the middle-range and long-range interactions do not "spoil" the advantageous energy conformation of the monopeptide. Therefore low-energy states of unbound residues should automatically be included in the composition of low-energy states of the fragments. In its turn, this suggests the existence of structural peculiarities in unbound fragments satisfying the middle-range and long-range interactions in the globule.

The conformational state of a residue is characterized by values of the dihedral rotation angles φ, ψ, and χ (see Fig. 9.4). At the next stage, a small number of low-energy states of oligopeptides (for example, tripeptide $n-1$, n, $n+1$) is taken into account based on known conformational states of dipeptide fragments which form it. Energy calculations show that interactions of residues in the dipeptide lead to noticeable differentiation of their optimal forms. At the same time, the most preferable conformations of the tripeptide turn to be the combinations of low-energy forms of adjacent dipeptides, which is the manifestation of the consistency of tri- and dipeptide interactions.

Thus, in the described method the search for the most stable conformation of protein molecules is reduced to the solution of the "block" task — the search for optimal conditions of packing rigid blocks upon variation of low-energy states of labile parts of the chain known from the analysis of short-range and middle-range interactions. As seen, here the idea about the determined and sequential character of self-organization of the protein globule is the same as that used in the empirical method. But instead of statistical correlations between the primary sequence and structure of the globule, direct calculations of intramolecular interactions and the principle of the minimum of energy as a criterion of optimal packing are used in the suggested method.

Specific calculations are based on the analysis of atom-to-atom interactions including non-valence, electrostatic, torsion interactions and hydrogen bonds. As an example, let us analyze the data of studying the three-dimensional structure and conformational possibilities of the *Arg*-1 – *Cys*-38 fragment of the molecule of bovine pancreatic trypsin inhibitor (BPTI) including 58 residues (E. M. Popov, 1981). At the first stage the structures of several tri- and tetrapeptides were determined that permitted obtaining low-energy conformations of larger fragments (see Fig. 9.9). Further analysis included the overcoming of steric hindrances, upon binding different fragments, and association of the *Phe*-22—*Tyr*-35 fragment with the *Arg*-1—*Tyr*-21 and *Tyr*-35—*Gly*-37 fragments (Fig. 9.9). The obtained structure was compatible with the experimentally determined geometry of this region with an accuracy of 0.1 nm. Thus such a semi-empirical conformational analysis can be successfully used for determining the three-dimensional structure of rather complex peptides (up to forty amino acids) basing on the initially chosen most advantageous conformational states of unbound peptides.

The "block" model of the polypeptide chain packing is used now in computer programs run on supercomputers or in distributed computing networks. For example, the Rosetta algorithm developed by the team of Prof. D. Baker (USA) generates an ensemble of models "collecting" the structure-conservative fragments in the specialized force field. These short (4–10 amino acid residues) overlapping fragments imitate "germ" structures in the block model of folding, and their conformation is chosen by the program from structures already obtained by experimental methods and by energy calculations.

Bioinformatics Methods. Comparative Modeling. A limited number of types of three-dimensional packing of protein molecules occur in nature, and different groups of evolutionary related proteins correspond to each type. According to the current structural classification, the known proteins are divided in $\approx 3\,500$ structural families

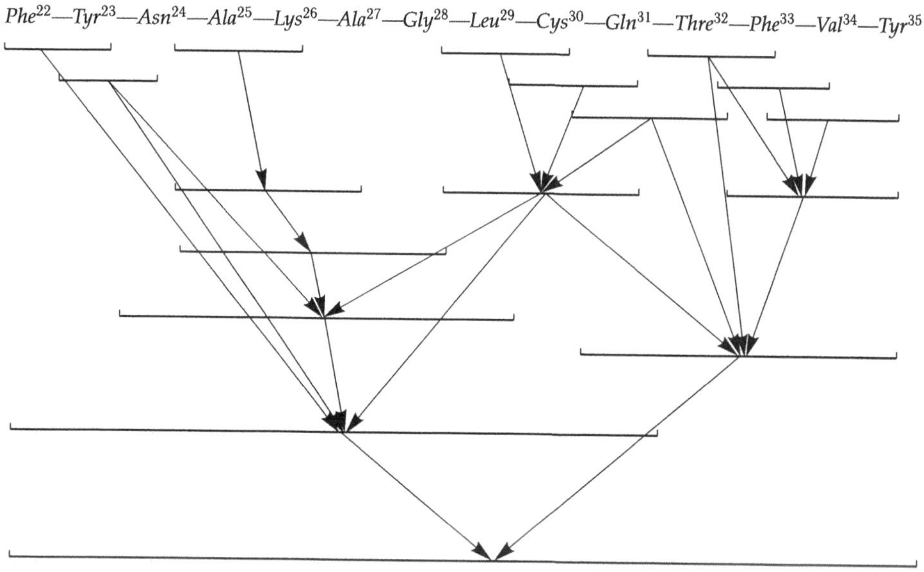

Figure 9.9. Sequence of taking into account interpeptide interactions when calculating the *phe*-22—*tyr*-35 fragment of BPTI.

forming ≈ 1 000 types of three-dimensional packing. For example, if a protein is known to belong to the family of integral cell membrane receptors, its structure should contain hydrophobic α-helices piercing the membrane approximately perpendicular to its plane (and polar and charged residues should be concentrated in "loop" regions (joining the transmembrane ones) and at the N- and C-termini). In some cases, the transmembrane domain may also consist of β-strands, but this is characteristic premodinantly of bacterial cell wall proteins.

It was found empirically that if the sequences of two proteins are identical by more than 30 %, with a high degree of probability the proteins are related to each other, and the extent of evolutionary divergence is not yet great enough for their structures to lose the inherent generality.

At present the homology modeling permits determining the structure of more than half of the proteins with yet unknown organization. The process of homology modeling includes several steps (Fig. 9.10) the main of them being the search for a pattern (a related protein with the known structure) and amino acid alignment.

Alignment of amino acid sequences of two or more proteins is one of the foremost tools of bioinformatics. It looks like a set of sequences written over each other in such a way that regions of maximal compatibility (homology) of the aligned proteins would be clearly seen (see Fig. 9.10).

Protein Self-organization. The α- and β-structural regions in an unfolded protein chain are formed mostly due to distribution of hydrophobic groups and local interactions between them. At the same time, long-range interactions with other regions of the protein chain provide its stabilization and folding in a globule. This provides the secondary structure elements formed in the unfolded chain to predetermine also the structure of the entire folding globule: the α-helical structure is dominating for α-helical proteins in the unfolded chain, and the β-structure is dominating in the unfolded chain of β-structural proteins.

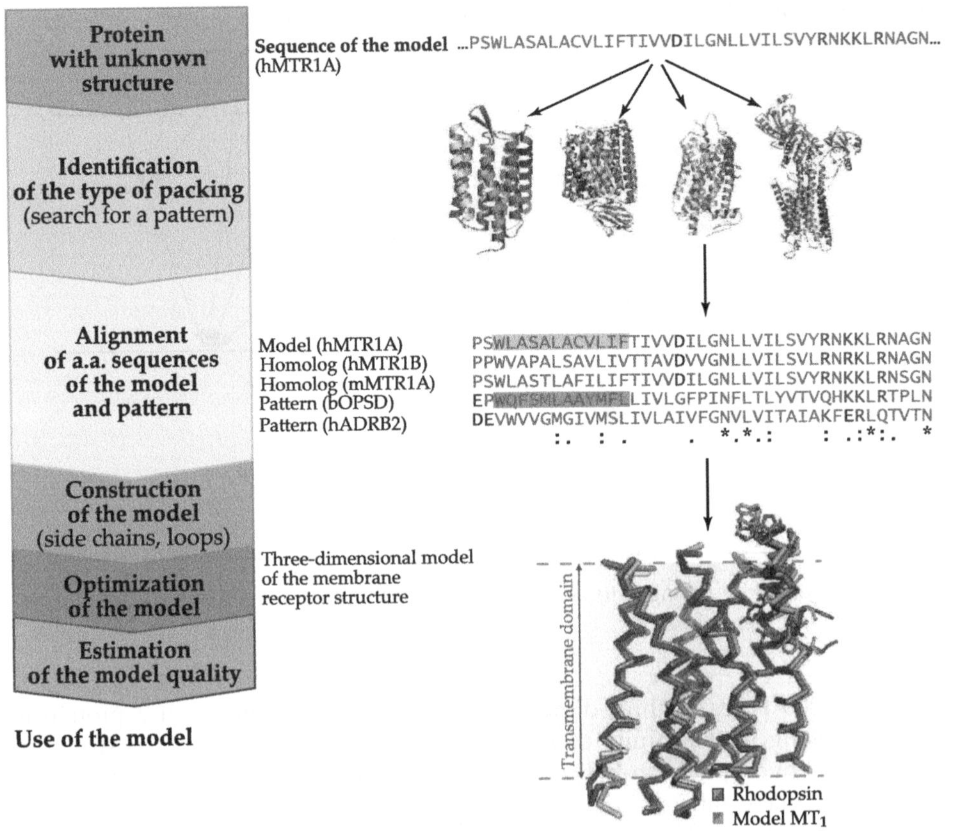

Figure 9.10. Diagram of homology modeling on the example of human melatonin MT1 receptor.

Identification of the structural pattern, i.e. protein with the known three-dimensional structure that is homologous to the modeled one (the identity of sequences exceeds 30 %).

Alignment of amino acid sequences pattern – model. Pairwise alignment serves to determine homologous residues in the pattern and model, which will have similar mutual positions in space. Multiple alignment can be useful for revealing conservative residues in the whole family (marked by an asterisk) or separate subfamilies of proteins (the three top sequences are melatonin receptors). Multiple alignment and profiles of sequences allow identifying weaker homology than pairwise alignment.

Model construction consists mainly in "conversion" of atomic coordinates of the pattern (in this case, visual rhodopsin) to the modeled protein (melatonin MT1 receptor) subject to the alignment. Regions having no homology with the pattern ("loops") should be completed independently. The first transmembrane segment of superposed structures of the model and the pattern shows side chains of residues "highlighted" in the alignment.

For example, for a predominantly α-helical calcium-binding protein from carp muscles the local maxima of the probability of formation of helical regions in an unfolded sequence correspond to hydrophobic regions of the chain with "helical" distribution of hydrophobic, initiating and terminate polar groups. Already at the first steps of secondary structure self-organization, α-helices "suppress" the β-structure almost over the whole length of the chain of this protein (Fig. 9.11). A similar

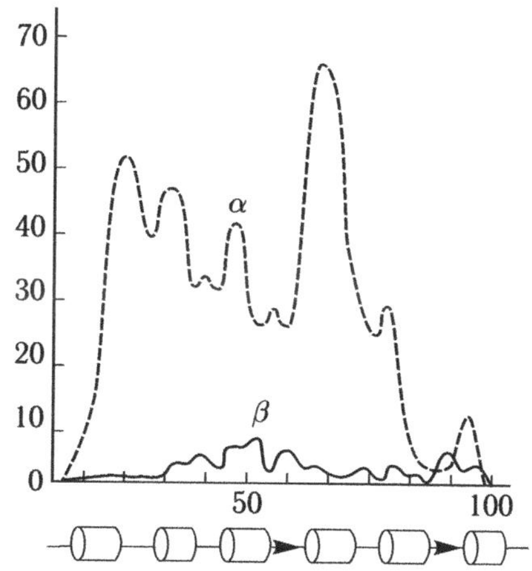

Figure 9.11. Secondary structure of an unfolded chain of calcium-binding protein from carp muscles at the first stage of self-organization (reproduced from Ptitsyn and Finkelstein, 1979). The abscissa axis shows numbers of residues in the protein chain, and the ordinate axis shows probabilities of α-helical (dashed line) and β-structural (solid line) states for each residue in the chain. The experimental localization of α-helices (cylinders) and β-structures (arrows) is shown under the abscissa.

situation is also observed in β-structural proteins (for example, ribonuclease A), where the β-structure predominates from the very beginning of self-organization. The final structure of a protein globule is a result of the directed process of chain folding with the pre-existing structural segments.

Still far from completeness, different molecular models of tertiary protein structure formation have been described, where stoichiometric peculiarities of α- and β-structures are taken into account. For example, α-helices should be packed so that side groups of one helix ("backbones" in positions 1–5 and 1–4) would get into the gaps ("gorges" in position 1–2) between side groups of another helix.

It can be considered that the process of protein self-organization consists of at least three stages (O. B. Ptitsyn, 1973). Formation of fluctuating germs of secondary structure, stabilized primarily by hydrogen bonds in the main chain, occurs at the first stage. At the second stage these germs form an intermediate compact structure of a protein molecule, where hydrophobic interactions of nonpolar side groups play the central stabilizing role. Finally, at the third stage a unique tertiary protein structure is formed stabilized by all basic types of voluminous interactions, including steric complementarity of atomic groups.

9.4 Peculiarities of Three-dimensional Organization of Nucleic Acids

As known, the foremost biological function of a DNA molecule is the storage and transfer of the information stored as a certain sequence of nucleotides in a double helix (Fig. 9.12). Accordingly, the chief requirement to the DNA structure associated with this circumstance — stability and preservation of genes — should be combined

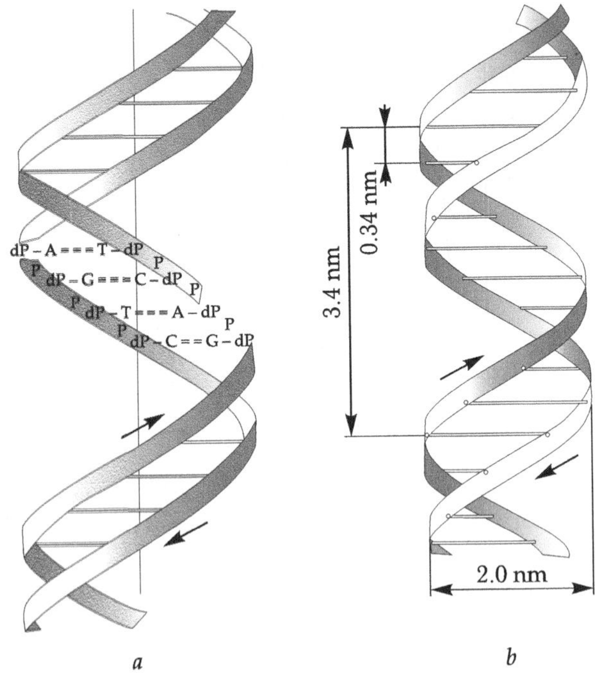

Figure 9.12. Structure of DNA. *a, b*: The double helix.

in a certain way with definite changes of its structure, in particular in the processes of interaction with proteins.[1)]

Physical Models of DNA. In the first group of methods of the greatest importance are the model of a flexible rod and the helix – coil model. □

In standard polymer chains, flexibility is determined by rotational isomerism (see Section 3 in Chapter 8). But in a model where DNA is represented as a rigid rod, its rigidity is characterized by two parameters: resistance to bending of the axis and resistance to twisting.

Resistance to twisting, or torsion rigidity, of DNA is determined as a change in the energy upon deviation of the angle between neighboring pairs of bases from its equilibrium value. According to the calculations, the numerical value of torsion rigidity for a circular polymer chain is $g_t = 0.036k_{\mathrm{B}}T$. At this resistance, the torsion rigidity has a special value in the case of closed circular DNA molecules consisting of two mutually hooked one-strand rings. Under usual conditions, the angle (36°) between the axes of base pairs adjacent in the length of the double DNA helix deviates by ±5° due to heat oscillations. At the same time, for transition between different forms of DNA (*B*- and *C*-forms) it is sufficient to change this angle by only 2.5°. Hence

[1)] *Assessment of the quality, optimization and use of the model.* The most complex stage in homology modeling is optimization of the model taking into account the available information on protein modeling. In general, the modeling of the structure by homology with the protein, performing a distinct function, cannot automatically produce a model suitable for important practical tasks. A necessary requirement is accurate optimization, turning the "workpiece" (which, as a matter of fact, is the model of "zero approximation") into an operating tool, which is the task depending rather on intuition and experience of the researcher than on certain computer methods.

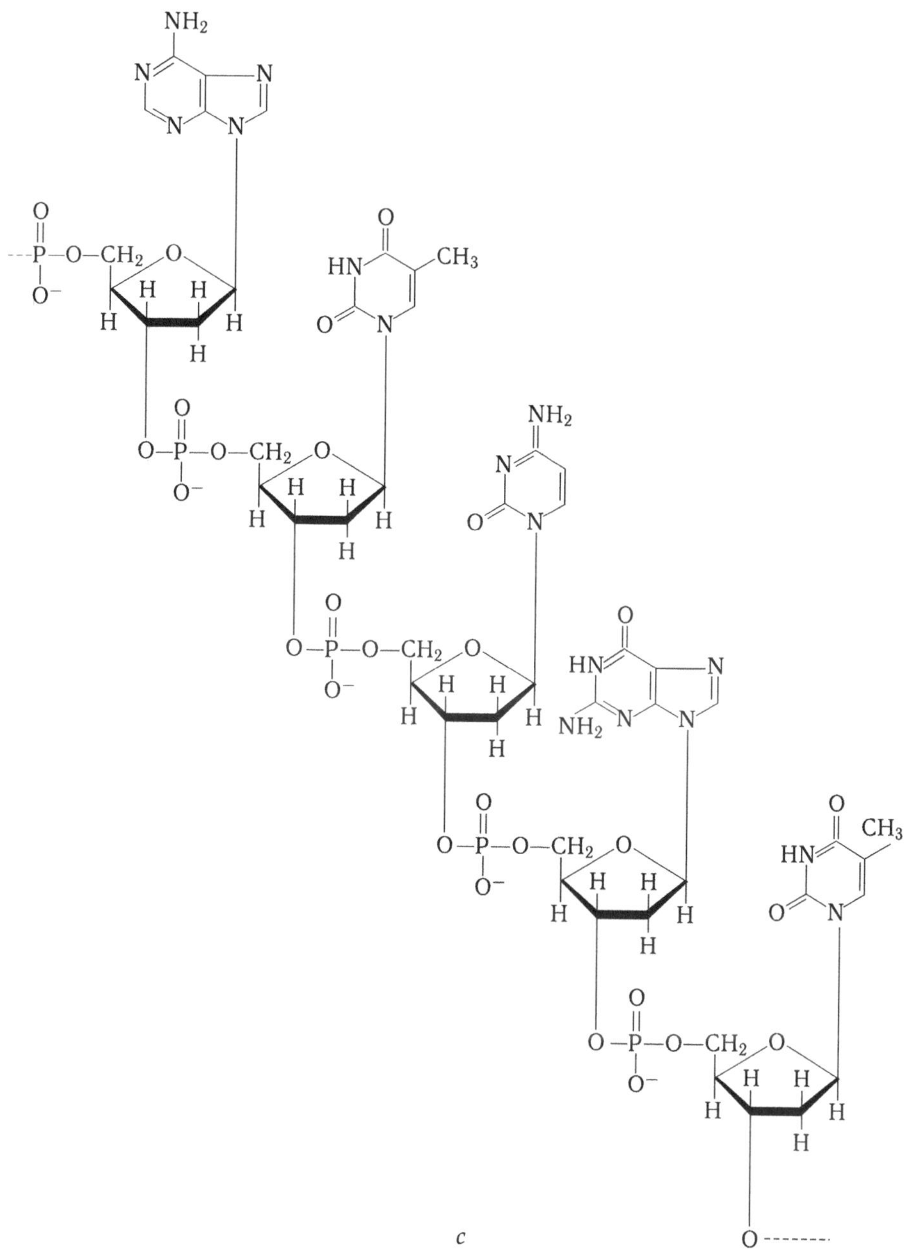

Figure 9.12 (*continued*). Structure of DNA. *c*: Region of the DNA chain.

it follows that continuous transitions from one conformation to another occur within a family of one type of forms (the *B*-form). As a matter of fact, different regions of the same DNA molecule can be concurrently in different conformations.

In the model of a flexible rod, the changes in the DNA conformation are associated with a small bend and twist of the double helix in each base pair, which accumulate in the large length. As a result, continuous bending of the DNA molecule takes place forming a ring with the diameter of about 10.0 nm. Subject to another point of view, the packing of the double helix upon DNA compactization can occur due to infrequent fractures at a large angle because of the disturbance of interplanar (stacking) interactions between base pairs. At usual physiological temperatures, DNAs undergo fluctuation openings of separate base pairs with probabilities not exceeding 10^{-5}. Probabilities of simultaneous opening of two and more base pairs are correspondingly one and two orders of magnitude lower.

Conformational Analysis of DNA. In contrast to well-studied helix – coil transitions, intensive studies of DNA transitions within a double-strand state without disruption of complementary pairs have been begun relatively recently. The aim of such studies is to reveal possible DNA forms and physical reasons for conformational transitions within the double-strand state. □

The significance of these problems is obvious from the biological point of view because in the process of functioning, the DNA can change its conformation due to interactions with other cell components, in the first place with proteins. The double helix of nucleic acid, having general rigidity over the helix length, is all together a kind of a hinge with a limited number of rotational degrees of freedom around singular chemical bonds (Figs. 9.13 and 9.14). The general geometry of a helical molecule is described by five parameters (Fig. 9.15), helical rotation τ (the rotation angle between adjacent pairs) and D (the distance of the pair from the helix axis) being the most important ones. The distinction between the *A*- and *B*-forms is the difference between values of the angles χ and σ (by 60°) and large positive D values and smaller τ values in the *A*-form.

The main reason for these differences is associated with the alternative geometry of the sugar ring in the forms of the *B*- and *A*-families.

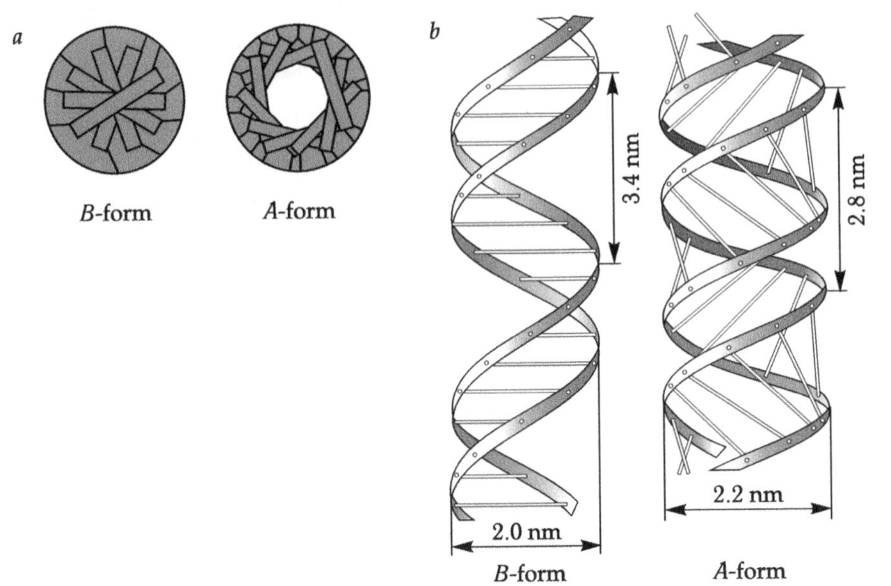

Figure 9.13. Polymorphism of the secondary structure of DNA: *a*, Top view; *b*, side view.

All known forms of DNA are divided in two families: the *A*- and *B*-forms. The axis of the helix passes across base pairs of the *B*-form near their centers of gravity; and in the *A*-form the pairs are displaced to the periphery of the molecule, so that the center has an orifice of about 4 nm.

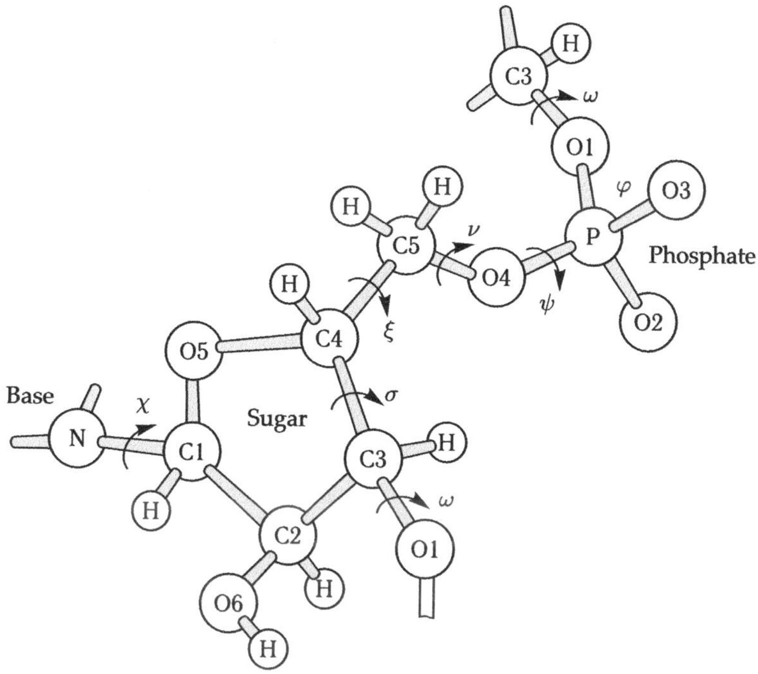

Figure 9.14. Designations of atoms and dihedral angles of the recurring unit in the sugar-phosphate chain.

The conformation of a nucleic acid monomer is determined by the conformation of the sugar ring, five angles of rotation around single bonds in the sugar-phosphate chain (ξ, ν, φ, ω, θ), and one χ angle (N–C1, the glycoside bond) which determines the orientation of the base relative to the sugar ring.

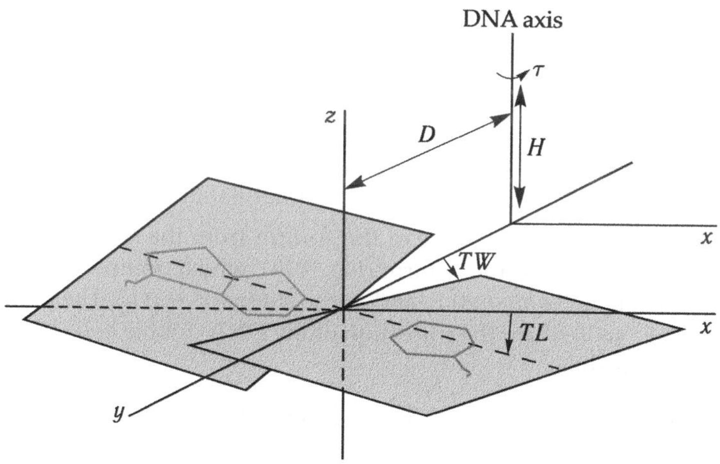

Figure 9.15. Five Arnott's parameters specifying the positions of bases in a regular double helix: rotation of pairs τ, distance from the axis of the helix D, inclination TL, "propeller" of bases in the pair TW, projection of the repeating unit to the helix axis H (reproduced from V. I. Ivanov, 1980).

The conformational map of complementary helices is given in Fig. 9.16 in the coordinates (D, τ). It represents the projection of 5-mer space of variable parameters to the plane (D, τ), the other parameters having values corresponding to the minimal helix energy.

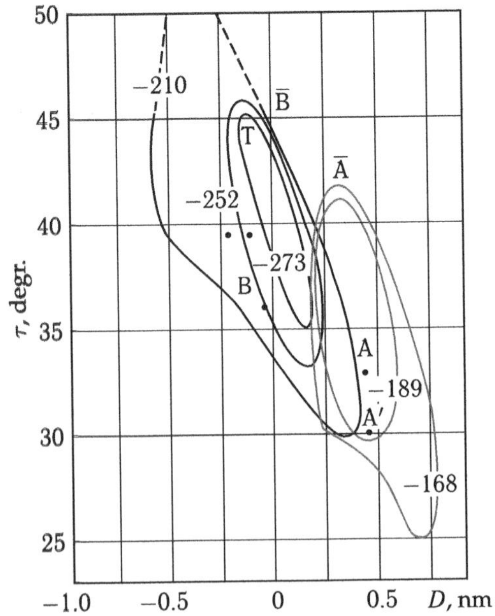

Figure 9.16. Map (D, τ) for canonical families of nucleic acid helices (reproduced from V. I. Ivanov, 1980).

Black lines show families of forms with C2-endo sugars (*B*-form), colored lines show families of forms with C3-edo sugars (*A*-form). Energy values in kJ/mol are marked on the lines of levels. Upper parts of some "ravines" are shown by dots to distinguish regions within the limits of which the helices have a very narrow sugar passage (to which glycoside bonds come in).

The map shows canonical families *B* and *A*, containing all known models of the double helix forms. The energy "ravines" restricted by equal-energy lines correspond to the regions of energetically advantageous structures. It is seen that the energy profile for the *B*-"ravine" is characterized by a rather flat bottom at $36° \leqslant \tau \leqslant 42°$.

The cardinal structural difference of the *A*-form from the *B*-one is a large aperture in the helix center in the *A*-form (Fig. 9.13), which occupies almost half the diameter of the whole helix ($d \sim 20$ E, and in the *B*-form $d \sim 9$ E). The pentose sugar ring is a molecular "switch" for the transition of the DNA double helix between families of the *B*- and *A*-forms. □

The biological role of the *A*-form is the subject of discussions.

Irregular Structures of DNA. At present, in addition to the *A*-form, much attention is paid to studying other irregular structures of DNA. They include cruciform structures, the left-handed *Z* form, different kinds of triple and quaternary helices, and the Slipped Loop Structure (SLS).

Under physiological conditions, like the *A*-form all irregular structures have higher energy as compared to that of the *B*-form.

Figure 9.17 shows a coiled-coil DNA molecule in which the double helix is right-handed, and the superspiralization is negative (the coiled-coil is left-handed).

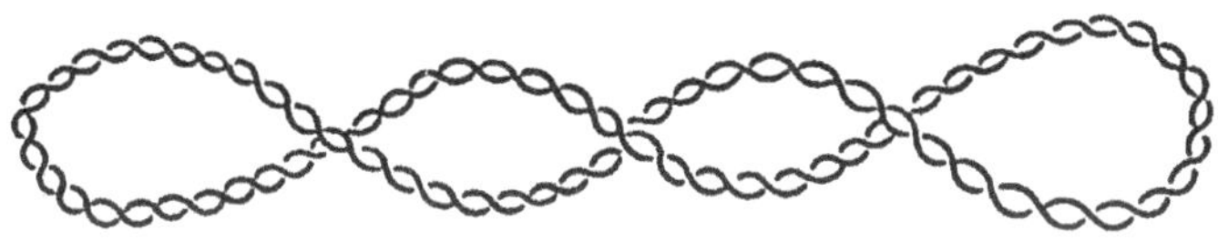

Figure 9.17. Superhelical DNA. Superhelicity is negative.

Different characteristics of DNA depend on the density of coiled-coils (the extent of superspiralization) σ, which is equal to the number of these coiled-coils τ per 10 base pairs in a state, when all pairs in the double helix are closed. For natural circular viral DNA molecules the σ value is mostly negative. The most typical value is $-\sigma = 0.05$ (~ 5 coiled-coils per 1000 base pairs). When the coiled-coil density $-\sigma$ grows, the probability of the opening of a pair increases, the most drastically this occurs from $\sim 9 \cdot 10^{-5}$ to 10^{-3} in the range of density changes from 0.06 to 0.12. In this case, not only separate pairs begin opening, but also stretched regions (Fig. 9.18) thus giving rise to cruciform structures. The density of superspiralization $-\sigma$ cannot exceed some limiting value ($-\sigma = 0.085$) because further increase of $-\sigma$ will be compensated by appropriate opening of base pairs in the double helix (Frank-Kamenetsky, 1983).

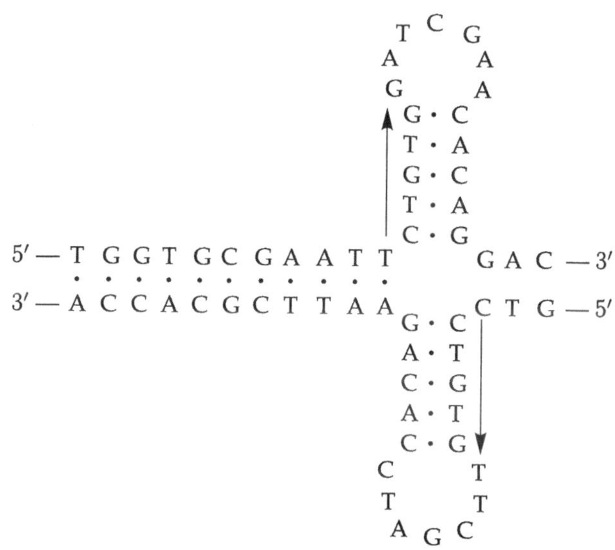

Figure 9.18. Two conformations of a region of the gene determining yeast phenylalanine tRNA. Cruciform structures appear at $-\sigma = 0.06 \div 0.08$ which compete with molten structures and are formed upon opening of intact DNA regions.

Pay attention to the fact that DNA sequences forming crosses which are located in different chains are themselves opposite transformations, or palindromes, i.e. are read alike both from the right to the left and from the left to the right. In Fig. 9.18, they are designated by arrows which indicate the direction of reading.

In studying the properties of irregular DNA structures, it is of the essence to find single-stranded regions as elements of these constructs. In this respect, the enzymatic method that allows localizing the place of attack of single-stranded nuclease turned out to be the most efficient. This method was used, for example, to reveal single-stranded regions in cruciform DNA and to demonstrate that upon the formation of the *Z*-form the boundaries between them and the *B*-form are attacked. It was found that single-stranded regions sensitive to single-stranded endonuclease are located in meaningful regulatory regions of the genome.

They are sequences which, as a rule, contain only purines (A and G) in one chain and only pyrimidines (C and T), i.e. homopurines — homopyridine sequences of the (G) · (C) or (GA) · (TC) type, in the other. The analysis of this problem has led to the disclosure of one more irregular form of DNA, the so-called *H*-form (Frank-Kamenetsky, 1986).

The *H*-form is connected with the existence of a triple complex consisting of two polypyrimidine (C – T) chains and only one polypurine (A – G) chain. □

9.5 State of Water and Hydrophobic Interactions in Biostructures

Recent data show that water is directly involved in formation of the structure of most important biopolymers and also in the processes of self-assembly of complex supramolecular structures. □

Peculiarities of Water Structure. As known, water properties are explained by the capacity of its molecules to form mutual hydrogen bonds between oxygen and hydrogen atoms O–H · · · O (see Section 2 in Chapter 8). A water molecule has such an electron structure that can form four hydrogen bonds at once which involve two protons and two lone-electron pairs of oxygen from each of them (Figs. 9.19 and 9.20).

Liquid water represents a system with well-developed hydrogen bonds, the properties of the system being thoroughly described by a continuous model, where almost all water molecules form on average hydrogen bonds. However, hydrogen bonds

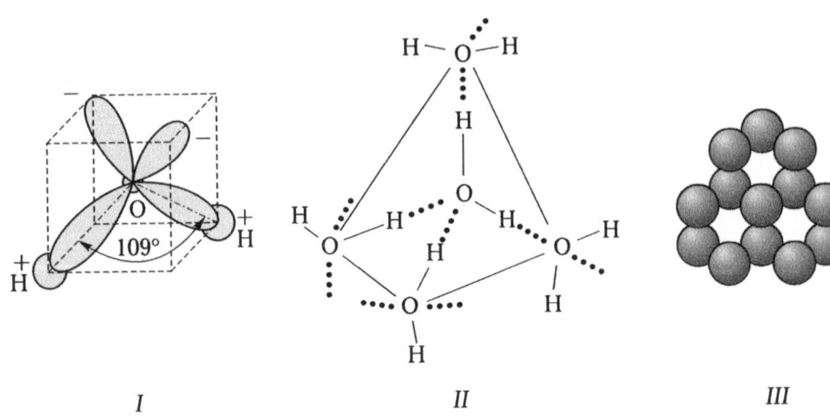

Figure 9.19. Structure of a water molecule (*I*) and its nearest environment in the structure of ice (*II*, *III*).

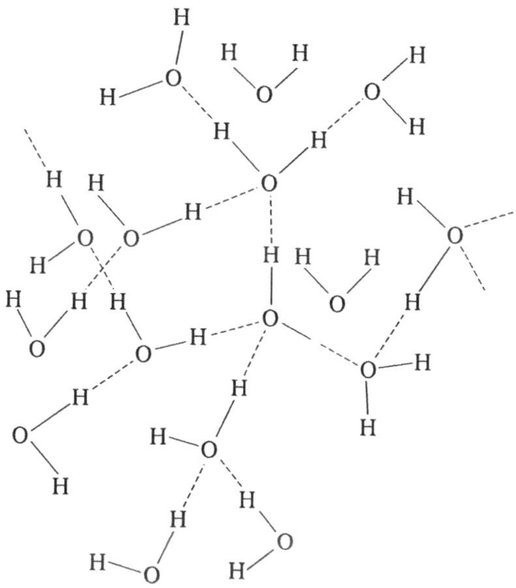

Figure 9.20. Structure of liquid water.

in water are characterized by a large set of angles and lengths. This means that the potential energy of a hydrogen bond is a continuous function of the angle between the intermolecular hydrogen bond and geometric parameters of water molecules. The maximum of the distribution function of the hydrogen bond energy corresponds to the distance between the oxygen atoms $R_{O\text{–}O} = 0.286$ nm. Hydrogen bonds in the ordered structure are linear and $R_{O\text{–}O}$ is reduced to 0.275 nm.

Usually, ice (*I*) has a tetrahedral lattice where each oxygen atom is located in the center of a tetrahedron, and adjacent oxygen atoms are at its apexes. Such a cellular structure with four molecules in each cell specifies the porosity and lower density of ice. The same structure, though with a large amount of disturbances, is also characteristic of liquid water. The framework of water molecules, bound quite firmly with each other by four hydrogen bonds, is responsible for a relatively high temperature of liquid water freezing, high heat of ice melting, high thermal capacity and heat of liquid water vaporization. D. Eisenberg and W. Kauzman introduced notions on the three types of states of liquid and solid water (Fig. 9.21). It is clear that thermodynamic parameters characterize largely the *D*-structure. The cluster model of the state of water is argued (H. Scheraga, 1962) according to which water is a mixture of monomer molecules and hydrogen-bound clusters. Clusters disintegrate and are formed again, which creates an average constant environment for each water molecule. According to the model of O. Ya. Samoilov, monomer water molecules fill in the voids in the cellular lattice of hydrogen bonds. When the temperature of ice rises, its ordered structure is broken and molecules are compacted with a decrease in the specific volume.

However, simultaneously the liquid is also expanding due to the increase of amplitudes of intermolecular oscillations. As a result of interaction of different factors, the minimum specific volume is attained at 4 °C. The volume decrease in the Samoilov model is associated with the activation processes of detachment of water molecules filling with the voids in the lattice.

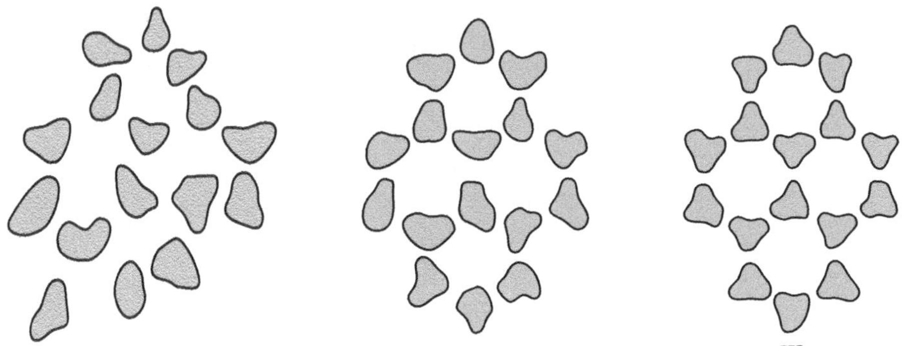

Figure 9.21. Instantaneous (*I*), oscillating averaged (*II*) and diffusion averaged (*III*) structures of water.

Water molecules oscillate near their average positions with the oscillation period $\tau_k \sim 2\cdot 10^{-13}$ s. At $t \ll \tau_k$ the instantaneous I-structure appears, where the molecule positions are not averaged (*I*) and may differ from equilibrium by no more than 15 %. At times exceeding τ_k but lower than the characteristic times τ_n of rotational and translational transfers ($\tau_n \sim 10^{-5}$ s for ice and $\tau_n \sim 10^{-11}$ s for water), am oscillatory averaged *K*-structure is obtained ($\tau_k < t < \tau_n$) (*II*). At $t \gg \tau_n$ diffusion shifts are averaged with the formation of the *D*-structure of water which has an ordered character (*III*).

From the thermodynamic point of view, the formation of the ordered structure of water causes a decrease in its entropy ($\Delta S < 0$), which is disadvantageous because leads to an increase in the total thermodynamic potential (free energy) $\Delta G = \Delta H - T\Delta S$ due to the positive member $(-T\Delta S) > 0$. But this increase is compensated in excess by the formation of four hydrogen bonds, which leads to a decrease in the energy (enthalpy) of water molecules $\Delta H < 0$. It is essential that in this case $|\Delta H| > |T\Delta S|$. □

Thus, on the whole upon spontaneous formation of the ordered structure of water the change in the free energy is negative: $\Delta F < 0$. This specifies the energetic efficiency of this process $\Delta F \simeq \Delta G$ at $\Delta V \simeq 0$ (see Eq. (5.10)). Water molecules undergo oscillatory motions near the equilibrium position with the characteristic oscillation time $\tau \sim 10^{-13}$ s. Moreover, due to fluctuation ruptures of hydrogen bonds (with the energy of rupture activation $\sim$20 kJ/mol at 298 K) separate water molecules can move into the voids of adjacent cells. In the liquid state, the characteristic averaged time of such self-diffusion is $\tau_D \sim 10^{-10} \div 10^{-11}$ s, whereas in ice at 0 °C it will essentially increase to $\tau_D \geqslant 10^{-5}$ s (for comparison, the time of valence oscillations of the O–H bond is $\tau_B \sim 10^{-14}$ s).

Just because of the efficient self-diffusion in the liquid state when water molecules occupy not only lattice points, but also a certain number of interstitial sites, the density of water exceeds that of ice. The most significant physicochemical properties of water, including ice modification, are evidently associated with the existence of a large number of transitional structure elements. The nature of transformations between the latter is not quite clear yet.

Hydrophobic Effect. Stabilization of macromolecular structure is closely related to the interactions of water with hydrophobic compounds. Studies of solubility of nonpolar hydrocarbons upon their transfer from nonpolar solvents to water demonstrated that dissolving of these compounds in water is an exothermal process.

It proceeds with a temperature increase and thermal energy release, which is evidence of a decrease in the enthalpy of the system upon mixing of hydrocarbons and water ($\Delta H < 0$). However, this decrease in enthalpy is compensated by a decrease in entropy ($\Delta S < 0$) and, consequently, an increase of the total thermodynamic potential ($\Delta G > 0$). On the whole, the decrease in enthalpy ($\Delta H < 0$) is smaller in its absolute value than the decrease in entropy and, accordingly, the dissolution of nonpolar compounds in water is a thermodynamically disadvantageous process:

$$\Delta G = \Delta H - T\Delta S > 0.$$

For example, upon ethane transfer from benzene and its solution in water at 298 K, the entropy decreases by $\Delta S = -84\ \mathrm{J/(mol \cdot K)}$, and the enthalpy increases by $\Delta H = -9240\ \mathrm{J/mol}$. The total increase in the thermodynamic potential of the ethane–water system as compared to the ethane–benzene system makes

$$\begin{aligned} \Delta G = \Delta H - T\Delta S &= -9\,240 + 298 \cdot 84 \simeq \\ &\simeq 25\,500 - 9\,240 \simeq 16\,000\ \mathrm{J/mol} = 16\ \mathrm{kJ/mol}. \end{aligned}$$

A direct consequence of this is that water repulses nonpolar substances, and hence for the latter, from the thermodynamic point of view, it is more advantageous to interact with each other than with water.

Detailed molecular rearrangements, reflecting the change in the water structure and mutual orientation of molecules of the dissolved substance and solvent, are rather complex.

First, the question arises what specific changes in the hydrocarbon–water system cause a decrease in its entropy. Apparently, solution of hydrocarbon in water should be associated with penetration of its molecules into the structured regions of water resulting in their destruction. Small molecules capable of embedding in the ice-like framework of water clusters dissolve better because they do not disrupt hydrogen bonds effectively. On the other hand, relatively large molecules of hydrophobic compounds usually disrupt hydrogen bonds in water.

The dissolved substance retards the translation (thermal) motion of water molecules dependent on dimensions of its particles since they create the "effect of an obstacle", making the space occupied by the particle inaccessible for water. As a result, a denser layer of water is formed around this particle. This phenomenon is called hydrophobic hydration. It leads to the compaction of water molecules adjacent to the hydrophobic surface and increases its averaged time of "settled" life to $\tau \sim 10^{-7}$ s.

The formation of the dense layer of water around nonpolar hydrocarbon radicals decreases the entropy of the system. This decrease is not compensated by weak interactions between hydrocarbons and water molecules. Consequently, when the concentration of nonelectrolyte increases, the molecules of the substance dissolved in water are united releasing concurrently part of the structured water around them with the resultant increase in the general system entropy. Thus the system becomes divided into two phases — the solvent and the hydrophobic compound. That is the reason for the entropic nature of hydrophobic interactions as a special type of binding of nonpolar particles with each other, which should reduce their common contacts with water. The forces conditioning these interactions are of a purely entropic nature. They are long-range ones because they emerge between hydrophobic groups at a distance exceeding the sum of their van-der-Waals radii. The energy of hydrophobic interactions varies from 2 to 40 kJ/mol and increases with an increase in the contact area of nonpolar particles. The hydrophobic effect is dependent on temperature. In the dense surface layer, the water molecules have to "wriggle" in order to avoid losing H-bonds. This allows partial compensation of the entropy decrease in the dense layer of water. But at a very low temperature these hydrophobic interactions, normally preserving the protein compact state, decline drastically that manifests itself also in a decrease of the protein melting heat. This decline may lead to the fact that at abnormally low temperatures (~ 10 °C) and lower, the energy of native protein (plus the energy of the water layer surrounding it) may become higher than that of denatured protein. As a result, the protein stability sometimes drops so that at physiologically abnormally low temperatures (approximately below zero degrees centigrade) the protein disintegrates. This phenomenon is known as "cold" denaturation (P. L. Privalov, 1988), which proceeds as a phase transition (cf. Section 3 in Chapter 7) similar to usual denaturation.

Hydrophobic Interactions in Proteins. The hydrophobic effect plays a significant role in the formation of biostructures. A real structure is ordered and compact and as a whole is largely determined by hydrophobic interactions. Individual amino acid residues differ in their hydrophobic properties and can behave as polar and nonpolar compounds. A thermodynamic estimate of the extent of hydrophobicity is made by the magnitude of the change of ΔG per a side group of amino acid upon its transfer from ethanol to water (C. Tanford, 1980).

Hydrophobic amino acids include such as *Try* ($\Delta G = +12{,}570$ J/mol), *Ile* (+12,440 J/mol), *Phe* (+11,100 J/mol), *His* (+5900 J/mol), *Met* (+5500 J/mol), and hydrophilic ones are represented by *Ala* (+3000 J/mol), *Glu* (+2300 J/mol), *Ser* (+170 J/mol), *Gly* (0 J/mol), *Asn* (−40 J/mol), *Gln* (−420 J/mol). The detailed calculations (D. F. Brandts, 1973), taking into account both the statistical character of the distribution of hydrophobic, hydrophilic and neutral groups within a macromolecule and the values of the free energy transfer of these groups to the surface or inside the globule, demonstrated that about 10–25 % of hydrophobic groups may be located on the globule surface. The experiments showed that in some cases the number of nonpolar groups on the surface even exceeds their number in the protein (in a lysozyme molecule their number are 32 and 16 respectively). Obviously, hydrophobic forces have a decisive meaning in determining the general character of the protein globule topography, but the detailed situation concerning interactions within a macromolecule is far more complicated.

Upon estimation of the role of different interactions in stabilization of globular proteins, it should be recognized that a native conformation state is dependent a joint well-balanced effect of a number of energetic and entropic factors. The hydrogen bonds formed between polar groups and water and inside the globule are the foremost factor in providing stability of separate regions of a protein molecule. They restrict local conformational changes inside the protein, determining the rigidity of the structure and the general character of potential barriers for internal motions of parts of the native structure. At the same time, hydrophobic interactions between side groups at separate regions of the main chain play a decisive role in the processes of folding of the globule from the main amino acid sequence and also in specifying its general form. In both cases, water as a solvent is of great importance facilitating polar interactions due to the formation of hydrogen bonds on the surface and inside the protein macromolecule. □

Methods of Quantitative Accounting of Hydrophobic Interactions. As has been stated, the hydrophobic effect is of a purely thermodynamic nature and has a cooperative character. Therefore, accurate quantitative accounting of hydrophobicity is complicated, and no physically correct method is available to calculate, for example, the contribution of the hydrophobic effect to the stability of a biological molecule or complex of molecules. Different scales of hydrophobicity of amino acid residues were obtained based on measurements in double-phase systems with various structures using different molecular fragments (isolated amino acids, amino acids in the composition of model peptides, etc.).

It is possible to take into account hydrophobic characteristics of a molecule in more detail using the approach of empirical molecular hydrophobic potential (MHP) in which constants of hydrophobicity are set already not for whole amino acid residues but for separate atoms (or small groups of atoms). As opposed to scales of hydrophobicity for amino acids in which corresponding values are directly measured, on the "atomic" level it is impossible to perform such measurements, because in standard conditions atoms are not free and form different compounds.

In the MHP method, "atomic" constants of hydrophobicity are calculated proceeding from the approximation of the additivity of hydrophobic properties. The base for such calculations is the large set of experimentally measured distribution coefficients of different compounds between nonpolar and polar phases. In this case, the octanol–water system and the corresponding coefficient $P_{o/w}$ (in the $\log P_{o/w}$ form) are regularly used. Another important feature of such approach is the determination of topological classification of "hydrophobic" types of atoms depending on the molecular environment: For a correct accounting of hydrophobic properties it is required to differentiate between, for example, the hydrogen atom included in the OH-fragment of the carboxyl group or the OH-group of alcohols and the hydrogen atom in the aliphatic or aromatic system. The molecular hydrophobic potential at any point A of space is determined as a superposition of monotonically decaying fields, created by each atom of the system, according to its atomic constant of hydrophobicity f_i:

$$\mathrm{MHP}_A = \sum_{i=1}^{N} f_i \times d(r_{iA}),$$

where N is the number of atoms in the system, r_{iA} is the distance between atom i and point A where the MHP is determined, and $d(r)$ is the function that determines the potential decay. As a rule, functions of the $d(r) = 1/(1 + r)$ and $d(r) = e^{-ar}$ types, where a equals 1 or 0.5, are used.

The environment of amino acid residues in membrane proteins is highly heterogeneous, varying from essentially nonpolar medium of the membrane itself to polar and charged domains "inside" the protein molecule. The MHP method, taking into account a detailed spatial distribution of hydrophobic properties, allows detailed mapping of hydrophobic interactions that play an important role in complex biomolecular systems (R. G. Efremov, 1995, 2007).

Heterogeneity of Distribution of Water Molecules. Heterogeneity of the macromolecular structure influences the distribution of water molecules. Apparently, water can exist as a bulk phase of free solvent and bound water, the state of water depending on the nature and localization sites of protein groups it interacts with. □

The amount of hydrated water determined by different methods makes 0.3–0.4 g H_2O/g of dry protein, while the total amount of water in crystals of globular proteins does not usually exceed 0.45–0.60 g H_2O/g of dry protein. So, the amount of free water in protein is not large. In particular, it can fill in the internal "cavities" free of the protein substance.

The general pattern of hydration of the protein surface is shown in Fig. 9.22. Separate centers are surrounded by water molecules which form also water bridges between hydration regions belonging to separate centers. As a result, an all-round hydration "lace-like" shell is formed which connects separate polar groups. The hydration shell on the globule surface includes water molecules belonging to local centers of hydration. The life-time of weakly bound water molecules in such a hydration shell is not high ($\tau \sim 10^{-11} \div 10^{-10}$ s), but close to a center it becomes much higher ($\tau \sim 10^{-6}$ s).

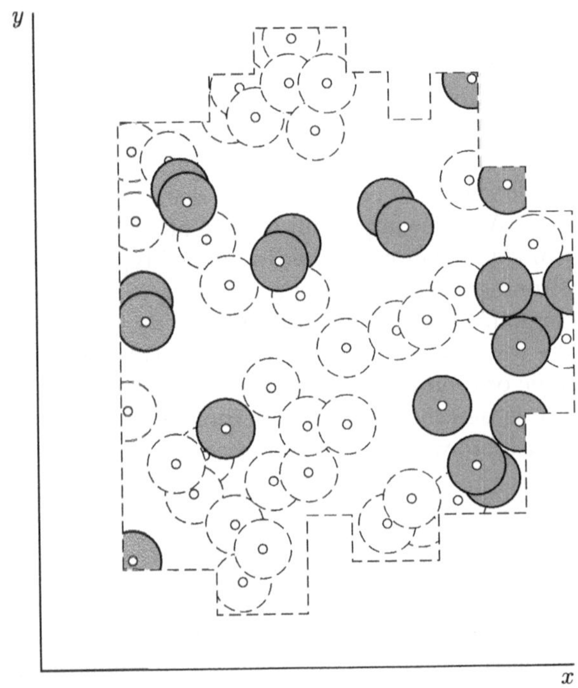

Figure 9.22. Map of hydration of the surface of an n-chymotrypsin molecule.

Solid lines designate the region of localization of water molecules connected by heteroatoms of surface side polar groups; dashed lines show the region of localization of water molecules on peptide groups.

In the strong primary layer, water molecules form with each other on the average no less than 2–3 hydrogen bonds. On the whole, two or three layers of water can be retained close to the protein surface, with no multilayer "fur coat" of crystallized water being formed. Upon water sorption, separate hydrophobic regions on the surface are "filled" with water only after its energetically more advantageous binding with neighboring polar groups is completed. Strongly bound water makes about 10 % of the mass of biopolymers and is characterized by high life-time values (10^{-3}–10^{-7} s).

Y The physicochemical nature of bound water is conditioned to a large extent by its interactions with structures of lower mobility. This determines the existence in solutions of biopolymers with several fractions of bound water, their life-times being between those of free (10^{-11} s) and strongly bound water (10^{-3}–10^{-5} s) (S. I. Aksenov, 1980). □

The functioning of proteins is closely associated not only with the character of their conformation, but, above all, with their conformation mobility depending on the presence of water. So, at a low extent of hydration of α-chymotrypsin preparations, the emerging additional contacts between surface dehydrated polar groups lead to an increase in the rigidity of the α-chymotrypsin globule and the loss of its enzymatic activity in dimethyl sulfoxide. In strongly dried preparations no activity is observed at all up to some critical value of the sample hydration. The activity is sharply restored with an increase in the extent of hydration in a narrow range of the increasing number of H_2O molecules from 170 to 180 per one protein molecule. It is likely that in this region disinhibition of certain degrees of freedom functionally essential for an enzymatic act takes place. Importantly that the amount of water required for this process is much less than would be required for accomplishing the formation of a complete hydration shell (Khurgin, 1980).

Detailed calculations and energetic maps plotted based on them (see Section 1 in this chapter, Fig. 9.5) demonstrated that upon hydration, there are ranges of values of the angles φ and ψ where stabilization and destabilization take place with corresponding changes in the depths of energy minima.

Y The influence of water on the conformational energy of peptides does not essentially change the localization sites of energy minima on the conformation map. But in this case, the relative stability of separate minima of conformational regions can change greatly, which, in its turn, creates prerequisites for characteristic changes of the protein conformation in different fluctuating processes. □

Dynamics of Protein Hydration. Let us analyze the main stages of changes in the state of water upon protein hydration on the example of lysozyme. At the beginning of hydration with the water content h, expressed in g per 1 g dry weight protein, $h = 0.01$–0.07, water interacts first of all with charged surface groups of protein, about 2 molecules of water per a group. When $h = 0.07$, transition from the dispersed (disconnected) state to the cluster one occurs that is associated with termination of hydration of charged groups. This transition can be detected by such methods and IR, EPR and neutron scattering.

At $h = 0.15$, a bond is formed connecting the molecules of hydrated water independent of the distances between them. In this range the electrical capacity of proteins is changed abruptly. A network of hydrogen bonds between molecules of hydrated water envelopes the globule surface. At $h = 0.25$, condensation of water begins, and water starts occupying regions on the protein surface, which are weakly interacting with it and yet not filled (the process of covering the hydrophobic surfaces of the protein globule).

When the extent of hydration is $h = 0.38$ (about 300 water molecules per globule), the lysozyme surface is covered by a monolayer of water. The interaction with charged and polar surface groups of protein results in locally ordered formations (structures) of hydrated water.

At the beginning of hydration when $h = 0.01$–0.07, the mobility of water is very low. When the hydration degree is $h = 0.25$, the correlation time of the rotational

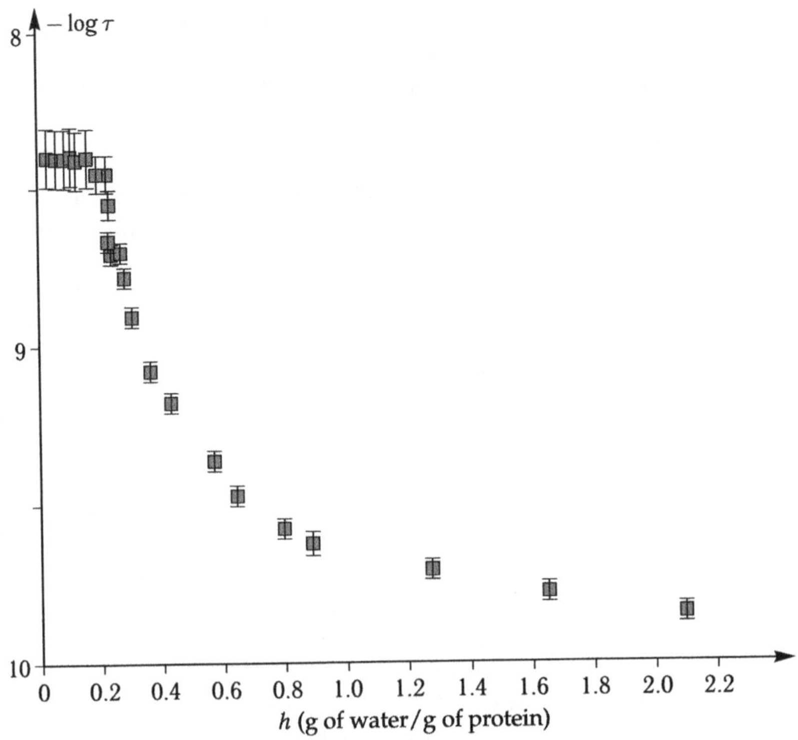

Figure 9.23. Dependence of correlation time τ for the TEMPONE spin label bound non-covalently to lysozyme dependent on the extent of its hydration (reproduced from J. A. Rupley et al., 1980).

diffusion τ_c decreases (see Section 2 in Chapter 11) from 5×10^{-8} s to 5×10^{-10} s, which suggests a large increase in the mobility of water with the increase in hydration (see Fig. 9.23).

At further addition of water and growth of the h values above the hydration value $h = 0.38$, the mobility of water increases so much that the τ_c values approach those for pure water. Subject to EPR, the rotational diffusion of the mobile bound ligand also increases greatly.

So, in the range between $h = 0.2$ and $h = 0.38$, hydrated water weakens the network of intramolecular hydrogen bonds, softens the structure of a lysozyme macromolecule which results in the emergence of noticeable conformational mobility. During subsequent hydration the protein conformational mobility does not grow. Figure 9.24 shows a general pattern of distribution of water over the surface of lysozyme.

9.6 Protein Folding

Because of a vast number of conformational substates in native protein ($\sim10^{100}$ for a chain of 100 residues provided each of them has about 10 low-energy conformations), the real process of folding should have a directed character. It cannot be

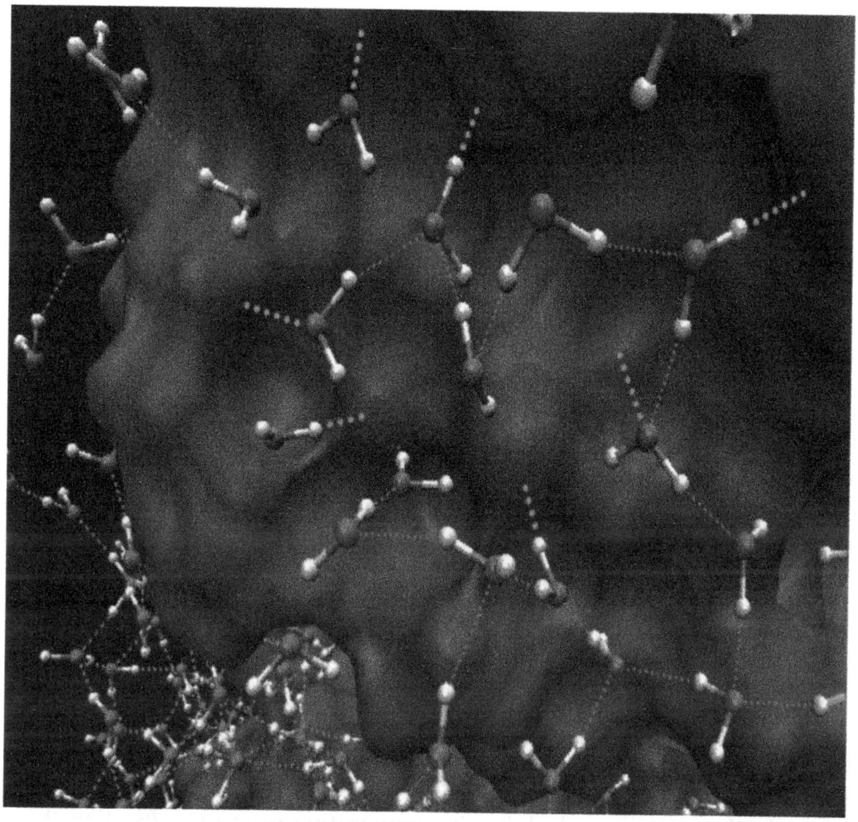

Figure 9.24. Hydrated protein. Hydrated water forms a network of hydrogen bonds between adjacent molecules of water and surface amino acid groups. Hydrogen bonds between water molecules are shown by red lines. Hydrogen bonds between water molecules and surface amino acid groups are shown by green lines (reproduced from Yu. F. Krupyansky et al. 2008).

realized by consecutive search for all versions to determine the energy minimum required for structure stabilization.

As a matter of fact, upon folding, of the protein chain the entropy drops due to the growth of ordering with a concurrent drop of energy due to the formation of contacts between the approaching links. If the folded structure is stable, its entropy loss ($T\Delta S < 0$) should be compensated by the energy gain ($\Delta U < 0$), so that altogether

$$\Delta F = \Delta U - T\Delta S < 0 \qquad (|\Delta U| > |T\Delta S|).$$

Suppose that the search for this energy minimum proceeds by selecting different types of packing. Then during folding, the chain should at first approach at random the native conformation, having lost almost all its entropy, and only after that an energy gain will take place as a result of the formation of corresponding contacts.

In this case, the folding process will be very slow, because at the first stage the entropy loss ($-T\Delta S$) is coupled to the growth of the free energy ΔG^* of the transient state formed during folding. In its turn, the duration of the process $\tau \sim \exp(\Delta G^*/RT)$ also depends exponentially on the ΔG^* value. Just this circumstance (the enthalpy drop prior to the beginning of the energy gain) underlies

"Levinthal's paradox", which states that the protein chain cannot form its most stable structure for a reasonable time by random selection.

Let the time of transition of one link from the coil to the growing globule be $\tau \sim 1$ ns, and the entropy drop by one amino acid residue having 10 conformations be about $R \ln 10$. Then the folding will take on the whole 10^N ns. By C. Levinthal's estimate, for proteins with $N = 58$ residues the time to form stable structure will be $\sim 10^N$ ns $\sim 10^{58}$ ns $\sim 10^{41}$ years.

Meanwhile, natural amino acid sequences fold spontaneously into a native globule in 10^{-2}–10^2 s. The stability of this globule is not too high (~ 0.1 kcal/mol per a chain link) and only slightly exceeds that of the unfolded form (a rather shallow energy minimum).

From the point of view of thermodynamics, a necessary condition for fast folding is compensation of the entropy decrease ($T\Delta C < 0$) by the energy decrease ($\Delta F < 0$) at each stage of folding (Finkelstein, 1996).

A consecutive process of the protein structure folding begins with the fixation of one link just in its final conformation, i.e. with the formation of the secondary structure nucleus. Then links adjacent in the chain are fixed in the final conformation and interactions stabilizing the structure are established.

In this situation, the insignificant value of the energy barrier ΔG^* is conditioned by the fact that the entropy decrease upon fixation of one link is compensated by the energy gain upon formation of the "required" contact, which is not disrupted later. This drastically decreases the total time of consecutive folding of the chain.

As a result, this mode of protein folding is ensured by the circumstance that the globular part of growing nuclei of protein structure is not rearranged during folding. This assumption complies with the principle of coordination of long- and far-range interactions in the protein underlying its cooperative properties (Section 2 in this chapter).

Calculations show that the maximal difference between the entropy decrease and energy increase upon fixation of n links

$$\Delta S(n)/\Delta S(N) - \Delta E(n)/\Delta E(N),$$

is only 0.19, so that the time of protein folding is $\sim \exp[(1 \pm 0.5)N^{2/3}]$ ns and at $N = 58$ is $10^{58 \cdot 0.19}$ ns $\sim 10^{11}$ ns ~ 1 min (Finkelstein, 1996), which approximately corresponds to the experiment.

It should be mentioned that at present theoretical studies in this field still do not allow taking into account explicitly the role of certain residues, the character of the amino acid sequence in the fast folding of the chain and native structure stabilization.

Modeling the Folding. The process of folding of amino acid sequences has been intensively studied by computer modeling methods. The folding chain should satisfy two different requirements.

As concerns thermodynamics, the native conformation should be unique and possess a clearly expressed energy minimum (stability) as compared to other close conformations.

From the kinetic point of view, a sequence should fold quite rapidly, and because of this it should be stable at not too low temperatures.

In other words, those temperatures are required, which are high enough to provide a high rate of the folding itself and at the same time rather low so that the formed native structure will be stable under these temperatures.

Three phase states of a polymer chain can be separated: coil, "molten" globule, and "frozen" globule with a small number of conformations. Transitions between these phase states can occur at definite temperatures:

$$\text{Coil} \xrightarrow{T_\theta} \text{Molten globule} \xrightarrow{T_c} \text{Folded conformation.}$$

The temperature T_θ specifies the boundaries of coil condensation to the molten globule state which has many conformations. □

At $T = T_c$ the molten globule undergoes freezing processes called "vitrification" by analogy with the overchilled liquid which is fixed in one of its numerous non-equilibrium states but has not yet formed its final equilibrium crystal structure.

It is at this stage that the native structure can form one specific conformation. But it is just here that controversy arises between kinetic and thermodynamic requirements to the folding process. If the real temperature of folding T_f is below the freezing point, $T_f < T_c$, the stability of the final conformation is provided, but the time of folding will be very large. On the other hand, at $T_f > T_c$ the folding is fast but the formed structure is unstable at these temperatures.

The controversy may be resolved by the shape of the energy profile of folding as shown in Fig. 9.25. Here we see a sharp slope of the energy profile and a deeper position of the native conformation level relative to the states closest to it. Namely,

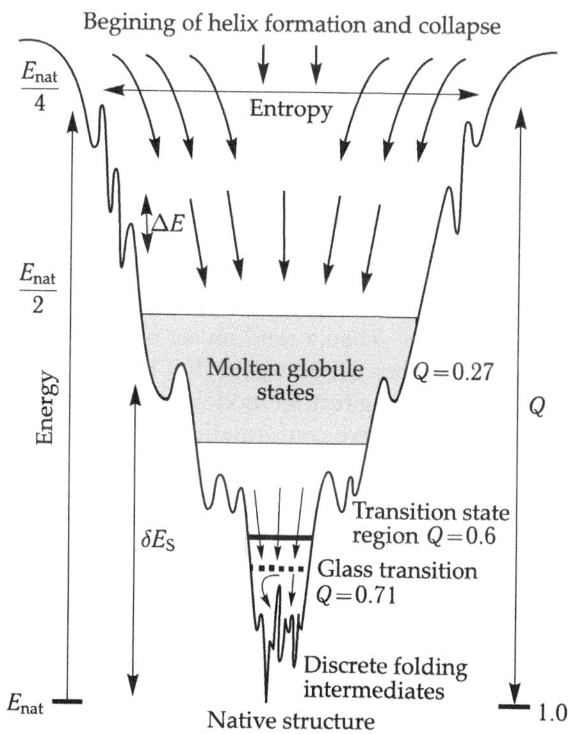

Figure 9.25. Scheme of the "folding funnel" for fast folding of a helical protein of 60 amino acid residues (reproduced from P. G. Wolynes, J. N. Onuchic and D. Thirmalai, 1995).

The width of the funnel is proportional to the entropy, and the depth is proportional to the energy. The portion of independent contacts Q is exactly determined and set for each of the states.

these characteristics specify not only a higher rate of folding, but also the stability of the native structure against thermal fluctuations at $T_f > T_c$.

M. Karplus and E. Shakhnovich with coauthors used a model of the lattice where the points contained beads interacting with each other as amino acid residues (Fig. 9.26). The model allowed studying a cubic lattice ($3 \times 3 \times 3$) with 27 elements (fragments of the chain). This model can imitate the molten globule state because it has no side chains whereas the compact "backbone" is preserved. It is proposed that the energy of this polymer chain depends only on the interaction of nearest fragments in the adjacent points of the lattice.

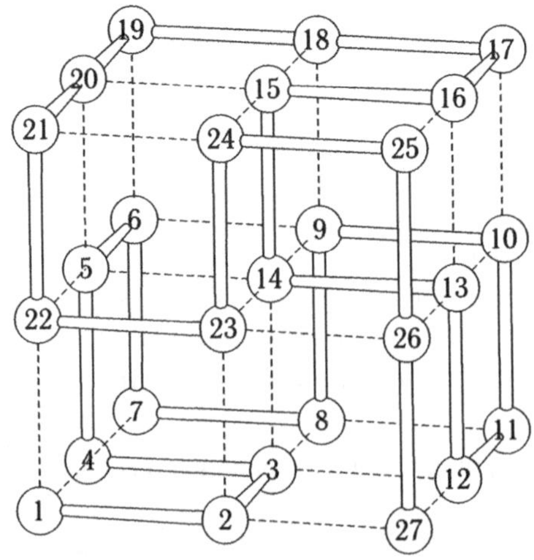

Figure 9.26. Lattice model of protein folding: an example of a compact chain structure of 27 monomers (numbered circles) with 28 contacts (dashed lines) (reproduced from Sali et al., 1994).

In such a system it is possible, by introducing the general condition of compactness (hydrophobic contraction) of the globule, to determine the minimal energy corresponding to the native state. Then a random set of interaction energy values is generated using a random number generator. In that way a set of different random model sequences is obtained. During further modeling the capacities of these random sequences to fold in low-energy native conformations are compared. It is required to choose those which can do this relatively quickly (for a limited number of steps) and after that compare the character of energy spectra of different sequences near the native state. □

The largest number of possible conformations in the model is $\sim 10^{16}$, but it was found that some selected sequences can find the native conformation in a far smaller number of steps ($\sim 10^7$). This testifies to the overcoming of Levinthl's paradox due to the "movement" over the "specified" pathway of folding.

In the range of 10^4 Monte Carlo steps a sharp transition to a compact globule is observed. At this stage, only $\sim 10^{10}$ coil conformations of the possible $\sim 10^{16}$ remain. This transition corresponds to condensation of the coil ($T < T_\theta$) and transition to the molten globule. Then (at $T < T_c$) globe "vitrification" and a search for one of 10^3 transient states take place. The transition to the native conformation occurs swiftly from any transient state and requires $\sim 10^5$ steps.

Figure 9.27 shows energy spectra of 10 random sequences, 4 of them are folding ones. The spectra include energy levels of 400 low-energy conformations. In folding sequences, the energy level of native conformation occupies the lowest position. Of fundamental importance is that it is separated from the nearest transient state by an energy gap ΔE, which is absent in spectra of non-folding sequences.

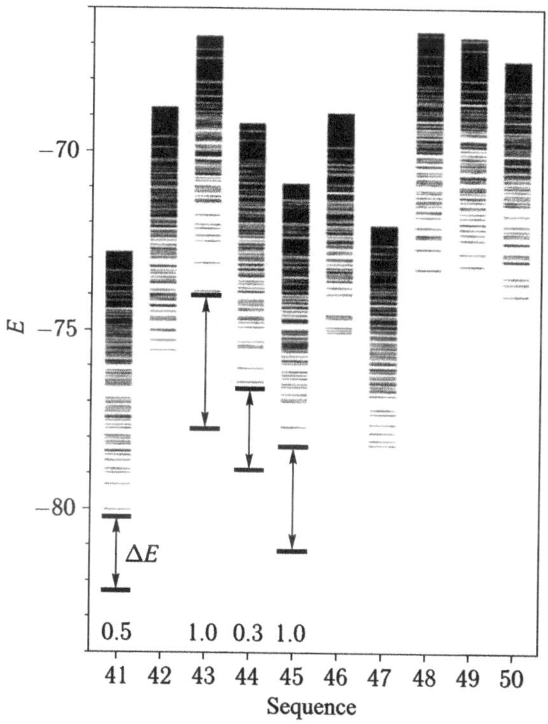

Figure 9.27. Energy range for 10 folding and non-folding sequences (reproduced from Sali et al., 1994).

The native state corresponds to the bottom band. The numbers under the spectra show the "tendency" to folding of corresponding sequences. If the number is absent, the "tendency" is zero. A sequence folds if its native conformation forms in 50×10^6 Monte Carlo steps.

We have already mentioned that temperature requirements of stability of the native state and the quickness of its establishment conflict with each other.

As seen in Fig. 9.27, the native conformation formed at relatively high temperatures nevertheless remains stable due to the energy gap ΔE, which separates it from the closest states (cf. Fig. 9.25). Accordingly, the folding sequences have such energy spectra that allow them to fold relatively fast at $T_f > T_c$ and maintain the stability of the native state under these conditions. All random sequences, stable only at $T < T_c$, are not the folding ones. If the folding occurs at $T_f \simeq T_c$, about 1 % of all sequences have an energy gap of several $k_B T$.

Recent results of model experiments have shown the role of structural factors in rapid folding. It has been demonstrated that the mechanism of rapid folding depends on the formation of the "folding nucleus", involving definite amino acids already at early stages. This folding nucleus is formed in rapidly folding sequences and is absent in slowly folding ones. Acceleration of folding is accompanied by stabilization of the nucleus due to internal contacts (E. I. Shakhnovich et al., 1996). It is interesting that residues in the folding nucleus of proteins CheY also play an essential role in protein functioning (phosphate binding). Thus, in this case, the folding nucleus makes also the site of the active center localization in the fully formed protein. □

Thus, a protein may have a definite number of specific positions of amino acid residues that are responsible for fast folding. In evolution the high rate of folding could be provided by such amino acids in the positions stabilizing the nucleus. This means that evolutionary selected natural proteins, the number of which ($\sim 10^5$) is much smaller than their total plausible number, are not a result of "the random choice storage". It is rather due to some fundamental properties that distinguish them from other unnatural sequences.

The general problem, which awaits not only its solution but also a correct physical formulation, is to understand the relationships between amino acid residues and their location in the main sequence and the structure and functional properties of the native protein.

Protein folding and realization of the above analyzed physical principles of folding occur in a living cell in a way differing from that *in vitro*. In a cell the microenvironment of the polypeptide chain includes ribosomal structures, enzymes, chaperone proteins and other factors absent in solution. All this allows concluding that in a living cell the sequence folds not from the state of a stochastic coil like upon renaturation in solution, but already on the ribosome without the release into the surrounding medium, i.e. co-translationally.

It was shown using the specially developed methods that the newly synthesized protein luciferase becomes enzymatically active already on the ribosome, when no less than 26–27 amino acid residues appear at its C-terminus (A. S. Spirin, 1995). So, the folding of luciferase into an active conformation takes place already on the ribosome, i.e. in a co-translational mode without the polypeptide chain leaving the ribosome.

In a cell special mechanisms (co-translational folding, chaperones) operate providing the implementation of the physical principles responsible for directed cooperative folding of the main sequence into a stable native protein conformation.

Returning to Anfinsen's dogma, it should be acknowledged that in a number of proteins the main amino acid sequence does determine the character of the stable secondary and tertiary protein conformations. However, it does not necessarily mean that specific pathways of protein folding in a cell and protein renaturation in solution are identical.

IV
Dynamic Properties of Globular Proteins

10

Protein Dynamics

10.1 Structural Changes in Proteins

The native three-dimensional structure is formed under the action of different energy and entropy factors. As well known in biochemistry, the change of the conformational state of a protein molecule under different conditions (pH, temperature, ionic composition) affects its functional activity. The questions arise: does the protein structure remain stable in the process of its functioning or does it change in the reaction, involving the protein molecule? In other words: does the structure of a biopolymer have internal conformational mobility, what is its nature and what is its functional meaning? This problem is of crucial significance for understanding molecular mechanisms of the functioning activity of biomacromolecules and their involvement in fundamental biochemical processes.

As known, characteristic times of many intramolecular transformations, including enzyme processes in biostructures, do not exceed 10^{-2}–10^{-3} s, and sometimes can be also several orders of magnitude shorter. Obviously, the conformational rearrangements accompanying them should also occur very fast. However, for such short times, they cannot just right away affect deeply the whole macromolecular structure and therefore at initial stages should have a local microconformational character, causing displacements of only separate groups of atoms. As such, local displacements spread on the other regions of the macromolecule it will lead to a general conformational change of the whole biopolymer structure. Recently, due to development and introduction of new methods it has become possible to overcome the experimental difficulties and obtain direct information on this type of intramolecular mobility of biopolymers.

Cooperative Rearrangements in Proteins. As an example, let us analyze conformational rearrangements in globular protein carriers of molecular oxygen — myoglobin (Mb) and hemoglobin (Hb). The structure of these proteins in a crystal form has been studied in detail using the X-ray analysis (Fig. 10.1).

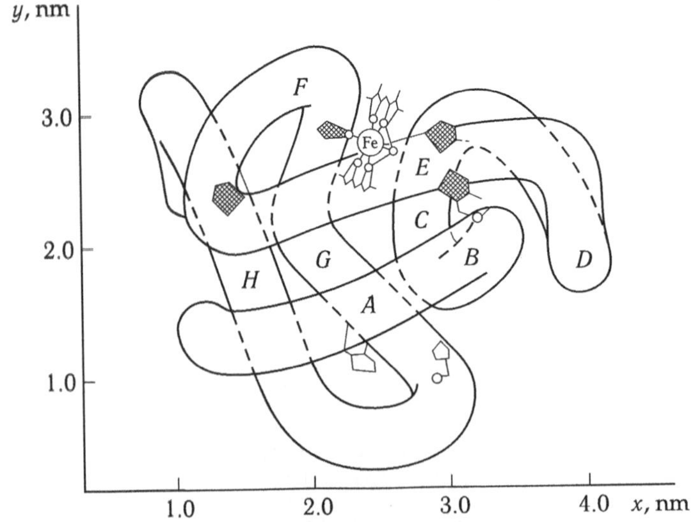

Figure 10.1. Three-dimensional structure of sperm whale myoglobin in projection to the xy-plane.

Myoglobin (oxygen carrier in muscles) contains one heme and one polypeptide chain including 153 residues, which are distributed mainly over eight α-helical regions (A–H). The heme, the center of which contains a Fe atom, is between helices E and F.

The bending of oxygen induces a number of conformational changes in a hemoglobin molecule, disclosed by M. Perutz. The association of oxygen with the transition of the Fe^{2+} ion to a low-spin state is accompanied by the concurrent 0.07 nm displacement of iron to the plane of the heme group. This displacement is transmitted by the histidine residue (F–8), and together with the histidine the helix (F) "pulls itself up" towards the heme to the molecule center, displacing the tyrosine residue from the cavity. After that a stage-by-stage disruption of salt bridges between α-subunits takes place as well as a displacement of subunits along contacts $\alpha_1-\beta_2$ and $\alpha_2-\beta_1$ by 0.07 nm (Fig. 10.2). The disruption of four salt bridges of the six ones upon oxygenation of the two α-subunits facilitates the disruption of the two other bridges and consequently facilitates the association of the following oxygen molecules to the remaining subunits, increasing their affinity to oxygen several hundred times. This is the cooperative character of O_2 attachment to Hb, whereby the beginning of Hb oxygenation facilitates the binding of the other O_2 molecules. □

Consider conversion of another enzyme — phosphofructokinase (PFK) which catalyzes phosphorylation of fructose-6-phosphaste (F6P) in glycolysis with the formation of 1,6-diphosphate (FDP) via the transfer of the ATP γ-phosphate group upon hydrolysis:

$$\text{F6P} \xrightarrow{\text{PFK}} \text{FDP}.$$

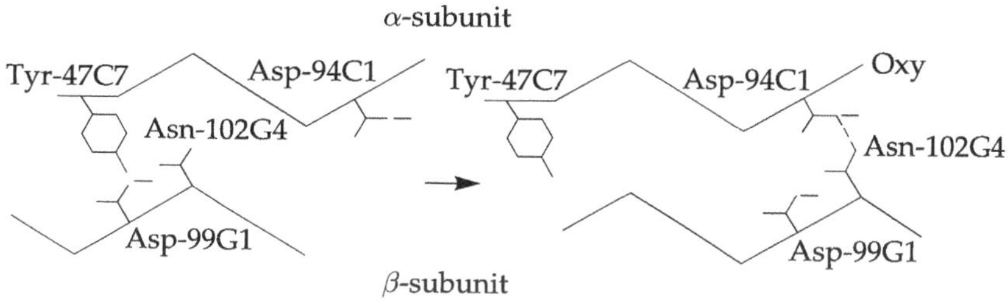

Figure 10.2. Structural changes occurring in hemoglobin upon oxygenation (see explanation in the text) (reproduced from M. F. Perutz, 1970; R. E. Dikerson, 1972).

As mentioned above (Section 2, Chapter 2), at this stage enzyme PFK is activated by the products of the reaction catalyzed by them (ADP, FDP), which causes autooscillations in glycolysis. Enzyme PFK has two states which are distinguished by the affinity to ADP. In the active *R*-form, the activator Mg-ADP is connected by hydrogen bonds to polar groups of protein subunits. The catalytic center of PFK lies on the border between subunits, where the allosteric transition takes place affecting the enzyme activity. The activation of enzyme PFK occurs here not by g the subunit rotation as in hemoglobin and phosphorylase, but by changes in its quaternary structure influencing the boundary realm between subunits.

Here the motions of quaternary structure elements are associated with closing and opening of polypeptide loops, which like levers "pull" subunits with them.

The described conformational rearrangements deeply affect the structure of proteins and represent a cascade of consecutively intercorrelated intramolecular acts proceeding directionally and closely associated with their functioning. These transformations occur spontaneously after the initial "push" (upon the attachment of atom Fe to O_2 and the following displacement of atom Fe to the cavity of the heme group). This local conformational change in the protein molecule is then spread also over its remaining part. The physical reason of the process consists in that the initial equilibrium of forces and, consequently, equilibrium deoxyconformation of the protein were irreversibly disturbed as a result of the oxygen attachment. The ensuing equilibrium oxyconformation and equilibrium of forces are established at another mutual arrangement of groups of atoms in the protein that is reflected in the induced conformational changes. The overall process of forming the equilibrium oxyconformation takes place in a row of consecutive non-equilibrium stages as relaxation of the non-equilibrium state, formed by fast attachment of O_2 to the initially equilibrium deoxyform. □

10.2 Conformational Mobility of Proteins by the Data of Different Methods

The above considered conformational transformations in proteins revealed by the X-ray analysis belong to the relatively large-scale macromolecular shifts, as the overall result of smaller local conformational changes. Obviously, a protein molecule has a number of different conformational substates which emerge fast due to short-living perturbations and fluctuations of the basic structure and between which equilibrium is achieved quickly. Hence, it becomes clear that different parts of the protein structure are involved in fast spontaneous motions varying from each other in a number of parameters. It has become possible to "watch" these fast internal motions, measure their characteristic times, and determine the localization sites in the protein globule mostly because of the introduction of advanced physical methods. Resonance methods of radiospectroscopy (electron paramagnetic, nuclear magnetic and nuclear gamma resonances), luminescence and hydrogen exchange methods play the cardinal role in studying the protein dynamics.

Luminescence and Its Application for Studying the Protein Dynamics

As known, upon absorption of the light quantum a molecule passes to the excited state as a result of the transfer of one electron to a higher energy level. The reverse transition from the excited state to the ground state may be accompanied by light quantum emission. Most frequently such a photoelectron in complex organic molecules is the π-electron involved in the delocalized conjugated double-bond system in the molecule. The transitions *per se* with light absorption and emission in the system of π-electrons are correspondingly transitions $\pi \rightarrow \pi^*$ and $\pi^* \rightarrow \pi$ between the ground and the so-called singlet excited state. They correspond to the transitions between bonding π^-- and antibonding π^*-orbitals.

Electron Levels in Molecules. Figure 10.3 shows curves of potential energies of the gound (S_0) and excited (S^*) states (cf. Fig. 8.1). The potential electron energy includes Coulomb interactions of all electrons and nuclei between themselves and with each other without taking into account the kinetic energy of nuclei. Since electrons move at rates by 2–3 orders of magnitude exceeding the rates of nuclear motion, it can be reckoned that electrons change their positions as nuclei remain practically immobile. Consequently, the electron energy $U(R)$ depends on the internuclear distance R as a parameter.

The separation of intramolecular motions into fast electron and slow nuclear underlies one of the fundamental laws of spectroscopy associated with the so-called

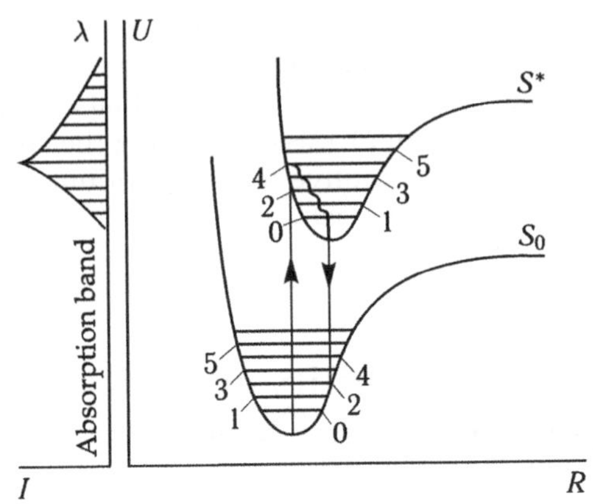

Figure 10.3. Curves of potential energy of the ground (S_0) and singlet excited (S^*) states of a two-atom molecule.

U is the potential energy, R is the internuclear distance, I is the absorption intensity, λ is the wavelength, 0–5 are vibrational sublevels of nuclear states.

Frank–Condon principle. Each electronic state S_0 and S^* is characterized by its potential curve $U^0(R)$ and $U^*(R)$. Transitions between them correspond to electron transitions between different energy levels. The motion (oscillation) of nuclei themselves occurs in the field created by quickly moving electrons, so that $U(R)$ is the potential energy of nuclear motion. The solution of the quantum-mechanical problem for nuclear motion shows that the energy of nuclear oscillations has discrete values, i.e., is quantized analogous to the electron energy. The appropriate vibrational sublevels (1, 2, 3) for nuclei are given on each curve of electron energy (Fig. 10.3). Light absorption and the transition $S_0 \rightarrow S^*$ occur during 10^{-15}–10^{-16} s, i.e. practically instantaneously, so that heavy nuclei have no time to change their positions and rates (the Frank–Condon principle). As seen from Fig. 10.3, during this transition (the vertical arrow) the reserve of internal vibrational and electron potential energy increases.

The reason for this is that the curves $U^0(R)$ and $U^*(R)$ are not identical. The initial internuclear distance for the ground state $U^0(R)$ corresponds to higher vibrational sublevels on the upper curve $U^*(R)$ of the excited state S^*. When the transition $S_0 \rightarrow S^*$ occurs the nuclear configuration still remains characteristic of the ground state S_0. Hence, the whole system becomes in an excited vibrational state. The vibrational relaxation occurs during the lifetime of the electron excited state S^*, the latter makes at average 10^{-9}–10^{-8} s for aromatic compounds. The excess vibrational energy dissipates into heat during 10^{-12}–10^{-11} s, and nuclei relax to lower sublevels (the wavy curve in Fig. 10.3) of the state S^*, from which the electron transition to the ground level S_0 occurs accompanied by the fluorescence light quantum emission.

Upon this transition the vibrational energy in the molecule increases again followed by its subsequent dissipation in 10^{-12}–10^{-11} s.

So, in accord with the Frank–Condon principle, "vertical" transitions between electronic states proceed at nearly immobile nuclei and are accompanied by a brief (for 10^{-12}–10^{-11} s) increase of vibrational energy in the molecule. "Non-vertical" transitions are much less probable. The position of the maximum of the spectral fluorescence band is determined by the most probable transition $S^* \rightarrow S_0$, while the magnitude of each fluorescence quantum is determined by losses of electron excitation energy to nuclei vibrations. □

In addition to fluorescence light emission, the state S^* can be deactivated otherwise. A nonradiative transition to the ground level $S^* \rightarrow S_0$ may take place when all the energy of electron excitation is converted to heat. Besides singlet (S) electron levels, the molecule has also triplet (T) electron levels located on the energy scale lower than singlet ones. The nonradiative transition from state S^* to state T is called intercombinational conversion $S^* \rightarrow T$. It is associated with the change in the spin orientation of the photoelectron to the opposite direction at the level S^*. As a result, the spins of the two initially coupled π-electrons, forming a chemical bond in the ground state S_0, now become parallel. Thus, the triplet molecule has two unpaired electrons and displays paramagnetic properties like biradicals.

The transition from the triplet state to the ground $T \rightarrow S_0$ also requires in its turn reorientation of the spin, thus lowering the probability of this transition. Therefore the lifetime of the triplet state is far greater than that of the singlet S^* state, being 10^{-6}–10^{-2} s and higher. Transitions $T \rightarrow S_0$ are accompanied by phosphorescence or can be nonradiative. A reverse transition from the triplet to the singlet excited state $T \rightarrow S^*$ may occur due to thermal activation, as a result delayed fluorescence is observed. Its spectra are similar to those of the $S^* \rightarrow S_0$ fluorescence while its lifetime is similar to that of phosphorescence. Inasmuch as the triplet molecule T lives rather long, it can have time enough to absorb the second quantum and pass to the second triplet level ($T_1 \rightarrow T_2$-absorption).

Figure 10.4 represents a general scheme of electron levels and transitions between them (Yablonsky's scheme).

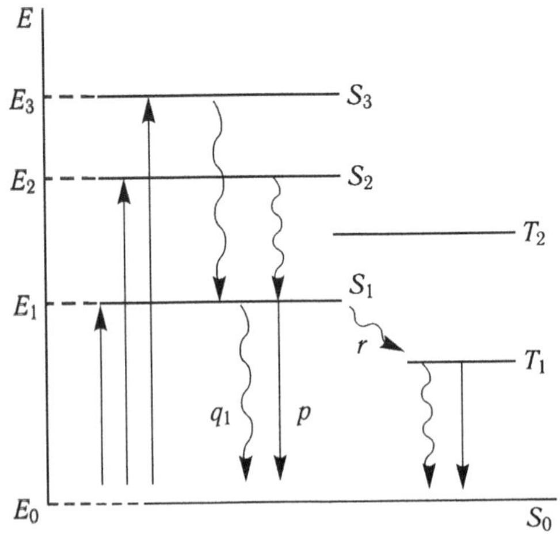

Figure 10.4. Electron levels in the organic molecule and transitions between them (Yablonsky's scheme).

p is the probability of transitions per a time unit to the ground level with fluorescent emission, q_1 is the same without emission, r is the probability of conversion to the triplet state.

The molecule has a system of triplet $T_1, T_2, \ldots$ and excited singlet levels $S_1, S_2, \ldots, S_n$. Upon transition to one of the highest excited singlet levels, the excess vibrational energy rapidly dissipates (10^{-12} s); the molecule gets to the lower singlet level S_1, from which the transition ($S_1 \rightarrow S_0$) or intramolecular conversion ($S_1 \rightarrow T_1$) take place. The overall probability (P_1) of deactivation is determined by the sum of values p, q_1, r: $P = p + q_1 + r$.

When the transition to the excited state considerably changes, the electric dipole moment of the molecule as compared to the dipole moment in the state S_0, the dipole field of the environment becomes also non-equilibrium. During the lifetime of the chromophore excited state, the dipoles of the environment can have time to reorient themselves in accordance with the new dipole field generated by the excited chromophore. Obviously, the relaxation of dipoles is restricted by dynamic properties of the environment. It is clear that if the time of the dipole relaxation τ_p of the environment is greater than the lifetime of the excited chromophore τ^* ($\tau_p \gg \tau^*$), the relaxation has no time to occur during τ^*, i.e. the chromophore environment is "rigid". On the contrary, at $\tau_p \ll \tau^*$ fast dipole relaxation in the surrounding occurs during τ^*. Due to the interaction between the dipole of the excited chromophore and oriented dipoles of the surrounding the decreases in the energy level of the excited electron causing a displacement of the fluorescent spectrum to the longer wave lengths occur.

Let us consider the diagram shown in Fig. 10.4. If $\tau_p \ll \tau^*$, during the lifetime τ^* the relaxation has time to pass on the molecule to lower sublevels of its excited state, wherefrom a luminescence quantum is emitted. On the contrary, at $\tau_p \gg \tau^*$

emission emerges from the top vibrational sublevels, which makes the emission spectrum maximum at the short-wave lengths. In a complex system (protein), where the change in the electronic state of the chromophore molecule causes rearrangement of its microenvironment, the relaxation time can exceed the lifetime of the electron excitation ($\tau_p > \tau^*$). This is not observed in individual chromophore molecules, where the intramolecular vibrational relaxation takes 10^{-12}–10^{-11} s, i.e. much less than time τ^* of states S_1 and T_1.

By changing the relaxation properties of the environment (with a temperature change), it is possible to influence the relationship of τ_p and τ^*. This is reflected in the position of the fluorescent spectrum maximum $\lambda_{\max}$ subject to the equation

$$\frac{1}{\lambda_{\max}} = \frac{1}{\lambda_\infty} + \left(\frac{1}{\lambda_0} - \frac{1}{\lambda_\infty}\right)\frac{\tau^*}{\tau_p + \tau^*}, \tag{10.1}$$

where λ_0 is the position of the spectrum maximum at $\tau_p \ll \tau^*$, and λ_∞ is its position at $\tau_p \gg \tau^*$. This method allows obtaining valuable information on the mobility of proteins upon studying the tryptophan self-fluorescence.

The indole ring of tryptophan is characterized by a large change of the transition moment upon excitation (up to $4D$). Because of this, the position of the maximum of its fluorescence spectrum is strongly dependent on the mobility of dipoles of the environment and the maximum can shift to 30 nm. As the time τ^* for the state S^* of tryptophan makes units of seconds, this permits studying structural rearrangements in proteins in a nanosecond time interval. It was demonstrated (Burshtein, 1977) that proteins have characteristic two-step curves of the fluorescence spectrum position on temperature in the range from $-90\,°C$ to $0\,°C$. The shifts of the fluorescence band by 4–9 nm in the range from $-90\,°C$ to $-20\,°C$ and by 5–12 nm in the range from $-20\,°C$ to $0\,°C$ (Fig. 10.5) are evidence of the freezing of motions in a protein matrix within a nanosecond range. Dried proteins have no such spectral shifts, which underlines the role of water as a necessary factor for providing the protein mobility (see Section 4 in Chapter 9). Apparently, the "freezing" of the fast relaxation mobility of proteins has a cooperative character, whereby the protein structures and the bound water freeze as a single microphase.

It was found that the kinetics of tryptophan fluorescence decay in different proteins is multiexponential. Hence it follows that there is a set of protein conformers,

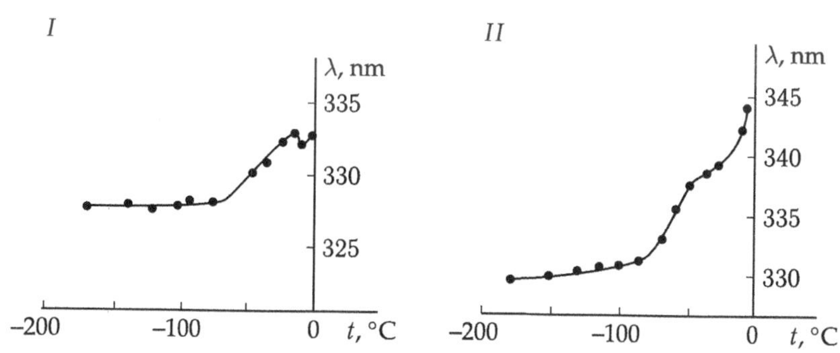

Figure 10.5. Dependence of the position of the fluorescence spectrum of aqueous β-lactoalbumin (*I*) and cobra neurotoxin (*II*) solutions on temperature at pH 6.5 (reproduced from E. A. Permaykov, 1977).

varying in their fluorescent properties, and that the time of establishing equilibrium between them is higher than the fluorescence lifetime.

Along with the change of parameters of the protein self-fluorescence (tryptophan residues), of great importance is the study of fluorescence of exogenous phosphorescent labels (for example, eosin). Since the lifetimes of triplet and singlet states can change widely, this allows widening the range of measurements of characteristic times of structural rearrangements from 10^{-9} to 1 s. For example, this method was used to record slower motions $\tau_p \sim 10^{-3}$ s in the surface layers and $\tau_p < 1$ s inside protein macromolecules in a photosynthetic reaction center.

Method of Spin Labels. Electron Paramagnetic Resonance (EPR). It is known that in a constant magnetic field of intensity $\mathbf{H}_0$ the electron spins will be oriented along the field. The energy level of an electron with the spin $s = 1/2$ is split in two levels according to the two spin directions parallel and antiparallel to field $\mathbf{H}_0$ (Fig. 10.6).

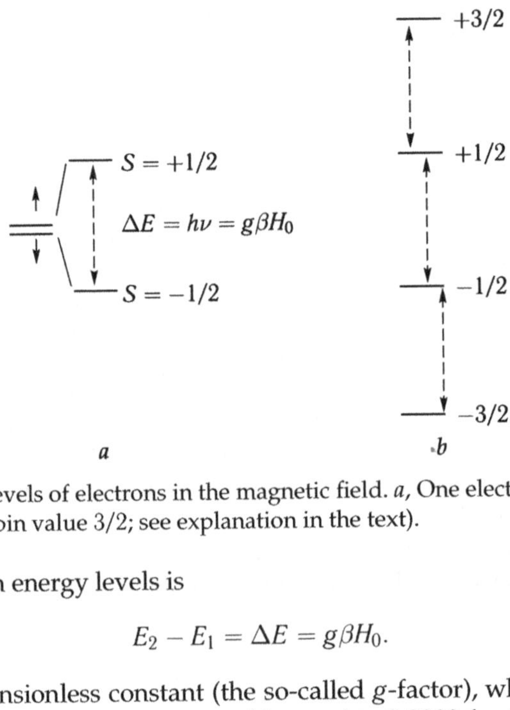

Figure 10.6. Energy levels of electrons in the magnetic field. *a*, One electron (spin 1/2); *b*, three electrons (maximal spin value 3/2; see explanation in the text).

The difference in energy levels is

$$E_2 - E_1 = \Delta E = g\beta H_0. \tag{10.2}$$

Here g is the dimensionless constant (the so-called g-factor), which depends on the electron structure of the environment and is equal to 2.0023 for the free electron; β is the Bohr magneton:

$$\beta = e\hbar/(2m_e c) = 0.93 \cdot 10^{-23}\ \text{J/T},$$

where e and m_e are the charge and electron mass respectively; c is the rate of light; $\hbar = h/(2\pi)$, h is the Planck constant ($6.62 \cdot 10^{-34}$ J $\cdot$ s). It is obvious that at a general number N of spins in the system the relation of spin populations at the bottom (N_1) and top (N_2) levels is determined by the difference of energies ΔE accordingto the Boltzmann distribution law:

$$N_1/N_2 = \exp\left[g\beta H_0/k_B T\right], \tag{10.3}$$

where $N_1 + N_2 = N$ and $N_1 > N_2$.

Now let us, in addition to constant $\mathbf{H}_0$, apply also alternative field of frequency ν, perpendicular to the constant field. Then forced transitions will be induced in the system between levels with the same probability for transitions $N_1 \rightarrow N_2$ and $N_2 \rightarrow N_1$. As from the start $N_1 > N_2$, at every time moment the number of transitions from the bottom level to the top one $N_1 \rightarrow N_2$ will exceed the number of reverse transitions $N_2 \rightarrow N_1$. Given

$$h\nu = g\beta H_0 \tag{10.4}$$

resonance absorption of the alternating field energy takes place. This phenomenon is called electron paramagnetic resonance (EPR).

From Eq. (10.4)we can find that at $H_0 = 1$ T, the resonance frequency for a free electron is $\nu_e = 2.8 \cdot 10^{10}\ \text{s}^{-1}$, i.e. is ascribed to the range of microwave emission. Upon increasing the power of microwave emission irradiating the sample, the initial difference between populations $n = N_1 - N_2$ the levels should obviously decrease. In its turn, this will lead to the disappearance of the resonance absorption of the alternative field, because

$$n(t) = n(0)\exp(-2tp),$$

where $n(0)$ is the difference of populations at $t = 0$, and p is the probability of the induced transitions between the levels.

Relaxation Processes in Spin Systems. An increase in the microwave power first leads to an increase in the amplitude of resonance absorption of the alternative magnetic field (the EPR signal), then some stationary value establishes and saturation takes place. The reason for this is associated with the interaction of the system of spins with their environment. This interaction causes reorientation of the spin and results in the transfer of the excess magnetic energy to other degrees of freedom, which in a solid includes oscillations of the crystal lattice. Nonradiative transitions between the two states of the spins, coupled to the interaction with the environment, are called the spin-lattice relaxation. These processes are the so-called spin-lattice relaxation time T_1. It reflects the average lifetime of the given spin state and corresponds to the time necessary to establish thermal equilibrium between the spin system and thermal vibrations of the lattice. Large T_1 time values indicate to a low rate of the spin-lattice interaction.

Figure 10.7 demonstrates the absorption line of the microwave field and its first derivative, which is usually recorded by EPR-spectrometers and, as a matter of fact, is an EPR signal.

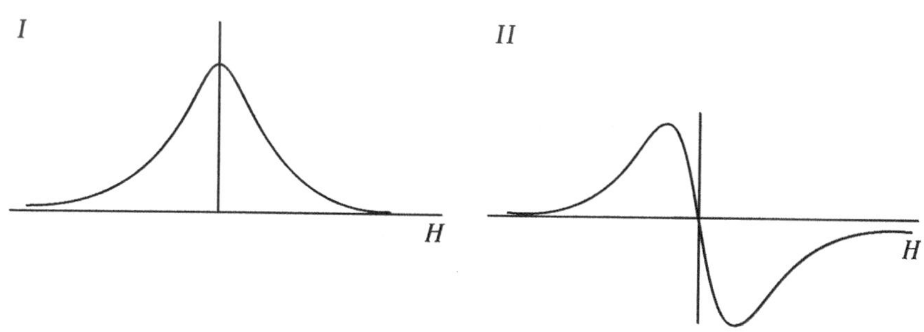

Figure 10.7. EPR signal. *I*, the absorption line of the microwave field; *II*, its first derivative. The abscissa axis is the magnitude of the constant magnetic field *H*, which changes gradually at a constant frequency of the microwave field up to attaining values corresponding to the resonance absorption.

Spin-Lattice Relaxation. How is the character of interactions in the spin system related to the EPR signal parameters? As seen from Fig. 10.7, the EPR signal has a wide absorption line. This is explained by the fact that due to relaxation processes, the state of the spin has a definite lifetime, which causes widening of the resonance signal because of the uncertainty relation for energy. Subject to this relationship, the energy E_0 of a system capable of disintegration during an average lifetime τ can be determined only with an accuracy of about the $\hbar/\tau$ value (generally called the Γ width of the energy level of the system).

Accordingly (cf. Eq. (8.4)),

$$\Gamma\tau \sim \hbar. \tag{10.5}$$

This is the uncertainty relation for energy. It means that the exact position of the energy level E_0 of the electron spin state in the system with the relaxation time T_1 can be determined by the position of the resonance line with an accuracy up to $\hbar/T_1$ ($E_0 \pm \hbar/T_1$), which just implies the widening of the resonance line E_0 as $\Delta H \sim \hbar/T_1$. □

The T_1 value is strongly dependent on temperature, increasing upon freezing. So, EPR signals of paramagnetic ions of metals are not frequently observed at room temperature due to a low value of T_1 and, as a consequence, a very large width. But at the temperature of liquid helium, the signals are recorded, because an increase in T_1 leads to the narrowing of the absorption line and the growth of its amplitude.

The spin-lattice relaxation is not the only process determining the line width. There are many other processes causing changes in relative energies of spin states due to the energy exchange in the spin system *per se*. They are the so-called processes of spin-spin interactions characterized by the time of the spin-spin relaxation T_2. For paramagnetic particles the T_2 value is practically independent of temperature and corresponds to the rate of equilibrium establishement between electron spins upon their interaction with each other. The contribution to the general line width of the spin-spin interaction is described by the value of about $1/T_2$. In a general case, the line width is determined as

$$\Gamma = \Delta H \sim \frac{1}{T_{exp}} \sim \frac{1}{(2T_1)} + \frac{1}{T_2}, \tag{10.6}$$

where T_{exp} is determined experimentally.

For free radicals $T_1 \gg T_2$, that is why in the absence of saturation their line width is determined mainly by T_2 ($\Delta H \sim 1/T_2$), which makes $\sim 10^{-8}$ s. Large values of T_1 (seconds, minutes) for organic free radicals are associated with the fact that the spin energy of electrons is transferred to oscillations of the lattice nuclei via the spin-orbital interaction, which is rather small for light nuclei. This conditions a relatively low rate of the spin-lattice relaxation of free radicals.

Physical Mechanisms of Spin-Spin Exchange. In solids and viscous solutions, the line width depends on dipole–dipole interactions of electron spins. The magnetic field, generated by the magnetic moment of the unpaired electron, induces an additional local field H_{loc} where the other electron is located equal to

$$H_{loc} \sim \mu \frac{3\cos^2\theta - 1}{R^3}, \tag{10.7}$$

Here μ is the magnetic moment component directed along the magnetic field of the magnet; R is the length of the position vector connecting the both electrons; θ is the angle between the directions of the magnetic field with the position vector.

Composition of this local magnetic field and the external magnetic field H_0 violates the resonance condition (Eq. (10.4)), which now is observed if

$$h\nu = g\beta(H_0 + H_{\text{loc}}).$$

Thus, instead of one resonance line from H_0, now we have a set of closely positioned lines which are a result of mutual influence of spins on each other due to local fields. This effect reduces the time T_2 for the whole spin system, widening the line subject to relation (10.5).

The resultant widened resonance line is the envelope of all adjacent lines of "local resonances". The dipole-dipole widening to about 0.1 mT is seen at distances of 3–4 nm between spins. In liquids of low viscosity, the dipole-dipole interactions are averaged due to fast molecular motions. In this case during the time inverse to the resonance frequency $t \sim 1/\nu$, multiple fluctuations of the position of radicals can occur. As a result, some lines of "local resonance" cannot be resolved and the resonance is observed at one averaged line with a decreased width. Freezing of the solution and a decrease in the viscosity again retard molecular motions and therefore a decrease in T_2 of spin states widens the resonance line.

At standard temperatures in liquids, other effects (chemical reactions, electron transfer, spin exchange), leading to changes in the magnetic surrounding of the unpaired electron lead to the widening of the line. For example, the exchange of spin states upon particle collision in solution decreases the lifetime T_2 of spins and widens the line width due to the uncertainty relation (Eq. (10.5)). The magnetic moments of nuclei in radical particles generate local fields, which are added to the external field H_0 and thereby have an effect on the EPR lines, causing their splitting manifested by a superfine structure (SFS).

Two types of interactions of electrons and nuclei are known — anisotropic and isotropic. Anisotropic interactions are analogous to the classical interaction of two magnetic fields and are described by the expression similar to (10.7), according to which the local magnetic field generated by the nucleus depends on the angle θ.

If the unpaired electron is localized on the asymmetric orbital p or d, the result of the spin-nucleus interaction will be dependent on the angle between the direction of the vector, connecting the electron and the nucleus, and the direction of the magnetic field.

For a symmetric s-orbital, the additional magnetic field created by the nucleus and determined by expression (10.7) is equal to 0 due to the averaging over all θ angles. In this case, as shown theoretically, the superfine splitting is determined by the final probability of the electron stay in the point of the nucleus location. Such an interaction is called contact interaction. The SFS emerging due to the contact interaction of the nucleus and the electron is independent of the direction and is called isotropic.

Local nuclear fields have different effects on the width of EPR lines because of different angels between the nucleus-electron line and the direction of the field H_0 (Fig. 10.8). Fast molecular motion of radicals in low viscosity media averages the acting local fields and averages the anisotropic component, thus narrowing the EPR line. If the interacting spins are simultaneously exposed to these local fields, responsible for SFS, the pattern becomes more complicated. Along with the widening of the line due to the exchange interactions between the spins, the averaging of different local fields will evidently take place that causes "smearing" of the SFS lines, their tightening to some averaged value and, consequently, some narrowing of the EPR lines.

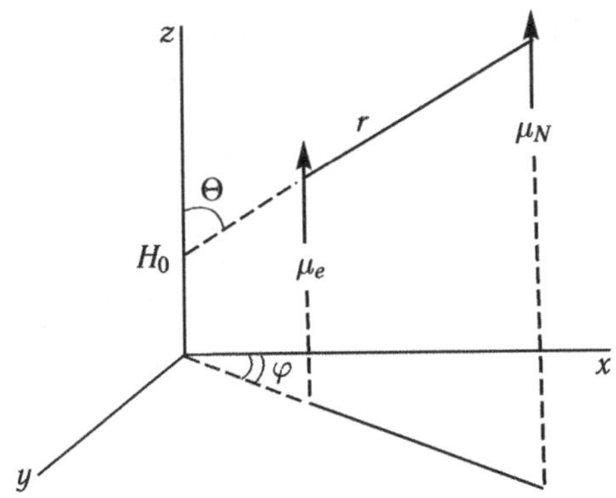

Figure 10.8. Anisotropic interaction of nuclear and electron magnetic resonances (reproduced from K. N. Timofeev, 1985).

μ_N is the nuclear magnetic moment; μ_e is the electron magnetic moment; r is the position vector; θ and φ are angles formed by the position vector r with axes z and x; H_0 is the external magnetic field directed along axis z.

The shape of the EPR spectrum depends also on the effect on the g-factor of the asymmetric anisotropic surrounding of the unpaired electron in the radical particle that determines the fine structure of the spectrum. Here the averaging leads to the removal of the g-factor anisotropy at fast intramolecular motions. Accordingly, changes in definite parameters of the EPR spectrum can yield important information on dynamic properties of the microsurrounding of the radical particle.

Method of Spin Labels. The main idea of the method is to bind a free radical to a functional protein group and to study characteristics of its EPR signals. In this respect, most convenient are nitroxyl radicals containing a free radical group $\dot{N}$–O.

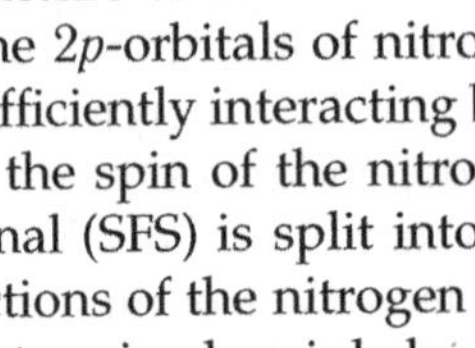

The scheme of the electron structure of the $\dot{N} \rightarrow O$ group is given in Fig. 10.9. The unpaired electron belongs to the $2p$-orbitals of nitrogen and oxygen and is delocalized between atoms N and O, efficiently interacting by the dipole-dipole mechanism with the magnetic moment of the spin of the nitrogen atom nucleus. Thereby, the absorption line of the EPR signal (SFS) is split into three components corresponding to the three different projections of the nitrogen nuclear spin to the H_0 direction. The shape of the spectrum is determined mainly by the anisotropic interaction.

Along with it, there is also a slight admixture of isotropic interaction. Figure 10.10 demonstrates an EPR spectrum of the paramagnetic label bound to lysozyme. As the SFS and the g-factor in EPR spectra of nitroxyl radicals are anisotropic, the Brownian diffusion affects the spectrum shape. The magnitude of the SFS splitting will depend on the angles between the axes of the paramagnetic fragment and the direction of the field. Rotation of the nitroxyl fragment relative to the field changes the position of lines in the field. In the case of maximally fast rotations $\nu_{rot} \geqslant 10^{11}\ s^{-1}$ all orientations

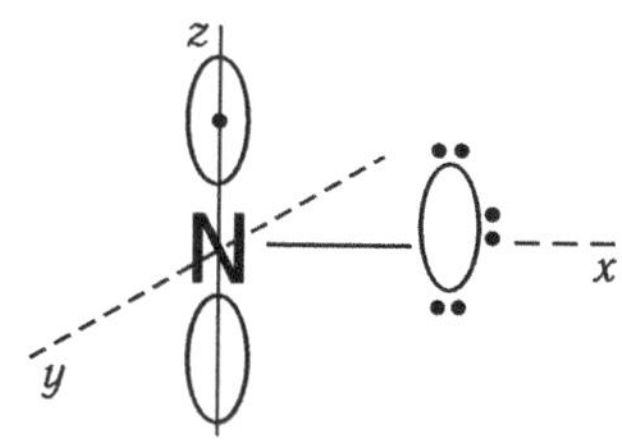

Figure 10.9. Scheme of a paramagnetic fragment of the nitroxyl radical.

The unpaired electron shown by a dot is localized on the R_z-orbital of the nitrogen atom. The efficient magnetic field acting on the electron is composed of the external magnetic field and the field generated by the nitrogen nucleus.

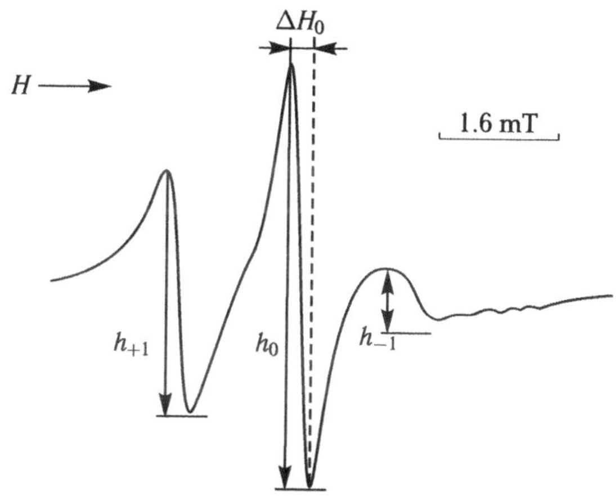

Figure 10.10. EPR spectrum of the paramagnetic label attached to *His*-15 of lysozyme (pH 7.0; $t = 26\,°C$) (courtesy of Lichtenstein).

h_{-1}, h_0, h_{+1} are intensities of components corresponding to magnetic moments of the nitrogen nucleus $M = -1, 0, +1$; ΔH_0 is the width of the central component.

are averaged, and the spectrum represents three equidistant lines of the same intensity, which in this case is due to the isotropic interaction of the unpaired electron with the nitrogen atom. But at the rotation frequencies of $10^{11}\ s^{-1} > \nu_{rot} \gg 10^9\ s^{-1}$, anisotropy of the SFS and g-factor is already displayed that changes the symmetric character of the spectrum (the high-field H_{-1}-component is most widened and has the lowest amplitude). Thus, based on the EPR spectrum parameters it is possible to estimate the correlation time of the radical motion (τ_c), i.e. the time interval enough for the particle to "turn" by the angle $\pi/2$. The $\nu = 1/\tau_c$ value is tentatively called the label rotation frequency.

In general, the rate of the nitroxyl label rotation is dependent on how deep it is localized in the protein globule layers (Lichtenstein, 1974).

When the label is localized in the external water layer, τ_c makes approximately $10^{-11} \div 10^{-10}$ s (Fig. 10.11). The surface layers, their "clefts" containing side protein groups and water molecules, are characterized by $\tau_c \sim 10^{-10} \div 10^{-8}$ s. In a deeper layer of the globule, containing external polypeptide chains, "cramped" side groups and tightly bound water, τ_c increases to $10^{-8} \div 10^{-7}$ s. This shows explicitly the slowing down of the label rotation rate upon its submersion that corresponds to the

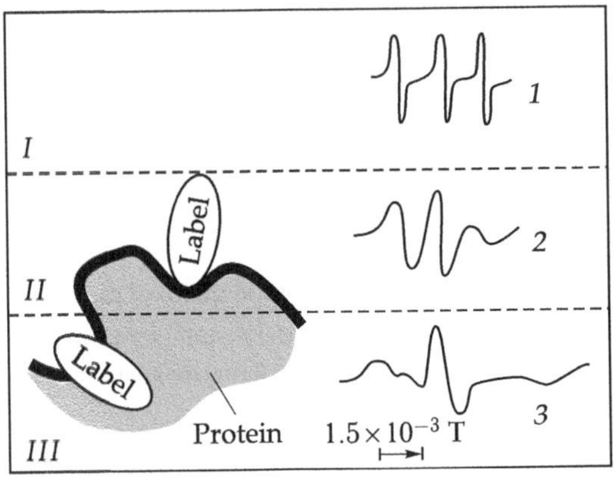

Figure 10.11. Mobility of different layers of the water-protein matrix (courtesy of Lichtenstein). *I*, The outer water layer of the matrix ($\nu \simeq 10^{11}$–10^{10}); *II*, the glycerol-like layer of the matrix ($\nu \simeq 10^{10}$–10^{8}); *III*, relatively solid layer of the matrix ($\nu \simeq 10^{8}$–10^{7}); ν is the correlation frequency of rotation diffusion of radicals (s^{-1}).

1, 2, 3 are typical EPR spectra of nitroxyl radicals submersed in corresponding layers of the matrix.

existence of a denser nucleus in the globule as compared to the soft "trimming" (see Section 2 in Chapter 7). On the whole, the correlation times that can be determined by the spin-label method are in the range of 0.1–300 ns, though the methods allowing determination of τ_c up to 10^{-3}–10^{-4} s are intensively developed.

If the radicals begin rotating, this makes a separate radical orientation relative to the external field to change randomly that in its turn changes the line positions in the EPR spectrum of separate radicals as well as changes the overall spectrum. □

On the other hand, the maximally fast rotation of the radical ($\nu > 10^{11}\ s^{-1}$) leads to complete averaging of EPR spectra anisotropy, so that an EPR spectrum represents three equidistant lines of the same intensity with the averaged SFS constant (1.6–1.7 mT) and averaged value of the g-factor (2.0059). It can be stated that in the case of maximally fast rotations the EPR spectrometer "has no time" to trace the changes in the radical orientation relative to the external field and "sees" only some averaged spectrum.

In the range of rotation frequencies of $10^{11}\ s^{-1} > \nu > 10^{6}\ s^{-1}$, anisotropy of the SFS and g-factor begins to show itself, thus changing the symmetric character of the spectrum. In the range of fast rotations this is manifested in the changed relationship between amplitudes and line widths of separate spectral components. Models are available for different types and frequencies of motion of nitroxyl radicals to determine τ_c.

For example, in the case of fast rotations $5 \cdot 10^{-11}\ s < \tau_c < 3 \cdot 10^{-9}$ s the time τ_c is determined by the formula

$$\tau_c = 6.65 \cdot \Delta H_{+1} \left(\sqrt{\frac{J_{+1}}{J_{-1}}} - 1 \right) \cdot 10^{-9}\ s,$$

where ΔH_{+1} is the width of a low-field line of the spectrum in mT; J_{-1} and J_{+1} are amplitudes of the high-field and low-field components of the spectrum.

In the range of slower motions, the methods are based on measuring the width of the lines and their displacements relative to the line positions in the spectra of radicals rotating maximally fast.

To retrieve information on the dynamics of biopolymers from EPR spectra of bound spin labels is a rather difficult task and can be solved using computer simulation of the spectra.

The experimentally found correlation times are compared to the calculated ones that allows us to draw conclusions on sphericity or nonsphericity of the protein molecule and the flexibility of separate segments of protein regions.

For example, data on correlation times for motions of hemoglobin, albumin, phosphorylase, lipase, and protein kinase demonstrated that they correspond to rigid globules with different ellipsoidal degrees.

A further development of the spin label method was the so-called EPR saturation transfer spectroscopy, which makes it possible to study rotation of spin labels, bound with macromolecules in the range of correlation times 10^{-7}–10^{-3} s.

The development of this method has enabled investigations of the dynamics of membrane proteins, protein–protein and protein–lipid interactions in biological membranes.

Thus, the saturation transfer spectroscopy was used to analyze the dynamics of Ca-ATPase in membranes of the sarcoplasmic reticulum and in proteoliposomes. As a result, the relation of conformational mobility and ATPase activity was revealed and intramolecular mobility and Ca-ATPase activity dependence on the nature of lipids associated with this protein was demonstrated.

Nuclear Magnetic Resonance (NMR) Method

One of the powerful tools for studying the dynamics of biopolymers is nuclear magnetic resonance. The essence of NMR is for the most part similar to that of electron paramagnetic resonance. Nuclei (besides nuclei with an even number of protons and neutrons — "even-odd" nuclei — such as basic carbon $^{12}_{6}C$ and oxygen $^{16}_{8}O$ isotopes) have spin values I (designation accepted for nuclei) and magnetic dipole moments μ different from zero. At the same time, magnetic moments of different nuclei also differ from each other.

The action of constant external magnetic field H_0 orients magnetic moments and in the case of spin $I = 1/2$ splits the initial level in two sublevels respectively (Fig. 10.12). The existence of two spin states with different energies in the external magnetic field

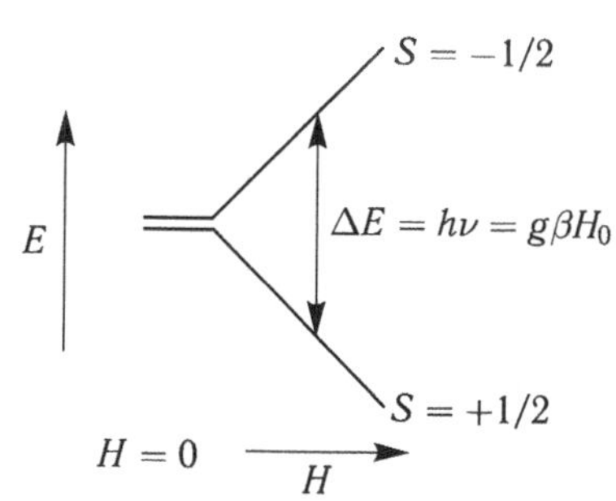

Figure 10.12. Resonance conditions of nuclear spin in the presence of the external magnetic field (splitting of the energy levels of the nucleus with spin 1/2).

leads to non-uniform distribution of spins in these states. In an equilibrium state, the distribution of spins over energy levels is described by the Boltzmann equation

$$N_{\text{upper}}/N_{\text{lower}} = \exp(-\Delta E/k_B T) = \exp(-hv/k_B T),$$

where N_{upper} and N_{lower} are populations of the nuclei of corresponding energy levels; k_B is the Boltzmann constant; T is the absolute temperature. For a field of intensity 18.8 T (v = 800 MHz) in the equilibrium state at room temperature, the ratio of populations is 0.999872, or for 1,000,128 nuclei in the lower state there are 1,000,000 nuclei in the upper energy state. Such a small difference is the reason for low sensitivity of the NMR spectroscopy method as compared to most part of other methods for studying biomacromolecules.

Upon applying an alternative field, the resonance conditions look as follows (cf. Eq. (10.4))

$$h\nu_N = g_N \beta_N H_0, \tag{10.8}$$

where g_N is the nuclear g-factor; $\beta_N = e\hbar/(2m_p c)$ is the nuclear magneton; m_p is the proton mass equal to $1836 m_e$. For the proton $g_N = 5.58$ and the nuclear magneton $\beta_N = 0.0505 \cdot 10^{-26}$ J/T.

In the case of NMR, the resonance conditions are more frequently written as

$$h\nu_0 = \gamma H_0,$$

where γ is the so-called gyromagnetic ratio or the ratio of the magnetic moment μ to the mechanical — spin I. In the field $H_0 = 1$ T, the resonance frequency for proton nuclear resonance will be $\nu_H = 4.26 \cdot 10^7$ Hz, which corresponds to the range of meter radio waves and is far smaller than the resonance frequency ν_e for the electron. Resonance conditions for nuclei, such as protons, included in the molecule structure, will differ from those for the free proton due to its screening by electron shells and the effect of nuclei from the chemical groups surrounding the proton. Thereby, the resonance magnetic field in (10.8) should be substituted by an efficient field which takes into account the surrounding. Moreover, magnetic moments of different nuclei interact with each other and with the electrons in the molecule, the character of this interaction depending also on the surrounding of the nuclei. These factors affect the parameters of the NMR spectrum and in this way yield information on the chemical properties and intramolecular dynamics.

Chemical Shift. The magnetic field near the nucleus is dependent on its electron surrounding, which changes with the formation of chemical bonds. This leads to the emergence of a chemical shift, proportional to the applied constant magnetic field. The physical reason for the shift is that circulation of electrons and corresponding electron currents in the molecule are induced by the external magnetic field. The direction of the electron currents is such that the additional magnetic field H_{ind} induced by them on the nuclei is proportional in its magnitude and opposite in its direction (subject to the law of induced current) to the external field H_0 ($H_{\text{ind}} = \sigma H_0$). Accordingly, the full efficient magnetic field which acts on the given nuclear moment looks like this

$$H_{\text{eff}} + H_0 - H_{\text{ind}} = (1 - \sigma)H_0, \tag{10.9}$$

where σ is the so-called screening constant (a dimensionless value) having the order of 10^{-6}. The σ value is anisotropic, but in liquids it is averaged. Taking into account

Eq. (10.9) the resonance condition is as follows:

$$h\nu_N = g_N \beta_N (1 - \sigma) H_0.$$

Consequently, now the resonance will be observed at another frequency.

Let us analyze two protons i and j with different chemical environments and, as a result, with different screening constants σ_i and σ_j. We can measure the relative chemical shift in their NMR spectra caused by their different positions in the chemical structure:

$$\nu = \nu_i - \nu_j = g_N \beta_N H_0(-\sigma_i + \sigma_j) = \nu_0(\sigma_j - \sigma_i),$$

or $\Delta\nu = \nu_0 \Delta\sigma$, where $\Delta\nu$ is the difference of resonance frequencies of the i-the and j-th protons, and $\Delta\sigma$ is the difference in screening constants of the same protons.

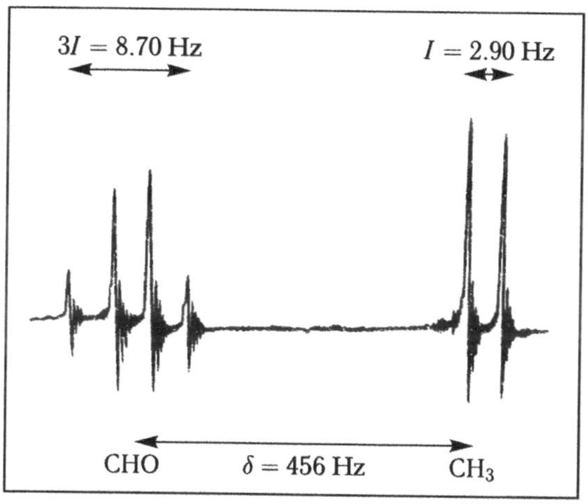

Figure 10.13. NMR spectrum of acetaldehyde CH_3CHO (reproduced from A. Carrington and A. Mac-Lachlan, 1970).

I is the magnitude of superfine splitting; the magnetic field increases from left to right.

The four lines for the aldehyde proton are caused by the additive effects of interaction with three protons of methyl groups. In this case, the total spin value 3/2 of the three chemically equivalent protons is achieved in one variant with the same direction of all three spins, 1/2 in three variants, −1/2 in three variants, and −3/2 in one variant.

The magnitude of spin-spin interaction is characterized by the spin-spin interaction constant (SSIC). It is designated by Latin symbol J, reflects the connection of nuclear spin precessions and is independent of the magnetic field intensity of the spectrometer. Spin-spin interaction is the result of interaction of electrons forming a chemical bond between nuclei. The SSIC value decreases rapidly with an increase in the number of covalent bonds connecting atoms. Direct SSIC (across one bond, ^1J) have the highest value ($^1J_{CH} \sim 140$ Hz, $^1J_{NH} \sim 90$ Hz, $^1J_{CC} \sim 40$ Hz, $^1J_{CN} \sim 9$–15 Hz). Geminate (^2J) and vicinal (^3J) SSIC (across 2 and 3 chemical bonds respectively) have values up to 20 Hz and are rather simply detected; long-range SSIC across 4 and more covalent bonds are vanishing (< 1 Hz). The vicinal SSIC play a significant role in contemporary structural studies because their values depend on the mutual orientation of the bonds connecting atoms, and permit finding torsion angles in the molecule, necessary to determine its structure. In addition, spin-spin interactions make it possible in a great number of NMR experiments to pass transfer magnetization between nuclei over the chain of covalent bonds.

The splitting occurs only as a result of interaction of spins with different electron clouds, i.e. which are localized in nonequivalent chemical positions and consequently, have different chemical shifts.

The value of constant J of the indirect spin-spin splitting is determined by the distance between the spins involved in the interaction. Along with the measurements of chemical shifts this enables direct studies of chemical structure, by identifying the types of radicals and measuring the distances and angles.

In contrast to spin-spin interaction, the dipole-dipole interaction of magnetic moments of nuclei is transferred directly in space. Due to fast isotopic rotation of molecules in solution, the dipole-dipole interactions are efficiently averaged and do not lead to additional splitting of signals in NMR spectra. However, the dipole-dipole interactions permit transferring magnetization from one nucleus to another during an NMR experiment. This phenomenon is known as the Nuclear Overhauser Effect (NOE). The interaction intensity and consequently the efficiency of the magnetization transfer are dependent on the distance between the interacting nuclei ($\sim 1/r^6$) that allows measuring the distances between atoms in the molecule. The maximally measurable distance is not large and does not exceed 6 E, but namely, the measurements of interatomic distances underlie the most part of structural studies.

The unique capacity of spin-spin and dipole-dipole interactions to transfer magnetization from one nucleus to another is the basis of multidimensional (2D, 3D etc.) NMR-spectroscopy, which at present is an integral part of all modern NMR studies.

Studies of the Dynamics of Protein Structure by NMR Method. These studies are based on measurements of spin-lattice (T_1) and spin-spin (T_2) relaxation times of nuclear magnetic moments and the dependence of these times on the characteristic correlation time τ_c of the motion.

The physical meaning of T_1 and T_2 times was considered earlier (cf. (10.6)). Nuclear spin-lattice relaxation is determined by the motion of molecules and alternative magnetic fields emerging upon such motion. The return of the system of nuclear spins to thermal equilibrium occurs due to the transfer of the energy absorbed from the radio frequency field to its surrounding, i.e. the "lattice". This requires that the spectrum of internal motion should have frequencies $\omega \sim 10^7$ Hz for nuclei H^+ equal to the resonance frequency (and its doubled value). At the same time, T_2 is affected by frequencies of internal motion which exceed the resonance width expressed in the frequency scale.

Ransom changes in energy levels of spins, causing no transitions between them and having no effect on T_1, however contribute to the line width $1/T_2$.

Consider a simplified case of two interacting protons, which are localized at distance r from each other and rotate isotropically. The following dependence of T_1 and T_2 on the resonance frequency ω_0 and correlation times τ_c of rotational motion of molecular groups including the considered protons are valid:

$$\frac{1}{T_1} = \frac{2}{5}\frac{\gamma^4\hbar^2 I(I+1)}{r^6}\left[\frac{\tau_c}{1+\omega_0^2\tau_c^2} + \frac{4\tau_c}{1+4\omega_0^2\tau_c^2}\right],$$
$$\frac{1}{T_2} = \frac{1}{5}\frac{\gamma^4\hbar^2 I(I+1)}{r^6}\left[\frac{5\tau_c}{1+\omega_0^2\tau_c^2} + 3\tau_c + \frac{2\tau_c}{1+4\omega_0^2\tau_c^2}\right], \quad (10.10)$$

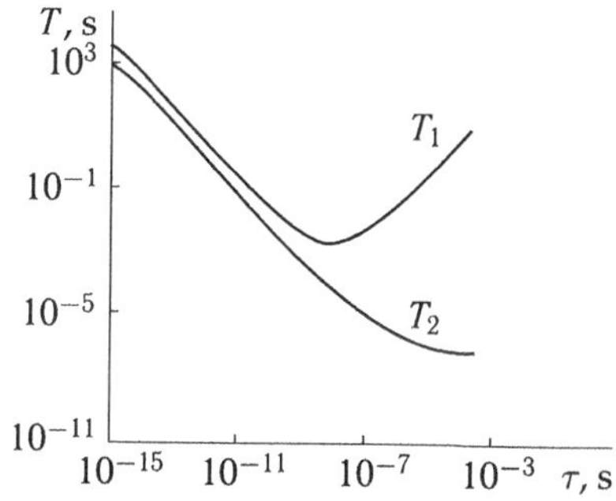

Figure 10.14. Theoretical dependence of the times of spin-lattice T_1 and spin-spin T_2 relaxation on the correlation time τ (the dependence was calculated for protons at the resonance frequency).

where γ is the gyromagnetic ration for protons; I is the nuclear spin. The dependencies of T_1 and T_2 on the correlation time τ_c, calculated from Eq. (10.10) can be represented graphically (Fig. 10.14).

For water $\tau_c \sim 2.7 \cdot 10^{-12}$ s at room temperature. In standard proton resonance spectrometers $\omega_0 = 1.9 \cdot 10^8$ rad/s at the resonance frequency of 60 MHz. Hence, it is clear that $\omega_0\tau_c$ is small, and in Eq. (10.10) this value can be neglected. Then for liquids T_1 is approximately equal to T_2. At $\tau_c = 2.7 \cdot 10^{-12}$ s and $r = 0.158$ nm, $T_1 = 6.7$ s. As seen from Eq. (10.10), T_2 grows linearly ($T_2 \sim 1/\tau_c$) with a decrease in the correlation time τ_c from 10^{-5} s (a solid, rigid lattice) to 10^{-12} s (a liquid).

With an increase in molecular mobility, the T_2 growth is a result of averaging the influence of different local fields on the spins of resonant nuclei. Consequently it leads to the T_2 growth related to their stay in the given spin state. The same dependence is also observed for T_1 in the range of values from $\tau_c \sim 1/\omega_0$ to $\tau_c \sim 10^{-12}$ s. In this range at low τ_c, both T_1 and T_2 are proportional to $1/\tau_c$. However at high τ_c values ($\tau_c \gg 1/\omega_0$) the curves for T_1 and T_2 diverge. At $\tau_c = 1/\omega_0$ the curve for T_1 reaches its minimum. After that T_1 at high relaxation times becomes proportional to τ_c rather than to $1/\tau_c$ and begins to grow again.

The drop of T_1 at the growth of τ_c upon transition from liquid to a solid state occurs due to the change in the spectrum of the lattice oscillations the increased contribution of the oscillations with the frequencies that coincide with resonance frequency ω_0. The growth of T_1 after reaching T_1 minimum at $\omega_0\tau_c = 1$ is associated with further changes in the oscillation spectrum namely a decrease in resonance oscillations ω_0 of the solid body lattice and correspondingly with a decrease in the spin-lattice relaxation. Accordingly, the minimum on the T_1 curve corresponds to the presence of such molecular motions in the τ_c spectrum of relaxation times that have $1/\tau_c$ values compatible with the resonance frequency ω_0. □

Estimations of the relaxation times of separate molecular groups of proteins by the band widths of NMR spectra are restricted by the possibilities of resolution of resonance bands. However, it permits obtaining information on low frequencies of internal motion which is almost inaccessible for other methods. For example, studies of the high resolution spectra of lysozyme and trypsin demonstrated that both (para- and meta-) protons of each tyrosine residue have nevertheless the same chemical shifts. This may indicate to the rotation of tyrosine rings relative to the peptide backbone at the average rate not lower than 10^4 s^{-1}.

Pulse Methods in NMR. In pulse methods, a system of spins oriented under a constant magnetic field is excited by a pulse of the radio frequency field and thereby is disturbed from equilibrium. This results in the deviation of the microscopic magnetization vector from its initial orientation along the direction of the H_0 field. As a result, the system of nuclear spins begins to gyrate around H_0, inducing voltage in the take-up reel that is recorded as a free induction signal after cessation of the radio frequency pulse.

In the spin-echo method, two and more pulses are used. One of them disturbs the spin system from the equilibrium, while the other (or others) corrects the alterations in the coherency of the spin movements brought about by different local fields associated either with the different chemical shifts of different groups or non-homogeneity of the constant magnetic field inside the sample. The use of different pulse sequences enables to record the return of the initial spin orientation and recovery of the so-called longitudinal magnetization characterized by spin-lattice relaxation time T_1. On the other hand, the decrease in the transverse magnetization with the characteristic spin-spin relaxation time, T_2 determined by the natural width of the resonance lines is also recorded. In more complicated cases, distribution of the T_2 values from groups with different mobility should be observed independent of their chemical shift. The faster decrease should correspond to protein globule regions of low mobility, and the slow decrease to relatively mobile regions. Consequently, the spin-echo method, which does not require adminitration of any additional labels, permits detecting the distribution of separate types of motions in the macromolecular structure.

The studies demonstrated that the T_1 value for protons depends on rotation of the protein globule as a whole as well as on the internal motion. For example, an increase in the protein molecular mass and a decrease in the rotational diffusion motion lead to a limited T_1 growth reaching maximal values of about 150 ms. The time of protein spin-lattice relaxation decreases with a temperature increase in the range from 0 to 40 °C. The decrease in correlation times τ_c of intramolecular motions with the temperature elevation shows unambiguously that the observed temperature dependence of T_1 corresponds to the right-hand part of the curve of the T_1 dependence on τ_c in Fig. 10.14. Here the conditions $T_1 > T_2$ and $\omega_0\tau_c > 1$ are fulfilled, where ω_0 is the resonance frequency of 10^8 s^{-1}.

It was found that temperature dependences of the time of spin-lattice relaxation of protons in the protein (serum albumin) have two minima: about −100 °C and about 0 °C (Fig. 10.15). In the range of the low-temperature minimum, where $T_1 \simeq 90$ ms, it is possible to determine the τ_c value. Assuming that the correlation time exponentially depends on temperature $\tau_c = \tau_0 \exp(E_{\text{act}}/RT)$, we find that $E_{\text{act}} = 10.5$ kJ/mol, $\tau_0 = 2.9 \cdot 10^{-12}$ s. At room temperature $\tau_c = 2.2 \cdot 10^{-10}$ s. This is well compatible with the τ_0 and E_{act} values obtained for rotation of methyl groups in polyalanine, polyvaline and polyleucine powders. The other minimum near 0 °C, dependent on the water content in proteins, is related to other types of internal motion. Its association with the internal motion follows directly from the fact that the curves coincide for SA and RNAse with molecular weights differing five times at negative temperatures. The correlation time of this motion at minimum T_1 is about 10^{-8} s and is evidently connected with oscillations of side groups in the polypeptide chain by angles exceeding 30 degrees.

Unlike recovery curves of longitudinal magnetization (T_1) close to monoexponential ones, the curves of the spin-echo signal decay in transverse magnetization from protons have a complex multicomponent character (see Fig. 10.16). The complex

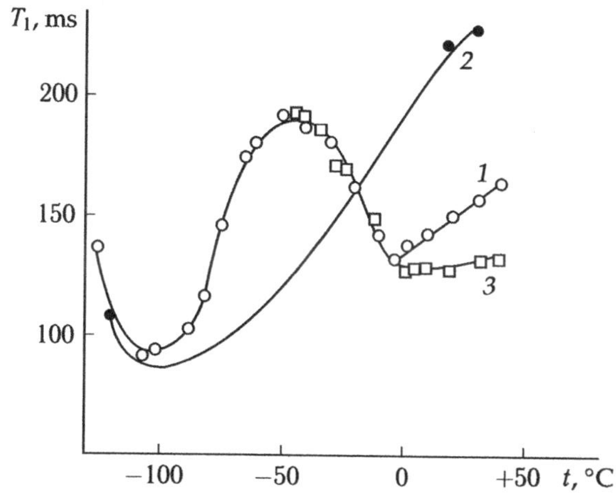

Figure 10.15. Temperature dependence of the time of spin-lattice relaxation of protons in human serum albumin (reproduced from S. I. Aksenov, 1978).

1, 50 % solution of serum albumin in D_2O; *2*, serum albumin powder lyophilized from D_2O; *3*, 50 % solution of RNAse in D_2O.

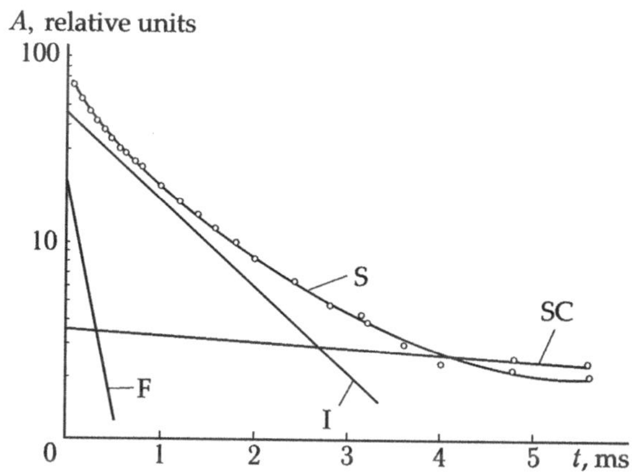

Figure 10.16. Curve of decay of transverse magnetization of immunoglobulin protons (4.5 % solution in D_2O) and its decomposition in components (from S. I. Aksenov, 1978).

S is the summation decay curve; SC is the slow component; I is the intermediate component obtained after subtraction of the slow component from the summation curve; F is the fast component obtained after subtraction of the slow and intermediate components from the summation curve (F is conditioned by rotation of the globule as a whole); I corresponds to relatively slow internal motions; SC characterizes the motion of external most mobile protein groups.

decrease is conditioned by the difference in the distances between protons in the CH_3–, CH_2–, CH– and aromatic groups, as well as by the effects of internal motion. Besides, unlike T_1 which is sensitive even to oscillations of chemical groups, the rotatory motion of separate groups or the globule as a whole has a noticeable effect on T_2. Here the motion of side groups of macromolecules is clearly pronounced. In some proteins other motions with frequencies exceeding the frequency of the protein globule rotation are also distinguished in much of the protein globule.

Studies of NMR Spectra of Protein Molecules. NMR spectroscopy methods have been widely used during the recent 20 years for studying the three-dimensional structure of proteins and nucleic acids in solution. An important advantage of this method is the possibility to study objects in the environment maximally close to the native conditions in aqueous solution, in lipid environment, and in a complex with other proteins or nucleic acids. Moreover, NMR spectroscopy methods permit studying the dynamic behavior of objects in a wide time range from pico- to milliseconds even at the level of separate residues. Unfortunately, the use of NMR spectroscopy methods is limited to studies of relatively small protein molecules with mass not usually exceeding 30–50 kDa.

Pulse NMR Spectroscopy. Let us consider physical bases of pulse NMR spectroscopy which now becomes the most powerful method for studying biopolymer structures. The direction of the external magnetic field and, consequently, of the magnetic moment of the sample are believed to be the positive direction of the axis z of the Cartesian coordinate system; this direction is called longitudinal. In the equilibrium, magnetization of the sample along the axes x and y (the transverse direction) is zero. In other words, the phases of different spins in this state are distributed randomly, so that no transverse magnetization exists. If the magnetization vector is somehow deflected from the axis z by the β angle, it will gyrate around the magnetic field direction with Larmor frequency circumscribing a cone with the β angle (Fig. 10.17). Accordingly, a transverse component of magnetization $M_0 \sin\beta$ will emerge in the xy plane and will rotate with Larmor frequency corresponding to the transition between the spin states (ω_0). The projections on the x and y axes will be correspondingly:

$$M_x = M_0 \sin\beta \cos\omega_0 t,$$
$$M_y = M_0 \sin\beta \sin\omega_0 t.$$

It is exactly the transverse magnetization that is recorded in the NMR experiment with the use of radio-frequency coils located in the direct vicinity of the sample. The precession of the magnetic field induces formation of an oscillating current in the coils which is recorded after amplification. This current is called free induction decay (FID).

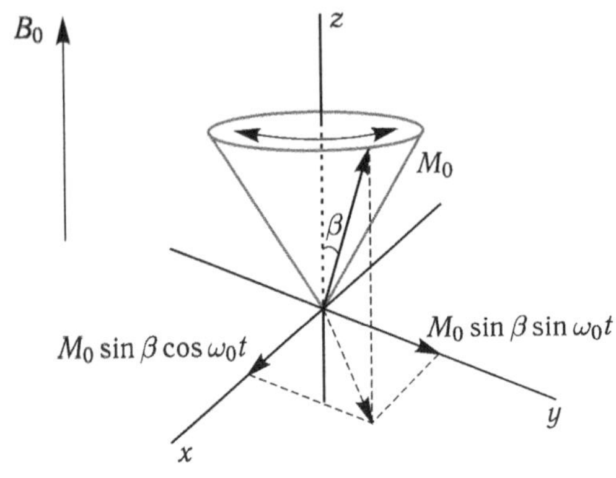

Figure 10.17. Precession of the magnetization vector M_0 in the external magnetic field B_0 generates the oscillating component of transverse magnetization.

NMR Experiments. Figure 10.18 shows the scheme of a simple NMR experiment. The general time of such an experiment can be tentatively divided in three periods. During the first period of duration d_1 the system establishes the state of the initial thermodynamic equilibrium. As a rule, the d_1 value makes from 1 to 5 magnitudes of the time of the longitudinal relaxation T_1. After the equilibrium state is established, one or several radio-frequency pulses generate non-equilibrium transverse magnetization of the sample. In the simplest case of a one-dimensional spectrum, the 90°_x pulse at the Larmor frequency of the observed nuclear spins turns the magnetization to the $-y$ axis. And finally, during the third period the magnetization gyrates in the xy plane, causing the emergence of the oscillating current in the radio-frequency coils of the sensor. The widely known in mathematics Fourier transform is used instead of time transformation to represent the signal in the frequency range.

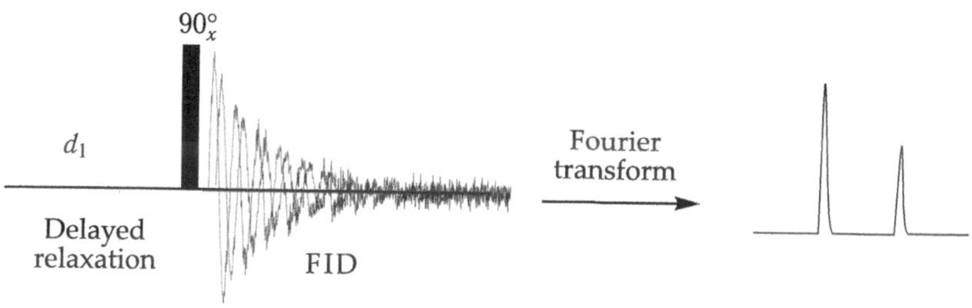

Figure 10.18. An elementary one-dimensional NMR experiment consists of three stages: (1) delayed relaxation; (2) one or several delayed pulses; (3) detection of FID. The Fourier transform of FID makes it possible to transfer the signal from the time range to the frequency range.

Multidimensional NMR Spectroscopy. Let us analyze the simplest two-dimensional variant of multidimensional NMR spectroscopy. If in a one-dimensional spectrum the signal intensity is measured versus one frequency, in a two-dimensional variant we get a spectrum of the intensity dependence on two frequencies, i.e. each peak in the 2D-spectrum is characterized by the intensity and two frequency coordinates. The type of the experiment determines to what the coordinates will correspond. For example, in the COSY (COrrrelation SpectroscopY) spectrum the peak frequency coordinates corresponds to chemical shifts of nuclei connected by the spin-spin interaction. In the ^{13}C-HMQC (Heteronuclear MultiQuantum Correlation) spectrum, one of the signal coordinates corresponds to the chemical shift of nuclei ^{13}C, and the second coordinate corresponds to protons ^{1}H bound to the nuclei. In NOESY (NOE SpectroscopY) experiments, the signal coordinates correspond to chemical shifts of the nuclei bound by dipole-dipole interactions.

A two-dimensional experiment may be roughly outlined as follows (Fig. 10.19). At the first stage (preparation) the equilibrium longitudinal magnetization is transformed into non-equilibrium or, in terms of quantum mechanics, we create a coherent spin state. At this stage, the initial pulse may be a simple 90° pulse, which creates transverse magnetization (single quantum coherence), or a group of pulses, which creates more complex magnetization (multiple quantum coherence) or transfers the magnetization from nuclei of one type to nuclei of another type.

Other two- and three-dimensional experiments are performed in an analogous way, for example, such as the pulse sequence NOESY. A central element of this experiment is the delay τ called the mixing time, during which magnetization is transferred

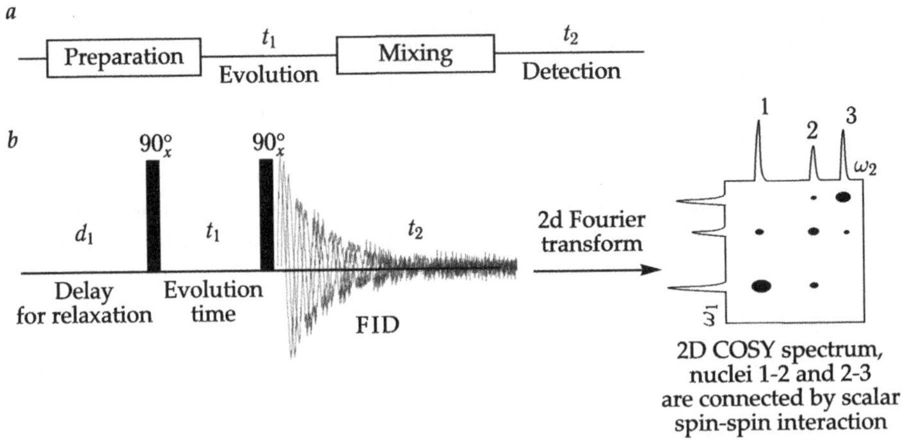

Figure 10.19. Mossbauer spectroscopy. *a*, Principal scheme of two/three-dimensional NMR consists of four stages. (1) Preparation: creation of nonequilibrium magnetization from equilibrium magnetization; (2) evolution of the created magnetization during the incremented time interval; (3) mixing — bringing the magnetization to the observed value; and (4) detection of FID. *b*, Diagram of two-dimensional COSY experiment. *c*, Diagram of two-dimensional NOESY experiment.

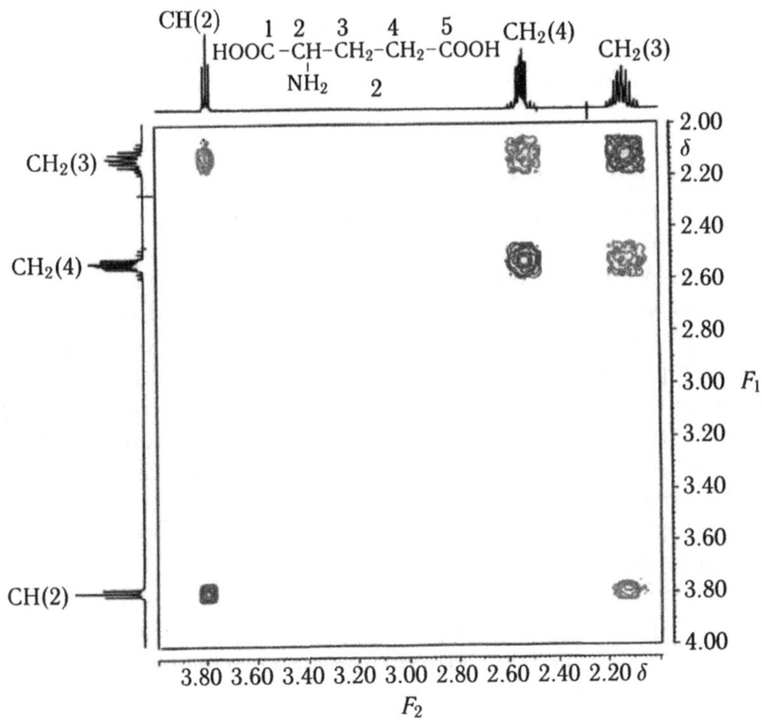

Figure 10.20. Two-dimensional COSY spectrum. At the top of the figure the chemical formula of the compound is shown; numbers indicate resonance protons and signal corresponding to them in a one-dimensional spectrum. Highest cross-peaks are observed for groups (3) and (4). Cross-peaks of groups CH(2) and $CH_2(4)$ are absent because the corresponding indirect constants of spin-spin interaction are close to zero.

between nuclei by the mechanism of dipole-dipole interaction. For this experiment, the intensity of the cross-peak will depend on the distance between the interacting nuclei subject following the law $\sim 1/r^6$ that makes it possible to measure distances between atoms and determine the three-dimensional structure of molecules.

Studies of Proteins by NMR Spectroscopy. Studies of the three-dimensional structure of a protein molecule using NMR spectroscopy can be tentatively divided in several stages. At the first stage, the signals are assigned by referring each signal to nuclei, involved in its formation. This stage can be divided in three phases: (1) assigning of signals of spin systems; (2) sequential assigning, i.e. linking spin systems in chains according to the main protein sequence; (3) assigning of signals in NOESY spectra in order to identify the adjacent nuclei. At the next phase, based on the obtained results different structural parameters are measured, for example, constants of spin-spin interaction (CSSI), the nuclear Overhauser effect (NOE). These data are used as spatial restrictions in calculating molecule structure. Algorithms of molecular dynamics and energy minimization realized in special programs are used in these calculations. As a result a set of structures is obtained which most of all satisfy the experimental data (Fig. 10.21).

The existing NMR methods permit studying structures of proteins with molecular mass to 100 kDa. Restrictions in the use of NMA spectroscopy methods for studying large protein molecules are associated with two basic factors. In the course of relaxation, the excited transverse magnetization of nuclear spins decreases rapidly to zero,

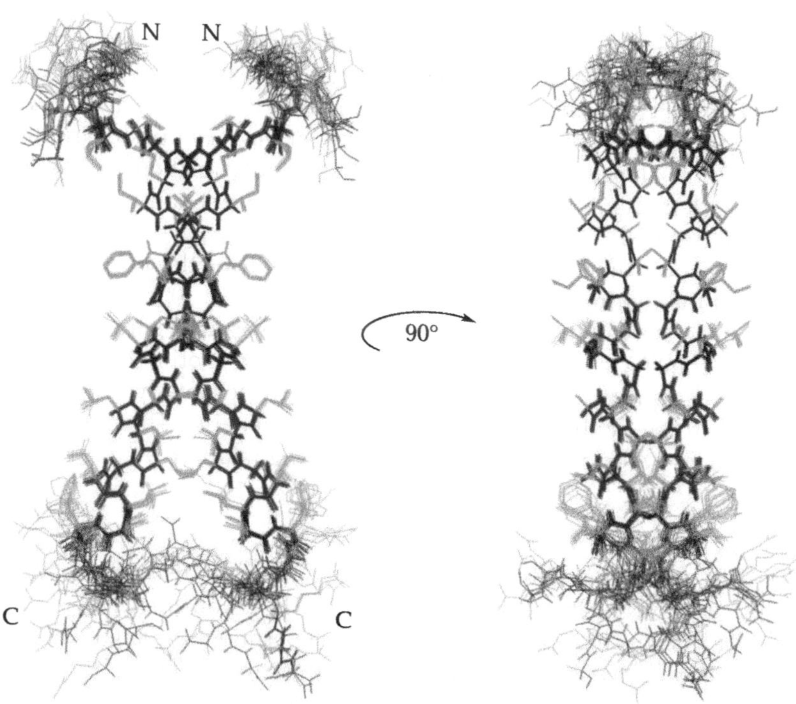

Figure 10.21. Three-dimensional structure of the dimer of transmembrane receptor domain EphA1. Ensemble of 12 calculated structures superposed by heavy atoms of the main chain. Atoms of the main and side chains are shown in black and gray, respectively (reproduced from M. L. Maisel, 2001).

which results in a decrease in the lifetime of excited NMR states and widening of resonance lines, i.e. a decrease in the sensitivity. The other essential circumstance is the overlapping of signals in NMR spectra.

In addition to obtaining structural data, another critical advantage of NMR spectroscopy over other methods is the possibility to investigate dynamic characteristics of molecules. To study the dynamic behavior of the protein main chain, ^{15}N nuclei are regularly used as well as ^{1}H protons covalently bound to them. In a number of experiments, the times of longitudinal (T_1) and transverse (T_2) relaxations of ^{15}N nuclei and the magnitude of heteronuclear NOE (dipole-dipole relaxation of ^{15}N nuclei on ^{1}H nuclei) are measured. The rate of different relaxation processes depends not only on the general rate of rotational motion of the molecule in solution, but also on different internal motions in the protein molecule. The analysis of relaxation data permits characterizing the internal motions in several time ranges. Those are fast motions occurring with characteristic times in the pico-subnanosecond range and slow motions with characteristic times in the micro-millisecond range. The possibility to analyze the slow internal motions of the protein molecule is a unique feature of NMR spectroscopy methods, because at present this range of times is inaccessible for studying with computational methods of molecular dynamics.

Method of Gamma-Resonance Spectroscopy and Its Physical Bases. This method also yields important information on the dynamics of proteins. It allows determining the amplitudes of atomic displacements in the protein structure at short times (10^{-7}–10^{-9} s). The method is based on that upon absorption of γ-quantum the nucleus is transferred from the ground state (E_1) to the excited state (E_2) according to the usual law $\Delta E = E_2 - E_1 = h\nu$, where $\Delta E = 10^3$–10^5 eV for nuclear levels. The absorption of γ-quanta is observed by nuclei of heavy atoms Fe, Cu and Pb. For the ^{57}Fe isotope with content in natural compounds of 2.2 %, the ΔE value for resonance absorption is 14.4 KeV, and the lifetime of the ^{57}Fe nucleus in the excited state is $\tau^* \sim 10^{-7}$ s. Hence, in accord with the uncertainty relation for energy (10.5), it is possible to determine that the natural width of the resonance line of γ-quanta absorption is very small $\Gamma \sim 10^{-8}$ eV.

Let the correlation time τ_c of the undirected motion of the nucleus is close to or smaller than the lifetime of its excited state ($\tau_c \sim \tau^* \sim 10^{-7}$ s). Then virtue of the momentum conservation law, upon γ-quantum absorption part of the quantum energy may be transferred to the translational energy of motion of the Mossbauer nucleus ^{57}Fe. Consequently, the γ-quantum energy used for internal excitation of the nucleus will decrease by the value equal to the recoil energy. Thus, the absorption resonance frequency ν_γ is determined by the following relation

$$h\nu_\gamma = \Delta E - mv^2/2,$$

where $\Delta E = E_2 - E_1$ (equal to 0.0144 MeV for ^{57}Fe), m is the mass of the Mossbauer nucleus, and v is the velocity of the recoil motion. The recoil energy ($E_{\text{loss}} = mv^2/2$) for the free ^{57}Fe nucleus makes $\sim$0.002 eV, which determines the small change in the frequency of γ-quantum absorption. However, as the natural line width is very small $\Gamma \ll E_{\text{loss}}$, the displacement of the absorption frequency ν_γ is quite noticeable.

As a rule, the sample is irradiated by γ-beams emitted from a radioactive γ-source where the excited ^{57}Fe nuclei are generated. In experiments, γ-quanta absorption spectra are recorded when the γ-source of $\pm$(0.1–10) cm/s moves at low velocities to the sample along the mutual axis connecting them. As a result, the γ-emission frequencies due to the Dopler effect change. The different velocities of the γ-source correspond to different recoil energy values, which reflect the statistical distribution of atoms in the sample over their self-motions and velocities and therefore resonance conditions of γ-quanta absorption.

The NGR (Nuclear Gamma-Resonance) spectra reflect the chemical and physical structures of the nucleus surrounding and are characterized by a chemical shift, quandrupole splitting, line form and superfine structure. At present, NGR is a powerful tool in decoding the atomic structure of protein active centers.

The Mossbauer Effect. The main source of information on the dynamic structure of proteins is the analysis of the shape of the spectrum relative to the widening of the NGR spectrum and the probability f' of resonance absorption of γ-quanta without recoil (the maximum of the absorption line, Fig. 10.22). In a solid, the recoilless absorption means that the recoil momentum is taken up by the whole crystal

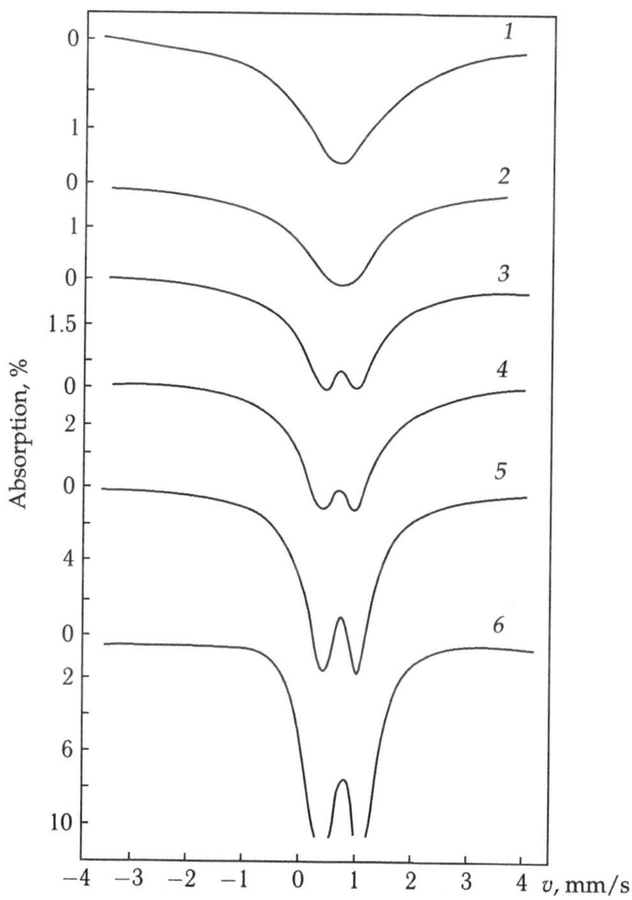

Figure 10.22. Mossbauer spectra of modified ^{57}Fe human serum albumin I in aqueous solution of glycerol, 100 mg/ml) (reproduced from G. I. Lichtenstein, 1974).

1–6, Temperature of 13.5, −4, −11, −22, −190 and −196 °C, respectively. The widening of the spectrum is caused by the diffusion of protein molecules. The change in the frequencies of widening is proportional to the γ-source velocity.

of a large mass M. In this case, the recoil energy of the crystal is M/m times smaller than the recoil energy of a single nucleus, i.e. is negligible and much smaller than the natural line width Γ. In other words, due to the absence of recoil lines not displaced by the energy appear in the spectra. No excitation of oscillation of separate atoms (phonons) in the lattice occurs, since the recoil momentum is accepted by the entire crystal as a whole structure. This is actually the Mossbauer effect.

The growth of the mean-square displacement $\langle x^2 \rangle$ of the Mossbauer nucleus decreases the probability of γ-quanta recoilless absorption f'. On the contrary, the f' value goes to unity when $\langle x^2 \rangle$ is small as compared to the λ wavelength of γ-quantum. The case of small $\langle x^2 \rangle$ values corresponds to the rigid connection of atoms in the lattice and correspondingly to the high value of the vibrational quantum, which significantly exceeds the recoil energy. Under such conditions, no transfer of recoil energy to the nucleus occurs. In the case of solids, the dependence of the f' value on $\langle x^2 \rangle$ is expressed by the Debye–Waller (Lamb–Mossbauer) formula

$$f' = \exp\left(-\langle x^2 \rangle / \overline{\lambda}^2\right), \tag{10.11}$$

where $\overline{\lambda} = \lambda/(2\pi)$, λ is the wavelength of γ-quantum ($\overline{\lambda} = 0.014$ nm for ^{57}Fe). This expression is valid when the relaxation time of the nucleus is rather small: $\tau_c \ll \tau^*$.

In the general case, the nucleus may be involved concurrently in several independent motions including atomic oscillations in chemical bonds, oscillation of the fragment of the system where the Mossbauer nucleus is tightly bound, and diffusion motions of the whole macromolecule. If these types of motion are independent of each other, the general dispersion of the nucleus positions $\langle x^2_{gen} \rangle$ comprises local intramolecular shifts $\langle x^2_{loc} \rangle$ and the total diffusion shift $\langle x^2_{dif} \rangle$ of the whole particle, where the Mossbauer atom is located:

$$\langle x^2_{gen} \rangle = \langle x^2_{loc} \rangle + \langle x^2_{dif} \rangle.$$

In their turn, local shifts may include nuclear oscillations and shifts with transitions between different conformational substates (see Section 1 in Chapter 9):

$$\langle x^2_{loc} \rangle = \langle x^2_{col} \rangle + \langle x^2_{conf} \rangle.$$

When the mobility of particles proceeds mainly as a whole, the decrease in the f' value is always accompanied by widening of the absorption line as in the case of a liquid. On the contrary, in systems where the shift is caused by intramolecular motions with small amplitudes of the whole region with the tightly fixed Mossbauer nucleus, the decrease in the f' value is not accompanied by widening. It rather reflects the decrease in the association of this region with the rest of the system (the crystal), for example, due to the decrease in viscosity upon local melting.

In model systems, this phenomenon is observed, for example, in "loose" particles of tin acids with internal dynamics. The Mossbauer atom shifts depend on the external conditions and may increase with the temperature rise or a decrease in the viscosity of the system. Thus, the melting of solids is always associated with widening of the spectrum and disappearance of the Mossbauer lines.

Temperature Dependences f' and $\Delta\Gamma$. Investigations of these dependences enable obtaining information on the character of mobility of Mossbauer nuclei and the properties of their environment. The most important advantage of this method is the possibility to determine also the amplitudes of atom motions. This is what makes its difference from other resonance methods which can determine only frequency characteristics of motions. Figure 10.23 shows curves of temperature dependence $f'(T)$ for ^{57}Fe-labeled protein preparations. For wet proteins, the probability of the f' effect is somewhat changed in the range of low temperatures, but decreases drastically at temperatures exceeding $-(60–30)$ °C without widening the gamma-resonance (GR) line. The Mossbauer spectra are widened only at temperatures above -20 °C at the final regions of the $f'(T)$ curve where the probability of the effect is already low. The dry protein is characterized by a weak temperature dependence of the factor f' and a constant width of the GR-line, which is characteristic of oscillation of ^{57}Fe ions in a solid matrix. The dependence of the f' value on the relative humidity of the sample (P/P_s) has a threshold character, which is the evidence of the cooperative nature of conformational mobility of the water–protein complex at the hydration degree $P/P_s \geqslant 0.4$ (see Section 4 in Chapter 9).

Therefore, hydrated proteins at temperatures 77–200 K do not differ in their dynamics from solids. Under these conditions the Mossbauer nuclei perform high-frequency oscillations with the amplitude $A < 0.01$ nm. At 200–260 K the motion with the average frequency $\omega > 10^7$ s^{-1} and the amplitude of displacements $A \sim 0.03 \div 0.06$ nm is released, and the GR-lines are slightly widened. Consequently, in its dynamic behavior the protein differs from solids and liquids. The temperature dependence $f'(T)$ as well as the increase in the effective value $\langle x^2 \rangle$ with T growth

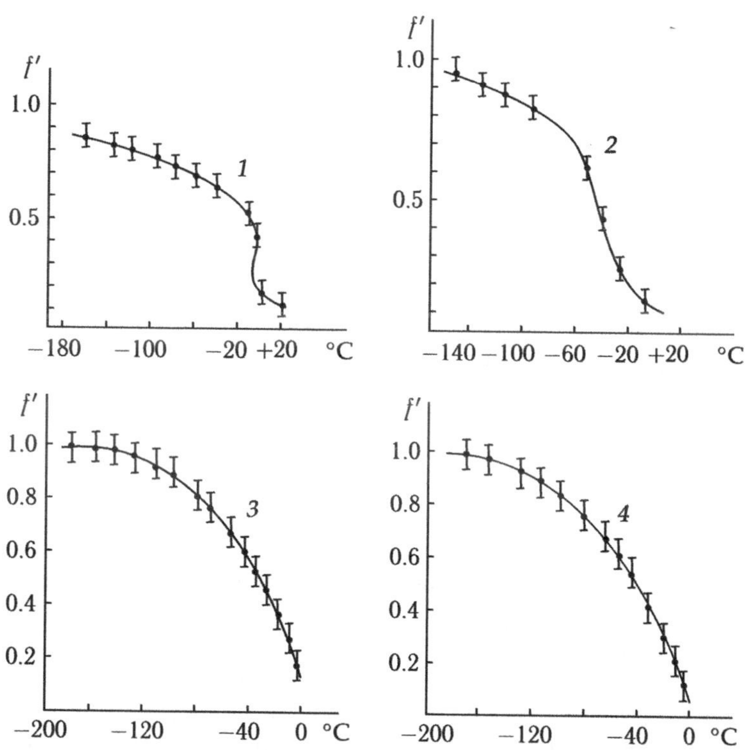

Figure 10.23. Dependence $f'(T)$ for powders (reproduced from E. N. Frolov, 1985) of: *1*, collagen at hydration extent $P/P_s = 0.97$; *2*, lysozyme at hydration extent $h = 0.16$ g H_2O/g protein; *3* and *4*, α-chymotrypsin of human serum albumin at hydration extent $P/P_s = 0.95$.

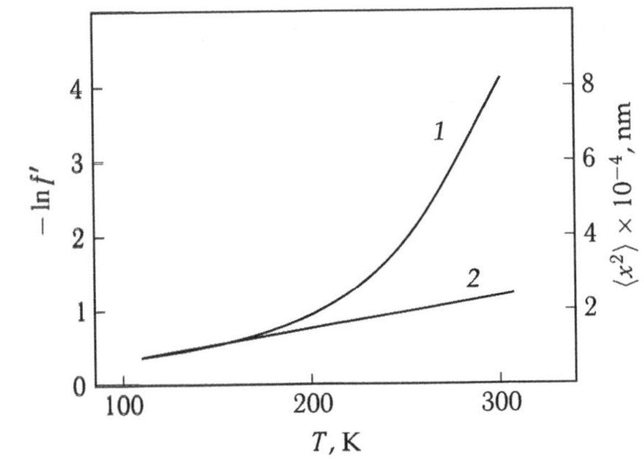

Figure 10.24. Temperature dependence of the effective mean-square displacement of ^{57}Fe nuclei for metmyoglobin (Met Mb) (reproduced from E. R. Bauminger et al., 1983).

1, Met Mb; 2 — $\langle x^2\rangle_v$; $\langle x^2\rangle \sim -\lambda^2 \ln f'$; $\langle x^2\rangle = \langle x^2\rangle_C + \langle x^2\rangle_v$; $\langle x^2\rangle_C$ is the contribution of conformational degrees of freedom; $\langle x^2\rangle_v$ is the contribution of oscillations of nuclei; $\langle x^2\rangle_v = k_B T/K$; K is the rigidity constant for oscillations of the ^{57}Fe.

for the protein have a sharp bend (Fig. 10.24, *1*) not typical for oscillations of atoms in solids (Fig. 10.24, *2*). Moreover, in this case the temperature increase does not result in large widening of the GR-line as it occurs in liquids. The latter reminds the properties of a solid-body matrix.

Studies of Internal Motions. For studying motions of different regions of the protein globule, the use of the NGR method is naturally restricted because of a relatively large number of Mossbauer atoms (^{57}Fe) in the protein. This disadvantage is to a certain extent compensated by the use of the GR spectroscopy method combined with X-ray dynamic analysis (XDA) based on diffraction of X-rays on different atoms. As compared to the characteristic frequencies of motions in the protein, the X-ray diffraction is an instantaneous process ($\tau \sim 10^{-15}$ s) and consequently the XDA method permits determining only the statistical disorder (defects in the lattice, irregularity in the protein crystal). As a matter of fact, here the values $\langle x^2_{\text{gen}}\rangle$ are determined which are caused by both statistical disorder and actual motions of atoms upon transitions between conformational substates rather than frequencies of motion:

$$\langle x^2_{\text{gen}}\rangle = \langle x^2_{\text{conf}}\rangle + \langle x^2_{\text{col}}\rangle + \langle x^2_{\text{stat}}\rangle,$$

where $\langle x^2_{\text{dif}}\rangle = 0$ is accepted for the protein, and $\langle x^2_{\text{stat}}\rangle$ is believed to be the same for all atoms.

The $\langle x^2_{\text{stat}}\rangle$ value is determined for the ^{57}Fe iron atoms as the difference of the $\langle x^2_{\text{gen}}\rangle$ value obtained by the XDA method and the $(\langle x^2_{\text{conf}}\rangle + \langle x^2_{\text{col}}\rangle)$ value determined by the GR spectroscopy method at $T > T_k$.

The general pattern of the dynamic structure of metmyoglobin obtained by the XDA method includes a "solid-body" nucleus ($A \lesssim 0.02$ nm) and the "liquid-like" external shell ($A \gtrsim 0.03 \div 0.04$ nm). At room temperature, the dimensions of the latter are much larger than those of the "solid-body" nucleus (see Section 2 in Chapter 7). The magnitudes of the amplitudes of displacements to the "liquid-like" regions, exceeding 0.03 nm even at 80 K, show that these motions of atoms

Proteins	
Primary events in photosynthesis and vision	10^{-13}–10^{-12} s
Local dynamics of atoms and small groups	10^{-12}–10^{-11} s
of side chains and segments of polypeptide chains	10^{-11}–10^{-7} s
Relaxation of polypeptide chain regions	$\sim 10^{-9}$ s
Motions of domains and subunits	10^{-8}–10^{-5} s
Reaction of transport of the proton of ionized groups	10^{-9}–10^{-7} s
Unfolding of α-helical regions	10^{-8}–10^{-6} s
Release of bound ligand molecules	10^{-6}–10^{-3} s
Kinetics of folding – unfolding	10^{-4}–10^{2} s
Exchange of tightly bound Ca ions	10^{-3}–10^{-2} s
Nucleic acids	
Vibrations of the DNA double helix	10^{-13}–10^{-11} s
Local motions of base pairs	10^{-10}–10^{-9} s
B–A-transitions to the DNA	10^{-9} s
Kinetics of intercalation	10^{-4}–10^{-3} s
Unpairing of base pairs	10^{-3}–10^{-1} s
Dissociation and formation of the double helix	10^{-3}–10 s
Biomembranes	
Relaxation of phospholipid heads	10^{-11}–10^{-9} s
Time of ion translocation through channels	10^{-9}–10^{-8} s
Phase transitions in phospholipid bilayers	10^{-2}–10^{-1} s
Flip-flop transitions in lipids	10^{4} s
Water	
Relaxation of volumetric water	10^{-12}–10^{-11} s
Relaxation of hydrated layers	10^{-11}–10^{-10} s
Lifetimes of bound water molecules	10^{-8}–10^{-2} s

correspond to transitions between different conformational substates. In the "solid-body" region where $A \lesssim 0.02$ nm, the conformational degrees of freedom are almost absent. The dimensions of the "solid-body" region are as a rule several per cent of the globular protein volume, and obviously even rigid α-helical regions move continuously, experiencing deformation shifts and oscillations or diffusing in the viscous environment.

11

Physical Models of Dynamic Mobility of Proteins

The protein dynamic behavior differs from the behavior of solids and liquids. Hence, the development of special physical models is required to study quantitative relationship between macroscopic parameters of proteins and their intramolecular mobility. Thereto, it is necessary to get to know the main physical ideas and approaches underlying the present-day concepts on the dynamic mobility of proteins.

11.1 General Characteristic of Molecular Dynamics of Biopolymers

From the physical point of view, biopolymers are extremely complex objects. The existence of linear memory, secondary and tertiary structures, as well as the essentially nonlinear character of interparticle interactions lead to rather diverse properties. □

As known, small oscillations of atoms in molecules and crystals with the amplitude $x_a \sim 0.001$–0.01 nm can be relatively well described in the context of harmonic approximation. The corresponding rigid constants for valence oscillations lie within the range $K = 10^5$–10^6 dyn/cm, and characteristic frequencies $\omega \sim 10^{13}\ \mathrm{s}^{-1}$. Effects of unharmonicity cause intermode interactions, leading to oscillatory relaxation with characteristic times $\tau \sim 10^{-12}$ s. However, the harmonic approximation is not valid at all for describing conformational motions in biomacromolecules with the amplitude $x_a > 0.01$ nm. The density of the protein globule is comparable with that of liquids and organic crystals. Therefore, here the displacement exceeding 0.01 nm, as in the case of liquids, requires formation of the density fluctuation or a "hole" characterized by the free energy of activation $\Delta G_0^{\#}$. In other words, conformational motions in biopolymers are restricted in the amplitudes, the same as in solids, but require energy activation as in the case of diffusion in liquids (Fig. 11.1). □

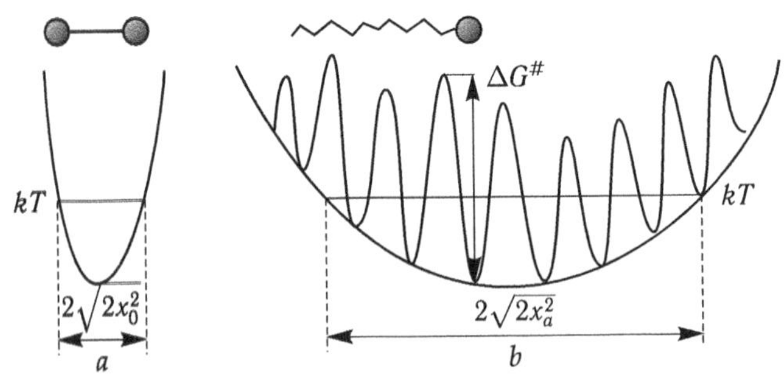

Figure 11.1. Characteristic quasi-one-dimensional profile of potential energy: a, for high-frequency (valence) oscillations; b, for conformational motions.

In local minima high-frequency oscillations occur practically in the same way as in case a. Energy fluctuations lead to migration from one of the local minima to the other and finally to limited diffusion within the whole classical accessible region. As a matter of fact, local minima and heights of barriers $\Delta G^{\#}$ are not fixed and fluctuate due to interactions with other degrees of freedom.

Secondary structure elements have a significant effect on the dynamic properties of biomacromolecules. Elements of the backbone formed by them experience limited Brownian motion with parameters determined by linear dimensions, rigidity, microviscosity etc. The motion of simple elements — side groups and bound molecules of the solvent — may be considered in the context of the model of the Brownian oscillator. Correlation times of these motions are in a nanosecond range, and overall amplitudes are $x_a \sim 0.03$–1 nm.

Proteins and media with a relatively rigid framework have a set of bound elements sharply differing in their correlation times. The fastest and small-scale fluctuations are typical of side groups. These groups form a liquid-like damping medium around helical regions of the polypeptide framework. The drastic difference in relaxation times of side groups and α-, β-elements simplifies the description of the fluctuating behavior of the system and allows using the hydrodynamic approximation and not taking into account correlations between the fluctuations of side groups and the bckbone. In other words, the existence of this hierarchy of relaxation times makes it is possible upon studying the fluctuation dynamics of the protein globule to model it as a liquid-like drop reinforced by a rigid backbone (Fig. 11.2). The examination of diffusion processes in structured protein media and the effect of the solvent viscosity on the protein dynamics is based on the dynamic model of the reinforced drop.

The Role of Solvent Viscosity and Correlation of Fluctuations and Protein Dynamics. The concept of protein viscosity as a measure of dissipative processes coupled to conformational motions is widely used in biophysics. Though at the first sight, the use of microviscosity does not seem difficult, the physical meaning of this

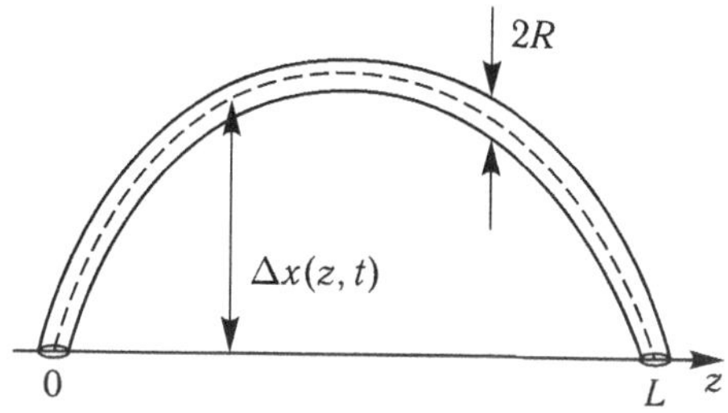

Figure 11.2. The most probable profile of bending/flexural fluctuations of a rigid element of the backbone.

notion is not so simple. This problem becomes of special urgency upon studying the effect of the solvent viscosity on the rate of intraprotein processes.

Conformational motions in the condensed phase (in proteins) may proceed only when a fluctuating cavity or a "hole" is formed in the appropriate position. Inside the globule the "hole" may form in two ways. First, at the expense of free volume of the globule, and second, due to the penetration of the "holes" formed in the solvent into the globule through its surface.

Conformational motions of surface groups in the external layer of the globule may occur only if the solvent has a corresponding "hole" (Fig. 11.3). The frequency of the "hole" formation is $k_0 \sim 10^{13}\ \mathrm{s}^{-1}$.

The formation of fluctuating cavities inside the globule proceeds due to a cascade of interconnected random processes. The overall process may be treated as diffusion

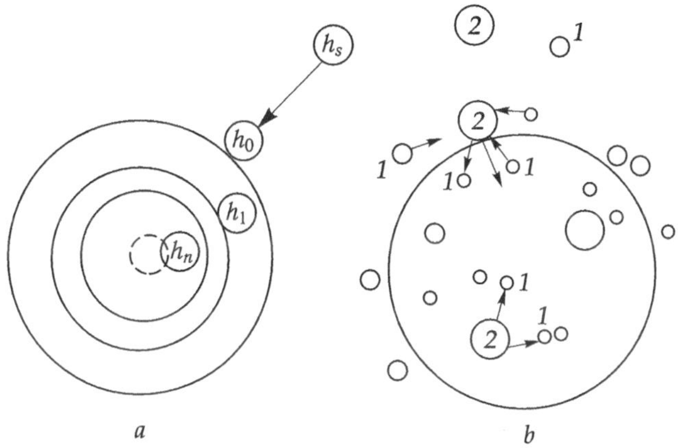

Figure 11.3. Hole-layer (*a*) and hole-recombination (*b*) models for the description of the effect of the solvent on the dynamics of the globule.

a, Holes h_s, which provide self-diffusion of solvent molecules, are absorbed on the globule surface and layer-by-layer penetrate inside the globule due to interactions with mobile groups; *b*, large holes of type *2* are formed due to recombination of small holes of type *1*, which diffuse freely in the globule. The hole of type *2* is required for conformational motions and penetrates into the globule only in exchange for an equivalent number of holes of type *1*. Diffusion of the hole of type *2* occurs inside the globule simultaneously with corresponding microconformational motions. After several diffusion acts, hole *2* may disintegrate in an equivalent number of holes of type *1*. The free volume of the globule is retained at all stages.

of the "hole" from the solvent to the globule. In such a system, microviscosity is determined by the concentration and mobility of the "holes", and the effect of the solvent on microviscosity decreases upon transition to deeper layers.

All conformational motions occur by the mechanism of limited diffusion. The effect of the solvent on relaxation times is caused by diffusion of the "holes" into the globule. The penetration of particles into the slit may take place only when its opening exceeds the van-der-Waals diameter of the group x_0. This leads to the fact that rigid elements of the structure impose additional spherical restrictions on the conformational mobility and thus retard selectively definite types of motions.

Diffusion in a structured medium is associated with the formation of definite conformations required, for example, in enzyme-substrate complexes (ESC). It is known that the surface of enzymes has, as a rule, a region formed by relatively rigid edges, and the penetration of the substrate to the active center is associated with the corresponding "fluctuating adjustment" of the forms and dimensions of this region and the substrate. Theoretical estimations of the rate of this process under substrate saturation and real protein rigidity and microviscosity are close to the observed values. When the diameter of the substrate increases more than by ~ 0.1 nm, the process is inhibited drastically (see Chapter 26).

11.2 Model of Limited Diffusion (Brownian Oscillator with Strong Damping)

Conformational Mobility of Proteins. Micromotions of protein fragments with amplitudes exceeding or equal to about 0.1 nm characterize transitions between separate conformational substates. The motion of each separate fragment is limited and affected by the nearest surrounding. As the length of the particle free path in a dense medium

$$l \simeq \sqrt{mk_B T}/\gamma \ll 0.1 \text{ nm}$$

(γ is the friction factor (11.6), and m is the particle mass of about 100 a.m.u.) is far less than the amplitude of conformational motions, the motion of the chain fragment has a diffusion (or stochastic) character. The displacement by 0.1 nm inside a condensed protein medium is necessarily coupled to a shift of individual molecular groups, surrounding the mobile fragment. This requires overcoming of potential barriers that actually corresponds to transitions between separate conformational substates. At the same time, according to the Frenkel kinetic theory of liquids the physical mechanism of such motion is equivalent to the motion in a viscous medium or, in the case of spontaneous motions, to diffusion.

The motion of a protein fragment is described by two parameters: the conformational potential $U(x)$ and the coefficient of conformational diffusion $D(x)$ dependent on the conformational coordinate x. The latter dependence reflects the microheterogeneous character of the protein medium, which is directly associated with inhomogeneity of potential barriers. Therefore, conformational motion is characterized by two spatial scales. On the one hand, there is a "comb" of potential barriers due to the interaction of the chain fragment with the dense environment, which prevents free oscillations of the fragment. On the other hand, the envelope of the minima of the potential "comb" represents the profile of a far more smoothly changing conformational potential $U(x)$, the motion along which is associated with the deformation and

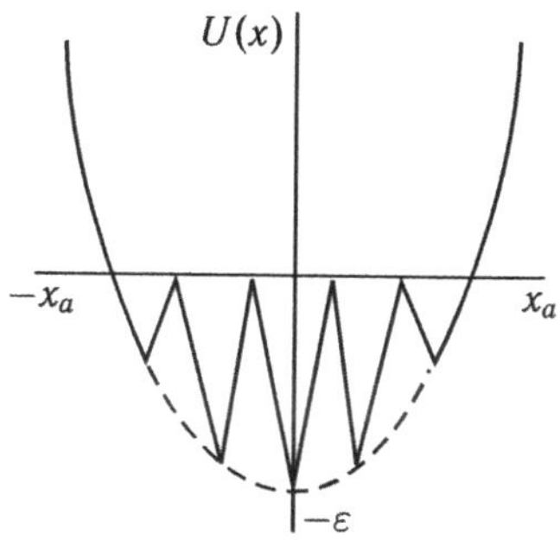

Figure 11.4. Conformational motion in viscous medium (the potential "comb" imitates viscosity).

Here x_a is the average amplitude of conformational fluctuations; $U(x)$ is the conformational potential; ε is the energy of activation of conformational motions.

change of the potential energy of the system (Fig. 11.4). To leave the boundaries of the conformational energy well is naturally forbidden because of the limited character of the fragment diffusion as a component of the protein structure.

Relaxation of the fragment inside the profile of the potential $U(x)$ resembles the aperiodic motion of a pendulum in a very viscous medium. An example of this can be the straightening of a compressed spring submersed in a very viscous liquid or resin. As for a definite type of the curve $U(x)$, in a harmonic approximation it is determined by the quadratic dependence of the oscillator energy on the amplitude of oscillations:

$$U = m\omega_0^2 x^2/2, \tag{11.1}$$

where ω_0 is the frequency of oscillations. As has been shown (see Section 2 in Chapter 9), at $T = 200$ K conformational motions of protein chains have relatively low average frequencies, $\omega \ll k_B T/\hbar \sim 3 \cdot 10^{13}\ \mathrm{s}^{-1}$. In a general form the dependence $U(x)$ can be represented as

$$U(x) \sim \alpha |x|^{1/\beta}, \tag{11.2}$$

where $\beta = 1/2$ for a harmonic oscillator.

Figure 11.5 shows different curves of the conformational potential depending on parameter β. In this case it determines the slope and dimensions of the potential well, within which diffusion transitions occur between separate conformational substates. We will consider only the case of a harmonic potential in analyzing how the mean-square shift (see Section 1 in Chapter 7) of the protein fragment changes with time depending on the temperature and its microenvironment.

Formulation of the Model. Let us consider the model (K. B. Shaitan and A. B. Rubin, 1981), where the behavior of a protein fragment in a highly structured medium is analyzed, and compare the obtained theoretical conclusions with the experimental data.

The main peculiarity of the problem is that at thermal equilibrium with the medium, each fragment experiences random thermal effects of the environment, i.e. from this point of view behaves as a Brownian particle. As known from the theorem on equidistribution of energy over degrees of freedom in thermal equilibrium, the average kinetic energy of one-dimensional motion of a free particle (per one

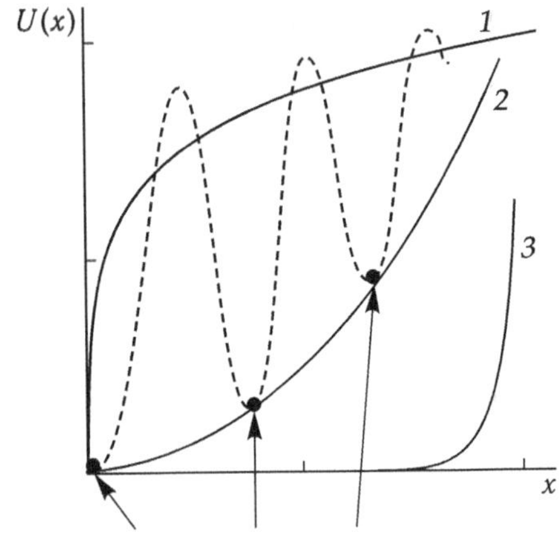

Figure 11.5. "Hopping" diffusion over local conformational substates in the potential $U(x) = \alpha|x|^{1/\beta}$ at different β.
1, $\beta \gg 1$; 2, $\beta = 1/2$; 3, $\beta \ll 1$.

degree of freedom) along the coordinate x is

$$\langle mv^2/2\rangle = k_B T/2, \tag{11.3}$$

where sign $\langle\ldots\rangle$ designates the thermal mean in ensemble ($v = dx/dt$ is the x component of the particle velocity). However, the fragment under discussion is in a dense medium and is not free: it is affected by forces, which can be tentatively divided in three types. □

1. The elastic force F_{elast} of a chain fragment caused by strong bonds with other molecular groups thus generating conformational potential $U(x)$

$$F_{\text{elast}} = -dU(x)/dx.$$

The existence of certain elasticity permits modeling the chain motion as a harmonic oscillator, for which $U(x)$ is determined by formula (11.1) and consequently

$$F_{\text{elast}} = -m\omega_0^2 x, \tag{11.4}$$

where m is the mass of the fragment (~ 100 a.m.u. $= 1.66\cdot 10^{-22}$ g); ω_0 is the frequency of conformational oscillations in the absence of damping ($\lesssim 10^{12}$ s^{-1}). Hence the elasticity constant is $m\omega_0^2 10^2$ g/s^2 at $\omega_0 \sim 10^{12}$ s^{-1}.

2. The friction force directed against the velocity $v = dx/dt$ and the force proportional to it

$$F_{\text{fr}} = -\gamma(dx/dt), \tag{11.5}$$

where γ is the friction factor.

By describing the resistance (friction) of the medium upon the particle motion in terms of microviscosity η, it is possible to estimate the friction factor by the known Stokes formula

$$\gamma = 6\pi b\eta, \tag{11.6}$$

where b ($\sim 0.1 \div 1$ nm) is the characteristic linear dimension of the moving fragment. At $b \sim 0.1$ nm and minimally allowable viscosity $\eta \sim 10^{-3}$ Pa·s (equal to the viscosity of water) the friction factor is $\gamma \sim 10^{15}$ a.m.u./s.

3. The force caused by random pushes from the environmental molecules due to the thermal motion and equal to

$$F_{\text{st}}(t) = \sum_i F_{\text{st}}(t_i), \tag{11.7}$$

where t_i are random instants of time. The amplitude of the force is also a random value.

As a result, we obtain the equation for the fragment motion in a dense medium (the Langevin equation) which looks like this

$$m\frac{d^2x}{dt^2} + \gamma\frac{dx}{dt} + \frac{\partial U}{\partial x} = F_{\text{st}}(t). \tag{11.8}$$

This is the model of the dynamics of motion of a chain region in a dense medium under the action of random thermal pushes from the microenvironment. A change in time of the coordinate $x(t)$ deviation from the mean value has a random character and is described by Equation (11.8). It is a stochastic equation, therefore, to solve it means to find the distribution functions x and $\dot{x}$ by the given distribution functions $F_{\text{st}}(t)$. Such equations are solved using the method of correlation functions. □

Correlation Function. The analysis of random deviations from the mean value of the variable permits characterizing the degree of dependency between the deviations. It is known that for statistically independent random values x and y and their deviations from the mean values $\overline{x}$ and $\overline{y}$ it is valid that

$$\langle (xy) \rangle = \langle x \rangle \langle y \rangle,$$

from which

$$\langle (x - \overline{x})(y - \overline{y}) \rangle = 0.$$

It is obvious that if random deviations are statistically dependent, these relations do not take place, i.e.

$$\langle (x - \overline{x})(y - \overline{y}) \rangle \neq 0.$$

Accordingly, the value

$$\langle (x - \overline{x})(y - \overline{y}) \rangle$$

can be regarded as the degree of dependence or correlation of random values. Let us see how the method of random values correlation can be applied to the analysis of protein fragment motions under the action of random thermal pushes, which correspondingly cause random deviations of the fragment coordinate $x(t)$ from the average value. We introduce function $\varphi(\tau)$ for random deviations $x(t)$, which is obtained as a result of: (1) multiplying the random value at the initial moment t, i.e. $x(t)$, by the value $x(t+\tau)$, which the variable x accepts in an additional time interval τ; (2) the following calculation of such products for a variety of t values; (3) their averaging over all t values:

$$\varphi(\tau) = \langle x(t)x(t+\tau) \rangle. \tag{11.9}$$

Apparently, the larger the time τ, the higher the reason to believe that $x(t)$ and $x(t+\tau)$ are statistically independent random values. In this case, at $\tau \to \infty$

$$\varphi(\tau) = \langle x(t)x(t+\tau) \rangle = \langle x(t) \rangle \langle x(t+\tau) \rangle = 0, \tag{11.10}$$

where it is taken into account that the average value is $\langle x(t) \rangle = 0$.

For a stationary random process, the correlation function $\varphi(\tau)$ is independent of the initial moment of time t. So, $\varphi(\tau)$ determines the degree of the statistical independence of random deviations $x(t)$ and $x(t+\tau)$. At $\tau = 0$ the correlation function is equal to the mean-square shift of the particle at the initial moment from the average position $\langle x(t) \rangle = 0$,

$$\varphi(0) = \langle x(t)^2 \rangle, \tag{11.11}$$

which is always positive independent of the sign of the deviation. Because the $\varphi(\tau)$ value does not depend on the choice of the moment of measuring t, and depends only on the time τ, then assuming the initial moment of time $t = 0$ and changing designations $\tau = t$, we can write

$$\langle x^2(t+\tau) \rangle = \langle x^2(t) \rangle \text{ or } \langle x^2(0) \rangle = \langle x^2(t) \rangle.$$

Due to this

$$\varphi(t) = \langle x(0)x(t) \rangle. \tag{11.12}$$

At high t values, the following asymptotic expression is valid

$$\varphi(t) = \langle x(t)x(0) \rangle \sim \langle x^2(0) \rangle \exp(-t/\tau_c), \tag{11.13}$$

where τ_c is the time of correlation of stochastic motion which determines the rate of damping of the random deviation x from the average value. For systems with many degrees of freedom $x_1, x_2, \ldots, x_n$ cross-correlation functions may be also introduced:

$$\varphi_{ik}(\tau) = \langle x_i(0)x_k(\tau) \rangle. \tag{11.14}$$

Because of the symmetry of equations of mechanics relative to the time inversion, it does not matter what value x_i or x_k should be taken upon averaging at an earlier instant of time and what at a later one:

$$\langle x_i(t)x_k(t+\tau) \rangle = \langle x_i(t+\tau)x_k(t) \rangle.$$

For the Brownian oscillator the average quadratic shift depends on t as follows:

$$\left\langle [x(0) - x(t)]^2 \right\rangle = \left\langle [\Delta x(t)]^2 \right\rangle = 2[\varphi(0) - \varphi(t)]. \tag{11.15}$$

It follows from the above expression that if at the beginning the coordinate was measured at instants of time differing from each other by value t, the average value of the squared difference of the coordinate is equal to the doubled difference between the values of correlation functions at these instants of time. The Fourier transform of the correlation functions yields the frequency spectrum of fluctuations, which permits us to distinguish frequencies making the main contribution to the fluctuation damping.

Determination of Time Correlation Functions. This is required for the description of thermal molecular motion in disordered media. Of great importance are correlation functions of velocity, angular velocity and orientation of molecules, which make possible to evaluate their time dependence on the character of thermal motion in the medium. For example, at a completely chaotic motion of rarefied gas molecules the velocity correlation function will decay exponentially with a characteristic time of about the average time between collisions of molecules, which lead to the loss of the interconnection between gas molecules. In crystals the free oscillations of molecules are correspondingly described by oscillatory correlation functions. But in dense liquids oscillations of these functions will be strongly damped. Finally, in a highly viscous medium (see below) they will damp with time proportionally to the viscosity of the medium. Thereby, the structure of disordered medium and the character of motions in it are displayed in the time dependence of correlation functions, the investigation of which permits determining important parameters of the medium.

Correlation functions make it possible to perform a quantitative comparative study of dynamic properties of macromolecular systems. To analyze these functions, methods of mathematical analysis and especially the Fourier analysis are used, which allow finding mode frequencies making the greatest contribution to the motion of the studied molecular group. The investigation of cross-correlation functions enables evaluating on the relationship between motions of different molecular groups in a protein.

Mean-Square Shift. Let us return to Eq. (11.8). Given $F_{st}(t)$, as a random force and white noise, is generated by surrounding molecules acting on the oscillator (11.8). It can be shown (K. Hiere, 1976) that the Fourier transform of the correlation function (11.13) is

$$\varphi(\omega) = 2\int_0^\infty \varphi(t)\cos\omega t\,dt = \frac{2k_B T_\gamma}{m^2(\omega_0^2-\omega^2)^2+\gamma^2\omega^2}. \tag{11.16}$$

By inverting the Fourier transform (11.16) we get formulas for the correlation function $\varphi(t)$ in the form $\varphi(t) = \left[k_B T/(m\omega_0^2)\right]\exp(-t/\tau_c)$, where $k_B T/(m\omega_0^2) = \langle x^2(0)\rangle$.

1) For weakly decaying oscillations in a solid at $\omega_0 \gg \gamma/(2m)$

$$\varphi(t) = \frac{k_B T}{m\omega_0^2}\exp(-t/\tau_c)\cos\omega_0 t, \quad \tau_c = 2m/\gamma. \tag{11.17}$$

2) For bounded diffusion at $\omega_0 \ll \gamma/(2m)$

$$\varphi(t) = \frac{k_B T}{m\omega_0^2}\exp(-t/\tau_c), \quad \tau_c = \gamma/(m\omega_0^2). \tag{11.18}$$

In both cases, the $k_B T/(m\omega_0^2)$ value (the mean square of the amplitude of thermal fluctuations of the particle position with the energy $k_B T$) is

$$\langle x_a^2\rangle = k_B T/(m\omega_0^2), \tag{11.19}$$

which can be readily found from (11.1) using the theorem on energy equidistribution over the degrees of freedom (11.3).

At $t = 0$ the value of the correlation function in (11.17) and (11.18) is also equal to the mean square of the fluctuation amplitude of the particle position:

$$\varphi(0) = k_B T/(m\omega_0^2). \tag{11.20}$$

Substituting (11.16), (11.17) and (11.18) in (11.15) we find that for limited diffusion

$$\langle [\Delta x(t)]^2 \rangle = \langle x_a^2 \rangle (1 - \exp(-t/\tau_c)), \quad \tau_c = \gamma/(m\omega_0^2), \tag{11.21}$$

and for oscillations in a solid

$$\langle [\Delta x(t)]^2 \rangle = \langle x_a^2 \rangle (1 - \exp(-t/\tau_c) \cos \omega_0 t), \quad \tau_c = 2m/\gamma. \tag{11.22}$$

In the case of small atomic oscillations in solids (11.17), the average correlation time $\tau_c = 2m/\gamma$ corresponds to the time τ_c of the damping of the amplitude of the oscillatory motion at ω_0 frequency. With the growth of γ the time τ_c decreases, that is equivalent to the decrease in the time interval, during which a noticeable change in the oscillation amplitude occurs due to energy dissipation.

At low τ_c values, the correlation between separate values of $x(t)$ and $x(t+\tau)$ vanishes rapidly. In this case the sequential values of $x(t)$ and $x(t+\tau)$ will be already independent, and this means that the correlation function turns to zero which follows from (11.17) at $\tau_c \to 0$.

Expression (11.18) describes another type of motion, and exactly the limited diffusion in the harmonic potential. It is valid at very high friction factors. Here the correlation time is proportional to the friction factor $\tau_c = \gamma/(m\omega_0^2)$. This is explained by the decrease in the diffusion rate caused by an increase in the resistance force or viscosity of the medium; delocalization of particles therewith slows down in diffusion.

In liquids due to limited diffusion $\langle [\Delta x(t)]^2 \rangle \sim 2Dt$, whereas in solids and proteins expressions (11.22) and (11.21) are correspondingly valid for $\langle [\Delta x(t)]^2 \rangle$. It can be demonstrated that under natural conditions f' for solids $\gamma/m \gg \Gamma$, but $\gamma/(2m) \ll \omega_0$ (i.e. low damping):

$$f'(T) = \exp(-a^2), \tag{11.23}$$

where $a^2 = \langle x_a^2 \rangle / \lambda^2$. Expression (11.23) is known as the Debye–Waller factor.

Temperature Dependence $f'(T)$. The $f'(T)$ value is determined by the dependence of $\langle x_a^2 \rangle$ on T and, as follows from relation (11.19), should have a smooth character,

$$f'(T) = \exp\left[-k_B T / m\omega_0^2 \pi^2\right], \tag{11.24}$$

reflecting a slow decrease of the factor f' with an increase of temperature. But this does not agree with the experimental data on the dependence of $f'(T)$ for proteins (see Fig. 10.23 and Fig. 10.24).

Let us analyze what determines the dependence of $f'(T)$ in the case of limited diffusion at strong damping $\gamma/(2m) \gg \omega_0$.

The temperature dependence of factor $f'(T)$ is closely associated with the value of the mean-square shift of the nucleus during the lifetime $\tau^* \simeq 10^{-7}$ s. In this case, the change of the factor f' versus temperature is determined not by the smooth dependence (11.19) of $\langle x_a^2 \rangle$ on T but by the strong temperature dependence of the correlation time $\tau_c = \gamma/(m\omega_0^2)$ due to the change of γ together with the temperature. As this takes place, the ratio of τ_c to the lifetime of the excited nucleus τ^* changes, and just this determines the magnitude of the nucleus shift (relative to λ) recorded during the lifetime τ^*. In fact, at a temperature increase the exponential decrease in $\gamma(T)$ depending on microviscosity η plays the main role rather than the slow increase in the parameter $a^2(T)$:

$$\tau_c = \gamma/(m\omega_0^2) = 6\pi b\eta/(m\omega_0^2).$$

If $\tau_c \gg \tau^*$, the viscosity of the medium is so high that the nucleus has no time to shift noticeably and reach the boundary of the potential well for time $t = \tau^*$: $\langle[\Delta x(t)]^2\rangle \sim \langle x_a^2\rangle t/\tau_c \ll \lambda^2$ and the factor $f' \simeq 1$.

An increase of temperature results in a decrease of viscosity:

$$\eta(T) \sim \exp(\epsilon/k_B T), \tag{11.25}$$

where ϵ is the energy of activation of the fluid microviscosity, and consequently as a result of the temperature growth the correlation time exponentially decreases:

$$\tau_c \sim \eta \sim \exp(\epsilon/k_B T). \tag{11.26}$$

At high temperatures where $\tau_c \ll \tau^*$, the standard Debye–Waller dependence (11.24) is observed, when during time τ^* the $\langle[\Delta x(t)]^2\rangle$ value reaches the limiting value $\lambda^2 a^2$. In the intermediate case when $\tau_c \sim \tau^*$, a harsh transition from $f' \sim 1$ values to $f' \sim \exp(-a^2)$ takes place. At $\tau_c \gg \tau^*$, the spectrum $g(\omega)$ consists of a narrow Mossbauer line. As the temperature increases, the spectrum widens gradually to the width magnitudes $\sim(\Gamma + a^2/\tau_c)$. Then, when the further temperature increase makes $\tau_c \lesssim \tau^*$, the "wings" of the spectrum line are rapidly widened together with the sharp decrease of the intensity in the central part of the spectrum and the drop of the f' values. At $\tau_c \ll \tau^*$, the widened part of the spectrum is seen as a smooth background against which a narrow line is clearly distinguished (Fig. 11.6).

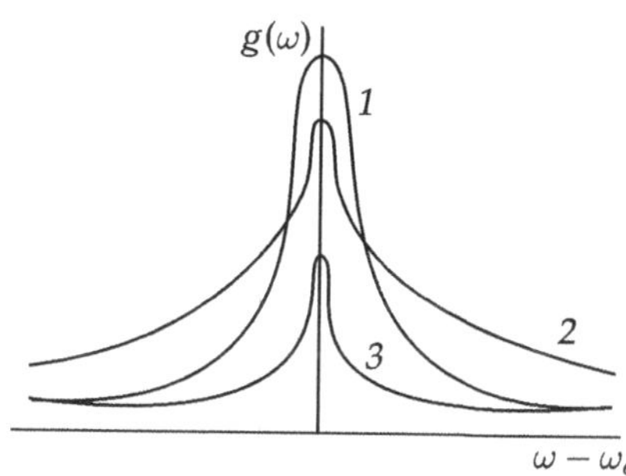

Figure 11.6. Changes in the shape of the Mossbauer spectrum $g(\omega)$ as a result of changes in the relaxation time τ, s (temperature increase).

1, $\tau \gg 10^{-7}$; *2*, $\tau \simeq 10^{-7}$; *3*, $\tau \ll 10^{-7}$.

Thus, a sharp decrease of the factor f' in a limited range of temperatures without widening of the spectral line is associated with the decrease of the correlation time below the critical value $\tau_c \simeq 10^{-7}$ s, rather than with the increase of the overall amplitude of motion x_a. The obtained formulas are applied in the analysis of experiments on the protein mobility which are performed using gamma-resonance spectroscopy. □

The processing of corresponding experimental data for myoglobin and chromatophores demonstrated that the microviscosity in the vicinity of the Mossbauer atom is $\sim$ 10 Pa · s at 300 K at linear dimensions $b \sim 0.5$ nm of a fragment. This is much higher than the viscosity of water (10^{-3} Pa · s), glycerol ($\sim$ 1 Pa · s) and the average viscosity of a biological membrane (0.1–1 Pa · s). The activation energy of the viscous flow ϵ determined by the dependence $f'(T)$ is 21 kJ/mol, and the conformational motion amplitude $x_a = 0.035$ nm (the amplitude of valence oscillations of the Fe atom is far less $x_0 \sim 0.001 \div 0.002$ nm). The mobility of the Fe atom is wholly determined by the motion of the corresponding fragment of the protein molecule. Dehydration of protein resulting in the increase of its viscosity and a drop in the water content should also lead to the growth of the factor f' that is actually observed in the experiments.

11.3 Numerical Modeling of Molecular Dynamics of Proteins

Atoms in the protein molecule are exposed to different actions from the microenvironment and are involved in various types of motion. Extremely fast thermal motions are closely associated with rare rearrangements, including functionally important ones. Internal motions of globular proteins are characterized by changes in the amplitude from 10^{-3} nm to 10 nm, energy from 0.4 J/mol to 420 kJ/mol and time from 10^{-15} s to 10^3 s. The motion of atoms and their groups in proteins occurs as if they proceed in a cage of the surrounding atoms. At low times ($\sim 10^{-13}$), such groups perform oscillatory motions relative to the small amplitude ($\lesssim 0.02$ nm). At higher times, collective shifts of the surrounding atoms occur that have either a local character or involve the whole molecule.

Method of Dynamic Modeling. This method permits obtaining detailed information on the processes in the range from 10^{-3} ps to 10^3 ps (1 ps $= 10^{-12}$ s) which involve several tens or hundreds of atoms.

As a model of the molecular system containing N atoms, the N set of interacting mass points is taken, the motion of which is described by the classical Newton equations:

$$m_i \frac{d^2\mathbf{r}}{dt^2} = \mathbf{F}_i \quad (i = 1, \ldots, N). \tag{11.27}$$

Mass points "model" atoms: m_i is their masses, and $\mathbf{F}_i$ is the forces acting on them. Given at the zero time instant the coordinates and rates of all particles determine their behavior in the subsequent instants of time. State-of-the-art computers make it possible to integrate numerically such equations for several hundreds or thousands of atoms at times $\gtrsim 10^{-9}$ s. The calculated trajectories of motion of all atoms give a complete pattern of structural changes in time.

The MD method suggests that the considered system is characterized by its energy, has the model Hamiltonian $\widehat{H}$ and possesses n degrees of freedom. We will accept that our ensemble obeys the laws of classical mechanics and can be completely described, if its energy is set as the function of generalized coordinates $(x_1, x_2, \ldots, x_N)$ and generalized impulses $(p_{x_1}, p_{x_2}, \ldots, p_{x_N})$ corresponding to them.

For an idealized ensemble, the energy can be written as the Hamiltonian function $H(x, p_x)$. In this case, the classical Hamilton equations of motion are as follows

$$\frac{dx}{dt} = \frac{\partial H}{\partial p_x}, \quad \frac{dP_x}{dt} = -\frac{\partial H}{\partial x}. \tag{11.28}$$

For N particles with mass m the coordinates will be an ensemble of Cartesian coordinates of these particles

$$x_\alpha, y_\alpha, z_\alpha \quad (\alpha = 1, 2, 3 \ldots, N),$$

and the corresponding pulses of $3N$ variables will be

$$m\dot{x}_\alpha,\ m\dot{y}_\alpha,\ m\dot{z}_\alpha \quad (\alpha = 1, 2, 3 \ldots, N).$$

From (11.28) without taking into account the energy dissipation, it follows that

$$\frac{dH}{dt} = 0, \tag{11.29}$$

and, as a result of integration of (11.29), the energy E of the ensemble

$$H(x, p_x) = E, \tag{11.30}$$

remains constant or is the integral of motion.

The behavior of the ensemble at any instant of time is set by the point in the phase space ($6N$ measurements) with coordinates x and p_x. The point travels along a phase trajectory in the phase space reflecting changes in the ensemble state in time. The main idea is to get characteristics of such a system in a computer experiment, which simulates the motion of the ensemble or its representative point in the phase space.

The characteristics of the ensemble or the observed values are obtained as the averaged ones over the overall space of the system states.

The set of these states comprises the phase space accessible to the system, and strictly speaking, the averaging should be conducted over all states. But in a real computer experiment, the averaging over the full ensemble is impossible. That is why the averaged characteristics of the system are calculated along the trajectories in the phase space. For example, the temperature of the system, as the average kinetic energy, per one degree of freedom of a particle is computed not by averaging over a large number of similar systems or the set of all possible microstates, but by averaging over time intervals.

In a real computer experiment the particle is driven over the trajectory in the phase space under the action of forces (11.4)–(11.7) and the kinetic energy is computed along the trajectory. This procedure represents the averaging not over all states, but the time averaging of the behavior of one particle. In principle, the particle should have time to "visit", while moving along its trajectory, all cells of the phase space. The equivalence of the values averaged over the ensemble to those averaged over time corresponds to the system characteristics called ergodicity.

The mean value $\langle A \rangle$ averaged over the ensemble is substituted for its mean value $\overline{A}_\infty$, averaged over time, i.e.

$$\langle A \rangle = \overline{A}_\infty.$$

It is evident that in a molecular system the time of observation t should several folds exceed the molecular times. There are obvious requirements to the conditions of motion along the phase trajectory in the computer experiment. The MD method makes use of the own dynamics of the model simulating the motion of the system in the phase space by setting equations of motion and finding their computer-aided numerical solution.

For a set of particles described by the equations of motion (11.28)–(11.30), this leads to trajectories ($x^N_{(t)}$, $p^N_{(t)}$) in the phase space with initially given coordinates $x_{1(0)}, x_{2(0)} \ldots, x_{N(0)}$ and impulses $p_{1(0)}, p_{2(0)} \ldots, p_{N(0)}$.

At other start conditions, the system will move in the phase space along another trajectory with constant energy. In this case, the ergodicity is taken into account, so that the averaging over time of the trajectory motion is equivalent to the averaging over the ensemble states. It is clear that when the time of the experiment is limited, the system cannot attend all phase space cells. It is namely the limited area of the phase space, over which the averaging is done, that is one the sources of the MD method errors.

In a computer simulation, the energy and temperature are calculated as the average kinetic energy per one degree of freedom. The obtained values can change at the random change of particle velocities at every time step, and also because of the limited accuracy of computer arithmetic operations and the rounding errors. Deviations from the real trajectory increase with an increase in the time step at computer-aided numerical solution of the equations, when the potential introduced in the equations instead of the real forces is set inaccurately.

In biomolecular systems, fast motions are the oscillations of the bonds connecting the heavy atom and the proton. They have a period of $\sim$ 10 fs. That is why the integration step to provide the stability should be of a much low value. As a matter of fact, the upper limit of the integration step, whereby the most frequently used Verlet integration scheme does not lose the stability, is $\sim$ 0.5 fs. An increase in the integration step permits more efficient filling of the phase space and, consequently, increases the estimation accuracy of the observed characteristics. To this end, it is required to eliminate fast motions in the system. This task is solved by constraining the most high-frequency oscillations of the bonds containing hydrogen atoms. This makes it possible to increase the integration step from 0.5 to 2 fs.

It is important when calculating the MD-trajectory that the "correct" way of filling the phase space would be realized spontaneously: transition between the two states is performed under the action of the force field. Only following such approach it is possible to reveal physically correctly relaxation characteristics to the equilibrium. Therefore, the MD method can be used to study not only equilibrium distributions, but also dynamic processes in biomolecular systems: either completely non-equilibrium or those which represent fluctuations in the equilibrium state.

Role of a Thermostat. Upon MD modeling, the average energy and impulse of the system should remain constant, provided by relatively fast relaxation of the system to the equilibrium distribution of Boltzmann velocities, for which, as a matter of fact, the very concept of a microstate is valid.

In numerical experiments, relaxation of the system to the equilibrium and to the assigned energy value experiencing random deviations is achieved by taking away or introducing additional energy into the system by the corresponding norming of particle velocities.

It is obvious that in a general case, the role of a thermostat is in maintaining the known relation between temperature T and the average kinetic energy E value of the system:

$$E = \frac{\sum_i^N m_i v_i^2}{2N}, \tag{11.31}$$

where N is the total number of atoms in the system.

The total kinetic energy of the particle is $E = \frac{3}{2}k_B T$ that gives the "instantaneous" temperature at each instant of time

$$T = \frac{\sum_{i=1}^{N} m_i v_i^2}{3Nk_B}. \tag{11.32}$$

Upon time averaging in the range from τ to t_0 we get the temperature in the real computer experiment:

$$\overline{T} = \frac{1}{3Nk_B \cdot \tau} \int_{t_0}^{t_0+\tau} \sum_{i=1}^{N} m_i v_i^2 dt.$$

As a rule, the MD method permits recording the constant thermostat temperature, which at equilibrium should evidently coincide with the average temperature of the molecular system. As the energy of the latter is much smaller than that of the thermostat, this condition should provide rapid establishment of the equilibrium. This is of special importance at the stage of relaxation, required at sharp deviations from the equilibrium temperature. These deviations are a consequence, in particular, of the large final step of integration, during which the system deviates from the true path that may cause a jump in energy and temperature. To accelerate the motion of the representative point in the phase space, numerical experiments are performed at relatively high temperatures. It provides an additional evidence of the importance of interactions with the thermostat to level the deviations from the equilibrium temperature value in dynamic experiments.

The effect of the thermostat, distorting real mechanical trajectories during counting, is revealed in the deviation between the calculated and the equilibrium Boltzmann distributions that should remain constant in the case of fast relaxation of the emerging fluctuations

Table 11.1 shows data on the time ranges of dynamic processes proceeding in biomolecular systems. As seen, proteins, nucleic acids and lipid membranes are very complex objects with motions in a large range of amplitudes and time scales.

In the classical description of the dynamics of biomolecules, the characteristic times of motions are in the range from tens of femtoseconds for oscillations of the bond lengths and angles to several seconds required for folding or complete denaturation of large protein molecules. The range of characteristic amplitudes from 0.1 to 100 E corresponds to these motions. The advanced technologies of calculating the molecular dynamics, based on the use of parallel computations with the utilization of highly efficient multiprocessor configurations, permits obtaining trajectories for different biomolecular systems with the length of no more than several microseconds (usually tens and hundreds of nanoseconds). It is clear that here the limiting factor is the dimensions of the simulated system (the number of atoms and molecules). In contemporary MD experiments, the objects of studies are frequently not separate molecules but whole molecular complexes, containing up to several million atoms. For example, they are proteins inserted in multicomponent cell membranes together with water molecules and different ions (D. L. Bostick and C. L. Brooks, 2007) and even such complex molecular machines as the ribosome (K. Y. Sanbonmatsu and C.-S. Tung, 2007). In addition to the "all-atom" models of molecules (the molecule contains all atoms), different simplified schemes are also available: models with

Table 11.1. Dynamic processes in biomolecular systems

Type of internal motions in biomolecules	Characteristic time scale
Oscillations of the bond lengths	10–50 fs
Oscillations of valence angles	20–110 fs
Global twisting of a helical DNA	1 ps
Deformation of hydrocarbon cycles in DNAs	1 ps–1 ns
Rotation of side chains in proteins exposed to the solvent	10–100 ps
Collective motions in protein molecules (motions of flexible loops, domain motions, allosteric conformational changes)	10 ps–1 μs
Global bend of a helical DNA	100 ps–100 ns
Lateral shift of lipids in the membrane at a distance of the lateral dimension of a molecule	1 μs–1 ms
Conformational changes in a superhelical DNA	1 μs–1 s
Rotation of side chains inside the protein globule	100 μs–1 s
Folding – protein denaturation	10 μs–10 s
"Flip-flop" motions of lipids in the membrane	1–105 s

only polar protons capable of forming hydrogen bonds, large-grain models in which particles describing groups of atoms are introduced, etc. The use of a simplified description of molecules in MD calculations makes it possible to remarkably increase their efficiency, but in this case it is necessary to sacrifice the extent of the detailing. On the whole, the use of the MD method for studying dynamic processes given in the lower part of Table 11.1 is yet strongly limited. For example, so far it is possible to describe the overall process of protein folding or denaturation only with the use of different artificial procedures, which accelerate these processes and unavoidably distort them, or only for very small protein molecules (e.g., 36-membered amino acid fragment of villin, see Fig. 9.3).

In long trajectories, very fast motions (up to 1 ps; see the upper part of Table 11.1) are well represented from the statistical point of view, but even here difficulties arise: the classical description of the dynamics of such motions is not quite correct. As a matter of fact, the classical approach is, strictly speaking, inapplicable for motions with relatively high frequencies, i.e. such that $h\nu/k_B \gg 1$. At the temperature of 300 K, this category includes all motions with characteristic times lower than 160 fs (the upper part of Table 11.1). Oscillations of the lengths of the bonds and valence angles in MD are described by the classical harmonic oscillator, the oscillatory energy of which at these frequencies is far larger than the energy of oscillations of the quantum oscillator. This discrepancy is partially resolved by constrains imposed on the most high-frequency motions — oscillations of the bonds containing a hydrogen atom (see above).

Nonetheless, at present the MD method can give a rather accurate description of dynamic processes represented in the middle part of the table. This is the pico- and nanosecond ranges which include an essential part of functionally important motions in biomolecular systems.

Molecular Dynamics of Trypsin Inhibitor (TI). Complete data of modeling the internal dynamics of TI are given in papers of M. Karplus and J. A. McCammon (1981). The initial coordinates of all heavy atoms were obtained from X-ray data by minimization of the potential energy of conformational interactions. The rates were adjusted to be equal in values but random in direction. The kinetic energy corresponded to the start temperature 300 K. In several picoseconds of integrating the equations of motion, artificial correction of the velocities is made because the increase in the average kinetic energy (temperature) of the molecule is observed due to atoms striking each other in the initial conformation. This procedure was repeated several times. As a result, in 35 ps the system demonstrates quite well its equilibrium behavior. At the end of the preparatory period, the average kinetic energy corresponded to the temperature of 306 K. This temperature is maintained during the basic period of modeling, when mechanical trajectories of atoms are directly calculated.

A detailed analysis of the dynamics of aromatic protein chains in the TI molecule demonstrated the key role of the microenvironment in relaxation processes of the emerging fluctuations. The aromatic ring *Tyr*-21 in the trypsin inhibitor is surrounded by other amino acid residues, so that the position of the potential energy minimum for it on the energetic map is strictly determined as compared to the free dipeptide, in which the aromatic ring is affected only by the main chain (see Sections 1 and 2 in Chapter 9).

The equation of the motion of the ring *Tyr*-21 is the Langevin equation (11.8) and in this case looks like this

$$I\frac{d^2\Phi}{dt^2} + \gamma\frac{d\Phi}{dt} + \alpha\Phi = F_{st}(t), \tag{11.33}$$

where $\Phi(t)$ describes fluctuations of the rotation angle; $I = 7.5 \cdot 10^{15}$ g · cm2/mol is the moment of inertia of the aromatic ring relative to the rotation axis; γ is the friction factor; α is the stiffness constant; $F_{st}(t)$ reflects random pushes experienced by the ring during microenvironment fluctuations. Figure 11.7 shows fluctuations $\Delta\Phi$ of rotation angle Φ of *Tyr*-21 expressed as deviations from the average value $\langle\Phi\rangle$.

As seen from the Figure, the rotational motion of the ring *Tyr*-21 in the protein has a less regular character as compared to the free fragment due to collisions with atoms

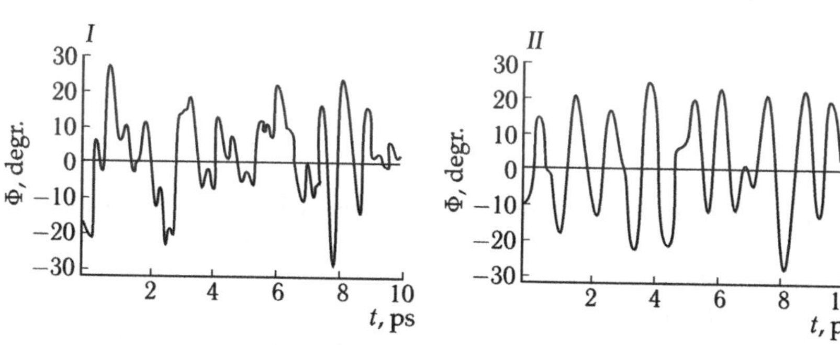

Figure 11.7. Changes in the rotation angle of the ring *Tyr*-21 (*I*) and isolated tyrosine fragments (*II*) upon dynamic modeling in the protein during 9.8 ps (according to M. Karplus, B. R. Celin and J. A. McCammon, 1979).

$\Delta\Phi = \Phi - \langle\Phi\rangle$. The rotatonal motion of the ring *Tyr*-21 in the protein has a less regular character as compared to the free fragments due to collisions with atoms of the microenvironment.

of the microenvironment, which slow down the ring motions. The time dependence of the normalized correlation function

$$\varphi(t) = \frac{\langle \Delta\Phi(t)\Delta\Phi(0)\rangle}{\langle [\Delta\Phi(0)]^2 \rangle}$$

is given in Fig. 11.8. It is seen that the curve of the function is rapidly damping in the protein compared to the isolated *Tyr*-21. The average time between noticeable changes of $F_{st}(t)$ is about 0.07 ps. Figure 11.7 (*I*) shows that during this short period the Φ value has no time to change noticeably. $F_{st}(t)$ is a random variable with the Gauss distribution, and the separate pushes corresponding to it affect the ring motions. As shown by the processing of the experimental data, the coefficient values in (11.33) are $\alpha = 5.5 \cdot 10^4$ J/(rad$^2 \cdot$ mol); $\gamma = 0.11$ g $\cdot$ cm/(s $\cdot$ mol), and the damping time $\tau = \gamma/\alpha = 0.2$ ps.

The example of rotation of *Tyr*-35 aromatic ring, deeply "immersed" in the protein molecule, permitted tracing molecular rearrangements necessary to overcome energy barriers.

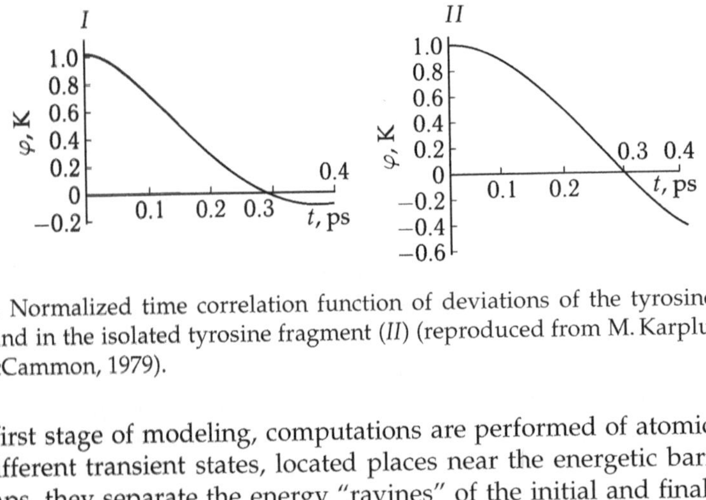

Figure 11.8. Normalized time correlation function of deviations of the tyrosine ring in the protein (*I*) and in the isolated tyrosine fragment (*II*) (reproduced from M. Karplus, B. R. Celin and J. A. McCammon, 1979).

At the first stage of modeling, computations are performed of atomic configurations in different transient states, located places near the energetic barrier. On the energy maps, they separate the energy "ravines" of the initial and final states. The modeling of the direct transition is performed by calculating the atomic mechanical trajectories passing through these transient states at the initial given velocities.

It was found that *Tyr*-35 ring crosses the potential barrier of rotation in 0.5–1.0 ps. The friction factors slow down the transition, but cannot turn it in the reverse direction.

Analysis of Trajectories with the Use of Maps of Free Conformational Energy of Molecules. In addition to the analysis of trajectories with the use of correlation functions, we will demonstrate also the results of studying the maps of free conformational energy of macromolecules (K. V. Shaitan, 2006). The form of these maps allows evaluating how the energetic characteristic of the molecule is related to the mobility of its separate groups. The free energy maps are plotted in the following way. The surfaces of the probability distribution over different energy conformations and their sections are determined by the MD method. On a logarithmic scale, these projections are actually the free energy maps of the molecule, expressed in $k_B T$ units. Figure 11.9 shows the map of free energy levels of dipeptides. It is seen that extended

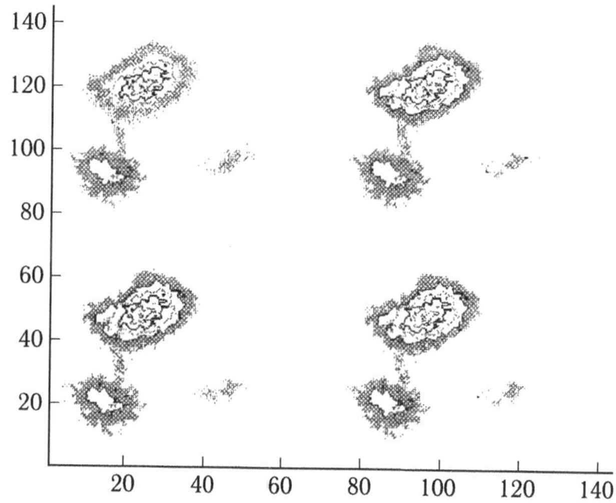

Figure 11.9. Map of probability levels (free energy) of a modified dipeptide *Tyr-Trp* in the space of torsion angles φ and ψ of the torsion residue (courtesy of Shaitan).

The regions of changes of the angles from $-360°$ to $360°$ are given on a proportional scale. The trajectory length is 5000 ps, $T = 1000$ K. The average frequency of collisions of heavy atoms with molecules of the medium is 10 ps^{-1}. The molecular mass of the particles of the medium is 18. A narrow neck for the transition in the configuration space between the two local minima is seen.

regions, where collective rearrangement of the dipeptide conformation occurs, are usually observed for correlated degrees of freedom.

For non-correlating pairs of variables, there is either a set of separate sharp local minima, the transition between the latter requires the overcoming of a high potential barrier, or vast areas of relatively free motion.

Behavior of Complex Biochemical Systems on a macroscopic level can be explained and predicted based on their microscopic characteristics. Some of their parameters can be determined by experimental methods, for example, NMR and X-ray analysis, which provide information on the structure and sometimes even on the processes occurring in molecular systems at the atomic level. However, not all microscopic properties that are of interest can be determined experimentally. The methods of molecular modeling make it possible to describe the macroscopic behavior of a molecular system in terms of its atomistic model. These methods permit the following: (1) understand and correctly interpret the experimental results, (2) give qualitative and sometimes quantitative estimations of macroscopic states, (3) interpolate or extrapolate experimental data to the hard-to-reach regions. A natural choice in constructing a realistic model is the quantum-mechanical approach, but it requires an extremely large calculation power — from $O(N_e^2)$ to $O(N_e^4)$, where N_e is the number of electrons in the molecule, and because of this it is used mainly for modeling small systems in spite of the essential advance in computational methods of quantum chemistry in the recent ten years. At present, relatively more accurate *ab initio* and more crude semi-empirical methods based on the solution of the Schrüdinger equation are used for objects containing from several tens to 10,000 atoms. The larger are the dimensions of the molecular system and the slower the processes under study proceeding in, the rougher approximations have to be used. At the same time, it is necessary to construct a model so that unavoidable assumptions, approximations and simplifications would have the least effect on the characteristics of interest. To model biopolymers the classical model in which a molecule is considered as a mechanical system of mutually interacting elements (atoms or groups of atoms) is usually used. Such a simplified representation is quite sufficient to solve a rather wide range of problems. Obviously, within the framework of a classical model it is impossible to study such effects quantum by their nature, as the formation – splitting of chemical bonds, tunneling of particles throughs potential barriers, the electron transfer processes, etc. Nevertheless, the current methods of molecular modeling, even in their classical version combined with the experimental methods, are the most powerful (and mostly the only) tool for the analysis of structure – dynamics – stability – function interrelations in biopolymers.

Potential of Forces. The MD method suggests that the effective potential energy (force field) of the molecule or a group of molecules can be represented as the sum of potentials (therms) emerging as a result of interactions of separate atoms or groups of atoms:

$$V_{\text{mol}} = \sum_{\substack{\text{Covalent}\\ \text{bonds}}} V_{\text{bond}} + \sum_{\substack{\text{Valence}\\ \text{angles}}} V_{\text{angle}} + \sum_{\substack{\text{Dihedral}\\ \text{angles}}} V_{\text{tors}} + \sum_{\substack{\text{Noncovalent}\\ \text{pairs}}} (V_{\text{vdw}} + V_{\text{el}}),$$

where V_{bond} and V_{angle} are energies of oscillations of the covalent bond and the valence angle, V_{tors} is the torsion energy (the energy of rotation of two chemical groups around the bond linking them), V_{vdw} and V_{el} are van-der-Waals and electrostatic energies of a pair of atoms not linked by a covalent bond (non-covalent potentials).

The formula contains a standard minimal set of therms used in most known programs for modeling biomolecules.

At present, the computer modeling of the dynamics of biomacromolecules employs several basic empirical potentials, developed in different research groups and varying in some details of the functional form, atomic representation and empirical parameters. □

The force fields, as a rule, have no universal character, and are developed for a definite type of objects. For example, for biopolymers the force fields CHARMM, AMBER, GROMOS AND OPLS can be used. They are parametrized on peptides, nucleotides, polysaccharides, lipids and different solvents. For low-molecular organic and inorganic compounds, force fields MM2/3/4 are more suitable.

On the whole, the up-to-date force fields reproduce well the equilibrium structural and dynamic properties of biosystems. Though as concerns non-equilibrium properties of proteins (time scales of conformational transitions, the processes of self-assembly), many questions still remain unclear.

11.4 Molecular Dynamics of Protein Myoglobin

We have analyzed the main experimental and hypothetical approaches to studying and physical modeling of intramolecular dynamics of biopolymers, and in the first place of proteins. The binding of low-molecular ligands with a heme protein is an elementary reaction, taking the latter as an example it is convenient to demonstrate how to use the above concepts.

Dynamics of Myoglobin. The most studied is the process of binding of carbon monoxide CO with the iron atom of the heme group of myoglobin. Figure 11.10 shows the section of the active center of myoglobin Mb with a plane drawn at a rectangular angle to the heme plane. The heme pocket is formed mainly by hydrophobic amino acids, and its size is about 0.5 nm. The iron atom forms covalent bonds with nitrogen atoms of the pyrrole ring and the nitrogen atom of the proximal histidine. From the side of the heme pocket it borders on the distal histidine. The nitrogen of the distal histidine is at a distance of 0.4 nm from the heme center. The inlet to the heme pocket is inhibited by propionic acid of the heme.

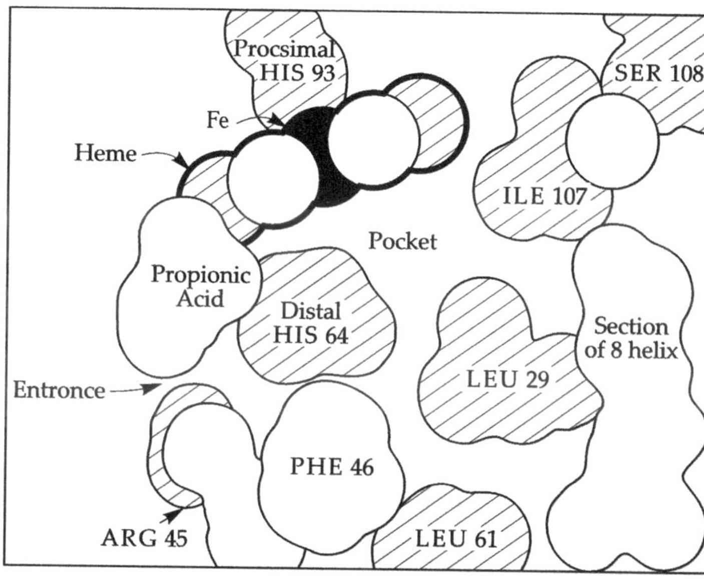

Figure 11.10. Cross-section across the region near the active center of myoglobin.
The inlet into the pocket of the heme is located between *His*-64 (Distal HIS 64) and propionic acid. Blackened groups are in the plane, and non-blackened ones are over the plane containing iron (reproduced from J. C. Kendrew et al., 1959).

The kinetics of reversible binding of the ligand CO to the heme iron and conformational changes of the protein interior of the active center, involved in the process, were studied in the experiments. In the initial state, MbCO is affected by the laser flash which breaks the bond between the ligand and the heme iron. As a result of photodissociation, myoglobin transits to the desoxy-form. This leads to a decrease in the absorption in the Soret band (for MbCO at 423 nm) and the emergence of an absorption band (band III) in the myoglobin spectrum at about 760 nm ($\sim$ 13,000 cm^{-1}), characteristic of its forms devoid of ligands.

After photodissociation the CO group can again bind to the iron atom that decreases absorption at 13,000 cm^{-1}. Because of this effect the kinetics of binding CO ligand to iron at different temperatures is studied by the laser flash induced changes of the absorption in the Soret band or at 760 nm. The shape of spectral band III and the change of its maximum position are the source of information on the conformational state of the heme protein.

Figure 11.11 shows the conformational energy map of Mb. The energy minimum is in the center. It corresponds to the hydrophobic pocket, the edge of which contains

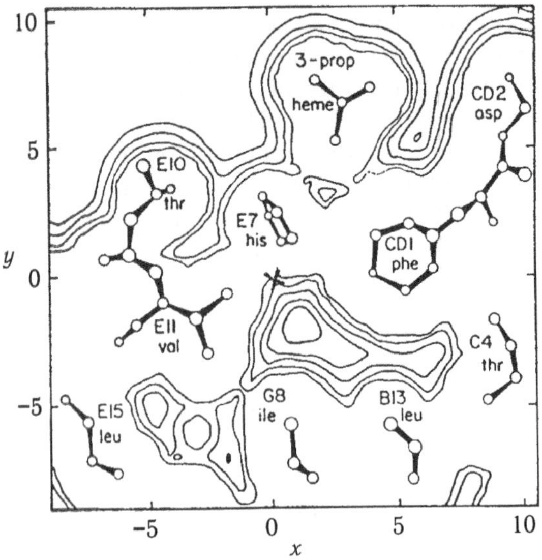

Figure 11.11. Contour map of the "myoglobin–ligand" interaction in the xy plane at $z = 3.2$ E (see the text).

The distances are given in angstrom units and the contour in kcal. The values correspond to 90, 45, 10, 0 and −3 kcal/mol relative to the ligand at infinity. The highest contour is the nearest to the atoms, the projections of which on the plane of the figure are shown by circles (reproduced from C. L. Brooks, M. Karplus and B. M. Pettitt, 1988).

the iron atom. The shortest way of the ligand from the pocket to the periphery of the heme molecule lies between *His* E7 and *Val* E11.

The calculations demonstrate that in the rigid protein structure the energy barrier that should be overcome by the ligand after photodissociation exceeds the value of 400 kJ/mol excluding its diffusion in the protein molecule at standard temperatures. This allows proposing that the transfer of the ligand in the heme cavity and its release to the outside, in the solution surrounding the protein, can occur only at intramolecular shifts of protein subgroups. In other words, the case in question is intramolecular rearrangements of the protein, which should have a functional character, enabling diffusion of ligands in the protein globule.

From the formal point of view, the recombination process (Fig. 11.12) includes several temperature-dependent stages, each of the latter having its own activation energy as the evidence of structural rearrangements in the protein.

The observed total rate of recombination $CO + Mb^* \to MbCO$ is characterized by a set of rate constants, corresponding to the distribution of Mb^* over conformational states, i.e. over values r and appropriate activation energy barriers.

At low temperatures ($T \sim 50$ K) the structure of the protein matrix is so rigid, that after splitting CO the iron atom Fe^{2+} cannot relax completely to its new equilibrium position in Mb. Thus, at low temperatures the recombination of CO and Mb does not affect relaxation processes in the protein. The position changes fundamentally at $T > 180$–200 K, when defrosting of conformational degrees of freedom of the protein takes place. Large-scale motions include the shift of the high-spin atom Fe^{2+} to the equilibrium position in the heme cavity of desoxymyoglobin. The overall relaxation of conformationally directed state of Mb^* to the equilibrium desoxy-state

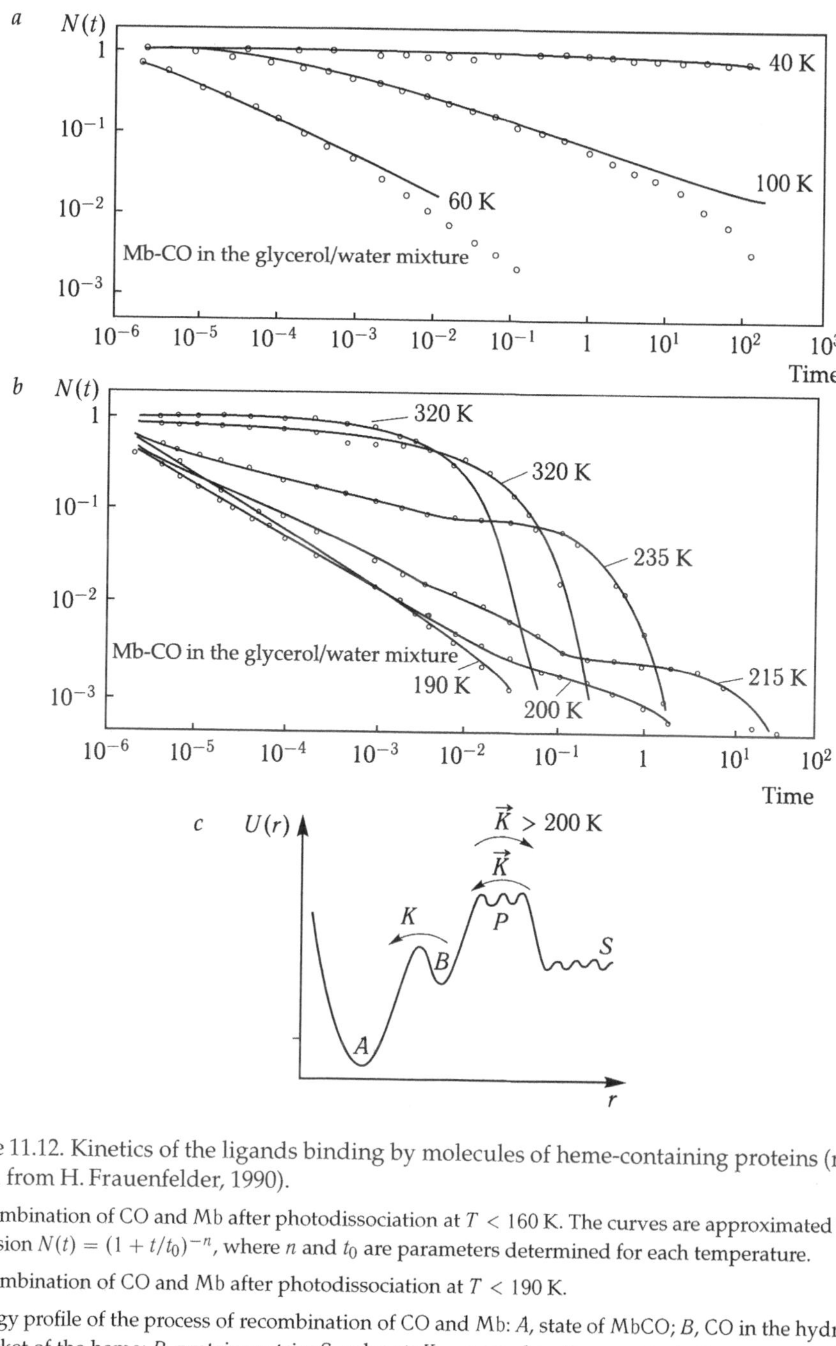

Figure 11.12. Kinetics of the ligands binding by molecules of heme-containing proteins (reproduced from H. Frauenfelder, 1990).

a, Recombination of CO and Mb after photodissociation at $T < 160$ K. The curves are approximated by the expression $N(t) = (1 + t/t_0)^{-n}$, where n and t_0 are parameters determined for each temperature.

b, Recombination of CO and Mb after photodissociation at $T < 190$ K.

c, Energy profile of the process of recombination of CO and Mb: *A*, state of MbCO; *B*, CO in the hydrophobic pocket of the heme; *P*, protein matrix; *S*, solvent; K_{BA}, recombination constant; *K*, constants of ligand migration in the protein matrix.

occurs, including the activation of motions of proximal *His*-93 on the F-helix. The system is characterized by distribution of activation barriers and rate constants, corresponding to different substates, over which the molecule fluctuates. The wavelength of the band III maximum increases again to 116 cm^{-1} at 300 K, suggesting an increase in the activation barrier, and the rate of recombination is decelerated at $T > 200$ K (Fig. 11.12). The reason for this is that at $T > 100$ K the CO ligand can leave the hydrophobic pocket moving away from Fe^{2+} into the solvent.

The problem of the CO ligand release from the hydrophobic pocket to the protein periphery is solved by methods of numerical modeling of intramolecular dynamics of Mb (Section 3 in this Chapter). The calculations have shown that the narrow place, determining the height of the barrier for the ligand diffusion is in the region of side chains of *His*-E7, *Val*-E11 and *Tyr*-E10. Torsional rotations of the three side chains by different angles require energy consumption not exceeding 8–10 kcal/mol. But at the same time the energy barrier decreases to 20 kJ/mol (M. Karplus, 2000) as compared to 400 kJ/mol in a rigid structure. This corresponds to the defrosting of the conformational dynamics in the protein. Figure 11.13 represents an energy map demonstrating the emergence of energy "valleys" for ligand diffusion in the protein.

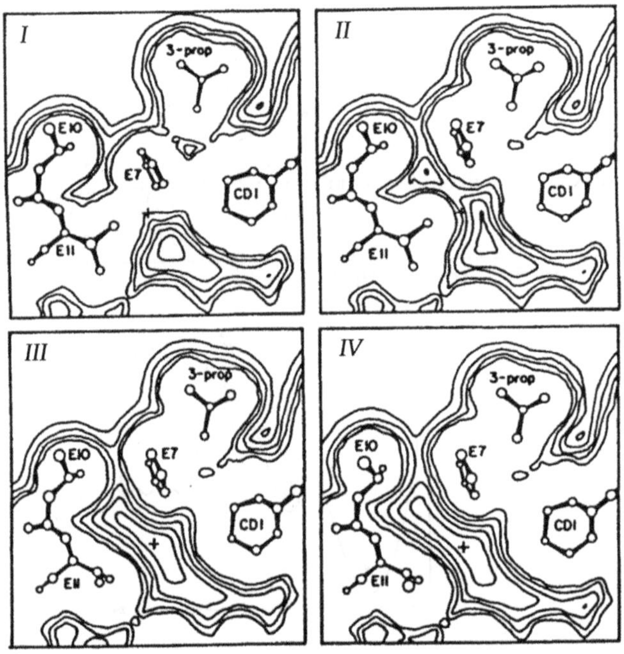

Figure 11.13. Contour map of myoglobin–ligand interaction in the *xy* plane (reproduced from C. L. Brooks, M. Karplus and B. M. Pettitt, 1987).

The distances are in angstrom units. The contours correspond to the values 90, 45, 19, 0 and −3 kcal/mol relative to the energy of the infinitely remote ligand. The higher energy levels are the closest to the atoms the projection of which on the plane of the figure is marked by circles. Parts *I–IV* show rotations of side chains.

11.5 Dynamic Models of DNA

In this section, we will analyze dynamic processes in DNA associated with its main biological function — storage of biological information and its transfer. The capacity of the double helix to untwisting is determined by complex dynamic properties of this molecule, consisting of a large number of atoms and having a quasi-one-dimensional structure with unusual symmetry, multiple degrees of freedom, many types of motions and a special distribution of internal interactions.

The classification of motions of the DNA molecule is given in Table 11.2.

The table shows that dynamic motions occur in a wide time range from femtoseconds to, at least, seconds. The general pattern of internal mobility is quite diverse. It contains motions of separate atoms, small groups of atoms, and whole fragments of the double helix. The motions include both fast and slow ones, motions with large and small amplitudes, elementary motions (oscillations of separate atoms or groups of atoms near the equilibrium positions with amplitudes not exceeding 0.1 E) and complex motions, such as local untwisting of the double helix or transition of some local DNA region to another conformational state. Frequencies associated with internal motions are in the range 1–100 s^{-1}, which is lower than the frequency values of internal oscillations in small isolated molecules.

Of special interest are motions with characteristic times living in nanosecond or microsecond ranges. Limited motions and oscillations of sugars, bases, phosphates and also rotational and bending motions of the double helix with the energy of activation of several kcal/mol correspond to the nanosecond range. They are characterized by the frequency of about several cm^{-1}. At small amplitudes of motions, linear models are suitable for their description. Provided the motions have relatively high amplitudes, the generation of collective motions becomes possible, which are associated with the formation of open states and untwisting of the double helix, characterized already by millisecond times. Their description requires nonlinear models.

Hierarchy of Structural and Dynamic Models. To construct a mathematical model of the DNA dynamics, the most fruitful is to use mechanical analogs — model systems well studied in mechanics with a similar set of structural elements, types of motions and interactions. In her monograph, L. V. Yakushevitch (2007) classified models of DNA motions in accord with the increasing level of their complexity. A scheme of the levels of the hierarchy of models is given in Fig. 11.14.

The most simple (level 1 in the figure) are elastic rod models with a circular section (in the micrograph the DNA resembles a fine elastic filament). The discrete analog is a chain of interconnected discs (or beads), one nucleotide pair corresponding to each disc. The dynamics of an elastic rod is well known in physics and is characterized by three types of internal motions: longitudinal shifts, rotational or torsional motions and transverse shifts.

Models of the second level take into consideration that a DNA molecule consists of two polynucleotide chains, interacting with each other across hydrogen bonds and twisted relative to each other in such a way that a double helix is formed. The model represents two elastic rods, weakly interacting with each other. The helical version of the model suggests that the rods are twisted into a double helix. In the direct (non-helical version) this is neglected. The two versions of the model have six types of internal motions: longitudinal, transverse and rotational motions in each of the rods. The internal structure of the chains is not taken into account. The discrete analog of this model represents two chains of discs, connected with each other by longitudinal and transverse springs, the rigidity of the longitudinal springs being much larger than that of the transverse ones.

The third level of the hierarchy takes into consideration the fact that each of the chains consists of three subunits: sugars, phosphates and bases. In the model the

Table 11.2. Classification of experimental and hypothetical data by the internal mobility of DNA

Time ranges	Main types of internal motions and their amplitudes	Activation energy, sources of energy	Experimental methods for studying motions	Hypothetical methods for studying motions
Picosecond	Oscillations of atoms with amplitudes $A \simeq 10^{-1}$ E	$E \simeq 0.6$ kcal/mol; the source is an external thermal reservoir	NMR, Raman spectroscopy, X-ray spectroscopy	Harmonic analysis, lattice model, methods of molecular dynamics
Nanosecond	Flexural and rotational motions of the double chain; restricted motions and oscillations of small groups of atoms: sugars, phosphates and bases ($A \simeq 5 \div 7$ E)	$E \simeq 5 \div 20$ kcal/mol; the source is collisions with "hot" molecules of the solvent	NMR, Raman spectroscopy, EPR, fluorescence	Harmonic analysis, methods of molecular dynamics, elastic rod model
Microsecond	Flexural motions; local twisting and untwisting of the double helix; opening of the base pairs	$E \simeq 5 \div 20$ kcal/mol; the sources are changes of pH, temperature, effect of denaturing agents	Optical anisotropy, reaction with formaldehyde, hydrogen exchange, NMR, EPR	Methods of conformational mechanics, topological models
Millisecond	Dissociation of the double helix; changes in supertwisting; rotation of the molecule as a whole	$E \simeq 10 \div 50$ kcal/mol; the sources are interactions with proteins and other ligands	Reaction with formaldehyde, hydrogen exchange, NMR, quasi-elastic scattering of light, hydrodynamic methods	
Second	Isomerization; division of bacteria, motions with amplitudes $A = 2 \div 3\ \mu$m			

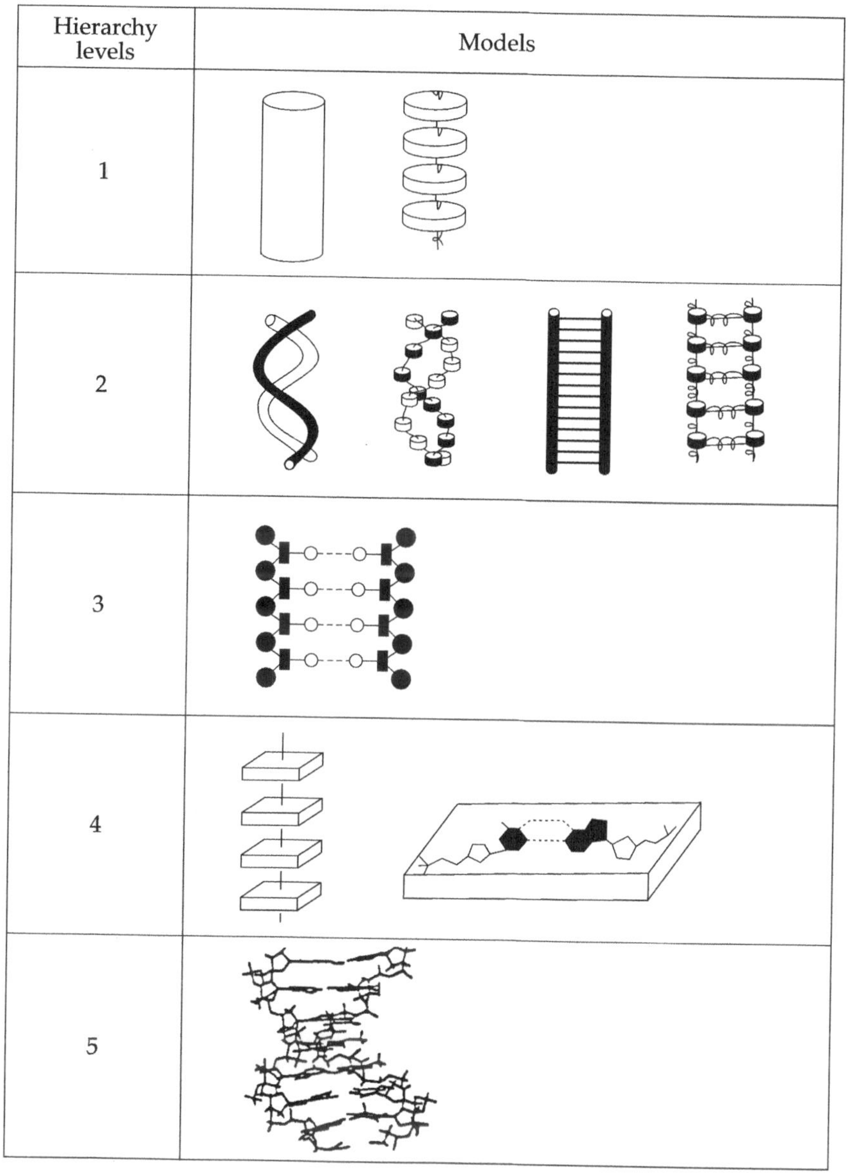

Figure 11.14. Hierarchy of dynamic models of DNA (reproduced from L. V. Yakushevitch, 2007).

motions of these groups of atoms are reproduced as motions of solids, weakly bound to each other. It is proposed that the bonds between separate subunits are weak and soft, while the bonds between atoms inside the subunits are strong and rigid.

The fourth level of the hierarchy describes internal motions in the "lattice" with the cell, formed by a bounded group of atoms (a pair of nucleotides) which recurs along the DNA molecule. In this approximation all displacements of nucleotide atoms are taken into account, but at the same time only homogeneous (homopolymer) chains are considered.

The models of the fifth level complete the hierarchy. They are models of molecular dynamics. These models reflect extremely precisely the structure and motions of the DNA. The complexity of their mathematical modeling is by an order of magnitude higher, and this modeling is possible only with the use of powerful computers.

Nonlinear Models of the Dynamics of DNA. A pioneering work in studying the internal dynamics of DNA was the research of S. W. Englander, N. R. Kallenbach, A. J. Heeger, J. A. Krumhansl, A. Litwin (1980). It consisted of two parts: experimental and theoretical ones. Using the method of hydrogen – tritium exchange the authors demonstrated in their experiment that open states may be supposedly formed in DNA. These states are determined as mobile local regions (with the length from one to several base pairs), inside of which the hydrogen bonds are disrupted (Fig. 11.15). The formation of such open states is associated with essential angular deviations of the bases from the equilibrium positions.

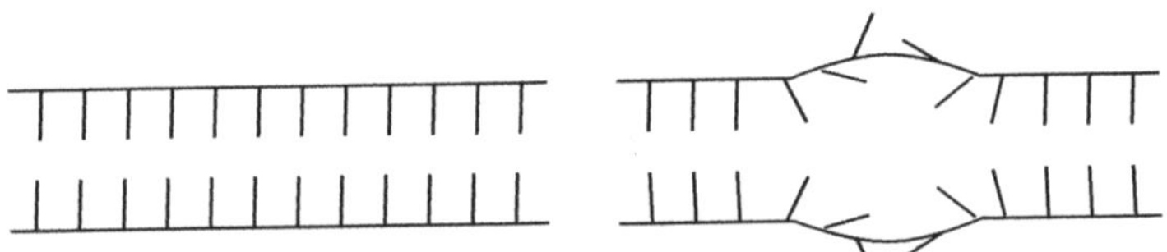

a *b*

Figure 11.15. Closed (*a*) and open (*b*) states of the double DNA chain.

Let us analyze in more detail the model of Englander and its modification (L. V. Yakushevitch, 2007) taking into account the peculiarities of DNA significant for explaining the mechanisms of DNA functioning.

A nonlinear model describes the processes of high amplitude rotational motions of the bases around the sugar-phosphate chains. These motions lead to the disruption of hydrogen bonds and formation of open states. In descriptions of the dynamic properties, the analogy between a DNA molecule and a chain of connected pendulums is used. The sugar-bound bases play the role of rotary pendulums in the DNA molecule, the sugar-phosphate chain plays the role of a horizontal chain, and the field induced by the second DNA filament weakly interacting with the first one across the hydrogen bonds between the bases plays the role of an external gravitation field (Fig. 11.16).

The dynamics of the chain of pendulums is well studied in mechanics and is described by a set of n nonlinear equations. For the n-th pendulum the equation looks like this

$$I\frac{d^2\varphi_n}{dt^2} = K(\varphi_{n+1} - 2\varphi_n + \varphi_{n-1}) - mgh\sin\varphi_n, \tag{11.34}$$

where φ_n is the angular deviation of the n-th pendulum from the equilibrium position, I is the moment of inertia of the pendulum, K is the coefficient of rigidity of the horizontal chain, m and h are the mass of the pendulum and its length respectively, g is the gravitation constant.

Let us consider the continuous approximation

$$z_n \to z, \quad \varphi_n(t) \to \varphi(z, t)$$

and take into account the analogy between the model of a chain of pendulums and DNA. Then the equation of rotational oscillations of DNA bases can be written

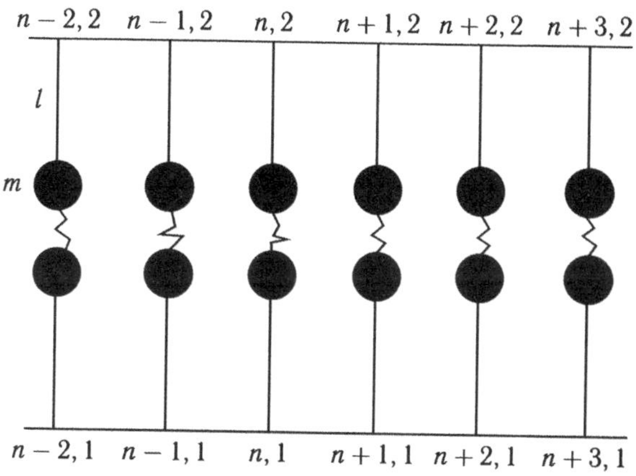

Figure 11.16. Mechanical analog of DNA used in the model of Englander. Two chains of pendulums are connected by springs.

as follows

$$I_0\varphi_{tt} - K_0\varphi_{zz} + V_0 \sin\varphi = 0. \tag{11.35}$$

Here I_0 is the moment of inertia of the base, K_0 is the coefficient of rigidity of the sugar-phosphate chain, $V_0 \sin\varphi$ is the force acting between the bases of base pairs.

This equation of the sine-Gordon type has an autowave solution of the "kink" type shown in Fig. 11.17.

A qualitative pattern corresponding to this solution is shown in Fig. 11.17, *b*. Two sugar-phosphate chains of DNA are represented here by two long lines and the bases by a variety of short lines. A local region with disrupted base pairs corresponds to the kink.

The described mode has a significant disadvantage — it takes into account the mobility of bases of only one of the two chains in the DNA. The second chain has a passive role of some average field. As a matter of fact, the both chains are

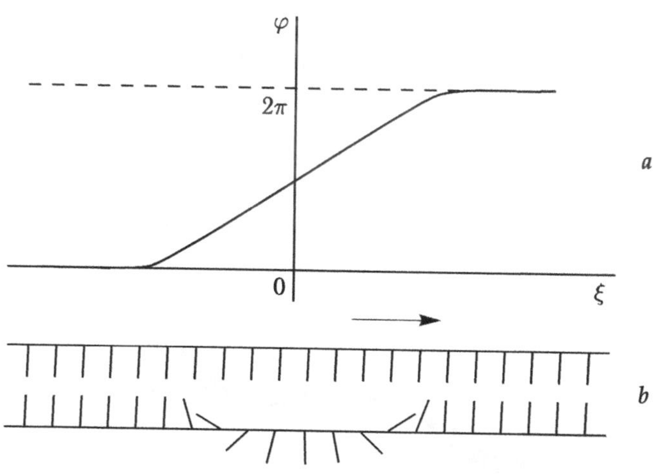

Figure 11.17. Solution of the sine-Gordon equation as a kink (*a*) and qualitative pattern corresponding to this solution (*b*) (reproduced from L. V. Yakushevitch, 2007).

equivalent. A mechanical analog is a model of two weakly bound elastic rods, and a discrete analog is the double chain of discs connected by longitudinal and transverse springs.

The solution for this model describes the local deformation (opening of base pairs), moving along the DNA molecule at a rate v. The model can be specified by taking into account the helical character of the DNA structure. In this case, instead of the straightened chain of pendulums oscillating in the "gravitation" field of the second chain, the chain of pendulums is considered folded around the axis in such a way that a helix can be formed. In the helical model, the suspension centers of the pendulums are positioned on the helix, and the pendulum threads are directed to the axis. Accordingly, in the equilibrium state the adjacent pendulums are turned to each other by an angle of 36°, and the direction of the "gravitation" field induced by the second chain, will also change upon transition from one pendulum to the other by an angle of 36° (Fig. 11.18).

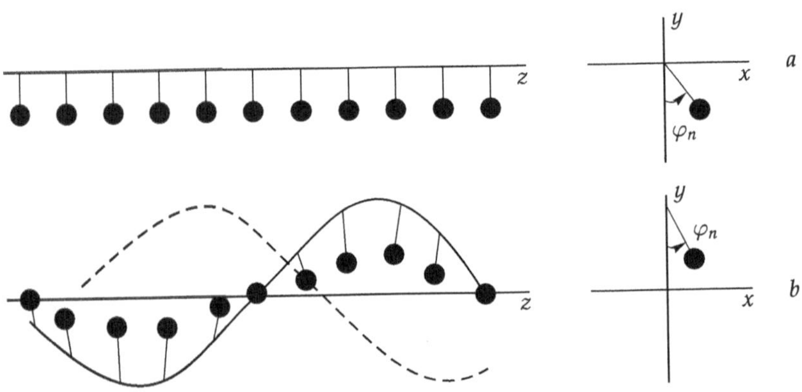

Figure 11.18. Mechanical DNA models: *a*, linear model; *b*, helical model (reproduced from L. V. Yakushevitch, 2007).

The above model allows a qualitative explanation of the effects of long-range interactions in the DNA molecule. The essence of these effects is clarified by the schemes in Figs. 11.19 and 11.20. They show a relatively long fragment of DNA with two specific sites (S_1 and S_2) and two protein molecules (P_1 and P_2). Numerous experiments demonstrate that the interaction of molecule P_1 and site S_1 can affect and change the character of interactions of molecule P_2 and site S_2, the distance between sites S_1 and S_2 therewith may be rather large (of about hundreds and even thousands of base pairs). It is because of this that these effects are called effects of long-range interaction. Within the framework of the considered model, the mechanism of this action may be as follows.

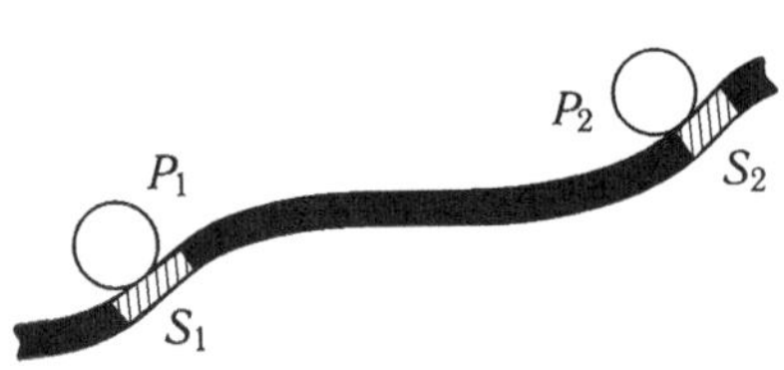

Figure 11.19. Schematic representation of DNA with isolated sites (S_1 and S_2) and two protein molecules (P_1 and P_2) (reproduced from L. V. Yakushevitch, 1998).

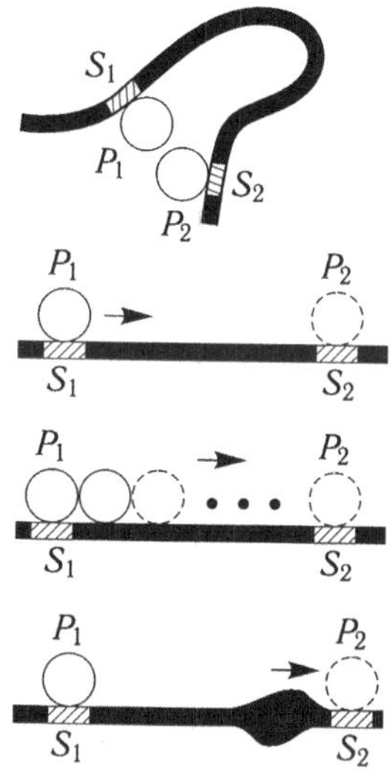

Figure 11.20. Models of interactions in DNA (reproduced from L. V. Yakushevitch, 1998).
a, The model explains the effect of long-range formation of a loop; *b*, the sliding model; *c*, the model of sequential binding; *d*, the model based on the possibility of propagation of conformational waves in DNA.

The binding of the molecule P_1 to the site S_1 leads to a local change in the vicinity of this site, and this local change (conformational wave) begins to propagate over the DNA. Having reached the site S_2, it will cause a change in the conformation of this site that in its turn will lead to changes in the coefficient of binding of the site S_2 with the molecule P_2.

Nonlinear Mechanisms of Regulation of Translation. Let us consider a significantly heterogeneous model of a fixed fragment of the DNA molecule, which contains basic functional units required for the DNA synthesis and its regulation: the promoter region P encoding region C, several regulatory regions R_1, R_2, R_3 and the terminal region T (Fig. 11.21).

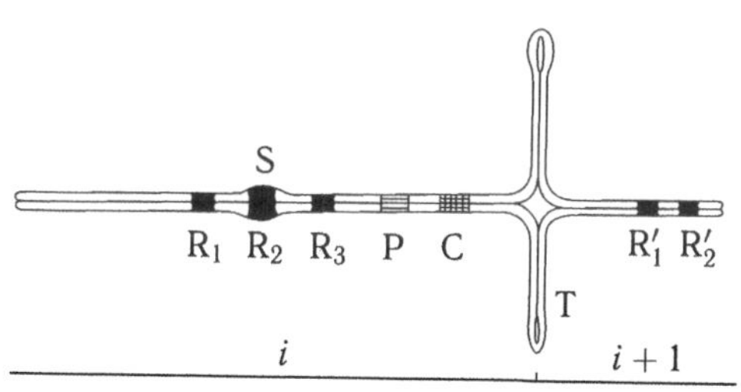

Figure 11.21. Schematic representation of a DNA fragment (reproduced from L. V. Yakushevitch, 1998).

At the first stage, the RNA-polymerase binds to promoter P, being accompanied by essential changes in the conformation of the binding region that can be transferred along the double DNA chain. This corresponds to the excitation and propagation over the DNA of the nonlinear conformational wave (the soliton). The excitation changes the conformation of the coding region C when advancing through it that leads to changes in the matrix properties of DNA and the rate of synthesis at the elongation stage. Within the framework of nonlinear dynamics, such excitation can be considered as a soliton, and the excitation motion represents the soliton propagation which resembles a kink.

Having passed the region C, the soliton reaches the terminator T which separates two genes — the i-th and the $(i+1)$-th. The experimental data show that this region has a heterogeneous structure. In Fig. 11.21, the region T looks like a cross. When advancing through this region, the wave may either be completely absorbed or overcome this barrier, but with the changed profile and rate. In the first case the result may be treated as termination, and in the second as a signal to initiate transcription in the second gene.

On the nonlinear model of Salerno (1992, 1994), it was shown that the direction of the initially stationary soliton depends on the sequence of bases in the vicinity of the starting point. It was demonstrated on the model of the $T7A_1$ promoter of the DNA molecule (plasmid B R322) that when the soliton was outside the promoter at the zero time, it also remained immobile later. But if the soliton was initially inside the promoter region, the wave started moving, the direction of its movement depending on the sequence of bases near the start point. These results show that the dynamically active regions of the promoter are in conformity with the functionally active regions.

11.6 Direct Modeling of Interactions of Proteins

The available models based on the method of Brownian dynamics make it possible to describe in detail the diffusion of a protein and its binding to another protein or a protein complex. However, for the description of processes occurring in subcellular compartments, containing many proteins and protein complexes, mobile carriers and hindrances for diffusion, this method is inapplicable.

This task can be solved using the method of a direct computer modeling of diffusion and interaction of proteins, based on the methods of Brownian dynamics.

The possibility of constructing direct models has appeared recently due to the advances in computer engineering, methods of object-oriented data bases and visualization methods.

In the method of direct computer modeling in virtual three-dimensional space, called a model scene, protein molecules are distributed. In the method of direct modeling, the law that governs in equations the mode of concentration changes in time (the law of mass action) is not given initially. Given are only the laws of diffusion of particles, their dimensions, forms and interaction of the two individual particles upon their collision. To describe the motion, the Langevin equation (11.8) is used which defines the change of each coordinate with time under the action of random and external forces:

$$\xi_x \frac{dx}{dt} = f_x(t) + F_x,$$

where x is the coordinate along which the motion is analyzed, ξ_x is the viscous friction factor along this coordinate, $f_x(t)$ and F_x are projections of random and electrostatic forces on the x axis, respectively. The random force $f_x(t)$ is distributed normally with the zero mean and dispersion, equal to

$$\frac{2k_B T \xi_x}{\Delta t}.$$

Here k_B is the Boltzmann constant, T is the temperature, Δt is the time step constant in this method. To estimate viscous friction factors, the protein molecule is approximated in the model by the ellipsoid of revolution rather than a sphere, as it is typically done in models of Brownian dynamics. With such an approach, the motion of each particle represents Brownian diffusion.

As an example, let us take the trajectory of diffusion of plastoquinone in a thylakoid membrane computed on a direct model (see Chapter 27), which contains complexes of photosystems 1 and 2 and the cytochrome complex. Visualization of the trajectories of motion of plastoquinone molecules shows that domains of plastoquinone diffusion (percolation clusters, Fig. 11.22) are formed in the membrane; their existence in the thylakoid membrane has been demonstrated experimentally. The coefficient of free diffusion of plastoquinone, calculated in a direct model with the assumption of free three-dimensional or two-dimensional motion, is an order of magnitude higher than the effective coefficient of diffusion of plastoquinone in the membrane filled with multienzyme complexes.

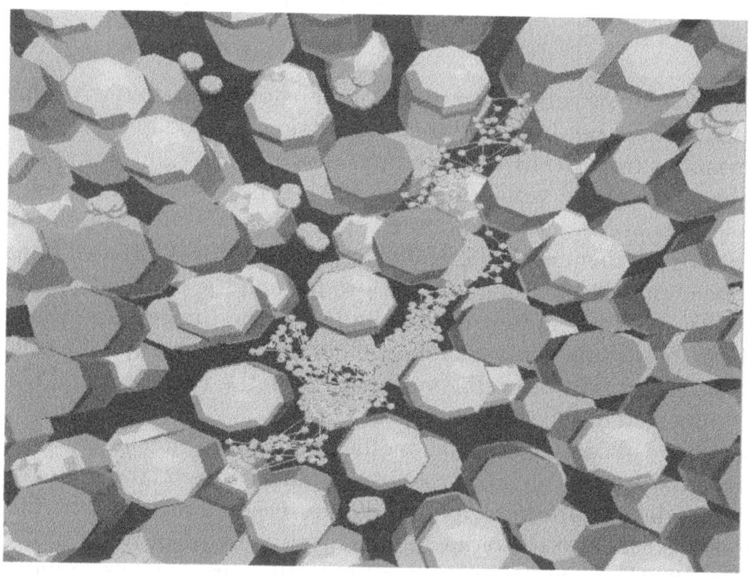

Figure 11.22. Visualization of the trajectory of the motion of a plastoquinone molecule in the membrane filled with complexes *PSI* and cyt bf (I. Kovalenko, G. Riznichenko, A. Rubin, 2002).

Diffusion of the other protein — plastocyanine in the lumen is greatly hindered by protruding parts of the transmembrane multienzyme complexes and other plastocyanine molecules, because in the native chloroplast the thickness of the lumen is comparable to the dimensions of plastocyanine.

Now let us analyze the interaction of proteins in the process of complex formation. This interaction can be tentatively divided into several stages: (1) Brownian diffusion of proteins to the docking site; (2) their approaching due to the action of electrostatic attractive forces between molecules, mutual orientation of molecules in space and formation of the preliminary complex; (3) formation of the final complex.

Consider the formation of a complex in a direct model. Upon diffusion of particles in a direct model, the two proteins may at random approach each other occupying an advantageous position for the binding (docking). This means that the distances between interacting parts of the molecules should be smaller than some distances (parameters of the model) or docking distances. At each time step, the attained distances between chosen amino acid residues are calculated. If at the given step these distances are smaller than the docking distances, the molecules can form a complex with the probability p or the docking probability, which is a parameter of the model. The random nature of the complex formation for the two proteins is realized in each definite case by comparing the initially set probability p of the complex formation with the number randomly chosen at the given time step. This comparison enables clarifying whether at the given time step a random event of complex formation took place.

The total number of complexes formed randomly at each step makes it possible to obtain a kinetic curve directly reflecting the time-dependent change in concentrations of initial agents (or formed complexes) during the reaction. The curve obtained in this way in the model may be interpreted within the framework of a mathematical model of bimolecular reaction initially based on the mass action law (MAL). Then, by approximating the kinetic curve obtained in the model based on the MAL, we get the rate constant of the second order reaction of complex formation consisting of the two proteins.

The mass action law is valid in the ideal solution approximation when the concentration of molecules is not high and their dimensions can be neglected. But in real systems, deviations from the MAL can occur.

Another example of the violation of the MAL conditions is observed when the reacting molecules are initially distributed non-uniformly. It is known that in the thylakoid chloroplast membrane the multyenzyme complexes ES1 are located only in the membrane regions protruding into the stroma, and the plastocyanine molecules interacting with them are uniformly distributed in the lumen. First plastocyanine molecules from the stromal regions of the membrane will enter into the reaction, and only later as a result of diffusion the molecules from the remote granal regions will be gradually involved in the reaction.

At the next stage of complex formation, the electrostatic interactions play a significant role upon the approaching of proteins. Partial charges on the proteins generate a heterogeneous electrostatic field. Its strength decreases noticeably with the distance from the protein surface and also due to the screening by polar water molecules and dissolved ions. If the given protein is far from other proteins, its motion is determined solely by Brownian diffusion. As soon as the protein comes close to the other proteins, its motion is influenced by the electrostatic field generated by the neighboring protein. As a result, a proper orientation for the complex formation can be achieved. As a rule, the electrostatic field of proteins is heterogeneous favors proper approaching and orientation of proteins.

The protein in the model can be represented as a region with a dielectric constant $\epsilon = 2$ and spatially distributed partial charges, while for the surrounding solution $\epsilon = 80$. Using the Poisson–Boltzmann equation, permitting us to take into account different values of dielectric permittivity of proteins and solution, it is possible to calculate the electrostatic field around each protein, generated by charges on the protein surface (Fig. 11.23).

The method for determining the protein binding regions and the structure of a preliminary complex, where electrostatic interactions of proteins upon their approaching are taken into account, makes it possible to simplify the task of the subsequent accurate modeling and prediction of the structure of the formed final

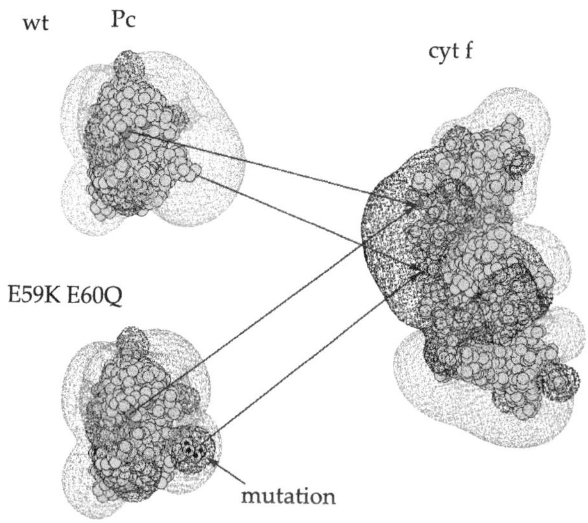

Figure 11.23. Equipotential surfaces −6.5 mV (light gray) and +6.5 mV (dark gray) of wild and mutant types of reduced plastocyanine and oxidized cytochrome f, calculated using the Poisson–Boltzmann equation; ionic force of 100 mM, $pH = 7.0$, $\epsilon_{\text{sol}} = 80$, $\epsilon_{\text{prot}} = 2$. The lines connect amino acid residues on the surface of plastocyanine and cytochrome f which are in direct contact with the complex (I. Kovalenko, G. Riznichenko, A. Rubin, 2002).

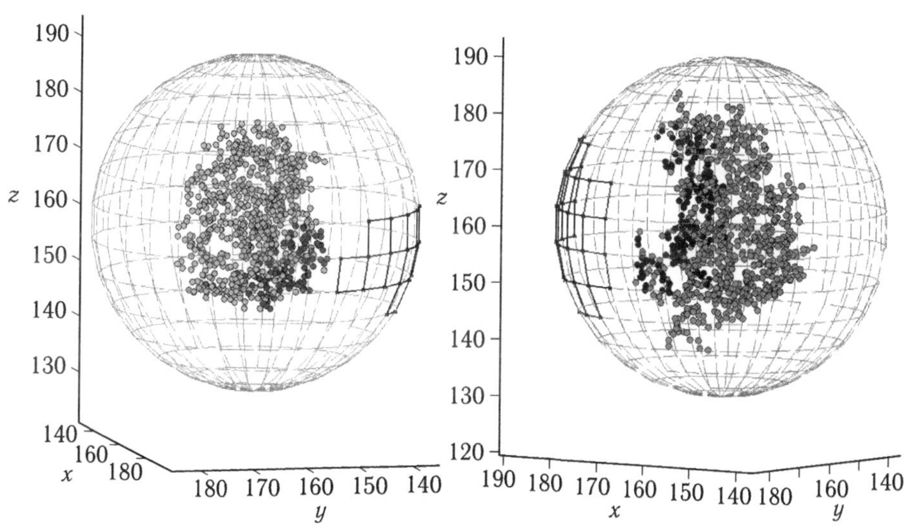

Figure 11.24. Representation of the probability of binding of barstar (left) and barnase (right) molecules, as a "globe" of probability distribution at the distance between the protein mass center ⩽ 40 E. Dark atoms belong to contacting amino acid residues of the protein complex barnase-barstar (according to the data of X-ray analysis). Dark cells correspond to protein regions with a high probability of binding (according to the calculations on the model) (I. Kovalenko, G. Riznichenko, A. Rubin, 2002).

complex. The set of mutual positions of the two proteins is separated into pair-wise nonintersecting subsets, corresponding to the interval of the rotation angle of 12° of one molecule relative to another molecule in the spherical coordinate system (Fig. 11.24). The sectors resulting upon such a rather detailed separation contain only one or two amino acid residues on the molecule surface. The task of the modeling is to find the sectors and corresponding amino acid residues which approach each other most frequently in the diffusion and electrostatic interactions of proteins. It means to estimate the probability of coming together of different amino acid residues and select those that correspond to the highest probability.

The figure shows the results of modeling for barnase and barstar molecules: proteins frequently approaching each other are dark gray in the corresponding sector and those rarely approaching each other are light gray. It is seen that the estimated regions with a high probability of binding (dark gray sectors) are opposite to the experimentally determined regions of binding (dark atoms), i.e. comply with the experimental data. In the model, the predicted structures of the preliminary complexes are formed as a result of electrostatic interactions and geometric complementarity of the binding regions. It is clear that in most cases analyzed the electrostatic interactions of proteins upon their approaching provide the highest probability for the mutual location of protein molecules favorable for their subsequent binding and formation of the preliminary complex which then turns into the final complex.

Accordingly, the possibility to study interactions of several protein molecules also permits modeling the formation of a large number of complexes, as this occurs in solution or a cell compartment, and observe the real kinetics of the process in time.

Further Reading

Part I

1) Eigen, M. and Schuster, P., *The Hypercycle. A Principle of Natural Self-Organization*, Springer, Berlin, 1979.
2) Peitgen, H.-O. and Richter, P.H., *The Beauty of Fractals*, Springer-Verlag, 1986.
3) Murray, J.D., *Mathematical Biology: I. An Introduction*, 3rd edition, Springer, 2002
4) Murray, J.D., *Mathematical Biology: II. Spatial Models and Biomedical Applications*, 3rd edition, Springer, 2003.

Part II

1) Prigogine, I. and Nicolis, G., *Self-Organization In Non-Equilibrium Systems*, Wiley, 1977.
2) Prigogine, I., *From Being to Becoming*, W.E. Freeman, San Francisco, 1980.
3) Haken, H., *Synergetics. Introduction and Advanced Topics*, Springer-Verlag, Berlin–Heidelberg, 2004.
4) Keplen, S.R. and Essig, E., *Bioenergetics and Linear Thermodynamics of Irreversible Processes*, N.Y., 1966.
5) Schuster, H.G., *Deterministic Chaos: An Introduction*, Physik, Weinheim, 1984.
6) Westerhoff, H.V. and Van Dam, K., *Thermodynamics and Control of Biological Free Energy Transduction*, Elsevier, Amsterdam, 1987.
7) Kondepudi, D. and Prigogine, I., *Modern Thermodynamics. From Heat Engines to Dissipative Structures*, John Wiley & Sons, 1998.

Part III

1) Cantor, C.R. and Schimmel, P.R., *Biophysical Chemistry*, W.H. Freeman and Company, San Francisco, 1980.
2) Schulz, G.E. and Schirmer, R.H., *Principles of protein structures*, Springer-Verlag, New York/Heidelberg/Berlin, 1979.
3) Serdyk, I.N., Zaccai, N.R. and Zaccai, J., *Methods in Molecular Biophysics: Structure, Dynamics, Function*, Cambridge University Press, Cambridge, 2007.

Part IV

1) De Gennes, P.G., *Scaling Concepts in Polymer Physics*, Cornell University Press, New York, 1979.
2) Cantor, C.R. and Schimmel, P.R., *Biophysical Chemistry*, W.H. Freeman and Company, San Francisco, 1980.
3) Hinchliffe, A., *Modeling Molecular Structures*, John Wiley & Sons, Ltd., Chichester, England, 2000.
4) Leach, A.R., *Molecular Modeling. Principles and Applications*, Pearson Education Limited, Edinburgh, England, 2001.
5) Serdyk, I.N., Zaccai, N.R. and Zaccai, J., *Methods in Molecular Biophysics: Structure, Dynamics, Function*, Cambridge University Press, Cambridge, 2007.
6) Jackson, M.B., *Molecular and Cellular Biophysics*, Cambridge University Press, Cambridge, 2006.

PART III

BIOPHYSICS OF MEMBRANE PROCESSES

V
Structure-functional Organization of Biological Membranes

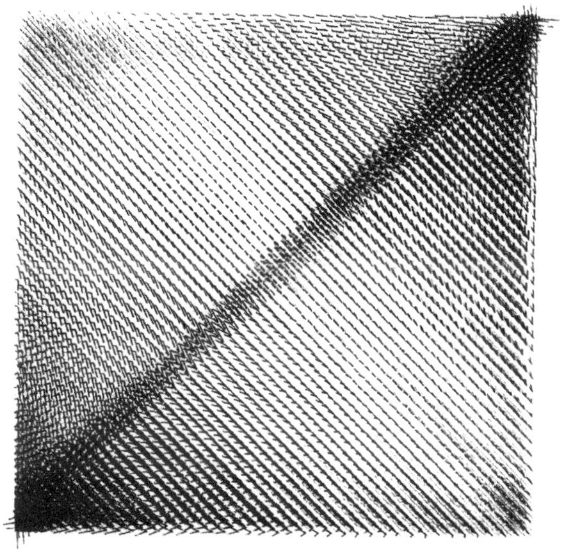

12

Molecular Organization
of Biological Membranes

13

Conformational Properties
of Membranes

12

Molecular Organization of Biological Membranes

12.1 Composition and Structure of Biological Membranes

Biological membranes are functional cell structures with the thickness of several molecular layers enclosing the cytoplasm and most of intracellular structures and forming a uniform intracellular system of small channels, folds and closed cavities. The thickness of biological membranes is occasionally more than 10.0 nm, but due to a relatively tight packing of their basic molecular components (proteins and lipids), as well as large overall area of cell membranes as a rule, they make more than half of the dry cell mass.

Biological membranes consist mostly of proteins, lipids and carbohydrates. Proteins and lipids make the most part of the mass of dry membranes. Usually, the portion of carbohydrates does not exceed 10–15 %, they being linked either to protein molecules (glycoproteins) or to lipid molecules (glycolipids). In membranes of different origin, the content of lipids varies from 25 to 75 % relative to protein mass (Table 12.1). □

Table 12.1. Content of Lipids in Mammalian Cell Membranes, % of the Mass of All Lipids

Lipids	Plasma membranes	Mito-chondrii	Lyso-somes	Nuclei	Endo-plasmic reticulum	Golgi apparatus
Phosphatidylcholine	18.5	37.5	23.0	44.0	48.0	24.5
Sphingomyelin	12.0	0	23.0	3.0	5.0	6.5
Phosphatidylethanol-amine	11.5	28.5	12.5	16.5	19.0	9.0
Phosphatidylserine	7.0	0	6.0	3.5	4.0	2.5
Phosphatidylinositol	3.0	2.5	6.0	6.0	7.5	5.0
Lysophosphatidylcholine	2.5	0	0	1.0	1.5	3.0
Diphosphatidyl glycerol	0	14.0	5.0	1.0	0	0
Other phopholipids	2.5	–	–	–	–	–
Cholesterol	19.5	–	14.0	10.0	5.5	7.5
Cholesterol esters	2.5	2.5	8.0	1.0	1.0	4.5
Free fatty acids	6.0	–	–	9.0	3.5	18.0
Other lipids	15.0	15.0	2.5	5.5	5.0	16.0

Biological membranes include lipids belonging mostly to three major classes: glycerophosphatides (phospholipids), sphingo- and glycolipids and also steroids.

Membrane lipids have a relatively small polar (charged) head and long uncharged (nonpolar) hydrocarbon chains. Polar heads of glycerophosphatides — phosphadylcholine, phsophatidylethanolamine and sphingomyelin — have positive and negative charges and at neutral pH are on the whole electrically neutral compounds (zwitterion lipids). Phosphatidylserine and phosphatidylinosite have one uncompensated negative charge each, while cardiolipin has two such charges.

The protein content of membranes is also exclusively diverse. With the rare exception, most membranes contain a large number of different proteins, their molecular mass varying from 10 to 240 kDa.

As regards their functioning, membrane proteins are subdivided in enzymatic, transport and regulatory ones. There are also structural proteins performing mainly the "supporting" functions. □

Investigations of the membrane structure are necessary for understanding their functioning. In 1935, J. F. Danielli and H. Davson proposed the first hypothesis on the structure of biological membranes, according to which a membrane consists of a double lipid layer, with both sides covered by layers of globular proteins.

At present, it is believed that proteins do not line the surface of the lipid layer of membranes but "float" on the surface as separate globular molecules or particles, more or less immersed in the membrane. This fluid mosaic model proposed by J. Lenard and S. J. Singer (1966) allows a reasonable explanation of a number of facts, in particular the dependence of many physiological functions of membranes and the activity of separate membrane enzymes on the phase state of lipids in a membrane, and its fluidity (viscosity). The later protein crystal model (G. Vanderkooi and

D. Green, 1970) differs from the fluid mosaic model only by postulating the existence of a rigid protein structure formed as a result of long-range protein–protein interactions in membranes.

Today various versions of the fluid mosaic model have become most popular. □

Interactions of polar and nonpolar groups of both the protein and lipid bilayer result in such peptide localization, where the maximally possible number of nonpolar amino acid residues are immersed in the bilayer. This obvious requirement determines the character of three main types of protein–membrane complexes including β-folded and α-helical structures having a different ratio of polar and nonpolar groups (Fig. 12.1).

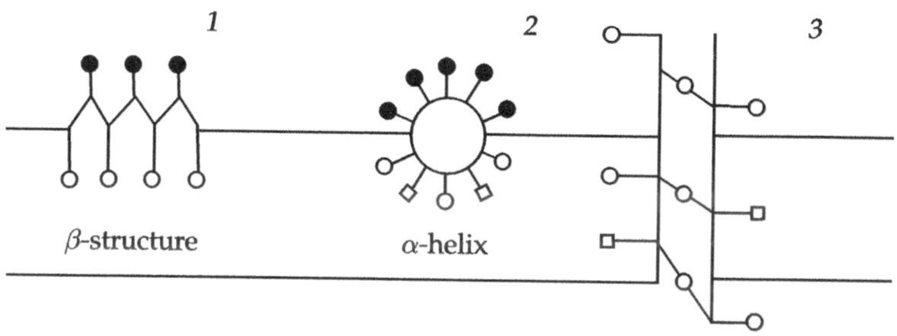

Figure 12.1. Three types of protein–membrane interactions.

1, Amphiphilic β-structure; *2*, amphiphilic α-helix; *3*, uniformly hydrophobic α-helix; black circles correspond to charged side chains; squares correspond to polar but uncharged side chains; white circles correspond to nonpolar side chains.

It is proposed that amphiphilic β-pleated sheets (type *1*) fold with the formation of pores for passive diffusion of substances across membraneS (see Chapter 14). Polar groups are located inside such a pore, while nonpolar groups are outside in contact with the bilayer.

Peptide hormones (Chapter 23) can form complexes (type *2*) with membranes. Type *3* structures are characteristic of integral membrane proteins, penetrating through membranes with their α-helices (for example, bacteriorhodopsins, Chapter 29; ionic channels, Chapter 18).

12.2 Formation of Membrane Structures

Monolayers. Liquid Crystals

Main membrane-forming lipids are compounds with an ideal combination of hydrophobic and hydrophilic properties. In a monomer form, they are relatively poorly water soluble, and the tendency of their polar heads to contact water provides their unique capacities to form variable relatively stable structures upon aggregation.
□

Insoluble Monolayers on the Water–Air Interface. Most of the membrane-forming lipids are not dissolved in the bulk water phase upon applying onto the water surface but rather promptly spread over the surface forming a monomolecular layer (I. Langmuir, 1933). Upon compression of a monolayer, intermolecular

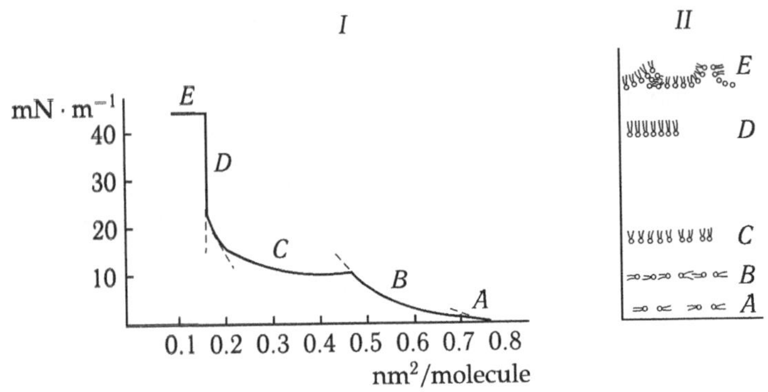

Figure 12.2. Area–pressure curve (isotherm) for a monolayer of long-chain lipids (*I*) and diagram of positions of lipid molecules on the boundary of the water–air interface (*II*).

A, State of two-dimensional gas; *B*, stretched liquid film; *C*, condensed liquid film; *D*, solid condensed state; *E*, state of collapse.

interactions are generated thus altering the state of ideal two-dimensional gas (Fig. 12.2, *A*).

Further increase in the pressure leads to the formation of the so-called stretched liquid film. In this state, molecules possess all the degrees of freedom as in the usual liquid, but their polar heads are fixed on the surface of the interface (Fig. 12.2, *B*). In this state, hydrocarbon chains of molecules demonstrate tendency to aggregation and mutual penetration.

The subsequent pressure increase will result in tighter packing of molecules, i.e. their vertical orientation in the monolayer. First the mobility of hydrocarbon chains will be higher than in polar heads, because the latter are strongly hydrated. In this state, the molecules retain the ability to move in the surface plane (lateral mobility) and the state can be regarded as a condensed liquid film (Fig. 12.2, *C*). Further condensation of the layer upon the pressure increase can occur only due to disruption of the hydrated shell of polar heads, following which the layer becomes practically incompressible and is in the solid condensed state (Fig. 12.2, *D*), where the area of the cross-section of molecules is independent of the length of the hydrocarbon chain. The area of the cross-section of the hydrocarbon chain is $\sim 0.2\ nm^2$, the area limit of saturated phosphatidylcholine molecules is $0.44\ nm^2$, and that of phosphtidylethanolamine is $\sim 0.4\ nm^2$. After exceeding some limiting pressure, the monolayer begins to break down (the state of collapse) (Fig. 12.2, *E*).

Artificial Bilayer Membranes

Multilamellar Liposomes. Planar lamellar structures have an inclination (at robust shaking) to form closed spherical particles. This process is energetically favorable, because in this case the minimum contact area of hydrophobic surfaces with water is attained. Simple mechanical actions lead to the formation of mainly multilayer particles (of several micrometers in diameter) with a concentric arrangement of the layers, i.e. multilamellar liposomes or vesicles (bags). Inside these particles, lipid bilayers separate the internal aqueous phase from the external solution. This arrangement allows the use of multilamellar liposomes for studying barrier properties of lipid bilayers and some other special tasks. Multilamellar liposomes are osmotically active: they change their volume upon changes in the osmotic properties of the external surrounding.

Proteoliposomes. Many membrane proteins and small fragments of biological membranes can be relatively easily included in the structure of artificial vesicular membranes. Such combined systems are called proteoliposomes. The efficiency of incorporation of most protein component in artificial membrane systems depends much on the lipid content of membranes, pH, salt composition, temperature etc. As a rule, the efficiency of incorporation of protein molecules or lipoprotein membrane fragments increases in the presence of small amounts of detergents. It is suggested that relatively small detergent molecules fill in the defects on the protein molecule boundary in the membrane, shielding the hydrophobic regions of the membrane from water.

The exclusive value of proteoliposomes as artificial membrane systems is determined first of all by wide possibilities provided by these systems for studying the functioning of many important components of biological membranes. In effect, the obtaining of proteoliposomes can be considered as the last stage preceding total reconstruction of functioning membrane systems from their components. □

Planar Bilayer Lipid Membranes. Lipids spontaneously forming lamellar layers can as a rule form bilayer structures (BLM or "black" films) on small apertures in thin hydrophobic materials. This phenomenon was first described by O. Mueller et al. (1962) who obtained BLM from brain phospholipids on small apertures (0.5–5.0 mm^2) in the Teflon partition separating two aqueous phases.

The general scheme of the cell for BLM formation and several sequential stages of the bilayer structure formation are shown in Fig. 12.3. The process of BLM formation begins with the application of a small amount of a proper lipid usually dissolved in liquid hydrocarbons (hexane, octane, decane etc.) on the aperture in a Teflon cup (Fig. 12.3, *I*). Then this thick film becomes spontaneously thinner down to the formation of a bilayer structure (Fig. 12.3, *III*).

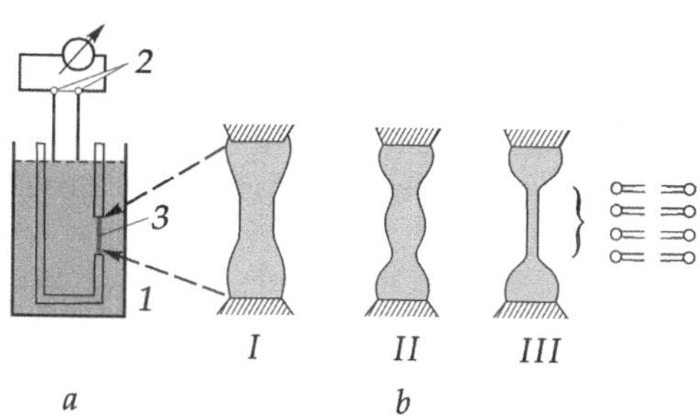

Figure 12.3. Formation of a bilayer lipid membrane (BLM).

a, Cell for obtaining BLM and studying its electric properties: *1*, Teflon glass; *2*, electrodes; *3*, a round hole in the wall of the Teflon glass.

b, Stages of the BLM formation (*I*, thick membrane; *II*, convex lens-like membrane; *III*, BLM).

The main forces determining the behavior of the film are the interphase (surface) tension σ and van-der-Waals forces of attraction between aqueous phases on the both sides of the film, squeezing it in the transverse direction — the disjoining pressure. □

Originally, the thickness of a lipid film exceeds 100 nm. At this thickness, the disjoining pressure is close to zero, and the behavior of the lipid mass is determined by the difference of the hydrostatic pressure in its peripheral curved part (torus) and the central relatively planar part. In accord with the law of Laplace, the difference in the hydrodynamic pressure Δp between the phase under the concave surface of the interface (in the torus) and the phase over this surface is determined by the expression

$$\Delta p = \sigma(1/R_1 + 1/R_2),$$

where R_1 and R_2 are internal and external radii of the surface curvature.

Apparently, in the central part of the film, where the radius of curvature is close to infinity, ($R_1 = R_2 \to \infty$), $\Delta p = 0$. This means that the hydrodynamic pressure in the torus is lower than in the planar part of the membrane, and therefore the liquid membrane-forming phase will be squeezed out from the central part in the radial direction. The rate of the generated flux will be maximal on the film periphery, in the regions adjacent of the torus, where the linear gradient of the hydrodynamic pressure is maximal. In accordance with this, the planar part of the film more swiftly becomes thinner along the periphery, forming a lens-like structure (see Fig. 12.3, *II*).

In the regions where the thickness of the film becomes thinner ~ 100 nm, the interactive forces of surface layers and aqueous phases interfere on both sides of the film. The combination of these forces either accelerates the film thinning — the negative disjoining pressure, or prevents this process — the positive disjoining pressure. The further thinning of the film causes either its disruption or abrupt formation of a stable, most frequently bilayer, structure. The formation of a bilayer structure is completed when is spreads over the whole area, directly contacting the torus. As a rule, the formation of a bilayer structure occurs for 5–20 min, but the duration of this process depends sharply on many parameters (the lipid composition, temperature, pH, ionic strength and composition of solutions and others, often unregulated parameters).

The problem of formation of closed vesicles, consisting of a lipid bilayer, in cells of organisms deserves special consideration. Apparently, this process in the cells, the same as in model systems, has the character of spontaneous self-organization. It is evidently not associated with the transport of lipids from the site of their biosynthesis and directed incorporation into the forming membrane. In any case, self-organization of the bilayer continues in the cell, where the processes of protein biosynthesis have been inhibited. However, it is not quite clear how far similar are the mechanisms of the bilayer self-organization in cells and in model systems. It is proposed that in both cases a common stage is self-organization of a monolayer and therewith the formation of large monolamellar vesicles, which later turn into multilamellar liposomes.

12.3 Thermodynamics of Membrane Formation and Stability

The formation of membranes and their stabilization are in many respects determined by surface phenomena and intermolecular interactions. Theoretical investigations of the properties of highly organized biological membranes are performed on hydrocarbon films and structures formed by lipids in aqueous solutions. □

Surface Activity, Work of Adsorption. The behavior of different compounds in solutions obeys the Gibbs adsorption equation:

$$-d\sigma = \sum_{i=1}^{m} \Gamma_i \, d\mu_i, \tag{12.1}$$

where σ is the interphase (surface) tension, Γ_i is the degree of adsorption of the i-th component in the surface layer, μ_i is the chemical potential of the i-th component.

As for diluted solutions

$$d\mu_1 \simeq RT \, d\ln a_1 \simeq RT \, d\ln c_1, \tag{12.2}$$

where a_1 and c_1 are the activity and concentration of the dissolved substance, the adsorption equation is as follows:

$$\Gamma_1 = -\frac{a_1 \, d\sigma}{RT \, da_1} \simeq -\frac{c_1 \, d\sigma}{RT \, dc_1}. \tag{12.3}$$

The $-(d\sigma/dc)_{c\to 0}$ value is called the surface activity. Surfactant substances are those which decrease the surface tension. For such substances, $d\sigma/dc < 0$ and $\Gamma > 0$. These compounds, well adsorbed in the surface layer and decreasing the surface tension, include also organic amphiphilic compounds, in particular, lipids.

Intermolecular Interactions in Thin Films and Membranes. Even a simple analysis of the action of factors, causing disintegration of thin hydrocarbon films and biological membranes, makes it possible to get a certain understanding of the peculiarities of different intermolecular (electrostatic and van-der-Waals) interactions forming these structures (see Chapter 8). In membrane systems, electrostatic interactions occur between anion lipids, amino- and SH-groups of amino acid residues of proteins (the positive charge), α-carboxyl groups of sialic acid (the negative charge) etc. Three types of electrostatic interactions are tentatively distinguished in membrane systems: lateral, or tangential, interaction of charged groups of molecules located in one half-layer of membranes; transmembrane interaction of charged groups located on different sides of the membrane; intermembrane interaction of charged groups located on the surface of two neighboring membranes.

By the moment, the theory of intermembrane electrostatic interactions has been developed further. The repulsive energy, for example, of erythrocytes turns to zero at distances exceeding 10 nm, and already at $\sim$ 2 nm it exceeds the energy of thermal motion by more than three orders of magnitude.

Dispersion forces are of greatest significance in van-der-Waals interactions in membranes (see Eq. (8.8)). They are predominant in the interaction of phases separated by the membrane.

When homogeneous cells having a not too high surface potential interact in solution, the dispersion component of the van-der-Waals interaction energy is displayed at larger distances than the electrostatic component (repulsion). This means that the total effect of such forces may lead to the emergence of the minimum of the cell interaction energy at distances between them from 3 to 8 nm. Hence it follows that under certain conditions, the cell suspension may contain cells, joined by mutual attraction. They should differ from other cells by a number of properties, including mobility. The existence of such cells is corroborated experimentally, for example, by the erythrocyte sedimentation or by the character of cell motion under the influence of the electric field. Naturally, the depth and width of the energy "well" is determined by a number of factors, including completely surface charges, pH of the medium, ionic strength etc. □

12.4 Mechanical Properties of Membranes

Mobility of Molecular Components in Membranes. The hydrophobic effect joining molecular components in membranes prevents their release to the aqueous phase outside the membrane. At the same time, the forces of intermolecular interaction do not usually impede the molecules in the membranes to exchange their locations, because in this case the contact area between water and hydrophobic regions of molecules does not change. As a consequence, molecular components in membrane systems retain individual mobility and can move within the membranes by force of diffusion. □

Lipid molecules most promptly perform rotations around their long axes. The correlation time of rotation motion τ_c (the time of rotation by an angle of 1 rad) of spin-labeled phospholipid molecules, sterols and fatty acids in different model and biological membranes in a liquid state, is $\sim 10^{-9}$ s. The rotational mobility remains relatively high ($\tau_c \sim 10^{-8}$ s) even at a temperature below the melting point of fatty acid chains of lipids in membranes.

The diffusional movement of lipid molecules along the layer, usually called lateral diffusion, also occurs very quickly. The lateral diffusion coefficient D of spin-labeled and fluorescent phospholipid probes in many synthetic and biological membranes is 10^{-7}–10^{-8} cm^2/s. The mean-square distance passed by the molecule along the membrane for time τ is

$$\bar{l} = \sqrt{4Dt}. \tag{12.4}$$

It can be readily computed that at $D = 6 \times 10^{-8}$ cm^2/s a lipid molecule moves along the membrane at a distance of about 5,000 nm for 1 s, i.e. can "run all over" the whole plasmatic membrane of such cells as erythrocyte, *E. coli* and others.

Another type of motion of lipid molecules in membrane systems is the transbilayer motion (the flip-flop-transition). It takes place in membranes at a relatively low rate because of a high barrier to be crossed by the lipid molecule polar head of the membrane hydrocarbon region. In model vesicular membranes, the rate of the flip-flop-transition is 10–20 h and higher, estimated by the transfer time of half of the numer of label molecules from one half-layer to the other. In biological membranes, this process can occur much faster. In the electric organs of the eel, this time makes 3–7 min, in erythrocyte membranes it is 20–30 min.

Protein molecules in membranes also have diffusional mobility. But due to the large mass, they move far slower. Accordingly, the correlation time of rhodopsin rotation in photoreceptor membranes is $\sim 10^{-6}$ s; for cytochromoxidase it is $\sim 10^{-4}$ s.

Lateral mobility of proteins as compared to that of lipids is also much lower. At room temperature, the coefficients of protein lateral diffusion in biological membranes are usually from 10^{-10} to 10^{-12} cm^2/s. Under physiological conditions, the membrane systems have properties of a two-dimensional liquid.

The currently obtained results of determination of membrane viscosity enable the conclusion that for normal functioning of most biological membranes the hydrocarbon area should be liquid-like, its viscosity being approximately equal to that of olive oil. □

Mechanisms of Lipid Bilayer Disruption. The above characteristics of elastic properties of membrane systems show that a membrane lipid bilayer is a relatively rigid structural formation. This is of great importance for maintaining the form of various cell structures and performing barrier functions by biological and synthetic membranes. However, it is known that the lifetime of a biological membrane is limited and depends on its composition and external conditions. Sometimes biological membranes have mechanical distortions and defects, these phenomena being conjugated with such substantial biological processes as cell fusion, lysis, secretion and hemolysis. □

At present, the most recognized mechanism of membrane disruption is related to the defects of the through pore type. As proposed, in this case the formation of a defect is accompanied by the reorientation of the lipid molecules located close to the defect boundary, with the formation of the so-called inverted pore. To give an approximate description, let us characterize the pore by radius r of an equivalent cylindrical defect. Energy E of the defect is equal to the work of generation of this defect minus the energy of the corresponding defect-free region of the membrane, i.e.

$$E = 2\pi\tilde{\sigma}r - \pi\sigma r^2, \tag{12.5}$$

where $\tilde{\sigma}$ is the linear tension of a length unity of the defect perimeter ($\tilde{\sigma} = h\sigma_n$); σ is the membrane tension; h is the membrane thickness; σ_n is the interphase tension of the "internal" surface of the defect (pore). □

This function has the shape of a curve with the maximum in point $r_{cr} = \tilde{\sigma}/\sigma$. Apparently the random defects at $r < r_{cr}$ will be "slammed" under the action of forces coupled to their formation. At the same time, a bilayer lipid membrane (BLM) may be disrupted by the emergence of only one defect with the supercritical value of radius $r > r_{cr}$.

Many biological membranes function in conditions when a high (25–300 mV) difference of electric potentials is applied to them (see Chapter 21) that sharply decreases the lifetime of BLM. However, a short-term action of an electric field on BLM increases the background conductivity and generates conductivity fluctuations (see Section 5 in Chapter 18). This shows the possibility of formation of elementary channels caused by the action of the fields.

Breakdown as a Consequence of Local Defects. The most complete experimental and theoretical analysis of an electrical breakdown of BLM was done in the studies of Yu. A. Chizmadzhev. It was found that the average lifetime $\bar{t}$ of BLM in an electric field decreases with an increase in the difference of the potentials on the membrane (Fig. 12.4). This demonstrates that the term "breakdown potential" φ_{cr}, used in the electromechanical model, has no clear sense and cannot be used as a parameter of membrane stability in the electric field.

Let us consider how the membrane energy will change upon the formation of a cylindrical pore of radius r. To calculate the energy of the defect, it is necessary to take into account the work, associated with the change in the interface (the increase

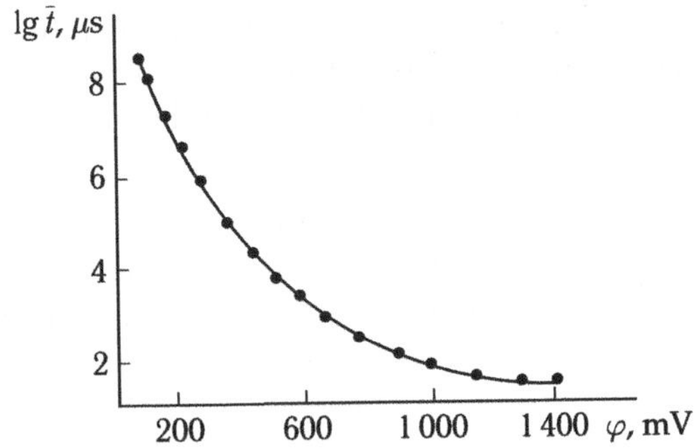

Figure 12.4. Dependence of the average lifetime $\bar{t}$ of BLM composed of lipids in decane on the difference of potentials φ on the membrane.

in the membrane–solution interface due to the formation of a side surface of the cylinder and the decrease in the interface due to the decrease in the number of regions corresponding to the cylinder ends), and in addition to estimate the change in the energy of the capacitor in the pore volume associated with the substitution of the broken dielectric (lipids with dielectric permittivity $\varepsilon_m = 2 \div 3$) for water ($\varepsilon_s \sim 80$).

The change in the membrane energy upon the formation t of a cylindrical pore is described by the following equation

$$E = 2\pi r\widetilde{\sigma} - \pi r^2\sigma - \Delta C\varphi^2/2, \tag{12.6}$$

where $\widetilde{\sigma}$ is the linear tension of the unit of the defect perimeter length ($\widetilde{\sigma} = h\sigma_n$); h is the membrane thickness; r is the pore radius; σ is the surface tension, i.e. the work required for the formation of 1 cm^2 lipid bilayer surface; ΔC is the change in the electric capacity of the membrane upon the emergence of a structural defect. In Eq. (12.6), the first two members correspond to the surface energy of the defect, and the last one reflects the change in the membrane electric energy upon the emergence of a through pore. The change in the electric capacity of the membrane upon the emergence of a pore can re represented as

$$\Delta C = \pi r^2 C_0(\varepsilon_s/\varepsilon_m - 1), \tag{12.7}$$

where C_0 is the capacity of the unit of area of the membrane. Consequently,

$$E = 2\pi\widetilde{\sigma} r - \pi r^2(\sigma + C\varphi^2/2), \tag{12.8}$$

where $C = C_0(\varepsilon_s/\varepsilon_m - 1)$. The dependence of the system energy on the defect radius described by Eq. (12.8) is presented by a curve with a maximum (Fig. 12.5). The figure shows that the defects of the small radius have a tendency to vanish. But the defects with the radius exceeding some critical value r_{cr} will irreversibly increase, thus resulting in the disruption of the membrane. The potential difference applied to the membrane leads to a decrease in the critical radius of the pore and the maximal energy value (the height of the energy barrier). Namely, this explains the increased probability of the membrane disruption in the electric field.

The dependence of the critical defect radius, corresponding to the maximum point on curve $E(r)$, is determined from condition $dE/dr = 0$ and is described

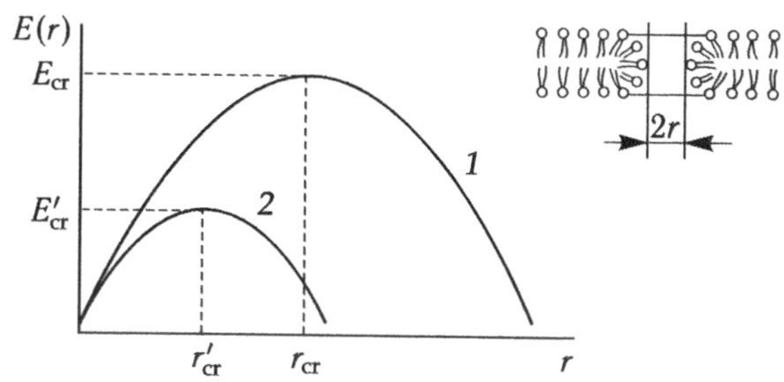

Figure 12.5. Dependence of energy $E(r)$ of the defect in a membrane on its radius r in the absence of electric field and upon application of potential difference φ (reproduced from V. F. Antonov, 1982).

1, $\varphi = 0$; 2, $\varphi > 0$; top right, representation of a pore in a bilayer.

by the relation

$$r_0 = \frac{\widetilde{\sigma}}{\sigma + C\varphi^2/2}. \tag{12.9}$$

By substituting (12.9) in (12.8) we find the dependence of the maximal value of energy E^* on the potential:

$$E^* = \frac{\pi\widetilde{\sigma}^2}{\sigma + C\varphi^2/2}. \tag{12.10}$$

Thus, under the action of the electric field the energy of formation of critical barrier E^* decreases, as a result of this the probability of the formation of the supercritical effect increases. With the emergence of this defect, the membrane disrupts spontaneously, because the increase in the size of the defect is accompanied by a decrease in the free energy of the system. The stability of the lipid bilayer is determined by the critical radius of the pore (Fig. 12.5). The larger value of the energy barrier corresponds to the larger critical radius of the pore. Destabilization of the membrane as a result of the phase transition of lipids or the electric breakdown goes in parallel with the decrease of the barrier. In this case, the decrease in the critical radius can lead to the situation when the existing pores will be on the descending branch of the curve (Fig. 12.5), thereby leading to unlimited growth of the pore and finally to disruption of the membrane. The magnitude of the linear tension of the pore perimeter plays a great role in stabilization of the membrane. The growth of the linear tension of the pore (from $5 \cdot 10^{-11}$ H to $6 \cdot 10^{-11}$ H) is accompanied by a sharp increase in the lifetime of the membrane (from 0.1 to $4 \cdot 10^6$ s), which is in good agreement with the average lifetime of the lipid bilayer of biological membranes.

Defects Identical to Through Pores upon Phase Transition of Lipids. In contrast to protein and peptide pores, lipid pores in biological membranes appear as a result of elastic deformations of the lipid matrix, the reasons for that being quite different. They are phase transitions in the lipid bilayer induced by the action of temperatures or chemical substances (ions of divalent metals) as well as electric and magnetic fields. In their turn, the pores generated in this way determine stability of the membranes. Formation of hydrophilic pores in bilayer lipid membranes upon phase transition from the liquid crystal state to gel was found in the laboratory of V. F. Antonov. The stress state of a lipid bilayer can be achieved by a change of temperature in the range of the main phase transition of the lipid forming the bilayer. The peculiarity of this process was first of all in that the generation of transmembrane pores occurred when the voltage on the membrane did not exceed 100 mV, which is relatively far from the voltage of the electric breakdown and corresponds to physiological values of membrane potentials (Section 1 of Chapter 19). The "freezing" of the lipid bilayer at physiological temperatures (23 °C for dimyristoyl phosphatidylcholine (DMPC) and 42 °C for dipalmytoyl phosphatidylcholine (DPPC)) is accompanied by a considerable decrease in the constant of lateral diffusion of lipid molecules and microviscosity of the lipid bilayer that makes it possible to drastically slow down the process of pore flowing and record reliably the stages of pore generation by the electric current fluctuations. Figure 12.6 shows current fluctuations corresponding to the conductance of hydrophilic pores in the bilayer membrane of distearoyl phosphatidylcholine (DSPC) at the phase

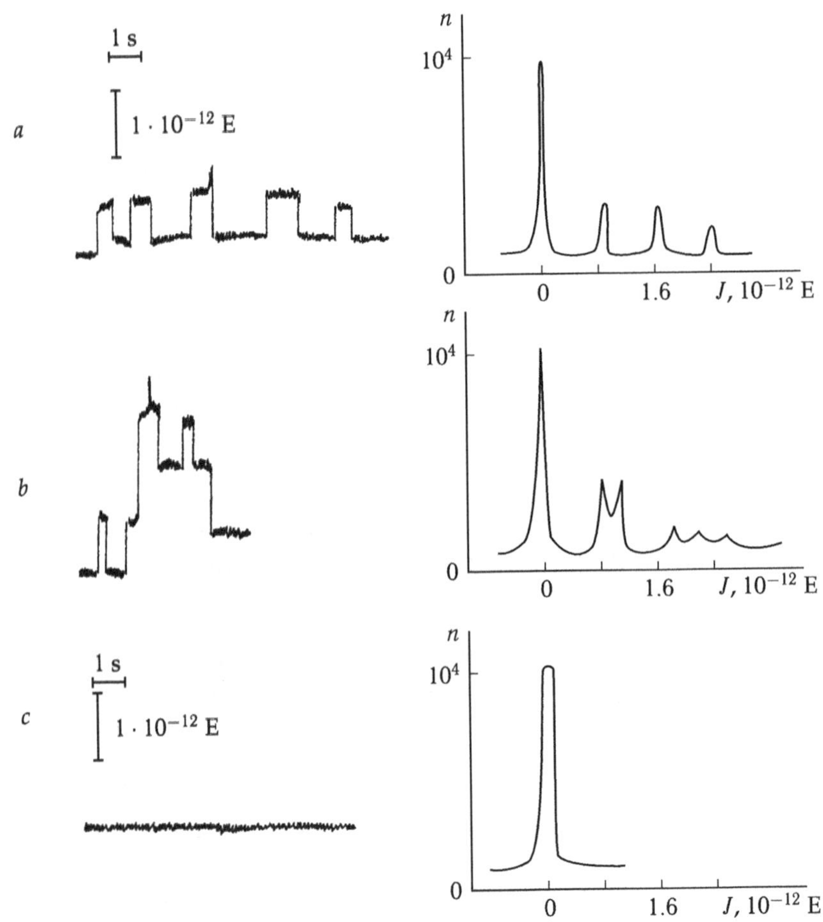

Figure 12.6. Current fluctuations across the bilayer lipid membrane.

Right, frequency distribution of currents plotted using a signal analyzer. The voltage on the membrane is 100 Mv.

a, The membrane is formed of 0.2% DSPC solution at 59 °C; *b*, current fluctuations in the gramicidin channel; *c*, bilayer membrane of egg lecithin in n-decane at 22 °C in the presence of 10^{-9} M gramicidin (reproduced from V. F. Antonov, 1980).

transition temperature (*a*). Current fluctuations in the gramicidin channel (*b*) are shown below for the comparison. As seen, in contract to peptide pores, lipid pores vary greatly in their dimensions — the histogram reveals at least 4 types of pores. Current fluctuations cannot be recorded in a liquid crystal state (*c*). The essential variability of lipid channels complies with the concept of random migration of pores.

Generation of pores in the lipid bilayer is caused by the decrease in its elastic energy as a result of freezing and emergence of local voids. The reason for generation of voids may be a change in the area due to the head of the lipid molecule which ensues from gosh-trans-transition (from 58 to 48 E^2 for DPPC).

The decrease in the elastic energy of the bilayer as a result of pore generation is compensated by the growth of the edge energy of the pore proportional to the pore perimeter. The increase in the pore radius suggests the appearance of a local energy minimum that is evidence of the energetically expedient structure of the membrane with the pore. A family of curves, corresponding to different dimensions of the cooperative unit, is shown in Fig. 12.7. As seen the first local minimum, corresponding to the stable pore, appears at the radius of 1.2 nm. The pore radius grows in parallel with the increase of the cooperative unit and reaches 2.5 nm. It is of interest that pores of such a radius are evidently adequate for the release of large drug molecules from liposomes and hemoglobin molecules from erythrocytes upon osmotic hemolysis.

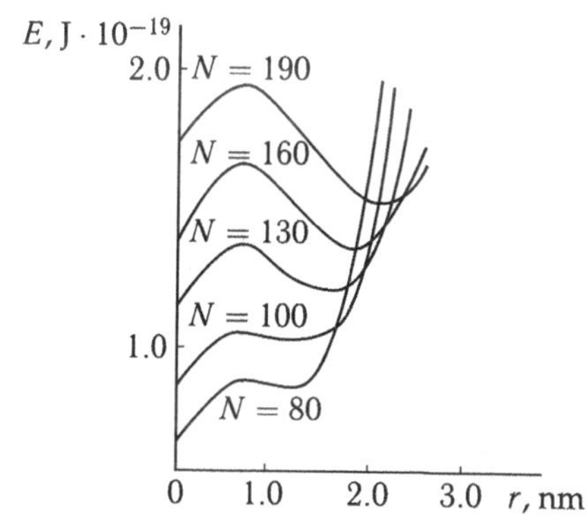

Figure 12.7. Dependence of changes in the Gibbs energy on the pore radius. N is the number of molecules, $\Delta a = 0.1$ nm^2 (reproduced from V. F. Antonov, 1992).

Initially, hydrophobic pores are formed spontaneously because of lateral heat fluctuations of lipid molecules. But when the radius value exceeds some critical value, reorientation of lipids with the formation of a hydrophilic pore becomes energetically more favorable. The heads of lipid molecules diverge so that hydrocarbon tails become accessible for the aqueous medium. This process may be called hydrophobic inversion of the pore. In this state the hydrophilic pore is a long-living hydrophobic pore filled with water. The pore size may change that is significant for its permeability. For example, the molecular dynamics modeling demonstrates that when the radius is less than 0.45 nm, the pore is impermeable for water molecules. However already at the radius of 0.65 nm, water and ions penetrate into it.

Figure 12.8 demonstrates a scheme of phospholipid pore evolution upon a phase transition of lipids from the liquid-crystalline state to gel retaining the bilayer liquid state. It is not excluded that the phase separation of lipids in the membrane can provide for enzyme concentration in liquid-crystalline domains, while the transport

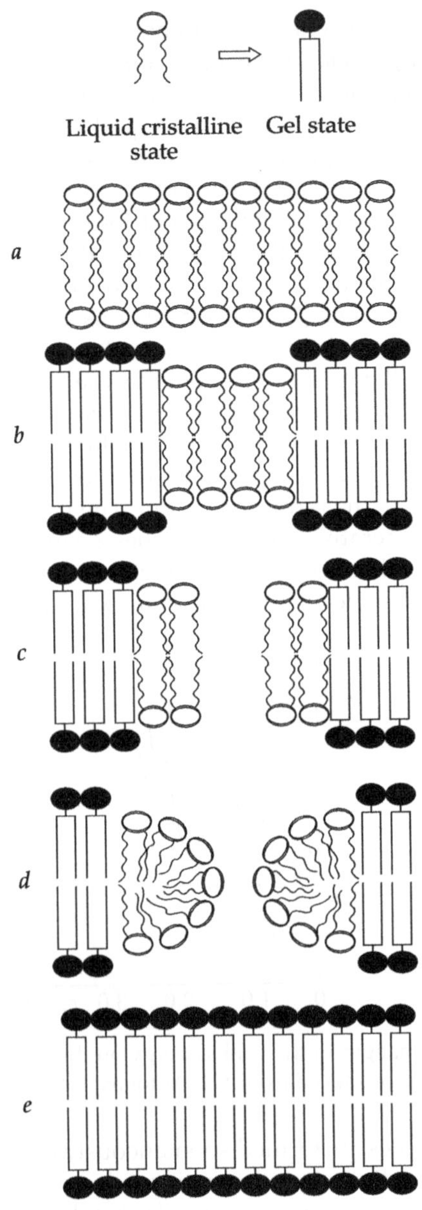

Figure 12.8. Scheme of evolution of a phospholipid pore upon phase transition from liquid-crystalline state to gel.

a, Double lipid bilayer in the liquid-crystalline state; *b*, coexistence of two phases in the bilayer; *c*, formation of a water-filled hydrophobic pore; *d*, inversion of the hydrophobic pore to the hydrophilic one with reorientation of phospholipid molecules at the pore edge; *e*, lipid bilayer in the gel-like state.

processes occur in gel-like domains or on their boundaries. This is probably associated with the existence of domain structures — rafts consisting of cholesterol, sphingomyelin, and some transport proteins in an ordered gel-like state under standard physiological conditions.

12.5 Effect of Electric Fields on Cells

Electric fields comprise a normal factor in the functioning of most biological membranes. At the same time, high voltage electric fields cause generation of qualitatively new phenomena. It was found in early studies on *Chara* cells that hyperpolarization of the cell membrane to some critical potential value increases greatly the transmembrane current — the phenomenon analogous to the electric breakdown of dielectrics. But in the case of cell membranes, the breakdown was completely reversible: upon repolarization of the cell the low conductivity of the membrane recovered, and the electric breakdown phenomenon itself could be observed repeatedly. The distortion of the membrane stability in strong electric fields was studied in detail on model systems — bilayer lipid membranes (see Section 4).

Distortions of the cell membrane permeability and cell death were detected upon the action of high voltage pulsed electric fields (pulses of durations from several microseconds to several milliseconds at the field voltage of several kV/cm) on cell suspensions. Cell lysis induced by an electric field was demonstrated for bacteria, yeasts, erythrocytes and protoplasts. Distortions of barrier properties of membranes are explained by the induction of the transmembrane potential up to 1 V. An accurate control of parameters of electrical treatment of cells makes it possible to induce a reversible increase in permeability of cell membranes (electroporation) in order to introduce biologically active substances and alien genes into the cells. It opens wide possibilities for the application of electric fields in molecular-genetic and biotechnological studies. Alongside with the membrane poration, electric fields can induce cell fusion, and this can be used for generation of heterocaryons and hybrids. Moreover, the action of electric fields can activate membrane proteins, stimulate cell processes, cause morphological changes such as formation of cell processes or local spherical swellings of the membrane (blebs) on the cell surface. That is why the effect of electric fields on living cells has become the subject of numerous studies.

Electroporation. As a rule, short pulses of a rectangular shape are used in studying electroporation. To calculate the transmembrane electric potential induced by the external electric field in a spherical cell, the Maxwell equation is used

$$\Delta\varphi = 1.5RE\cos\Theta, \tag{12.11}$$

where R is the cell radius, E is the voltage of the external electric field, Θ is the angle between the orientation of the electric field and the radius vector. The uneven distribution of the induced potential over the cell surface was corroborated in experiments with the use of potential-sensitive dyes.

After the action of the electric pulse resulting in the formation of pores in the membrane, permeable for ions and sucrose, the cells swell, as a rule, for a period from several seconds to several minutes. Subsequent to electroporation due to the formation of through pores in the membrane, the salt and sucrose concentrations in the cell and the medium rapidly equalize. However, the membrane remains impenetrable for cytoplasm macromolecules, which support the excess osmotic pressure in the cell. This enables water movement into the cell accompanied by its swelling. Thus, the cell swelling after electroporation has a colloid-osmotic nature.

Electrofusion of Membranes. This phenomenon is evidently closely associated with electroporation of membranes. Some hypotheses are based on relatively simple physicochemical models and the theory of colloid stability. As a matter of fact, the membrane fusion between contacting cells may be compared to coagulation of hydrophobic colloidal particles. In the existing stability theory developed by B. V. Deryagin in collaboration with L. D. Landau and independently by E. Fareway and J. Overback, the process of adhesion is considered as a result of the joint action of van-der-Waals attractive forces and electrostatic repulsive forces between the particles having the surface charge (cf. Sections 2 and 3). Depending on the balance of forces in the fine water film separating particles, there emerges either positive pressure preventing the particle association or negative pressure causing the thinning of the water interlayer and formation of contacts between particles.

For intermembrane distances of about 1–10 nm, only the electrostatic repulsion and van-der-Waals attractive forces can be taken into account. The resultant force of membrane interaction is represented in Fig. 12.9.

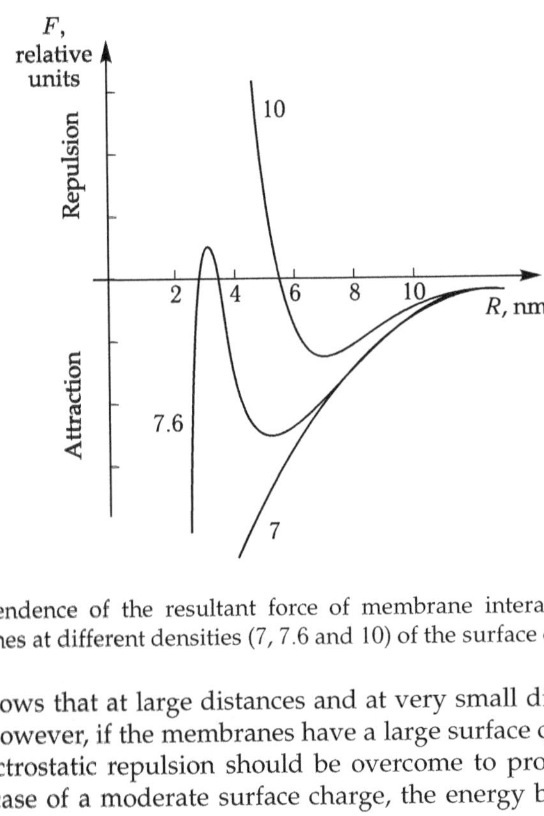

Figure 12.9. Dependence of the resultant force of membrane interaction on the distance between membranes at different densities (7, 7.6 and 10) of the surface charge.

The figure shows that at large distances and at very small distances attraction is predominant. However, if the membranes have a large surface charge, a very essential force of electrostatic repulsion should be overcome to provide the membrane joining. In the case of a moderate surface charge, the energy barrier of membrane fusion is surmountable.

The experiments performed on bilayer lipid membranes demonstrated that the fusion process in the given model system has a stage of formation of a bilayer membrane interlayer and the so-called trilaminar structure (Fig. 12.10). The initial contact between the membranes is formed due to the existence of the defects — local swellings. The probability of emergence of swellings is dependent on the molecular geometry of phospholipids and is the highest for phospholipids, where the dimensions of the hydrophobic tail of the molecule exceed those of the polar head. The flat bilayer, formed from cone-shaped molecules, is in a stress state and has defects in the form of swellings. A bridge is formed in a random contact zone of the defects (Fig. 12.10, *b*). After that, the inner monolayers move away from the contact area,

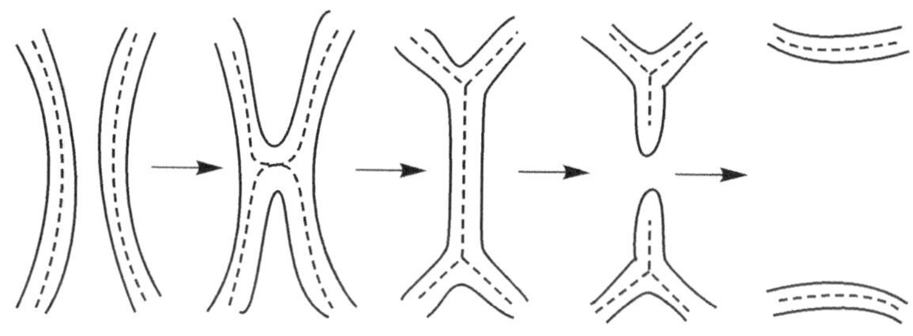

Figure 12.10. Putative mechanism of fusion of phospholipid membranes.
a, Plane-parallel contact; *b*, formation of a bridge between local swellings of closely located (inner) monolayers; *c*, trilaminar structure; *d*, formation of a through pore in the bilayer partition; *e*, formation of a tube (complete fusion).

and the outer monolayers form a bilayer partition (Fig. 12.10, *c*). Complete fusion of membranes occurs subsequent to the formation of a through pore in the lipid bilayer. The probability of pore formation is the highest for phospholipids having the shape of an inverted cone (a large polar head and a narrow hydrophobic tail).

The mechanism of electrofusion of unbroken cells is represented as follows. Under the action of the external electric field, coaxial pores are formed with the current flows through them. The electric force lines are distributed in such a way that pressure is generated bringing together the membranes and locking the pore edges. Upon locking the pore edges of contact membranes, a membrane tube connecting the cytoplasm of the fused cells is formed. The swelling of the cells caused by electroporation induces membrane tension, due to which the perimeter of the intercellular bridge increases and the overall mixing of the cell cytoplasm takes place.

Cell Motion. Two types of cell motion exist in electric fields. A constant field causes movement of cells having a surface charge (the phenomenon of electrophoresis). When a variable inhomogeneous field affects cell suspensions, cell motion called dielectrophoresis takes place. Upon dielectrophoresis, the surface charge of the cells plays no essential role; the mechanism of motion consists in the interaction of the induced dipole moment with the external field.

The difference between directions of the dielectrophoretic forces at low (kilohertz) and high (megahertz) frequencies is explained by different orientations of the induced dipole moment relative to the external electric field. From courses on electricity the orientations of dipole moments of weakly conductive dielectric particles in the conductive medium and that is of highly conductive particles surrounded by a low-conductive medium are well known. In the case of low-frequency field action, the membrane is a good insulator, and thus the current flows bypass the cells via the conductive medium. The induced charges are distributed as shown in Fig. 12.11, *a* and amplify the field strength inside the particle. This distribution of the induced charges corresponds to the antiparallel orientation of the particle dipole moment relative to the external field.

Under the action of the high-frequency field, the membrane conductivity is high; and since the electroconductivity of the intracellular medium exceeds the electroconductivity of the salt-free extracellular medium, the current flows predominantly across the cell (Fig. 12.11). Under these conditions, the distribution of induced

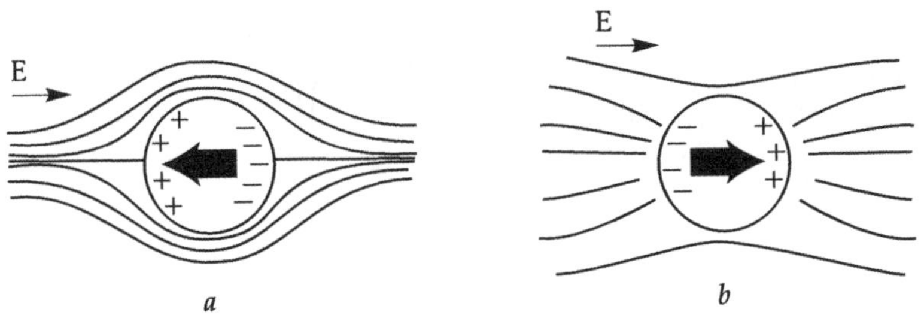

Figure 12.11. Distribution of induced charges on the particle surface and orientation of the induced dipole moment.

a, Non-conducting particle in the conductive medium (the cell in the low-frequency field); *b*, conducting particle in the weakly conductive medium. Vector **E** shows the orientation of the external electric field.

charges on the surface of particles corresponds to the parallel orientation of the dipole moment relative to the external field.

The forces acting on cell membranes in the electric field play a significant role in the approaching of the membranes upon electroinduced cell fusion. The mechanism of electroinduced fusion is evidently somewhat different from the mechanism of fusion of planar BLM (see above). According to one of the hypotheses, electrofusion begins from that stage of formation of coaxial pores in contact of membranes of adjacent cells (Fig. 12.12). In the area of coaxial pores, the electric force lines are positioned tangentially to the membrane surface and thus create the force which makes the membrane regions near the pore approach each other. As a result of action of these forces, the probability of locking the pore edges increases (Fig. 12.12, *b*). Subsequent to the formation of the through pore, the field directed tangentially to the pore edge tries to stretch the pore (Fig. 12.12, *c*) that results in complete fusion of the cells.

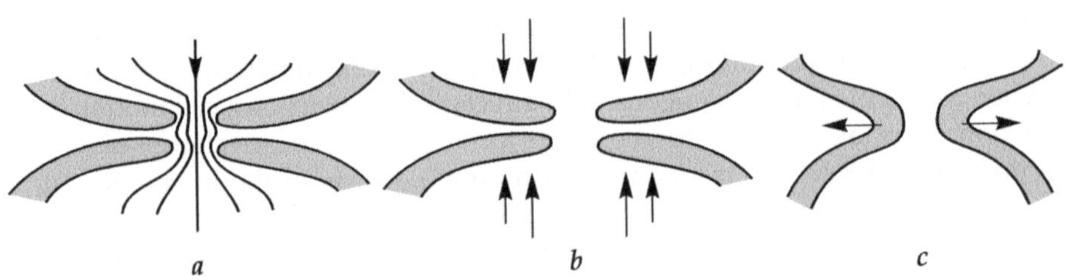

Figure 12.12. Mechanism of cell fusion.

a, Positions of electric force lines; *b*, forces acting on the membranes at pore edges, the length of arrows corresponds to the magnitude of the acting force; *c*, the field, directed tangentially to the surface of the membrane in the pore edge, stretches the pore and leads to the complete cell fusion.

13

Conformational Properties of Membranes

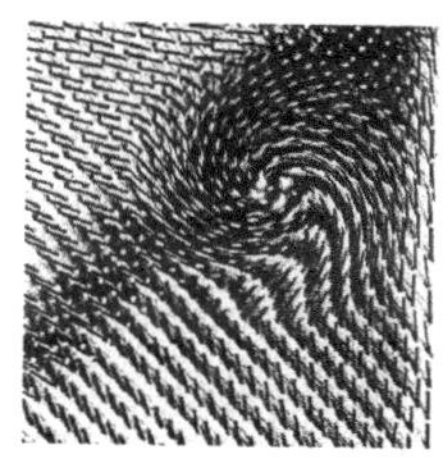

13.1 Phase Transitions in Membrane Systems

Biological and synthetic membrane systems can have two main phase states: a solid two-dimensional crystal (gel) or a liquid-crystalline (molten) state.

The temperature point of the basic phase transition, or parameter T_{pt} of the systems, formed from individual lipids, depends greatly on the length and degree of unsaturation of hydrogen chains of lipids. The temperature of the phase transition grows with the increase in the length of the chains and decreases with increased unsaturation of fatty acid residues.

Fatty acid chains of natural lipids usually contain unsaturated bonds. Therefore, in most biological membranes, the T_{pt} value is negative, i.e. under natural conditions the membranes are in the "liquid" state. The basic phase transition in these membranes occurs only when they are "frozen" in the aqueous phase.

The Loop (Kink) Hypothesis. The melting of fatty-acid chains upon phase transition is caused by rotational isomerism (see Section 3 in Chapter 8). The *trans*-configuration has the lowest energy and the *cys*-configuration has the highest energy.

The energy of *gosh*-conformations (*gosh* (+) and *gosh* (−), rotation by $\pm 120°$ relative to the *trans*-configuration) rather little exceeds that of the *trans*-configuration (by 2–3 kJ/mol), but these states are separated by an energy barrier of about 12–17 kJ/mol high. While hydrocarbon chains in a complete *trans*-configuration are linear structures, the emergence of a single *gosh*-conformation in the chain leads to bending of the spatial configuration of the chain by an angle of about 120° (see Fig. 8.4). This bending creates serious steric impediments in tightly packed membrane systems with a complete *trans*-conformation of hydrocarbon chains that inhibits the emergence of single *gosh*-conformations.

Upon melting of hydrocarbon chains in membranes, steric impediments diminish with synchronous emergence in the chain of two *gosh*-configurations (*gosh* (+) and *gosh* (−)) separated by the C−C bond in the *trans*-conformation. Though the energy of such a state is twice higher than that of a single *gosh*-conformation, the emerging rotational state of the chain does not cause extensive widening of the lattice, because upon consecutive rotations of the chain by +120° and −120° the spatial configuration of the chain remains rectangular on the whole (Fig. 13.1). The chain region which is in the *gosh* (+)-*trans*-*gosh* (−)-conformation forms a kink or a loop in the hydrocarbon chain. □

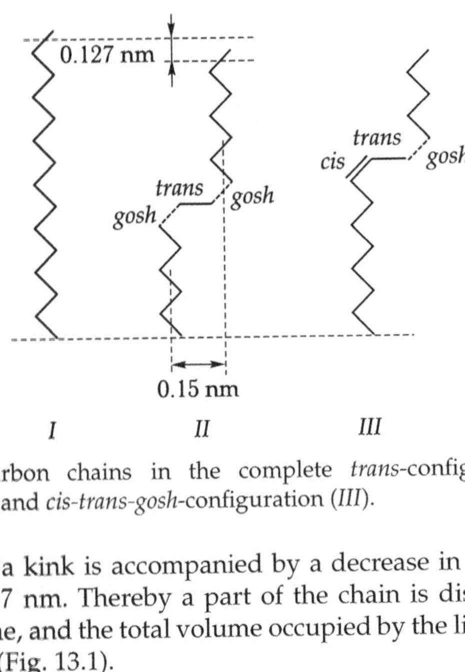

Figure 13.1. Hydrocarbon chains in the complete *trans*-configuration (*I*), *gosh-trans–gosh*-configuration (*II*), and *cis-trans-gosh*-configuration (*III*).

The formation of a kink is accompanied by a decrease in the effective length of the chain by ~ 0.127 nm. Thereby a part of the chain is displaced by ~ 0.15 nm forming a free volume, and the total volume occupied by the lipid molecule increases by 0.025–0.050 nm^3 (Fig. 13.1).

The formation of one kink in the hydrocarbon chain is insufficient for its melting, but single kinks facilitate the emergence of kinks in neighboring hydrocarbon chains, forming alternating kink blocks (Fig. 13.2). Such blocks can appear either in one half-layer of the membrane, or in two hydrocarbon loops located opposite to each other. With the growth of the number of kinks in hydrocarbon chains the disorder in the hydrocarbon area in the membranes increases drastically.

The low viscosity (high fluidity) of the hydrocarbon area in membranes in the liquid-crystalline state (at $T > T_{pt}$) is supposed to be associated with the increase in the amplitude of torsional oscillations around C−C-bonds, the emergence of *gosh*-conformations (kinks) and their rapid isomerization to neighboring positions. □

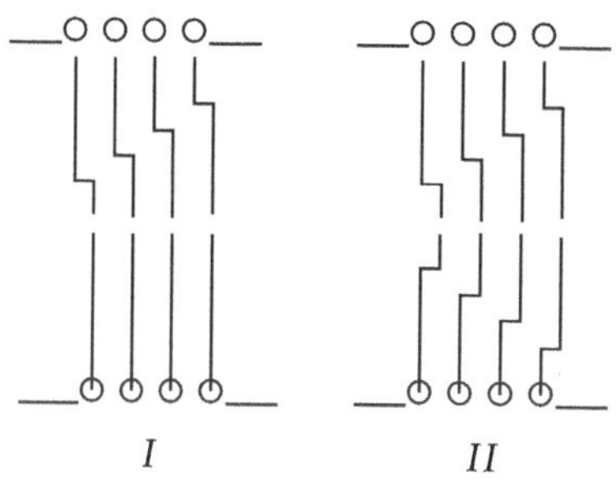

Figure 13.2. Kink-blocks in hydrocarbon chains of membranes.
I, in one half-layer; *II*, in two monolayers of the lipid bilayer.

Under conditions of high fluidity of the hydrophobic area of membranes, a kink can be isomerized, moving along the hydrocarbon chain because of the synchronous rotation by 120° of the corresponding sequence of the C–C-bonds. Upon displacement to an adjacent position the kink moves by ~ 0.13 nm.

Molecular Order Parameter. The high mobility of hydrocarbon chains in membranes allows the assessment of the dynamic conformation of these structures or of the most probable (average) form of the chains.

The molecular order parameter S_n is used to describe the dynamic deviation of the conformation of such chains from the normal to the surface:

$$S_n = \frac{3}{2}\,\overline{\cos^2 \beta_n} - \frac{1}{2}, \tag{13.1}$$

where β_n is the angle between the normal to the bilayer and the normal to the plane, formed by two vectors of the C–H-bonds in the n-th segment of the chains (Fig. 13.3). Apparently, at ideal order $S_n = 1$, and for the isotropic medium $S_n = 0$. □

Figure 13.4 shows how S_n changes for each subsequent deuterated link in palmitic acid incorporated biosynthetically into the *A. laidlawii* mycoplasma membranes. An analogous distribution of the order parameter S_n along the chain is observed also in bilayers of egg lecithin or some synthetic lipids.

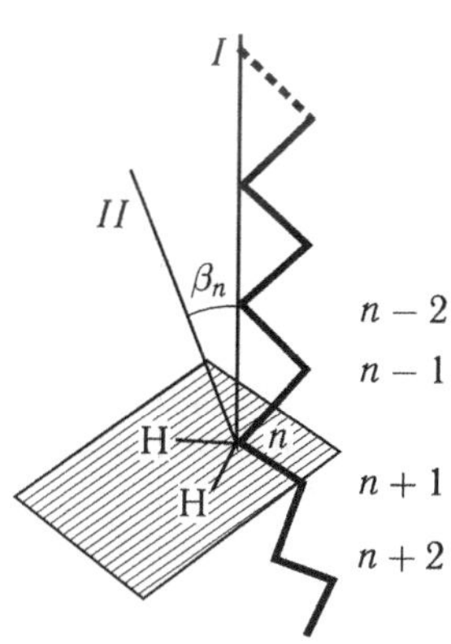

Figure 13.3. Determination of angle β between normal *I* to the membrane surface and normal *II* to the plane formed by vectors of the C–H-bonds in the n-th segment of the hydrocarbon chain.

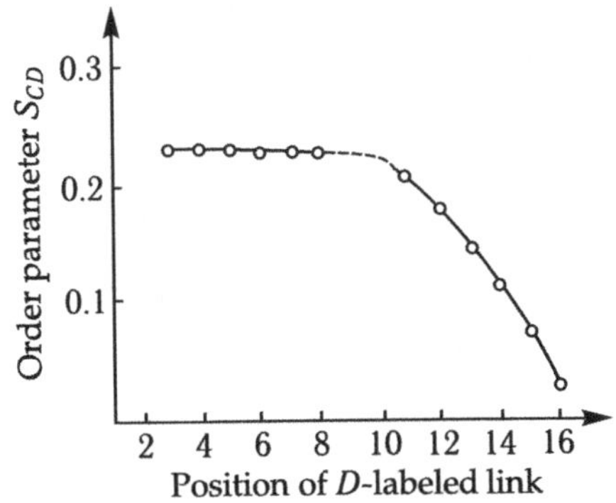

Figure 13.4. Dependence of the order parameter of the S_{CD}-bond on the position of the deuterated link in the palmytine chain incorporated biosynthetically into the *A. laidlawii* mycoplasma membranes at 42 °C ($S_{CD} = 0.5S_{CH}$).

As seen, after the 8-th–9-th methylene link, the disorder increases abruptly approaching the isotropic motion near the terminal methyl groups.

Thermodynamic Parameters of Phase Transitions. From the thermodynamic point of view, phase transitions in membrane systems can be regarded as transformations of aggregate states ($1 \leftrightarrow 2$).

Phase transitions in lipid bilayers occur in a relatively wide temperature range ($\Delta t \sim 0.2 \div 1.0\,°C$). This is conditioned by the fact that one phase in a bilayer is generated necessarily in the matrix of another phase ("liquid" or "solid" bilayers) with the formation of a large number of domains of a new phase of microscopic dimensions, but with a large specific surface area.

On the boundary of the domain, the molecules have an unusual packing that is reflected in the internal surface tension σ. Because of the changes in the system volume during phase transitions, mechanical forces χ will have an effect on the interface of the phases (domains). As a result, the liquid-crystalline phases of the domains will be affected by the compressive forces, and the gel volumes will be extended. Consequently, this generates an additional dependence of their energies (G_1 and G_2) on the internal surface tension and mechanical forces for the two phases (1 and 2) coexisting in the lipid bilayer:

$$G_1 = f(T, p, \sigma_{1,2}, \chi_{1,2}) \quad \text{and} \quad G_2 = f(T, p, \sigma_{2,1}, \chi_{2,1}).$$

Accordingly, the complex equilibrium between the phases is established in a relatively wide temperature range, and the steepness of the phase transition decreases, indicating to the decrease in the extent of cooperativity of transition (see Section 6 in Chapter 9). A sharp transition can occur only in the case of an infinite cooperative unit formed by interacting molecules that corresponds to the melting of an ideal crystal. In lipids of membrane structures, the number of molecules N joined in cooperative units does not exceed several hundreds.

Several approximate methods are proposed for the evaluation of the van't Hoff enthalpy of transition, for example, by the transition half-width $\Delta T_{1/2}$:

$$\Delta H_{\mathrm{VH}} \simeq \frac{7T_{\mathrm{pt}}^2}{\Delta T_{1/2}}. \tag{13.2}$$

To evaluate the extent of the process cooperativity, the calculated ΔH_{VH} value and the measured calorimetric heat of transition ΔQ are compared. In the absence of cooperativity $\Delta Q/\Delta H_{VH} = 1$, and for a cooperative process this value is much less than a unity. In general terms, the extent of cooperativity is determined by the relation $\Delta H_{VH} = N\Delta Q$ where N is the number of molecules in the cooperative unit. To characterize the extent of cooperativity of the process, the cooperativity parameter $\sigma = (\Delta Q/\Delta H_{VH})^2$ is also used. It is clear that the lower σ, the higher the cooperativity of the process. The relationship of the cooperativity parameter and the size of the cooperative unit can be found from the expression

$$N = 1/\sqrt{\sigma}. \quad \square \tag{13.3}$$

Theories of Phase Transition. Attempts to describe theoretically phase transitions of lipids in membranes have been made repeatedly. In general theories, attempts are made to describe the process completely and compute the temperature of the phase transition and its heat effect. Possible configurations of lipids determining the phase transition are found based on the analysis of the system energy representing the sum of energies of intramolecular and intermolecular interactions. In the simplest model proposed by J. F. Nigel (1973), the phase transition of a bilayer is considered in the framework of the theory of rotational isomerism (see Section 3 in Chapter 8) as an order–disorder transition with the assumption of an optional combination of *trans*- and *gosh*-conformations in each chain. Upon counting the number of possible configurations, it is suggested that the neighboring chains are in one plane and their hydrocarbon atoms can occupy the cells of a planar hexagonal lattice. Steric restrictions are taken into account by assuming that one cell of the lattice cannot contain more than one C-atom. The heat of transition calculated based on such a rough two-dimensional approximation was relatively close to the experimentally measured values.

The models of phase transition, taking into consideration a larger number of factors, are based on the molecular field theory with the use of the "order parameter"concept. In constructing these models, it is suggested that the configuration of a system of lipid molecules, the same as of nematic liquid crystals, is determined by the mutual orientation of adjacent molecules.

The effect of cholesterol and proteins on phase transitions was analyzed and the dependence of the order parameter of lipids in the nearest surrounding of protein and cholesterol molecules in membranes was studied. The results imply that lipids can form several molecular layers with a gradually changed order around the protein molecules.

13.2 Lipid–Lipid Interactions in Membranes

Inhomogeneity of Membrane Composition, Lipid Rafts. Under real conditions, not only biological but frequently also synthetic membranes are remarkably inhomogeneous in their lipid and protein composition. Specific interactions of membrane components take place due to the inhomogeneity of the lipid composition.

In 1992 D. Brown and J. Rose put forward a concept of lipid rafts augmenting the heterogeneous fluid mosaic model of the membrane composition. The membrane regions consisting of cholesterol and sphingolipids and including up to 10–15 accompanying protein molecules were called lipid rafts.

Hydration of Lipids. Among different factors determining the state of lipids in membranes, of great importance are electrostatic forces of attraction and repulsion between charged polar heads, steric factors, taking into account the shape of lipid molecules and the character of location of their heads and hydrophobic hydrocarbon tails, "the hydration force" as well as hydrogen bonds between lipid heads.

The nature of repulsive hydration forces is not of an electrostatic character but is revealed against the background of Coulomb interactions, increasing sharply non-exponentially at short distances. For example, upon approximation of dihexadecyl dimethylaminoacetate bilayers this occurs at a distance of ~ 11 E between the surfaces (V. Parsedjan, 1989). But an addition of Ca^{2+} ions to the system can induce their interaction with polar groups, due to which hydration repulsion is altered and, as a result, the bilayers stick together in a structure having no water in its composition.

Hydration of lipids depends on their nature and in many respects determines their physical properties. As a rule, lesser hydration is observed in lipids with donor and acceptor groups, involved in the formation of hydrogen bonds. Their decreased hydration is caused by the participation of the lipid polar head groups in the formation of hydrogen bonds between each other rather than with the surrounding water molecules (J. M. Boggs, 1987).

The release of water molecules from the bilayer surface, accompanying this process, results in the growth of the system entropy. The latter contributes to the compensation of energy consumed in the disruption of hydrogen bonds between lipids and water. Such hydrogen bonds are easily disrupted and formed again between other lipids for periods of ~ 10^{-11}–10^{-12} s. The unified system of labile hydrogen bonds enhances cooperative properties of the system and, in particular, raises temperature T of phase transitions from gel to liquid crystal contrary to the

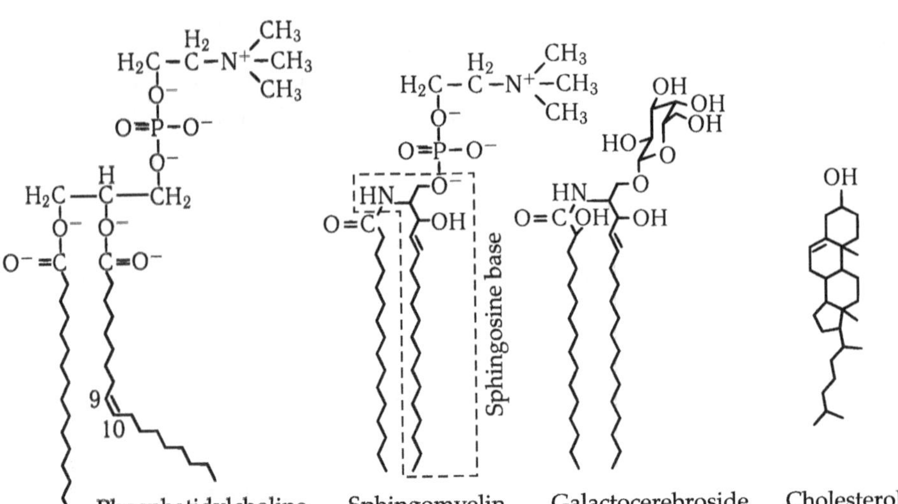

Figure 13.5. Example of the structure of phospholipids (phospatidylcholines), sphingolipids (sphingomyelin and galactocerebroside) and cholesterol.

The scheme shows the *cys*-conformation of the double bond of fatty acid in the structure of phosphatidylcholine and the *trans*-conformation of the double bond in the structure of sphingolipids. The sphingosine base is enclosed by a dashed line.

destabilizing effect of repulsive electrostatic forces of polar heads that in contrast decreases T.

Generation of Lipid Rafts. Let us consider a typical mechanism of lipid raft formation, associated with the difference in the geometry of membrane lipids and with the interlipid hydrogen bonds. Double bonds in fatty acids of most *phospholipids* are, as a rule, in the *cys*-conformation, whereas double bonds in fatty acids of *sphingolipids* are in the *trans*-conformation. Because of such a conformation, sphingolipids can have tighter and higher ordered packing. The tight packing of many sphingolipids stabilizes the hydrogen bonds between OH-groups of the sphingosine base and α-OH-groups of lipid-based fatty acid (see Fig. 13.5). Accordingly, at physiological temperatures these lipids are in the highly ordered gel-like phase. On the contrary, the *cys*-conformation of double bonds in most phospholipids prevents their tight packing, and thus at physiological temperatures they are in a relatively inadequately ordered liquid crystalline phase.

Cholesterol, another important component of lipid rafts, interacts predominantly with sphingolipids, facilitating segregation of membrane lipids into structured microdomains with sphingolipids and unstructured microdomains with usual phospholipids. It is believed that cholesterol is also located on the boundary between sphingolipids and phospholipids, decreasing the energy of interaction of the two phases. The interaction of cholesterol and sphingolipids changes the phase of the raft from the gel-like one to the liquid crystalline (see Fig. 13.6).

13.3 Lipid – Protein and Protein – Protein Interactions in Membranes

Lipid – Protein Interactions. These interactions are based on intermolecular dispersive and electrostatic forces, hydrogen bonds or other effects of binding. Lipid-protein interactions and phenomena caused by them are tentatively classified as follows: protein – lipid monolayer interactions, protein – lipid bilayer interactions, and lipid – protein interactions in membranes including lipid-depending enzymes.

The lipid dependence of the activity of membrane enzymes can be clearly displayed under selective extraction of membrane lipids and at subsequent addition

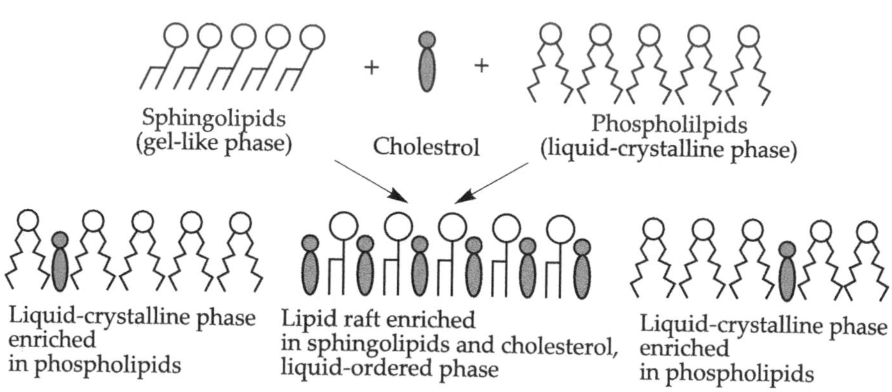

Figure 13.6. Phase separation in membrane lipids.

Cholesterol facilitates the formation of lipid rafts and stabilizes them in the liquid ordered state.

of definite lipids to delipidized membranes. Moreover, during the recent 5–8 years it has been demonstrated that the activity of membrane proteins often depends on the presence of sphingolipids and cholesterol in the membrane and increases with the growing viscosity of the membrane. In most cases, such proteins are continuously or briefly localized in lipid rafts, and the growth of the protein activity is evidently facilitated by conformational rearrangements, induced by their interaction with the lipids. Mild extraction of cholesterol results in a decrease in the activity of such proteins or its complete loss.

Protein – Protein Interactions occur in membranes as a reversible intramembrane aggregation of membrane proteins, frequently accompanied by a change in the functional and enzyme activity of the system. For example, protein intramembrane particles reversibly aggregating at pH < 5.5 are uniformly distributed in erythrocyte membranes. The aggregation is sensitive to the structure of the aqueous phase; it is inhibited at an increasing concentration of electrolytes and low pH values. The intramembrane aggregation of protein particles in erythrocytes correlates with the change in the distribution of surface receptors.

The theory of "membrane microdomain structure" describing compartmentalization in the structure of the functioning cell membrane has become widespread. According to this theory, oligomer structuring of lipid domains and protein clusters of the membrane makes it possible to give a more comprehensive explanation of the mechanism of the cell signal (CS) transfer as compared with the mechanisms based on the role of individual molecules. Thus a link with lipid rafts is known for two signal systems of the cell: receptor tyrosine kinases and receptors mediated by proteins G. The role of lipid rafts is not the same for different types of receptors.

It is regarded that in most cases the localization of the receptor, helper proteins and effector enzymes in one lipid raft facilitates and accelerates the process of signal transfer from the stage of receptor binding with the hormone to the direct action or generation of secondary messengers such as inositol-3-phosphate.

By now it has been clarified that cycles of aggregation – disaggregation of proteins in cell membranes are a common phenomenon manifested upon pinocytosis, at some stages of cell cycle, upon interaction and fusion of membranes etc. It is supposed that aggregation-associated interactions may be based on the forces of electrostatic character or more complex interactions, mediated by the peculiarities of the lipid surrounding of proteins, as well as the local crystallization of lipids in membranes.

To study the processes of diffusion and formation of a complex of two proteins, the method of Brownian dynamics is effectively exploited. The Brownian dynamics approach is used to describe the motion of protein molecules, each molecule being considered as a separate rigid body performing chaotic motion under the action of the solvent molecules (Chapter 11).

The interaction of two proteins upon formation of a temporary complex can be tentatively divided into several stages: (1) Brownian diffusion of proteins in the space of the corresponding compartments; (2) their approaching due to the action of electrostatic forces of attraction between the molecules, mutual orientation of molecules in space and formation of a preliminary complex; (3) formation of the final complex; (4) performing a function (for example, electron transfer) in the complex.

If proteins are considered as solids, the kinetics of their interaction can be described in the framework of a simple model, where two protein molecules collide due to diffusion. If they have collided in an appropriate mutual orientation, they form

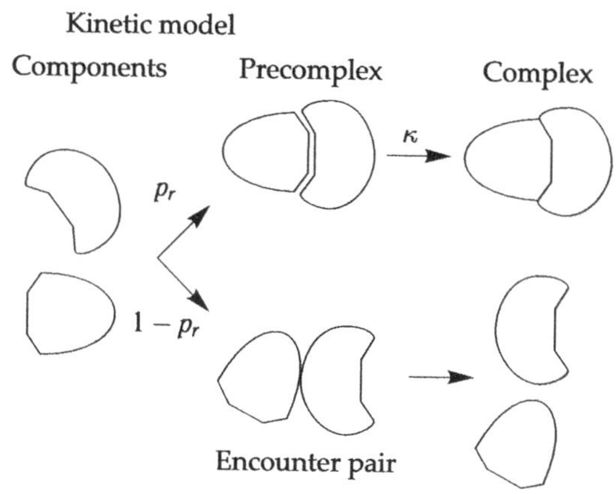

Figure 13.7. Kinetic model of the protein–protein interaction.

Components of the complex collide chaotically thus leading to the formation of a short-lived random protein–protein pair. With low probability p_r their mutual orientation appears to be suitable for the subsequent complex formation. Then they form a preliminary complex — a transient state upon binding which turns into a stable complex with high probability κ.

a preliminary complex — a transient state which rapidly turns into a stable complex. Otherwise the collided pair immediately dissociates (Fig. 13.7). The scheme has a simple biomolecular stage and one transient state after which only very fast events take place: water molecules leave the protein–protein interface and the short-lived interactions optimize the structure of the formed complex.

Provided the protein is located far from other proteins, at first the motion is determined only by the free Brownian diffusion. When the proteins approach each other, the total electrostatic charge of each of them manifests its effect. For example, barnase and barstar proteins have a common charge of an opposite sign that increases the rate of their approaching. With further approaching, non-homogeneities in the electrostatic field become meaningful, the proteins are oriented in the electrostatic field, even more increasing the rate of preliminary complex formation. If the electrostatic interaction does not occur (for example, at high ionic strengths), the magnitude of the bimolecular binding constant is approximately equal to $k_a = 10^5\ M^{-1}s^{-1}$. The electrostatic interaction at average ionic forces (50–100 mM) can increase the binding rate by three orders of magnitude (for the barnase–barstar pair $k_a = 10^8\ M^{-1}s^{-1}$). Protein plastocyanine and *cytochrome f* molecules both have generl negative common electric charges. However the binding site on *cytochrome f* has a region of a positive potential that facilitates an appropriate spatial orientation of *cytochrome f* favorable for the formation of the preliminary complex. This leads to the increase in the rate constant by ten-fold of the binding with plasticyanin at average ionic strength values as compared to the rate constant at high ionic strengths, when the electrostatic field is completely screened.

13.4 Peroxidation of Biomembrane Lipids

The process of peroxidation of lipids is observed upon regular functioning of biomembranes. It becomes more intensive at pathological states of organisms.

An important role in studying the mechanism of this phenomenon belongs to the Bach–Engler theory of peroxide oxidation (1897), the Semenov theory of branched chain reactions (1934), as well as to the works of Prof. N. M. Emanuel's school on liquid-phase oxidation of hydrocarbons and the works of Prof. B. N. Tarusov's school on the function of chain free radical oxidation of lipids in cell damage.

The process of lipid peroxidation begins with the formation of free radical products as a result of disruption of the C–H- or C–C-bonds in the hydrocarbon molecule. It occurs more readily in hydrocarbons with unsaturated double bonds. The lowest disruption energy is inherent to the CH-bond that is adjacent to the double bond in the α-position. But even in this case, more than 300 kJ/mol is needs to break the bond. That is why the overall rate of lipid peroxidation is determined primarily by the initiation stage of the free radical process.

Lipid peroxide ROOH is formed in the reaction of the free radical intermediate, namely alkyl radical $R^{\cdot}$ with oxygen. Firstly, peroxide radical $ROO^{\cdot}$ is formed, and then via its interaction with substrate molecule RH the peroxide molecule emerges:

$$R^{\cdot} + O_2 \rightarrow ROO^{\cdot} + RH \rightarrow ROOH + R^{\cdot}.$$

Reactions, in the course of which the original hydrocarbon radical R is regenerated from the nonoxidized substrate molecule RH, are called chain reactions. The number of cycles that the free radical center manages to complete prior to its destruction regulates the chain length.

Activated Forms of Oxygen. In an oxygen molecule, two of the twelve electrons are unpaired. Due to this the oxygen molecule has the total spin 1 and the properties of a biradical. It is paramagnetic and its basic state is triplet $^3O_2$ ($\uparrow\uparrow$). In the triplet state $^3\Sigma_g$, oxygen readily reacts with iron ions and other paramagnetic radicals. At the same time, molecular oxygen has relatively low activity upon interaction with organic molecules which are in the ground (singlet) state. Sequential single-electron reduction of oxygen to water leads to generation of active intermediate products (Fig. 13.8).

At the first stage of oxygen reduction, superoxide anion radical $O_2^{\overline{\cdot}}$ is formed. Its oxidation–reduction potential is −0.32 V, which is close to the potentials of reduced pyridine nucleotides. In proton-rich solvents $O_2^{\overline{\cdot}}$ is protonated to perhydroxyl radical $HO_2^{\cdot}$. The logarithm of the dissociation constant of the reaction is 4.8, and therefore

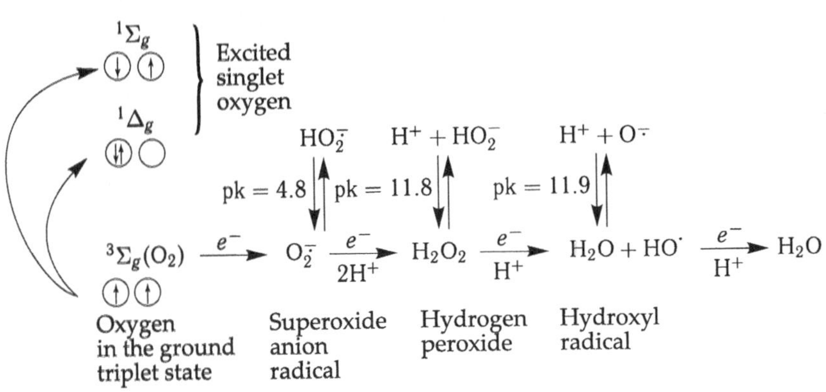

Figure 13.8. Generation of activated oxygen forms in the processes of its single-electron reduction and excitation.

the processes with the participation of $HO_2^{\cdot}$ occur mainly in the acidic medium. Further reduction of superoxide anion-radical results in the formation of hydrogen peroxide. In the presence of ions of variable valence this compound disintegrates to hydroxyl radicals $HO^{\cdot}$ (see below). Dissociation of hydrogen peroxide into ions occurs only in the alkaline medium (pH = 11.8).

At the next stage of single-electron reduction of hydrogen peroxide, the bond between oxygen atoms is broken and a water molecule and hydroxyl radical $HO^{\cdot}$ are formed. The process of oxygen reduction is terminated by the formation of a second water molecule.

The hydroxyl radical has extremely high reactivity. The rate constant of hydrogen detachment from the hydrocarbon chain by radical $HO^{\cdot}$ is close to the diffusion constant 10^9–10^{10} $m^{-1}s^{-1}$. The interaction of the hydroxyl radical with the oxidation substrates is practically nonselective.

Oxygen can perform transition from the initial triplet state to one of the two excited states differing by the localization of outer electrons (Fig. 13.8). Such a transition occurs, when oxygen gains energy in both photochemical and dark reactions. The lower excited state Δg has an excess energy of 94.5 $kJ \cdot mol^{-1}$ compatible with the energy of infrared emission with the wavelength 1270 nm. The higher energy state $^1\Sigma_g$ (156.7 $kJ \cdot mol^{-1}$) has a very short lifetime in water systems (about 10^{-10} s) and its involvement in biological processes is hardly probable. Singlet oxygen in water has the lifetime of about 2 μs. It actively enters into reactions with many unsaturated compounds forming specific products, for example, dioxitanes and internal transannular peroxides. The rate constant of interaction of singlet oxygen and polynonsaturated fatty acids is about 10^4–10^5 $m^{-1}s^{-1}$. Figure 13.9 demonstrates the main transformation reactions of active oxygen forms.

The central position in the above scheme is occupied by superoxide anion radical $O_2^{\bar{\cdot}}$. Its spontaneous dismutation leads to the generation of hydrogen peroxide and, as suggested by some data, singlet oxygen. The interaction of hydrogen peroxide and superoxide anion radical (the Haber–Weiss reaction) results in the formation

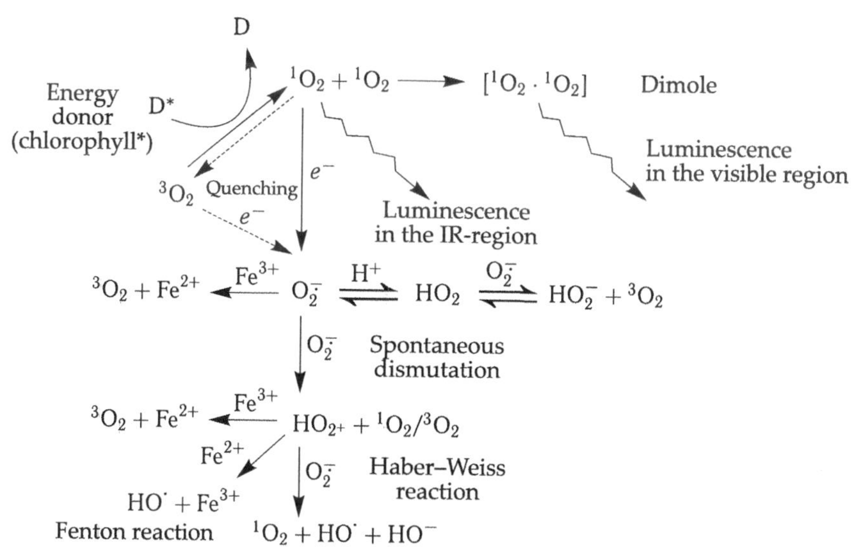

Figure 13.9. Main reactions of active oxygen forms.

of hydroxyl radical $HO^{\cdot}$ and possibly singlet oxygen. Trivalent iron or another oxidized ion of variable valence can be reduced by hydrogen peroxide or superoxide anion radical. Homolytic disintegration of hydrogen peroxide upon its interaction with bivalent iron (the Fenton reaction) produces hydroxyl radical $HO^{\cdot}$.

Activated Oxygen in Biological Systems. Generation of activated oxygen in cells occurs during aerobic respiration and photosynthesis. It is probable that the primary product of oxygen reduction — superoxide anion radical — would leave the boundaries of the respiratory or photosynthetic chain and would enter the cytoplasm. The efficiency of this process usually increases when an electron transport chain is "overloaded", and the degree of its carrier reduction enhances. In the respiratory chain, the electron donor for oxygen can be the semiquinone form of ubiquinone. In the photosynthetic electron transport chain (see Chapter 27), a significant portion of $O_2^{\bar{\cdot}}$ is generated at the level of photosystem 1 electron acceptors and ferredoxin. In photosystem 2 plastoquinone and plastocyanine can accomplish single-electron reduction.

Activated oxygen is also involved in biosynthetic processes, for example, upon formation of prostaglandins in animals or ethylene phytohormone in plants.

It is proposed that active oxygen forms both under physiological conditions and upon oxidative stress can play the role of initiators of DNA fragmentation with endonuclease (apoptosis, i.e. programmed cell death) or even directly attack DNA (hydroxyl radical). The biological role of cell apoptosis is related to the protection of organisms from malignant cell regeneration.

Different assumptions were made on the role of activated oxygen as a messenger in regulation of cell homeostasis. Unsaturated lipids of biological membranes are one of the most vulnerable targets for activate oxygen.

Lipid peroxidation is the factor capable to change chemical composition and physical parameters, ultrastructural organization and functional characteristics of biological membranes. The process of peroxidation involves mostly unsaturated phospholipids of biological membranes, mainly polyunsaturated phosphatidylethanolamine. Peroxidation increases membrane viscosity due to the decreased number of liquid hydrophobic lipids in bilayer regions, generation of intermolecular cross-links and the increased number of ordered lipids with limited mobility in the bilayer.

In biological membranes (membranes of erythrocytes, mitochondria, sarcoplasmic and endoplasmic reticulum, lysosomes), permeability for different ions, nonelectrolytes and macromolecules is induced because of lipid peroxidation. This effect when membranes lose their barrier functions underlies the pathogenesis of many diseases.

As a rule, the scheme of membrane lipid peroxidation is compiled similar to reactions of liquid-phase oxidation of hydrocarbons.

$$
\begin{array}{llll}
0 & RH \rightarrow R^{\cdot} & k_0 & \text{Generation of the chain} \\
1 & R^{\cdot} + O_2 \rightarrow RO_2^{\cdot} & k_1 & \left.\right\} \text{Prolongation of the chain} \\
2 & RH + RO_2^{\cdot} \rightarrow R^{\cdot} + ROOH & k_2 & \\
3 & ROOH \rightarrow RO^{\cdot} + {}^{\cdot}OH & k_3 & \text{Degenerated branching} \\
4 & R^{\cdot} + R^{\cdot} \rightarrow & k_4 & \\
5 & R^{\cdot} + RO_2^{\cdot} \rightarrow \left.\right\} \text{Molecular products} & k_5 & \left.\right\} \text{Disruption of the chains} \\
6 & RO_2^{\cdot} + RO_2^{\cdot} \rightarrow & k_6 &
\end{array}
\qquad (13.4)
$$

Here RH is a hydrogen molecule (a fatty acid residue); $R^{\cdot}$ and $RO_2^{\cdot}$ are alkyl and peroxide free radicals, ROOH is hydroperoxide; $k_0, k_1, \ldots$ are constants of corresponding reactions.

Inasmuch as reaction 1 occurs at a high rate ($k_1 \simeq 10^7 \div 10^8$ $M^{-1}s^{-1}$) and the concentration of oxygen in liquid-phase systems is usually relatively high, it is assumed that the condition $[R] \ll [RO_2]$ is fulfilled in liquid-phase systems. This means that in liquid-phase systems radical RO_2 is the cardinal one, and it is the only one that participates in the prolongation of the chain (reaction 2) and enters the reaction of chain disruption (reaction 6).

Defense Mechanisms. Biological systems are adapted to aerobic conditions. They are protected from toxic action of activated oxygen by enzyme systems, antioxidants and quenchers of singlet oxygen.

Antioxidants are substances protecting biological systems from oxidation due to their interactions with free radicals. The most typical antioxidants are phenolic compounds capable to efficiently interact with active free radicals of lipids forming low-activity inhibitor radicals. In different biological membranes, those are tocopherols and a number of hydroquinones.

The other group of antioxidants is represented by compounds efficiently interacting with active forms of oxygen. They sharply decrease the concentration of superoxide radicals in the system catalyzing the dismutation reaction (superoxide dismutase — SOD) and generating hydrogen and peroxide oxygen in the triplet state:

$$2O_2^{\overline{\cdot}} + 2H^+ \rightarrow H_2O_2 + O_2.$$

The rate constant of this reaction is very high $2 \cdot 10^9$ $M^{-1}s^{-1}$.

In the systems where $^1O_2$ is generated, quenchers of singlet oxygen act as antioxidants. In this respect, the most efficient is β-carotene, the rate of its $^1O_2$ quenching being regulated by diffusion ($k \sim 10^{10}$ $M^{-1}s^{-1}$). α-Tocopherol is also an effective quencher of $^1O_2$ ($k \sim 10^6 \div 10^7$ $M^{-1}s^{-1}$). This process occurs mostly by the mechanism of physical quenching when the singlet state of oxygen is deactivated. The chemical quenching of $^1O_2$ is accompanied by oxidation of the inhibitor and, as a rule, proceeds at a lower rate (for tocopherol $k \sim 10^6 \div 10^7$ $M^{-1}s^{-1}$). The overall constant of physical and chemical interaction with $^1O_2$ is about $(3 \div 5) \cdot 10^7$ $M^{-1}s^{-1}$ for histidine and tryptophan. In photosynthetic membranes, singlet oxygen is formed in the reaction of oxygen $^3O_2$ with monomer chlorophyll or bacteriochlorophyll molecules P (see Chapters 27 and 28) which are in the triplet state (see Fig. 13.9):

$$^3P + {}^3O_2 \rightarrow {}^1O_2 + P.$$

As a result of subsequent reactions, the pigment is oxidized and chemiluminescence arises in the system (see below):

$$^1O_2 + P \rightarrow P^*_{\text{oxid}} \rightarrow P^*_{\text{oxid}} + h\nu$$

due to the recombination of radicals upon decomposition of labile pigment peroxides.

The protective function of carotinoids (see Chapter 27) is realized by the mechanism of physical quenching. □

Chemiluminescence. Chemiluminescence (CL) is the emission of light, generated as a result of a chemical reaction. Two stages (chemical and physical) are distinguished in the mechanism of the CL reaction. At the chemical stage, electron-excited products are generated. The probability of their formation is equal to the ratio of the number of excited molecules to the number of reacted molecules and determines quantum yield of excitation Y_{exc}. At the next physical stage of CL, the generated excited molecule undergoes transition from singlet or triplet state to the ground state accompanied by light quantum emission. The probability of light quantum emission, i.e. proportion of the number of quanta to the number of excited molecules is called quantum yield of luminescence Y_{lum}. Quantum yields of excitation and luminescence are always smaller than unity.

The emission spectrum of CL reaction is usually compatible with the luminescence characteristics of the final reaction product or the activator substance. An excited molecule emits light either from the singlet (fluorescence) or triplet (phosphorescence) level. The short-wave boundary of the CL spectrum is determined by the following condition:

$$h\nu \leqslant E_a + \Delta H,$$

where $h\nu$ is the magnitude of a short-wave quantum on the boundary of the CL spectrum, E_a is the activation energy of the chemical reaction, and ΔH is the enthalpy change.

To emit a light quantum in the visible range of the spectrum in a single chemical act, the energy portion of 1.8 to 3.1 electron-volt is required (41–71 kcal/mol or 171–298 kJ/mol). The most exothermal reactions are the recombination of free radicals or ionic pairs and also the electron transfer from the anion radical to the oxidizer. These reactions can be accompanied by CL under corresponding conditions.

Generation of singlet excited oxygen with its subsequent transition to the ground state can also initiate CL. Of the greatest importance is the transition from the singlet state $^1\Delta g$ to the groundone with the emission of infrared quanta $\lambda = 1270$ nm at a very low quantum yield.

Apparently, CL observed in biological systems occurs with the involvement of activated oxygen molecules is mainly due to the decay of oxidized products of organic compounds rather than the result of direct emission of quantum by singlet excited oxygen.

As a rule, spontaneous CL of animal and plant tissues proceeds in parallel with their normal life activity. This luminescence usually enhances under stress effects that suggests activation of oxidizing processes under such conditions. The research school of Prof. B. N. Tarusov has contributed greatly to the investigation of the nature of luminescence of living organisms.

Chemiluminescence is one of the most sensitive methods for revealing and assessment of the level of reactions involving activated oxygen forms in biological systems. A chemiluminescence indicator (luminol or lucinogenin) is frequently added to the system studied to determine and measure the concentration of activated oxygen. These compounds continue to exist as activated oxygen forms and at the same time chemiluminesce with a high quantum yield.

VI
Transport of Substances and Bioelectrogenesis

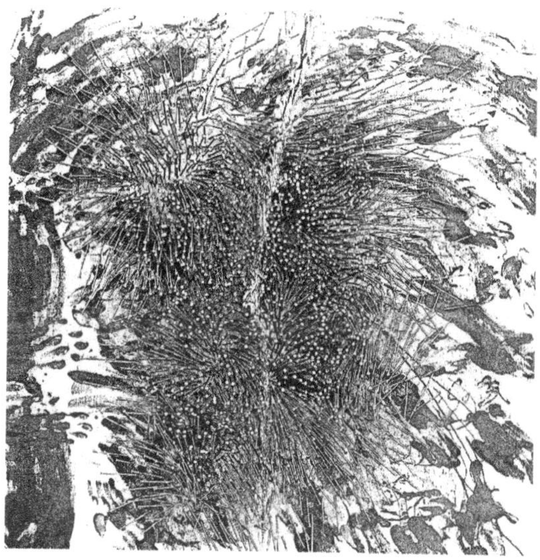

14

Nonelectrolyte Transport

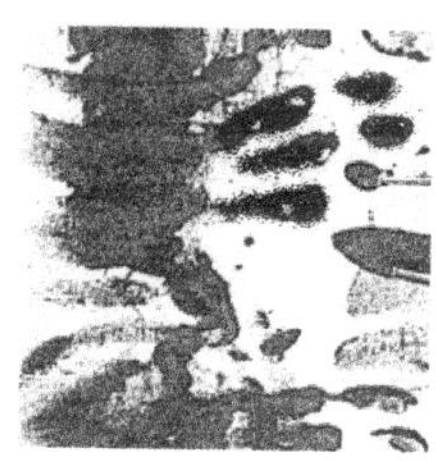

14.1 Diffusion

The movement of many uncharged substances across membranes obeys the diffusion laws. The diffusion process was first described quantitatively by A. Fick. The first Fick law reflects the simple fact that substance flow J along axis x is proportional to the driving force, i.e. concentration gradient dc/dx:

$$J = -D\,dc/dx, \tag{14.1}$$

where D is the diffusion coefficient in $cm^2 \cdot s^{-1}$; the dimension of the flow is $mol \cdot cm^{-2} \cdot s^{-1}$.

In the general case of non-stationary diffusion, the substance concentration changes with time at any point along axis x. Let us consider in detail one-dimensional diffusion of the substance towards axis x (Fig. 14.1). Imagine an elementary volume of area $S = 1\ cm^2$ and thickness dx. Suppose that in point x the flow is equal to J and in point $(x + dx)$ the flow is equal to $J + (\partial J/\partial x)\,dx$, where $\partial J/\partial x$ is the flow gradient in the direction of axis x. The amount of the substance accumulating in the considred volume $V = S\,dx$ per time unit is equal to $-(\partial J/\partial x)\,dx$. The amount of the substance accumulating in the elementary volume may be expressed as the result of changes in the concentration in time $\partial c/\partial t$ multiplied by the volume. By equating these two different expressions describing the accumulation of the substance in the volume, we obtain the so-called equation of continuity:

$$-\partial J/\partial x = \partial c/\partial t. \tag{14.2}$$

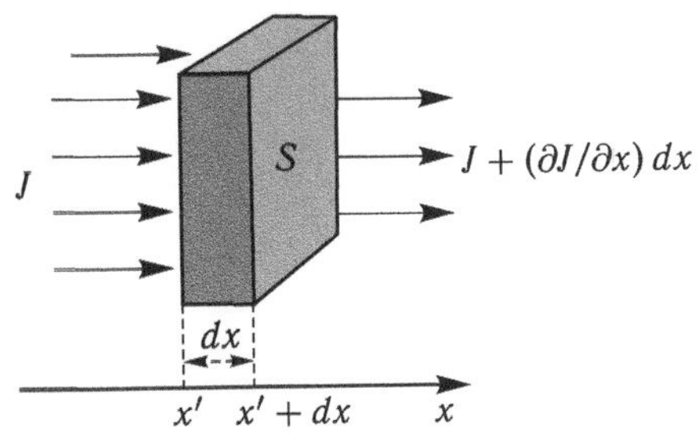

Figure 14.1. Gradient of flow J upon one-dimensional diffusion across an elementary volume of area S and thickness dx.

Substituting Eq. (14.1) in Eq. (14.2) we get the expression for the second Fick law:

$$\partial c/\partial t = D\partial^2 c/\partial x^2. \tag{14.3}$$

Equation (14.3) describes the change in the concentration of the solved substance dependent on coordinate x and time. Using the method of substitution we can obtain the solution of differential equation (14.3) as follows:

$$c = \frac{c_0}{2(\pi D t)^{1/2}} \exp(-x^2/4Dt). \tag{14.4}$$

In this equation, c_0 is the total amount of the substance at zero time $t = 0$ in point $x = 0$, and c is the concentration in point x at any moment of time t. Figure 14.2 shows profiles of the substance concentration by coordinate x for two different moments of time described by Eq. (14.4). For curve regions convex upwards $\partial^2 c/\partial x^2 < 0$ and, according to Equation (14.2), $\partial c/\partial t < 0$, i.e. the substance concentration decreases with time. At curve regions remote from the center, $\partial^2 c/\partial x^2 > 0$. Consequently, $\partial c/\partial t > 0$, i.e. the substance concentration in these regions grows with time. The result obtained reflects the known fact that diffusion is always directed to equalize the concentration gradients.

The concentration profile obtained at time t (see Fig. 14.2) is characterized by parameter $\overline{x}$. The substance concentration at distance $\overline{x}$ decreases by e times as compared to the concentration in the initial point if the exponent index in Eq. (14.4) is unity: $\overline{x}^2/4Dt = 1$.

Consequently,

$$\overline{x}^2 = 4Dt. \tag{14.5}$$

The above formula characterizes the important property of the diffusion process: the time required for diffusion increases proportionally to the squared distance. □

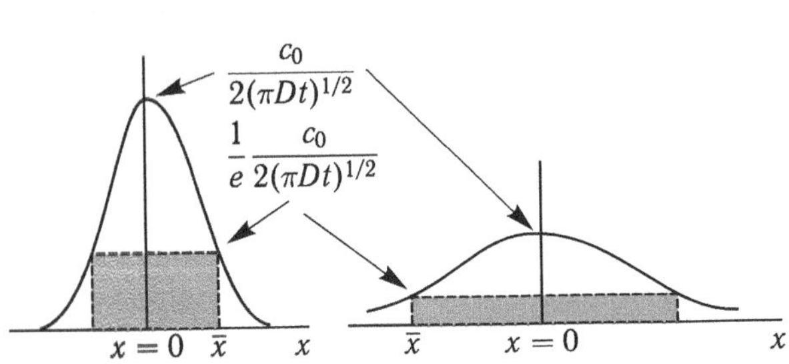

Figure 14.2. Concentration profiles of the diffusing substance for two instants of time t.

The diffusion coefficient is defined by the size and shape of molecules. The diffusion coefficient D is about 10^{-5} $cm^2 \cdot s^{-1}$ for small molecules in water. For spherical particles diffusing in liquids, it is possible to use the Stokes–Einstein equation relating the diffusion coefficient to radius of particles (r) and medium viscosity (η):

$$D = \frac{k_B T}{6\pi r \eta}. \tag{14.6}$$

Assuming that at constant η and T the molecular mass of substance M is proportional to the volume of spherical molecules $\left(M \sim \frac{4}{3}\pi r^3\right)$, we get the following relation

$$DM^{1/3} = \text{const}, \quad \text{or} \quad \lg D + \frac{1}{3}\lg M = \text{const}. \tag{14.7}$$

In a diagram form, relation (14.7) between $\lg D$ and $\lg M$ is shown by a straight line with the angular coefficient of −0.33, obtained in the case of diffusion in water of large protein molecules having almost a spherical shape (ribonuclease molecules). For small molecules from hydrogen to trisaccharides, the experiment gives a steeper straight line:

$$DM^{1/2} = \text{const}. \tag{14.8}$$

This deviation is explained by the fact that molecules of small dimensions cannot be considered as spherical. Moreover, Eq. (14.6) is valid for particles diffusing in an ideal homogeneous liquid. For small molecules, comparable in dimensions with water molecules, this condition is not evidently fulfilled.

To provide penetration of nonelectrolytes from water to the hydrophobic portion of the membrane or the narrow membrane pore, partial or complete dehydration of molecules is required, i.e. it requires energy to overcome interactions of polar groups of the molecule (–COOH, –OH, $-NH_2$) with water dipoles. For example, the activation energy values for penetration of ethylene glycol, glycerol and erythritol across synthetic phospholipid membranes, as well as across membranes of isolated cells, are close to that of dehydration of these compounds. The requirement for dehydration of molecules is the reason for strong temperature dependence of the membrane permeability coefficient for a number of nonelectrolytes. Although a variety of compounds diffuse across biological membranes, at the same time even relatively small molecules of amino acids and monosaccharides do not practically penetrate across membranes of most cells due to simple diffusion. □

14.2 Facilitated Diffusion

The mechanism of transport involving a carrier can be represented as a kinetic scheme:

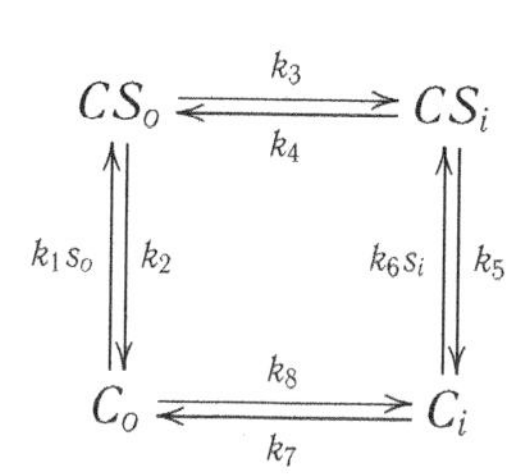

,

where C and S are the carrier and substrate molecules.

Let us introduce the following designations: c_o and c_i are concentrations of carriers in the free state on the outer and inner sides of the membrane, and cs_o and cs_i are carriers bound to the substrate; s_o and s_i are concentrations of substrates in the external and internal solutions, respectively; and k_1 to k_8 are rate constants of separate stages. The expression for the rate of the substrate flow may be obtained by solving the system of equations describing the stationary state at $dc_o/dt = d[cs_i]/dt = dc_o/dt = 0$. Hence, the substrate flow J is equal to

$$J = k_1 c_o s_o - k_2 cs_0 = k_3 cs_o - k_4 cs_i = k_5 cs_i - k_6 c_i s_i = k_7 c_i - k_8 c_o. \quad (14.9)$$

As the total concentration of the carrier c_i is invariable, we can write

$$c_t = c_o + cs_o + c_i + cs_i. \quad (14.10)$$

The solution of the above system of equations (14.9) and (14.10) in general terms includes different combinations of constants from k_1 to k_8 as well as the substrate concentrations s_o and s_i on both sides of the membrane. The initial rate of the substance transfer $s_i = 0$, is described by the expression analogous to the Michaelis–Menten equation

$$J = \frac{J_{\max} s_o}{K_m + s_o}. \quad (14.11)$$

At a very low substrate concentration s_o, Eq. (14.11) is reduced to the formula for the case of simple diffusion (linear dependence on concentration s_o). The K_m constant is equal to the substrate concentration at which $J = J_{\max}/2$ and is an important parameter characterizing the rate of the whole process. In a real system, the total flow across the membrane always includes a component specified by simple diffusion (Fig. 14.3).

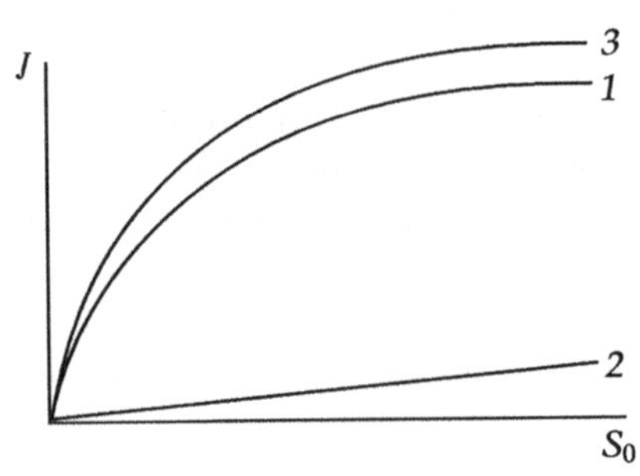

Figure 14.3. Dependence of initial rate J of the substance supply into the cells on the concentration in the outer medium S_0.

1, Transport with the involvement of a carrier; *2*, simple diffusion; *3*, total rate of the transfer.

The driving force for transport with the involvement of a carrier is the gradient of chemical or electrochemical potential of the substance. The functioning of systems with facilitated diffusion, as well as with simple diffusion is directed to equalize the gradients and establish equilibrium in the system. But the gradients of substances can be maintained for a prolonged time due to the processes in which the penetrating molecules are consumed or formed in biochemical reactions on one side of the membrane. □

14.3 Water Transport. Aquaporins

Water transport across biological membranes is a fundamental process for all living organisms. In general terms, the driving forces for water transport are membrane gradients of hydrostatic (ΔP) and osmotic pressure ($\Delta\pi$). If a membrane separates two aqueous solutions of some substance, the bulk flow J_V, determined as the water volume transferred across a unit of the membrane area per unit of time, is

$$J_V = L_P \cdot (\Delta P - \sigma\Delta\pi) = L_P \cdot (\Delta P - \sigma RT\Delta c). \tag{14.12}$$

Here L_P is the hydraulic permeability of the membrane in $m/(Pa \cdot s)$; Δc is the difference in the concentrations of the solved substance in compartments separated by the membrane. The dimensionless coefficient of selectivity (often called an osmotic reflection coefficient) is introduced to Eq. (14.12) to take into account the effect of the solved substance flow on the water flow. In the case of ideal semipermeability of the membrane (permeable only for water but not for the solved substance), $\sigma = 1$.

It has been found that two pathways of transmembrane water transport exist. The first pathway is simple physicochemical diffusion of water across the lipid bilayer. It is characterized by a relatively low rate and high energy of activation ($P_f \leqslant 5 \cdot 10^3$ m/s, $E_a > 10$ kcal/mol).

The second pathway was first found upon studying the water permeability of human erythrocyte membranes. The experiments performed in the 50s of the last century demonstrated that the erythrocyte membrane has water-filled pores. Because of this the water permeability of the erythrocyte membrane is much higher than the permeability of the lipid bilayer ($P_f > 10_4$ m/s). The energy of activation of water transport through the pores is lower ($E_a < 5$ kcal/mol) as compared to the energy of activation of physicochemical diffusion.

The protein forming a water pore in the erythrocyte membrane was isolated only twenty years later. This small membrane protein (the molecular mass of 28 kDa) was afterwards called aquaporin 1 (AQP1).

Aquaporins guarantee highly efficient transfer of water molecules across biological membranes: $3 \cdot 10^9$ water molecules are transferred across an AQP1 pore per 1 s. Water transport across aquaporins is passive and occurs with the same efficiency in both directions.

Aquaporins are integral membrane proteins forming homotetramers in the membrane; a water pore is present in each of the four monomers. The polypeptide chain of each of the subunits crosses the membrane six times, and N- and C-termini of the chain are in the cytosol. The chain regions immersed in the lipid bilayer form α-helical domains (Fig. 14.4).

Loops B and E carry short α-helical domains immersed in the membrane. These domains take part in the formation of a channel for water.

The water channel for aquaporins has a dumbbell-like shape with two wide mouths on the extracellular and cytoplasmic sides. The mouths are bound by a narrow and long selective pore. At the junction of the mouth and the selective pore on the outer side of the cell, the channel has the smallest diameter. In aquaporin AQP1 the channel diameter on this level is 2.8 E that is approximately equal to the diameter of the water molecule. Water molecules in the selective pore are positioned as a linear sequence and form hydrogen bonds with definite amino acid residues of the protein (Fig. 14.5).

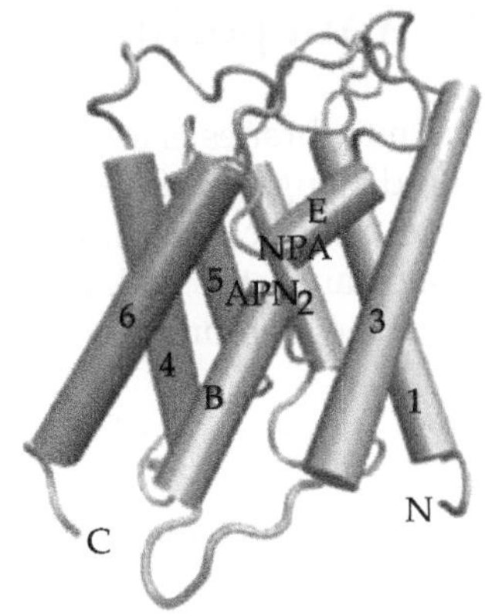

Figure 14.4. Structure of the aquaporin AQP1 monomer. α-Helices crossing the membrane are designated by numerals 1–6; connecting loops containing short α-helical domains are designated by letters B and E. The structure is taken from the Protein Data Bank (1J4N).

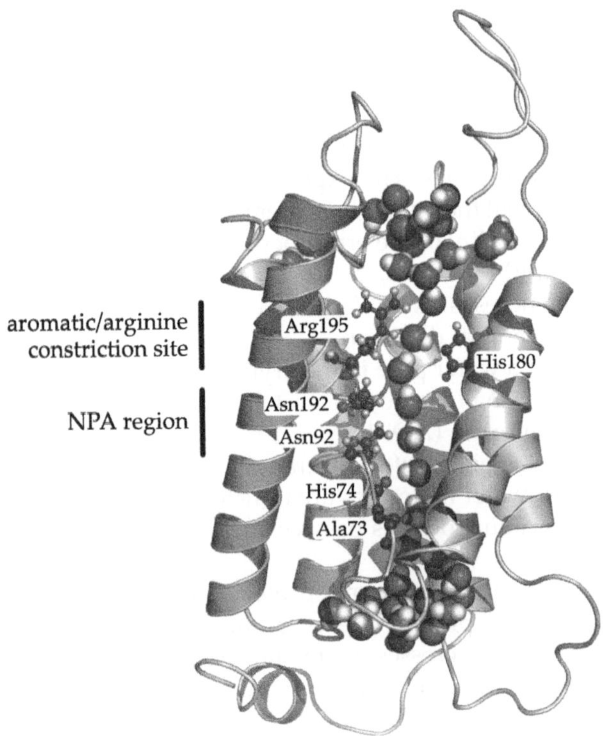

Figure 14.5. Movement of water molecules through the aquaporin AQP1 pore (the results of molecular dynamic calculations). Spherical representation of water molecules is shown; some amino acid residues interacting with water are shown as balls and rods (reproduced from J. S. Hub et al., 2009).

These residues form hydrophilic centers on the surface of the pore which is hydrophobic on the whole. Water molecules move along the selective pore, "hopping" from one hydrophilic center to another under the action of heat fluctuations.

15

Ion Transport. Ionic Equilibria

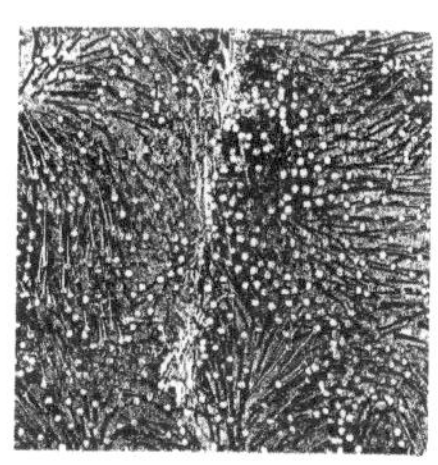

The rate of ion permeability across the membrane is determined by its properties such as the thickness, the value of its dielectric permeability, the presence of fixed electric charges on its surface, the sign and density of fixed charge localization in the membrane, the dimensions and number of pores in the membrane, the presence of fixed charges in pores. In this chapter, we analyze the driving forces of ion passive transport and ionic equilibria in membrane systems. □

15.1 Electrochemical Potential

The driving force of diffusion is the difference of chemical potential values of the given substance in the two areas between which the diffusion occurs. The chemical potential of the solved substance (μ) for conditions, when the contribution of hydrostatic pressure can be neglected, is

$$\mu = \mu_0 + RT \ln c, \tag{15.1}$$

where μ_0 is the standard chemical potential dependent on the nature of the solvent, and c is the concentration. At large concentration values, c is substituted for the activity in Eq. (15.1). The electrochemical potential is a corresponding notion for ions, the movement of which depends not only on the concentration but also on the electric potential. The electrochemical potential ($\overline{\mu}$) of an ion of type i for conditions, when the activity of the ion is equal to its concentration, and the contribution of the hydrostatic pressure is negligibly small, is

$$\overline{\mu} = \mu_0 + RT \ln c + zF\varphi, \quad (15.2)$$

where φ is the electric potential; z is the valence; T is the absolute temperature, K; R is the gas constant equal to about 8.31 J · mol^{-1} · K^{-1}; F is the Faraday constant ($\approx 9.65 \cdot 10^4$ C · mol^{-1}). The electrochemical potential is the measure of the work required to transfer 1 mol (1 gram-equivalent) of ions from the solution of the given concentration and electric potential to the infinitely remote point in vacuum. This work is summed up of the energy needed to overcome the forces of chemical interaction $\mu_0 + RT \ln c$ and the work to transfer charges in the electric field $zF\varphi$.

The passive movement of ions takes place from the region with high electrochemical potential to the region with lower electrochemical potential. The driving force of the ion transfer is the gradient of electrochemical potential $d\mu/dx$.

15.2 Ion Hydration

The ion interaction with the solvent molecules is determined by the standard chemical potential μ_0. Ion hydration in water leads to changes in their effective radius. The radius of hydrated ions is usually assessed using the Stokes–Einstein relation (see Eq. (14.6)):

$$r = \frac{k_B T}{6\pi D\eta}. \quad (15.3)$$

At 25 °C, this radius in nanometers is $0.245 \cdot 10^{-5}/D$, where D is in cm^2·s^{-1}. The radius of the Na^+ ion calculated in this way is 0.184 nm. This value is twofold higher than the van-der-Waals radius (see Section 1 in Chapter 8), measured by X-ray diffraction on crystals, and is caused by the presence of water molecules around the ion.

The application of the above relation (15.3) to particles of atomic dimensions is limited because the medium cannot be regarded as homogeneous at dimensions of diffusing particles close to the dimensions of water. In addition, the diffusing particle affects the local structure of water and as a result the effective local viscosity can differ from the macroscopic viscosity.

According to the contemporary theory, the moving ion is affected not only by the hydrodynamic friction force proportional to the radius, but also by the so-called dielectric friction force. The hydrodynamic friction force F acting on the spherical particles moving in liquids at rate v, is described by the Stokes law:

$$F = 6\pi\eta r v. \quad (15.4)$$

This force is proportional to the particle radius and is independent of the charge. In contrast to it, the dielectric friction force is proportional to the squared charge and inversely proportional to the ionic radius cube. The resultant of the two forces has the minimum at the radius of about 0.2 nm. The existence of the dielectric friction force explains also the low mobility of multivalent ions.

The origin of dielectric friction can be imagined having considered a homogeneous dielectric, flowing around a charged sphere. The dielectric is polarized upon approaching to the charge and loses polarization upon moving away from it.

Provided the change in polarization proceeded as quickly as the change in the electric field upon the charge approaching or removing, polarization could be spherically symmetrical and would not create the total force acting on the ion. However, in real dielectrics one of the polarization vector components relaxes relatively slowly (for water the relaxation time is $\sim 10^{-11}$ s). Because of the delay, polarization of the medium approaching the charged sphere, is smaller than polarization of the removing medium, where polarization has already terminated. Polarization is not spherically symmetrical, and the movement of the ion is more retarded the faster it moves.

□

In accord with the Born electrostatic theory, the energy of the ion of radius r in the medium with dielectric permeability ε is determined by the formula

$$W = \frac{z^2 e^2}{8\pi r \varepsilon \varepsilon_0}, \tag{15.5}$$

where ε_0 is the absolute dielectric permeability of vacuum, equal to about $8.85 \cdot 10^{-12}$ F/m. The change in the free energy upon ion transfer from water to the membrane can be approximately calculated using the formula

$$\Delta W = \frac{Z^2 e^2}{8\pi \varepsilon_0 r}\left(\frac{1}{\varepsilon_2} - \frac{1}{\varepsilon_1}\right), \tag{15.6}$$

where ε_1 and ε_2 are relative dielectric permeability of water and nonpolar solvent, respectively ($\varepsilon_1 \approx 80$ and $\varepsilon_2 \approx 2$–3).

The values of free energy calculated for transfer of Cs, K and Na ions from vacuum to water are correspondingly about −412, −517 and −727 kJ/mol that exceeds the experimental values. The electric fields near the ions are so high (10^6–10^7 V/cm), that the molecules which are the nearest to the water molecule ion are almost completely oriented. This state is called dielectric saturation. The ordered structure of water in the hydration shell is most complete near the ions smallest in dimensions, which create the most intensive local electric fields. As soon as the order becomes complete, further polarization is impossible. This is equivalent to the decrease in the effective dielectric permeability, and the total energy of polarization in real systems is smaller than in the case of an ideally polarized dielectric.

15.3 Ionic Equilibrium on the Phase Interface

Ionic equilibrium between two aqueous solutions separated by a membrane is described in the simplest way. In the case of equilibrium of one-type ions, the ion electrochemical potentials are the same in both solutions:

$$RT \ln c_1 + zF\varphi_1 = RT \ln c_2 + zF\varphi_2,$$

from where

$$\Delta\varphi = \varphi_2 - \varphi_1 = \frac{RT}{zF} \ln \frac{c_1}{c_2}. \tag{15.7}$$

The Nernst equation (15.7) shows that under electrochemical equilibrium the difference of potentials on the membrane is determined by the ratio of concentrations of the given ion in the two contiguous solutions. After the transition to decimal logarithms in Eq. (15.7) and substitution of $2.3RT/F \approx 58$ mV (at 20 °C), the Nernst equation for univalent ions is as follows:

$$\Delta\varphi = 58 \lg(c_1/c_2). \tag{15.8}$$

In biological systems, the distribution of K^+ ions between the cytoplasm of animal cells and the environment is in relatively good agreement with the Nernst equation. However, the distribution of Na^+ ions in most cells varies greatly from the equilibrium one.

Let us analyze the ionic equilibrium on the interface of water and the nonpolar solvent. Such a system can be regarded as a model of the interface of water and the lipid membrane with dielectric permeability from 2 to 3.

Imagine a system of two immiscible liquids, for example, oil and water, in which electrolyte A^+B^- is dissolved. Because of dissimilar lipophilicity of A^+ and B^- ions, the so-called interphase difference of potentials emerges between phase volumes.

Figure 15.1 shows a scheme of profiles of an equilibrium distribution of concentrations of A^+ and B^- ions and the profile of the electric potential near the interface of two phases with different dielectric permeability ($\varepsilon_1 > \varepsilon_2$) for the case when the solubility of anions in the nonpolar medium is higher than the solubility of cations. The plane $x = 0$ corresponds to the phase interface, on the left there is the medium with dielectric constant ε_1 and on the right — with the dielectric constant ε_2. If the anions are of high lipophilicity, their total content in the organic phase is somewhat higher than the content of cations. Hence the nonpolar phase has an excess negative charge and a lower potential than the aqueous phase. The excess of one-sign ions in each of the contiguous phases is so small, that cannot be detected with the available analytical methods. In the regions remote from the interface, the concentrations of ions and cations are the same due to the condition of electroneutrality of bulk phase volumes ($c_A = c_B$). The electric potential in the phase volume is constant and has the values φ_1 and φ_2 in the aqueous phase and organic solvent, respectively. The concentration of ions in the phase with low dielectric permeability c_2 is lower than the concentration c_1 in the aqueous phase, because the transition of charged particles from the phase with high permeability ε to the phase with low permeability is associated with the growth of the electrostatic energy of the charged particle.

Near the interface, on the right and left of the boundary, diffusion layers are located (see Fig. 15.1), in each of them the concentrations of ions and cations being dissimilar. The condition of electroneutrality for these regions is not fulfilled: one phase has a small excess of negative charges, and the other — of positive charges. In this region, the potential is not constant but depends on the spatial coordinate x. Thus, the potential near the interface of the two phases changes not unevenly but gradually at some distance from the boundary.

The formulas describing the difference of electric potentials and distribution of concentrations between bulk phase volumes can be obtained from the condition of electrochemical equilibrium in the system. Upon equilibrium, the electrochemical potential for each ion is the same at any point of the system (does not depend on coordinate x). Accordingly, $\overline{\mu}_A^{(1)} = \overline{\mu}_A^{(2)}$ and $\overline{\mu}_B^{(1)} = \overline{\mu}_B^{(2)}$, i.e.

$$\begin{aligned} \mu_{0A}^{(1)} &= RT \ln c_1^A + F\varphi_1 = \mu_{0A}^{(2)} + RT \ln c_2^A + F\varphi_2 \qquad (z = +1), \\ \mu_{0B}^{(1)} &= RT \ln c_1^B - F\varphi_1 = \mu_{0B}^{(2)} + RT \ln c_2^B - F\varphi_2 \qquad (z = -1), \end{aligned} \tag{15.9}$$

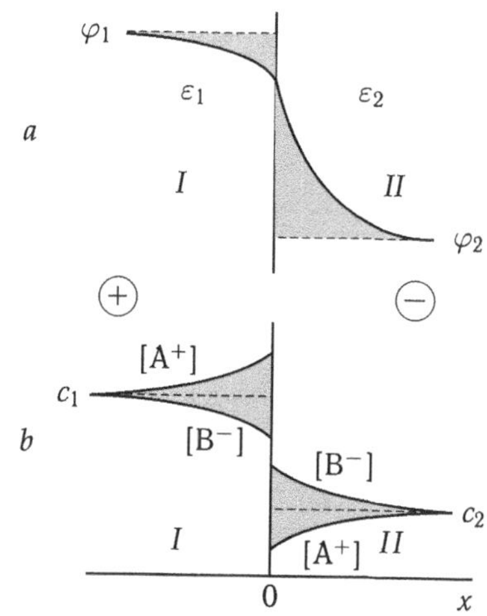

Figure 15.1. Distribution profiles of potential and ion (A^+ and B^-) concentrations on the phase interface (*I* and *II*).

I, Aqueous solution ($\varepsilon_1 = 80$); *II*, oil ($\varepsilon_2 = 2 \div 3$); *a*, potential (φ) distribution; *b*, distribution of concentrations (c) of cations and anions; c_1 and c_2, concentrations of electrolytes in phase volumes *I* and *II*. The plots reflect the case when anions are solved better in the nonpolar phase than cations.

where indices 1 and 2 apply to the first and second phases, and indices A and B — to cation A^+ and anion B^-. □

After regrouping Equations (15.9) look as follows

$$\begin{aligned} -F\Delta\varphi &= -\Delta\mu_{0B} + RT\ln(c_2^B/c_1^B), \\ F\Delta\varphi &= -\Delta\mu_{0A} + RT\ln(c_2^A/c_1^A), \end{aligned} \quad (15.10)$$

where $\Delta\varphi = \varphi_1 - \varphi_2$; $\Delta\mu_{0A} = \mu_{0A}^{(1)} - \mu_{0A}^{(2)}$. In the phase volume, the concentrations of ions and cations are the same: $c_1^A = c_1^B = c_1$ and $c_2^A = c_2^B = c_2$. Solving equations relative to c_2/c_1 or $\Delta\varphi$, we get formulas reflecting dependences of the differences of potentials between phase volumes and distribution of ions between phase volumes on standard chemical potentials of ions A^+ and B^- in aqueous and organic media:

$$\ln(c_2/c_1) = \frac{\Delta\mu_{0A} + \Delta\mu_{0B}}{2RT}, \qquad \Delta\varphi = \frac{\Delta\mu_{0B} - \Delta\mu_{0A}}{2F}. \quad (15.11)$$

Relations (15.11) become more obvious, if the difference of standard chemical potentials for each ion $\Delta\mu_0$ is expressed through the interphase distribution coefficient γ, which by definition is equal to the ratio of ion concentrations in the contiguous phases at the very interface:

$$\gamma = c_0^{(2)}/c_0^{(1)}. \quad (15.12)$$

In this formula $c_0^{(1)}$ and $c_0^{(2)}$ are ion concentrations at the interface itself (plane $x = 0$) on its leftand right sides, respectively.

To express $\Delta\mu_0$ through the distribution coefficients, let us write the condition of equilibrium for the phase interface:

$$\mu_{0A}^{(1)} + RT\ln c_{0A}^{(1)} = \mu_{0A}^{(2)} + RT\ln c_{0A}^{(2)}. \quad (15.13)$$

Members $F\varphi$ in the left-hand and right-hand sides of this equation are absent, because the electric potential on the interface is continuous and the $\varphi_0^{(1)}$ and $\varphi_0^{(2)}$ values are the same near plane $x = 0$. Taking into account that $c_0^{(2)}/c_0^{(1)} = \gamma$, we get from Eq. (15.13)

$$\ln \gamma_A = \frac{\Delta\mu_{0\text{A}}}{RT} \quad \text{and} \quad \ln \gamma_B = \frac{\Delta\mu_{0\text{B}}}{RT}. \tag{15.14}$$

As for Equations (15.11), the interphase difference of potentials and the ratio of electrolyte concentrations in the two phases can be expressed as:

$$\ln(c_2/c_1) = \frac{1}{2}(\ln \gamma_A + \ln \gamma_B), \quad \text{or} \quad c_2/c_1 = \sqrt{\gamma_A \gamma_B}; \tag{15.15}$$

$$\Delta\varphi = \frac{RT}{2F} \ln(\gamma_B/\ln \gamma_A). \tag{15.16}$$

As follows from Eq. (15.16), the interphase jump change of the potential occurs only when the distribution coefficients of cations and anions are different. With equal coefficients, the interphase difference of potentials is zero. □

The coefficient of ion distribution can be estimated in the first approximation from Eq. (15.14), showing that the coefficient is exponentially dependent on the work of the ion transfer from aqueous phase 1 to nonpolar phase 2. This work is equal to the energy of ion dehydration and can be estimated by the Born formula (see Eq. (15.5)). If the ion radius is accepted to be 0.2 nm and the dielectric constant of the membrane $\varepsilon = 3$, the distribution coefficient will be 10^{-20}. In this case, the permeability of bilayer lipid membranes would be essentially lower than the measured values ($\sim 10^{-8}$ ohm $\cdot$ cm^{-2} for non-modified membranes).

A substantial decrease in the ion energy in the membrane may be specified by other factors: (1) ions can form ion pairs within the membrane; (2) the membrane may have pores with a high dielectric constant the passage through which requires a lower energy of activation than the ion transfer from the solution to the lipid; (3) the ion may be "hidden" in a neutral "carrier" molecule solvating it (increases the effective radius r) thus facilitating its dissolution in the membrane.

The interaction of ions with membrane fixed charges leads also to a change in the profile of ion energy in the membrane. For a homogeneous uncharged membrane (see Fig. 15.2, *III*), there is a barrier the magnitude of which reflects the work to transfer the charged particle to the layer with low dielectric permeability. For phospholipid membranes with fixed charges on their surfaces, in the energy profile on the membrane edges there appear additional barriers corresponding to the energy of activation of ion adsorption–desorption in the regions with fixed charges (see Fig. 15.2, *IV*).

15.4 Profiles of Potential and Concentrations at the Interface

Equations (15.15) and (15.16) apply to the jump change of the potential and the ratio of concentrations between phase volumes. The distribution of ion concentrations in the bilayer near the interface is determined by the Boltzmann statistical equation which can be obtained from the condition of equality of the electrochemical potential

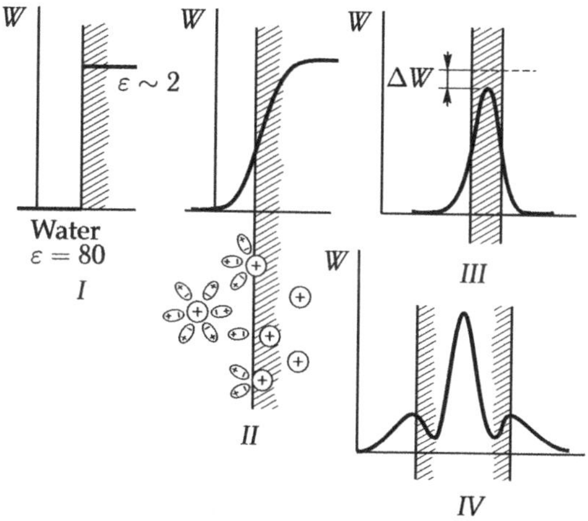

Figure 15.2. Profiles of ion free energy on the interface of immiscible liquids and in membranes. *I*, Edge effects are not taken into account; *II*, at small distances near the interface; *III*, in a thin uncharged membrane; *IV*, in a membrane with fixed charges: $W = \frac{\varepsilon\varepsilon_0}{2}\int_r^{\infty} E^2\,dV$, where W is the ion energy, E is the field strength, dV is the elementary volume, ε is the dielectric permeability of the medium, ε_0 is the absolute vacuum dielectric permeability, and r is the ion radius.

in any point on the x axis (see Fig. 15.1). For arbitrarily plane x in phase 1 we have (assuming $\overline{\mu}_x = \overline{\mu}_0$)

$$RT \ln c_x + zF\varphi_x = RT \ln c_0 + zF\varphi_0,$$

where c_x and φ_x are the ion concentration and electric potential at distance x; c_0 and φ_0 are the concentration and potential in phase 1. Hence we have

$$c_x = c_0 \exp(-zF\varphi/RT), \tag{15.17}$$

where $\varphi = \varphi_x - \varphi_0$ is the potential in plane x calculated from the potential level in the solution volume. Equation (15.17) shows that upon approaching to the negatively charged phase (see Fig. 15.1) the local concentration of cations increases, while the local concentration of anions decreases. A ten-fold change in the concentration of a univalent ion is attained in the region where the potential changes by 58 mV.

The interphase difference of potentials determined by Eq. (15.16) is associated with the charge separation in the vicinity of the interface and with the formation of a double electric layer. The characteristic of electric potential distribution in the bilayer, in particular near the water–nonpolar solvent interface, is solved based on one of the principal equations of electrostatics — the Poisson equation. For the analyzed one-dimensional case (one spatial coordinate x), this equation looks like

$$\frac{d^2\varphi}{dx^2} = -\frac{\rho(x)}{\varepsilon\varepsilon_0}, \tag{15.18}$$

where $\rho(x)$ is the density of electric charges in plane x

$$\rho(x) = F\sum zc_x, \tag{15.19}$$

ε is the absolute dielectric constant having the ε_1 value at $x < 0$ and ε_2 at $x > 0$. The Poisson equation connects the second potential derivative in plane x with the charge density in this plane. If we assume that the electrolyte concentration in the phase volume is c, then with account of the Boltzmann equation (15.17) the charge density may be expressed as

$$\rho = zF(c_{A^+} - c_{B^-}) = zFc\left(\exp\frac{-zF\varphi}{RT} - \exp\frac{zF\varphi}{RT}\right), \tag{15.20}$$

where $\varphi = \varphi(x) - \varphi(\infty)$ is the potential in plane x relative to the potential in the phase volume $\varphi(\infty)$, and $z = z_+ = -z_-$. In the case when the potential value on the interface $\varphi(0)$ is not high relative to the potential in the phase volume, the exponential members in Eq. (15.20) can be expanded into a series. □

After such simplification, the Poisson – Boltzmann equation is written as

$$\frac{d^2\varphi}{dx^2} = \kappa^2\varphi, \tag{15.21}$$

where

$$\kappa^2 = \frac{2cz^2F^2}{\varepsilon\varepsilon_0 RT}. \tag{15.22}$$

The solution of Eq. (15.21) is as follows (the Debye approximation or the Gouy model)

$$\varphi = \varphi_0 \exp(-\kappa|x|), \tag{15.23}$$

where φ_0 and φ are the electric potential values on the interface and in plane x, measured relative to the potential in the phase volume; $|x|$ is the absolute value of coordinate x. Equation (15.23) shows that the electric potential changes exponentially with a change in the distance from the interface (Fig. 15.3). The double electric layer spreads into the depth of the both phases on the both sides of the interface. Depending on the distance, the steepness of the potential change is determined by the κ value which depends on the ion concentration in the given phase

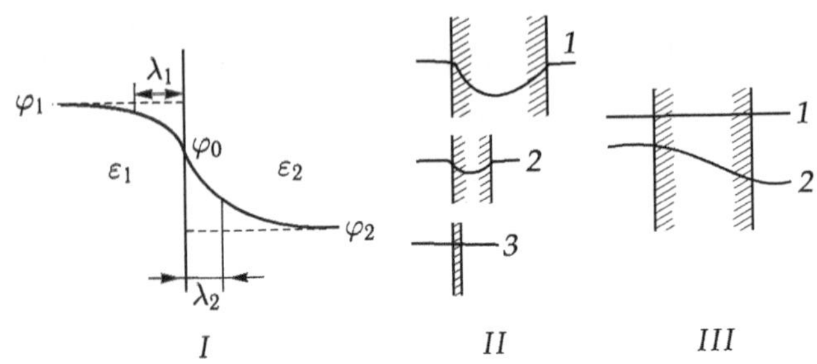

Figure 15.3. Profiles of the potential distribution on the interface (*I*), in membranes of different thickness (*II*), in a thin membrane in the absence of the external field (*III*, *1*) and upon applying the electric potentials difference (*III*, *2*).

φ_1 and φ_2 are electric potentials in the phase volume; φ_0 is the electric potential on the interface; ε_1 and ε_2 are correspondingly dielectric permeability of the two phases; λ_1 and λ_2 are the screening length in the first and second phases.

and on the dielectric permeability. The κ value changes to κ_1 and κ_2 correspondingly for phases 1 and 2. The main change in the potential near the interface occurs at a distance equal to $1/\kappa$. The value $\lambda = 1/\kappa$ characterizes the thickness of each plate of the diffusion double layer and is called the screening length or the Debye length.

As seen from formula (15.23), at $x = 1/\kappa$ the potential φ has the value φ_0/e, i.e. approximately by 63 % lower than the potential difference between the interface and the phase volume. The regions with the lengths λ_1 and λ_2 on the right and left sides of the interface carry an excess number of ions of different signs.

The screening length depends substantially on the electrolyte concentration in the given phase:

$$\lambda = \kappa^{-1} = \left(\frac{\varepsilon \varepsilon_0 RT}{2cz^2F^2} \right)^{1/2}. \tag{15.24}$$

In diluted aqueous solutions, the screening length, i.e. the region of the diffusion double layer, reaches tens of nanometers, and in the concentrated solutions it makes tenth parts of a nanometer. When divalent ions are present in the solution, the double electric layer becomes mote tight, i.e. the screening length decreases. □

The electric potential profile in membranes surrounded by the electrolyte is essentially dependent on the membrane thickness. If the thickness exceeds greatly the screening length, the situation does not differ from the case of infinitely extended phases. The potential profile in such a membrane is shown on curve *1* (Fig. 15.3, *II*). The main part of the interphase difference of potentials decreases in the membrane, the screening length in the membrane being much larger than the screening length in the electrolyte solution. Since the values of dielectric permeability in the aqueous solution and the membrane differ greatly, the curve of changes in the potential has a bump on the interface.

In the other limiting case, the membrane thickness is much smaller than the screening length. Namely, such a situation occurs in cell and bilayer lipid membranes with the thickness of 7–10 nm. In this case the sudden change of the potential within the membrane does not occur (curve *3* in 15.3, *II*). It can be thought that in the absence of the external applied field, the potential inside the membrane is constant over its entire thickness. The distribution of potential in such a thin membrane upon superposition of an external field is demonstrated in Fig. 15.3, *III*. The potential in the thin membrane changes linearly, and the potential gradient is constant over its entire thickness: $d\varphi/dx = \text{const}$. The suggestion on the constant electric field in biomembranes underlies the equations used for the description of ion transport across biomembranes and for calculation of the membrane potential value (the Goldman equation; see (16.7)). □

15.5 Double Electric Layer

The presence of fixed surface charges on cell membranes results in the formation of a double electric layer near the membrane that is a key factor in many bioelectrochemical processes.

According to the concepts developed by G. L. Gouy and D. L. Chapman (1910), the distribution of ions on the charged surface is dependent on two factors: electrostatic attraction retaining counterions near the surface, and thermal motion

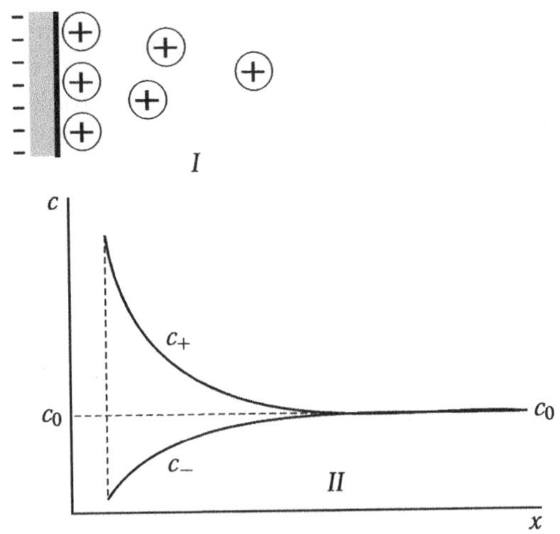

Figure 15.4. Ion distribution in a double electric layer.

I, Excess cations; *II*, ion concentration at different distances from the surface; c_0 is the electrolyte concentration in the volume, c_+ and c_- are concentrations of cations and anions; the abscissa shows the distance from the interface, the ordinate shows the electrolyte concentration.

of these ions, which on the contrary tend to equalize their concentrations in the surface layer and volume. In the established equilibrium, the density of ion distribution near the charged surface decreases as the distance from the surface increases. Equilibrium concentrations of cations c^+ and anions c^- in the surface layer and the solution volume are schematically shown in Fig. 15.4. The layer of the solution with changed concentrations near the surface is called a diffuse layer. Dependent on the conditions, its thickness changes by several orders of magnitude — from tenth parts of a nanometer to micrometers. □

The ion concentration near the charged surface is determined by the Boltzmann law (see Eq. (15.17)). The profile of potential in the double layer region (Fig. 15.5) is determined by the Poisson – Boltzmann equation obtained by combining the Poisson equation (see Eq. (15.18)) with the account of the charge density in plane x (see Eq. (15.19)) and the Boltzmann equation (see Eq. (15.17)). The Poisson – Boltzmann equation is as follows

$$\frac{d^2\varphi}{dx^2} = -\frac{F}{\varepsilon\varepsilon_0}\sum zc_0 \exp(-zF\varphi/RT). \tag{15.25}$$

According to the Gouy – Chapman theory, the potential decreases with an increase in the distance from the membrane in conformity with the equation

$$\varphi(x) = \frac{2RT}{F}\ln\frac{1+\alpha\exp(-\kappa x)}{1-\alpha\exp(-\kappa x)}, \tag{15.26}$$

where x is the distance from the interface, κ is the reciprocal Debye length (see Eq. (15.22)). The α value is set by the expression

$$\alpha = \frac{\exp\dfrac{F\varphi_s}{2RT} - 1}{\exp\dfrac{F\varphi_s}{2RT} + 1},$$

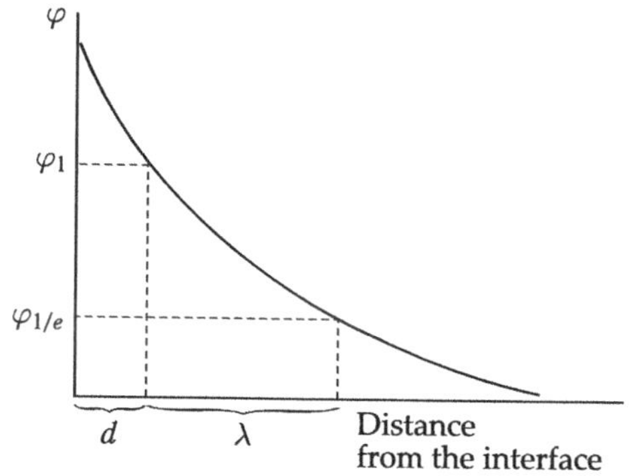

Figure 15.5. Distribution of potential φ in the double electric layer.

Here d is the thickness of the dense layer; λ is the screening length; φ_1 is the potential in the plane of maximal approximation of free cations.

where φ_s is the surface potential, i.e. the potential on the interface of the membrane and the solution. In the case of low surface potentials, equation (15.26) is simplified and is the same as Equation (15.23).

Alkaline metal cations and bivalent cations affect the surface potential mainly by screening the membrane fixed charges. In contrast to this, H^+ ions have their effect predominantly by binding — protonation of the corresponding groups, i.e. by affecting the density σ of fixed charges. Because of this, H^+ and other ions, capable of specific adsorption on membranes, are called potential-determining ions as opposed to indifferent ions involved only in the screening of fixed charges. □

The membrane surface potential cannot be measured experimentally using direct methods. However, it is possible to change the other value close to φ — the electrokinetic potential ζ called the zeta potential. The zeta potential is determined as the potential of the sliding boundary also counted from the potential level in the solution volume. The sliding boundary depends on the first layer of ions with their hydration shells and the first layer of water molecules surrounding the solid phase. This layer does not move relative to the charged surface and is involved together with the membrane particle upon its movement relative to the liquid. Consequently it is usually proposed that $\zeta \leqslant \varphi_1$ (Fig. 15.5). The difference between ζ and φ should be the smaller the less the $d\varphi/dx$ ratio decreases in diluted solutions.

The value of the zeta potential can be calculated by measuring the rate of motion of membrane charged particles in the electric field using the Smoluchowski formula:

$$\zeta = \frac{\eta v}{\varepsilon\varepsilon_0 E}, \tag{15.27}$$

where η is the medium viscosity, v is the rate of particle motion, and E is the potential gradient of the external electric field. The v/E ratio is called electrophoretic mobility.

16

Electrodiffusion Theory of Ion Transport Across Membranes

Theories of passive membrane transport of ions were developed with regard to physicochemical membranes. There are two approaches for the description of a direct flow of charged particles across a membrane: discrete and continuous ones. In the discrete approach, based on the Eyring theory of absolute reaction rates, it is suggested that the particle passes across the membrane in several discrete jumps over the activation barriers. The continuous approach is based on the concept of diffusion of particles in a homogeneous membrane.

A continuous approach based on the electrodiffusion model is most expedient for description of ion transport across synthetic lipid membranes. To describe ion transport across selective ion channels it is more logical to use a discrete approach (see Chapter 18).

16.1 The Nernst–Planck Equation of Electrodiffusion

In the electrodiffusion model, the membrane is considered as a continuous homogeneous medium, where diffusion of point non-interacting particles takes place. The total flux of ions j of an arbitrary type, moving passively and independently in such a homogeneous medium towards the x axis, is proportional to the ion concentration, their mobility and the force acting on the ion. The general transport equation is as follows:

Flux = Concentration × Acting force × Mobility. □

Thus, flux J of ions j with concentrations in the x plane c and mobility u, is determined as

$$J = cu(-d\overline{\mu}/dx). \tag{16.1}$$

Taking into account the definition of electrochemical potential (see Eq. (15.2)) and substituting the derivative $d\overline{\mu}/dx$ in Eq. (16.1), we obtain

$$J = -uRT\frac{dc}{dx} - uczF\frac{d\varphi}{dx}, \tag{16.2}$$

where z is the ion valence, T is absolute temperature, R is the gas constant, and F is Faraday number. Equation (16.2) is the electrodiffusion equation or the Nernst–Planck equation. It describes ion diffusion in solution or in a homogeneous uncharged membrane. The first member in the right-hand side of the equation describes free diffusion (the diffusion component of the total flux), and the second member expresses the migration of ions in the electric field (the migration component). The dimension of the flux is in $\text{mol} \cdot \text{cm}^{-2} \cdot \text{s}^{-1}$.

The Planck–Henderson approach to the solution of electrodiffujsion equation (16.2) suggests that the condition of neutrality is fulfilled not only for the volume of phases separated by a membrane, but also for the membrane itself.

We have

$$\Delta\varphi = -\frac{u_+ - u_-}{u_+ + u_-}\frac{RT}{F}\ln\frac{c_2}{c_1}, \tag{16.3}$$

where $\Delta\varphi = \varphi_2 - \varphi_1$; u_+ and u_- are correspondingly cation and anion mobility in the membrane. Relation (16.3) is the Henderson equation. It allows the calculation of the diffusion potential generated between two electrolyte solutions with different ion concentrations.

The Henderson equation (16.3) is applicable for membranes of microscopic thickness (about micrometers and higher) but is not valid for the case of thin lipid and cell membranes, where the conditions of local electroneutrality are not observed over the overall thickness of the membrane. It is essential also that the thickness of biomembranes is lower than the Debye screening radius.

16.2 Constant Field Approximation

The second approach to the solution of the Nernst–Planck equation known as constant field approximation, is based on the assumption that changes in the electric potential are linear or that the field intensity over the entire thickness of the membrane is constant ($d\varphi/dx$ = const). This condition is valid for thin membranes, where the concentration of charge carriers is low and the thickness of the double electric layer is very high (see Section 4 in Chapter 15), i.e. in the case of ion diffusion across lipid bilayers and cell membranes. □

If the profile of the electric potential in the membrane with thickness h is linear ($d\varphi/dx = \varphi/h$ = const), the Nernst–Planck equation (16.2) is similar to the inhomogeneous linear differential equation

$$dc/dx + Ac = -B, \tag{16.4}$$

where $A = zF\varphi/RTh$, $B = J/uRT$, φ is the transmembrane potential difference. The solution of Eq. (16.4) shows the dependence of the total passive flux J_j of ions j on the difference of potentials φ on the membrane and the concentrations of ions at the membrane edges:

$$J_j = \frac{zF\varphi u}{h} \cdot \frac{c' - c'' \exp(zF\varphi/RT)}{1 - \exp(zF\varphi/RT)}, \tag{16.5}$$

where c' and c'' are ion concentrations j at the membrane edges correspondingly in the planes $x = 0$ and $x = h$. For practical purposes, in Eq. (16.5) it is necessary to substitute concentrations c' and c'', relating to the membrane phase, for ion concentrations in solutions on both sides of the membrane. A. Hodgkin and B. Katz proposed that ion concentrations at the membrane edges (c' and c'') are proportional to the concentrations in the outer and inner washing solutions (correspondingly c_o and c_i):

$$c' = \gamma c_o, \qquad c'' = \gamma c_i, \tag{16.6}$$

where γ is the distribution coefficient dependent on the ion lipophilicity. Then Eq. (16.5) for the passive ion flux across the membrane will be as follows

$$J = \frac{zF\varphi P}{RT} \cdot \frac{c_o - c_i \exp(zF\varphi/RT)}{1 - \exp(zF\varphi/RT)}. \tag{16.7}$$

The dimension of the P ($P = uRT\gamma/h$) value, which is the permeability coefficient, is in cm/s. It is directly proportional to the ion mobility in the membrane and the distribution coefficient and inversely proportional to the membrane thickness. The magnitude of current I transferred by the given ion is $I = zFJ$. □

Equation (16.7), derived by D. E. Goldman (1943) and further complimented by A. Hodgkin and B. Katz (1949). makes it possible to compute the passive ion flux provided we know its concentration in solutions on both sides of the membrane, the potential difference across the membrane and the membrane permeability for the given ion. To calculate the membrane permeability coefficient, it is required to determine the one-row fluxes and the total ion flux with the use of a radioactive label and also to obtain evidence that active transport systems are not involved in the ion transport. On plant cell membranes, the permeability coefficients for K^+ vary from 10^{-8} to 10^{-6} cm/s; the P_{Na} values are typically by an order of magnitude lower than the P_K values, while the P_{Cl} values do not exceed 10^{-8} cm/s.

Volt-ampere Characteristics. The Goldman equation (16.7) predicts a nonlinear dependence of the transmembrane ion flux on the potential difference across the membrane.

Upon a drastic difference of ion concentrations on both sides of the membrane, the reasons for nonlinearity of volt-ampere characteristics are due to the electrostatic field effect on the profile of ion concentration in the membrane (Fig. 16.1). Upon applying differently oriented electric fields, the concentration of ion transferring electric current in the membrane either increases or decreases, leading correspondingly to an increase or decrease in the membrane electropermeability — the rectifying effect. □

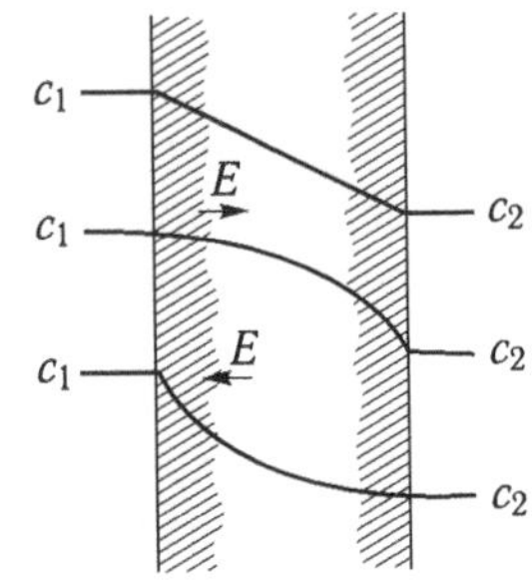

Figure 16.1. Profiles of cation concentration in the membrane in the absence of electric field and upon application of the external electric field E of different orientations.

The expression for the potential difference, formed on the membrane as a result of ion diffusion, is derived from the equation for ion fluxes (16.7). At equilibrium, when there is no electric current across the membrane, the sum of currents transferred by ions is zero: $\sum_j I_j = 0$. With account for only univalent ions

$$I_{\mathrm{K}} + I_{\mathrm{Na}} + I_{\mathrm{Cl}} = 0. \tag{16.8}$$

After substitution and reduction of all members by the common factor, we get

$$P_{\mathrm{K}}[\mathrm{K}_o] - P_{\mathrm{K}}[\mathrm{K}_i]\exp(F\varphi/RT) + P_{\mathrm{Na}}[\mathrm{Na}_o] - P_{\mathrm{Na}}[\mathrm{Na}_i]\exp(F\varphi/RT) + \\ + P_{\mathrm{Cl}}[\mathrm{Cl}_o] - P_{\mathrm{Cl}}[\mathrm{Cl}_i]\exp(F\varphi/RT) = 0. \tag{16.9}$$

When solving the obtained equation relative to φ we obtain the expression for the membrane potential:

$$\varphi = \frac{RT}{F}\ln\frac{P_{\mathrm{K}}[\mathrm{K}_o] + P_{\mathrm{Na}}[\mathrm{Na}_o] + P_{\mathrm{Cl}}[\mathrm{Cl}_i]}{P_{\mathrm{K}}[\mathrm{K}_i] + P_{\mathrm{Na}}[\mathrm{Na}_i] + P_{\mathrm{Cl}}[\mathrm{Cl}_o]}, \tag{16.10}$$

where φ is the difference of potentials on the membrane, P_j is permeability of corresponding ions; indices o and i refer to ion concentrations in the outer and inner phases. The expression (16.10) is the Goldman equation for the membrane potential. It demonstrates that the electric potential on the membrane is determined by the difference in stationary ion concentrations on both sides of the membrane and by different coefficients of their permeability. Consequently, the strongest effect on the membrane potential is exerted by the ion having the highest coefficient of permeability that is observed in experiments. The relation (16.10) describes adequately the experimental dependences of the potential on the ion composition if the range of concentrations is not rather wide. This reveals serious restrictions of the electrodiffusion theory in describing electric characteristics of a living cell.

Ussing Relation. The total ion flux determined by Eq. (16.7) is equal to the algebraic sum of two one-row fluxes: $J = J_i - J_o$, where J_i is the flux to the inner phase, separated by the membrane, and J_o is the flux to the outer solution. Apparently, in Eq. (16.7) describing the total ion flux, the component members correspond to the incoming flux and the output flux. The expressions for one-row fluxes are as follows:

$$\overrightarrow{J} = J_i = \frac{zF\varphi P}{RT} \cdot \frac{c_o}{1 - \exp(zF\varphi/RT)},$$
$$\overleftarrow{J} = J_o = \frac{zF\varphi P}{RT} \cdot \frac{c_i \exp(zF\varphi/RT)}{1 - \exp(zF\varphi/RT)}, \tag{16.11}$$

where $\overrightarrow{J}$ and $\overleftarrow{J}$ are ion fluxes directed to the inner and outer solutions. As seen from the above equations, the incoming flux is linearly dependent on the ion concentration in the outer solution and is not dependent on the concentration in the inner medium. In its turn, the outcoming flux depends only on the concentration in the inner solution. This means that the incoming and outcoming fluxes are independent in the constant field approximation.

It follows from eq. (16.11) that unidirectional fluxes are interrelated by the Ussing–Theorell correlation:

$$\frac{J_i}{J_o} = \frac{c_o}{c_i \exp(zF\varphi/RT)}. \tag{16.12}$$

The implementation of the Ussing–Theorell equation is a criterion of passive transport, when the ion flux is determined only by the concentration gradient and the electric field action. Deviation from the equality (16.12) can suggest the involvement of the active transport system in the ion transport across the membrane or can be a consequence of the violation of the independence condition of ion fluxes. For example, it may happen upon diffusion through narrow channels in the membrane (see Chapter 17). In this case, the usual electrodiffusion description (see equations (16.2), (16.7) and (16.12)) based on the concept of diffusion of point non-interacting particles is not valid. □

Unidirectional fluxes are determined experimentally using radioactive isotopes by adding them to one of the solutions and measuring the rate of the isotope inflow into the other solution separated by the membrane.

17

Induced Ion Transport

17.1 Bilayer Lipid Membranes

The main regularities of ion transport across membranes were studied in experiments with different model systems, the most closely related to biomembranes being the bilayer lipid membrane (BLM) (see Section 2 in Chapter 12).

The energy consumption required for ion penetration into the nonpolar phase can be estimated using the Born formula (see Eq. (15.5)). According to the latter, the energy needed for ion to transfer from water to a membrane depends on ion radius r and dielectric permeability of water ε_w and the membrane ε_m.

In the SI system, this formula has the simplest and obvious form:

$$W = \frac{z^2 e^2}{2r} (1/\varepsilon_m - 1/\varepsilon_w), \tag{17.1}$$

where z is the ion valence, and e is the elementary charge. The free energy values calculated by this equation for K^+ transition from water to the nonpolar solvent with $\varepsilon_m = 2$ has a high value (250–350 kJ/mol). That is why there is a barrier preventing the transition of alkaline metal ions across the hydrophobic part of the lipid bilayer in the non-hydrated form.

As seen from Eq. (17.1), the energy of ion transfer to a membrane decreases with the increase in the ion radius. Consequently large organic ions penetrate across the BLM more easily than alkaline metal cations. Such large lipophilic ions as dipicrylamine and tetraphenyl boron may directly pass across membranes. With a change in the concentration gradient in the case of one penetrating ion, the difference in electric potentials on the BLM changes in accord with the Nernst equation. This method made it possible to demonstrate passive movement of penetrating ions across mitochondrial membranes, submitochondrial particles, and bacterial chromatophores upon energy-dependent generation of membrane potential (see Chapter 21).

As seen from Eq. (17.1), the magnitude of the energy barrier in the membrane decreases, and accordingly the membrane permeability for the ion increases not only upon the growth of its radius, but also as the ε_m value becomes closer to ε_w. Namely, these physical principles underlie the ion transfer by ionophores. Ionophores can form a large-size complex with the ion (the carriers) or form a water-filled pore in the membrane (the channels). These transfer mechanisms (with the involvement of carriers and through the ion channels) have been studied in more detail in experiments with BLM. □

17.2 Mobile Carriers

The ion transfer across the membrane with the involvement of a carrier proceeds via stages of the ion–ionophore complex formation on one side of the membrane, complex transport across the membrane, ion release on the other side, and the return of ionophore. Two schemes of the carrier performance can be proposed: a small "carrousel" when the ionophore is not released from the membrane and a large "carrousel" when the ionophore is transported across the membrane, and the formation and disintegration of complexes occur in the immiscible layers near the membrane. The mechanism of a small carrousel is activated, for example, upon the transfer of K^+ in the presence of valinomycin.

In the membrane with electric permeability ε_m, the energy of the complex of an ion with radius r and a carrier of radius b may be written based on the Born formula (Fig. 17.1):

$$W = \frac{e^2}{2\varepsilon_m b} + \frac{e^2}{2\varepsilon_c}(1/r - 1/b), \tag{17.2}$$

where ε_c is the effective dielectric permeability of the inner sphere of the complexon. The energy of the ion–carrier complex is much lower than the energy of the dehydrated ion. For example, for a charged complex of radius 1 nm, the free energy of the transfer from water to the membrane is about 15 kJ/mol versus the values of 250–350 kJ/mol corresponding to the dehydration energy of free ions of radius 0.1–0.15 nm.

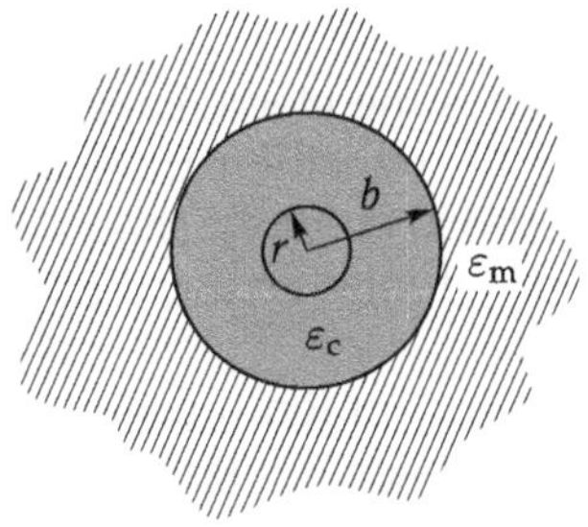

Figure 17.1. To the calculation of the electrostatic energy of the ion–carrier complex. ε_m is the dielectric constant of the membrane, ε_c is the effective dielectric constant of the inner cavity of the complexon, r is the ion radius, b is the radius of the ion–carrier complex.

X-ray analyses of such complexes demonstrate that the cation enters the inner cavity of the ionophore molecule, therewith the formed structure is stabilized due to the ion interaction with 6–8 polar (ether, amide or ester) groups. In such complexes, the metal ion is screened from the interaction with the solvent. The chemical structure of a valinomycin molecule is shown in Fig. 17.2, *I*. The high selectivity of valinomycin in the formation of complexes with K^+ is achieved due to the good steric correspondence of K^+ and the inner cavity of valinomycin formed by CO groups. In complex with K^+, valinomycin dehydrates K^+ completely, whereas the carbonyl (CO–) dipoles substitute the hydration shell. The Na ion can fit in the valinomycin cavity, but due to a smaller (as compared to that of K^+) radius it is not able to interact with oxygen of carbonyl groups. Consequently, the energy of the Na ion binding to water of the hydration shell is higher than the energy of its binding to the valinomycin molecule. If the valinomycin molecule had been relatively flexible and the cavity dimensions could have been "adjusted" for Na^+, valinomycin would not have had its good selectivity. A valinomycin molecule transfers no more than 10^4 ions per second across the BLM. □

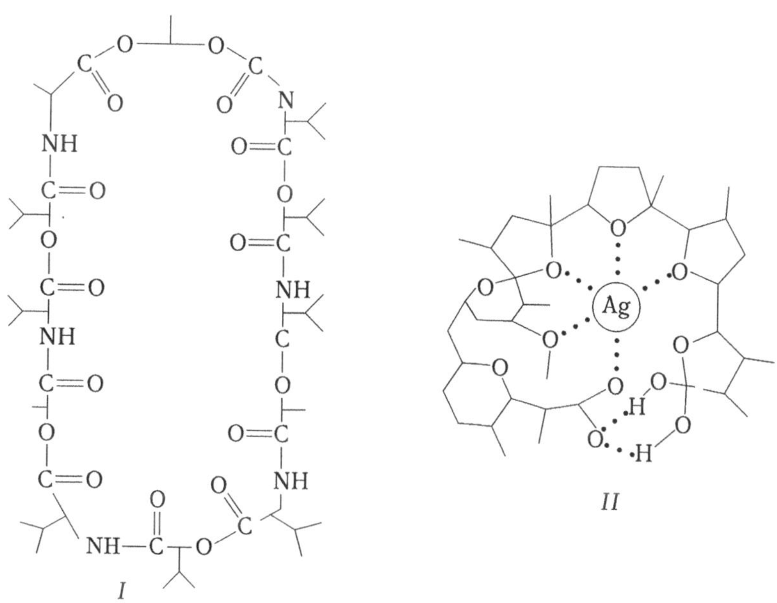

Figure 17.2. Chemical structure of some ionophores.

I, valinomycin molecule; *II*, system of metal–oxygen and hydrogen bonds in nigericin silver salt.

Some ionophores such as nigericin and grisorixin have formally a noncyclic structure, but they are also capable of binding metal ions, thus forming complexes in which the molecule is in the folded, pseudocyclic conformation. Nigericin and its analogs have the COOH-group ionized at neutral pH values. The hypothetical structure of the cation–nigericin complex is also shown in Fig. 17.2, *II*.

The analyzed modes of carrier operation are demonstrated in Fig. 17.3. The figure also shows the mechanism of H^+ transport by lipophilic carriers of protons. The proton transport across the membrane is realized mainly by weak acids, and their dissociated form attaches H^+ with the formation of a neutral molecule. The non-dissociated form of protonophore crosses the membrane moving along

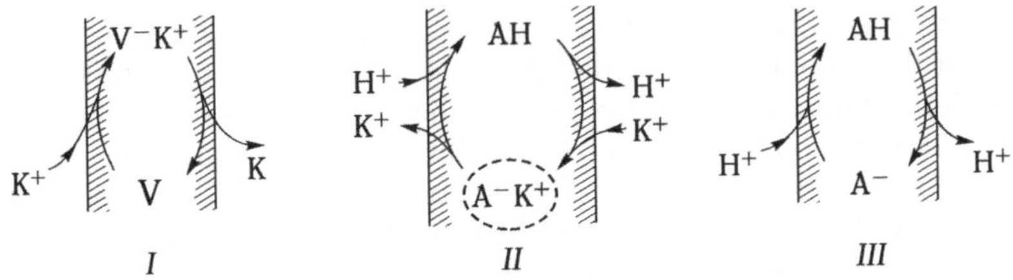

Figure 17.3. Induced ion transport with involvement of different carries.

I, neutral ionophores (valinomycin); *II*, nigericin (exchange transport of H^+ and K^+); *III*, lipophilic disconnectors of phosphorylation (proton transfer).

A^- is the dissociated form of the carrier; AH is the protonated form of the carrier; V is valinomycin.

the concentration gradient; the negatively charged anion form is transferred in the opposite direction under the action of the electric field. This mechanism is characteristic of uncouplers of oxidative and photosynthetic phosphorylation such as 2,4-dinitrophenol, carbonyl cyanide chlorophenylhydrazine, tetrachlorotrifluormethyl bensymidasole etc. (Fig. 17.4), which increase the permeability of BLP and energy-conjugated membranes for H^+.

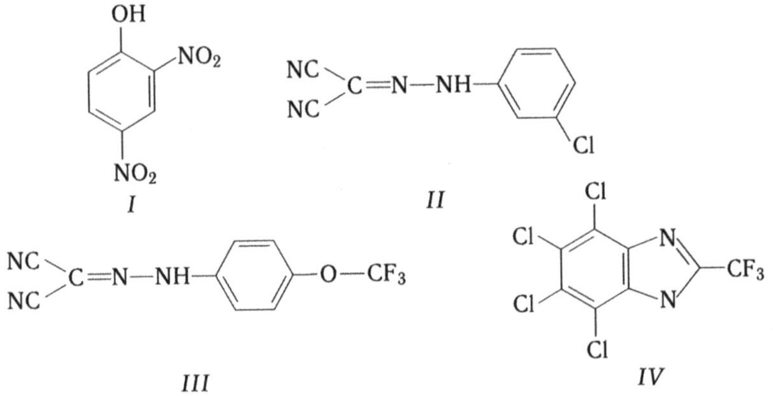

Figure 17.4. Chemical structure of some lipophilic carriers of proteins.

I, 2,4-dinitrophenol; *II*, carbonyl cyanide chlorophenylhydrazine; *III*, carbonyl cyanide trifluoromethoxyphenyl hydrazine; *IV*, tetrachlorotrifluormethyl bensymidasole (TTFB).

17.3 Channel-forming Reagents

Ion motion is facilitated greatly, if ionophore molecules form a complex with an aqueous pore — a channel. The most known channel-forming compounds are gramicidin A, alamethycin, amphothericin, monazomycin, and polyene antibiotics. □

Molecules of these compounds have affinity to the aqueous and organic phases that, on the one hand, permits them to form a water pore, and on the other hand, leads to strong sorption of the antibiotic on the membrane. The outer part of the molecules in the pore is hydrophobic, and well polarized groups are inverted inside the channel.

The charged or strongly polarized groups can be located on one end of the molecule. Such groups are an "anchor" retaining the polar terminus on one side of the membrane and allowing the molecule to pierce through the hydrophobic part of the membrane.

Ion channels in the BLM can be found experimentally by the specific conductance noise, generated due to their opening and closing. This feature, peculiar to the channel functioning, permits to distinguish this conductance mechanism from the channel transport via carriers.

Channels Formed by Gramicidin A. The peptide chain of a gramicidin A molecule is formed by fifteen hydrophobic amino acids (Fig. 17.5), the characteristic feature here is the alternation of L- and D-configurations of amino acids as well as the presence of a formyl group at the N-termiunus of the molecule (the head) and ethanolamine at the C-terminus (the tail). The molecule is completely hydrophobic except for carbonyl and imino groups of the peptide chain and terminal polar groups.

CHO – L-*Val* – *Gly* – L-*Ala* – D-*Leu*

L-*Ala* – D-*Val* – L-*Val* – D-*Val* – L-*Try* – D-*Leu*

L-*Try* – D-*Leu* – L-*Try* – D-*Leu* – L-*Try* – NHC_2H_4OH

Figure 17.5. Structure of gramicidin A.

Each horizontal row of amino acids corresponds to approximately one turn of the $\pi^6_{L,D}$-helix; diagonal lines correspond to peptide bonds connecting separate turns of the helix.

When in the membrane, the gramicidin A molecule folds in a helix-like structure — a semi-pore stabilized by hydrophobic bonds and forms a hollow cylinder of about 3 nm in length and nearly 0.5–0.8 nm in diameter. This structure is called the $\pi^6_{L,D}$-helix. The name is determined by that each helix turn is formed by six amino acid residues where the CO-groups of L-amino acid residues are oriented along the helix axis towards the C-terminus, whereas the corresponding groups of D-residues and glycine towards the N-terminus.

Upon addition of small amounts of gramicidin A to the BLM, jump type changes in the current across the membrane are observed. At fixed voltage on the BLM, discrete changes of about the same magnitude take place in the membrane conductance (Fig. 17.6). They may be caused, for example, by the formation and disintegration of the channels upon association of the two monomers and dissociation of the gramicidin dimer. The dimer structure forms an open-end transmembrane channel with conductance of about 40 pS · m^{-1} in 0.1 M solution of KCl. Roughly 10^7–10^8 ions/s can be transported through such a channel.

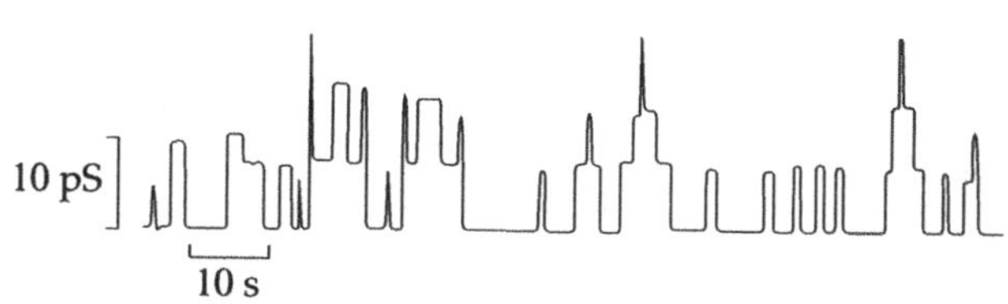

Figure 17.6. Records of fluxes through single channels formed by gramicidin A in BLM.

The reasons for transitions between conducting and non-conducting states of the channel may be diverse. An alternative opening mechanism of gramicidin channels is shown in Fig. 17.7.

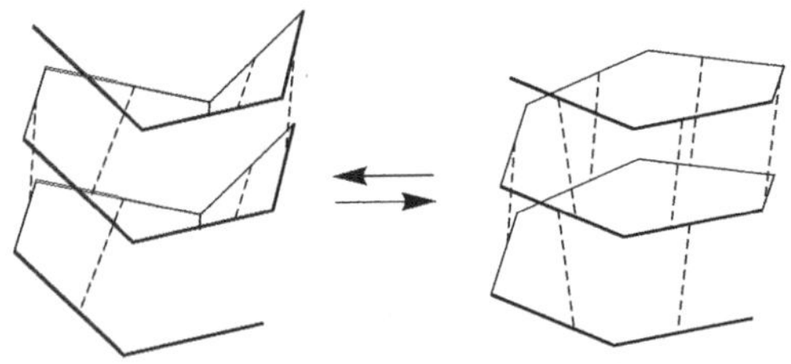

Figure 17.7. Equilibrium of anti-β_2^6- (*I*) and $\pi_{\rm L,D}^6$-helices (*II*) of gramicidin A; dashed line show hydrogen bonds (reproduced from D. W. Urry, 1972).

It is believed that the $\pi_{L,D}^6$-helix can be in equilibrium with another structure — the so-called anti-β_2^6-helix. The anti-β_2^6-helix has no internal cavity and therefore it cannot accommodate metal ions. Since the anti-β_2^6-helix has a smaller dipole moment than that of the $\pi_{\rm L,D}^6$-helix, equilibrium between these forms can shift to either side under the action of the external electric field.

Gramicidin is an extremely effective modifier of BLM. Its submicromolecular concentrations increase the conductance by 10^6–10^8 times relative to the conductance level of non-modified membranes.

17.4 Effect of Surface and Dipole Potentials on the Ion Transport Rate

Membrane conductance induced by neutral carrier ionophores changes substantially upon addition of charged surface active substances and dipole modifiers to aqueous solutions. This is explained by the fact that the movement of charged ion–carrier complexes across BLM and cell membranes depends greatly on the jumps of the boundary potential on the boundary between the membrane and electrolyte.

The boundary potential φ_b (Fig. 17.8) is summed up of two components — the surface potential φ_s and the dipole potential φ_d:

$$\varphi_b = \varphi_s + \varphi_d. \quad \square \tag{17.3}$$

The surface potential is specified by membrane fixed charges formed by dissociated groups of lipid polar heads as well as by ionized groups of amino acids of structural proteins of the membrane. The charges fixed on the membrane surface with the counterions attracted to them form a double electric layer (see Section 5 in Chapter 15).

The dipole jump of the potential is created by oriented dipoles of both water and lipid polar groups. Based on the measurements of the boundary potential of monolayers of different lipids, it was concluded that the hydrophobic region of BLM is charged positively at several hundreds of millivolts relative to the potential of aqueous solutions near the boundary with the membrane. It is accepted that in contrast to the surface potential, the dipole potential is independent of pH and ionic strength of the aqueous solution.

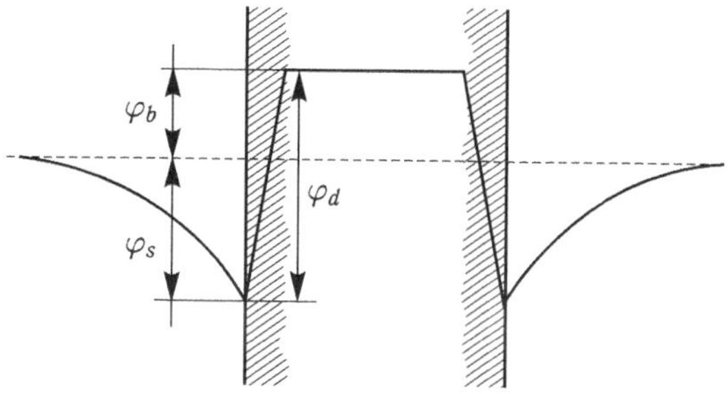

Figure 17.8. Boundary potential (φ_b) as the sum of the surface (φ_s) and dipole (φ_d) potentials. The membrane is in the symmetric ion surrounding in the absence of the externally applied difference of potentials.

The surface and dipole potentials are among the key factors determining the permeability of lipid bilayers for penetrating ions and charged ion–carrier complexes. The surface potential value determines the equilibrium ion concentration in the membrane on the boundary with the solution:

$$c_M = \gamma c \exp\left(-\frac{zF\varphi_s}{RT}\right), \tag{17.4}$$

where γ is the coefficient of the ion or charged complex distribution between the membrane and water. If the surface layer is negative, which is characteristic of most cell membranes and many BLM, the concentration of cations near the membrane is increased, and accordingly increased is the cation concentration in the membrane on the boundary with the solution. □

The dipole potential value determines is one of the components, which determines the height of the potential barrier for the transition of charged particles to the membrane.

The increase in the negative surface potential (at a constant value of the dipole potential) should go in parallel with the increase in the BLM conductance, induced by positively charged ion–carrier complexes. The experimental study of the dependence of BLM conductance, induced by positively charged ionophore nonactin–potassium complex, supports this assumption explicitly. Inasmuch as the BLM charge is formed as a result of dissociation of acidic and basic groups of phospholipid heads, it changes depending on pH, thereby inducing the conductance rise with an increase in pH in the presence of nonactin.

If the distribution density n of dipole groups oriented perpendicular to the membrane plane is known as well as the value of dipole moment μ, the dipole potential can be calculated:

$$\varphi_d = \frac{n\mu}{\varepsilon_0}. \tag{17.5}$$

The change in the BLM conductance upon absorption of charged detergents and multivalent metal ions on it is also explained by an alteration in the surface BLM potential.

18

Ion Transport in Channels

18.1 Discrete Description of Transport

The classical description of ion transport across biological membranes is based on the concept of diffusion of charged particles not interacting between each other in the homogeneous continuous medium. Electrodiffusion equations obtained in the approximation of a constant field (the Goldman – Hodgkin – Katz approximation) are the theoretical basis of electrophysiology. They include Equations (16.7) and (16.10) describing the flux of one type of ions across the membrane and also the membrane potential under conditions of zero electric current:

$$
\begin{aligned}
J &= \frac{zF\varphi P}{RT}\frac{c_o - c_i \exp(zF\varphi/RT)}{1 - \exp(zF\varphi/RT)}, \\
\varphi &= \frac{RT}{F}\ln\frac{P_K[K_o] + P_{Na}[Na_o] + P_{Cl}[Cl_i]}{P_K[K_i] + P_{Na}[Na_i] + P_{Cl}[Cl_o]}.
\end{aligned}
\tag{18.1}
$$

The assumption on the linear dependence of the potential on the distance is justified in the first approximation in thin membranes of $\leqslant$ 10 nm thick. However, the other prerequisites of the electrodiffusion theory are violated in the case of ion transport. □

The membrane is not a homogeneous phase for the ion motion because ion transport across the membrane is realized through specialized regions — ion channels which are lipoprotein complexes of a complex structure. These molecular complexes generate pores in the membrane. At the same time, ion channels can open and close i.e. perform spontaneous transitions between conducting and non-conducting states.

In narrow channels (the sodium channel of 0.31×0.51 nm and the potassium channel of 0.45×0.45 nm) independent diffusion is impossible, and only one-row ion motion is allowable. Due to the narrowness of the pore, strong ion – ion interactions can occur in the channel.

The differences in surface potentials at the ends of the pore can result also in diverse values of the distribution coefficient. Local non-homogeneities of the electric field may also exist, associated with the protein molecule charges in the channel.

Consequently, there are phenomena which cannot be explained in the framework of the free diffusion concepts. They include interaction of fluxes and blocking the channel at high electrolyte concentrations.

At present, the theory based on general physical principles is accepted.

1) The ion entrance to the channel is accompanied by substitution of water in the hydration shell for polar groups lining the channel cavity.
2) Peculiarities of ion interaction with molecular groups of the channel are taken into account by corresponding profiles of potential ion energy, presented as a number of sequential potential wells and barriers.
3) The ion is retained in each potential well for a rather long period (as compared to the frequencies of heat fluctuations). The hopping is possible only into an empty well. Another electron cannot get into the occupied potential well due to the electrostatic repulsion by the ion that is already in the well.
4) Hopping from one well to another occurs under the action of thermal fluctuations. The probability of hopping depends on the applied electric field. □

Based on these proposals, the equation for the ion flux can be derived using the theory of absolute rates of Eyring reactions, provided the ion transport proceeds as a series of sequential jumps over energy barriers in the channel. Side barriers correspond to "inlet" regions of the channel, where the primary process of dehydration takes place, while the central barrier corresponds to the selective region (a selective filter). If the rate of ion transfer through the channel is limited by the central energy barrier (fig. 18.1), unidirectional ion fluxes are determined by the equations

$$\overrightarrow{J} = c_0 A \exp\left[-\left(\frac{E}{RT} + \frac{\varphi F z}{2RT}\right)\right] = c_0 \nu \exp\left(-\frac{zF\varphi}{2RT}\right),$$
$$\overleftarrow{J} = c_i A \exp\left[-\left(\frac{E}{RT} - \frac{\varphi F z}{2RT}\right)\right] = c_i \nu \exp\left(\frac{zF\varphi}{2RT}\right), \tag{18.2}$$

where A is the constant, E is the barrier height in the absence of the electric field, ν is the rate constant of transition through the central barrier in the absence of the electric field.

The form of the volt-ampere characteristic of the ion channel can be used to reveal what stage (the boundary or inner one) determines the overall ion transfer through the channel. Figure 18.2 illustrates the relationship between the energy profile (I) and

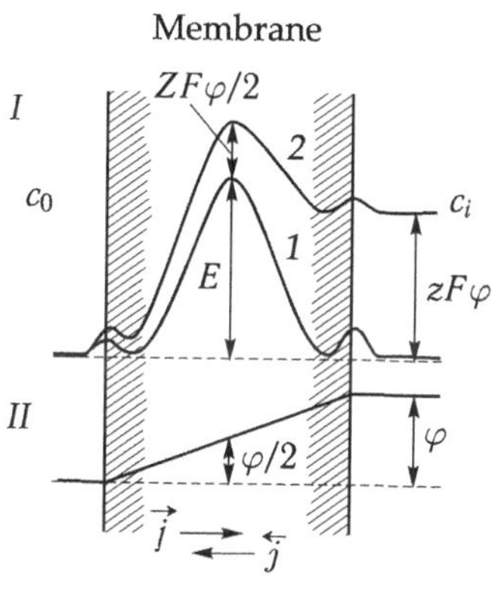

Figure 18.1. Energy profile in the three-barrier model of the channel (*I*) in the absence of the field (*1*) and upon superposition of the difference of electric potentials on the membrane (*2*) and profile of the electric potential φ on the membrane (*II*).

c_o and c_i are the electrolyte concentrations in the solution surrounding the membrane; E is the height of the central barrier; z is the ionic valence; F is the Faraday number; $\vec{j}$ and $\overleftarrow{j}$ are one-row ion motions through the channel.

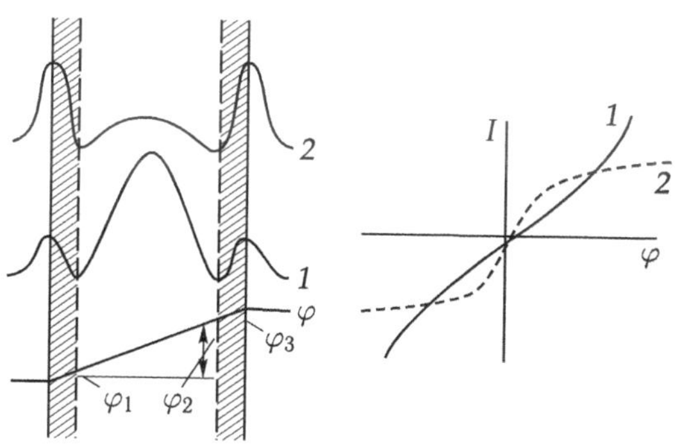

Figure 18.2. Energy profiles in channels with the inner limiting stage (*I*, *1*) and the interfacial limiting stage (*I*, *2*), and corresponding volt-ampere characteristics of the channels (*II*).

I, current; φ, electric potential; φ_1, φ_2 and φ_3, voltage drop in the membrane regions corresponding to three energy barriers.

the form of the volt-ampere characteristic of the channel (*II*). If the current through the channel is determined by the transfer over the central barrier, the volt-ampere characteristic will bend to the current axis (curve *1*). But if the current is determined by the rate of ion entering into the channel, then at high voltage on the membrane all entering ions will be transferred through so that the current will be saturated (curves *2*). Rate constants of ion transfer over side barriers depend little on the electric field, because the difference of potentials applied to the membrane almost completely decreases in the inner part of the membrane (voltage drop φ_2) and the potential decrease on the boundary itself is inessential (voltage drops φ_1 and φ_3). Figure 18.2 demonstrates that the total jump of the potential on the membrane is summed up of small-value potential jumps on the boundary (φ_1 and φ_3) and the foremost decrease of the potential on the inner part of the membrane.

As seen from Eq. (18.2), the relationship of unidirectional fluxes is

$$\overrightarrow{J}/\overleftarrow{J} = \frac{c_o}{c_i} \exp\left(-\frac{zF\varphi}{RT}\right), \tag{18.3}$$

which is compatible with the Ussing formula (see (16.12)) based on the elecrodiffusion model. It is because in the case of diluted ion solutions, when the channel is mostly empty, the ions do not "interfere" with each other. However, even in this case, the symmetry of the volt-ampere characteristic is substantially dependent on the position of the central barrier relative to the middle of the membrane. When the cntral energy barrier is positioned near the one side of the membrane, the ion flux in one direction depends essentially on the potential, whereas the counter flux is weakly sensitive to the membrane potential changes. As a result, general rectifying effect is observed.

18.2 Channel Blocking and Saturation

At high ion concentrations, the both binding sites at the "inlet" and "outlet" change their states not only in the process of the transport itself through the channel, but also due to the interaction with the surrounding solutions. In other words, the transport rate in the channel should depend on the probability of filling the binding sites on the membrane boundary. It may lead to the filling of the both binding sites and thus to the channel blocking at high electrolyte concentrations. The probability of filling the inlet region of the channel is dependent on the ion concentration and can be determined from conditions of the established equilibrium between fluxes directed from the solution to the channel and backwards. A unidirectional flux depends on the ion concentration in both solutions surrounding the membrane. Accordingly, the principle of independence of ion fluxes applied to channels at high ion concentrations is not valid here (see (16.12) and (18.3)).

Hence for a channel with two binding sites (a "two-site channel"), a decrease in conductance at high ion concentrations in surrounding solutions is observed when the both binding centers are filled (the channel is in the 1 1 state). For example, this is observed for gramicidin channels formed in BLM in the presence of growing Cs^+ concentrations.

Ion channel as a dynamic structure. If one of the binding sites in the channel is filled, for the second ion entering the free site will be hindered due to additional energy consumption necessary to overcome the Coulomb interaction with the ion that is already in the channel. In other words, the entrapment of an ion to the left potential well will lead to an increase in the energy barrier for another ion to enter the right potential well from the solution (Fig. 18.3).

Thus, upon taking into account Coulomb interactions the form of the energy profile depends on the position and number of ions in the channel.

When the number of particles in the channel changes, the energy profiles change in such a way that the access of the second ion to the channel, as compared to that of the first ion, is impeded, and on the contrary the second ion is promptly released from the channel. A detailed analysis of the model demonstrates that the Coloumb ion – ion interaction accelerates greatly the total ion trasnport through the channel.

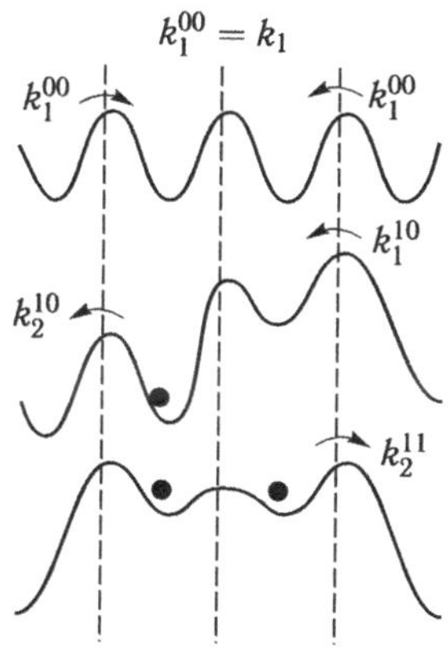

Figure 18.3. Dependence of the energy profile on the channel filling with account for Coulomb interactions.

k_1 are constants of ion release from the solution; k_2 are constants of ion entering the solution. Top indices correspond to different states of the channel at different filling of the binding sites.

The height of barriers can change dynamically not only because of the Coloumb interactions of ions in the channel, but also as a result of conformational rearragmenents in the protein, forming the ionic channel, which can be initated by hopping of the ion between separate binding sites. □

Such periodic conformational rearrangements may occur in the ATPase of energy-congugating membranes in the process of its functioning, in cytochromes upon changes in their redox state, in bacteriorhodopsin of halophilic bacteria upon excitation of the chromophore group (see Chapters 21 and 30).

Transport in an open channel. Theory of selectivity.

The penetration of an ion into the pore is facilitated by that the free energy growth caused by the loss of the hydration shell (500–700 kJ/mol) is compensated by the energy decrease upon interaction with dipole groups of the channel.

Another important factor facilitating the penetration of an ion into the channel is the Coloumb interaction of the transported cation with the fixed anion center in the channel lumen. The existence of dipole groups and fixed charges in the channel does not allow the energy of ion transfer from the solution into the channel to exceed $\sim$ 30.5 kJ/mol. □

The dramatic dependence of ion penetrability on their geometry and dimensions, observed in experiments, suggests that it is determined by the topography of the selective filter corresponding to the narrowest place . When the ion dimension reaches its critical value, its penetrability drops to zero. On the whole, however, the structural factors, determining the penetrability, are not reduced only to the ion geometry and the selective filter, but should include details of the chemical structure of channel macromolecules as well as movements of water molecules in the pore associated with the ion motions. The molecular dynamics method for intramolecular mobility (see Section 3 in Chapter 11) allows studies of the dynamics of ion motion at short time intervals 10^{-8}–10^{-9} s close to the resolving power of the method.

Movements of water molecules were simulated in the narrow part of the K^+ channel pore ($\sim$ 0.3–0.5 nm) through which ions are trasnported (M. Green, 1991). It was found that the presence of charged protein groups (COO^- and NH_2^+) in channel walls leads to the transition of water molecules inside the pore to the bound state, thus closing the narrow place in the channel and blocking the passage of ions.

18.3 General Properties of Ion Channels in Nervous Fibers

Among different ion channels in biological membranes two main types of channels in nervous fibers (Na^+- and K^+-channels) are characterized in most detail.

The experimental data on ion transport in the Na^+-channel are adequately described based on different models with distinct energy profiles (Fig. 18.4, curves *1* and *2*). The input and output Na^+ fluxes in the Na^+-channel are independent; this corresponds to the energy profile with a high central barrier (curve *1*). However, such a form of the energy profile does not agree with the data on the existence of a fixed anion group in the selective center of the Na^+-channel that is supported by the dependence of the Na^+-channel conductance on pH of the medium. Protonation of the acidic group (pK 5.2) leads to blocking the Na^+-channel. At physiological values of pH, the profile, where the central energy barrier is split in two, corresponds to a greater extent to the channel state with the deprotonated acidic group (Fig. 18.4, curve *2*). In such a channel, the fixed anion center facilitates the access of cations to the channel by attracting them. But at the same time, this impedes the ion release from the main potential well that can be facilitated when the second cation appears in the lateral binding site. Electrostatic repulsion promotes the release of the first ion, and the second ion may fill in the central well.

According to the data of B. Hille, different organic cations with dimensions not exceeding 0.3×0.5 nm, can be transported through Na^+-channels, these dimensions complying with the minimal cross-section of the pore. The dimensions of many penetrating ions are similar to those of a partially hydrated Na ion (Fig. 18.5). It is obvious that the Na ion can pass through the selective filter retaining from 1 to 3 water molecules in its shell. Figure 18.5, *b* shows the energy profile of the Na-channel.

The proposed energy profile of a potassium channel is shown in Fig. 18.4. This profile corresponds to conditions when the slowest stage of transport is the transition of particles across the membrane boundary. The form of the energy profile suggests also that ion desorption is decelerated relative to the adsorption from the solution. A detailed analysis of the model with account for the above suggestions demonstrates that the channel is constantly in a filled double-particle state. Because of this

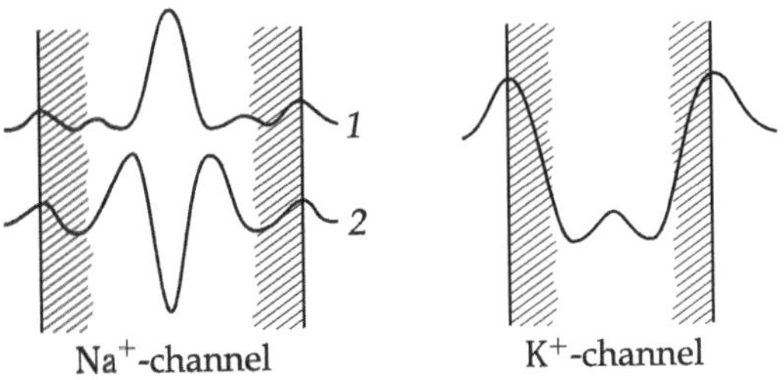

Figure 18.4. Energy profiles of the sodium and potassium channels in excitable membranes. *1* and *2* are alternative models of the Na^+-channel.

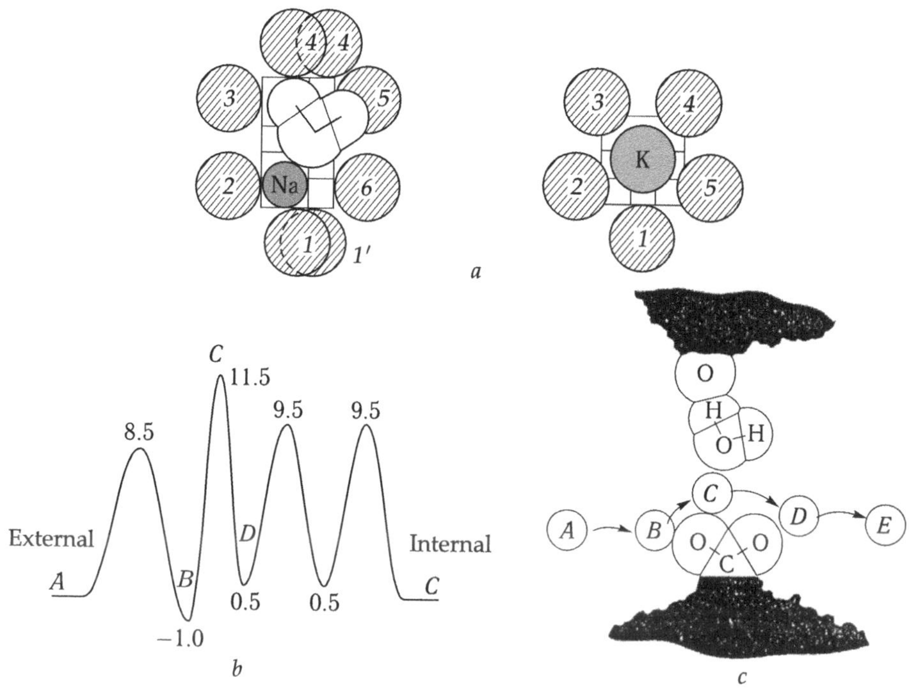

Figure 18.5. Hypothetical structure of selective filters in Na- and K-ion channels (reproduced from B. Hille, 1992).

a, Front projection of a selective filter; *1, 1′* is the COO group; *2–6* are oxygen atoms; the graticule is in E; the Na-channel contains a water molecule.

b, Energy profile of Na-channel. Designations *A–E* correspond to ion energies at different stages of its transition through the channel. The energy levels of potential barriers for the Na ion are given in *RT* units. For ions with lower penetrability, the *C* peak should be higher; ions with stronger binding have a deeper potential well *B* in the channel.

c, View of a selective filter of Na-channel in the direction parallel to the membrane plane. *A–E* designate different positions of ions in the channel. In position *C* the ion passes through the narrowest place of the channel — the selective filter with the charged COO^--group (bottom) and the oxygen atom (top) as well as the water molecule. The dark areas at the top and bottom show the channel orifice.

the particle release from the channel into the solution is accompanied by a sequential shift of ions over the whole channel. And as there are two wells in the channel, the "order" of the reaction increases twofold.

Method of Local Voltage Fixation (Patch-Clamp). Ion channels of biological membranes are studied primarily by the local patch-clamp method. In this method, voltage is fixed on a microregion of a cell surface that makes it possible to record currents in single ion channels. In 1976 E. Neher and B. Sakmann were the first to perform such measurements. A glass pipette with the orifice diameter from 0.5 to 1.0 μm and electric resistance of about 5 MOhm is used as a measuring electrode. Low pressure is created in the pipette and a fragment of a cell membrane is drawn into its tip. The cell membrane forms a tight and mechanically stable contact with the pipette tip and thereby provides reliable electric insulation of the membrane fragment from the remaining cell surface and the extracellular solution. The insulation resistance reaches 100 GOhm (giga-Ohm contact) thus specifying low values of leakage current and a low level of current noise. These conditions make it possible to record changes in the current associated with the opening and closing of separate ion channels.

Measurements of currents in single channels at different voltage magnitudes and different ion composition of the surrounding solutions yield data on the channel conductance, average lifetime of the open and closed states, provide information on potential dependence and selectivity of the channels and give grounds for constructing kinetic models of ion channel functioning. This method allows also investigation of modifications in the structure-functional characteristics of the channels upon the action of various pharmacological preparations.

An essential peculiarity of most channels is spontaneous jump-like changes in conductance observed upon measuring the current in single channels. Upon transitions between the open and closed states, the current emergence and disappearance in each channel taken separately occur in a jump-like manner. However, the total changes of the current in the ensemble of channels, each changing its stage with definite probability, are averaged statistically and on the whole have a gradual character (Fig. 18.6).

18.4 Molecular Structure of Channels

The opening and closing of channels is controlled by a special molecular "gating" apparatus that opens the pore in response to the external action. Two types of channels are distinguished: *potential-dependent* channels, where the gating apparatus is controlled by the transmembrane potential, and *ligand-operated* channels, where one protein molecule combines a receptor part binding the neural mediator, and a channel part where the gating apparatus is controlled first of all by the state of the receptor part.

Figure 18.7 shows a structure-functional scheme of a channel where the gating apparatus is opened with a controlling device sensitive to the external electric field. The channel "body" consists of a transmembrane protein immersed in the lipid bilayer. The macromolecule includes 1900–4000 amino acid residues arranged in one or several polypeptide chains, as well as several hundreds of sugar residues covalently bound to amino acids in the outer space. To open and close the channel gates, a reversible change of the protein conformation is required, thereby moving the gates in the appropriate position. In its turn, the probability of the channel opening and closing is determined by the state of the sensor, containing several charged groups moving under the influence of the applied membrane electric field in the channel of electric excitable membranes.

The charge movements induced by the electric field bring about conformation rearrangement of the channel protein required for performing the work.

Suprafamily of Potential-Dependent Channels. The data on the amino acid sequences of potential-dependent Na-, K- and Ca-channels demonstrate a remarkable similarity in the structure of their transmembrane domains. The most studied are the structure and gating mechanisms of K^+-channels. The knowledge of the molecular structure obtained using the methods of genetic engineering and X-ray analysis of crystallized protein molecules has provided a deeper understanding of the basics of

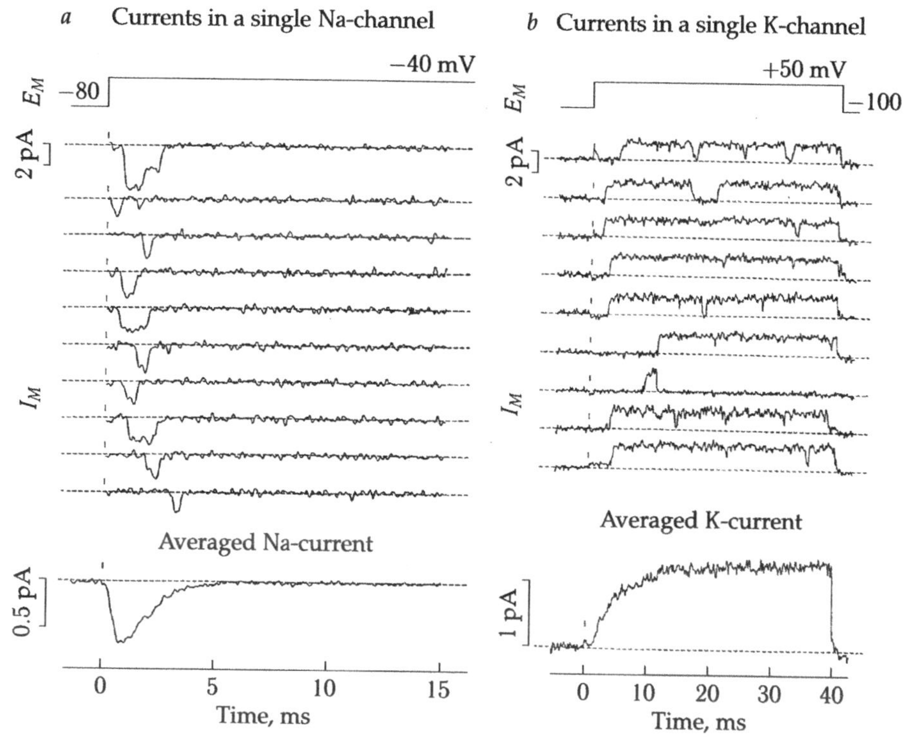

Figure 18.6. Currents in single channels and integral ion current (reproduced from B. Hille, 1992).

a, The gating function of single Na-channels. Patch-clamp records of currents of single Na-channels in murine muscle fiber at a stepwise voltage shift from −80 to −40 mV. Recording was done in the "attached cell" configuration under depolarization of the fiber with Cs ions. The upper part of the figure shows 10 sequential records of currents at the filter transmission band of 3 kHz. The top record of the current reflects a simultaneous opening of the two channels, and the other records correspond to the opening of only one channel. The studied membrane fragment (patch) may contain more than 10 Na-channels. The lower part of the figure shows the current curve obtained as a result of current averaging over the ensemble.

b, The gating function of single K-channels. Patch-clamp records of currents of single K-channels in a gigantic squid axon at a stepwise voltage shift from −100 to +50 mV. To avoid contact with the surrounding Schwann cells, the axon was incised and the micropipette was pressed to the cytoplasmic surface of the membrane. The top part of the figure shows also 9 sequential records of currents through the channels with conductance of 20 pS. The bottom curve is the result of averaging the currents of single channels over the ensemble (averaging of 20 records).

electrostatics of ion interaction with the channel pore and the mechanics of the gating mechanism.

Bacterial K^+-channels (KcsA) and channels of inward rectification of eukaryotes (designated K_{ir}) have the simplest structure of transmembrane subunits. Such a channel is formed by four α-subunits, each of them consisting of two transmembrane α-helical domains (S5 and S6) as well as of a short α-helical domain P and the sequence of the selective filter.

In addition to domains S5, S6 and P, each of the four channel-forming α-subunits from the family of potential-dependent K^+-channels (K_v) contains also four additional transmembrane domains (S1–S4) (Fig. 18.8). Each domain S4 carries four

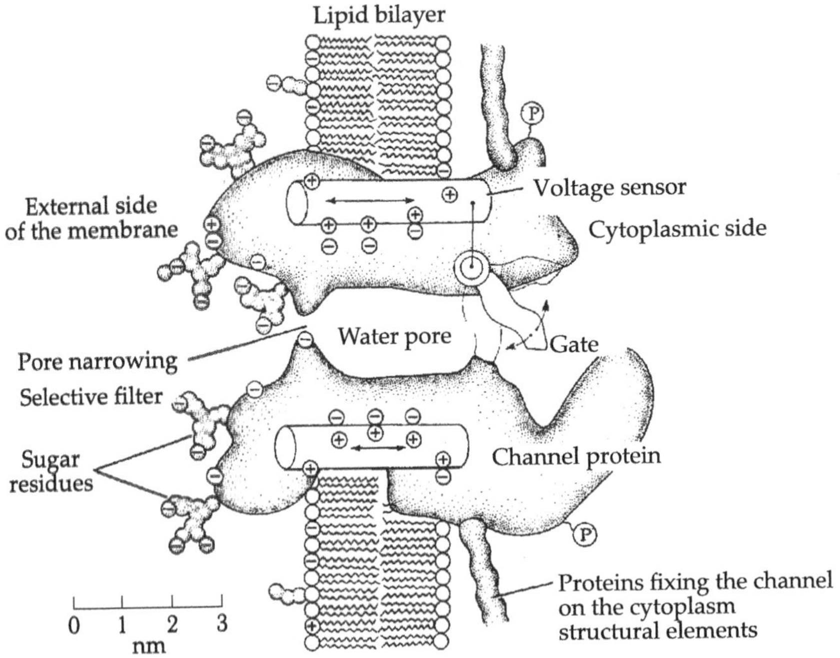

Figure 18.7. Schematic representation of a potential-dependent ion channel (reproduced from B. Hille, 1992).

The channel is shown as a macromolecule, positioned in the membrane, with a through cavity in the middle. The part of the molecule directed to the outer solution carries glycoside groups.

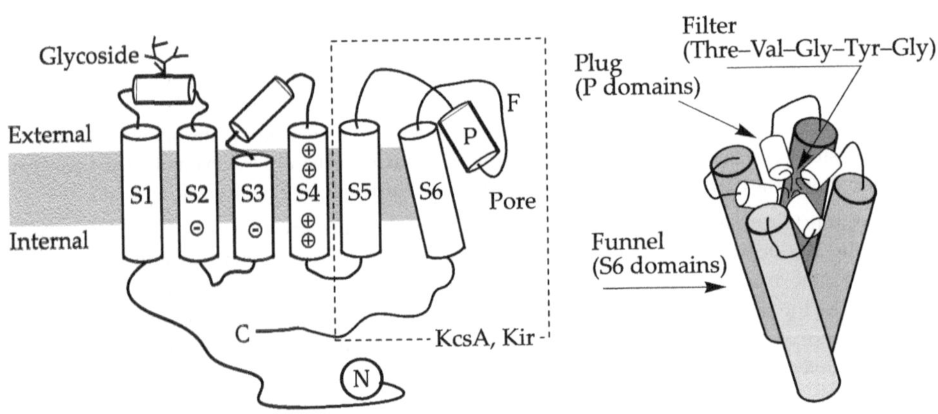

Figure 18.8. Schematic representation of α-subunit of K-channels (*I*) and pore formation with domains S6 and P upon tetramerization (*II*).

I, The dashed line encloses the most conservative domains S5, S6 and P that are the basic transmembrane domains of all K-channels. Domain S4 carries charged amino acids, determining the potential sensitivity. *II*, Four domains S6 form a funnel with a plug of domains P and the selective filter. The bending of the funnel edges results in opening of its lower (gating) part.

positively charged amino acids. The experiments with channels mutant by these amino acids demonstrated that the existence of these charges on domains S4 provides the dependence of the channel opening and closing on the membrane potential.

Consequently, domain S4 of α-subunit represents a potential sensor. The other transmembrane domains are also involved in regulation of the channel activity and provide the binding to cytoplasmic subunits and ligands. □

The membrane part of the channel can be represented as a four-edged "funnel" in the membrane with its wide part directed to the external side of the membrane being closed with a "plug" that has a narrow orifice of the selective filter. In a closed state of the channel, the thin part of the "funnel" directed to the internal side of the membrane is substantially narrowed, and in an open state it is widened permitting the ions to pass through the channel (Fig. 18.8, *II*; Fig. 18.9).

The middle part of the channel accommodates the "central cavity" — an opening of about 10 E in diameter, filled by several water molecules and formed by nonpolar amino acids of α-helices of domain S6. The existence of a water-filled cavity (cf. Section 4 in Chapter 12) decreases considerably the potential energy of the channel inside the pore thus providing high conductance. It should be noted that α-helices of domains P have an essential total dipole moment with the negative charges directed inside the central cavity. The helices of domain P are inclined in the membrane in such a way that the superposition of the dipoles inside the central cavity forms a site with a low potential energy for cations, and therefore several potassium ions are constantly in the cavity.

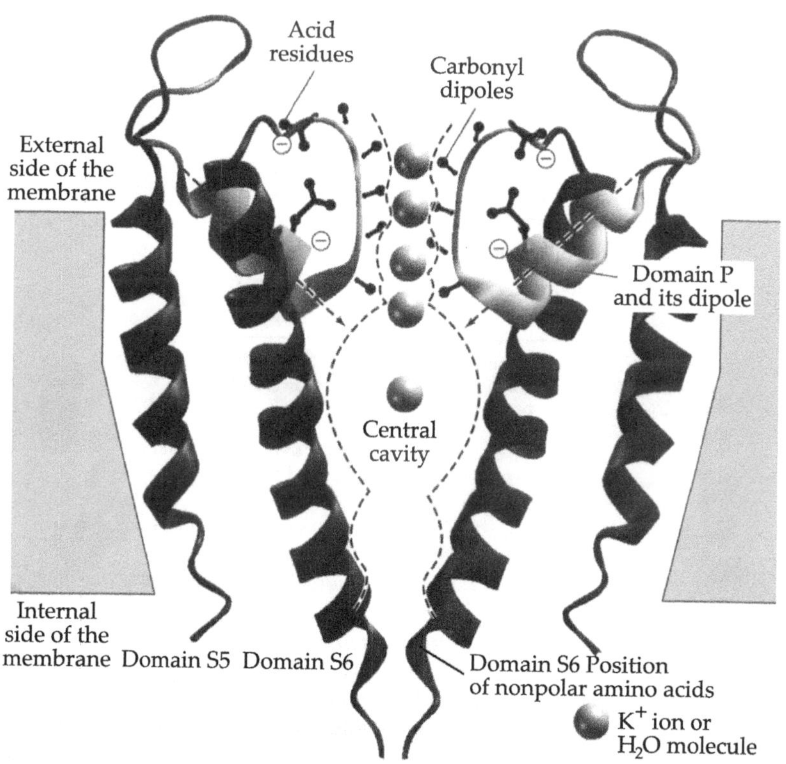

Figure 18.9. Electrostatics of ion motion inside K-channels (reproduced from B. Hille, 2000).

The pore part of the KcsA channel is represented schematically (two monomers of the four are shown). Light circles designate probable positions of K$^+$or water ions. The dashed line encircles the water pore of the channel. Dotted arrows indicate dipole moments of domains P.

Gating Mechanism. In a closed state, the C-termini of α-helices of each of the four domains S6 are positioned so that they do not allow the ions to pass through the channel. The ion movement is also prevented by the presence of nonpolar amino acids with aromatic side groups (Phe) at the C-ends of S6 that creates electrostatic and steric impediments for ion transport. A comparison of the structures of crystallized channels in the open and closed states and also the use of the molecular dynamics method allowed the conclusion that the channel opening occurs due to cooperative bends and turns of domains S6. Glycine residues facilitating the bending of α-helices are located in the bending points (Fig. 18.10). It is noteworthy that, as demonstrated by crystalline structures and molecular dynamics on simplified models, a functionally closed state can be established without complete closing of the pore, but its narrowing to < 4 E and formation of hydrophobic surrounding in it are quite sufficient for this. Consequently, a turn of domains S6 leads also to the widening of the gating pore and concurrent creation of hydrophilic surrounding in it. The position of the gate depends on the conformational changes in the other subunits of the channels.

a

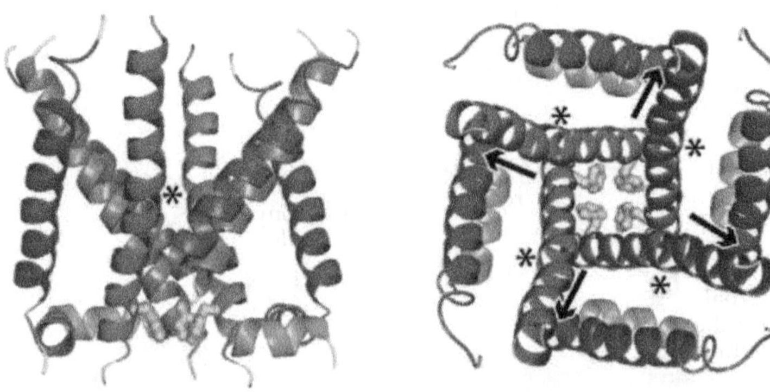

b

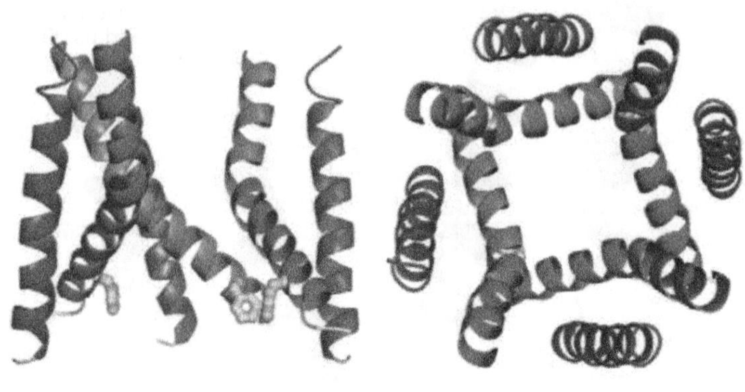

Figure 18.10. Scheme of the gating mechanism in K-channels.

Side view (on the left) and top view (on the right) of the main transmembrane domains. Domains S5 are colored red, domains S6 are violet, and aromatic amino acids of the gating apparatus are yellow.

(*a*) The channel is in the closed state. The N-terminus of domain S5 and the C-terminus of domain S6 are deleted to show the hydrophobic gate. (*b*) The channel is in the open state. For the sake of clarity the subunit closest to the viewer is deleted in the left figure. Arrows show the direction of movement of α-helices; asterisks show the positions of Gly residues facilitating the bending of α-helices (reproduced from Doye, 2004).

Ligand-operated Channels. In the nervous system and neuromuscular junctions, the process of information transfer depends on the operation of postsynaptic receptors, binding to neuromediators and changing the potential of the postsynaptic membrane. Channels, carrying the receptor domain, with the probability of opening dependent on the binding of definite ligands are called ligand-operated channels (receptor channels).

All receptor channels of this family are formed through pentamerization, as a rule, by several different subunits. The generation of such heteropentamers combined with a large number of subunit versions provides a variety of properties of these channels in different tissues.

Because of the accessibility of the natural source, the most studied is the structure of the acetyl choline receptor, found in large amounts in the electric organ of the *Torpedo* ray. This receptor is a heteropentameric complex including up to five different types of subunits (α, β, γ, δ, and ε).

Each of the five subunits forming the pentamer spreads over the whole length of the channel and contributes to the formation of both the receptor and the channel parts of the protein, being located along the normal to the membrane (Fig. 18.11, *a*). □

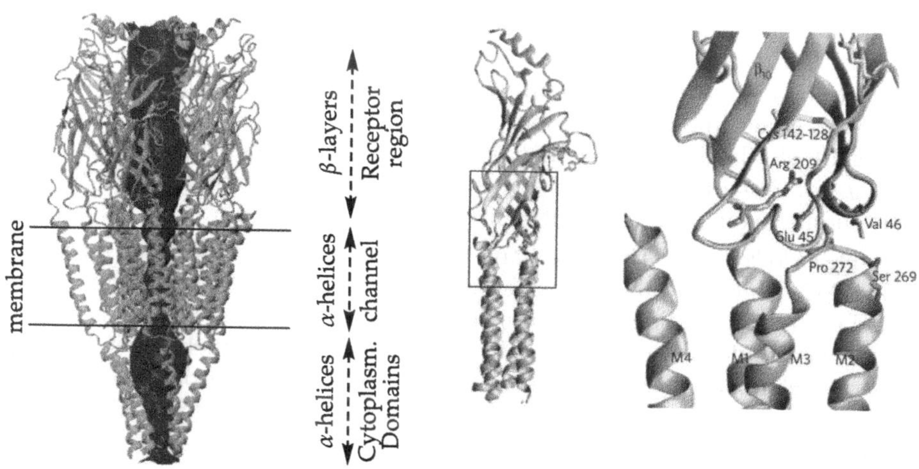

Figure 18.11. (*a*) Model of the structure of the acetyl choline receptor in the *Torpedo* ray electric organ. (*b* and *c*) Coupling of binding between acetyl choline and the gating mechanism (reproduced from M. Sine and G. Engle, 2006; structure from PDB No.2BG9).

(*a*) Ribbon representation of the structure. α-Subunits are colored orange, the other subunits are colored blue. Horizontal lines show the cell membrane boundaries. Dark blue designates the water cavity in the channel. The receptor part formed by β-layers is located over the membrane plane, whereas α-helical domains pierce through the membrane and form the cytoplasmic part of the protein.

(*b*) Side view of the α-subunit. Loop C and β-layer 10 are colored green, showing the pathway of information transfer from the binding site of acetyl choline to the upper part of the channel.

(*c*) Enlarged representation of the coupling between the receptor and the channel parts of the protein. Key amino acids involved in the coupling are indicated. The Cys loop is formed by cysteine residues 142–128, the cysteine bridge is colored yellow. Gly45 is located on the β_1–β_2 linker and Pro272 is located on the M1-M2 linker.

Ligand binding. Five aromatic (Tyr and Trp) are concentrated in the middle of the ligand binding; four of them belong to the α-subunit, and one either to subunit ε or δ. The structures of acetylcholine-binding proteins from shellfish crystallized in complex with ligands showed that the ligands are wholly surrounded by the protein. The ligands are stabilized by $\pi-$cation and dipole–cation interactions, hydrogen bonds and van-der-Waals forces (see Chapter 8). Trp143 generates a strong $\pi-$cation interaction with the ligand, and the other conserved amino acids of the binding center stabilize this bond. The hydrophobic surrounding of the binding center stabilizes the ligand by van-der-Waals interactions. Finally, the N-terminal peripheral loop of the peptide of the binding center (loop C) at the α-subunit plays the role of a "cover" that closes the entrance to the binding center cavity (Fig. 18.11, *b*). The molecular dynamics method, measurements of both Trp fluorescence and crystalline structures of receptors made it possible to establish that the acetylcholine binding results in the transition of loop C to the state preventing the ligand release from the binding center and thereby decreasing the dissociation constant of acetylcholine. For example, the molecular dynamics starting from the crystalline structure of the acetylcholine receptor shows that loop C transits from the closed state to the open one, this movement being prevented by the bound acetylcholine.

The acetylcholine binding leads to the movement of loop C that causes the shift of β10 transferred to the channel part via the Arg209–Gly45 bond. These amino acids are present in all channels of the Cys loop. A change of the charge in one of these amino acids to an opposite one by the site-directed mutagenesis disturbs the operation of the channel, whereas the simultaneous reversion of the charge in both amino acids results in an almost normally operating receptor.

The shift of Gly45 following that of Arg209 leads to the transition of Pro272 to the *cys*-conformation and the bending of loop M2–M3, shifting M2 to the state corresponding to the open gate of the channel.

Gating Mechanism. This mechanism includes first of all the motion of domains M2, while the other transmembrane domains form the necessary framework. The barrier impeding the ion flux in the closed state of the channel is closer to the cytoplasmic side of the membrane. In a closed state, the channel pore in this place has the diameter of about 3 E, and in an open state it is 8 E. It was demonstrated that upon opening of the channel the receptor is twisted relative to its long axis. The data on the channel structure evidence that the gate opening is accompanied by a turn of domain M2. The binding constant of the channel is nearly 20 times higher than the constant of the channel closing: the channel opening takes tens of microseconds and its closing occurs for about hundreds of microseconds.

When the receptors of the Cys loop are activated, the following sequence of events takes place on the molecular level. The ligand binds to the receptor in the part of the channel above the membrane; loop C moves thus closing the release of the agonist from the binding center pocket; domain β10 moves that through the coupling of conservative Arg and Gly leads to conformational rearrangements of the interface between the above-membrane and channel parts including *trans–cys*-isomerization of proline in loop M2–M3, and as a result transfers domain M2 to the state inherent to the open gate. The scheme of all stages of the information transfer from receptor to the channel is given in Fig. 18.12. □

Blocking of Channels. Blocking of channels with different pharmacological reagents is used for studying the structure of channels and the mechanism of their

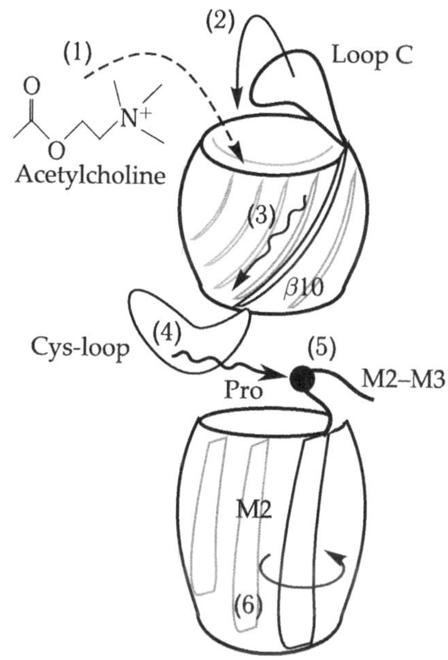

Figure 18.12. Principal scheme of events accompanying the ligand binding to the receptor and the coupling with the gating mechanism. Only receptor and channel parts of the protein are shown. Key structures engaged in the process are indicated.

(1) Ligand binding; (2) transition of loop C to the closed state; (3) change in the position of the β-layer 10; (4) change in the mutual arrangement of amino acids in the Cys loop; (5) transition of Pro from the *trans*- to *cys*-conformation; (6) turning and rolling of domain M2 leading to opening of the channel gate.

functioning. Two types of blocking are known: (1) the binding directly in the channel pore thus preventing the ion transfer though it, and (2) the allosteric binding of the reagent by macromolecular structures in the channel that stabilizes the pore conformation in the closed state and thereby impedes the channel opening. The both types of blocking can take place concurrently under the action of chemical substances.

Pay attention to the obvious analogy with the competitive (for the place in the active center) and allosteric types of the enzymatic process inhibition (Chapter 26).

Such substances as tetrodotoxin TTX, saxitoxin STX and tetraethylammonium TEA played the greatest role in studying the selective inhibition of definite types of ion conductance (Fig. 18.13).

It is accepted that a molecule of blocker T binds reversibly to the receptor. Figure 18.14 shows the scheme of action of anesthetics, displaced from the complex with the receptor by Na^+ and H^+ ions, penetrating to the channel from the outside. In its turn, the formation of the complex affects the efficiency of the gating mechanism, leaving the gate open and strongly slowing down the channel inactivation. For example, the blocker *N*-methyl strychnine added to the open channels blocks them and fixes the gate in the open state, so that it cannot be closed until the blocker leaves the channel. Thus, the blocker access to the receptor is allowed only when the gate is open. Upon closing of the gate, after the formation of the TRec-complex, the anesthetic appears to be locked in the channel. The effect of anesthetics TTX and STX on the gating mechanism is not direct and is conditioned by their association with other molecules contributing to the gate closing (the allosteric mechanism).

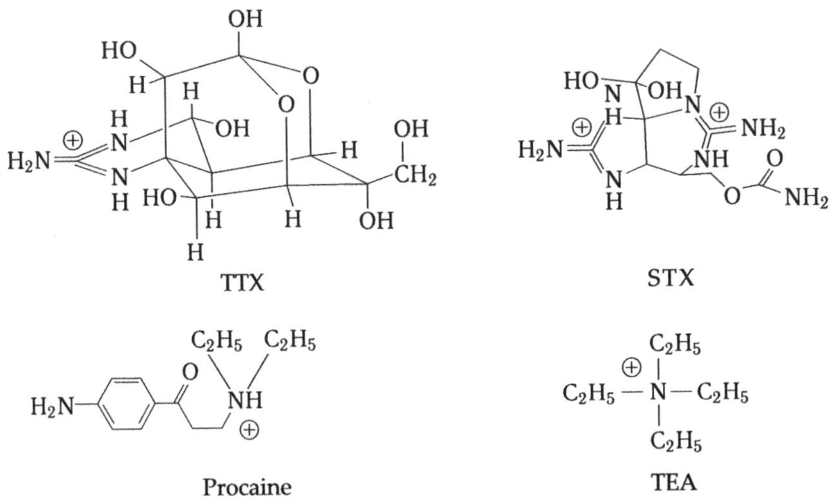

Figure 18.13. Chemical structure of blockers of ion channels (reproduced from B. Hille, 1992).

Tetrodoxin (TTX) and saxitoxin (STX) are natural toxins blocking specifically Na-channels. The ion of tetraentyammonium (TEA) is a simple quaternary derivative of ammonium that blocks K-channels. The action of all these reagents is reversible.

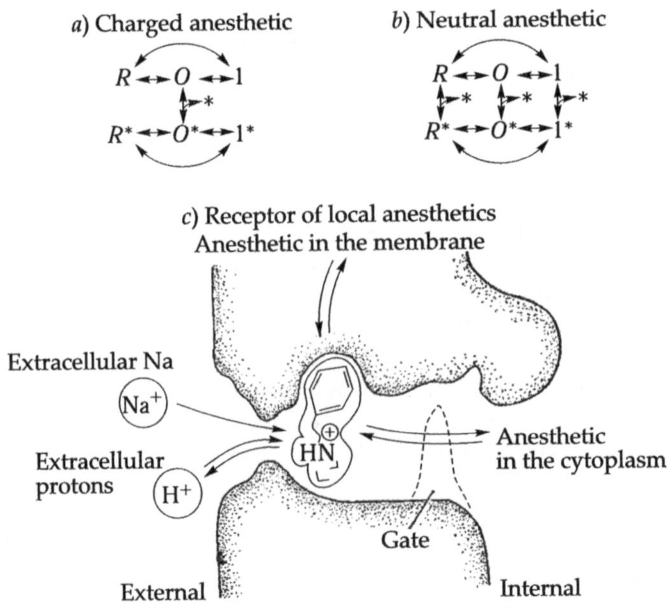

Figure 18.14. Hypothetical scheme of blocking the ion channel with local anesthetics (reproduced from B. Hille, 1992).

(*a*) States of the Na-channel and transitions caused by the interaction with charged anesthetic molecules: R, resting state; O, open state; I, inactivated state. Charged (hydrophilic) anesthetic molecules can penetrate through the channel only when the gate is open.

(*b*) Charged anesthetic molecules can bind and unbind even when the gate is closed.

(*c*) Two ways through which the anesthetics can penetrate to the receptor in the channel orifice. The hydrophilic way is blocked by closing the gate. Extracellular Na^+ and H^+ can interact with the bound anesthetic molecules via the selective filter.

Inhibition of the Gating Mechanism. The inactivation process is easily inhibited by the action of chemical reagents penetrating to channels from the side of the axoplasm. The slowing down of the activation may result in that the duration of the potential action (an open Na-channel) would be prolonged to several minutes. The effect is observed upon introduction of pronase (a mixture of proteolytic endopeptidases) when Na-channels remain open under the action of extended depolarizing pulse losing their capacity for inactivation and closing.

Apparently, the sites of action of inactivation inhibitors of the gating mechanism in Na-channels are peptide loops between repeats III and IV with accessible amino acid residues — arginine, tyrosine, tryptophan and histidine. The receptor for their binding differs from that for TTX or STX. Local anesthetics compete with them for the binding site.

Under standard conditions, inactivation of Na-channels not only promotes the closing of the gate but also is accompanied by changes in the external part of the channel structure where the receptor is located.

Different substances can modify the gating mechanism of channel closing and opening. For example, enzymes (pronase) has an irreversible effect from the side of the cytoplasm and irrevocably damage the structures responsible for inactivation of the channel. Other inhibitors (peptide toxins) act from the outside slowing down inactivation of the channels. Toxins soluble in lipids suppress activation (opening of the channels), slow down inactivation and decrease the ionic selectivity of the channel. At last, changes in pH, intracellular concentration of divalent ions and ionic strength affect the functioning of the gating mechanism. Thus, the inactivation involves structural elements located outside the channel and in the region of the gate. At the same time, activation is dependent on the structures deeply immersed in the lipid part of the membrane and inaccessible to external chemical agents. Figure 18.15 shows the sites of action of different inhibitors on Na-channels.

18.5 Electric Fluctuations of Membrane Properties

Electric fluctuations in membranes are conditioned by thermal noise, movement of single ions through channels and changes in the number of open channels due to the processes of their opening and closing. The analysis of spontaneous fluctuations on the level of a whole membrane or single molecules makes it possible to disclose mechanisms of channel functioning. □

The analysis of fluctuations is based on the determination of their statistical and spectral characteristics. An important parameter of a stochastic process $\{x\}$ is the standard deviation σ and dispersion σ^2 of fluctuations relative to average value $\overline{x}$:

$$\sigma^2 = \frac{1}{n}\sum_{k=1}^{n}[x_k - \overline{x}]^2 . \tag{18.4}$$

Time interval τ when $x(t+\tau)$ does not correlate with $x(t)$, is the measure of the rate of correlation decay. The fluctuations may be characterized by the so-called autocorrelation function (see Section 2 in Chapter 11) determined as the mean of the product of random value x at instant of time t by its value after time interval τ:

$$C(\tau) = \langle g(t)g(t+\tau)\rangle . \tag{18.5}$$

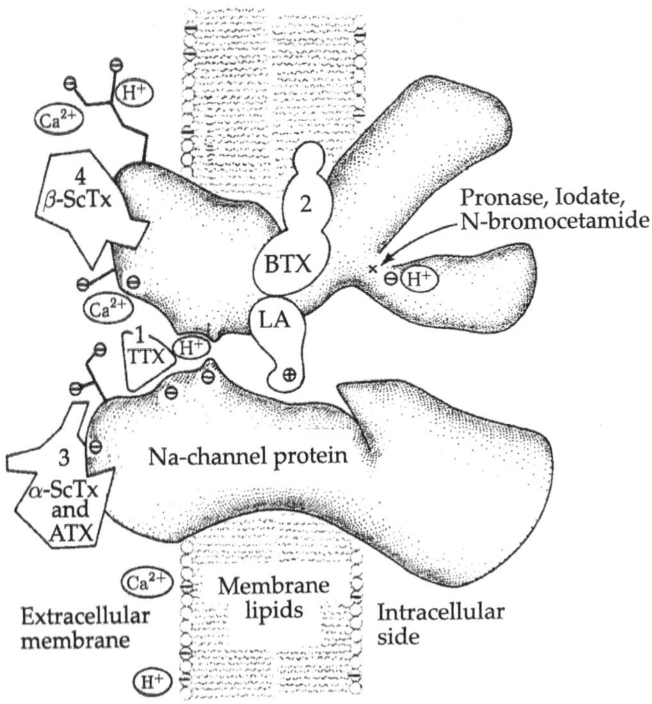

Figure 18.15. Toxin receptors in the sodium channel (reproduced from B. Hille, 1992).

Hypothetical scheme of the structure of Na-channel. 1, Binding site of tetrodotoxin and saxitoxin; 2, binding site of lipophilic modifiers of the gating function (batrachotoxin BTX); 3, binding site of peptide toxins from scorpion ScTx and actinia ATX (inactivation modifiers); 4, binding site of β-toxins from scorpion (β-ScTx) (activation modifiers). Internal parts of the membrane subjected to the action of chemical modifiers of inactivation (pronase, iodate) are shown: the binding site of local anesthetics (LA) inside the hydrophilic pore and negative surface charges attracting bivalent cations and Na ions to the mouth of the channel.

If the random value fluctuates near 0, i.e. the mean has been subtracted, the value of the autocorrelation function at $\tau = 0$ coincides with the signal dispersion: $C(0) = \sigma^2$. As the time shift τ increases, the overlapping of the initial signal with its shifted copy becomes smaller and smaller and for a random process decreases monotonously, tending to zero for relatively high τ values.

Another essential parameter of the signal is its spectral density or the power spectrum. It characterizes the distribution of the fluctuation intensity of the signal as a frequency function and can be measured in experiments.

According to the physical nature, the following noise sources exist in a membrane: thermal noise caused by thermal perturbations of charge carriers, shot noise that is a consequence of discreteness of charge carriers, in this case of ions, and also fluctuations of conductance of single channels reflecting their transitions between open and closed states. The recorded fluctuations of electric properties of the membrane are overlapping of all sources of noise. Separate components can be isolated if spectral properties of each source of noise is known. □

Thermal noise characterizes passive properties of the membrane — its capacity and resistance. The characteristic feature of thermal noise is its dependence on temperature and independence on the frequency for current fluctuations. When shot noise is described, it is assumed that ions move across the membrane independently at a constant rate, remaining in the constant field (see Section 2 in Chapter 16), and the membrane dielectric permittivity is stable.

Noise of Channel Opening and Closing. The integral current across the membrane represents superposition of currents passing through single ion channels. Each channel transits from the open (conducting) state to the closed (non-conducting) state. Jumps of the conductivity of single channels have a stochastic nature, and the probability of transitions depends on the external parameters such as the membrane potential, concentration of neuromediators etc.

At normal temperature, spontaneous opening and closing of the channels take place that have an effect on equilibrium fluctuations of the membrane conductance. According to the fluctuation-dissipation theorem, the system in thermodynamic equilibrium responds to the weak applied perturbations in the same way as to spontaneous fluctuations. Consequently, there is a direct connection between reversible spontaneous fluctuations in the system and its non-equilibrium (for example, irreversible energy dissipation) linear response. In practice, the system response frequently has an exponential form ($f(x) \propto \exp\{-x/\tau\}$). Hence it follows that spectral properties of the dynamics of single molecules define the system behavior on the macroscopic level. Thus, proceeding from the concepts on aurocorrelation properties of molecular variables, it is possible to predict spectral properties of spontaneous fluctuations of the macroscopic system and its linear response to the external action. In its turn, by approximating the experimentally measured macroscopic characteristics by the predicted dependences it is possible to estimate the dynamic parameters of single molecules. For channels such characteristics are conductance of a single channel γ, channel average lifetime in an open state τ_o and also rates of channel transition between separate states.

The simplest representation suggests only two states of conductivity — open (O) and closed (C), and in the kinetic respect the transitions between them are the first order reaction:

$$C \underset{q}{\overset{p}{\rightleftharpoons}} O. \tag{18.6}$$

In this case, the channels perform transitions quite randomly and independently. The probability to find a channel in an open state is $P_o = p/(p+q)$, and $P_c = q/(p+q)$ in a closed state. Then the time-averaged current through a single channel is equal to iP_o, and the squared current is equal to i^2P_o. The expression for dispersion (18.4) can be written as follows:

$$\sigma^2 = \frac{1}{n}\sum_{k=1}^{n} x_k^2 - \bar{x}^2.$$

Then dispersion of microscopic fluctuations of the current is

$$\sigma_i^2 = i^2P_o - i^2P_o^2 = i^2P_o(1-P_o) = i^2P_oP_c. \tag{18.7}$$

Under conditions of independent channel transitions, the macroscopic current is $I = NiP_o$, where N is the number of channels. Dispersion of macroscopic current fluctuations has the form:

$$\sigma_I^2 = Ni^2P_o - Ni^2P_o^2 = iI(1-P_o), \tag{18.8}$$

or, which is equivalent,

$$\sigma_I^2 = iI - I^2/N. \tag{18.9}$$

Hence it follows that dispersion of macroscopic current fluctuations in the given approximation is associated by a quadratic relation with the average current. When the probability of the channel opening is low, for example, at low potentials for potential-dependent channels or at low concentrations of the ligand, $p \ll 1$ and the measured dispersion of current fluctuations is proportional to the average current:

$$\sigma_I^2 \approx iI, \tag{18.10}$$

this dependence can be used to estimate the current through a single channel. The conductivity of a single channel can be estimated as a derivative of the dependence of the current level on the membrane potential:

$$\gamma = \frac{di}{d\varphi} = \frac{d}{d\varphi}\left(\frac{\sigma^2}{I}\right).$$

Kinetic scheme (18.6) suggests the following behavior of the macroscopic system from n_o open and n_c closed channels:

$$\begin{aligned} \dot{n}_o &= pn_c - qn_o, \\ \dot{n}_c &= -pn_c + qn_o, \end{aligned} \qquad n_o + n_c = N, \tag{18.11}$$

where N is the total number of channels. In such a system, relaxation to the stationary value after perturbations occurs exponentially:

$$n_o(t) = \bar{n}_o - (\bar{n}_o - n_o^0)\exp[-(p+q)t], \tag{18.12}$$

where $\bar{n}_o$ is the stationary number of open channels, and n_o^0 is their number at the zero moment. The characteristic time scale of the system, i.e. the average time of reaching the stationary state, is $\tau = 1/(p+q)$. The same time scale is also characteristic of spontaneous fluctuations. To simplify the subsequent analysis, it is traditional to study the situation when the constant of the channel opening is low: $p \ll q$. In this case, the time scale associated with the constant of the channel closing is dominating. In experiments such a situation is observed, for example, for ligand-operated channels when the neuromediator concentration is small.

In this approximation, channel openings are a stationary stochastic process, when the distribution of channel lifetimes in an open state is described by the monoexponential function of the type $P_o(t) = (1/\tau_o)\exp(-t/\tau_o)$ with the time constant τ_o. Consequently, the autocorrelation function of this process is also a simple exponent

$$C(\tau) = \sigma_I^2 \exp[-\tau/\tau_o] \tag{18.13}$$

with the same time constant (Fig. 18.16).

If the autocorrelation function is defined by equation (18.13), the power spectrum of this process is as follows:

$$S(f) = 2\sigma_I^2 \int_0^\infty \exp[-\tau/\tau_o]\cos(2\pi f\tau)d\tau; \quad S(f) = \frac{2\sigma_I^2\tau_o}{1+(2\pi f\tau_o)^2}. \tag{18.14}$$

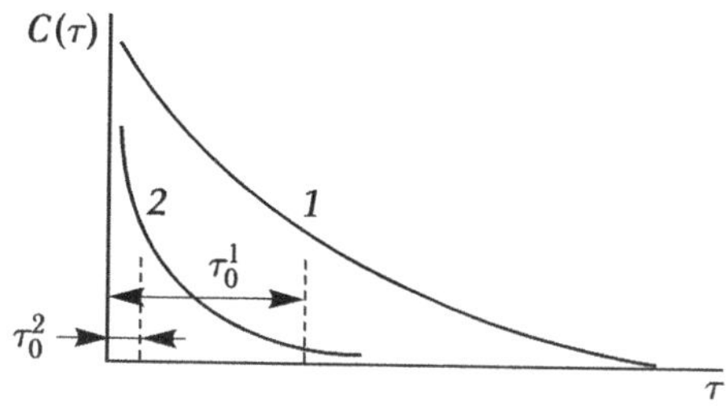

Figure 18.16. Autocorrelation functions corresponding to slow (*1*) and fast (*2*) fluctuations. τ_o is the characteristic time of the exponent coinciding with the average lifetime in an open state.

The general form of the dependence $S(f) = S_0/(1 + (f/f_c)^2)$ is called the Lorentz curve and is characteristic of stochastic processes with linear responses to weak perturbations in the form of damped exponentials. At low frequencies, the fluctuation intensity is independent of the frequency, and at high frequencies it decreases inversely to the square frequency $(1/f^2)$. At frequency $f = f_c$ called the cutoff frequency, the spectrum amplitude decreases twofold. In this representation the cutoff frequency is associated with the characteristic lifetime of the channel: $f_c = 1/2\pi\tau_o$. When the kinetic scheme includes several stages, the distribution of lifetimes of the channel is approximated by the sum of several exponentials, and the power spectra by the sum of several Lorentz functions with different f_c values. □

At low average lifetime of the open state, the spectrum parameters are associated with characteristics of a single channel (detailed conclusion see in C. R. Anderson and C. F. Stevens (1973)):

$$S(f) = \frac{2\gamma I(V - V_{eq})/q}{1 + (2\pi f/q)^2}, \tag{18.15}$$

where γ is the conductivity of a single channel, and V_{eq} is the equilibrium potential for the given ion. Relationship (18.15) is deduced based on the above correlations: $\sigma_i^2 = iI$, $q = 1/\tau_o$ and $\gamma = i/(V - V_{eq})$. Upon approximation of the experimentally obtained power spectrum of the Lorentz curve fluctuations, we can calculate the average lifetime of the channel ($\tau_o = 1/2\pi f_c$), and the conductivity of a single channel, $\gamma = S_0/2\tau_o I(V - V_{eq})$, can be calculated by the spectrum plateau level.

One of the experimental objects with the power spectrum corresponding to the Lorentz function is the current passing through the end-plate of the neuromuscular preparation at low acetylcholine concentrations. An example of fluctuations of the current through the frog end-plate membrane is demonstrated in Fig. 18.17. At rest the fluctuations are small upon fixing the potential at −100 mV. Activation of acetylcholine receptors with low acetylcholine concentrations results in an increase of both the current and the fluctuation intensity. As seen from Fig. 18.18, power spectra of these fluctuations are well described by Lorentz functions. Upon cooling, the shot frequency of the object f_c decreases, suggesting an increase in the average lifetime of the channel in an open state τ_o and the decrease in the constant of the channel closing. In addition to the currents in the end-plate, fluctuations of current and conductivity, whose spectrum is proportional to $1/f^2$, are observed also in the membrane of the squid axon and in the Ranvier's constrictions of frog myelin nerve fibers.

The channel conductivity is estimated as the tangent of the angle of inclination of the linear portion of the volt-ampere dependence of the channel. An example of such dependence is shown in Fig. 18.19.

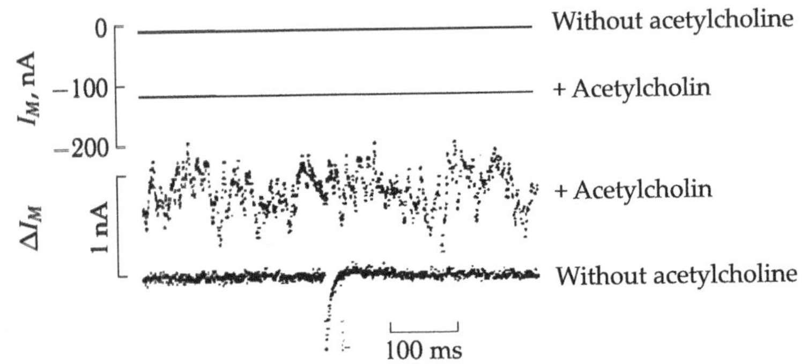

Figure 18.17. Current fluctuations in the synaptic membrane (reproduced from B. Hille, 1992).

Currents were measured on the frog muscle fiber upon fixation of the potential. Top curves represent measurements of the current at a low amplification factor; bottom curves are measurements at a high amplification factor. In the stationary neuromuscular ending the current recorded upon weak amplification is equal to zero. The record made at a high amplification factor reflects a low level of noise and a single pulse of the inward current, caused by spontaneous isolation of the neutrotransmitter from the single presynaptic bead. When low-concentrated acetylcholine is added, a large constant inward current is recorded at small amplification. Records made upon high amplification, reveal fluctuations caused by stochastic opening of a large number of channels. The temperature is 8 °C.

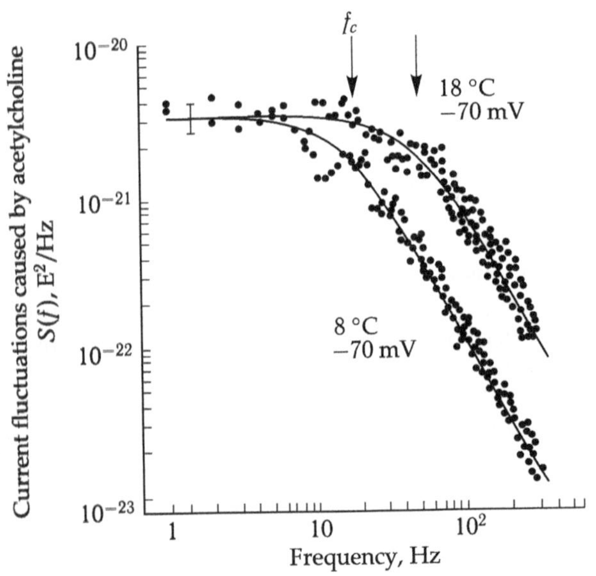

Figure 18.18. Spectrum of current fluctuations caused by the action of acetylcholine on the frog muscle fiber (reproduced from C. R. Anderson and C. F. Stevens, 1973).

The spectrum of fluctuation density $S(f)$ is represented as a frequency function on a double logarithmic scale. Arrows show the cutoff frequency of Lorentz curves (solid lines). A temperature increase causes the growth of the cutoff frequency.

A random process, the behavior of which is defined only by the state at the current instant of time rather than by the prehistory, i.e. the process without memory of the preceding states, is calles a markovian process. Just this process is generated upon modeling the dynamics of ion channels using kinetic diagrams,

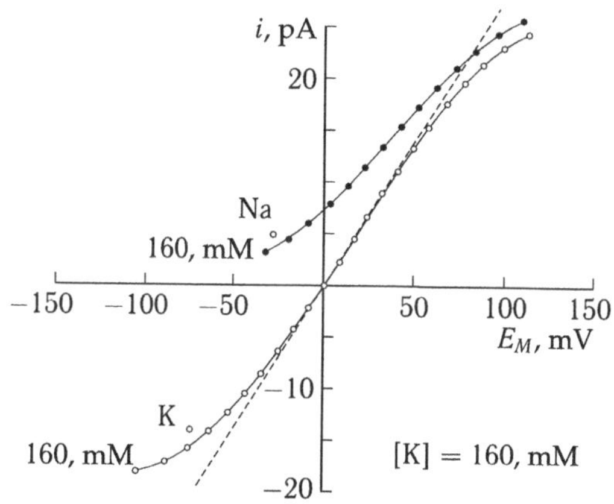

Figure 18.19. Measurements of the single channel conductivity (reproduced from G. Yellen, 1984).

The volt-ampere characteristic of a single Ca-dependent K-channel in bovine chromaffin cells. An isolated mebrane fragment in the "outside–out" configuration was placed in 160 mM KCl or NaCl, and the micropipette contained 160 mM KCl. In symmetrical KCl solutions the slope of the dashed line corresponds to the conductivity of 265 pS.

where random transitions between the fixed set of states take place in accord with the fixed set of trasnition constants. □

Non-markovian Kinetics of Channels. Upon studying the currents through single ion channels with the method of local fixation of potentials, it was found that for a number of channels the spectral density of current fluctuations depends on the frequency not as the Lorentz function but as $1/f^{\beta}$ where $\beta \approx 1$. The existence of fluctuations characterized by $1/f$-spectra of power demonstrates that the molecular dynamics has special properties differing from the above simple notion. The $1/f$-noise is widely spread in nature, but there are no simple and universal models explaining its mechanism. For example, it cannot be produced by integrating or differentiating the signal from some noise source or by using simple stochastic differential equations.

Y The power spectrum dependent on frequency as $1/f$ indicates to the existence of long-term correlations (memory) in the studied fluctuating process. In other words, $1/f$-noise characterizes the non-markovian process. □

Most likely, the macromolecule in the channel does not perform transitions between the fixed set of conformational states; the conformational transitions are cooperative and are characterized by a large number of substates (see Chapter 11). The data of molecular dynamics show the existence of $1/f$-noise even in conformational transformation of chains (A. Takano et al., 1998). It is believed that the key role in generation of fractal $1/f$-noise in current fluctuations of the channel belongs to the conformational dynamics of its gating mechanism.

Y It is obvious that the main source of $1/f$-noise and long-term correlations in the current through separate types of ion channels is their transitions between open and closed states. These conformational rearrangements of the channel cannot be described in the framework of a completely random, stochastic process. □

19

Ion Transport in Excitable Membranes

19.1 Action Potential

Nerve impulses generated in separate fibers have a constant amplitude and shape. A single nerve impulse — the action potential — lasts for about 1 ms and spreads at a rate from 1 to 100 m/s.

The notions on the process of generation of nerve impulses are based on the studies performed by A. Hodgkin, A. Huxley and B. Katz on the giant axon of the squid nerve fibers. To measure the action potential inside the axon, a fine glass capillary with the ending diameter less than 0.5 μm filled with KCl solution is inserted without causing any noticeable effect on the axon activity.

When hypepolarizing or subthreshold depolarizing current pulses are passing, the membrane behaves like a passive parallel RC-chain, charging exponentially from the initial to the new level with characteristic time τ equal to the product of the membrane resistance and capacity (Fig. 19.1). Under the action of subthreshold depolarizing pulses, potential values in the internal part of the fiber shift rapidly to the level +(40–50) mV and then return with a short-term stage of residual hyperpolarization. The amplitude of the action potential is 100–120 mV.

Like the protoplasm of many other cells, the axoplasm contains high concentrations of K^+ and relatively low concentrations of Na^+ and Cl^-. Potassium ions are free inside the fiber and not bound to proteins or other large molecules.

At rest, the membrane is more permeable for K^+ than for Na^+. Due to K^+ diffusion from the axoplasm to the outside, at rest the internal content of the nerve fiber is negative relative to the outer solution. This constant difference of potentials — the rest potential — is 50–70 mV. What factors define this value? If the membrane were

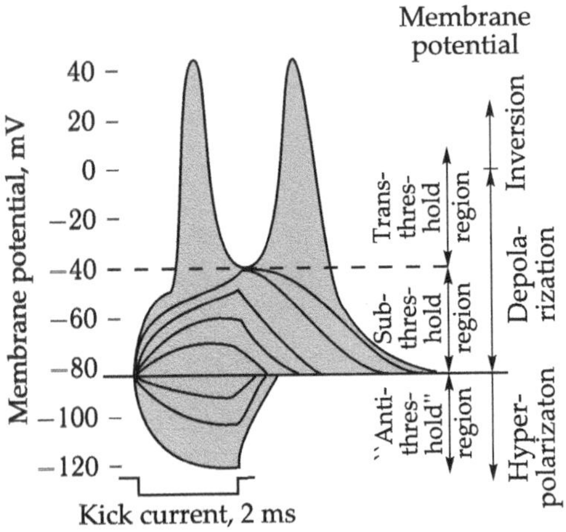

Figure 19.1. Impulse generation as a result of local depolarization (reproduced from B. Katz, 1968).

Under the action of the kick current of fixed duration (2 ms) but of different strength and direction, changes in the memrbane potential appear, represented as a family of curves in the plot.

permeable only for K^+, the equilibrium difference of potentials would have reached the value determined by the Nernst equation (see (15.7)) for the potassium electrode:

$$\varphi_K = \frac{RT}{F} \ln \frac{[K_o]}{[K_i]}, \tag{19.1}$$

where φ_K is the equilibrium potential for K^+. When the concentration of K^+ is 20 mM in squid blood and 400 mM in its axoplasm, the calculated φ_K value is −75 mV. As a matter of fact, in intact axons rest potentials are found to be about −70 mV. Consequently, under such conditions the membrane behaves as the K^+-electrode. But at lower concentrations of K^+ in the medium, the rest potential differs greatly from the calculated values of the equilibrium potential φ_K.

It was found that the dependence of the rest potential on the K^+ concentration in the medium is much better described by the Goldman equation (see (16.10)), based on the electrodiffusion theory in the constant field approximation.

At rest, the distribution of Cl^- is close to equilibrium, i.e. the Cl^- flows from the cell and inside it are the same. In this case, the condition of the zero current through the membrane (see (16.8)) $I_K + I_{Na} + I_{Cl} = 0$ is simplified to $I_K + I_{Na} = 0$ and the rest potential is calculated by the formula:

$$\varphi = \frac{RT}{F} \ln \frac{P_K[K_o] + P_{Na}[Na_o]}{P_K[K_i] + P_{Na}[Na_i]}. \tag{19.2}$$

At rest, the membrane potential (19.2) approaches the equilibrium potential for K^+, because under such conditions the permeability coefficient for K^+ is far larger than the permeability coefficient for Na^+ ($P_K : P_{Na} = 1 : 0.04$). □

Impulse Generation. The reversion of membrane potential during impulse generation is caused by that at the crest of the action potential spike the membrane is selectively penetrable for Na^+, so that the relationship of ion permeability values becomes $P_K : P_{Na} = 1 : 20$. As seen from formula (19.2), under these conditions ($P_{Na} \gg P_K$) the potential on the membrane approaches to the equilibrium sodium potential φ_{Na}. The difference of potentials on the membrane penetrable only for Na^+ is expressed by the Nernst formula:

$$\varphi_{Na} = \frac{RT}{F} \ln \frac{[Na_o]}{[Na_i]}. \tag{19.3}$$

The potential limiting value estimated using the above formula is +55 mV, exceeding only slightly the experimentally observed values of the inverse potential. The same conclusions are drawn if a more complete formula with account for Cl^- flows is used for expressing the membrane potential:

$$\varphi = \frac{RT}{F} \ln \frac{P_K[K_o] + P_{Na}[Na_o] + P_{Cl}[Cl_i]}{P_K[K_i] + P_{Na}[Na_i] + P_{Cl}[Cl_o]}. \tag{19.4}$$

At rest $P_K : P_{Na} : P_{Cl} = 1 : 0.04 : 0.45$, and in the excited state at the peak of the action potential when the condition of zero current through the membrane is again valid, $P_K : P_{Na} : P_{Cl} = 1 : 20 : 0.45$. □

It is postulated that different types of ions pass across the cell membrane via spatially separated pathways. This complies with the current notions on the existence of separate types of ionic channels selectively permeable for Na^+ and K^+.

The excitation process develops due to the dependence of membrane permeability for ions on the membrane potential. When critical depolarization is achieved, and the membrane permeability for Na^+ increases, these ions run inside and cause further depolarization of the membrane. The process continues until the potential shifts to the equilibrium sodium potential. Under such conditions, the outside and inside Na^+ fluxes are matching. After that permeability for K^+ increases and K ions begin leaving the cell following the gradient of their electrochemical potential. During this process the membrane is repolarized. The release of K^+ stops when the membrane potential approaches the value of the equilibrium potassium potential. □

The amount of the incoming Na^+ is quite sufficient for repolarization of the membrane in the period of the action potential.

When one spike transfers along the axon, the system returns to the initial state, except some excess of Na ions that have appeared in the cell during depolarization together with some deficiency of K ions that left the cell during repolarization. Changes in the intracellular concentration of Na^+ and K^+ caused by a single action potential are extremely small, but could be substantial upon multiple advances of nerve impulses. To maintain the constant level of intracellular concentrations of K^+ and Na^+ in the cell membrane, the membrane has a Na^+/K^+ pump system (see Section 2 in Chapter 20), providing active release of Na^+ in exchange for the entry of K^+ to the cell and operating at the expense of ATP energy.

19.2 Ion Currents in the Axon Membrane

Excitation mechanisms were deciphered as a result of membrane current measurements by the voltage clamp method (K. Cole, 1949). In standard measurements of action potential, the permeability values for Na^+ and K^+ are a function of two variables: the membrane potential value and the time. In the voltage clamp method, shifts of the membrane potential are regulated which makes it possible to discriminate the effect of these two factors due to the inhibition of autocatalytic development of the action potential.

There is no current in a resting membrane. If the membrane potential is shifted from the resting potential level towards depolarization, permeability of the Na^+- and K^+-channels change in a certain way from one state to the other. In this case, in each of the parallel coupled regions there are currents proportional to the difference between the equilibrium ion potential φ_i and the maintained membrane potential:

$$I_i = g_i(\varphi - \varphi_i), \tag{19.5}$$

where g_i is the conductance of the membrane for this type of ions. Total current I_{tot} across the membrane is equal to the sum of the capacitive current and the ionic current caused by the movement of ions along separate pathways:

$$I_{\text{tot}} = C_{\text{m}}\, d\varphi/dt + I, \tag{19.6}$$

where C_{m} is the membrane capacitance, $d\varphi/dt$ is the rate of changes in the membrane potential, and $I = \sum I_i$ is the transmembrane ionic current.

If the potential on the cell membrane is not maintained at the fixed level, then during the period of the action potential $d\varphi/dt \neq 0$ sizeable capacitive current also passes across the membrane. An important advantage of the voltage clamp method is that the capacitive current flows only for a very short time following the membrane potential shift to the designated level. After a short-term spike the capacitive current terminates because the membrane potential does not change ($d\varphi/dt = 0$). Since the duration of the action potential is several milliseconds, it is important that the potential formation on the new level and the drop in the capacitive current should occur much faster, for example, for about ten microseconds. In this case, the current, developed after termination of the capacitive current, will be caused only by the ion transport across the membrane.

In response to the stepwise depolarization of the squid axon from the resting potential to 0 mV, the membrane current consisting of four constituents is recorded (Fig. 19.2). Capacitive current I_C represents a short-lived impulse generally terminating to the moment when the voltage reaches the plateau. Some dipole molecules in the membrane slowly change their orientation after imposing step voltage that results in the emergence of a small-amplitude "tail" of the capacitive current, decreasing relatively slowly. After the drop of I_C the membrane current is completely defined by the movement of ions through the membrane channels.

Two basic constituents (potassium and sodium) and one secondary constituent can be distinguished in the ionic current. The latter is a low-value leakage current that does not change with time. The Na^+-current is caused by a short-lived Na^+ flow through Na^+-channels. This current decreases rapidly and in about 1 ms is changed for the inward K^+-current.

If the membrane is depolarized so that the difference of potentials is equal to the voltage of a sodium battery (φ_{Na} is the equilibrium Na^+-potential), the Na^+-current is absent, and it is possible to observe only K^+-current emerging with a typical delay. When the potential shifts to positive values exceeding the φ_{Na} level, the Na^+-current has an outward direction (Fig. 19.3).

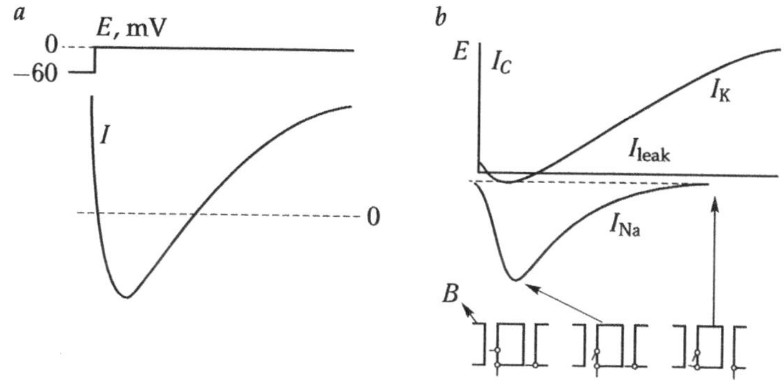

Figure 19.2. Current flowing across the squid axon membrane upon stepwise depolarization from the resting potential level (−60 mV) to zero (*a*); representation of the total current as a sum of four constituents (*b*) and different states of the gating system of Na- and K-channels (reproduced from C. M. Amstrong, 1975).

I_C is the capacitive current; I_{Na} is the sodium current; I_K is the potassium current; I_{leak} is the leakage current; downward deflections correspond to the inward current, upward deflections correspond to the outward current.

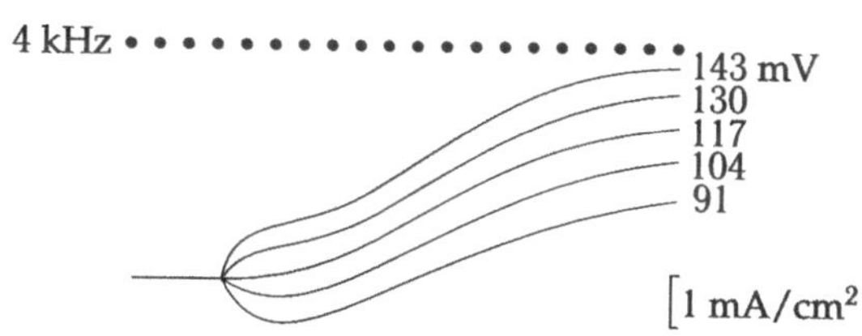

Figure 19.3. Membrane currents at different shifts of the membrane potential (the temperature is 3.5°). Upward deflections correspond to the outward current; numerals designate the magnitude of the membrane potential shift (reproduced from A. L. Hodgkin, 1964).

The ionic current can be separated into constituents by substituting Na^+ ions for non-penetrating choline cations. In this case, the kinetic curve includes only the K^+ constituent (Fig. 19.4).

The separation of the current into constituents is clearly illustrated by the experiments with the use of a number of pharmacological regents. An addition of small amounts of tetrodoxin (TTX) to the axon suppresses completely the Na^+-current (Fig. 19.5). Provided tetraethyl ammonium (TEA) was injected or introduced to the axon by means of perfusion up to the concentration of 15 mM, the K^+-current is suppressed almost completely. The experiments with the use of TTX and TEA demonstrate that Na^+- and K^+-channels are separate structures as suggested in the diagram proposed by A. Hodgkin and A. Huxley. □

Further verification of the existence of two types of channels was obtained in experiments with the use of pronase. It was found that perfusion of the axon with the solution containing pronase does not affect the kinetic curves of the K^+-current, but essentially changes the kinetics of the Na^+-current. The treatment with pronase leads to the suppression of the inactivation stage of the Na^+-current (Fig. 19.5, *III*), i.e. the stage of the rapid decrease in the Na^+-current disappears when the peak value is achieved. The magnitude of the current remains high during the whole voltage impulse.

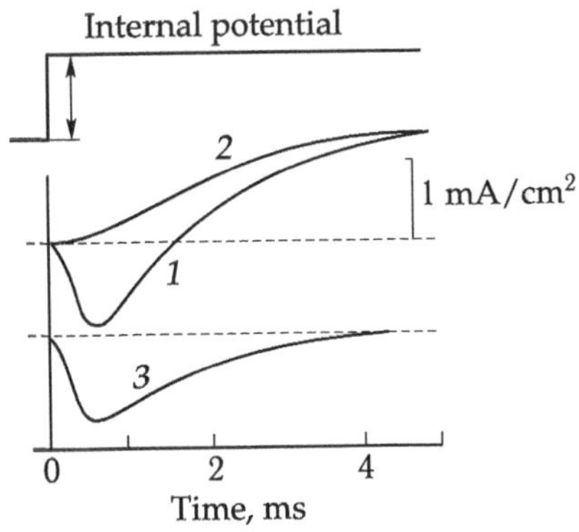

Figure 19.4. Experimental separation of membrane current I in potassium and sodium constituents (reproduced from A. L. Hodgkin, 1964).

1, The axon is in the sea water, $I = I_{Na} + I_K$; *2*, the greater portion of sodium is substituted for choline, $I = I_K$; *3*, the difference between *1* and *2*, $I = I_{Na}$, the temperature is 8.5 °C; the upward deflection corresponds to the outward current.

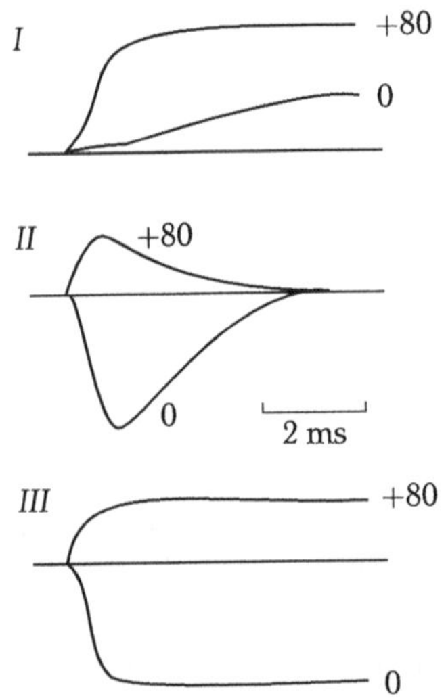

Figure 19.5. Effect of TTX, TEA and pronase on ionic currents in the axon membrane caused by the membrane potential shift to 0 and to +80 mV.

I, Potassium current in axons treated with TTX; *II*, sodium current in axons perfused with a solution containing 15 mM TEA; *III*, the same as *II* after addition of pronase to the perfused solution (2 mg/ml); numerals at the curves are the values of the potential of the internal solution.

Figure 19.6 shows calculated changes in the conductance caused by the rapid shift of the potential inside the fiber from the resting potential level (−56 mV) to zero. As seen, at stepwise depolarization of the membrane, the Na^+ conductance starts changing from an extremely low value, grows rapidly to 25 mSm/cm^2 and then decreases exponentially. The K^+ conductance increases following an S-shaped curve and reaches a constant level in 5–6 ms. Changes in the K^+ and Na^+ conductance are gradual and reversible. The sodium conductance is restored by about 10-fold faster than the potassium conductance.

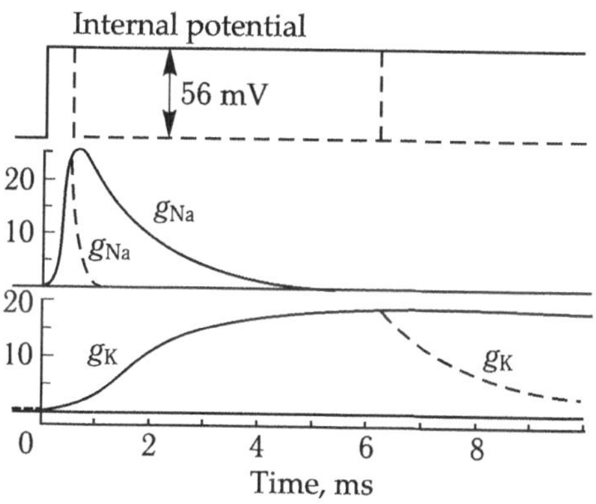

Figure 19.6. Time-dependent changes in sodium and potassium conductance upon depolarization of the membrane at 56 mV (reproduced from A. L. Hodgkin, 1964).

Solid lines show conductance upon prolonged depolarization, dashed lines show conductance upon repolarization of the membrane in 0.6 and 6.3 ms.

19.3 Description of Ionic Currents in the Hodgkin–Huxley Model

Using the data of experiments on voltage clamp in the squid axon, A. Hodgkin and A. Huxley formulated a number of fundamental principles:

1) Na^+ and K^+ transfer is realized by various noninteracting structures.
2) Currents change as a result of the change in the membrane permeability (for Na^+ and K^+).
3) The membrane transmission capacity is controlled by the electric field. □

A. Hodgkin and A. Huxley proposed an adequate mathematical description of the experimentally observed kinetics of changes in the sodium and potassium conductance at different shifts of the membrane potential (Fig. 19.7). The model suggests that the membrane permeability for Na and K ions is controlled by some gating particles that move in the membrane upon changes in the electric field.

Such a displacement may be represented as jumps of positively charged particles between two potential wells over the energy barrier at rate constants α and β (Fig. 19.8) dependent on the height of the energy barrier and consequently on the potential. A rise of the electric potential on the left side of the membrane leads to an increase in constant α and a drop of constant β.

At a stepwise change in the difference of membrane potentials, the particles are redistributed between the two states. This is accompanied by changes in the newly established equilibrium values of probable filling of the right and left potential wells (x and $(1-x)$), respectively.

The process of changes in probability (x) of the right well filling after the stepwise potential shift can be described by the differential equation:

$$dx/dt = \alpha(1-x) - \beta x. \tag{19.7}$$

Setting boundary values $x = x_0$ at $t = 0$ and $x = x_\infty$ at $t = \infty$, we get

$$x = x_\infty - (x_\infty - x_0)\exp[-t(\alpha + \beta)]. \tag{19.8}$$

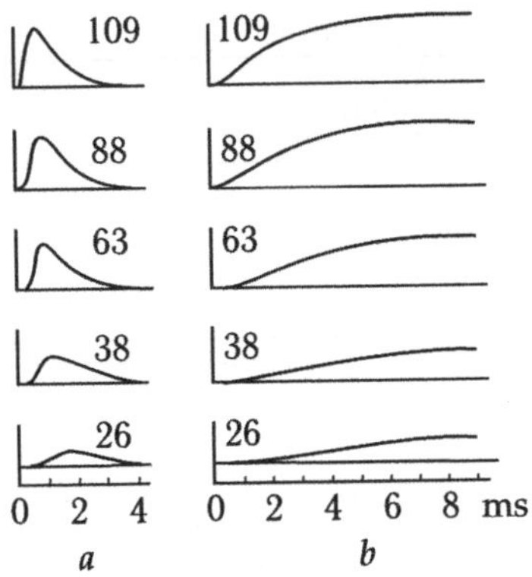

Figure 19.7. Changes in the sodium (*a*) and potassium (*b*) conductance of the axon membrane at different displacements of the membrane potential from the resting state.

Numbers designate the magnitude of depolarization (mV).

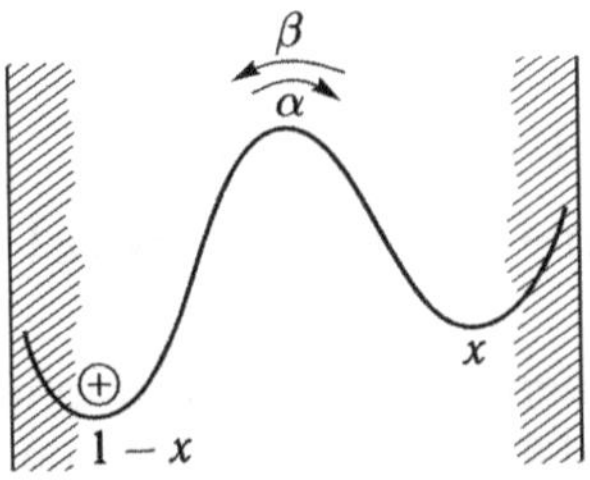

Figure 19.8. Model demonstrating displacement of charged particles in the membrane upon changes in the electric field.

x is the probability for the particle to be on the right side of the membrane; $(1 - x)$ is the probability of the particle to be on the left side of the membrane; α and β are potential-dependent constants of transition of charged particles across the membrane.

The equation shows that accumulation of particles on the right side of the membrane occurs according to the exponential law at time constant τ where $\tau = 1/(\alpha + \beta)$.

Changes in Potassium Conductance. In case the channel opened as a result of transition of one gating particle from left to right, the conductance of ion channels would change in time also exponentially, i.e., it could be written that $g_i = x\overline{g}_i$ where $\overline{g}_i$ is the maximal conductance for type i ions. However, the optimal correspondence of the kinetic curve for potassium conductance in the squid axon is attained on the assumption that four gating particles should arrive to a definite region of the membrane in order to open the potassium channels.

Let n designate the probability that one particle controlling the K^+-conductance is in the right-hand part of the membrane. Then probability p that there are four particles simultaneously is n^4. Therefore, at any instant of time, the K^+-conductance of the membrane is

$$g_K = \overline{g}_K n^4, \tag{19.9}$$

where n is defined by an equation like (19.7) and can vary from 0 to 1:

$$dn/dt = \alpha_n(1 - n) - \beta_n n. \tag{19.10}$$

Particles controlling the permeability of K^+-channels are called n-particles.

Let us analyze a simple case when in the initial state all n-particles are on the left side of the membrane ($n = 0$) and K^+-channels are closed. After imposing the large difference of potentials with the plus sign on the left side of the membrane, all n-particles flow to the right well. The kinetics of accumulation of n-particles on the right side of the membrane is described by the exponent (see Eq. (19.8)):

$$n(t) = n_\infty - (n_\infty - n_0)\exp(-t/\tau_n), \tag{19.11}$$

where $\tau_n = 1/(\alpha_n + \beta_n)$. Here $n_0 = 0$ and $n_\infty = 1$, so

$$n(t) = 1 - \exp(-t/\tau_n).$$

The magnitude of K^+-conductance proportional to the fourth power of n grows not exponentially but with a large delay period. The n^4 value starts changing rapidly only when the n values exceed 0.7.

The above example explains also the way of determining kinetic parameters of activation of K^+ conductance. To define the constant of activation time τ_n it is necessary to normalize the kinetic curves of conductance, i.e., to plot dependences $g_K(t)/\overline{g}_K = n^4$. Then, by extracting the fourth root, to plot the $n(t)$ dependence and determine the time constant of the exponent $\tau_n = 1/(\alpha_n + \beta_n)$ for the given level of the membrane potential.

If the difference of potentials applied to the membrane is not high enough, not all of the gating particles would move from the left side of the membrane; this means that parameter n would not reach its limiting value ($n = 1$) and K^+ conductance would be established at the level lower than the maximal $\overline{g}$. The n_∞ value is determined from experimental data using the formula:

$$n_\infty = (g_\infty/\overline{g})^{1/4}, \tag{19.12}$$

where g_∞ is the stationary level of K^+ conductance at the given potential, and $\overline{g}$ is the maximal K^+ conductance. It follows from Eq. (19.10) that at equilibrium ($dn/dt = 0$)

$$n_\infty = \alpha_n/(\alpha_n + \beta_n).$$

Consequently, in the family of kinetic curves of K^+ conductance (see Fig. 19.7, *b*) it is possible to find rate constants α and β at different potentials using the system of equations:

$$\tau = 1/(\alpha + \beta), \qquad n_\infty = \alpha/(\alpha + \beta). \tag{19.13}$$

As seen from Eq. (19.13), $\alpha = n_\infty/\tau$.

Changes in Sodium Conductance. Kinetic curves of Na^+ conductance have a more complex shape (see Fig. 19.7, *a*): the conductance grows to the maximum (activation) and then decreases (inactivation). The change in Na^+ conductance was described assuming the presence of activating m-particles and inactivating h-particles. The model of channel opening suggests that three m-particles enter a certain region of the membrane. The transition across the membrane of one inactivating particle causes the channel closing. As a result, changes in Na^+ conductance are described by the equation:

$$g_{Na} = \overline{g}_{Na} m^3 h, \tag{19.14}$$

where $\bar{g}_{Na}$ is the maximal Na^+ conductance, m is the probability of one m-particle to be in the given region of the membrane, h is the probability that the inactivating particle is absent in this region. Upon applying the field that moves charged particles from the left side of the membrane (Fig. 19.8) to the right side, m grows in the range from 0 to 1, and h decreases (from 1 to 0).

The kinetics of redistribution of particles across the membrane upon applying the electric field is described by single-type equations:

$$dm/dt = \alpha_m(1-m) - \beta_m m, \qquad dh/dt = \alpha_h(1-h) - \beta_h h, \qquad (19.15)$$

where α and β are rate constants depending only on the membrane potential (at constant temperature and concentration of Ca^{2+}). The solution of these equations is as follows:

$$\begin{aligned} m(t) &= m_\infty - (m_\infty - m_0)\exp(-t/\tau_m), \quad &\text{where } \tau_m = 1/(\alpha_m + \beta_m); \\ h(t) &= h_\infty - (h_\infty - h_0)\exp(-t/\tau_h), \quad &\text{where } \tau_h = 1/(\alpha_h + \beta_h). \end{aligned} \qquad (19.16)$$

The m_∞ and h_∞ values are determined from the boundary conditions:

$$m_\infty = \alpha_m/(\alpha_m + \beta_m), \qquad h_\infty = \alpha_h/(\alpha_h + \beta_h). \qquad (19.17)$$

As a result of the analysis of the kinetic curves of activation an inactivation of Na^+ conductance, A. Hodgkin and A. Huxley plotted graphic dependences of parameters $\alpha_m, \beta_m, \alpha_h, \beta_h, \alpha_n, \beta_n$, as well as m_∞, n_∞ and h_∞ on the membrane potential (Fig. 19.9 and 19.10) and determined empirical relations describing the variation of the above dependences.

Based on the empirical dependences represented in Fig. 19.9, it is possible to demonstrate that the ratio of transition rate constants of m-particles across the membrane depends on the potential as described by the formula:

$$\frac{\beta_m}{\alpha_m} \simeq \exp\left(-\frac{3.9F\varphi}{RT}\right) \sim \exp\left(-\frac{4F\varphi}{RT}\right). \qquad (19.18)$$

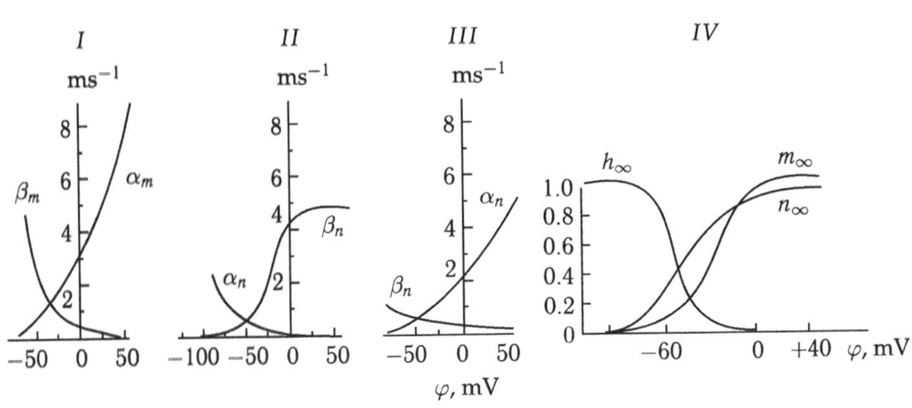

Figure 19.9. Dependence of constants α and β on the membrane potential (I–III) and of stationary values m, h and n on the membrane potential (IV) (reproduced from V. S. Markin et al., 1981).

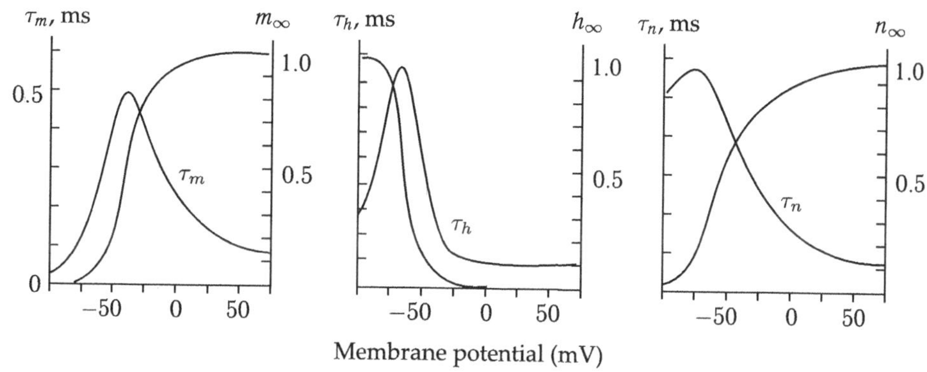

Figure 19.10. Potential-dependent parameters of the Hodgkin–Huxley model.

Time constants τ_m, τ_h, and τ_n and stationary states m_∞, h_∞ and n_∞ calculated from empirical equations of the Hodgkin–Huxley model for the squid giant axon membrane at 6.3 °C. Membrane depolarization results in the rise of m_∞ and n_∞ and a decrease of h_∞. The constants of relaxation time have the maximal values in the region of the resting potential and become shorter at any shift of the resting potential (reproduced from B. Hille, 1970).

Table 19.1. Dependences of transition rate constants for particles m, h and n on transmembrane potential φ in the Hodgkin–Huxley model

Particle	$\alpha(\varphi)$	$\beta(\varphi)$
m	$0.1\dfrac{25-\varphi}{\exp\left(\dfrac{25-\varphi}{10}\right)-1}$	$4\exp\left(\dfrac{-\varphi}{18}\right)$
h	$0.007\exp\left(\dfrac{-\varphi}{20}\right)$	$\dfrac{1}{\exp\left(\dfrac{30-\varphi}{10}\right)+1}$
n	$0.001\dfrac{10-\varphi}{\exp\left(\dfrac{10-\varphi}{10}\right)-1}$	$0.125\exp\left(\dfrac{-\varphi}{80}\right)$

The dependences of rate constants α and β on the potential used by Hodgkin and Huxley for the squid axon are given in Table 19.1.

The curves in Fig. 19.9 and 19.10 show that the stationary K^+ current occurs in a wide range of potentials, whereas the stationary Na^+ current is limited to a narrow region, where m^3 and h are concurrently not equal to zero. Using the above mathematical model, Hodgkin and Huxley reproduced with high accuracy the nerve impulse generated by the axon membrane under various experimental conditions. After slight modification of the model, the process of electric excitation in the shellfish soma neuron was also simulated (P. G. Kostyuk and O. A. Kryshtal, 1981).

Separation of transmembrane currents into separate components played an extremely important role for understanding the mechanisms of electrical excitability of membranes: characteristics were obtained for selectivity, potential dependence, turn on and turn off kinetics of intramembrane pathways — ionic channels. On the whole, as a result of these studies general notions on the structure-functional organization of Na^+- and K^+ channels controlled by changes in the electric field (Fig. 18.7) were formulated, and functional elements of ionic channels responsible for selection mechanisms (selective filter), activation ("activation gate") and inactivation ("inactivation gate") were indicated (Sections 6 and 7 in Chapter 18). □

In the Hodgkin–Huxley model, the activation and inactivation of channels are considered as completely independent processes defined solely by the displacement of charged particles. The total form of the model can be represented by four differential equations:

$$\frac{dV}{dt} = -\frac{1}{C}\left[\overline{g}_{\text{Na}} m^3 h(\varphi - \varphi_{\text{Na}}) + \overline{g}_{\text{K}} n^4(\varphi - \varphi_{\text{K}}) + \overline{g}_L(\varphi - \varphi_L)\right] + I_{\text{app}}, \tag{19.19}$$

$$\frac{dm}{dt} = \alpha_m(1 - m) - \beta_m m, \tag{19.20}$$

$$\frac{dh}{dt} = \alpha_h(1 - h) - \beta_h h, \tag{19.21}$$

$$\frac{dn}{dt} = \alpha_n(1 - n) - \beta_n n, \tag{19.22}$$

where $\overline{g}_L$ is the conductance of the potential-dependent leakage, and I_{app} is the applied external current acting as a controlling parameter. Consequently, all basic properties of membrane electrical excitability are described by corresponding dependencies of coefficients α and β on potentials at h, m and n. It should be noted that the form of nonlinearities n^4 and m^3h is specific namely for the description of K^+- and Na^+ channels of the squid axon. The dynamics of gating variables of potential-dependent channels of other animal species can be described using other powers for n and m. For example, nonlinearity m^2h corresponds to the Na^+current in leach neurons, and nonlinearity n^2 corresponds to the K^+current in mammalian nerve fibers.

Though the above models are rather effective, they do not take intoo account the molecular mechanisms of channel opening and closing, only formally describing them by displacement of hypothetical charged particles. In the light of concepts (Section 4 in Chapter 18) on macromolecular structure of channels, these processes should be directly affected by conformational rearrangements in the channel that have been revealed in the observed experimental temperature dependencies, but yet are not analyzed theoretically.

19.4 Gating Currents

The simplest interpretation of Hodgkin–Huxley equations describing the dependence of the conductance of excitable membranes on the potential suggests the presence in the membrane of mobile charged particles m, n and h redistributed in the membrane under the action of the electric field and transferring the channel in the conductive or nonconductive state.

If charged particles capable of moving in the electric field are present in the membrane, it is evidently possible to record the current caused by movement of these gating particles. The motion of the gating charge or the gating current can be observed experimentally. □

Such an experiment consists in the following. If the membrane represents an ideal dielectric, upon imposing a step-function voltage impulse the capacitive current,

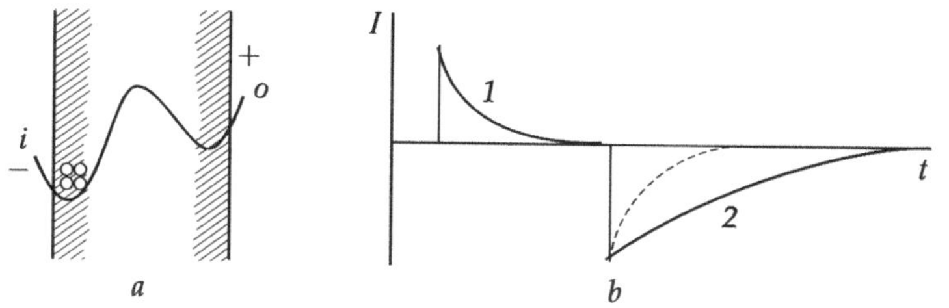

Figure 19.11. Asymmetric displacement current in a membrane containing mobile charged particles.

A, Asymmetric distribution of particles in the membrane in the initial state; *B*, kinetics of the capacitive current upon imposing hyperpolarizing (*1*) and depolarizing (*2*) voltage pulses.

The dashed line shows changes in the capacitive current when mobile charges are absent in the membrane; *i* and *o* are internal and external sides of the membrane.

or the displacement current, would flow through it during the initial instants of time, its magnitude and kinetics being independent of the polarity of the applied impulse. Another situation occurs when the membrane contains charged particles capable of moving in the electric field, the distribution of these mobile particles being asymmetric in the initial state. Let us propose for certainty that in the initial state all gating particles are on the left (internal) side of the membrane having a lower potential (Fig. 19.11). If the membrane is hyperpolarized by applying the step-function voltage impulse of the same polarity as that of the existing potential, the capacitive current would flow through it similar to that in the absence of gating particles. However, upon imposing the voltage of the opposite polarity, other than the main component of the capacitive current, an additional current is generated caused by the displacement of gating particles from left to right. The magnitude of this current may be rather small as compared to the main component of the current.

Combining the methods of intracellular perfusion, potential fixation and synchronous signal accumulation, on the squid giant axon, K. Armstrong and F. Besanilla found a small asymmetric displacement current supposedly associated with activation of Na^+-currents.

The gating current is shown in Fig. 19.12 as compared to the Na^+ current. In the above experiment, 95 % of Na^+ in the outer solution was substituted for nonpenetrating ions of trishydroxymethylaminomethane (tris). Figure 19.12 shows that the Na^+ current of the magnitude which under these conditions is much lower than usual, is preceded by the short-time outward current (gating current). If all Na ions are removed from the solution, it becomes clear that the gating current slowly drops after the initial peak and for the most part terminates to the moment when the peak of the Na current is reached in the standard solution (Fig. 19.12, *B*).

A characteristic peculiarity of asymmetric displacement currents is their exponential dependence on time that corresponds to the Hodgkin–Huxley model. The analysis of the characteristic relaxation time of the asymmetric current versus the potential demonstrated that this time is approximately comparable with time $\tau_m(\varphi)$ in the Hodgkin–Huxley model. The coincidence suggests the association of asymmetric displacement currents with conformational rearrangements accompanying the opening of Na^+channels.

An essential corroboration that the asymmetric displacement current is associated with the activation of inward Na^+ current was provided by the comparison of the

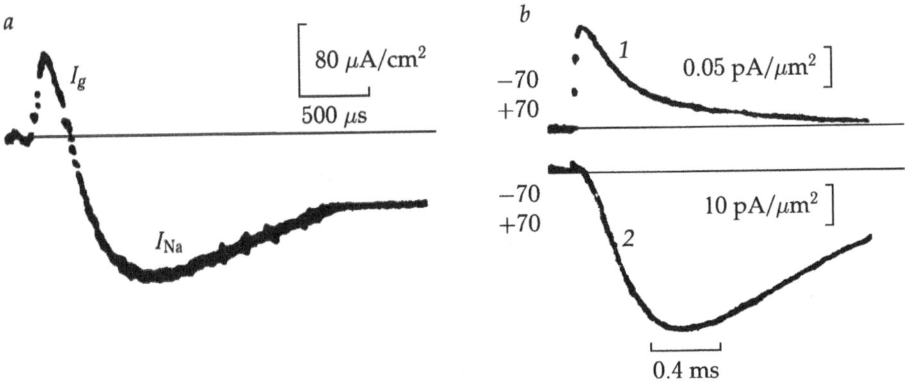

Figure 19.12. Gating current I_g and sodium current I_{Na} in the squid axon (reproduced from C. M. Armstrong, 1975).

a, 95 % of Na^+ are substituted for tris ions; *b*, gating current under conditions of complete substitution of Na^+ for tris (*1*), Na^+ current recorded on the same axon under standard conditions in seawater (*2*).

experimentally measured and theoretical dependences of the displaced charge on the value of the membrane potential. Based on the experiment, the value of the displaced charges can be found integrating the gating current in time:

$$Q = \int_0^\infty i(t)\,dt. \tag{19.23}$$

An example of such dependence is given in Fig. 19.13. Here the ordinate axis shows values of the displaced charge in nC/cm^2. In some cases the dependence of the $\Theta = Q/Q_{\text{max}}$ value on the potential is measured. This value shows the portion of displaced charges relative to the maximally possible transferred charge.

The theoretical dependence of Θ on the membrane potential value can be obtained by writing down the condition for the equilibrium distribution of charged gating particles between two possible states. The distribution of particles in the presence of the potential difference φ on the membrane is described by the Boltzmann equation:

$$1 - \Theta = \Theta \exp\left(-\frac{zF\varphi}{RT}\right), \tag{19.24}$$

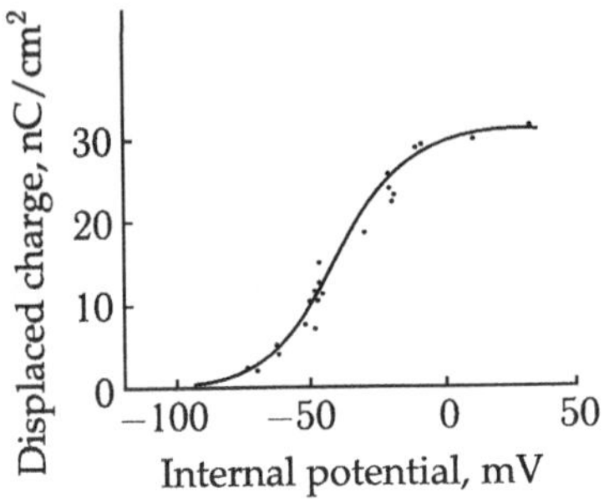

Figure 19.13. Dependence of the displaced charge in the frog muscle membrane on the applied potential (reproduced from W. Armstrong, 1978).

where $(1-\Theta)$ is the filling of the left potential well, and Θ is the filling of the right well. From Eq. (19.24) it follows that the number of particles displaced in the membrane from left to right is

$$\Theta = \left[1 + \exp\left(-\frac{zF\varphi}{RT}\right)\right]^{-1}. \tag{19.25}$$

The same value defines the portion of open channels as dependent on the applied potential φ and valence z of the displaced particle. Figure 19.14 shows curves of $\Theta(\varphi)$ at different z values.

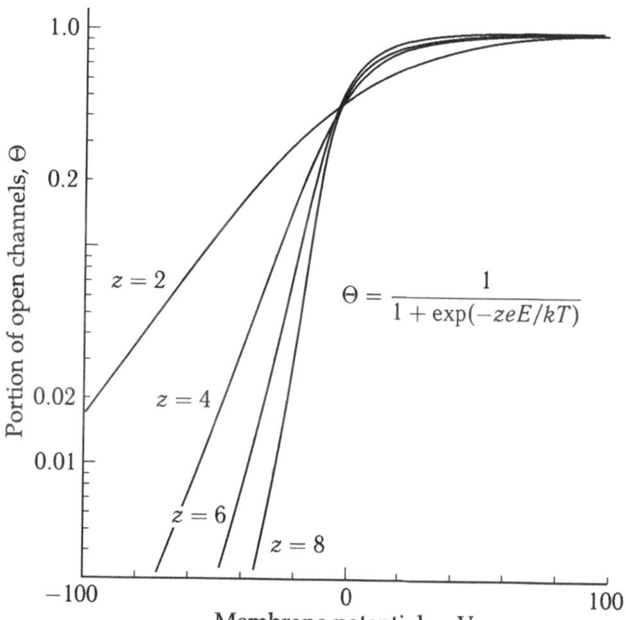

Figure 19.14. Boltzmann's description of potential-dependent channels.

Upon description of potential-dependent channels having two (open and closed) states, it is supposed that the channel opening is regulated by motion of particles with total charge z between opposite sides of the membrane. For particles with a higher charge z the theoretical dependence of the probability for an open state is characterized by higher steepness (reproduced from B. Hille, 1992).

The coincidence of the theoretical (Fig. 19.14) and experimental (Fig. 19.13) curves of such dependence for i_K and i_{Na} is observed at $z = 4, 5$ for g_K and $z = 6$ for g_{Na}. Consequently, the opening of the Na^+ channel occurs upon the displacement of the gating charge equivalent to 6 elementary charges. When the value of the displaced charge Θ_g in units of elementary charge e is known, dividing it by $z = 6$ it is possible to find also the density of Na^+ channels.

For Na^+ channels the value of Θ_g varies in different objects within a wide range from 260 $e/\mu m^2$ (rat cardiac muscle) to 1500–2200 $e/\mu m^2$ (squid giant axon), and the density of Na^+ channels changes correspondingly from 40 to 250–350 μm^2. For K^+ channels, $\Theta_g \sim 500\ e/\mu m^2$, and for Ca^{2+} channel, $\Theta_g \sim$ 1500–2000 $e/\mu m^2$.

The simplest interpretation of the gating currents is that they are a result of transition of m particles between two states upon activation of Na^+-channels. However the skip of several charged groups over the whole width of the membrane is hardly probable. It is most likely that the displacement currents are caused by relatively small displacements of a large number of charged groups, associated with the opening of the gating mechanism of the channel.

Peculiarities of gating currents. The determination of the complete quaternary structure, numerical experiments with molecular dynamics simulations and site-directed mutagenesis provide a possibility to understand the mechanisms underlying the kinetic regularities of opening, closing and inactivation of ion channels. A cardinal task is identification of certain molecular groups, their motion resulting in generation of the gating current in the channel under the action of the external electric field. We have already seen the role of charges in these processes in the channel (Section 4 in Chapter 18). Note that the opening – closing cycle is not connected with the disruption and formation of some covalent chemical bonds, and may include a cascade of conformational changes in the channel protein. The threshold character and high velocity of opening of separate closed channels exhibit also the concerted cooperative character of these processes.

19.5 Impulse Propagation

The principal property of the nerve impulse is its capability to propagate along the fiber at a constant rate without decay. Propagation of excitation is associated with the local currents between resting and active (excited) regions (Fig. 19.15). For the simplicity let us accept that the outer surface of the axon is equipotential, i.e., the external environment has high electrical conductivity. In the region of generation of the action potential, the internal part of the fiber is charged positively, while in the neighboring unexcited regions it has a negative charge. As a result, local current emerges between the excited and resting sites of the nerve that depolarizes the membrane just before the active region. Upon reaching critical depolarization, this region is also excited. In such a way excitation is transferred further. The unidirectional impulse travel along the nerve fiber is provided by that the regions, where the action potential has terminated, lose their ability to be excited for some time (refractoriness). □

In fibers of vertebrates a large portion of their surface is coated by myelin which is an insulator. Therefore the local current flows between Ranvier's constrictions

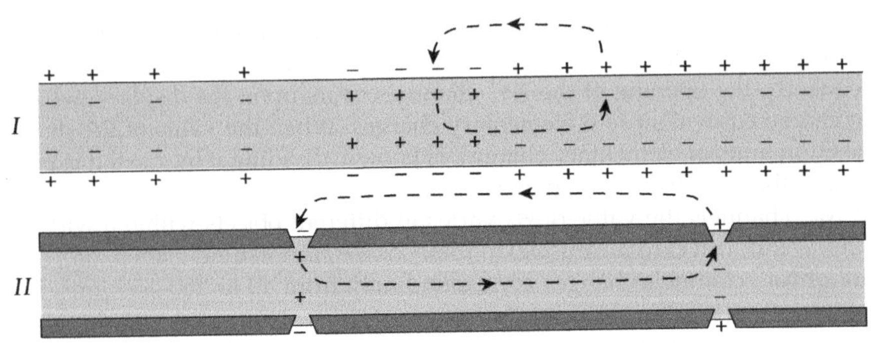

Figure 19.15. Illustration of the theory of local currents (reproduced from A. L. Hodgkin, 1964). *I*, unmyelinated nerve fiber; *II*, myelinated nerve fiber.

(Fig. 19.15, *II*). As a result, the excitation jumps from one Ranvier's constriction to another, this occurring at a higher rate and with less energy losses than in unmyelinated fibers of a comparable diameter. Low energy losses are explained by the fact that the electric capacity of the myelin coating is very small and its charging requires far less energy than the charging of the axon membrane of invertebrates with high electric capacity. Below we will consider only the process of impulse propagation along unmyelinated fibers.

The fibers, the nerve impulse propagates along, are usually compared to a cable having imperfect electric properties (low electrical conduction of the axoplasm, insufficient insulating properties of the membrane, high membrane capacitance).

The distribution of the membrane potential over the fiber $\varphi(x,t)$ is in general described by the cable equation ensuing from the condition of equality of the total current across the membrane i_M to the sum of the capacitive and ion currents (see Eq. (19.6)):

$$\frac{1}{r_i}\frac{\partial^2 \varphi}{\partial x^2} = I + C\frac{\partial \varphi}{\partial t}, \tag{19.26}$$

where C is the membrane capacitance per unity of the fiber length, r_i is the resistance of the axoplasm per length unity, I is ion current flowing across the membrane; the potential value φ is counted off the resting potential level. Current I is summed up of partial ion currents: sodium, potassium and leakage currents (see Section 2 in this Chapter). The dynamics of changes in each of the partial currents is described by the empirical Hodgkin–Huxley equation.

Accordingly, the task of determining the rate of impulse propagation and its form is reduced to the solution of Eq. (19.26) in the system with Hodgkin–Huxley equations. As a result of computer-based numerical solutions, it is possible to calculate the form and rate of the action potential propagation (Fig. 19.16).

The rate of a stable impulse is described by a simple formula if the conductance of the resting membrane is neglected:

$$v_1 = \left(\frac{i_1}{\varphi^* r_i C^2}\right)^{1/2}. \tag{19.27}$$

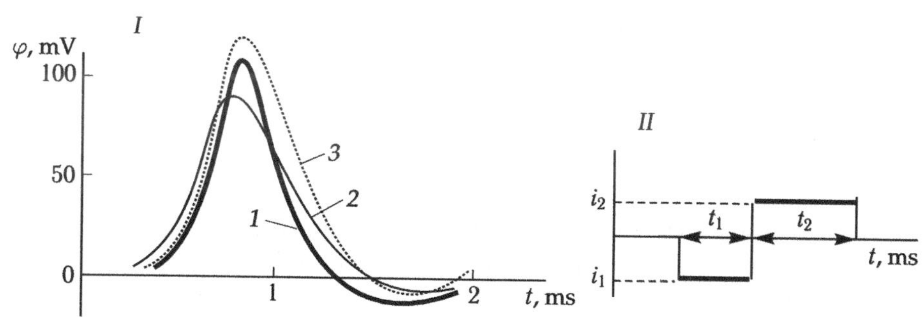

Figure 19.16. Propagating impulse. *I*, Shape of the impulse; *II*, approximation of the ion current upon excitation (downward impulse corresponds to the inward Na^+ current; upward impulse corresponds to the outward K^+ current) (reproduced from V. S. Markin et al., 1981).

1, Squid axon at 18.5 °C; *2*, numerical solution of the system of Hodgkin–Huxley equations; *3*, shape of the impulse upon approximation of ion current with two steps; t_1 and t_2, duration of the inward and outward currents; i_1 and i_2, amplitudes of the inward and outward currents.

The expression includes only parameters of the first phase of the excitation current, i.e. the propagation rate is determined by the leading edge of the travelling pulse.

Studies of the dependence of the impulse propagation rate on the diameter of the nerve fiber in experiments with unmylienated fibers demonstrated that the rate is approximately proportional to the square root of the fiber diameter. The same result is achieved using the above formula. For the squid giant fiber, the rate is about 21 m/s. □

Based on the model of the threshold membrane generator of ion current, one can examine propagation of impulses on nonhomogeneous and branching fibers as well as analyze the interaction of impulses in parallel fibers. For example, the analysis of impulse travelling along an extending fiber (Fig. 19.17) shows that as the extension site is approached, the impulse rate decreases, and after the extension it begins growing until a new stationary value, exceeding the initial one, is reached. Consequently, the transition to higher rate values occurs not monotonously but with a delay, which is the larger the greater the difference in the fiber section. At a rather large extension the impulse can terminate entirely.

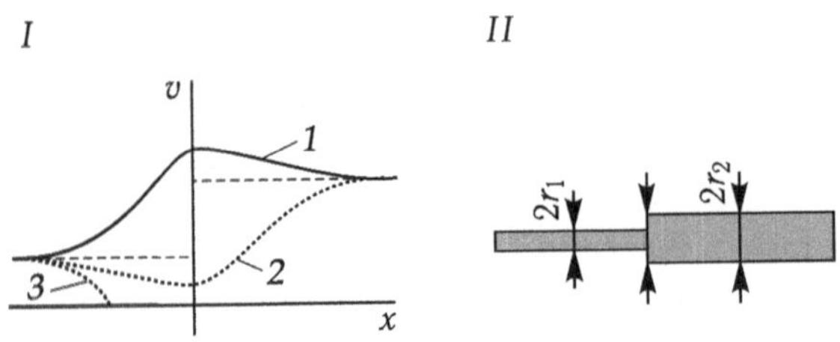

Figure 19.17. Travelling of the impulse along the extended fiber. *I*, Change in impulse rate *v* dependent on its direction; *II*, scheme of the extending fiber.

1 and *2*, Impulses are travelling; *3*, impulse are blocked; the dashed line shows the rate of impulse propagation far from the site of changes in the fiber diameter.

When an impulse moves backwards, no blocking occurs. The impulse can at all times travel from the wide fiber to the narrow one. However, under conditions of inhomogeneity, the change of the rate has an opposite character. When approaching the narrowing, the impulse rate grows, and then it begins decreasing to a new stationary value. Thus, the plot of the rate shows a peculiar hysteresis loop the origin of which is quite clear from the physical point of view. The wide fiber of a high capacitance located in front represents a powerful drain for the charge therefore the potential value grows slower to the threshold. On the contrary, a high charge cannot be absorbed in the narrowed part of the fiber, and this part is a reflecting barrier, close to it the potential grows faster and as a result the rate of impulse propagation increases.

In organisms, nerve fibers are usually combined in bundles or nerve trunks, where each fiber represents an independent communication link. But when a nerve impulse is travelling along a fiber, it creates an electric field in the surrounding intercellular fluid that can affect the membrane potential of nearby fibers. As a matter of fact, upon processing the nerve trunks with special chemical substances it is possible to see not only reciprocal disturbances but also transmission of excitation between neighboring fibers.

Investigations of simultaneous travelling of impulses along different fibers revealed that interaction of impulses may lead to their synchronization. As a result,

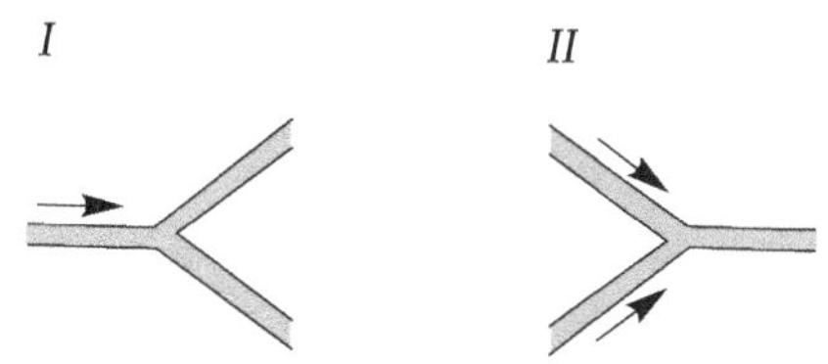

Figure 19.18. Impulse propagation through the branching center.
I, impulse travels along one fiber; *II*, impulse travels along two fibers simultaneously.

a collective impulse may be generated upon concurrent excitation of fibers. However, only impulses, the rates of which differ no more than by 10% can be synchronized, their interactions increasing greatly with the growth of resistance of the environment. When a fiber is excited, a sign-changing membrane potential is induced in the neighboring fiber: first the fiber is hyperpolarized (decreased excitability), then depolarized (increased excitability) and, finally, hyperpolarized once again. Direct excitation transfer to the neighboring fiber grows with an increase in its radius.

20

Active Transport

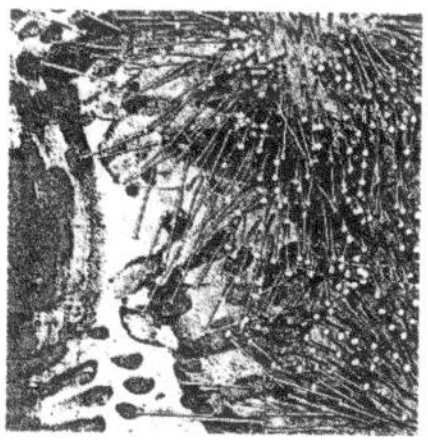

The existence of concentration gradients of Na, K and Ca ions between the intra- and intercellular media is a crucial condition for maintaining the resting potential (see Eq. (16.10)) and providing electric excitability of nerve and muscle cells.

The distribution of Na^+ and Ca^{2+} differs significantly from the equilibrium distribution: the both components of the passive flow (by the concentration gradient and in the direction of the electric field) are directed inside the cells. To maintain the low non-equilibrium level of Na^+ and Ca^{2+} concentration in the cell, a mechanism of their active release into the environment or into the intracellular compartments is required. Such systems of active transport provide ion transfer against electrochemical potential gradient at the expense of ATP energy (the Na^+-pump of plasmatic membranes, the Ca^{2+}-pump of membranes of the sarcoplasmic reticulum; see Chapter 22) or the energy of redox reactions (the H^+-pump of mitochondria, chloroplasts and other energy-coupled membranes; see Chapter 21). □

Active transport of most cations in membranes is carried out by the P-type ATPases — specialized enzymes constituting a family of homologous proteins. All of them are characterized by the formation of covalently bound aspartyl phosphate upon ATP hydrolysis closely associated with the ion transfer. The P-type ATPases are quite diverse and occur both in prokaryotes and eukaryotes. This family also includes Ca^{2+}-ATPases of the sarcoplasmic reticulum and plasmatic membrane, Na^+/K^+-ATPase, H^+/K^+-ATPase and others. All of them are combined by a similar structure and the enzyme working cycle, which can be divided in the binding of ions of one type on one side of the membrane, phosphorylation of the enzyme, ion transfer across the membrane, hydrolysis of phosphate and transfer of second-type ions across the membrane in the inverse direction. Thus, in the case of Ca^{2+}-ATPases, Ca^{2+} ions are ions of the first type, and protons are ions of the second type, while in the case of Na^+/K^+-ATPase such a pair is formed by Na^+ and K^+, respectively.

20.1 Calcium Pump

Ca^{2+}-ATPases located in membranes of the sarcoplasmic reticulum have been studied in more detail. Different tissues contain several varieties of Ca^{2+}-ATPases; all of them exchange Ca^{2+} ions for protons, but in dissimilar stoichiometry. For example, Ca^{2+}-ATPase of the sarcoplasmic reticulum exchanges two Ca^{2+} ions for 2–3 H^+, thus representing an electrogenic pump, though the H^+ gradient on the sarcoplasmic membrane is not generated because protons can is freely penetrate it.

General Structure. The calcium pump is a monomer, i.e. is formed by one subunit. It has three cytoplasmic domains (phosphorylating domain P, nucleotide-binding domain N and transmission domain A), ten transmembrane α-helices and several small loops in the lumen. Domain A, connected to helices M1–M3 with rather long linkers, serves as a transfer link between the gating mechanism controlling the association and dissociation of Ca^{2+} and transformation of ATP with domains N and P. Domain P contains a phosphorylating residue Asp351 coordinating magnesium ions in Asp703 and other residues critical for the given family of enzymes. Domain P has a wedge-like form with a flat upper surface that allows domain A to perform large-scale rotation sliding over its surface. Domain N contains amino acids responsible for the adenosine binding (for example, Phe487) and responsible for the overlapping of ATP and domain P, for example Arg560. Of the three cytoplasmic domains, this domain displays the highest variability among P-type ATPases. The three domains are well distinguished for state $E1 \cdot 2Ca^{2+}$, whereas in other structures the cytoplasmic part is compact.

There are ten transmembrane helices, many of them (M2–M10) protrude deep into the cytoplasm (Fig. 20.1). Two helices (M4 and M6) have a proline residue in the middle and during the working cycle are partially untwisted. Helices M1–M6 perform considerable movements during the working cycle, whereas helices M7–M10 are relatively static and may serve for docking the pump in the membrane.

The scheme of Ca^{2+}-ATPase can be represented as shown in Fig. 20.2. The first stage is characterized by the binding of Ca^{2+} ions in the membrane part of the protein, and ATP is bound in its cytoplasmic part. The binding constant of Ca^{2+} ions is about 10^7 M^{-1}. The nucleotide binding is accompanied by occlusion of Ca^{2+} ions from the cytoplasmic side of the membrane. At the second stage, ATP is hydrolyzed with the formation of enzyme $E1{\sim}P$.

The phosphorylated form of enzyme $E1{\sim}P$ is unstable from the conformational point of view, and at the next stage of the cycle the affinity of Ca^{2+}-binding centers to calcium ions changes concurrently with the change in the character of the phosphate group binding to the enzyme. The energy, previously concentrated in the high-energy phosphate bond of the $E1{\sim}P$ complex, is expended for the change in the binding constant of calcium ions and the enzyme. Due to the change in the three-dimensional structure of the enzyme, calcium ions get access to the lumen of the sarcoplasmic reticulum. The binding constant of Ca^{2+} decreases upon formation of a stable phosphorylated form of the enzyme from 10^7 to 10^4 M^{-1}.

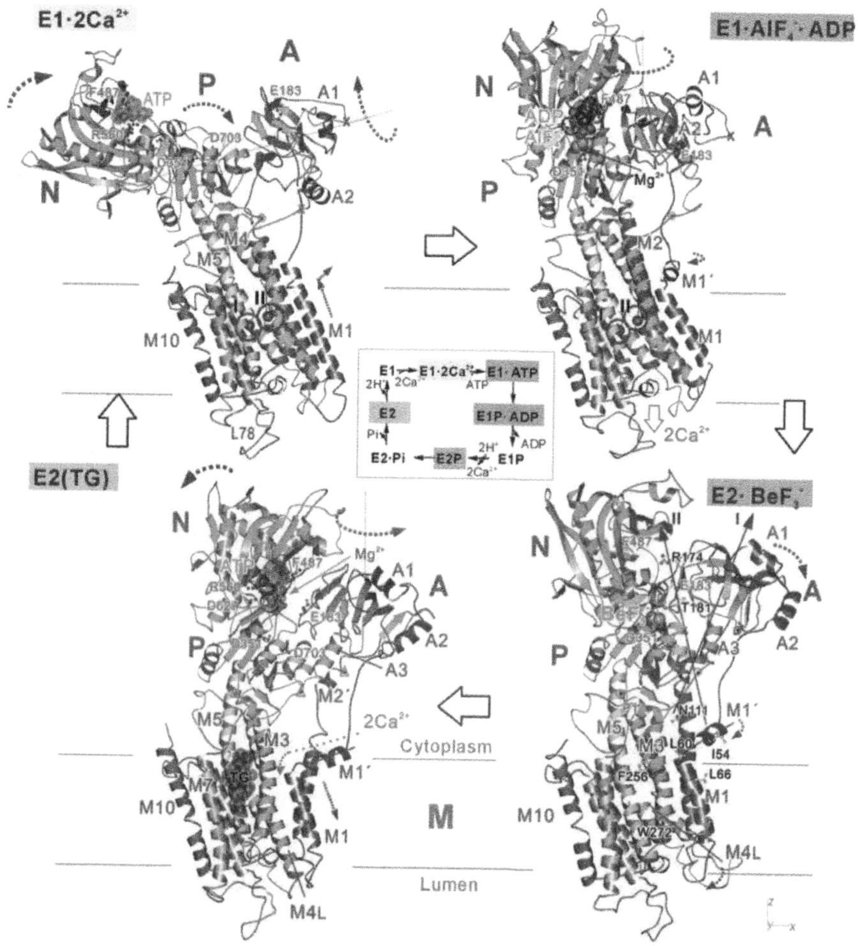

Figure 20.1. Four main conformational states and the kinetic scheme of the Ca^{2+}-pump cycle. The coloring of the protein structures changes gradually from the N-terminus (blue) to the C-terminus (red). Violet circles with Roman numerals show Ca^{2+} ions. Cytoplasmic domains A, N and P are indicated; three α-helices of domain A (A1–A3) and ten transmembrane domains (M1–M10) are shown. Thin orange lines show the rotation axis of domain A, excluding the $E2 \cdot BeF_4^-$ structure where long arrows designate rotation axes upon $E1P \rightarrow E2P$ (*I*) and $E2P \rightarrow E2 \cdot Pi$ (*II*) (reproduced from C. Toyoshima, 2008).

Using relation (5.12) where the constant of chemical equilibrium is associated with the change in the free energy during the reaction ΔG_0, i.e. $\Delta G_0 = -RT \ln K$, we can find that the change in the free energy upon a decrease in the binding constant from 10^7 to 10^4 M^{-1} makes 17.8 kJ/mol. Thus, the total energy losses for pumping Ca^{2+} across the membrane are about twice lower than the energy of ATP hydrolysis (around 40 kJ/mol at pH 9) that is enough for the transfer of two calcium ions.

After dissociation of Ca^{2+} ions and phosphate, the enzyme recovers its initial state E1.

The working cycle of the enzyme is accompanied by essential conformational transformations, when large protein domains perform large scale mutual movements interacting with different surfaces at separate stages of the cycle.

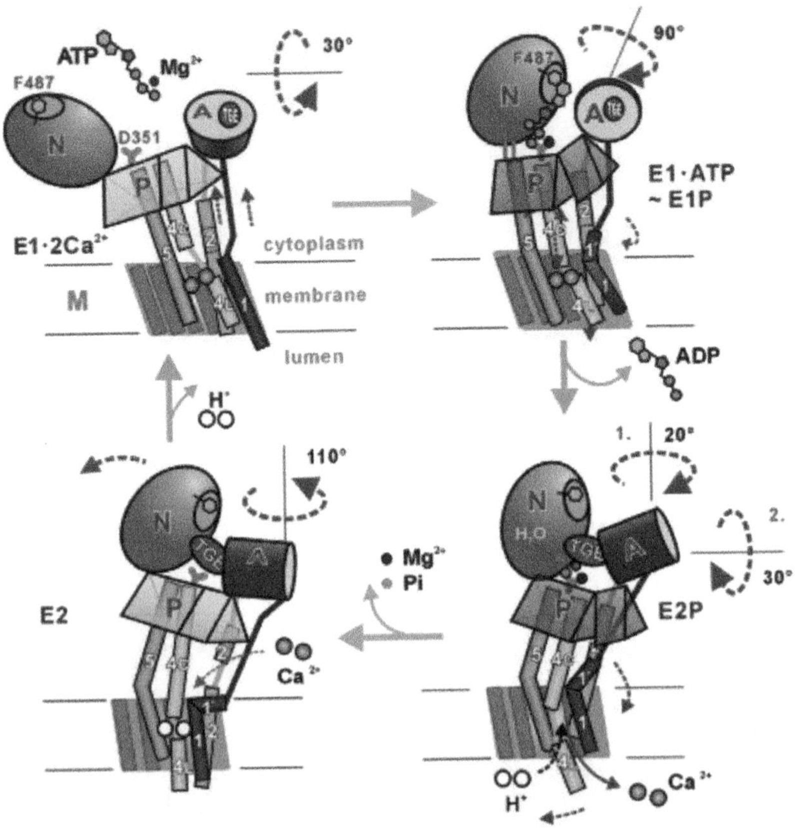

Figure 20.2. Scheme of the reaction cycle of the Ca^{2+}-pump and accompanying motions of protein domains (reproduced from C. Toyoshima, 2008). See explanations in the text.

Transition $E1 \cdot 2Ca^{2+} \rightarrow E1P$**: Phosphorylation of the Enzyme and Occlusion of** Ca^{2+} **Ions.** The central moment of the working cycle is the formation of a high-energy aspartylphosphoanhydride complex (enzyme phosphorylation): $E1-ATP \rightarrow E1{\sim}P$. This stage is followed by transition to the state with a lower energy $E2-P$ coupled to the ion transfer across the membrane. The generation of the high-energy intermediate is associated with occlusion of Ca^{2+} ions from the cytoplasmic side that does not permit Ca^{2+} to be released back into the cytoplasm.

ATP is bound close to the loop between domains P and N and thereby joins them so that adenosine binds to domain N, and γ-phosphate together with the Mg^{2+} ion binds to domain P causing its bending (Fig. 20.2). The abundance of hydrogen bonds, formed around ATP and the metal, suggests the strength of this configuration.

The binding of ATP and transfer of γ-phosphate lead to the rotation of domain N by about 90°, and the head end of the domain acquires a compact configuration. The bending of domain P causes rotation of domain A by about 30°. Due to this helix M1 linked to domain A stretches upwards and bends so that its amphiphilic N-terminal part (M1′) is packed up on the membrane surface (Fig. 20.1 and 20.2). This bending makes the top region of the transmembrane part containing conservative amino acids occupy the space close to Gly309, forming a hydrophobic cluster that like a plug closes the access to the binding sites of Ca^{2+} from the side of the cytoplasm.

As a result, the conformation of the side chain of Gly309 is fixed and the both Ca^{2+} ions are locked in their binding sites. During this, domains M1 and M2 form a V-like structure moving similar to a rigid body in the course of the remaining reaction cycle and transfers motions of domain A to other transmembrane helices, in the first place to the lumen of M4.

The transfer of γ-phosphate to Asp 351 also immobilizes domain N in a strongly tilted position that is accompanied by the emergence of mechanical conjugation between domain A and N and prepares domain A to be rotated by 90° at the next stage.

Transition E1P $\longrightarrow$ E2P **and Release of** Ca^{2+} **into the Lumen.** At this stage, transition from high-energy state E1 to state E2 takes place, the latter is characterized by lower energy and lower affinity to Ca^{2+}. Disorganization of Ca^{2+} binding sites and the opening of the lumenal gate results in dissociation of Ca^{2+} ions into the lumen. Consequently, the energy is expended not only for the displacement of Ca^{2+} ions within the membrane, but also for the change of the binding constant of these ions, proceeding in parallel with substantial conformational changes in the protein molecule.

The rotation of domain A causes a large-scale rearrangement of transmembrane helices M1–M6, including the considerable shift of M4, the bending of M5 towards M1 (Fig. 20.1) and the rotation of helix M that brings about disruption of the Ca^{2+} binding sites.

Transition E2P $\longrightarrow$ E2: **Hydrolysis of Aspartylphosphate and Closing of the Lumenal Gate.** Upon transition E2P $\longrightarrow$ E2 $\cdot$ P_i, domain A makes an additional turn relative to the rotation axis near Thre181, differing from the rotation axis of the preceding turn. The rotation terminates when domains A and P form an electrostatic bond. During this transformation the luminal gate is closed again. In state E2, the head fragment of the enzyme is compact and domains N and A are closely linked by several hydrogen bonds.

Cycle Closure: E2 $\longrightarrow$ E1. The charge unbalance in empty binding sites of Ca^{2+} should be compensated by corresponding conformational changes, inclusion of water molecules and protonation. The calculation of electrostatic interactions demonstrates that all carboxyl groups of the binding sites should be protonated in state E2.

However, as a result of thermal perturbations, the head fragment of the enzyme opens and protons are released into the cytoplasm and dissipate across the membrane of the sarcoplasmic reticulum permeable to protons. Thus, at pH 7, E2 is a transient state and the major part of the enzyme molecule is in state E1, when carboxyl residues in the binding cavity of Ca^{2+} are not protonated and therefore have a high affinity to Ca^{2+} ions. Consequently, the ATP energy is most likely expended for an ordered change of the affinity of ion-binding centers to cations, and the domain movement itself occurs already due to thermal vibrations.

20.2 Sodium-Potassium Pump

J. C. Skou (1957) was the first to achieve success in studying Na^+/K^+-ATPase. He studied ATP splitting in preparations of squid nerve membranes and demonstrated that hydrolysis of a substantial part of ATP depends on the coupled action of Na and K ions. The ATPase activity stimulated by Na^+ and K^+ ions is inhibited by ouabain. An increase in the ouabain concentration had the same inhibitory effect on Na^+ transport and ATPase activity. The results of quantitative studies showed that the hydrolysis of one ATP molecule is accompanied by import of two K ions into the cell and export of three (and sometimes two) Na ions from the cell. Thus, it was recognized that the Na^+-pump is an enzyme system that is localized in the plasmatic membrane of the cell, pumping Na^+ from the cell to the extracellular environment and K^+ backwards. The pump transports Na ions against electrochemical potential gradients due to coupling ion-exchange processes with hydrolytic splitting of ATP by Mg^{2+}-dependent Na^+/K^+-stimulated ATPase.

The Na^+/K^+-ATPase enzyme consists of two polypeptide chains with molecular masses of 84,000 and 5,700 forming its large and small subunits. The protein is strongly linked to phospholipids, their complete removal from the membrane causes disappearance of the ATPase activity. The larger subunit is homologous to the protein of Ca^{2+}-ATPase and provides analogous binding sites to Na^+, K^+ and ATP, while the β-subunit is unique for P-type ATPases, transporting K^+ ions; it is involved in occlusion of K^+ ions and forwards the α-subunit to the plasmatic membrane.

On the whole, the scheme of pump functioning reproduces that of the Ca^{2+} pump given Ca^{2+} substituted for Na^+ and protons for K^+:

$$E_1{-}\mathrm{ATP} + 3\mathrm{Na}_i \rightarrow E_1{-}\mathrm{ATP}{\cdot}3\mathrm{Na}_i \rightarrow E_1{\sim}\mathrm{P}{\cdot}3\mathrm{Na}_i \rightarrow E_2{-}\mathrm{P}{\cdot}3\mathrm{Na}_e \rightarrow E_2{-}\mathrm{P} + 3\mathrm{Na}_e,$$
$$2\mathrm{K}_e + E_2{-}\mathrm{P} \rightarrow E_2{\cdot}2\mathrm{K}_i \rightarrow E_1{-}\mathrm{ATP} + 2\mathrm{K}_i. \qquad (20.1)$$

Na^+ ions bind to the enzyme from the cytoplasmic side of the membrane and catalyze the protein phosphorylation by the previously bound ATP molecule. Phosphorylation moves the protein from state E_1 with a high affinity to Na^+ and ATP ($K_M = 0.2\ \mu$mol/l) into state $E_1 \sim$ P. Therewith the access to the ion-binding cavity from the side of the cytoplasm is terminated, i.e. occlusion of Na^+ ions takes place. After that, transition into the $E_2{-}P$ state occurs with a lower energy accompanied by the opening of the extracellular gate into the binding cavity of Na^+ ions and by a decrease in the enzyme affinity to Na^+. Na^+ dissociates from the extracellular side of the plasmatic membrane. The enzyme in the form of $E_2{-}P$ binds K^+ ions from the extracellular side triggering dephosphorylation of the protein that, in its turn, results in the closing of the extracellular gate in the ion binding cavity and generation of the $E_2{-}P$ form. The release of K^+ ions into the cytoplasm is catalyzed by the binding of ATP to the site having a relatively low affinity to ATP ($K_M \approx 150\ \mu$mol). The conformational changes occurring due to the ATP binding switch the pump from the state with conformation E_2 to the state with conformation E_1 having a low binding constant for K^+. It results in the release of these ions. Consequently, the rise of the enzyme affinity to ATP provides energy for the release of K^+ from the cytoplasmic side of the cell, and phosphorylation — for the release of Na^+ from the extracellular side.

The ejection of three moles of Na^+ requires the energy of about 19 kJ under normal physiological conditions, and the energy released upon hydrolysis of the high-energy bond in the Na_1E_1 ATP complex makes 23 kJ/mol. The transport of two moles of K^+ into the cell requires $\sim$ 16 kJ, and the energy released upon disruption of the $K_e{-}E_2{-}P \rightarrow K_e + E_2 + P$ bond makes almost 18 kJ/mol. The released energy is used for the transfer of Na^+ and K^+ from ionophore centers into the corresponding near-membrane space.

20.3 Electrogenic Ion Transport

Active ion transport can be electrically neutral or electrogenic. A transport system is neutral if its functioning is accompanied by the exchange of intracellular ions for extracellular ones in the "charge-for-charge" manner. In this situation, the system of active transport is only a means for maintaining concentration gradients and is not involved directly in generation of potential difference across the cell membrane. Membrane potentials are generated then only due to ion diffusion by concentration gradients.

In the case when the number of charges transferred in one direction per unit of time is not compensated by the sum of the charges transported in the opposite direction, the transport mechanism directly generates additional difference of potentials across the membrane. For example, when in the electrogenic mode, the Na^+-pump operates exchanging three Na ions for two K ions. □

Systems of electrogenic active transport were first revealed and studied in detail on frog skin and later also on other epithelial tissues. If two chambers containing Ringer's solution are divided by frog skin, the difference of potentials reaching 100 mV is generated between the external and internal surfaces of the skin (the internal surface is positive with respect to the external surface). By supplying voltage in the opposite direction (Fig. 20.3), we can compensate the potential difference on the skin to zero (the so-called short-circuit of the skin). Under such conditions, the passive transport of ions should terminate (due to the absence of either concentration gradients or potential difference), and unidirectional flows of passively transported ions should equalize. However, in reality the flow of labeled Na^+ directed from the external skin surface to the internal one, is much higher than the Na^+ flow in the opposite direction. The total Na^+ flow across the frog skin is proportional to the electric current magnitude. Therefore, the magnitude of the short-circuit current is proportional to the rate of Na^+ active transport directed from the external skin surface to the internal one.

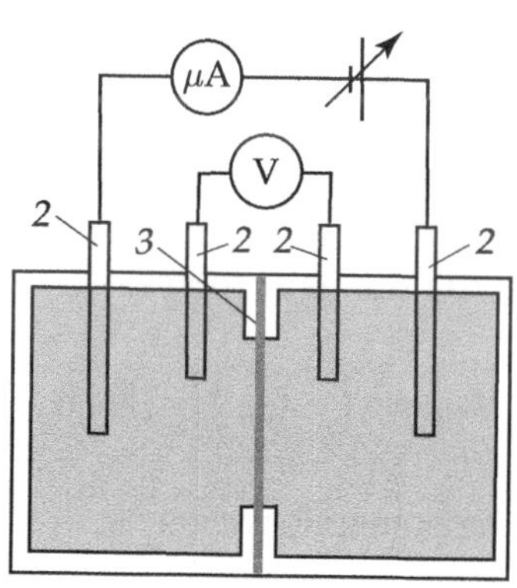

Figure 20.3. Principal scheme of measuring the short-circuit current on the frog skin.
1, Measuring electrodes; *2*, tone electrodes; *3*, frog skin.

The mechanism of active transport of Na^+ on the frog skin from the side of the mucous membrane (the external surface of the skin) towards the serous membrane (the internal surface of the skin) is close to the mechanism of ion transfer across the

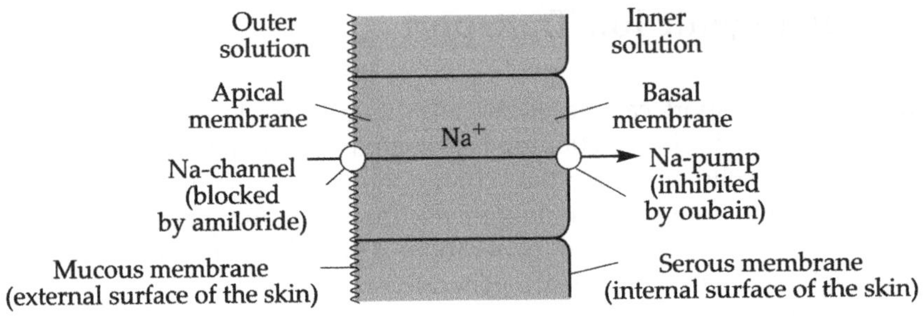

Figure 20.4. Sodium active transport in epithelial tissues.

epithelium in the intestine and renal ion channels. Figure 20.4 shows the general scheme of Na^+ transfer across the epithelial layer.

The system of active electrogenic transport of Cl^- has been disclosed in many epithelial tissues. The contribution of these systems to the short-circuit current is typically much lower than the current of active transport of Na^+.

The expression for a membrane potential can be obtained using the kinetics of chemical reactions. For example, if the cell membrane has a system of active transport of Na^+ outwards, and K, Na and Cl ions are transferred passively, the stationary flows of K^+, Na^+ and Cl^- directed into the cell and to the external surrounding can be written as given below with account of the effect of membrane potential φ:

Inward flow	*Outward flow*
$[\mathrm{K}_o]k_\mathrm{K}\exp(-\varphi F/RT)$	$[\mathrm{K}_i]k_\mathrm{K}$
$[\mathrm{Na}_o]k_\mathrm{Na}\exp(-\varphi F/RT)$	$[\mathrm{Na}_i]k_\mathrm{Na} + [\mathrm{Na}_i]k_a$
$[\mathrm{Cl}_o]k_\mathrm{Cl}\exp(\varphi F/RT)$	$[\mathrm{Cl}_i]k_\mathrm{Cl}$

Here k_K, k_Na and k_Cl are rate constants characterizing passive penetration of K, Na and Cl ions across the membrane; k_a is the rate constant of active transport; indices o and i refer to ion concentrations in the external surrounding and the cytoplasm of the cells.

In a steady state, the sum of all inward flows is equal to the sum of the outward flows (Cl^- flows in opposite directions are usually equal). The condition for the equality of inward and outward flows is as follows:

$$[\mathrm{K}_o]k_\mathrm{K}\exp\left(-\frac{\varphi F}{RT}\right) + [\mathrm{Na}_o]k_\mathrm{Na}\exp\left(-\frac{\varphi F}{RT}\right) = [\mathrm{K}_i]k_\mathrm{K} + [\mathrm{Na}_i]k_\mathrm{Na} + [\mathrm{Na}_i]k_a. \quad (20.2)$$

Solving Eq.(20.2) relative to the potential, we get

$$\varphi = -\frac{RT}{F}\ln\frac{k_a\,[\mathrm{Na}_i] + k_\mathrm{K}[\mathrm{K}_i] + k_\mathrm{Na}[\mathrm{Na}_i]}{k_\mathrm{K}[\mathrm{K}_o] + k_\mathrm{Na}[\mathrm{Na}_o]}. \quad (20.3)$$

The first member in the numerator of the sublogarithmic expression reflects the activity of the electrogenic ion pump. If the active transport is nonelectrogenic, this member is absent and expression (20.3) becomes identical to the Goldman equation (see (18.1)) for the membrane potential conditioned by passive ion flows.

20.4 Proton Transport

Proton transport plays an important role in the functioning of bioenergetics systems.

Transfer of H^+ across membranes can be performed by three types of mechanisms. Some membranes have mobile proton carriers (plastoquinone in the photosynthetic membrane of chloroplasts). Membrane protein can also perform conformational transitions upon proton binding on one side of the membrane and protein deprotonation on the other side of the membrane, coupled to the turning of the macromolecule in the membrane, when the attached proton crosses the membrane. Finally, protons can be transported across the membrane via special structures — H^+ channels. A proton channel represents a narrow cavity formed by protein polar groups. □

A proton channel is an essential component of all H^+-ATPases. It is formed by the hydrophobic part of the subunit of the coupling factor — the CF_0 complex. Through the H^+ channel, protons get to the catalytic region of the transport system where the transfer of H^+ is coupled to the reactions of synthesis — ATP hydrolysis. Consequently, the key parts of HH^+-pumps are the proton channel and the active center.

Proton transport via H^+-channels is frequently compared to the transfer of H^+ along a regular lattice formed by the system of hydrogen bonds, similar to that in ice. In the ice structure, protons can move from one water molecule to the neighboring molecule (see Section 5 in Chapter 9). As a result, an ion pair is formed that consists of a positively charged ion of hydroxonium H_3^+O and hydroxyl OH^-. When there is no electric field, the proton exchange occurs at random, however upon generation of the potential difference, directed migration of defects takes place. As a result, protons are rapidly displaced along the ordered chain of hydrogen bonds.

Proton transfer relays may take place only upon alternation of proton jumps and turns of chemical bond in the defect region (Fig. 20.5). The figure shows that translocation of one proton along the chain occurs in two stages. At first, the proton jumps from one group to another (Fig. 20.5, *I*) and then the group turns (Fig. 20.5, *II* this is a turn around the C–O bond). Thus, the charge is additionally displaced and the system of bonds returns to the initial state. As a result of the turning of a chemical

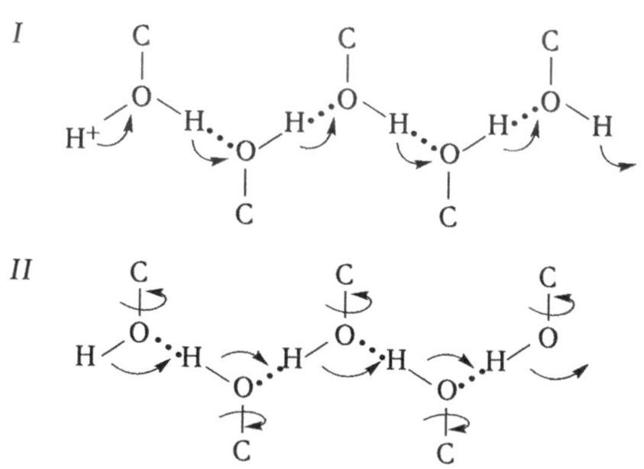

Figure 20.5. Proton transfer along a chain of H-bonds.

I, Stage of a proton jump; *II*, stage of rotation of groups (migration of the rotation defect).

group or a water molecule, a configuration is formed that does not allow the proton to return to its initial state. In the α-helical rigid structure, such turns of groups are rather hampered; that is why the α-helical system of H-bonds cannot transfer protons.

The magnitude of the activation barrier depends greatly on the distance between donor – acceptor pairs. In most cases, the distance between the atomic nuclei in the system of hydrogen bonds is 0.23–0.27 nm. An increase in the distance by 0.1 nm brings about a proportional growth of the barrier by 125–250 kJ/mol. It means that a change by only 0.1–0.2 nm in the distance between oriented pairs in proteins may inhibit the proton conductivity of proteins. Factors affecting the proton conductivity of proteins can be external (transmembrane) electric fields and the presence of charged groups. The time of a proton transfer in the elementary act is comparable to the time of vibrations and fluctuations of bond lengths due to atomic vibrations, i.e. is in the range $(10–50) \cdot 10^{-15}$ s. The molecular dynamics simulation (Section 3, Chapter 11) makes it possible to calculate the trajectories of proton movement over the barrier upon transition between the two groups. For example, the trajectory of the proton transition from Glu35 in lysozyme to the oxygen atom in the substrate molecule was determined (A. Warshell, 1991). As a result of random fluctuations of charges and dipoles of interacting groups for $\sim$ 16 ps, a transient configuration with a decreased barrier is formed that is favorable for the proton transfer. Inasmuch as the following jump of the proton takes several femtoseconds (10^{-15} s) the limiting stage of the proton transfer is namely the formation of an active configuration of atoms at random fluctuations ($\sim$ ps).

VII
Energy Transformation in Biomembranes

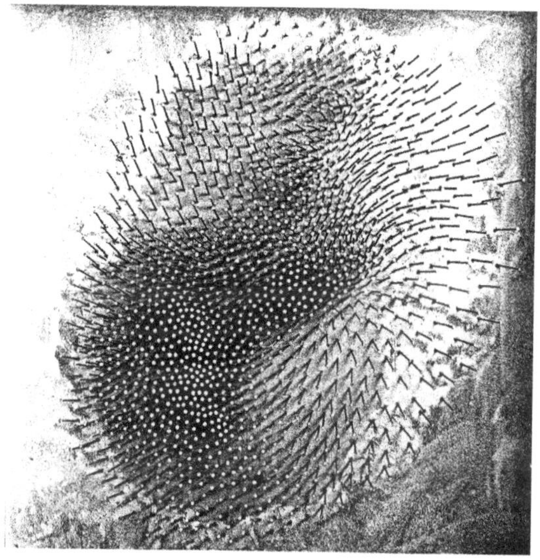

21

Electron Transport
and Energy Transformation
in Biomembranes

22

Physics of Muscle Contraction,
Actin-Myosin Molecular Motor

23

Biophysics of Intracellular
Signaling Processes

21

Electron Transport and Energy Transformation in Biomembranes

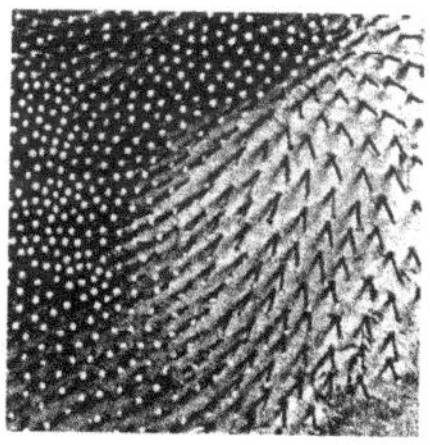

21.1 General Description of Energy Transformation in Biomembranes

The processes of energy transformation upon electron transport in membranes of some bacteria, mitochondria, chloroplasts and chromatophores are similar fundamentally. In all such systems, the energy of the electron flux provides the energy for the synthesis of APT molecules.

A variety of hypotheses have been proposed on the mechanism of membrane phosphorylation that are distinct from each other by different redox reactions and coupling components taking part in energy transformation and the ATP synthesis. Earlier hypotheses of chemical coupling proposed generation of a high-energy compound carrying electron A and intermediate X involved in energy transfer at the stage of electron transport from A to B:

$$AH_2 + B + X \rightleftarrows A \sim X + BH_2. \tag{21.1}$$

At the next stage, the energy is transferred from $A \sim X$ to $X \sim Y$, where Y as X is an intermediate only in the processes of energy transfer and does not contain electron transport chain (ETC) components:

$$A \sim X + Y \rightleftarrows X \sim Y + A. \tag{21.2}$$

Finally, at the third stage, phosphorolysis of the non-phosphorylated high-energy molecule $X \sim Y$ and transphosphorylation to a chemically stable ATP molecule take place:

$$X \sim Y + P_H + ADP \rightleftarrows ATP + X + Y. \tag{21.3}$$

As seen from reactions (21.1)–(21.3), the energy transfer from pair (AH_2, B) to ATP proceeds by the shift of the chemical equilibrium to the right resulting in the formation of excess ATP.

Is this principle per se possible in biological membranes? As calculations show, the high-energy character of the phosphate bond in ATP is explained by the energy of repulsion of atoms due to the excess negative electron density on oxygen atoms in phosphor residues. Therefore, the high-energy bond $\sim$ P itself can be formed because of the directed induced movement of electron density in intermediate compounds, when the electron energy excess does not dissipate to heat during hydrolysis. Consequently, to provide effective energetic coupling by the shift of chemical equilibrium, the electron energy excess concentrated on the chemical bonds of intermediate unstable compounds should not dissipate to heat in the reaction, but be rather preserved, and transformed without energy losses to the energy of the final stable high-energy compound (ATP). □

As known (see Chapters 10 and 11), dissipation of the excess electron energy in processes of electron-vibrational relaxation in a condensed phase (membrane proteins) occurs for 10^{-11}–10^{-12} s. It means that free of dissipation energy transfer along the chemical pathway is possible if the complete process of energy transformation occurs in one elementary act of 10^{-11}–10^{-12} s. Only in such a case the electron energy excess concentrated on certain bonds of atomic groups of unstable intermediate compounds will have no time to dissipate to heat during electron-vibrational relaxation. It is clear how difficult it is to perform all transformations (21.1)–(21.3) for such a short period. It is especially obvious that the process should involve several low-molecular compounds which are suggested to be spatially separated in the membrane and then diffuse to the reaction site. Hence an important conclusion can be made that the highly effective energy coupling, based on the shift of chemical equilibrium in the chain of reactions involving unstable compounds, cannot be realized in the condensed phase. To this end, a unique complex of molecules is required where energy transformation occurs in one act avoiding thermal losses. Namely, such a complex is the enzyme ATP-synthetase. Unsuccessful searches for chemical compounds among components of the electron transport chain and energy carriers have led to the necessity to scrutinize other principles of coupling. The most recognized is Mitchell's chemiosmotic theory, according to which the electron transfer along the electron transport chain results in proton transfer across the hydrophobic barrier of the membrane. Mitchell's chemiosmotic coupling principle, formulated in a general form, suggests that electron transfer is associated with the ATP synthesis via the formation of transmembrane difference of electrochemical potentials of hydrogen ions $\Delta\bar{\mu}H^+$:

$$\Delta\bar{\mu}H^+ = F\Delta\varphi + 2.3\,RT\Delta pH, \tag{21.4}$$

where $\Delta\varphi$ is the difference of electric potentials, ΔpH is the difference of concentrations of hydrogen ions on both sides of the membrane, F is the Faraday constant, R is the gas constant, and T is the temperature. It is accepted that the energy $\Delta\bar{\mu}H^+$ is consumed for the ATP synthesis from ADP and P using a special enzyme —

membrane ATP-synthetase (H^+-ATPase). However, the molecular mechanisms realizing the transmembrane proton transfer upon electron transfer as well as the mechanism of ATP formation by means of ATPase remain disputable.

21.2 Mechanisms of Proton Translocation and Generation of $\Delta\bar{\mu}H^+$ in Respiratory and Photosynthetic Chains of Electron Transport

The formation of ΔpH can proceed by different mechanisms (see Section 4 in Chapter 20). First, it is proposed that there exist asymmetrically located special membrane proteins transporting the proton. Their affinity to the proton and the proton-binding properties are changed upon electron transfer that leads to corresponding changes in the concentration of protons on both sides of the membrane. Second, a redox-dependent proton channel may be formed in the protein portion of the electron carrier. In this case, the electron carrier should perform concurrently the function of the proton pump. It is proposed that proton translocation upon redox transformations of carriers occurs either due to conformational changes in the carriers per se (or a special protonophore protein), or due to the redox-dependent protonation — deprotonation.

Third, proton transfer can be specified by the functioning of mobile carriers that, being recovered on one side of the membrane, attach the proton simultaneously with the electron and then diffuse to the other side of the membrane, where the carrier is oxidized and the proton is released into the near-membrane region due to dissociation. Therefore, P. Mitchell proposed that in each proton-transferring segment of the ETC, hydrogen carriers ($2H^+ + 2e^-$) interchange with electron carriers ($2e^-$) (Fig. 21.1) forming a redox-loop. As seen from the diagram of a redox-loop, hydrogen carrier *D* transfers hydrogen to carrier *A* on the opposite side of the membrane. Upon oxidation of *D* and recovery of *A*, two protons are released into the surrounding. Electrons are transferred from *A* transmembraneously to the other electron carrier *B*. One can be assume that the hydrogen carrier, as an uncharged one like quinone, diffuses by the concentration gradient to *A*. At the same time, electron transfer from *A* to *B* occurs opposite to the electric field generated upon the transmembrane transfer

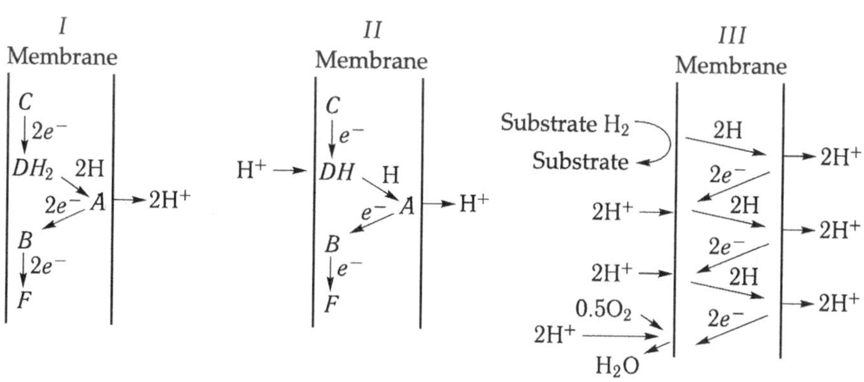

Figure 21.1. Transfer of charges across the membrane.

I and *II*, separate redox-loops; *III*, mitochondrial respiratory chain.

of protons. At such organization of the electron transport, the oxidation of electron donors DH with the final acceptor B on the same side of the membrane is coupled to the transmembrane transfer of protons and, consequently, generates $\Delta\bar{\mu}H^+$ on the membrane. The transfer of the two electrons in the redox-loop is accompanied by translocation of two protons.

21.3 ATPase Complex

At present, proton ATPases have been isolated practically in all types of coupled membranes: mitochondria, chloroplasts, chromatophores. The ATP-synthase complex, or H^+-ATPase, is a reversible enzyme having both ATP-synthase and ATPase functions. The ATP synthesis is realized due to $\Delta\bar{\mu}H^+$, and ATP hydrolysis results in that proton ATPase generates interfacially the transmembrane difference of electrochemical potentials H^+. Thus, in H^+-ATPase processes occur according to the general scheme:

$$\Delta\bar{\mu}H^+ \underset{\text{ATPase}}{\overset{\text{ATP-synthase}}{\rightleftarrows}} \text{ATP}. \tag{21.5}$$

The ATP-synthase complex (F_0F_1-ATPase) consists of soluble ATPase (factor F_1) and membrane components (complex F_0). The general topography of the ATPase complex is given in Fig. 21.3.

The coupling factor of ATPase (factors F_1 for mitochondria or CF_1 for chloroplasts) is a polyfunctional protein with a complicated quaternary structure. It consists of three types of large subunits (α, β, γ) with molecular mass of 30–60 kDa and two types of minor subunits (δ, ε) with molecular mass 11–20 kDa. The stoichiometry of the complex is as follows: $\alpha_3\beta_3\gamma\delta\varepsilon$. Decomposition of the complex into separate subunits leads to the loss of the enzyme activity. The "cap" with the height of 80 E and the width of 100 E of a mushroom-like outgrowth of H^+-ATPase corresponds to factor F_1 partially immersed in the membrane, while its base corresponds to hydrophobic proteins of complex F_0, including three types of polypeptides (a, b, c) with molecular masses from 6.5 kDa to 30 kDa, and provides the binding of factor F_1 to the membrane and proton transfer upon the enzyme functioning. In addition to α-β-subunits, each complex contains one polypeptide a, two proteins b and 9–12 copies of protein c of the water-soluble complex. The general stoichiometry of the whole complex is as follows: $\alpha_3\beta_3\gamma ab_2c_n$, where $n = 9 \div 12$. Subunits α and β are homologous and are packed in protein globules forming a single complex, where α- and β-subunits are arranged in turn around the γ-subunit looking as a slightly curved rod of 90 E long. Kinetic and structural data evidence for the presence of three interacting hydrolytic sites, each β-subunit containing one such site located in the slot between middle α- and β-subunits, though the major portion of amino acid residues in the active center belongs to the β-subunit. The γ-subunit looks like protruding from globule F_1 playing the role of a connecting link between membrane and water-soluble F_0 fragments of ATPase (Fig. 21.2).

F_0 forms a channel arranged in a particular way (see below) that allows proton transfer across the whole membrane from the aqueous phase to the hydrophobic fragment of the membrane and after that to water to the other side of the lipoprotein barrier. The most probable mechanism of protein transfer seems to be the token passing over proton-donor and proton-acceptor groups of amino acids including Arg, Tyr and Glu residues.

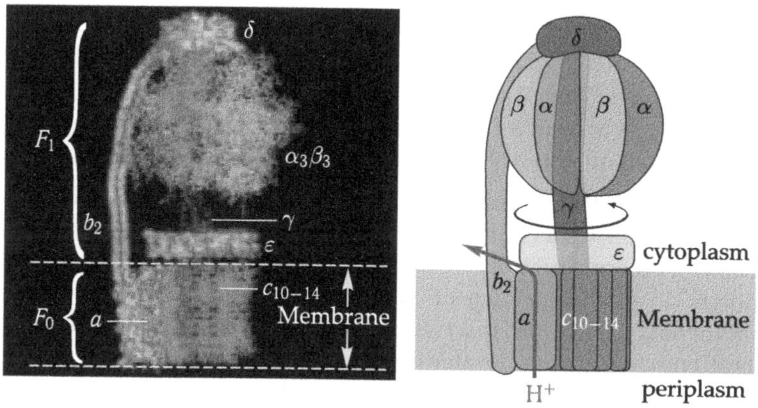

Figure 21.2. Scheme of subunits. ATP hydrolysis (reproduced from G. Oster, H. Wang, 2002).

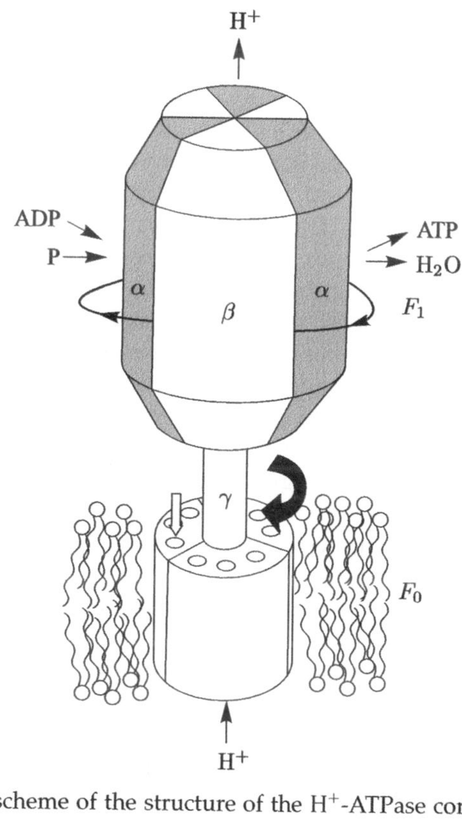

Figure 21.3. Simplified scheme of the structure of the H^+-ATPase complex (reproduced from J. K. Lanyi, 1995).

Suggesting that individual transfer of H^+ (either along hydrogen bonds or by tunneling between the centers) proceeds at distances not exceeding 0.2–0.3 nm, we can calculate that the number of groups should be no less than 15–20 to cross the entire membrane (4–5 nm).

The basic catalytic properties of factor F_1 are provided by α- and β-subunits containing binding sites of nucleotides of different affinity. The enzyme α-subunit has non-catalytic centers of high affinity to ADP and high specificity of its binding. The complex localizes active centers of ATPase reaction, the substrate for it is the Mg^{2+}-ATP complex (free Mg^{2+} and ATP are competitive enzyme inhibitors). The ATPase functioning is accompanied by cooperative conformational rearrangements involving the enzyme quaternary structure that requires energy supply.

The problem of ATP synthesis in the H^+-ATPase-synthase should be considered based on the general concepts on the role of electron-conformational interactions in mechanisms of enzyme catalysis (Chapter 26).

As known, an elementary act of catalysis is performed in the enzyme active center spontaneously when the reactive configuration is achieved between reacting substrate and enzyme groups located in the active enzyme center at distances of about chemical bond lengths. At stages when the substrate and the enzyme interact to produce an active configuration and then release the formed product, conformational changes occur in the enzyme. Such external factors as temperature, ionic strength of the solution, viscosity can affect these relaxation stages. However, the direct act of catalysis in the formed active configuration does not require further heat energy activation. □

It is with these concepts that P. Boer's results are compatible. An elementary act of ATP generation may proceed in the active center of H^+-ATPase when a membrane is de-energized and $\Delta\bar{\mu}H^+ = 0$ and even in the active center of isolated factor F_1 in solution. However, the total ATP synthesis does not practically take place upon de-energizing. Studies of isotope exchange reactions ^{18}O (ATP $\rightleftarrows$ $H_2{}^{18}O$)) demonstrated that the stage determining the extremely low rate of total ATP synthesis in the absence of $\Delta\bar{\mu}H^+$ is the release of the synthesized ATP from the active center (P. D. Boyer, 1974). Namely, this process is accelerated (by 1000-fold) upon membrane energizing. Consequently, the energy of $\Delta\bar{\mu}H^+$ is consumed mainly for expelling the tightly bound ATP from the enzyme catalytic center. The energy is also consumed in binding phosphate and ADP in the enzyme the catalytic centers that function in a definite order. It means that the energy of $\Delta\bar{\mu}H^+$ is consumed not in the elementary act of ADP $\sim$ P covalent bond formation in the enzyme active center, but upon the substrate binding and release of the reaction products (ATP or ADP and P) from the active center.

The complex contains active centers, which can be either open or closed. The binding of ADP and P in one center correlates with conformational changes in another center that facilitates the ATP release. Similar to this, during ATP hydrolysis the ATP binding and hydrolysis correlate with fast dissociation of ADP and P from the alternative center. The role of $\Delta\bar{\mu}H^+$ components is to protonate definite molecular groups due to ΔpH or to change positions of polar groups under the action of $\Delta\varphi$ electric field. Therewith (see Section 3 in Chapter 26) separate protein groups change their motions in the conformational potential. Just this can directly affect the formation of the enzyme-substrate complex upon the substrate binding and the release of the reaction product, i.e. affect the processes occurring in the course of conformational relaxation.

It was shown that subunit γ can turn around the axis coinciding in its direction with subunit γ. There is direct experimental evidence for such rotation of subunit γ inside the coupling factor F_1 (H. Noji et al., 1997) at the expense of ATP hydrolysis energy. Using a fluorescent microscope, the Japanese researchers observed the rotation of the γ-subunit upon ATP hydrolysis in F_1 molecule. Three β-subunits of the F_1

complex were immobilized on a stable substrate, and a fluorescent fragment of an actin filament of ~ 1 micron length was chemically "cross-linked" to the γ-subunit terminus. Upon functioning of the enzyme and ATP hydrolysis, rotation of the actin filament took place that terminated in the absence of ATP.

On the whole, the cooperative rearrangement of the H^+-ATPase structure and rotation of γ-subunits are related to the directed character of electron conformational interactions in the F_1 complex. As conformational relaxation processes proceed the quaternary structure of the F_1 complex changes providing both the substrate binding and the removal of the reaction products formed in the active center.

Because of the γ-subunit structure asymmetry, the rotation of γ-subunit in the F_1 complex makes its β-subunits contact different sites of the γ-subunit. This changes the states and brings about conformational rearrangements in active centers. As γ-subunit turns by 120°, each of the α- and β-subunits occupies the site of another such α- and β-subunit with respect to the γ-subunit site they contact (see Fig. 21.3).

Thus, each of the three active centers proceeds via the three conformational states differing by the degree of their affinity to ATP, ADP and P molecules in their catalytic centers.

Figure 21.4 shows a scheme of changes in the states of a center during ATP synthesis. In state *1*, the β-subunit center is open and ATP and P molecules are relatively weakly bound in it. As rotation of γ-subunits proceeds accompanied by the conformational rearrangements of the F_1 complex, the center transfers to state *2* where the ATP synthesis directly occurs.

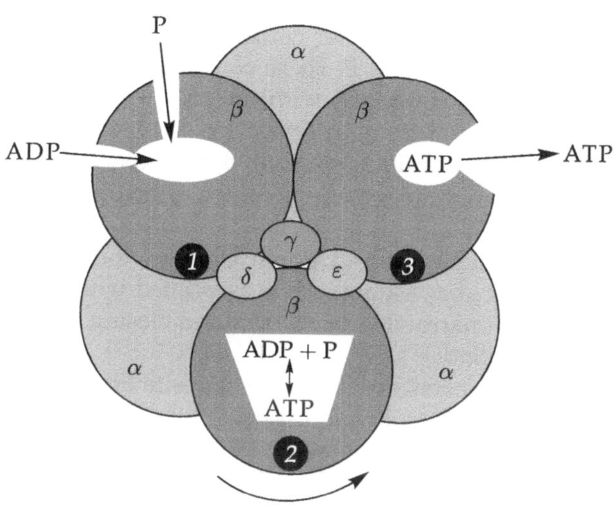

Figure 21.4. Scheme illustrating energy-dependent changes in the state of catalytic centers (reproduced from P. D. Boyer, 1974).

Here ADP and P are strongly fixed in the catalytic center in an active configuration necessary for the formation of a covalent bond between phosphate groups of ADP and P. As mentioned above, this stage does not require additional energy (P. D. Boyer, 1974). In this state, spontaneous acts of formation and splitting of the ADP–P covalent bond take place because the equilibrium constant of the reaction

$$ATP \leftrightarrows ADP + P_I$$

is close to unity.

At the next stage, the center passes to state *3* where due to the structural rearrangement the strong bond between the formed ATP molecule and the center is now weakened and the ATP molecule is released. The vacant site is occupied by new ADP and P molecules from the solution. The described rearrangements are cooperative and involve all three catalytic centers of β-subunits in the F_1 center. As the whole cycle includes three stages and H^+-ATPase contains three subunits, a new ATP molecule is generated after each structural transition.

The rotation has a step-wise character and proceeds in parallel with discrete turns by a definite angle. This conclusion has been verified by experimental data on the rapid recording of the rotation of γ-subunits with an attached actin filament upon ATP hydrolysis. Upon ATP hydrolysis the rotation of γ-subunits will proceed in the opposite direction. It was found that a turn by 120° is divided by two angle steps of 90° and 30°. A turn by 90° occurs with a delay of about 0.1 ms, especially at low ATP concentrations. It is caused by the binding of the ATP molecule in the open center rather than by its hydrolysis. The delay before the following turn by 30° is independent of the ATP concentration and consists of two stages of $\sim$ 1 ms each.

The system returns to the initial state with an open center where the ATP binding may take place again. Apparently, the turn by 30° is a consequence of the release of hydrolysis products (ADP or P) from the center.

Complex F_0 can be considered as a rotating ensemble of subunits having 9–12 pauses due to the proton flux with time binding of protons on each subunit. Each turn of the γ-subunit inside the complex (or the ensemble of $\alpha_3\beta_3$-subunits around subunit γ) has the result in every center corresponding to whether its hydrolytic site is occupied by ATP or ADP and P. These conformational changes (mechanical changes in the structure) are connected with the transfer of three or four protons across the membrane at each third stage of the cycle.

Recently, molecular dynamics simulation and direct experiments have revealed in many details the structural rearrangements accompanying the rotation of subunits in the APTase complex.

Let us briefly consider these results. Upon hydrolysis, the ATP binding proceeds in an open catalytic center of a β-subunit. It is associated with the formation of 15–20 H-bonds that results in narrowing of the slot and closing of the β-center where ATP is already strongly bound.

Upon ATP synthesis, the events occur in the opposite direction. In ATP-synthase for ATP release from the β-center it is necessary to weaken the H-bonds. This takes place due to the γ-subunit rotation, when the P-loop moves from the α/β-surface with an increase in the lengths of the H-bonds and their weakening. The tension, weakening and loss of H-bonds between ATP and P-loop are accompanied then by the breaking of the remaining H-bonds of ATP with β-Arg-189 (helix C) and α-Arg-373 (α-subunit).

Upon ATP binding, the energy is stored as elastic deformation of the protein. After hydrolysis and ATP disintegration, the energy of elastic deformation is consumed for the release of ADP and $P^\bullet$ products.

The analysis of the molecular dynamics of hydrolysis in F_1 showed that ATP decay includes its interaction not only with water but also with amino acid residues of the protein (K. Schulten, 2003).

On the whole, the process of ATP hydrolysis in protein F_1 has an endothermal character in contrast to that in solution.

It should be underlined once again that the synthesis of the phosphate bond itself in ATP does not require activation energy and occurs spontaneously in the active center, while the energy is consumed for the release of the protein products.

Upon hydrolysis, the ATP binding causes bending of the upper part of the β-subunit by an angle of about 30° that is accompanied by the movement of helices B and C near the active center.

The bended β-subunit "pushes" the part of the γ-subunit, protruding stray of the central axis, and makes it turn like the crankshaft rotation in a motor. This rotation is the key moment in the ATP hydrolysis and synthesis. The rotation of asymmetric γ-rod consecutively moves the catalytic β-centers to different microsurrounding, changing their conformational states and affinity to ATP and ADP or P•. After the release of ADP and P, the β-subunit straightens up and the center becomes open and ready to accept a new ATP molecule.

Thus, cyclic conformational "bending/straightening" of β-subunits push and rotate the γ-shaft coupled to the ATP fixation, hydrolysis and release of ADP and P•.

Upon ATP synthesis, the γ-shaft rotates in the opposite direction due to the energy of proton gradient (see Section 3).

It is believed that upon rotation of γ-subunits and substrate binding (ADP and P), each β-center passes through three states: substrate-free open, then "loose" with a weakly bound substrate, and finally compact with a strongly bound ADP molecule, where the product (an ATP molecule) is formed. In these states, the structure of F_1 and its changes were revealed in a series of studies by J. Walker (1994–2000) employing the methods of X-ray structure analysis. The obtained results were used in molecular dynamics simulation of the F_1 complex upon γ-subunit rotation. The motion was simulated starting from some "starting" structure F_1 to the pre-assigned structure corresponding to the location of atoms in β-centers as established by the X-ray method (M. Karplus, 2002). In this simplified model the trajectory length did not exceed 5×10^{-10} s and the shifts of atom positions from the initial locations made $\sim$ 10 E.

It turned out that electrostatic interactions play a great role in the coupling of mutual motion of γ- and three β-subunits. On the γ-subunit there are a great number of positively charged amino acid residues (^{16})Arg and (^{22})Lys localized on the surface of its "protruding" and helical parts.

These positive side chains on the γ-subunit interact with negative residues Glu and Asp on the three β-subunits (by 2–3 on each of them) surrounding the γ-subunit. The rotation of this subunit is facilitated because of electrostatic consecutive interactions of negative (on the β-subunits) and positive (on the γ-subunit) charges, all together forming a kind of an "ion path". The "sliding" of subunits along this "ion path" as a low-energy trajectory is facilitated due to the attraction of their surfaces, "lowering" the mechanical resistance.

21.4 ATPase Complex as a Molecular Motor

As has been mentioned above, the functioning of the ATPase complex is accompanied by mutual transformation of chemical energy of ATP hydrolysis/synthesis and transmembrane gradient of protons into the mechanical work of rotation of protein subunits.

Such transformation of chemical and mechanical energy occurs also in other proteins performing motor functions. Examples are kinesin directed motions along the microtubule, DNA transfer by the bacteriophage portal protein, motions of flagella, directed motion of DNA polymerase along the DNA strand.

In all these cases, separate displacements of motor parts, due to their low energy and space scales, experience significant effect of Brownian motions of the surrounding — water molecules, other atomic groups. Moreover, the Brownian motion itself and thermal fluctuations are required for the motor work. This is what distinguishes molecular motors from macroscopic motors. The latter, generally speaking, are affected by temperature but not by thermal fluctuations.

The problem is as follows: in what way can stochastic thermal fluctuations similar by the energy magnitude to motion energy of the molecular motor part lead to directed mechanical motions along the selected degrees of freedom? Among the degrees of freedom of the molecular motor, there are purely stochastic ones reflecting the effect of thermal fluctuations and friction forces (see Chapter 11) and affecting the motions by the distinguished mechanical degrees. The force potential of this motion depends on the interaction of motor parts with each other and with the surrounding solvent. Mechanical motion is coupled to the "chemical" degree of freedom, or the coordinate of the chemical reaction, involving molecules — the source of chemical energy. Such systems have been considered in the theory of active media (Chapter 4). Here we will make some general conclusions from the currently developed theory of molecular motors (G. Oster, 2000).

In the simplest case, a motor possesses only one chemical and one mechanical degree of freedom. The motion along the chemical coordinate proceeds with a change in the reagent concentration, while the motion along the mechanical coordinate characterizes the conformational shifts under the action of applied forces. The region of "intersection" of the two trajectories (more exactly, two cross-sections of the potential energy surface of the motor) is the site of conjugation of the chemical reaction and the mechanical motion. The molecular motor motion is described by the Langevin or Fokker–Planck stochastic diffusion equations including random (Brownian) force, friction forces and the action of conservative forces of the potential field (Chapter 11). On the whole, the motion has a stochastic diffusion character; and in contrast to the macromolecular motor, the friction forces exceed the inertia forces, and conjugation of chemical and mechanical processes is achieved in definite areas of the motor microstates.

The direction of structural rearrangements in the motor is a consequence of its definite molecular configuration and its changes due to the protein electronic state alteration.

A simple example of such mechanochemical coupling in protein is considered in Chapters 10 and 26, where chemical binding of a substrate molecule in the enzyme active center caused directed conformational rearrangements in a protein. Non-directional mechanical (conformational) displacements of a protein occur spontaneously due to the stochastic Brownian forces as a consequence of a shallow energy minimum of the equilibrium conformational potential of the protein (Chapter 9). However, only coupling to the chemical reaction that changes the electronic conformational equilibrium imparts a directed character to these displacements. This principle is the basis for a molecular motor model — the so-called the Brownian ratchet mechanism. As known, in a ratchet mechanism the gear wheel rotates only in one direction under the action of an external force. Its rotation in the reverse direction is prevented by a "locking pawl" which slides under one of the teeth. In the Brownian ratchet, the energy source for the motor movement is thermal fluctuations which can, in principle, cause a displacement of the protein subunit in two opposite directions. Due to the action of the thermal fluctuations, the motor can switch to one of the two possible states, one of which is "necessary" and the other "parasitic". The chemical reaction, the products of which are disabling the "parasitic" state at every step, plays the role of a ratchet ("locking pawl") (Fig. 21.5). This happens upon dissociation of proton-containing groups on the rod in the left-hand cell at high pH. As a result the rod acquires a negative charge and thus cannot move now to the right-hand cell across the dividing hydrophobic membrane. In the right-hand cell the rod is neutral at low pH and can move to the left-hand cell due to thermal fluctuations. Given the frequency of thermal fluctuations lower than that of chemical acts, at each step the blocking has time to happen that results in the protein transition only to the "necessary" state (G. Oster, 2000).

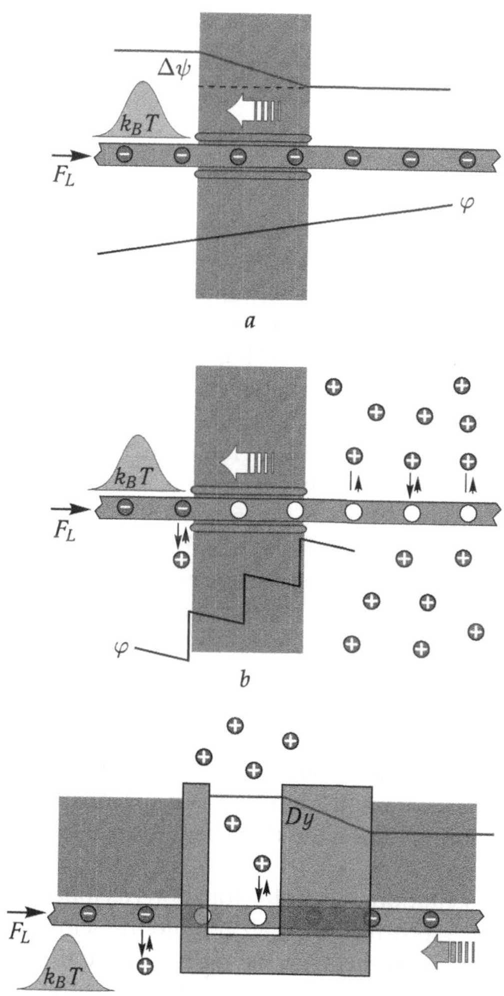

Figure 21.5. The Brownian ratchet mechanism (modified from G. Oster).

So, because of the chemical work only fluctuations acting in the "required" direction are "selected" and thereby the process although stochastic by its nature acquires a vector character. Naturally, this does not violate the second law of thermodynamics since this selection is provided by an independent source of chemical energy.

The motor behavior can be in general described by kinetic schemes, the analysis of which allows obtaining the values of energy efficiency, maximal velocity and rate of chemical energy consumption per one tact or in the whole cycle, as well as the efficiency of mechanical coupling (see Chapter 5).

Due to small molecular dimensions of the motor, it moves under the action of the stochastic Brownian force. Instantaneous fluctuations change greatly the magnitude and direction of the motion velocity. Consequently, force f is already not constant at each instant of time. The motor overcomes the friction forces which depend on the viscosity of the medium and are proportional to the motion velocity (11.5).

However, due to small inertia forces, at each instant of time the motor with a ratchet chemical mechanism can use thermal fluctuations to increase its thermodynamic potential (free energy). It is clear that this is equivalent to cooling the nearest surrounding of the molecular motor, so that in general the process has an endothermal character and the work is performed due to the thermal energy of the surrounding.

Work of the ATPase Complex Motor. Let us consider the models available in biophysics demonstrating the work of the ATPase complex motor. Figure 21.6 shows a flow diagram of location of the subunits in the ATPase complex corresponding to Fig. 21.2. The complex consists of two parts. One part is called a starter and includes subunits α, β, δ, i.e. hexamer F_1 with α_3- and β_3-subunits. The other part of the complex (a rotor) consists of subunits γ, ϵ and 9–12 copies of protein C. The boundary between the starter (a) and the rotor (c) has two water-filled proton semichannels that provide the movement of protons between the rotor and the internal and external regions separated by a cell hydrophobic membrane. In the diagram, the internal region (the cytoplasm) has low pH, while pH of the external region (the periplasm) is high, and the whole complex works in the mode of ATP synthesis due to the energy of the transmembrane proton gradient. Asp61 residues, carrying negative charges in the deprotonated state, are located on the C subunits in the rotor, whereas a positively charged Arg210 residue is on the α-subunit of the starter. The value of its pK_a is rather high, so that it is always protonated and carries positive charge $+1$.

The action of thermal fluctuations can cause only stochastic undirected displacements of the rotor. As a result, additional interactions playing the role of a ratchet mechanism in the motor are required to provide the directed rotation of the rotor at the expense of thermal energy.

In the ATPase complex, this is provided by electrostatic interactions of amino acid residues of the rotor and starter with simultaneous protonation and deprotonation of Asp61 residues on C-subunits upon their entering the proton semichannels (G. Oster, 2000).

In a simplified representation, the motor functions as follows. Due to thermal fluctuations the protonated Asp61 residues can get into the semichannels connecting the rotor to the cytoplasm and periplasm. In the right-hand semichannel, Asp61 will stay in the protonated state because of its contact with the region of low pH in the cytoplasm. In the left-hand semichannel, Asp61 will have a negative charge due to proton dissociation into the periplasm region with high pH. Deprotonation of Asp61 in the

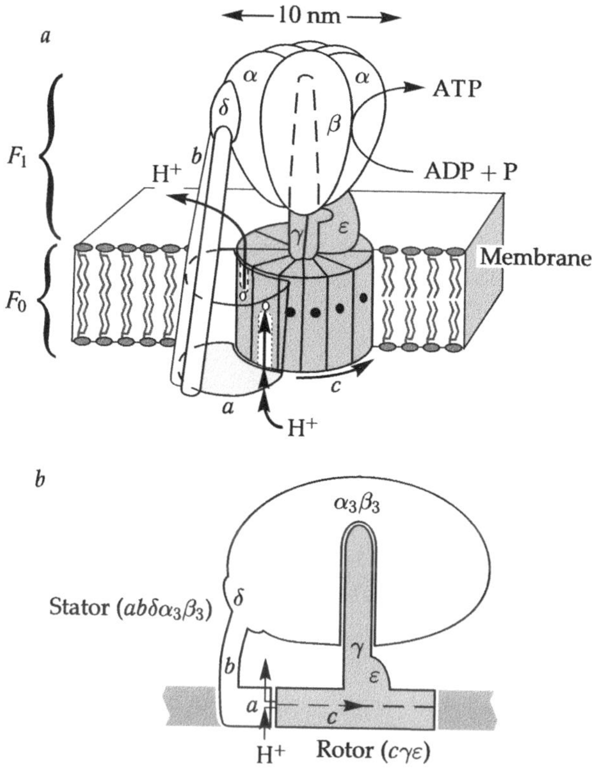

Figure 21.6. Flow diagram of location of subunits in the ATPase complex (modified from G. Oster).

left-hand semichannel is additionally facilitated because of the decrease in its pK_a caused by positively charged Arg210 located between the semichannels in the starter.

In the right-hand semichannel, neutral protonated Asp61 can diffuse only to the right into the hydrophobic membrane. Its displacement to the left is prevented by the Asp61$^-$ ion in the left-hand semichannel, which interacts with Arg210. When protonated Asp61 has moved to the right, its place in the right-hand semichannel is occupied by Asp61$^-$ from the left-hand semichannel that cannot move to the left into the hydrophobic region. Arg210$^+$ retains Asp61, however not strongly enough to prevent its movement to the place of protonated Asp61 that has moved to the right. New Asp61 gets into the left-hand semichannel and is again deprotonated there. Thus, the general displacement of the rotor to the right takes place, i.e., one tact of the motor rotational cycle is performed.

The presence of a positive charge on the starter prevents rotor diffusion in the reverse direction, tightly coupling the proton transfer and the rotor movement. The rates of protonation/deprotonation are much higher than the velocity of the rotor rotation. The electric interactions naturally decelerate the rotor rotation, but this "drawback" is compensated by that at high pK_a of Asp61 residues ($pK_a \approx 7.5$) random acts of protonation/deprotonation are prevented, thereby they are closely coupled to the instants of the rotor passing the site of positive Arg219 localization in the starter.

22

Physics of Muscle Contraction, Actin-Myosin Molecular Motor

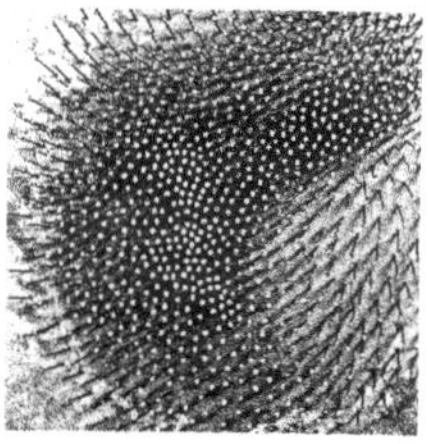

22.1 General Description of Energy Transformation in Systems of Biological Motility

The moving ability is a characteristic property of living organisms from protozoa to most complex ones. Contraction of skeletal, cardiac and smooth muscles and plant leaf movement, ciliary beating and flagella rotation, cell division and protoplasm movement are diverse types of motility having a common feature — transformation of chemical energy, released upon ATP hydrolysis, into mechanical work. Among linear molecular motors is myosin, moving along actin filaments, as well as kinesin and dinein, moving along microtubules formed by tubulin.

Upon muscle contraction, the mechanical work is performed by the heads of myosin molecules, arranged in supramolecular structures, when they interact with polymerized actin. The main regulator of the motility in all muscles is calcium. Elucidation of molecular mechanisms for generation of force, transformation of chemical energy of ATP hydrolysis into mechanical work and mechanisms controlling these processes is the chief task of biophysics of biological motility.

22.2 Basic Information on Properties of Cross-striated Muscles

Skeletal and cardiac muscles as well as flight muscles in insects are ascribed to cross-striated muscles. A skeletal muscle consists of longitudinal fascicles of muscle fibers. A fiber is a long multinuclear cell with the cross-section from 10 to 100 μm. The fiber length corresponds frequently to the muscle length reaching 12 cm.

Mechanics and Energetics of Muscle Contraction. A skeletal muscle starts to offer noticeable resistance to extension only of the length much exceeding its natural value in the organism. In response to electric excitation, this muscle develops an active force and becomes much more rigid. The muscle mechanical response to electric excitation depends on both the stimulating signal and the mechanical restrictions imposed on the muscle. Given the ends of the muscle or fiber fixed, the muscle generates an active force or mechanical tension. Tension is the force per cross-sectional area, i.e. a value having the dimension of pressure. The mode of muscle contraction at a constant length is called isometric. Figure 22.1 shows how the tension in the muscle fiber changes upon its electric stimulation in an isometric mode.

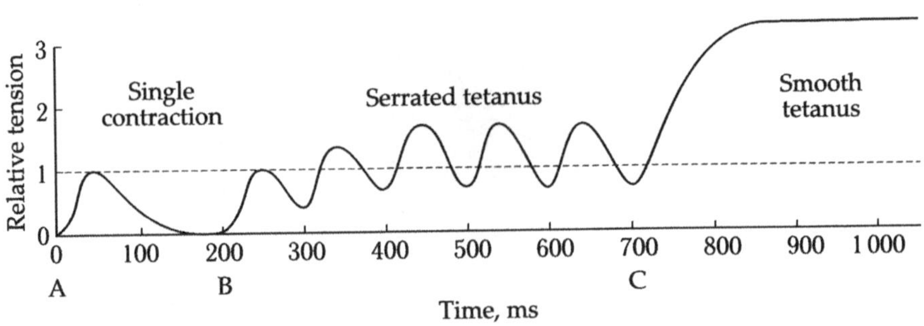

Figure 22.1. Response of mechanical tension generated by a muscle fiber to the increase in the frequency of exciting electric stimuli.

A, B and C are single contraction, serrated and smooth tetanic contractions, respectively.

When the stimulation frequency is increased above some limit, single contractions are partially combined — the muscle has no time to relax before the next stimulus comes. This contraction mode is called a serrated tetanus. Upon further frequency increase, the generated force continues growing until it reaches some threshold after which the tension remains constant even in between the stimuli. This contraction mode is called a smooth tetanus. Upon smooth tetanus contraction, a muscle develops the maximally possible tension which we will designate as P_0. In some cases the magnitude of the external load on the muscle, i.e. the load lifted by it, is maintained constant and the changes in the muscle lengths are noted. This is an isotonic contraction. An essential part of information on mechanical and energy properties of muscles was obtained by A. V. Hill when he studied the frog tailor's muscle in the mode of smooth tetanus contraction. If a muscle, which contracted in an isometric mode, is relieved and allowed to shorten when lifting the load, the rate of shortening is constant in time. The relation of this velocity of shortening v to the generated force P (at a constant rate it is equal to the weight of the load lifted by the muscle) is called the characteristic Hill equation and is described by a hyperbola with displaced

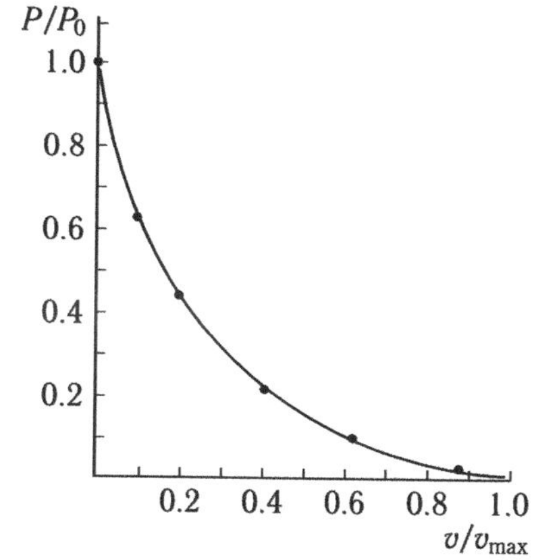

Figure 22.2. Hill's relation is the dependence of the rate of isotonic shortening v and the developed force P for the frog tailor's muscle at 0 °C (reproduced from J. Squire, 1981).

Solid line, experimental dependence; circles, calculations using E. Huxley model (1957, see lower). P_0 is the maximal, i.e. isometric force; v_{max} is the maximal rate of shortening.

axes (Fig. 22.2):

$$(P + aP_0)v = (P_o - P)v_{max}, \tag{22.1}$$

where P_0 is the isometric force ($P = P_0$ at $v = 0$), a is the dimensionless parameter, and v_{max} is the maximal velocity of muscle contraction, i.e. the velocity of its stress-free shortening. Parameter a for the frog tailor's muscle is 0.25 at 0 °C. For other muscles it can differ somewhat. Under these conditions in the frog muscle $P_0 = (2.5\text{–}3.0)^5$ Pa that corresponds to 2.5–3.0 atm, and $v_{max} = 2.0\text{–}2.5 l_0\ \text{s}^{-1}$, where l_0 is the muscle or fiber length at rest.

Let us now consider the energetics of muscle contraction. Power W, i.e. work performed by a muscle per time unit, is equal to the product of shortening velocity v and force P. By expressing the velocity via force using Hill's relation, we find that the dependence of W on P is described by the formula:

$$W = \frac{Pv(P_0 - P)}{P + aP_0}. \tag{22.2}$$

As seen, at $P = 0$ and $P = P_0$ the muscle power turns to zero because it either shortens without stress or generates force without shortening.

Upon contraction, the muscle not only performs work, but also evolves heat. According to the first law of thermodynamics, the sum of the work of external forces over the muscle, $\Delta A'$, and the heat flow to it, $\Delta Q'$, is equal to the change in the internal energy of the system, ΔU. The work A, performed by the muscle, is equal to the work over the muscle with an opposite sign, i.e. $A = -A'$, and the produced heat Q is the heat flow with an opposite sign, $Q = -Q'$, and therefore

$$-\frac{dU}{dt} = W + \frac{dQ}{dt}, \tag{22.3}$$

where U is the internal energy.

If the activated muscle shortens and performs work, more heat is produced as compared to that upon isometric contraction (Fenn's effect). According to the first law of thermodynamics, the rate of the total excess energy production equal to the rate of the internal energy changes in the muscle will be as follows:

$$-\frac{dU}{dt} = \frac{dQ}{dt} + W = (A + P)v. \tag{22.4}$$

At constant temperature, the change in the internal energy of the muscle is conditioned mainly by the change in the concentrations of different substances. As demonstrated by V. A. Engelgardt and M. N. Lyubimova (1939), interaction of actin with myosin underlying muscle contraction, is accompanied by hydrolysis of adenosine-triphosphate (ATP). Therefore, it was natural to suggest that just this reaction determines the rate of changes in the internal energy of the muscle upon its contraction.

At physiological concentrations of ATP, ADP and P_I in a muscle, the free energy of hydrolysis of one ATP molecule is about $(8 \div 10)10^{-20}$ J which makes 48–60 kJ/mol. At physiological temperature, heat energy $k_B T$, where T is the absolute temperature and k_B is the Boltzmann constant, is about $4 \cdot 10^{-21}$ J. Accordingly, the free energy of ATP hydrolysis is 20–25$k_B T$. Remember that we should distinguish the standard free energy of ATP hydrolysis, ΔG^0_{ATP}, determined under standard conditions (the pressure of 1 atm, temperature of 25 °C, and 1 mol/l ATP, ADP and phosphate concentrations), and the real free energy determined at those concentrations of ATP, ADP and inorganic phosphate P_I which are in the muscle at the given instant of time. The standard free energy of hydrolysis, ΔG^0_{ATP}, is about −30 kJ/mol, while the real free energy is determined from the relation

$$\Delta G_{ATP} = \Delta G^0_{ATP} + RT \ln \frac{[ADP][P_I]}{[ATP]}, \tag{22.5}$$

where [ATP], [ADP] and $[P_I]$ are concentrations of the corresponding substances in mol/l, R is the universal gas constant, and T is the absolute temperature. Concentrations of ATP, ADP and phosphate in muscles are approximately $5 \cdot 10^{-3}$ M, $5 \cdot 10^{-4}$ M and 10^{-3} M, respectively; when substituted in the relation for the free energy, this gives 50–60 kJ/mol.

Structure of Muscle Cells. A muscle fiber is a complicated multinuclear cell containing thousands of thinner elongated formations — myofibrils — of 1–2 μm in diameter, which stretch from one end of the fiber to the other. A microfibril consists of elementary contractile units — sarcomeres (Fig. 22.3).

Under a light microscope, sarcomeres look like recurring dark and light bands, and their boundaries in neighboring microfibrils coincide with each other. This order of neighboring microfibrils imparts characteristic cross-striation to the muscle, and this has given the name to cross-striated muscles. Dark bands in the middle of a sarcomere are anisotropic — their refraction index is dependent on light polarization. Therefore, they are called A-bands. Light bands having no anisotropic features are called isotropic bands (I-bands, or I-discs). An A-band is formed by filaments of 15 nm in diameter and about 1.6 μm in length. They were called thick filaments. Thinner filaments of about 10 nm in diameter and 1 μm in length branch out from Z-discs to the sarcomere center. In accord with the sliding filament theory, muscle contraction takes place as a result of relative displacement (sliding) of thick and thin filaments in a sarcomere towards each other without changes in their

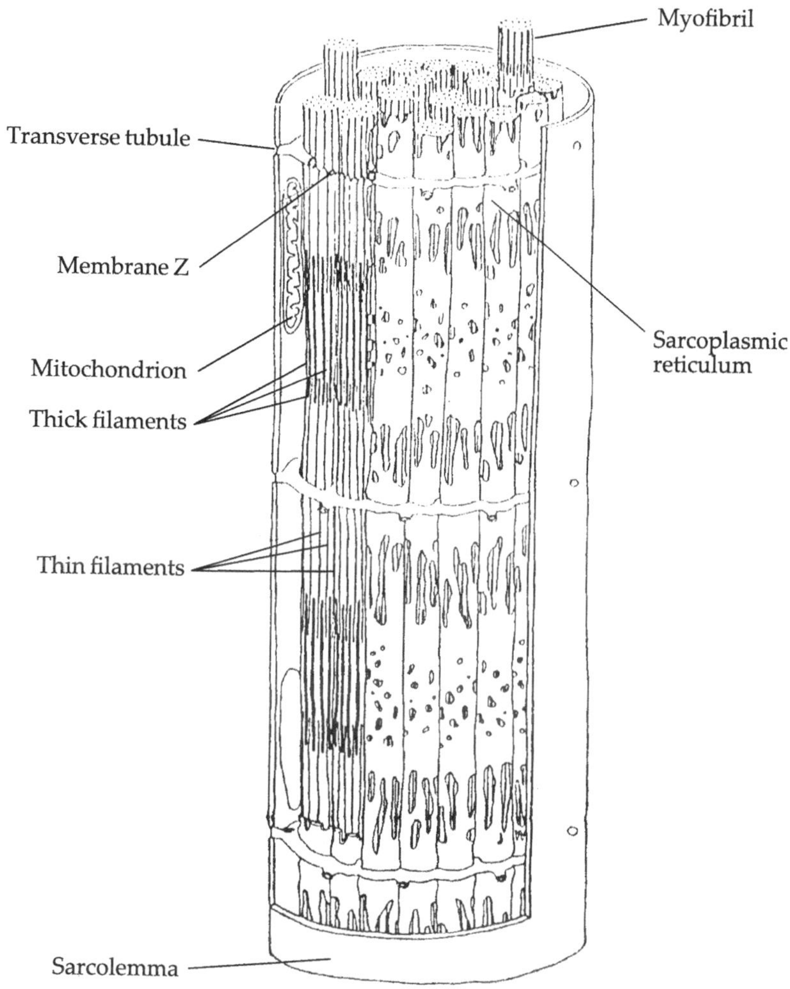

Figure 22.3. Schematic representation of the structure of a muscle cell (reproduced from R. Woledge et al., 1992). In one region the transverse tubule is removed to show one of membranes Z separating the myofibril into sarcomeres.

lengths. The reason for relative sliding of filaments is cross-bridges — structures interconnecting thick and thin filaments.

The bridge hypothesis of force generation. By 1957, the bridge hypothesis had been formulated. According to it, the sliding of filaments in a contractile muscle occurs as a result of action of independent force generators uniformly distributed in a sarcomere between actin and myosin filaments. Myosin bridges, protruding from thick filaments are linked to thin actin filaments, generate an active force or mutual displacement of the filaments, and after that detach; and the cycle is repeated again. This hypothesis was experimentally proved by A. M. Gordon, A. F. Huxley and F. J. Julian (1966), who determined the dependence of isometric force P_0 on the length of sarcomeres, i.e. on the length of the area of overlapping of thick and thin filaments in a sarcomere.

It was found that when the sarcomere length changes, the maximal tension of fiber P_0 changes in line with the change in the length of the overlapping area of actin and myosin filaments, i.e. it is proportional to the number of the functioning bridges. In addition, upon contraction of an unloaded muscle the maximal velocity of isotonic shortening v_{max} remains constants in a considerable range of sarcomere lengths. This fact supports the basic postulate of the bridge hypothesis that bridges work independent of each other, i.e. both the force generated by a separate bridge and kinetic parameters of the bridge cycle are independent of the number of functioning bridges and are completely determined by the properties of the bridge itself.

22.3 Structural Organization of Muscle Contractile and Regulatory Proteins

To understand how cross bridges generate force and realize mutual sliding of thick and thin sarcomere filaments, let us consider in more detail their composition and structure. Corresponding data were obtained as a result of electron microscopic, X-ray diffraction and biochemical studies of many research groups. Thick filaments consist mainly of myosin.

Thick Filaments. A molecule of muscle myosin II (Fig. 22.4) with molecular mass of $\sim$ 500 kDa consists of two heavy and four light chains with molecular masses of $\sim$ 200 kDa and $\sim$ 20 kDa, respectively. Thin alpha-helical regions of the two heavy chains, intertwining into a coiled coil, form a long rod of $\sim$ 2 nm in diameter and $\sim$ 150 nm in length. The general view and designation of different myosin sites are given in Fig. 22.4.

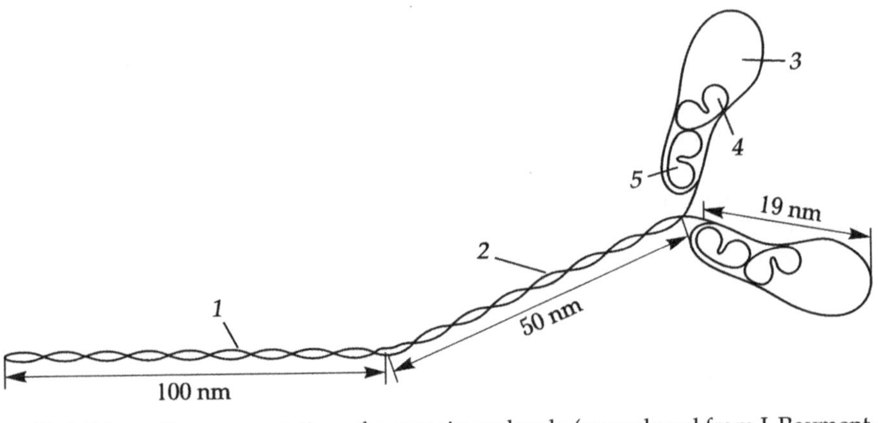

Figure 22.4. Schematic representation of a myosin molecule (reproduced from I. Rayment and H. Holden, 1994).

1, Light meromyosin; *2*, subfragment II; *3*, globular region of a heavy chain (subfragment I); *4*, major light chain; *5*, regulatory light chain.

N-termini of heavy chains diverge, forming two massive globular heads, or subfragments I, of $\sim$ 16 $\times$ 5 nm in size, which are connected via a flexible hinge to the rod region of the molecule. Close to the hinge in each subfragment I, there are the major and regulatory light chains located pairwise. The regulatory light chain is in the direct vicinity from the site where the heads are linked to the rod. In cross-striated muscles, the stems of a thick filament are formed by tightly packed backbone regions of myosin molecules, whereas subfragments II and globular heads are located on their surface and are involved in the interaction with actin filaments.

In a quiescent muscle, subfragments II are secured to the backbone of the thick filament so that myosin globular heads on the surface of the thick filament form a kind of a three-stranded helix (the Squire model) with subunit and axial pitches of the helix of 14.3 nm and 42.9 nm, respectively (Fig. 22.5). The figure shows also the thick filament structure obtained using electron microscopy in the group of R. Craig (M. E. Zoghbi et al., 2008).

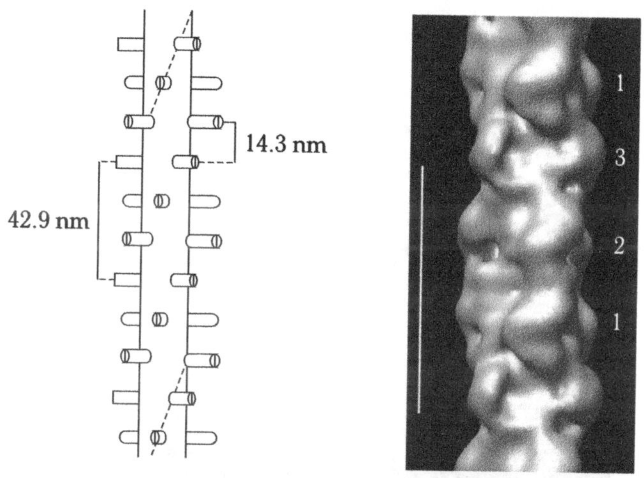

Figure 22.5. Three-stranded model of a thick filament. Left: the Squire model (reproduced from J. Offer, 1974). Right: contemporary electron microscopy reconstruction of the thick filament structure of murine cardiac muscle (reproduced from M. E. Zoghbi et al., 2008). The while vertical line shows the 42.9 nm span of the myosin helix in the thick filament range, where C protein is present.

The model was constructed assuming that myosin subfragments II are secured to the filament backbone; for convenience myosin heads are shown as cylindrical protuberances.

Thin filaments (Fig. 22.6) of 8–10 nm in diameter and about 1 μm in length consist of actin, regulatory proteins (coiled-coil tropomyosin and troponin) and also contain structural protein nebulin.

Each tropomyosin molecule interacts with 7 actin monomers and one troponin complex. Polarity of the thin filament structure is seen when it is decorated with myosin heads: the attachment of these molecules to actin globules in filaments creates a specific structure of "arrowheads", inserted in each other and oriented in the same direction towards the sarcomere center.

The binding of Ca^{2+} with troponin C rearranges the whole troponin complex. Troponin I stops binding to actin, as a result tropomyosin can relatively easily rotate around the actin filament axis. Therewith myosin binding regions on the actin filament open, and myosin heads become able to bind to actin and form cross-bridges — thus muscle contraction takes place. Formation of bridges causes further rotation of the troponin–tropomyosin complex around the thin filament axis and further, more complete, opening of interaction regions with myosin on actin.

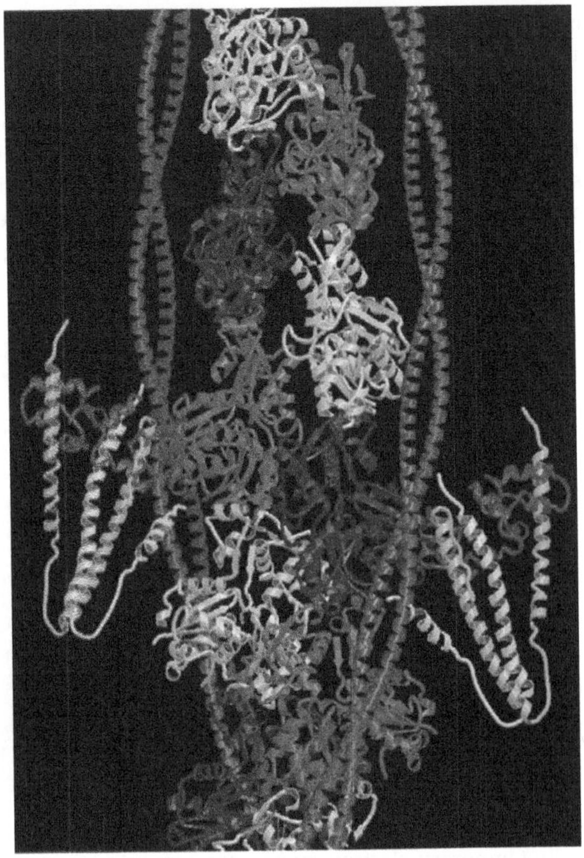

Figure 22.6. Scheme of a thin filament structure (reproduced from A. Pirani et al., 2006).
From white to blue: actin globules; yellow, light blue and lavender: troponin complex; red: tropomyosin strand.

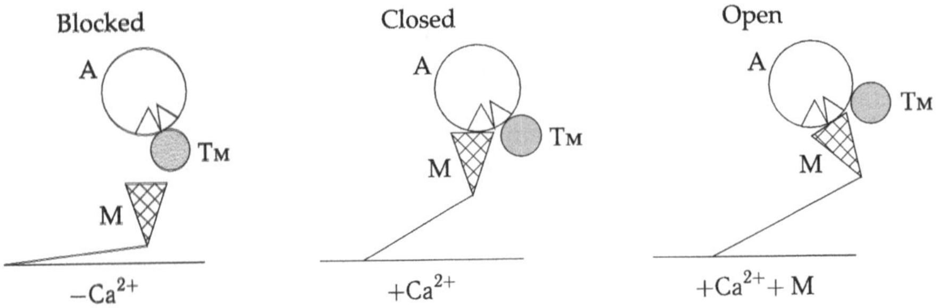

Figure 22.7. Scheme of movement of the troponin–tropomyosin complex under the actin of Ca^{2+} ions and myosin heads (reproduced from D. F. McKillop, M. A. Geeves, 1993).

Troponin of skeletal and cardiac muscles has two functional centers and one functional center of Ca^{2+} binding, respectively. Schematic representation of calcium regulation of thin filaments in cross-striated muscles is shown in Fig. 22.7.

In the "blocked" state (Ca^{2+}), myosin heads (M) cannot bind to actin monomers (A), because the myosin-binding region of actin is blocked by tropomyosin (Tm).

In the closed state, the binding of Ca^{2+} ions to troponin causes displacement of Tm and release of the system that makes possible weak association of M and A. Only after transformation of the weak actin-myosin bond into a strong one does further movement of Tm take place and an "open state" of regulatory complex is formed.

22.4 Mechanochemical Transformation of Energy in Muscles. Lymn – Taylor Scheme

In 1939, V. A. Engelgardt and M. N. Lyubimova discovered that myosin is an ATPase. They were the first to recognize that namely this reaction is the source for muscle contraction energy and initiated mechanochemical studies of muscle contraction. In the early 1970s, cardinal details of the mechanism of ATP hydrolysis by myosin were clarified by R. W. Lymn and E. W. Taylor (1971), and soon defined more exactly by D. R. Trentham and C. R. Bagshaw (1973) and other researchers using methods of nonstationary biochemical kinetics. A very notable property of myosin ATPase was that hydrolysis itself, i.e. disruption of the bond between ADP and gamma-phosphate, as a result of the water molecule attack, is reversible, and consequently it is not associated with the great change in the energy of the system. The free energy of an ATP molecule is consumed mainly at the moment of its binding in the active center of myosin as well as upon discharge of products, i.e. is consumed for performance of mechanical work and for uncoupling of the myosin head from actin. Another important peculiarity of muscle contractile proteins is the dependence of the stability of the actin-myosin complex on the nucleotide (ATP) bound in the active center of myosin. In the absence of the nucleotide or in the presence of ADP, the affinity of myosin to actin is very high and the equilibrium constant makes $< 1\ \mu M$. It is just because of this that after ATP removal from the sarcoplasm the muscle fiber becomes exceptionally rigid, because all myosin heads are strongly bound to thin filaments independently of the state of the regulatory system. This state is called a rigor state (from Latin *rigor mortis* meaning postmortem rigidity). An addition of ATP relaxes the fiber (in the absence of Ca^{2+}) or causes its active contraction in the presence of Ca^{2+}. The affinity of myosin-ATP and myosin-ADP-phosphate complexes to actin is far lower, and the equilibrium constant is $100\ \mu M$. Therefore, in a contractile muscle at every instant of time only part of myosin bridges are bound to actin, and these bridges can detach from actin and again bind to it at a rather high rate.

As a result of such studies, a scheme was proposed including cycles of events accompanying ATP hydrolysis and force generation upon actin-myosin interaction, the so-called Lymn – Taylor scheme (Fig. 22.8).

Let us consider this cycle from the moment when the myosin head, having discharged products of ATP hydrolysis, is strongly bound to actin in the rigor state. However under physiological conditions when the ATP concentration is about 5 mM, this state is short-lived, because ATP rapidly (at a rate of $\sim 10^6\ M^{-1}s^{-1}$) and strongly binds to the actin center of the myosin head. After ATP binding, the myosin head rapidly (for about 1 ms) rearranges that results in a decrease in the strength of its link with actin d 100–1000 times so that it detaches from actin. Hydrolysis of an ATP molecule proceeds on the myosin head uncoupled from myosin. The head with the hydrolysis products ADP and P_I, still remaining in its active center, again binds to actin, forming a weakly bound complex with it. This process is closely associated

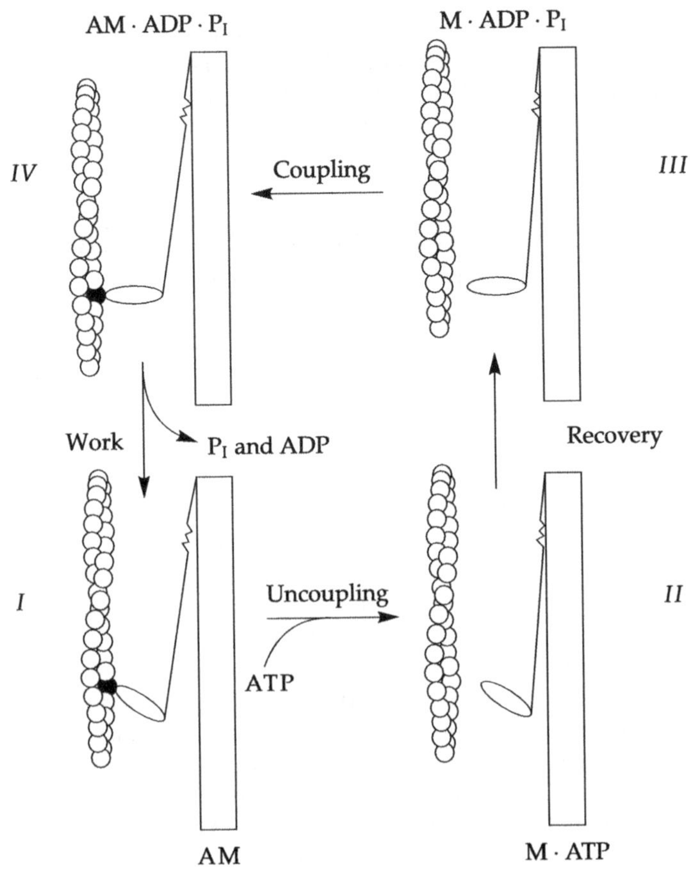

Figure 22.8. Lymn–Taylor scheme of coupling of ATPase and mechanical cycles in the work of a myosin bridge.

A, Actin filament (actin globules are not shown); M, myosin filament; *I–IV*, consecutive stages of the working cycle. State *I* is the rigor state of the bridge; transition is accompanied by force generation (bridge working stroke); double arrows show the direction of forces acting on the thick and thin filaments. Conformational changes in the myosin head are shown as changes in its geometrical shape.

with force generation occurring due to isomerization of the actin-myosin complex, as a result of that the subfragment I region most remote from actin is displaced by 5–10 nm relative to the actin-binding center. This leads to a relative displacement of the thick and thin filaments. According to the contenporary concepts, the force-generating step takes place before phosphate is discharged. After discharge of phosphate, the stability of the actin-myosin complex increases. The following discharge of ADP terminates the cycle and prepares the myosin head to the binding with the next ATP molecule. The discharge of ADP is slowed down upon the bridge tension, i.e. upon the muscle stretching, and is accelerated upon its shortening. This explains the Fenn effect, i.e. increase in the rate of energy production and ATP hydrolysis upon muscle shortening. Under isometric conditions when a muscle does not perform mechanical work, the rate of ATP hydrolysis is rather low. It increases several times upon muscle shortening caused by moderate load, corresponding to maximal power.

22.5 Three-dimensional Structure of Actin and the Myosin Head

The current progress in understanding the molecular mechanism of operation of the actin-myosin motor is to a great degree on the data on the atomic structure of actin and the myosin head.

Actin Filament Model. To decipher the three-dimensional structure of protein, it is required to obtain high-quality monocrystals suitable for X-ray analysis. Since globular actin at its high concentration in solution had propensity for polymerization, i.e. the formation of long filaments, it can be crystallized only in complex with other proteins or after chemical modification that prevents filament formation. This was done, for the first time, by W. Kabsh et al. (1990) (Fig. 22.9).

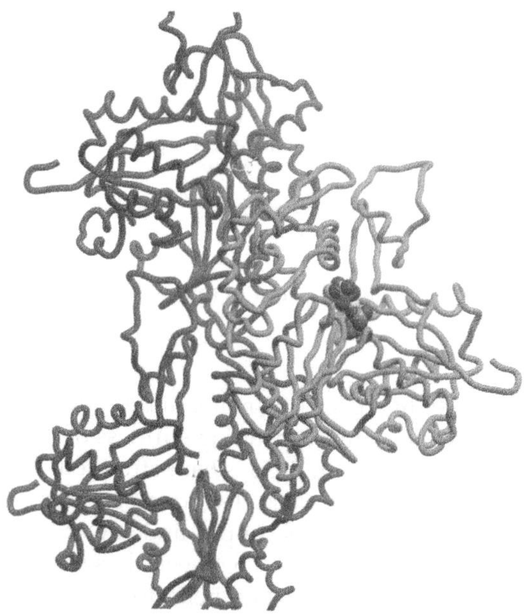

Figure 22.9. Actin filament structure (reproduced from T. Oda et al., 2009).

Three actin monomers are shown by different colors. A filament is a helix with the axial pitch of 2.75 nm and the rotation angle of neighboring monomers of 167°. The balls in the central actin monomer designate an ATP molecule.

Structure of the Myosin Head. Due to the presence of long coiled-coil regions (subfragment II and light meromyosin), the attempts to crystallize a whole myosin molecule were also unsuccessful.

Figure 22.10 shows the structure of subfragment I from a shellfish muscle in two different states — in the presence of ADP and vanadate and in the absence of the nucleotide in the ATPase center. The myosin – ADP – vanadate complex is a long-lived analog of the myosin – ADP – P_I complex. The two structures correspond to the beginning and end of the working stroke of the myosin head in the Lymn – Taylor cycle.

Subfragment I consists of a massive (about 80 kDa) "motor" domain and a long "lever". The basis of this lever is a long alpha-helix (8.5 nm). The long alpha-helix

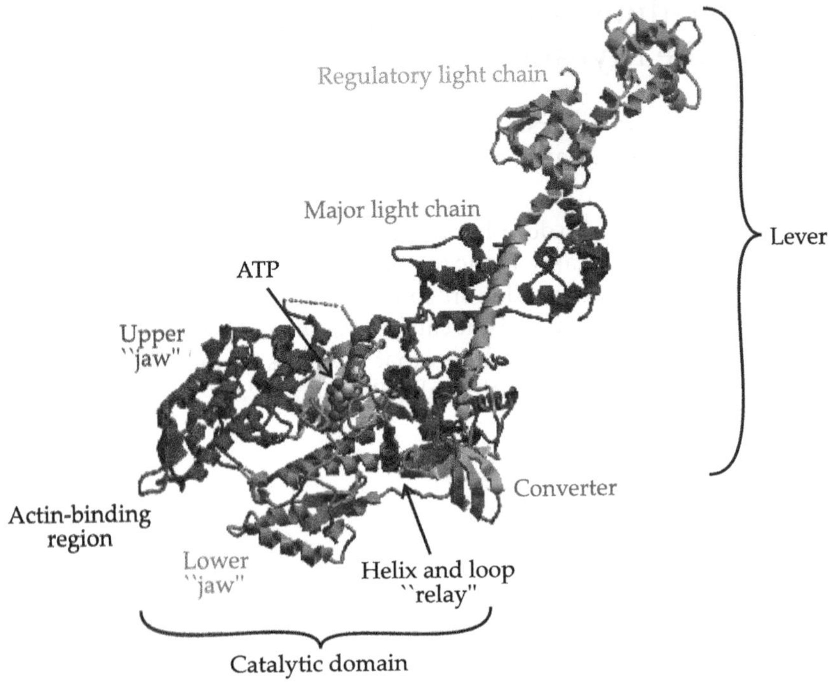

Figure 22.10. Atomic structure of subfragment I in myosin (reproduced from C. Cohen et al., 2000, 2006).

allocates sequentially two light chains (the major and regulatory ones) stabilizing its structure and apparently increasing the bending stiffness of the lever.

The nucleus of a motor domain is a seven-stranded beta-sheet, containing a nucleotide-binding center on its surface. Two compact domains called the upper and lower actin-binding regions, or the upper and lower "jaws", are attached to the beta-sheet. There is a gap, or "mouth" between these domains, which is closed upon strong, "rigor" binding of the myosin head to actin. The closing of this gap is accompanied by the movement of the upper domain, as a result of that the nucleotide-binding "pocket" opens. This direct mechanical relationship explains the decrease in the stability of nucleotide binding by the myosin head upon its strong binding to actin and also the weakening of the actin-myosin bond in the presence of ATP or ADP and phosphate in the active center of the head, though the actin and nucleotide binding centers are at a considerable distance from each other.

22.6 Mechanism of the Work Cycle of the Actin-Myosin Motor

The both closed "jaws" of the motor domain are involved in the formation of the bond between the head and actin. Therewith the nucleotide-binding "pocket" of the active center is open. The ATP binding by the myosin head is accompanied by untwisting of the beta-sheet in the motor domain center and consequently it becomes flat. As an equilibrium "natural" state of the beta-sheet is just the twisted state, the untwisting is accompanied by storage of elastic mechanical energy in the myosin head. The source for this energy is ATP molecules. Upon binding of the ATP molecule head, or more exactly its complex with Mg^{2+} in the nucleotide "pocket", the stability of the ADP bond with gamma-phosphate diminishes so much that the hydrolysis splitting of the bond are accompanied only by an insignificant change in the free energy of the system. The equilibrium constant of the hydrolysis is about 10, making only approximately 10 % of the total energy released during a hydrolysis cycle of the ATP molecule. A far larger portion of this energy is stored in the myosin head upon the ATP molecule binding.

If the nucleotide-free myosin head has firmly attached to actin, thus forming a "rigor" complex with it, the binding of Mg to ATP causes a changed configuration of the nucleotide "pocket". Therewith the upper actin-binding domain rotates relative to the remainder part of the molecule and partially closes the nucleotide "pocket" with a concurrent opening of the "mouth" between the upper and lower actin-binding "jaws". This conformational rearrangement prevents simultaneous formation of stable bonds with actin by both "jaws". As a result, the actin-myosin bond weakens so that the head detaches from actin.

At the next stage of the cycle, the nucleotide "pocket" is closed. This is accompanied by formation of a complete set of bonds between ADP, Mg^{2+} and phosphate with a number of amino acid residues of the upper and lower actin-binding regions, including the motor domain attached to the beta-sheet. Finally, the "pocket" closing when gamma-phosphate can detach from ADP, is accompanied by a large-scale rearrangement of the head. The most remarkable manifestation of this rearrangement is the lever rotation. The active center of myosin has two loops called "switches". One of them ("switch 1") is connected to the upper actin-binding "jaw". It is involved in information transfer by the active center and the upper actin-binding domain ("jaw") and controls the closing of the "mouth" between the upper and lower "jaws". The closing of this "switch" is coupled to the opening of the "mouth" and weakening of the following disruption of the bond between the myosin head and actin. The second "switch" is connected to the lower "jaw". The closing of "switch 2" causes slight ($\sim 20°$) untwisting of the alpha-helix stretching from the actin-binding center to the motor lever. It is called a "relay helix". The end of the "relay helix" contains a "relay loop" contacting a small but rigid domain, which consists of a three-stranded beta-sheet and is called a "converter". In its turn, the "converter" is a part of a long motor lever. Due to such mechanical contacts, the untwisting of the "relay helix" upon the closing of "switch 2" leads to rotation of the "relay helix", which turns over the "converter" together with the whole lever. The rotation angle of the "converter" from one stable state to the other, caused by the relay rotation by a relatively small angle, is far more considerable and reaches 70°. Such rotation, occurring under physiological conditions after detachment of the myosin head from actin, "cocks" the motor lever and prepares it to performing the next work stroke. In 2005–2007 S. Fisher with coauthors created a chain of events accompanying the closing of the ATPase "pocket" and the lever "cocking" using the methods of nonlinear molecular mechanics. They discovered a number of molecular "springs" and "small levers", due to the coordinated work of which small (< 0.5 nm) movements of "switch 2" turn into 10-nm movement of the lever end.

The ATP hydrolysis proceeds in the closed active center of myosin when the lever is in position corresponding to the beginning of the work cycle. The teams of S. Fisher (2006) and B. Grigorenko (2007) described the results of molecular dynamic modeling of the MgATP hydrolysis in the close nucleotide "pocket" of myosin, revealing the role of several water molecules and a number of conservative amino acid residues in decreasing the activation barrier upon disruption of the gamma-phosphate bond and the subsequent formation of inorganic phosphate.

It is not clear yet how the formation of a stable bond between myosin subfragment I and actin and opening of the nucleotide "pocket" coupled with this process trigger the lever turning, i.e. performance of a work stroke of the motor. According to the lever hypothesis, it is at this stage that the myosin motor performs mechanical work which can make up to 40% of the ATP hydrolysis energy, i.e. $10k_BT$, or 4×10^{-20} J. Along with the current concepts, this proceeds before the release of inorganic phosphate. It is proposed that the energy for this process, at least partially, is drawn from the elastic energy of the untwisted beta-sheet of the motor domain. It is also possible that the movement of "switch 1" upon closing of the "jaws" facilitates in some way the movement of "switch 2" and the turn of the lever associated with it.

The rate of ADP release and stability of its binding to actin depend on the bridge strain. The stretching of the bridge upon muscle stretching slows down the ADP release, while the removal of the load upon muscle shortening, on the contrary, facilitates the ADP release and preparation of the myosin head to binding with a new ATP molecule and performance of the next cycle. This may be connected with the mechanical deformation of the active center pocket when the load is applied to the motor lever. In its turn, the dependence of the rate of the ADP release on the load explains the increase in the rate of energy generation in the muscle upon its shortening under moderate load (the Fenn effect known since the 20-s of the last century). After the ADP release, the myosin head forms a stable, "rigor" complex with actin that may be considered as the final phase of the bridge cycle, characterized by minimal free energy. The head is ready to perform a new cycle of work triggered by the binding of a new ATP molecule.

23

Biophysics of Processes of Intracellular Signaling

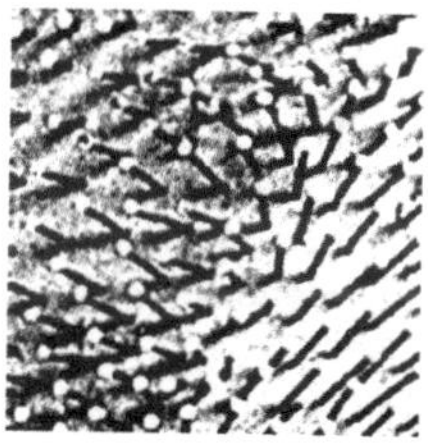

23.1 General Regularities of Intracellular Signaling

Cellular Signaling. Generation of cellular signal (CS), information acquisition and processing obey a number of general principles. The process begins with detection of extracellular events (chemical reagent binding, light quantum absorption, mechanical or temperature effects, etc.) by membrane proteins (receptors) followed by their transfer via secondary intermediaries (SI, messengers). On a plasmatic membrane, signal generation can be mediated not only by modification of membrane proteins, but also by changes in the membrane potential. This is valid for the functioning of nervous, hormonal and immune systems in animals (Fig. 23.1).

Usually signaling molecules (SM) do not penetrate into the cell and interact specifically with receptors localized in the plasmatic membrane. Note that hydrophobic steroid and thyroid hormones can penetrate across the plasmatic membrane into the cell, where they interact with receptor proteins, localized in the cyto- and/or nucleoplasm. Upon CS (chemical substance, quantum of light, mechanical action) realization, the SM stimulates the release of SI molecules by interacting with membrane proteins. In general, signal transformation across the membrane is reduced to three main stages. The first stage is the SM interaction with the receptor; the second is conformational rearrangement of the receptor molecule and a change in the function of specialized membrane mediator proteins; and the third is the formation of secondary messengers-intermediaries (SI), which are small molecules (of ions) diffusing in cells into definite subcellular structures and thereby providing signal

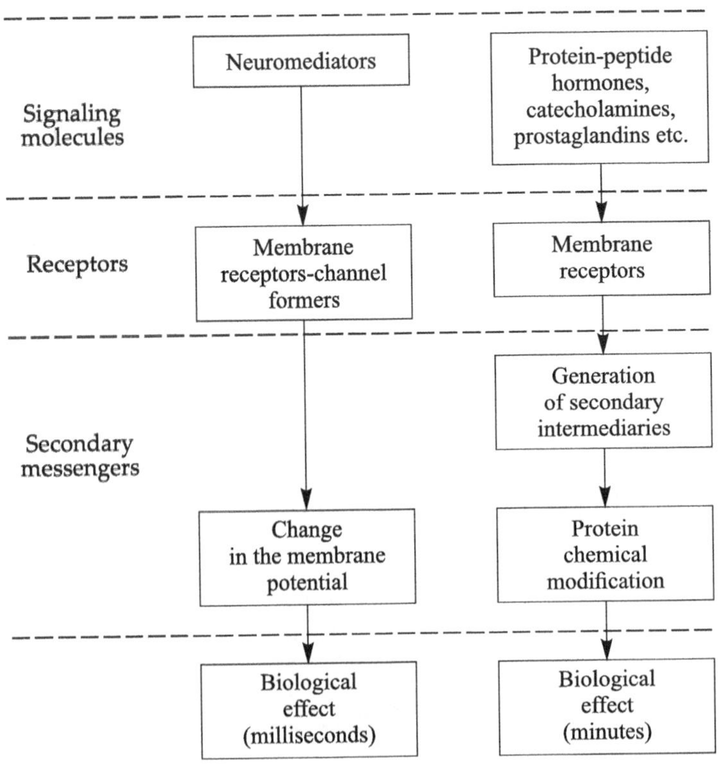

Figure 23.1. Main processes of cell regulation.

propagation. Every cell has a complex converting all external signals into intracellular ones and then transforming these signals into intracellular functions. As a rule, a signal is transmitted into the nucleus by means of translocation into it of cytosol protein kinase or activated transcription factor (phosphorylated by cytosol protein kinase or released from the complex with another protein). In the mitochondria, the signal is transmitted otherwise — by translocation from cytosol of secondary agents (Ca^{2+} or cAMP).

Types of Receptors. Receptor proteins are divided into two classes: globular (cytoplasmic) proteins and membrane (integral) proteins. Globular proteins are freely floating or fixed in the cytoplasm high-affinity compounds for certain types of molecules. An example is regulation of cell protein synthesis with aldosterone. In blood this hydrophobic hormone readily passes across the cell membrane and forms a complex with the receptor in the cytoplasm. This complex can penetrate into the nucleus, where is disintegrates and the released receptor serves as the initiation factor for expression of corresponding genes. Integral proteins bind SM on the external side of the membrane and, due to the change in the receptor protein conformation, generate a signal on the internal side of the membrane. The receptor, independently of the SM nature, has a general structure: a region located outside the cell, an intramembrane region, and a region immersed in the cytoplasm. The external receptor region is specific to extracellular changes in the surrounding, whereas the internal region is specific to the receptor-associated intracellular protein. Namely, the properties of the latter determine which is the intracellular signaling systems to interact with.

At present, three types of receptor membrane proteins are distinguished.

a) Proteins consisting of one transmembrane polypeptide chain. They are allosteric enzymes with the active center located on the internal side of the membrane. This type includes insulin, growth factors and cytokine receptors. The SM binding provokes a change in the receptor conformation and phosphorylation of tyrosine residues (in the first place, the protein molecule of the receptor itself is phosphorylated — autophosphorylation).

b) Oligomeric protein – receptor complexes with ion channels. The ligand binding results in activation of the channel for Na, K or Cl ions. A classic example of such type of intracellular signaling is cation acetylcholine receptors (AcR).

c) Receptors coupled to GTP-binding proteins (GTPases). The polypeptide chain of these proteins includes seven transmembrane segments. The signal transformation is realized from GTP-binding proteins to effector proteins located on the adjacent membrane regions or in the cytoplasm. The function of these proteins is to modify cell metabolism.

Now the classification of cell signaling systems is made similar to secondary messengers and metabolic processes regulated by them.

The so-called G proteins are used in most of these signaling processes. Receptors coupled to G proteins (R) transmit the signal from primary messengers to intracellular targets as follows: R → G protein → effector protein. Effector proteins in signaling systems like R → G protein → effector can be adenylate kinase catalyzing the cAMP synthesis from ATP, phospholipase C hydrolyzing phosphatidylinositol with the formation of diacylglycerol (DAG) and inosytoltriphosphate (IP_3), phosphodiesterase cleaving cGMP into GMP, some potassium and calcium channels.

It is important that upon CS transmission in the sequence 'receptor → G protein → effector protein', the initial external signal can be amplified multiply. This occurs because one receptor molecule, while staying in the activated state (R^*), has time to convert several G protein molecules into an activated form (G^*). For example, in the rhodopsin (see Chapter 29) → G → cGMP-phosphodiestarase visual cascade, several hundreds or thousands of G^* molecules can be formed per each R^* molecule, and this means that at the first stage of the $R^* \rightarrow G^*$ cascade the external signal amplification factor is 10^2–10^3. At the next stage of the cascade (G^* → effector protein), each G^* molecule interacts only with one effector protein molecule, but even here the signal is amplified, because in the cytoplasm a large number of secondary signal molecules emerge (disappear) per each G^* molecule and, correspondingly, activated effector protein. In photoreceptor cells at the second stage of the cascade, one molecule of activated cGMP-phophodiesterase can cleave up to 3000 cGMP (IM) molecules per one second. Since the external signal amplification occurring at both stages is summed, the signal amplification factor can be as high as 10^5–10^6 in visual cells, when it passes along the cascade.

Receptors of Excitable Cells. Upon generation of excitation, several types of different receptors are activated in a cell, and each cell is sensitive to several types of SM (neuromediators, hormones, prostaglandins, growth factors). Note that CS transmission is not always coupled to inclusion of G proteins, because many receptors have properties of an independent channel. The most characteristic example of reception of a fast cell response (several milliseconds) is the acetylcholine receptor (ACR) (Fig. 23.2).

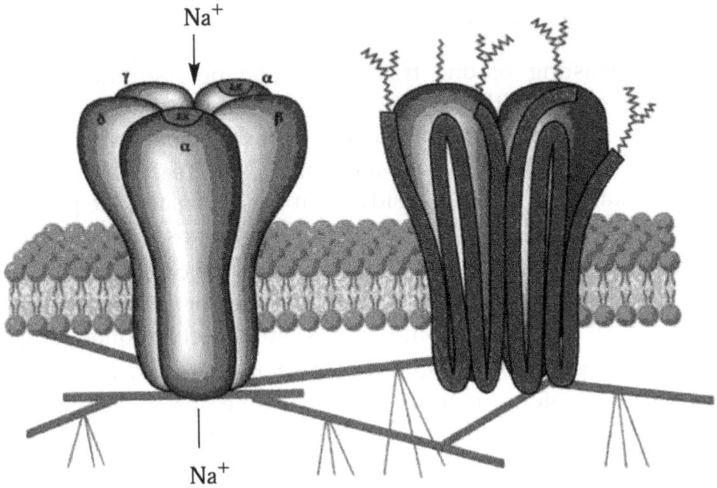

Figure 23.2. Structure of a nicotine cholinergic receptor forming an ion channel.

Upon acetylcholine binding, the choline receptor of a postsynaptic membrane changes its conformation so that Na and Ca ions pass inside the cell through the mouth formed by the subunits. Membrane depolarization results in the release of K^+ ions from the cell. The flux of K^+ ions brings the membrane potential back to its initial value. As a rule, excitation of the axon causes the release of a large amount of acetylcholine into the synaptic slot; instantaneously the neuromediator molecules saturate all receptors and cause intensive inflow of Na^+ ions into the cell (action potential generation). However, hormonal regulation is realized further in a nervous fiber via the secretion of SM from the Schwann cell (SC) into the axon.

23.2 Biophysics of Cellular Signaling

Peculiarities of Intramembrane Reception. Receptors, G proteins and effector enzymes are present in the plasmatic membrane as individual and frequently independent integral proteins. Therefore, their interaction is related to the characteristics and velocity of their movements that is affected considerably by the membrane structure, particularly by its lipid composition. Lateral diffusion of receptors (LDR) in the cell plasmatic membrane can be detected if a fluorescent compound is linked to the receptor. A laser light pulse induces photooxidation and "bleaching" of the fluorophore (the loss of its ability to fluoresce) at a small cell region. After that, fluorescence of the bleached region becomes removed due to diffusion into this region of "undamaged" receptors is recovered. The parameters of the fluorescence recovery allow calculation of LDR coefficients. For a two-dimensional case (motions in the membrane plane), $D = (x^2)/4t$, while for a three-dimensional case, $-D = (x^2)/6t$. The D values for receptors obtained by the method of recovered fluorescence after photobleaching (RFAP, see Section 3) are about 10^{-10} cm^2/s and less. Since no complete recovery of fluorescence of labeled proteins in the "bleached" regions takes place, the method makes it possible to determine fractions of mobile and "immobile" proteins.

Movements in a medium with hindrances are described using the percolation (leakage) theory. Lateral diffusion of receptors in membranes, frequently having no expressed discrete character, are considered as continuum percolation. In the percolation theory, the cluster (the oligomer complex) in a biological membrane is considered as an uninterrupted pathway for the diffusion of membrane proteins. In such a case, when the part of the space, accessible for protein movement is decreased below the threshold, the cluster is divided into smaller disconnected clusters, and as a consequence, diffusion over large distances becomes impossible. Diffusion coefficients and the mobile lipid fraction in membranes with changed composition of lipids and immobilized oligomer proteins were determined experimentally. It was found that as the area accessible for the movements decreased, both the coefficients of protein lateral diffusion and the portion of their mobile fraction also decreased. As known proteins are able of lateral diffusion only in membrane regions occupied by fluid (liquid-crystalline) lipids, and consequently, the percolation cluster should consist of fluid lipids.

Role of Oligomer Protein and Lipid Complexes in Cell Signal Transformation. As noted above, in addition to the direct interaction of a single SM molecule with a membrane protein, there is also another mechanism of cell signaling: the interaction of SM with a complex of cluster proteins (CP). CP are regarded as the main protein structure in membranes.

Peculiarities of CS Regulation by Intracellular Mediators. As mentioned above, after CS transformation in the plasmatic membrane — in the case of receptors coupled to G proteins — concentrations of intracellular mediators, like cAMP or Ca^{2+}, affecting the activity of protein kinases (PK) in the cytoplasm, change. As a result, cascade signal amplification takes place: the number of molecules increases at each stage, so that the total number of intracellular mediator molecules considerably exceeds the initial number of SM bound to receptors, while the number of enzyme molecules, phosphorylated by PK, by orders of magnitude exceeds the number of mediator molecules. Such a system has peculiar dynamic behavior described by a number of mathematical models. For example, Ca^{2+} ions are characterized by regular oscillatory changes upon cell signaling. Oscillations of the intracellular ion concentration with periods from < 1 min to 30 min are observed in cells. Under the effect of SM and R with G proteins, phosphatidylinosite-specific phospholipase C, hydrolyzing phosphatidylinosite-4,5-biphosphate with the formation of inosite-1,4,5-triphosphate (ITP) and diacylglycerol (DAG), is activated. ITP affects the intracellular Ca^{2+} pool via its receptors (Fig. 23.3). Thus, ITP regulates the release of Ca ions into the cytosol, and, in their turn, they trigger oscillation cycles of Ca^{2+} discharge from another Ca^{2+} pool not sensitive to IPT.

Variables Z (concentration of free Ca^{2+} in the cytosol) and Y (concentration in the IP_3-insensitive pool) are considered in the above model. The kinetics of changes in these concentrations can be written as follows:

$$\begin{aligned} \frac{dZ}{dt} &= v_0 + v_1\beta - v_2 + v_3 + k_f Y - kZ, \\ \frac{dY}{dt} &= v_2 - v_3 - k_f Y, \end{aligned} \tag{23.1}$$

where v_0 and kZ correspond to the rates of Ca^{2+} inflow and release from a cell without hormonal stimulation; v_2 is the rte of ATP-dependent Ca^{2+} pumping from the cytosol

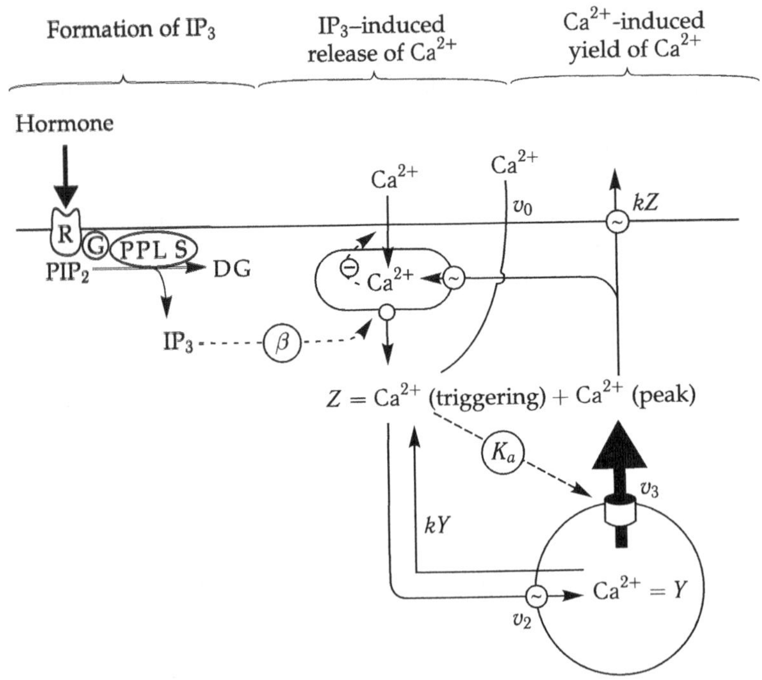

Figure 23.3. Diagram of regulation of intracellular concentration of Ca^{2+} ions.
PPL S is phosphatidylinosite-specific phospholipase S; IP_3 is inosite-1,4,5-triphosphate; PIP_2 is phosphatidylinosite-4,5-biphosphate; DG is diacylglycerol.

into the IP_3-insensitive pool; v_3 is the rate of Ca^{2+} release from this pool into the cytosol; $k_f Y$ is the rate of passive leakage from Y into Z; kZ is the rate of cytosol Ca^{2+} release from the cell. Pumping of Ca^{2+} into the IP_3-insensitive pool and Ca^{2+} release from it into the cytosol are of cooperative nature, and cytosol Ca^{2+} itself activates (autocatalysis) the release of Ca^{2+} from the IP_3-insensitive pool.

Taking this into account, the expressions for rte v_2 of Ca^{2+} access and v_3 of its release from the IP_3-insensitive pool into the cytosol will be the following

$$v_2 = v_{M2} \frac{Z^n}{K_2^n + Z^n},$$
$$v_3 = v_{M3} \frac{Y^m}{K_R^m + R^m} \cdot \frac{Z^p}{K_A^p + Z^p}. \tag{23.2}$$

In these equations, v_{M2} and v_{M3} are maximal rates of Ca^{2+} pumping and leakage, respectively. These processs are described by the Hill equations, and cooperativity coefficients are n and m; p is the degree of cooperativity of activation, and K_2, K_R, and K_A are threshold constants for pumping, leakage and activation. Equations (23.1) have the only steady state solution, though not always stable. As known (see Chapter 2), when non-equilibrium stationary state becomes unstable, continuous oscillations may emerge following the limit cycle trajectory. In the considered system, they appear at $29.1\,\% < \beta < 77.5\,\%$ provided other parameters v_0, k, k_f, v_1, K_2, K_R, v_{M2}, v_{M3} have physiological values. Beyond this range of parameters, the stationary state is stable where oscillations do not take place. A further theoretical analysis demonstrated that the efficiency of phosphorylation of substrates by Ca^{2+}-dependent protein kinase is related to oscillations of the Ca^{2+} ion concentration and increases with the increase in their frequency.

Figure 23.4 shows simulated oscillations of the cytosol ring of Ca^{2+} upon varying the model parameter β to 30%. The change in the shape of peaks is a consequence of the additional inhibition of agonists affecting the processes of Ca^{2+} entering and release. There is also another scheme is used to describe cell signaling. In this scheme, processes regulating intracellular oscillations of Ca^{2+} (the endoplasmatic reticulum (ER), mitochondria, Ca^{2+}-binding proteins in the cytoplasm) are more detailed (Fig. 23.5).

In this case, a cell was considered as a closed system not interacting with other cells and the intercellular space. The dynamics of changes in Ca^{2+} was described by a system of three differential equations (for cytosol Ca^{2+}, mitochondrial Ca^{2+} and Ca^{2+} in ER) and two conservation laws concerning the total amount of Ca^{2+} in the cell and the total number of Ca^{2+}-binding proteins in the cytosol. Since Ca^{2+} flows through the plasmatic membrane are included into the model, the cell is not a closed system any more, and the total amount of Ca^{2+} in the cell is not constant. Therefore, an additional equation, describing the dynamics of Ca^{2+} binding to buffer proteins in the cytosol, is added to the system. Simulation allows revealing in more detail the functional role of cytoplasmatic Ca^{2+} oscillations. The latter provide additional coding of information, communicated to the cell, not only by the amplitude, but also by the frequency and shape of oscillations of intracellular SM concentrations.

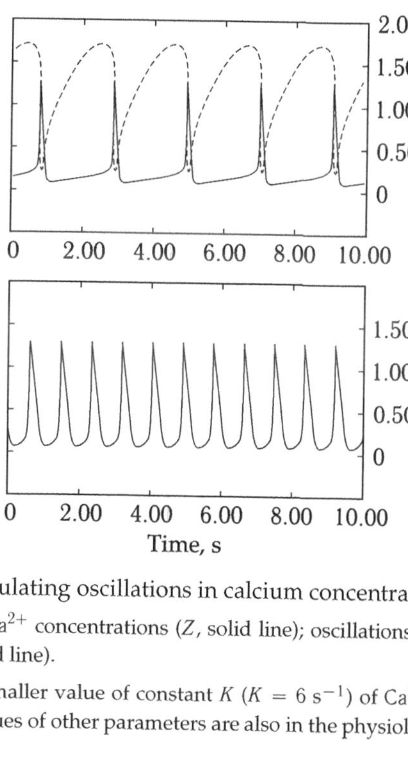

Figure 23.4. Results of simulating oscillations in calcium concentrations in a cell.

Top: Oscillations in cytosol Ca^{2+} concentrations (Z, solid line); oscillations in Ca^{2+} concentrations in the IP_3-insensitve pool (Y, dashed line).

Bottom: Oscillations Z at a smaller value of constant K ($K = 6\ s^{-1}$) of Ca^{2+} release from the cell (in the top part $K = 10\ s^{-1}$). The values of other parameters are also in the physiological range (reproduced from A. Goldbeter et al., 1990).

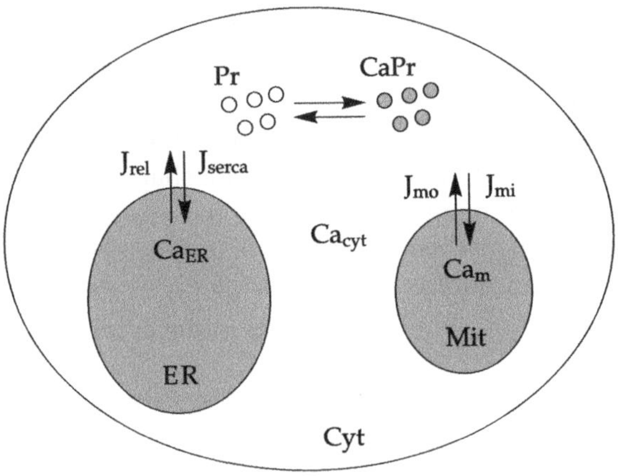

Figure 23.5. Diagram of Ca^{2+} flows in a single-point model.

ER, the endoplasmic reticulum; Mit, mitochondria; Pr, Ca^{2+}-binding proteins in the cytoplasm; J_{rel}/J_{serca}, ion fluxes from the endoplasmic reticulum and into the endoplasmic reticulum; J_{mo}/J_{mi}, ion fluxes from mitochondria and into mitochondria; Ca_{cyt}, cytoplasmic calcium.

23.3 Methods for Studying Cell Signaling

At present, the methods of choice for studying intercellular signaling are microscopic methods (including different versions of electron, optical and probe microscopy), as well as a number of other methods (for example, Raman spectroscopy of conformational scattering). These methods allow not only evaluating the external form and morphology of objects, but also the functional state of both the whole cell and its separate organelles in particular. Below we will consider only some methods, allowing visualization and evaluation of not only morphological state of tissues and cells but also characterization of membranes, in particular their viscosity affecting transportation and diffusion of membrane proteins. These methods provide studies of membrane protein diffusion and mapping their localization in the native cell without distortion of its functional state.

Scanning Probe Microscopy (SPM). SPM is a set of methods based on different principles to study surfaces. The main peculiarity of scanning probe microscopy is the use of a microscopic probe, which moves over the chosen region of the studied sample (the process of scanning) and interacts with each surface point of the region. The result of the sample scanning is a topographic image of the studied region where the value of coordinate z depends on the interaction of the probe with the surface in each point of the sample. Moreover, it becomes possible to evaluate the change in the magnitude of the probe interaction in each point of the surface within time, i.e. evaluate the dynamics of changes in the recorded parameter in the given point. The principal scheme of a scanning probe microscope is given in Fig. 23.6.

A wide variety of SPM includes the following: scanning tunneling microscopy (STM) based on the evaluation of the magnitude of the tunnel electric current between the probe and the sample; atomic force microscopy (AFM) most widely used in biology for measuring the attractive and repulsive forces of macromolecules; electrochemical microscopy (ECM) for evaluation of the electric current between the probe and the point on the surface; magnetic force microscopy (MFM) for evaluation

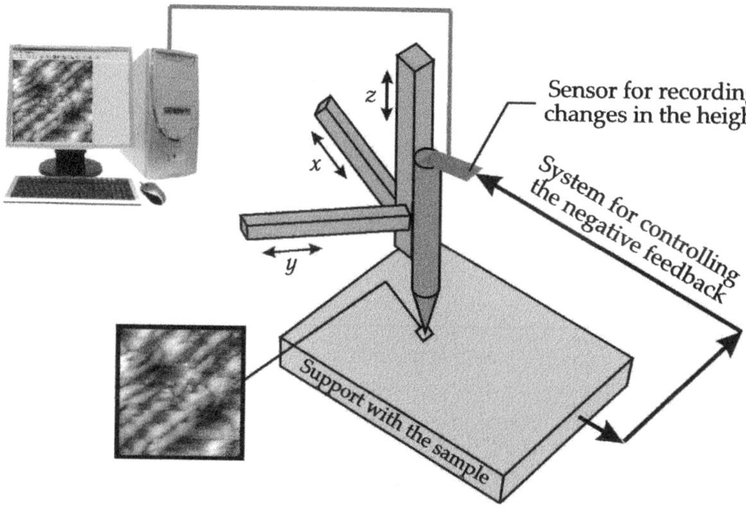

Figure 23.6. Principal scheme of a scanning probe microscope.

A scanning tunneling microscope image of horseradish peroxidase layered on a support of highly oriented pirografene is given as an example. The size of the side of the square is 30 nm.

of the interaction between the probe and the surface; and near field microscopy (NFM) that is a combination of optical and probe microscopy where the use of an optical probe allows a resolution of about several tens of nanometers.

Scanning Tunneling Microscopy (STM). A thin metal needle serves as a probe in STM. It is fitted on an electromechanical drive. When the drive is brought to contact the sample surface at a distance less than 10 E, the tunnel current is running in the gap between the needle point and the sample. As known, upon a change in the probe-to-sample distance by one angstrom the magnitude of the tunneling current changes by an order of magnitude. This makes it possible to determine with high accuracy the distance between the probe and the surface. To bring the probe needle point to the sample and to scan the surface, a piezoengine of piezoelectric ceramics (the material changing its dimensions under the effect of the control voltage) is used that allows high-accuracy probe movements thus reaching atomic resolution. It has become possible to use STM to obtain images and volt-ampere characteristics of different macromolecules, proteins (for example, enzymes and channels) as well as cell membranes.

Atomic Force Microscopy (AFM) allows obtaining images of objects from tens of nanometers to tens of microns. AFM is successfully used for obtaining images and evaluating properties of different biological objects: crystals of amino acids, proteins, cell membranes, DNA molecules and other macromolecules, biopolymer films, viruses, different cells, tissues etc. An elastic macroscopic console (cantilever) with a sharp needle is used as a probe. As a result of nano- and microscopic interactions of the needle with the sample, the cantilever can be bent by a rather large value which is measured. The deviation of the cantilever is proportional to the interaction force of atoms (van der Waals forces) of the probe and the sample, as well as to the effect of electric and magnetic fields, friction etc. At small distances (of about 1 E), repulsive forces emerge between the probe needle point (in an ideal case, it is one

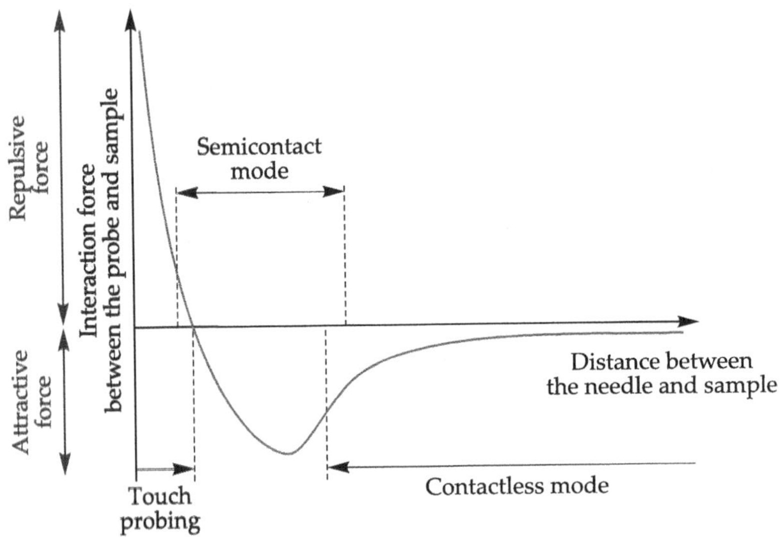

Figure 23.7. Typical dependence of the interaction force on the distance between the probe needle point and the sample surface. Indicated are portions of the curve which correspond to different modes of performance of an atomic force microscope.

atom) and the sample atoms, while attractive forces appear at large distances. Several sets of techniques are available for AFM measurements: touch probing, semicontact probes (with intermittent contact) and contactless (Fig. 23.7).

Upon the contact mode of the microscope operation, the repulsive force between the sample and the probe needle point is controlled, and deviations of the cantilever reflect the surface relief and also the resistance (if the object is conductive) or the distribution of friction forces over the sample surface. When the semicontact mode is used, the interaction of the probe with the sample surface is not constant, and it is only for a short period that the probe "senses" the forces repulsing it. In the contactless mode, attractive forces make the main contribution to the magnitude of the probe and sample interaction. As compared to the contact mode, semicontact and contactless modes have certain advantages because the cantilever pressure on the sample surface is much smaller or even absent that allows working with an intact biological object.

Near Field Microscopy, or Near-Field Scanning Optical Microscopy (NFSOM). This type of microscopy is a combination of scanning probe and traditional optical microscopy. Lateral resolution of classical optical microscopes is restricted by the Abbe diffraction limit which can be described by the relation

$$\Delta x = \frac{0.61\lambda}{NA}, \tag{23.3}$$

where λ is the wavelength of the light source, and NA is the digital aperture.

However, the use of NFSOM makes it possible to overcome this restriction. A NFSOM microscope is a scanning probe microscope where a pointed optical fiber coated with a metal film on each side is used as a probe. The diameter of the aperture at the probe end is so small and the aperture is located so close to the surface, that diffraction does not affect the resolution. In this case the lateral resolution

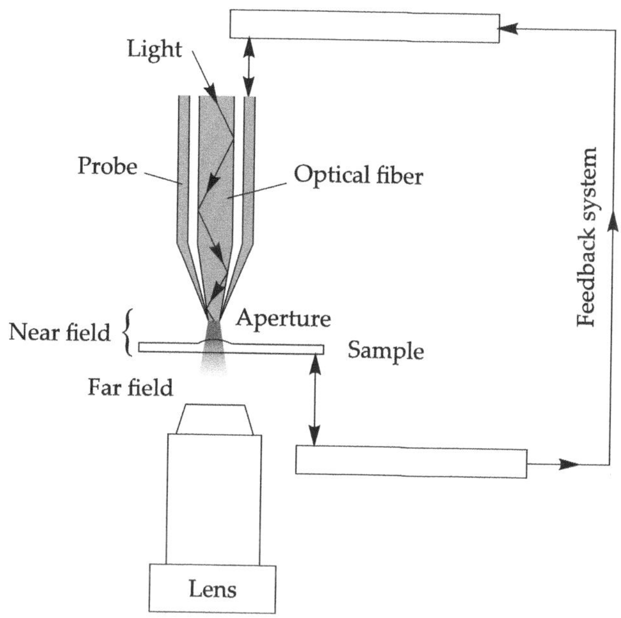

Figure 23.8. Principal scheme of a scanning NFSOM microscope.

of the method depends on the aperture diameter, which can be less than 50 nm that is far smaller than the wavelength of the light source. The probe is located in the direct vicinity of the studied sample surface (5–10 nm), thus providing sample excitation if the size of the light spot on the surface is much smaller than the wavelength. In this case, the distance of the evaluated point on the sample surface from the light source (the probe tip) is much smaller than the wavelength in the so-called "near field" area. The distance between the probe and the sample is maintained due to the negative feedback system (Fig. 23.8). Diffuse light from the area of the probe tip and the sample surface gets into the microscope objective. Recording can be performed either in transmitted light (the objective is underneath, see Fig. 23.8), or in reflected light, then it is scattered at an angle and the recording objective is located on the top. The measuring accuracy of the objective depends on the aperture size: at high values this may make tens or hundreds of nm, but usually are 10–50 nm. Therefore, the method is commonly used for surface mapping.

Interference Microscopy (IM) makes it possible to record changes in the sample phase that allows high contrasting. An image obtained by the IM method is conditioned by combination of two coherent waves moving along different optical pathways. The condition of coherency is achieved by splitting the beam into two ones. One of the obtained beams is directed through the observed object. The other beam is a reference one and does not pass through the object. When the reflected and reference beams are superimposed, an interference image (phase image of the object) is formed, and it is recorded by the analyzer. Upon phase imaging, the signal is normalized along the wavelengths and the value of the difference in the optical pathways of the two beams, or phase height Φ, is determined:

$$\Phi = \frac{(\varphi_0 - \varphi_{obj})}{2\pi}\frac{\lambda}{2}, \tag{23.4}$$

where φ_0 is the initial phase (rad); φ_{obj} is the phase shift by the object (rad); λ is the wavelength of the source (nm).

Generally, at each point of the object, the height (Φ) is connected with its geometrical parameters (thickness) (z):

$$\Phi(x, y) = \int_0^{z_{\max}} (n(x, y, z) - n_1)\, dz,$$

where n_1 is the refractive index of the buffer solution, its value is constant; $n(x, y, z)$ is the value of the refractive index at the cell point with coordinates x, y and height z.

The phase portrait (PP) of the cell is the distribution of the phase height over different regions of the object. The phase height value at each point (i) represents a sum of heights of different optical media (z) where n is refractive indices of the corresponding medium:

$$\Phi_i = \left(n_1^i z_1^i + n_2^i z_2^i + \ldots + n_n^i z_n^i\right) - n_1 z.$$

Thus, in addition to the enhanced contrast of biological objects, interference microscopy makes it possible to evaluate quantitatively changes in the phase height. Therewith the phase height depends both on the refractive index of the object and on its thickness.

The method allows observation of transparent and colorless objects without the use of stains. The main difference of IM from the method of phase contrast is the possibility to measure quantitatively with a high accuracy (to 1/2000) the optical path differences, introduced by microobjects. IM makes it possible to evaluate not only the refractive index and dimensions of the object, but also the total mass and concentration of dry substance in the object (for example, in a plant or animal cell), because the refractive index depends linearly on its concentration in the volume. The method of interference contrast is often used in combination with other microscopy methods, in particular, observation in polarized light. Its use together with ultraviolet microscopy allows, for example, determination of the content of nucleic acids in the total dry mass of the object. Interference microscopy includes also methods which use microinterferometers. Examples of erythrocyte images measured by the method of interference microscopy are given in Fig. 23.9.

Fluorescence Correlation Spectroscopy (FCS) is based on the analysis of fluorescence induced by laser emission. The FCS method has a high spatial resolution and high sensitivity. This is explained by a small number of fluorescent molecules required for measurements and by that the laser irradiated volume can be reduced to minimally allowable diffraction limit (23.3). Present-day objectives provide light focusing into a spot with transverse (radial) dimensions of about fractions of a micrometer. Moreover, to decrease the studied volume, a confocal principle of signal recording is used analogous to the recording used in a confocal microscope (see below Fig. 23.12). As a rule, this is achieved by means of a diaphragm mounted in front of the photodetector and cutting off parallel or weakly focused beams. This allows researchers to record the change in the state of even one molecule. The value of the solution volume whose signal, induced by the laser beam, is recorded (effective focal volume, V_{eff}) is determined by the formula

$$V_{eff} = \pi^{\frac{3}{2}} r_0^2 z_0, \tag{23.5}$$

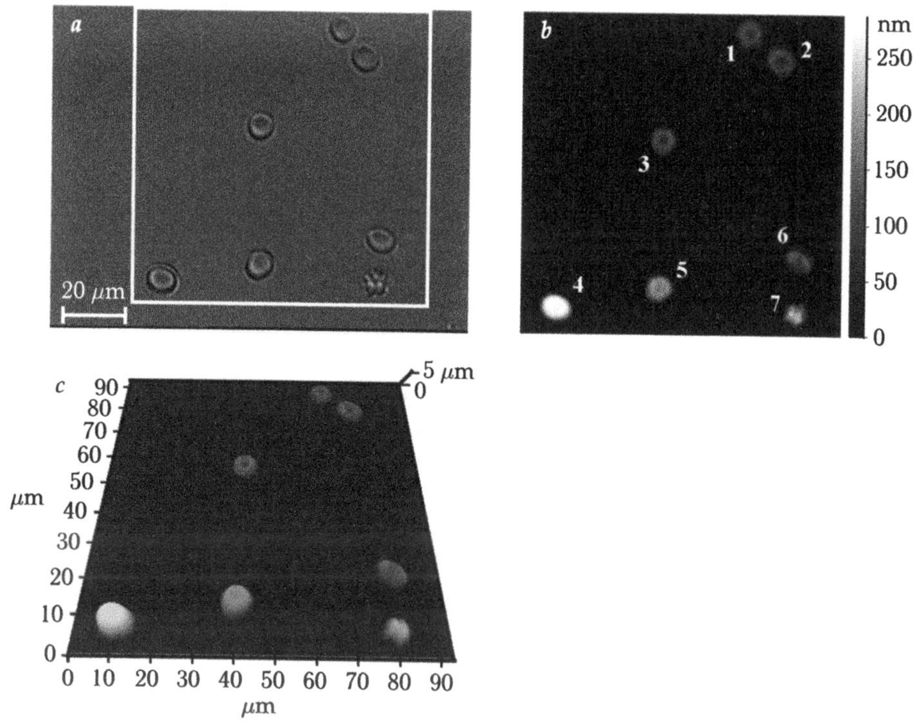

Figure 23.9. Human erythrocytes in blood plasma measured by the method of laser interference microscopy.

The figure shows images of different forms of erythrocytes obtained by the traditional method (*a*) and phase images of erythrocytes (*b* and *c*). The following forms of erythrocytes are shown: 1–3, discocytes; 4, an aggregate of three cells; 5, an aggregate of two cells; 6, stomatocyte; 7, echinocyte. The figure gives the phase height of cells (in nm) (*b*), the three-dimensional image of an object, the phase height of which is recalculated into the real thickness of objects with the use of the refractive index of erythrocytes — 1405 (*c*).

where r_0 and z_0 are lateral and axial radii of the studied object depending on lateral (determined by expression (23.3)) and vertical resolutions. Vertical resolution depends on the digital aperture of the used objective and is determined by the formula as $2.33n/NA$. Usually, the magnitude of vertical resolution makes a value from 2 to 6 times higher than the lateral resolution.

The principle of the FCS is based on recording the correlation function (Chapter 11) of fluorescence fluctuations in the laser-illuminated volume (Fig. 23.10). Fluctuations in the intensity of the recorded signal are caused by fluctuations in the time-dependent number or quantum yield of fluorescent molecules in the studied volume.

A change in the concentration of fluorophores is observed in the volume illuminated by the light beam (illuminated volume). A change in the concentration is a stochastic process conditioned by chaotic motions of fluorophores in solution. A change in the concentration of fluorophores in the illuminated volume can be described by the following correlation function:

$$C\left(\vec{r},t\right) = \langle C\rangle + \delta C\left(\vec{r},t\right), \tag{23.6}$$

where $\langle C\rangle$ is the average concentration, and expression $\delta C\left(\vec{r},t\right)$ is determined by stochastic fluctuations in time conditioned by substance diffusion. By substituting

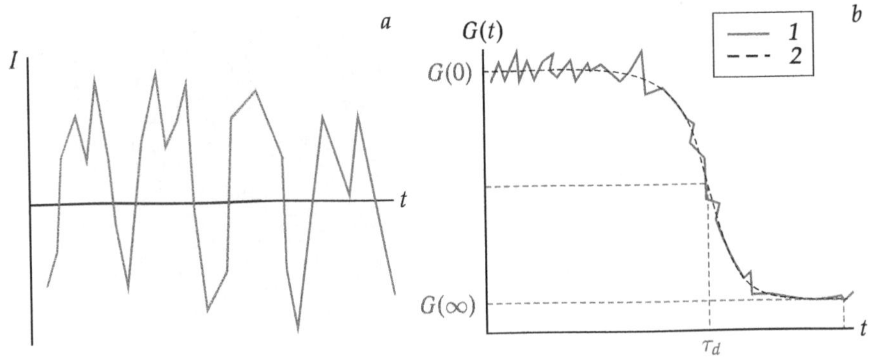

Figure 23.10. Example of a correlation function used in the method of FCS.
a, The change in fluorescence intensity in the studied volume; *b*, the correlation function for this change in the intensity. *1*, Experimental value of the correlation function; *2*, approximation of the linear function. Intersections of dashed lines with axes t and $G(t)$ indicate experimentally determined values τ_d, $G(0)$ and $G(\infty)$, respectively.

Expression (23.6) in the formula describing the concentration change due to diffusion, it is possible to obtain the following expression:

$$\frac{\partial \delta C\left(\vec{r},t\right)}{\partial t} = DV^2 \delta C\left(\vec{r},t\right). \tag{23.7}$$

The number of photons, determining the fluorescence intensity in the volume with time at the time period Δt, depends not only on the fluorophore concentration but also on the magnitude of the quantum yield and the distribution of the illumination profile in the studied volume:

$$n\left(t\right) = \Delta t \int d^3\vec{r}\,\rho\left(r\right) qC\left(\vec{r},t\right), \tag{23.8}$$

where q is the quantum yield coefficient; ρ is the illumination profile of the studied volume. The latter is described in the simplest case by the Gaussian distribution near the point:

$$\rho\left(x,y,z\right) = \rho_0 e^{-2\left(x^2+y^2+z^2/\left(\frac{z_0}{r_0}\right)\right)/r_0^2}. \tag{23.9}$$

If function $n(t)$ fluctuates near some average value $\langle n \rangle$, expression (23.9) can be rewritten as

$$\langle n \rangle = \Delta t q^2 \langle C \rangle \int d^3\vec{r}\,\rho\left(r\right)$$

or by substituting expression (23.5) for the effective focal volume:

$$\langle n \rangle = \Delta t q^2 \langle C \rangle \rho_0 \pi^{\frac{3}{2}} r_0^2 z_0. \tag{23.10}$$

It follows from the above that fluctuations in the number of photons near $\langle n \rangle$ will be determined by the expression:

$$\delta n\left(t\right) = n\left(t\right) - \langle n \rangle = \Delta t \int d^3\vec{r}\,\rho\left(r\right) q\delta C\left(\vec{r},t\right). \tag{23.11}$$

To analyze such processes, correlation functions are used (see Chapter 11). Note that determination and calculation of correlation functions upon evaluation of the fluctuation intensity for the FCS method remind in many aspects the situation with fluctuations in membrane conductivity described in Section 8 of Chapter 18. The correlation function $G(t)$ describing the change in the fluorescence intensity can be represented by the following expression:

$$G(t) = \frac{1}{\langle n \rangle^2} \sum_{i=0}^{M} \frac{\delta n(t)\, \delta n(t+\tau)}{M}, \tag{23.12}$$

where M is the number of time intervals Δt.

If for simplicity we accept that the initial instant of time is equal to 0 and $t = \tau$, expression (23.12) can be written as

$$G(t) = \frac{1}{\langle n \rangle^2} \langle \delta n(0)\, \delta n(t) \rangle \tag{23.13}$$

or by substituting in Expression (23.11):

$$G(t) = \frac{\Delta t}{\langle n \rangle^2} \iint d^3\vec{r} d^3\vec{r'} \rho(r) \rho(r') q^2 \langle \delta C(\vec{r}, 0)\, \delta C(\vec{r}, \tau) \rangle. \tag{23.14}$$

A standard way to solve similar equations is to use the Fourier transformation. In the simplest case, the following correlation function is used for description of 3D-diffusion:

$$G(t) = G(0) \frac{1}{\left(1 + \frac{t}{\tau_D}\right) \sqrt{1 + \frac{z_0}{r_0} \left(\frac{t}{\tau_D}\right)}} + G(\infty). \tag{23.15}$$

$G(0)$ and $G(\infty)$ are the function values at the instant of time equal to 0 and at infinity, τ_D is the characteristic time of the function decay. Parameters $G(0)$, $G(\infty)$ and τ_D are determined experimentally by analyzing this function (see Fig. 23.10, *b*). Using parameter $G(0)$ it is possible to evaluate the mean value $\langle N_p \rangle$ and the mean concentration $\langle C \rangle$ of fluorescent particles in the studied volume:

$$G(0) = \frac{1}{\langle N_p \rangle} = \frac{1}{V_{eff} \langle C \rangle}. \tag{23.16}$$

The diffusion coefficient of particle (D) can be determined using the τ_D value:

$$D = \frac{r_0^2}{4\tau_D}. \tag{23.17}$$

In addition, the analysis of parameters and the form of the correlation function makes it possible to determine the hydrodynamic radius of particles (if their form is known), velocities of intramolecular conformational transitions and also kinetic constants of reactions for these molecules.

As a rule, FCS is used in combination with different methods of optical microscopy, particularly confocal and two-photon microscopy. The method can be used for working not only with fluorescent objects, but also for evaluation of the state of luminescent objects of different types (for example, for evaluation of luminescence of gold particles, or quantum dots, or different phosphorescent objects). So, specifically the FRET method (see below) is also referred as a variation of FCS.

FCS has been actively used in different fields of physics, chemistry and biology for more than three decades. At present, the method is actively used in biological research. This is due connected both to its high sensitivity that allows working with nano- and subnanomolecular concentrations of studied objects in volumes, hundreds of times smaller than the cell volume, and to the possibility to determine mobility of fluorescent particles in a wide range from microseconds to seconds (see Fig. 23.11). Moreover, the FCS method is characterized by fast measuring time: only tens of seconds are needed to estimate changes in fluorescence and their nature in the studied volume. However, it should be noted that the use of the FCS method for studying living cells is limited because of the high total substance concentration and rather complex internal structure of the cell.

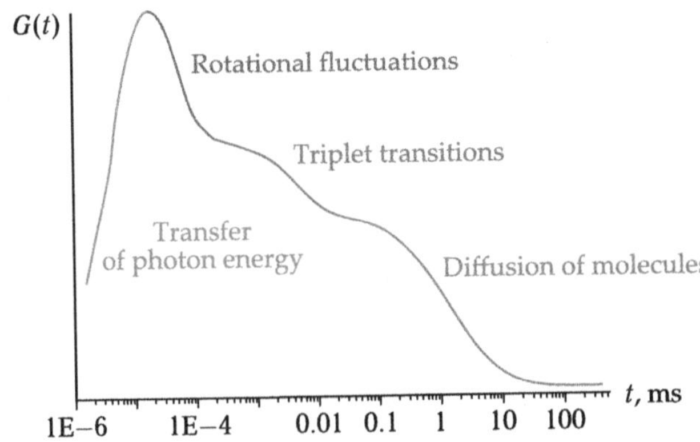

Figure 23.11. The use of the correlation function for evaluation of different states.

Another peculiarity of this method is that signals from immobile molecules, for example, those connected with internal structures of the cell or the cell surface, do not contribute to the correlation function and therefore are not considered.

FCS is used in immunology for highly specific detection of the antigen–receptor complex at very low concentrations, for studying the ligand–receptor interaction, conformational rearrangements of DNA, spontaneous protonation–deprotonation of the fluorescence center in proteins possessing green fluorescence (for example, GFP), for analysis of DNA restriction at nanomolecular concentration (of about 1 μl) in very small volumes and for other model and intracellular studies.

Two-photon Laser Microscopy (TLM). This type of fluorescence microscopy has made it possible to observe living tissues at the depth of more than one millimeter. TLM is a variant of a multiphoton fluorescence microscope. As compared to a confocal microscope, a two-photon fluorescence microscope has the following advantages: high penetrating capacity, low phototoxicity due to a much less amount of the stain used. A two-photon laser microscope is based on the following principle of two-photon excitation. Two low-energy photons excite the fluorophore (the molecule or its part capable of fluorescence) in one quantum act. The result of such excitation is that thereafter the excited molecules emit a fluorescent photon then recorded by the analyzer. The emitted fluorescence photon energy is higher than the energy of one exciting photon. The probability that the both exciting photons will be absorbed by one molecule is very low. Therefore an extremely intensive flux of exciting photons is needed. It can be obtained using a laser source emitting pulse radiation upon high frequency of pulse repetition (more than 80 MHz).

In a two-photon microscope, an infrared laser beam is focused by a convergent lens of the objective. Normally, a high-frequency (80 MHz) sapphire laser is used emitting pulses of 100-femtosecond length that provides a high photon flux density required for two-photon absorption. The most frequently used fluorophores have an excitation spectrum in the range from 400 to 500 nm, while the excitation wavelength is 700–1000 nm (the range of infrared waves). If a fluorophore absorbs two photons simultaneously, it will get enough energy to transfer to the excited state. Then the excited fluorophore will emit one photon (in the visible region of the spectrum) with wavelengths depending on the fluorophore type.

The maximal fluorescence is observed in the focal volume (where a laser beam is focused) and reduces sharply in the area outside the focus. The light emitted by the fluorescent specimen is enhanced by means of a highly sensitive photomultiplier. With a single-channel light detector, light in this focal volume generates one pixel of image. To obtain a two-dimensional pixel image, scanning over the specimen plane is performed.

Method of Confocal Laser Scanning Microscopy (CLSM). Confocal microscopy is a non-damaging method for obtaining two-dimensional optical sections of a specimen that allows investigations of the three-dimensional topology or internal structure of semitransparent objects. The specimen thickness, as a rule, can reach 100 μm, while the thickness of a microscopic section under corresponding conditions is less than 500 nm. Confocal microscopy allows investigations of the structure of objects of 5–200 μm in size at a resolution of about ± 0.2 μm.

The difference of a confocal microscope from a standard optical fluorescence microscope is that at each instant of time the image of one point of the object is recorded, and the image of the whole object is constructed afterwards by scanning. To record the light emitted only by one point, a small-size diaphragm is mounted behind the objective lens so that the light emitted by the analyzed point, containing the fluorescent substance (see Fig. 23.12), passes across the diaphragm and will be recorded, while the light coming from other points is blocked by the diaphragm. This requires fulfilling the conditions of confocality (coincidence of focuses that gives

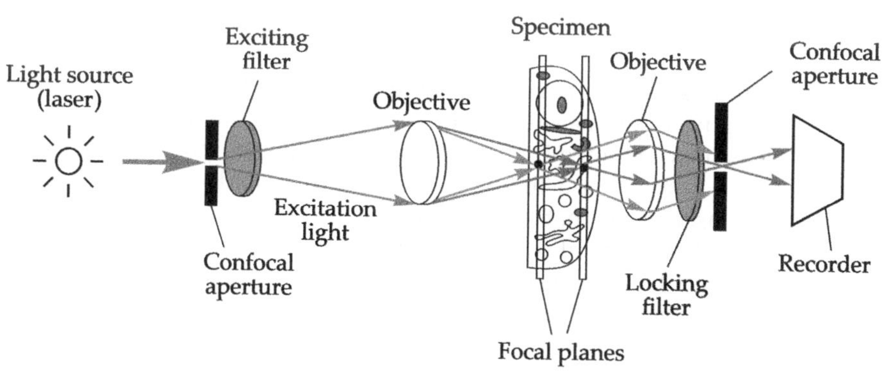

Figure 23.12. Principal design scheme of a confocal laser scanning microscope.

the name to the method). It consists in that the detector diaphragm projection on the object should just coincide with the focus of the light illuminating the object.

An essential point in recording is a conceivably smaller diameter of the detector diaphragm that can be provided using rather powerful light sources (lasers). A combination of these peculiarities makes it possible to increase remarkably the resolution and contrast of specimens, as compared to the usual light microscopy, due to the fact that "parasitic" light from the points, adjacent to the analyzed one, does not reach the detector, while the illumination of the recorded image point increases concurrently.

Fluorescence generated in the part of a specimen, on which the objective is focused, reaches the detector and is recorded. In usual light luminescence microscopy, fluorescence of the entire illuminated specimen part is detected, though the regions outside the focus are visible at low-contrast. The method of CLSM enables "layer-by-layer" inspection of living cells (similar to tomography) and quantitative evaluation of their fluorescence imaging. The objective table moves in the $x-y$ plane and also along the z axis. Each obtained image is fixed by the computer and after that the image of each separate vertical optical section is reconstructed. This procedure is impossible for usual optical light microscopy. Quantitative characteristics of an object with CLSM are obtained determining the fluorescence intensity by means of a photoelectronic multiplier.

Fluorescence Resonance Energy Transfer (FRET). The resonance energy transfer mechanism (Chapter 25) is a radiationless process occurring between two dye molecules, when the excited donor transfers its excitation energy to the nearest acceptor due to the dipole-dipole interaction. Energy transfer necessitates coincidence of vector directions of dipole moments of the donor and acceptor and partial overlapping of the donor fluorescence and the acceptor absorption spectra. In this case, the efficiency of energy migration is inversely proportional to the sixth power of the distance between the donor and acceptor. This method has become a method of choice upon studying both the dependence of excitation energy migration between molecules and the changes in the molecular configuration and molecular interactions. Combined with the method of fluorescence labeling (for example, GFP), this method is a sensitive tool for determining internal and external intermolecular distances of about 1–10 nm. The use of the method allows investigators to accomplish subcellular mapping of protein–protein interactions, visualize compartments and organelles where these interactions take place. However, the method has some disadvantages. For example, to obtain a reliable result, a large selection of cells (hundreds and thousands of cells obtained using flow cytophotometry) is required.

Method of Fluorescence Recovery after Photobleaching (FRAP). This method is used in cell biophysics for measuring the coefficient of lateral diffusion of proteins and lipids in plasmatic cell membranes. Fluorescence-labeled proteins or phospholipids are inserted in a cell and the fluorescence is recorded using a fluorescence microscope. The laser beam that excites fluorescence is focused on the plasmatic membrane site (of about several square microns). After that, the beam is attenuated to cause only the label fluorescence rather than photooxidation of the membrane components. At the beginning of this experiment, the excitation beam intensity is enhanced for a short period (tens of ms) so that photooxidation (photobleaching) of the fluorescent label is specially generated. Then the emission intensity value is recovered to the initial low level and the cell fluorescence intensity after

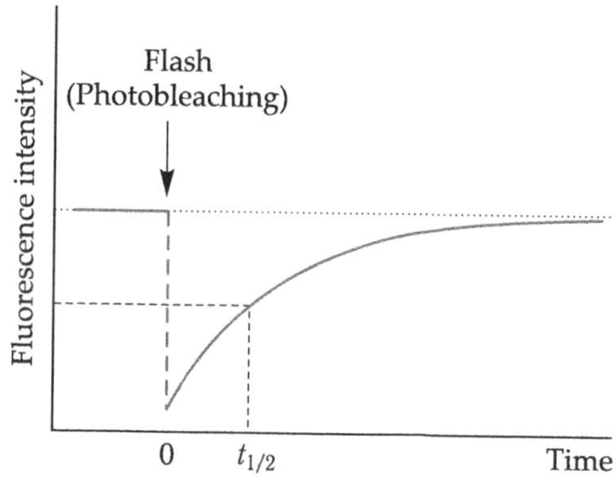

Figure 23.13. Diagram of changes in fluorescence intensity upon their measurements by the FRAP method.

photobleaching of the label is recorded. As the photooxidation product does not fluoresce, the recovery of the initial fluorescence level takes place due to lateral diffusion of labeled protein or lipid molecules from neighboring cell regions. A schematic change in the fluorescence intensity in experiments with the use of the FRAP method is shown in Fig. 23.13. Parameters of fluorescence recovery enable the estimation of coefficients of lateral diffusion of receptors (see Section 2). Using the FRAP method, it is possible to evaluate the lateral diffusion coefficient, D_L, of labeled substances:

$$D_L = \frac{w^2}{4t_{1/2}} \gamma_D, \tag{23.18}$$

where $t_{1/2}$ is the time of half-recovery (the time when the fluorescence intensity value becomes equal to the half of initial fluorescence intensity prior to photobleaching), w is the radius of the bleached region, and γ_D is the estimated constant dependent on the beam profile and the degree of photobleaching.

Thus, recording of the recovery rate of the initial fluorescence level in the cell membrane region after photobleaching allows determination of diffusion coefficients of membrane proteins and lipids and quantitative characterization of the protein mobility in living cell membranes.

An important factor, determining the mobility of receptors and proteins interacting with them (and thereby transfer of the hormonal signal into the cell), is the membrane macrostructure. Membrane proteins form aggregates which can be retained by the cytoskeleton and submembrane structures, the latter including proteins fodrin, spectrin or TW260/240. Therefore, receptors move not in an open lipid "sea" of the membrane, but between the "islands" of immobile protein aggregates. The protein movements in the membrane are impeded also by the presence of lipids in the gel phase ("solid" lipids). Consequently, the velocity of motion will be affected considerably by the portion of the membrane area accessible to protein motion.

Diffusion in a medium with barriers can be described by the percolation theory — a part of theoretical physics. The main concepts of this theory can be illustrated on the example of lattice percolation. Imagine a lattice consisting of knots and links where each knot can exist with a certain probability. A combination of knots connected with each other by the nearest-neighbor distances forms a cluster. There is a critical concentration of knots (the percolation threshold) above which a percolation cluster penetrating the whole lattice is generated in the system for the first time.

Movements in the lattice are possible in the cluster from knot to knot following the links connecting them. Accordingly, the presence of a percolation cluster conditions the possibility of movements over the whole lattice.

The FRAP method was used by A. S. Sobolev (1994, 1995) to determine diffusion coefficients and the mobile fraction of proteins in native plasmatic membranes with changed portions of liquid lipids and immobilized protein conglomerates.

As known, proteins are able of lateral diffusion only in regions of the membrane occupied by liquid (liquid-crystalline) lipids; consequently, a percolation cluster should represent a cluster of liquid lipids. It was found that a decrease in the space for free movement caused a decrease in both the lateral diffusion coefficient of proteins and the portion of their mobile fraction. This change was compatible with the percolation theory predictions, i.e. had a threshold. Experimentally, determined critical indices were in good agreement with the theoretical ones.

The applicability of the percolation approach to the description of motions and interactions of membrane proteins, interacting with adenylate cyclase, made it possible to evaluate the dimensions of the membrane space occupied by these proteins, characterize peculiarities of hormonal effects at different portions of the membrane accessible for protein motion, evaluate the influence of the submembrane cytoskeleton development in the process of cell differentiation to the hormonal response.

Combined Scattering Spectroscopy (CS, Raman Spectroscopy). The combined scattering effect was discovered in 1928 by L. Mandelstam and in the same year by C. Raman and K. Krishnan. If monochromatic radiation of intensity I is falling onto a transparent liquid, a portion of the light is scattered and spreads in all directions. For atoms, molecules and their complexes with dimensions much less than the radiation wavelength, the magnitude of light scattering is not large but can be recorded. Light scattering is frequently accompanied by a change in the vibrational energy of the scattering particles. This type of scattering is called combined (Raman) scattering. In this case, in addition to the basic line of the incident light a spectrum of the scattered light has several new lines, their frequency being somewhat higher and lower than the incident light frequency. Thus, evaluating the intensity and relation of intensities of these lines to each other we can determine the change in the vibrational state of the molecules, and consequently evaluate the state of the medium (substance) with respect to its molecular and interatomic bonds.

The emergence of these lines can be explained using the phenomenon of substance polarization P that depends on the substance polarization factor (α) and the electric field intensity (A) of the incident light wave of frequency ν_0:

$$P = \alpha A_0 \cos(2\pi\nu_0 t). \tag{23.19}$$

In the substance, charges oscillate at frequency ν_m near the equilibrium position:

$$q = q_0 \cos(2\pi\nu_m t). \tag{23.20}$$

As a rule, the electric field amplitude A_0 is constant within the period of these variations. However, oscillations of charges change the polarization of the material with time that can be described by the Taylor series:

$$\alpha = \alpha_0 + \left(\frac{\partial\alpha}{\partial q}\right) q + O\left(q^2\right), \tag{23.21}$$

where $O(q^2)$ means the value "of about q^2". As the q^2 value is much smaller than q, it can be neglected together with subsequent terms in this series. Substitution of expressions (23.20) and (23.21) in (23.19) provides the following expression for the dependence of the substance polarization on the light wave:

$$P = \alpha_0 A_0 \cos(2\pi\nu_0 t) + \\ + \frac{1}{2} q_0 \left(\frac{\partial\alpha}{\partial q}\right) A_0 \left[\cos\{2\pi(\nu_0 - \nu_m) t\} + \cos\{2\pi(\nu_0 + \nu_m) t\}\right]. \tag{23.22}$$

The first term corresponds to the typical Rayleigh light scattering, while the second one describes the combined scattering. The second component contains frequencies of scattered light $(\nu_0 - \nu_m)$ and $(\nu_0 + \nu_m)$ differing from the Rayleigh frequencies and corresponding to vibrational transitions in the complex. The probability of combined scattering is dependent on the derivative value $\partial\alpha/\partial q$. If upon a vibrational transition the polarization does not change, $\partial\alpha/\partial q = 0$, the transition is not active. In three-dimensional space, the $\partial\alpha/\partial q$ value is determined by the polarization tensor. Vibrations are Raman-active if at least one component of the polarization tensor is changed. Therefore, Raman-scattered light has frequencies $(\nu_0 - \nu_m)$ and $(\nu_0 + \nu_m)$. The frequency of the first line is less than that of the incident light, and such lines are called Stokes lines. Meanwhile, the frequency of the second line is higher, and these lines are called anti-Stokes lines.

Intensities of anti-Stokes lines in Raman scattering are much lower than those of Stokes lines. Raman scattering spectra are generally described by wave numbers, the wave number of excitation light is accepted to be the origin of coordinates. Raman scattering lines are characterized by parameter $1\lambda_1 = -1\lambda_2$, whereas wavelengths λ_1 and λ_2 are expressed in centimeters. Stokes lines have a lower frequency than the excitation radiation, whereas anti-Stokes lines are positioned symmetrically and have a higher frequency.

For anti-Stokes scattering, a photon should obtaint an additional energy (due to thermal energy of scattering molecules). Therefore, anti-Stokes lines disappear completely at absolute zero (−273 °C), when it is impossible to obtain thermal energy.

In a quantum-mechanical description, the case in question is "transitions between vibrational levels appearing due to interactions with the radiation field". Because of changes in the substance caused by the light quantum, energies of the excitation and scattered photons are different and their difference corresponds to the energy of the considered vibration (E_{col}) in the particle. It can be written as

$$E_{col} = h\nu_e - h\nu_s, \tag{23.23}$$

where $h\nu_e$ is the energy of the excitation quantum, and $h\nu_s$ is the energy of the scattered quantum. The $(h\nu_e - h\nu_s)$ difference is called a "frequency shift" and corresponds to the frequency of vibrations of the studied bond. Upon recording CS spectra, the intensity of scattering radiation is measured at different wavelengths or frequencies. When plotting the spectra, CS intensities are plotted on the ordinate axis, while the calculated frequency shift is plotted on the abscissa axis; ν_e is the laser radiation frequency, and ν_s is all the frequencies at which the spectrum was recorded. Peaks in the CS spectra are the result of transitions between vibrational levels of the molecule. If a peak is associated with the change in one bond vibration, the frequency shift corresponding to the peak maximum is equal to the vibration frequency of this bond. But if the peak is caused by vibrations of a group of bonds that as a rule occurs in complex molecules, it is impossible to find an accurate correspondence between the frequency shift and vibration frequencies of the bonds. In this case, the change in parameters of the group as a whole can be characterized. In contrast to other optical methods, CS reveals even insignificant changes in the structures of studied substances. The analysis of the state of living cells with use of CS can yield information on conformational transitions of molecules occurring as a result of their direct functioning or under the influence of natural intracellular surrounding.

Revolutionary changes in Raman spectroscopy occurred at the beginning of the 1960s after the onset of lasers that made it possible to obtain Raman scattering spectra on small-volume samples. The following improvement of optical filters and the advent of more sensitive light sensors widened considerably the range of Raman scattering application in microscopy.

Nowadays Raman spectroscopy is commonly used in biology to study polymer biological macromolecules — polypeptides, polynucleotides and polysaccharides. The analysis of protein structures using Raman spectroscopy makes it possible to evaluate spectrum bands characteristic of amide A (vibrations of N–H-bonds), amide 1, amide 2 and amide 3. When nucleic acid structures are analyzed, studied are the bands of the Raman spectrum characteristic of water, sugars, nucleotides (vibrations of C=O-, C=N- and C=C-bonds), symmetrical and asymmetrical vibrations of PO_2 bonds, vibrations of phosphate PO-bonds and sugar CO-bonds, as well as skeletal, deformation vibrations of all bonds in polynucleotides. When structural changes in polysaccharides are analyzed using Raman spectroscopy, vibrations of OH- and CH-bonds in sugar, cellulose and chitin are evaluated. In cell biophysics, Raman spectroscopy is of choice for studying conformational changes in proteins and phospholipids in biological and model membranes. For example, investigations of vibrations of C–C- and C–H-bonds made it possible to reveal the dependence of conformational changes in hydrocarbon chains of fatty acids on the order of membrane phospholipids in cell membranes (see Fig. 23.14).

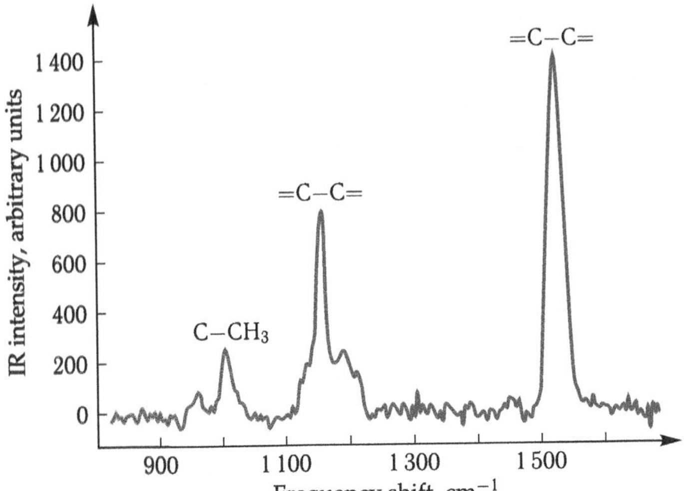

Figure 23.14. Typical CS spectrum of carotenoids obtained from a myelinated nerve. The size and positions of the spectral peaks characterize the state of the corresponding bonds in a molecule.

There are a variety of types of Raman spectroscopy. Thus, if the excitation radiation is in the absorption band of a molecule (bond), resonance Raman scattering of the substance is observed. In this case the recorded peaks can be higher than non-resonance peaks by 2–3 orders of magnitude. Resonance Raman scattering is caused by coherent vibrations of molecules excited by the electric field of a laser beam. Therefore, in a mixture containing many different types of molecules it is possible to evaluate the state of only one type of molecules that is of interest.

Another interesting type of Raman spectroscopy is surface-enhanced Raman spectroscopy (SERS). The method is based on the ability of some molecules to increase the intensity of Raman peaks by 10^5–10^6 times on specially prepared metal surfaces.

Raman microscopes allowing visualization of local changes in CS spectra in the cell volume have become tools of choice lately. The spatial resolution of such tools is about 1 μm, while the spectral resolution is about 1 cm^{-1}. It is frequent that such a microscope combines CS and confocal microscopes or CS and atomic force microscopes in one setup.

Further Reading

Part V

1) Damjanovich, S. (Ed.), *Biophysical Aspects of Transmembrane Signaling*, Springer-Verlag, Berlin, 2005.

Part VI

1) Alberts, B., Bray, D., Lewis, J., Raff, M., Roberts, K. and Watson, J.D., *The Molecular Biology of the Cell*, 4th edn., Garland Science, New York, 2002.
2) Hille, B., *Ionic Channels of Excitable Membranes*, Sinauer Associates, Sunderland, Mass, 1992.
3) Huguenard, J. and McCormik, D.A., *Electrophysiology of the Neuron*, Interactive Tutorial, Oxford, New York, 1994.
4) Jackson, M.B., *Molecular and Cell Biophysics*, Cambridge University Press, 2006.

Part VII

1) Bagshaw, C.R., *Muscle Contraction*, 2nd edn., Chapman & Hall, London, 1993.
2) Nicholls, D.G., *Bioenergetics. To the Chemiosmotic Theory*, Academic Press, London, New York, Paris, San Diego, Sao Paulo, Sydney, Tokyo, Toronto, 1982.
3) Racker, E., *Mechanisms in Bioenergetics*, Academic Press, New York, London, Toronto, Sydney, San Francisco, 2014.
4) Hill, A.V., *The Mechanism of Muscular Contraction*, Nobel Lecture, 12 December 1923.
5) Lehninger, A., Nelson, D. and Cox, M., *Principles of Biochemistry*, Worth Publishers, New York, 1997.
6) Ernster, L. (Ed.), *Molecular Mechanisms in Bioenergetics*, Elsevier, Amsterdam, 1992.
7) Damjanovich, S. (Ed.), *Biophysical Aspects of Transmembrane Singnaling*, Springer-Verlag, Berlin, 2005.

VIII
Electronic Properties of Biopolymers

24

Fundamentals of Quantum Description of Molecules

25

Mechanisms of Charge Transfer and Energy Migration in Biomolecular Structures

26

Mechanisms of Enzyme Catalysis

24

Fundamentals of Quantum Description of Molecules

Intramolecular mobility of biopolymers (proteins) described in previous chapters has a spontaneous character and is conditioned by thermal motion. However, a change in the electronic state of proteins induces definite conformational transitions in them. They achieve to achieve a new equilibrium state and energy minimum corresponding to the new electronic state of macromolecules.

In this chapter, we will shortly outline quantum mechanical concepts used in biophysics to describe electronic states of molecules and charge transfer processes. The obtained results will be used for model calculations of certain processes in biological systems.

24.1 Introduction

The main parameter, characterizing the state of any physical system, is energy that in physics is determined via the work of forces.

For a molecule as a system, consisting of a number of interconnected atoms and isolated from all possible external factors (a molecule in vacuum), the total energy is its internal energy. It is convenient to divide this energy in parts and represent it as the sum of energies of electronic (E_{el}), vibrational (E_v) and rotational motions (E_{rot}):

$$E = E_{el} + E_v + E_{rot}. \tag{24.1}$$

It is notable that this differentiation of energy is approximate, because generally speaking, there are interactions proceeding between these motions with energies designated as E_{el-v}, E_{el-rot}, E_{v-rot}. In the next chapter, we will consider the role of electron-vibrational interactions in the evolution of the state of a molecular system. The magnitudes of these energies are related as follows:

$$E_{el} \gg E_{v} \gg E_{rot}. \tag{24.2}$$

As known, the key dynamic equation in classical mechanics is the Newton equation.

In distinction to classical mechanics, quantum mechanics uses operators corresponding to dynamic variables rather than the variables themselves.

In quantum mechanics, the key equation describing the evolution of a physical system is the Schrodinger equation

$$i\hbar\frac{\partial\psi(q,t)}{\partial t} = \widehat{H}\psi(q,t), \tag{24.3}$$

where $\widehat{H}(\widehat{q},\widehat{p},t)$ is the operator of the system total energy, or the Hamiltonian; $\widehat{q}$, $\widehat{p}$, t are operators of generalized coordinates, generalized pulses and the time, respectively; $\psi(q,t)$ is the wave function describing the state of the system; and $\hbar = h/2\pi$ where h is the Planck constant.

Wave function (ψ), its type being determined from the solution of the Schrodinger equation for a certain system, contains complete information on the system. Generally, the wave function is a complex value.

The physical meaning of the wave function is disclosed in the interpretation proposed by M. Born: $|\psi(q,t)|^2 = \psi^*(q,t)\psi(q,t)$ is the probability density of the particle to be in the point with coordinate q. Here function $\psi^*(q,t)$ is a complex conjugate wave function. If the probability density $|\psi(q,t)|^2$ is multiplied by volume element dv, containing point q, probability dw is obtained for detecting the particle in this range of the space: $dw = |\psi(q,t)|^2\,dv$.

The physical content of operators can be explained with the use of the following determinations. Let operator $\widehat{F}$ correspond to some real (or observed) physical value F. By definition, the average value of operator $\widehat{F}$ of this value in state ψ is calculated according to the rule (the so-called quantum mechanical average)

$$\langle F\rangle = \int \psi^*\widehat{F}\psi\,dv, \tag{24.4}$$

where integration is done over the whole space. In order the average value to correspond to the physically observed value, it should be *real*, i.e. the next equality should be valid:

$$\langle F\rangle = \langle F\rangle^*.$$

Or else, in a general case, the equality should be effective:

$$\int \psi_2^*\widehat{F}\psi_1\,dv = \int \psi_1\widehat{F}^*\psi_2^*\,dv, \tag{24.5}$$

where wave functions ψ_1 and ψ_2 characterize different states corresponding to physical value F (or superposition of such functions). Integrals like those in

Equation (24.5), often called matrix elements, generally determine the probability of transition of a quantum system from state (ψ_1) to the other state (ψ_2) caused by some physical action presented by operator $\widehat{F}$ (see below).

Introduction of average values (Equation (24.4)) means that in reality we can observe deviations from average values. The magnitude of probable deviations is estimated, using the root-mean-square deviation well known in statistics. We can put the question whether such states of quantum systems, where the root-mean-square deviation of physical value F is zero, i.e. the condition $\langle \Delta F \rangle^2 = 0$ takes place. The answer to this question leads to the equation

$$\widehat{F}\psi_F = F\psi_F, \tag{24.6}$$

where functions ψ_F satisfying this equation are called eigenfunctions, and values of F are eigenvalues of this operator. The totality of the operator eigenvalues is its spectrum.

In the state described by eigenfunction ψ_F of operator F, physical quantity F has a definite value equal to the eigenvalue of this operator. The spectrum of operator eigenvalues indicates to possible results of measuring the F value.

In some cases, one eigenvalue of an operator has several corresponding linearly independent eigenfunctions. Such states of a system are called degenerated, and the number of wave functions corresponding to this eigenvalue is called the degeneracy multiplicity.

It should be reminded that the physical meaning of wave function ψ is that $|\psi|^2 \cdot dv$ determines the probability of a quantum system in state ψ to be in volume element dv of the space. The $|\psi|^2$ value is termed the probability density or the probability distribution function. As a system is necessarily in some point of the space, the integral of the distribution function over the total space should be equal to zero, i.e.

$$\int_{-\infty}^{\infty} |\psi|^2 \, dv = 1. \tag{24.7}$$

24.2 Stationary and Non-stationary States of Quantum Systems. Principle of Superposition of States

Stationary States of Quantum Systems. Let us consider solutions of the Schrodinger equation for the case when the potential energy of a quantum system does not explicitly depend on time. Write the complete (or temporal) Schrodinger equation (24.3):

$$i\hbar \frac{\partial \psi(r,t)}{\partial t} = \widehat{H}\psi(r,t), \tag{24.8}$$

where $\widehat{H}$ is the energy operator of a quantum system (the Hamiltonian); $\hbar = h/2\pi$ where h is the Planck constant; ψ is the wave function of the system which is the function of time t and spatial coordinates, their totality will be designated as r.

The total energy H of a system, called the Hamiltonian function, is the sum of kinetic and potential energies. For a single particle, the Hamiltonian function is $H = \dfrac{p^2}{2m} + U(r,t)$. If in this expression we turn to operators of corresponding values, we will get the Hamiltonian operator, or the Hamiltonian, representing the sum of operators of kinetic and potential energies. For a single particle of mass m, moving in the potential force field $U(r, t)$, this operator is

$$\widehat{H} = -\frac{\hbar^2}{2m}\Delta + U(r, t), \tag{24.9}$$

where Δ is the Laplace operator that in Cartesian coordinates is

$$\nabla 2 \equiv \Delta = \frac{\partial^2}{\partial x^2} + \frac{\partial^2}{\partial y^2} + \frac{\partial^2}{\partial z^2}. \tag{24.10}$$

Wave function $\psi(r, t)$ will be the product of two functions, one of them depending only on time and the other only on the coordinates:

$$\psi(r, t) = \chi(t)\varphi(r).$$

Consequently, the complete wave function of the *stationary* state of a quantum system is as follows:

$$\psi(r, t) = A\varphi(r)\exp\left(-i\frac{E}{\hbar}t\right). \tag{24.11}$$

As seen, the complete wave function (Equation (24.11)) of the *stationary* state of a quantum system is *time-dependent*, where value E corresponds to the energy of the given stationary state of the system.

Wave functions, obtained from the solution of the stationary Schrodinger equation for a hydrogen atom, are of principal importance in quantum mechanics.

The wave function of the ground state of a hydrogen atom most frequently used for estimate calculations is

$$\varphi_0 = \frac{1}{\sqrt{\pi a_0^3}}\exp(-r/a_0). \tag{24.12}$$

Principle of Superposition. An important condition underlying many quantum effects is the principle of superposition of states. This principle is a consequence of the Schrodinger equation as a linear differential equation. According to the theory of differential equations, any sum of particular solutions of a linear differential equation is also its solution. This sum is called a superposition, or linear combination, of partial solutions.

In quantum mechanics, the principle of superposition of states is one of the fundamental principles. If a system can be found in states described by wave functions $\psi_1, \psi_2, \psi_3, \ldots$, it can be as well in other states described by wave functions that are a linear combination of the initial ones, i.e.

$$\psi = a_1\psi_1 + a_2\psi_2 + a_3\psi_3 + \ldots.$$

For example, if we have two states ψ_1 and ψ_2, the wave function

$$\psi = a_1\psi_1 + a_2\psi_2 \tag{24.13}$$

also describes some state of the system.

The principle of superposition reflects the most cardinal property of quantum systems that does not take place in classical physics. This property is that state (24.13) is to some extent intermediate between initial states ψ_1 and ψ_2. The larger the contribution of one of the initial states to the sum, the more the state (24.13) is "similar" to it. For example, while considering the physical nature of a chemical bond or an electron transfer between two localization centers, it is summed that the electron is present simultaneously in both centers, as if "spreading" over the space between these centers.

24.2.1 Harmonic Oscillator

In addition to the issue on a hydrogen atom, an accurate solution can be obtained for a quantum harmonic oscillator.

Quantum Harmonic Oscillator. Let us consider a one-dimensional harmonic oscillator that in a classical version represents a small ball of mass m fixed on a small spring of rigidity k. As known from textbooks on physics, the potential energy of a strained spring is $U = \frac{1}{2}kx^2$, where $x = l - l_0$ is the deformation of the spring, i.e. the difference between the strained spring length l and its length l_0 in the absence of deformation (the equilibrium length). The eigenfrequency of such an oscillator is $\omega = \sqrt{k/m}$. Using this definition, it is possible to represent the potential energy of the oscillator as $U = \frac{1}{2}m\omega^2x^2$.

Let us write the stationary Schrodinger equation:

$$H\psi(x) = E\psi(x), \tag{24.14}$$

where the Hamiltonian is

$$H = -\frac{\hbar^2}{2m}\frac{\partial^2}{\partial x^2} + \frac{1}{2}m\omega^2x^2. \tag{24.15}$$

Introduce dimensionless variables:

$$\xi = \frac{x}{x_0}, \quad x_0 = \sqrt{\frac{\hbar}{m\omega}}, \quad \varepsilon = \frac{2E}{\hbar\omega}. \tag{24.16}$$

Formally, herein the introduced parameter x_0, as will be seen below, has the meaning of an amplitude of the so-called zero vibrations of the quantum oscillator. In dimensionless variables, the Schrodinger equation (24.14) is

$$\left\{\frac{\partial^2}{\partial\xi^2} - \xi^2 + \varepsilon\right\}\psi(\xi) = 0. \tag{24.17}$$

Solving the Schrodinger equation, we get the energy values for the oscillator:

$$E_n = \hbar\omega\left(n + \frac{1}{2}\right). \tag{24.18}$$

The above expression determines the energy spectrum of the harmonic oscillator. It is easily seen that this spectrum is equidistant, i.e. the difference in energies between any two neighboring energy levels is $\hbar\omega$. The ground state of the harmonic oscillator, i.e. the state with the minimal energy, is the state at $n = 0$. From expression (24.18) it follows that the energy of the ground state is $E_0 = \frac{\hbar\omega}{2} \neq 0$. In contrast to this, for the classical harmonic oscillator zero values of total energy characterize its ground state. The fact that the energy of the ground state of the harmonic oscillator is not zeero is a purely quantum effect, having no classical analog. It means that in the ground state the harmonic oscillator performs the so-called zero oscillations of amplitude x_0.

The following values can be found for the amplitude of zero oscillations:

$$x_0 = \sqrt{\frac{\hbar}{m\omega}}. \tag{24.19}$$

The strict derivation of this relation is based on the argument that zero oscillations of the quantum oscillator are a consequence of Heisenberg's uncertainty principle.

24.2.2 Normal Coordinates and Normal Frequencies of a System of Coupled Oscillators

Let us consider a system of bound harmonic oscillators. The total vibration motion of such a system is much more complicated than vibrations of a single oscillator. This is due to the links between oscillators that make their displacements relative to the equilibrium positions independent. During the general motion of bound oscillators, the amplitude and frequency of oscillations of each separate oscillator change. Therefore, the general pattern of the system vibration motion is very complicated. However, this complicated motion is quite easily described, if we introduce some generalized coordinates (called normal coordinates).

The notion of normal coordinates can be explained by the example of two coupled oscillators.

Apparently, there exists a system of equations describing the vibration process performed by a system of coupled oscillators. At the same time, one should understand that this system represents a whole unified vibration system. Consequently, there should be such solutions of equations that correspond to vibrations of this unified system. In this case, it is possible to find two frequencies of the system (Ω_1, Ω_2) called eigenfrequencies of the vibration system of coupled oscillators

$$\Omega_{1,2}^2 = \frac{1}{2}\left[\omega_1^2 + \omega_2^2 \pm \sqrt{(\omega_1^2 - \omega_2^2)^2 + 4\nu_1^2\nu_2^2}\right], \tag{24.20}$$

where the following designations are introduced: $\omega_1^2 = (k_1 + \gamma)/m_1$, $\omega_2^2 = (k_2 + \gamma)/m_2$, $v_1^2 = \gamma/m_1$, $v_2^2 = \gamma/m_2$. Frequencies ω_1 and ω_2 are called partial frequencies. A partial oscillator is the oscillator where all oscillators of a unified system, except for the isolated one, are fixed in their equilibrium positions. Parameters ν_1 and ν_2, having the frequency dimension, are called the coupling coefficients because their value is determined by the rigidity (γ) of the bond. Consequently, free vibrations of a system of two bound oscillators can occur only at these frequencies.

The physical meaning of normal vibrations is as follows. If in the system of bound oscillators vibration is excited with a frequency corresponding to one of eigenfrequencies, for example, Ω_1, all oscillators of this interconnected system will vibrate at this *normal frequency*, though the vibration amplitude of each of them is individual. It depends on initial conditions and on coefficients of amplitude distribution. Figure 24.1,*b* illustrates this.

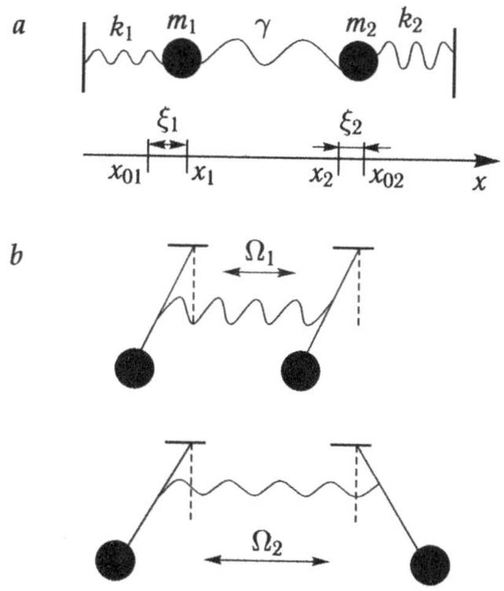

Figure 24.1. Schematic representation of the vibration system of two coupled oscillators.

a, Two bodies of masses m_1 and m_2 are linked to immobile walls by small springs with rigidity k_1 and k_2, and coupled to each other by a spring of rigidity γ. Coordinates x_{01} and x_{02} denote equilibrium positions of these bodies, whereas x_1 and x_2 denote correspondingly their current positions. To displace the bodies from equilibrium positions, designations $\xi_j = |x_j - x_{0j}|, j = 1, 2$ are introduced.

b, Normal vibration modes are shown for a system of two coupled oscillators — a low-frequency (soft) mode (Ω_1) corresponding to cophased, or asymmetric, vibrations of individual oscillators and high-frequency (rigid) mode (Ω_2), corresponding to antiphased, or symmetrical, vibrations of individual oscillators of the system.

Let us note also that displacement of each individual oscillator of the system is a linear combination (superposition) of normal coordinates, and this makes it possible to develop relatively simply a theory of electron-vibrational interactions. It is clear that if the system of bound oscillators is excited in an arbitrary way, all possible vibrations will be excited in it. For an external observer such a vibration process will look like a complicated pattern of apparently random motions. However, it is possible to represent this complicated random (or thermal) motion as a superposition of normal vibrations. This makes the fundamental physical meaning of normal vibrations — phonons as collective states of complex molecular systems called elementary excitations.

An interesting effect, caused by interaction of partial systems, is energy transfer from one oscillator to the other. It is seen most explicitly for systems with strong connectedness. If both oscillators are pulsating, energy is transferred periodically from one oscillator to the other. Namely, the complete energy transfer is the physical meaning of the concept of strong connectedness of systems. The time τ of energy transfer from one oscillator to the other depends on the coupling factor value:

$$\tau = \frac{\pi}{\Omega_2 - \Omega_1} = \frac{\pi\omega}{\nu}. \qquad (24.21)$$

Thus, the time of energy transfer is inversely proportional to the coupling factor. In a system of N coupled oscillators, such as in particular polyatomic molecules, the total number of atoms included in a molecule, is n, and a molecule is known to have $N = 3n - 6$ vibrational degrees of freedom. By introducing normal coordinates it is possible to represent small vibrations of such a molecule as a superposition of vibrations (N) of independent (normal) harmonic oscillators, each of them vibrating with its own (normal) frequency (Ω_n). The operator of the vibration energy of the molecular system is expressed as the sum of energies of normal oscillators.

24.3 Theory of Non-stationary Excitations — Theory of Transitions

Two-level System. We will consider a two-level system with nondegenerate states and illustrate methods of nonstationary theory of excitations describing transitions in a system between its states under the effect of external excitation.

Let us have a system that can be found in two states φ_1 and φ_2. The Hamiltonian of the system will be designated H_0. Functions φ_1 and φ_2 are eigenfunctions of this Hamiltonian, i.e. they satisfy the Schrodinger equation:

$$H_0\varphi_n = E_n^0\varphi_n, \quad n = 1, 2. \tag{24.22}$$

Functions φ_n are orthonormal. Eigenvalues of the Hamiltonian (E_1^0 and E_2^0) determine energy levels of the system. We will suppose states φ_j to be nondegenerate and for the sake of certainty will suppose that $E_2^0 > E_1^0$. Stationary states φ_1 and φ_2 of a system with Hamiltonian H_0 are states with definite energy E_1^0 or E_2^0. This means that if the system is, for example, in state φ_1, it will remain in it for ever. In order to attract the system to state φ_2, an external action is required. Assume that at some instant of time such action (V) took place. Let us call it perturbation. With account for this excitation, the Hamiltonian of the system will be:

$$\widehat{H} = \widehat{H}_0 + \widehat{V}.$$

For this Hamiltonian, states φ_1 and φ_2 are not eigen ones; therefore, in principle transition of the system from one state to the other becomes possible. To describe this process it is necessary to analyze the non-stationary Schrodinger equation with Hamiltonian $\widehat{H}$:

$$i\hbar\frac{\partial\psi}{\partial t} = \left(\widehat{H}_0 + \widehat{V}\right)\psi. \tag{24.23}$$

To solve the problem, we will expand the desired function ψ over the complete system of orthonormal Hamiltonian functions H_0, which in this case is presented by only two functions. Since wave functions of φ_1 and φ_2 are not eigenfunctions of Hamiltonian $\widehat{H}$, expansion coefficients should be time-dependent:

$$\psi = a_1(t)\varphi_1(r)e^{-i\frac{E_1^0}{\hbar}t} + a_2(t)\varphi_2(r)e^{-i\frac{E_2^0}{\hbar}t}. \tag{24.24}$$

Here we have explicitly singled out the time-dependent part of complete wave functions of stationary states of the unexcited Hamiltonian (see Eq. (24.11)).

As a result, we get the following system of equations:

$$\begin{cases} i\hbar\partial a_1/\partial t = a_1V_{11} + a_2V_{12}e^{i\omega_{12}t}, \\ i\hbar\partial a_2/\partial t = a_2V_{22} + a_1V_{21}e^{i\omega_{21}t}, \end{cases} \tag{24.25}$$

where matrix elements V_{ij} are calculated by the general rules (24.5),

$$V_{ij} = \int \varphi_i^*(r) V \varphi_j(r)\, dv,$$

and integration is performed over the entire space. Besides, the following designations are introduced

$$\omega_{ij} = \frac{E_i^0 - E_j^0}{\hbar}.$$

System of equations (24.25) is equivalent to equation (24.23). Now we will use the theory of perturbations. Suppose that perturbation V, which is responsible for the system to pass from one state to the other, is a small value as compared to the energy of the initial system H_0, i.e. $\frac{V}{H_0} = \alpha \ll 1$. In addition, we will introduce initial conditions and namely suppose that prior to perturbation the system was, say, in state φ_1. Taking into account decomposition (24.24) this means that $a_1(0) = 1, a_2(0) = 0$. Let us take system of equations (24.25) and use the method of successive approximations. Represent the desired coefficients $a_j(t)$ as the following series

$$a_j(t) = a_j(0) + a_j^{(1)} + a_j^{(2)} + \dots. \tag{24.26}$$

Here the following condition is valid: the ratio of a series member to the previous member is of the order of α, i.e. $a_j^{(1)}/a_j(0) = \alpha \ll 1$. We are interested in the system transition from state φ_1 to state φ_2, or, in other words, temporal evolution of coefficient $a_2(t)$.

Restricting ourselves to the first order of the theory of perturbations, we get that the probability to find the system in state φ_2 in time τ after the beginning of perturbation is expressed as follows:

$$p_2(\tau) = \frac{1}{\hbar^2} \left| \int_0^{\tau} V_{21}(t) e^{i\Omega_{21}t}\, dt \right|^2, \tag{24.27}$$

where $\Omega_{21} = \frac{1}{\hbar}(E_2^0 + V_{22} - (E_1^0 + V_{11}))$. Generally, if $V_{11} \neq V_{22}$, the frequency is $\Omega_{21} \neq \omega_{21}$ (remember $\omega_{21} = (E_2^0 - E_1^0)/\hbar$).

The Case of Time-Independent Perturbation. Generally, perturbation $V = V(t)$ is a time function. But now let us consider an essential particular case, when perturbation is explicitly independent of time. Thus, from Equation (24.27) we get

$$p_2(t) = \frac{2V_{21}^2}{\hbar^2} \cdot \frac{1 - \cos \Omega_{21} t}{\Omega_{21}^2}, \tag{24.28}$$

where the integral upper limit in Equation (24.27) is designated as t.

Differentiating Equation (24.28) with respect to time, we get the probability of transition per time unit:

$$\frac{dp_2(t)}{dt} = \frac{2V_{21}^2}{\hbar^2} \cdot \frac{\sin \Omega_{21} t}{\Omega_{21}}. \tag{24.29}$$

The energy conservation law should be valid upon transition of a system from one state to the other. Since for a perturbed system states φ_j are not stationary, the energy of these states changes with time. Consequently, transition from state φ_1 to state φ_2 may occur, when their energies become equal. This means that transition will be possible at $\Omega_{21} \to 0$. It follows from Equation (24.29) that to realize this transition the limit transition should take place:

$$\lim_{\Omega_{21}\to 0} \frac{\sin \Omega_{21} t}{\Omega_{21}} = \pi\delta(\Omega_{21}) = \pi\hbar \cdot \delta(E_2 - E_1). \qquad (24.30)$$

The above condition reflects energy conservation in the system upon its transition from one state to the other. Substitution of Equation (24.30) into Equation (24.29) gives us the expression called the "Fermi's golden rule":

$$\frac{dp_2(t)}{dt} = \frac{2\pi}{\hbar} |V_{21}|^2 \delta(E_2 - E_1). \qquad (24.31)$$

To satisfy condition (24.30) in real systems, some processes compensating for the difference in energies of states $(E_2 - E_1)$ should take place. This is determined by interaction of the system with the surrounding medium or external fields. The mechanism of this interaction determines peculiarities of such a process as, for example, electron transfer in different molecular systems including biological ones.

24.4 Model of a Hydrogen Molecule Ion. Nature of Chemical Bonds

The nature of the tunneling effect, determining charge transfer in quantum systems, as well the nature of chemical bonds, determining all structure-dynamic properties of molecules, will be explained using a simple model of a two-level system. In quantum mechanics, it plays approximately the same fundamental role as the model of the harmonic oscillator in classical physics.

Let us consider a two-level system taking the example of a model that will be called the model of a hydrogen molecule ion, i.e. system H_2^+. Model H_2^+ represents a system consisting of a hydrogen atom and a proton, located at distance R from each other (Fig. 24.2, *a*, *b*). In this case, we suppose the nuclei of the system (i.e. protons) to be immobile. This is the so-called zero adiabatic approximation.

Formulation of the Problem. Let us introduce a spherical system of coordinates $\{r, \theta, \varphi\}$, combining its origin with one of the immobile protons and directing polar axis z along the straight line connecting the protons (Fig. 24.2, *a*). Radius vector $\vec{R}$ is directed along the polar axis and determines the position of the second proton, $|\vec{R}| = R$. Let, for certainty, the initial configuration is such that the electron is localized near the first proton at the origin of the coordinates. Thus, the initial state of the system will be the state when a hydrogen atom is at the origin of the coordinates, and the proton is in point $\vec{R}$ (Fig. 24.2, *a*).

If the hydrogen and proton atoms are interchanged, we will get the state when the proton is at the origin of coordinates and the hydrogen atom is in point $\vec{R}$ (Fig. 24.2, *b*). We will call this state the final one. Evidently, as a result of this rearrangement the system energy will not change, though, as seen, the states of the system will be different. Different wave functions correspond to these states. Consequently, our system is degenerate.

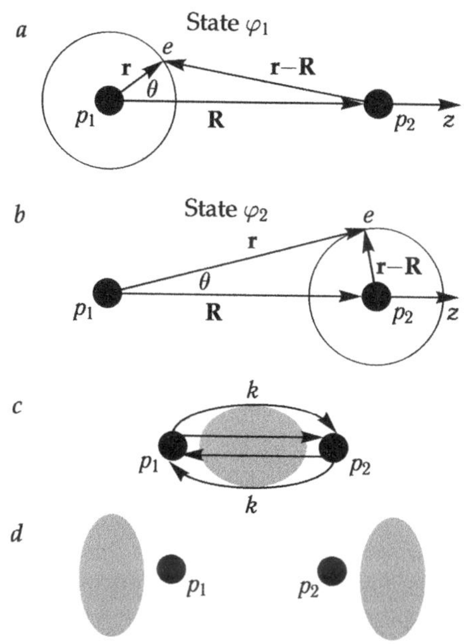

Figure 24.2. Schematic representation of electronic states in the model of a hydrogen molecule ion H_2^+.

State φ_1 corresponds to the electron localization on the first proton p_1 in system H_2^+ (*a*), and state φ_2 is the electron localization on the second proton p_2 (*b*). The two protons are at fixed distance R from each other. The origin of coordinates is combined with the position of first proton p_1. Radius vector **R**, linking the two protons (p_1 and p_2) of the system, is directed along polar axis z. Polar angle θ is the angle between the polar axis and radius vector **r**, drawn from first proton p_1 to electron e. Radius vector **r** − **R** is drawn from second proton p_2 to the electron. Stationary states of system H_2^+: stable state (connecting orbital ψ_1) (*c*) and unstable state (antibinding orbital ψ_2) (*d*). Gray color shows schematically the electron cloud, i.e. the region of the most probable electron location. Symbols k and k_- are nominal indicators for rate constants of direct and reverse transfer between protons upon solution of a non-stationary version of the task.

Let us write the Hamiltonian of a complete system taking into account that protons are immobile and, hence their kinetic energy is zero:

$$H = -\frac{\hbar^2}{2m}\nabla_r^2 - \frac{e^2}{r} - \frac{e^2}{|\vec{r}-\vec{R}|} + \frac{e^2}{R}. \tag{24.32}$$

To determine eigenfunctions of this Hamiltonian, we will use the Schrodinger equation. Because our system is degenerate, i.e. the difference in energies of these states is zero, already in the zero approximation we will search for solution in the form of a linear combination, and as basic functions will choose wave functions of unperturbed Hamiltonian

$$\psi = a_1(t)\varphi_1 + a_2(t)\varphi_2, \tag{24.33}$$

where coefficients $a_1(t)$ and $a_2(t)$ are time functions, and the desired function itself (24.33) obeys the non-stationary Schrodinger equation

$$i\hbar\frac{\partial\psi}{\partial t} = \left(H_0^{(1,2)} + V_{1,2}\right)\psi. \tag{24.34}$$

Here either first or second indices are chosen.

Then, following a standard procedure, we will get the system of equations

$$\begin{cases} i\hbar\dot{a}_1 + i\hbar S\dot{a}_2 = (E_0 + V_{11})a_1 + (E_0 S + V_{12})a_2, \\ i\hbar S\dot{a}_1 + i\hbar\dot{a}_2 = (E_0 S + V_{21})a_1 + (E_0 + V_{22})a_2, \end{cases} \tag{24.35}$$

where designations are introduced for the overlap integral of wave functions S and matrix elements V_{ij}

$$S = \int \varphi_1^*\varphi_2\,dv = \int \varphi_2^*\varphi_1\,dv,$$

$$V_{11} = \int \varphi_1^* V_1 \varphi_1\,dv, \qquad V_{12} = \int \varphi_1^* V_2 \varphi_2\,dv,$$

$$V_{21} = \int \varphi_2^* V_1 \varphi_1\,dv, \qquad V_{22} = \int \varphi_2^* V_2 \varphi_2\,dv,$$

and integration is done over the whole space.

System of equations (24.35) is equivalent to Equation (24.34).

Stationary Solutions. Let us determine stationary states of the system. In a stationary state, coefficients a_j, as stated above in Equation (24.11), should be represented as

$$a_1(t) = A_1 \exp\left\{-i\frac{E}{\hbar}t\right\}, \qquad a_2(t) = A_2 \exp\left\{-i\frac{E}{\hbar}t\right\},$$

where coefficients A_j are already independent of time. Energy values E of stationary states should be determined. Substituting this a_j value in Equation (24.35) and making necessary transformations, we obtain an algebraic system of equations

$$\begin{cases} (E_0 + V_{11} - E)A_1 + (E_0 S + V_{12} - ES)A_2 = 0, \\ (E_0 S + V_{21} - ES)A_1 + (E_0 + V_{22} - E)A_2 = 0. \end{cases} \tag{24.36}$$

As known from linear algebra, a homogeneous system of linear equations has a nontrivial solution when its main determinant is equal to zero. Making up this determinant and equating it to zero, we obtain a quadratic equation for determining energy eigenvalues E of the Hamiltonian (see Equation (24.32)). It is seen, that this procedure is quite analogous to that for determining natural frequencies of the system of two coupled harmonic oscillators (24.20).

Prior to searching for solution of this system, let us pay attention to matrix elements. It is seen that $V_{11} = V_{22}$ and $V_{12} = V_{21}$. This circumstance simplifies considerably the search for solutions, and we get two energy values:

$$\begin{aligned} E_1 &= E_0 + \frac{V_{11} + V_{12}}{1 + S}, \\ E_2 &= E_0 + \frac{V_{11} - V_{12}}{1 - S}. \end{aligned} \tag{24.37}$$

Further, coefficients A_j can be determined for energy values E_1 and E_2 from system (24.36). It is seen that for $E = E_1$ the equality $A_1 = A_2$ is valid, and for $E = E_2$

we have $A_1 = -A_2$. The wave function,corresponding to the stationary state of the system with energy E_1, is as follows:

$$\psi_1 = \frac{1}{\sqrt{2(1+S)}}(\varphi_1 + \varphi_2)\exp\left(-i\frac{E_1}{\hbar}t\right). \quad (24.38)$$

And the wave function corresponding to the state with energy E_2 is

$$\psi_2 = \frac{1}{\sqrt{2(1-S)}}(\varphi_1 - \varphi_2)\exp\left(-i\frac{E_2}{\hbar}t\right). \quad (24.39)$$

Functions (24.38) and (24.39) describe the stationary states of complete system H_2^+ and represent, as seen from these expressions, a superposition of states of an unperturbed (initial) system. Remember that the states φ_1 and φ_2 of the initial system correspond to localized states of the electron (of the hydrogen atom) in points $\vec{r} = 0$ and $\vec{r} = \vec{R}$, respectively. Mixing of these states means that in a complete system the electron is delocalized — it is simultaneously (in part) near the first proton and near the second proton. This phenomenon was called interference of states and is a consequence of the principle of superposition. Now we can state only that the electron is localized in the space, where the both protons are located, i.e. in the space with an effective radius of about internuclear distance R. Detailed information is contained in wave functions. We can indicate a region, where, for example, $\int |\psi|^2\, dv = 0.9$. This region will correspond to the electron cloud — the region of the electron localization. The electron is, as they say, "smeared" between two centers, and we cannot determine more exactly where the electron is, because it is delocalized.

The region of electron localization is determined by wave functions in Equations (24.38) and (24.39) of the stationary states. It is possible to obtain explicit expressions for system energies $E_{1,2}(R)$ in these states. The analysis of these energies shows that $E_1(R)$ has a minimum that corresponds to the stable state of system H_2^+. In this state the system represents a real ion of a hydrogen molecule, and therefore wave function (24.38) is called a binding orbital. This consideration clarifies the basic ideas underlying the quantum description of the chemical bond. Contrary to this, energy $E_2(R)$ has no minimum and is called an antibinding, or loosening orbital for the hydrogen molecule ion. In state (24.38), the electron density is concentrated mainly between protons (Fig. 24.2, *b*), and in state (24.39), it is outside this range (Fig. 24.2, *d*).

Figure 24.3 shows plots of dependencies $E(R)$ for the two stationary sates of model system H_2^+ minus the intrinsic energy E_0 of the hydrogen atom. It is seen that energy $E_1(R)$ has a minimum (stable state) achieved at $R = 2.5a$, i.e. $R \approx 1.32$ E, and energy $E_2(R)$ has no minimum (unstable state), growing monotonically with a decrease in the internuclear distance. Notice also that at $\rho \approx 6$, or, which is the same, at $R \approx 3$ E the molecular hydrogen ion dissociates.

24.5 Method of Molecular Orbitals

Using the hydrogen ion model, we have learnt the main idea underlying the method of linear combination of atomic orbitals (LCAO). Now we will consider briefly the method of molecular orbitals (MO).

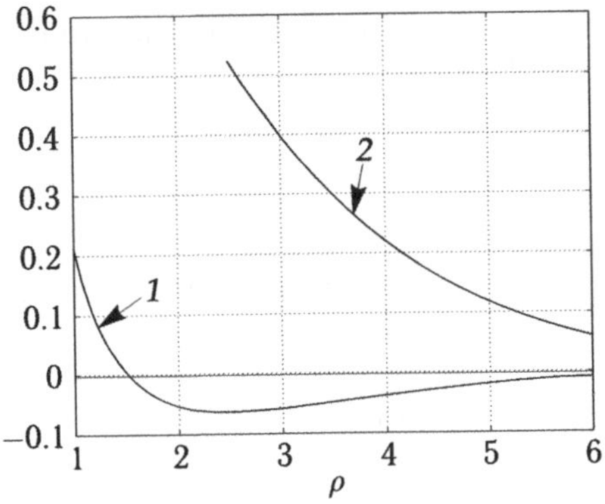

Figure 24.3. Dependence of energy $E(\rho)$ of two stationary states of system H_2^+ minus the intrinsic energy E_0 of hydrogen atom which is independent of R.

State ψ_1 is characterized by energy $E(\rho)$ (curve 1), state ψ_2 is characterized by energy $E_2(\rho)$ (curve 2). Parameter $\rho = R/a$, where $a \approx 0.5$ E is the Bohr radius. On the ordinate axis, the energy is measured in Hartley units.

The complete wave function (ψ_{mol}) of a molecule in the orbital approximation is constructed from molecular orbitals. Molecular orbitals themselves are determined as a linear combination of atomic orbitals. Therefore, in an abbreviated form this method is called LCAO MO. Let us enumerate main characteristics of this method. It is supposed in this method that the electron belongs to a molecule on the whole and moves in the field of all nuclei of this molecule. The state of the electron is described by a single-electron wave function ψ_k, where k is a definite set of quantum numbers. Namely this function is called a molecular orbital. Apparently, this function ψ_k is multicentric (according to the number of nuclei in the molecule). A complete state of the electron is described by the so-called spin-orbital that is the product of coordinate function ψ_k by spin function of the electron (χ_k), i.e. $\Psi_k = \chi_k \psi_k$. A molecular orbital describes the stationary state of the molecule with corresponding energy E_k. This energy is approximately equal to the energy of ionization from the given molecular orbital. The electron configuration of a molecule, i.e. the combination of its molecular orbitals, is constructed based on the two fundamental principles — the principle of the lowest energy and the Pauli principle. It means that for description of the electron configuration of a molecule with $2n$ or $2n - 1$ electrons, n molecular orbitals (or n energy levels) are required, because each orbital can be occupied by two electrons. If orbitals are degenerate (i.e. correspond to the same energy value), they are filled according to the first Hund's rule[1)]. The orbital containing two electrons is called closed, otherwise it is open. The coordinate part of a full wave function of the molecule is set as the product of single-electron wave functions:

$$\psi_{mol} = \prod_{j=1}^{n} \psi_j.$$

And, finally, the molecule energy can be represented as

$$E_{mol} = E_{MO} + E_{QQ} + E_{ee},$$

1) According to the Hund's first rule, the ground state for this electron configuration of an atom with equivalent electrons will be that having the maximal number of unpaired electrons on degenerate orbitals, i.e. the state with the maximal total spin.

where E_{MO} is the sum of orbital energies, E_{QQ} is the energy of internuclear repulsion, and E_{ee} is the energy of interelectron repulsion. The formation of a chemical bond is possible upon socialization of electrons and charge transfer that depends essentially on overlapping of atomic functions. For example, for atoms A and B the overlap integral is

$$S_{AB} = \int \varphi_A \varphi_B \, dv.$$

If this integral is equal to zero, the chemical bond is not formed. The equality of this integral to zero can be quite easily established based on the properties of molecule symmetry and the knowledge what atomic orbitals are involved in the formation of the bond. Symmetry of the overlap cloud is an obvious case for classification of diatomic bonds σ, π, δ etc. A stricter basis for such classification is values of projection of the electron angular momentum to the molecule axis characterized by quantum numbers $\Lambda = 0, 1, 2, \ldots$.

For the σ-bond, the overlap cloud is symmetrical relative to the line between atoms, i.e. has an axial symmetry. The wave function with such a bond is independent of the angle of rotation around the molecule axis and is characterized by $\Lambda = 0$. For the π-bond, the overlap cloud is symmetrical only relative to the plane passing through the molecule axis, therewith $\Lambda = 1$. The δ-bond is characterized by the existence of two mutually perpendicular planes passing through the line between atoms, relative to which the overlap cloud is symmetrical, therewith $\Lambda = 2$ etc. Notice that these ideas have a strict meaning for a diatomic bond, while in a polyatomic system they become rather relative.

Multiple bonds are formed with involvement of σ- and π-electrons (an example is an acetylene molecule). At the same time, the bond becomes rigid relative to rotation around it. Conjugated bonds represent an alteration of double and single bonds in the molecular chain herewith the localization of the chemical bond is partially disturbed. A typical example is benzene molecule C_6H_6. It is a planar molecule — six carbon atoms are located in apexes of a regular hexagon. Three valence electrons of each carbon atom are involved in the formation of three σ-bonds (two with neighboring carbon atoms and one with its hydrogen atom). The fourth valence electron of the carbon atom is a π-electron. Each such π-electron is involved in the formation of the bond concurrently with two adjacent carbon atoms rather than with one atom. This delocalization of π-electron (π-bond) leads to a possibility to move six π-electrons of the benzene molecule along the ring from one carbon atom to another. Thus, a ring current is generated. If a benzene molecule is placed in the magnetic field, perpendicular to the benzene ring, a ring current is generated in the molecule leading to the emergence of a magnetic moment of the molecule (diamagnetism).

24.6 Manifestation of Electronic Properties of Biopolymers

Chromophore Groups. In spite of high complexity of macromolecules, they contain separate constituents — chromophores with definite individual structural and spectral characteristics. The peptide group as the main characteristic chromophore of polypeptide chains has an absorption band at 190 nm caused by $\pi \rightarrow \pi^*$-transitions. Another chromophore group is carbonyl group >C=O existing in all amino acids. Not only $\pi \rightarrow \pi^*$-transitions contribute to absorption in the ultraviolet region of the spectrum, but also the so-called $n \rightarrow \pi^*$-transitions. They exist because the p-orbital (n-level) of oxygen contains a lone-electron pair not involved in the formation of the bond with the carbon atom. Light absorption at 225 nm causes $n \rightarrow \pi^*$-transitions, as a result of this the electron from the oxygen lone-electron pair gets to the antibinding π^*-orbital (Fig. 24.4). This transition is rather weak, because it is forbidden due to the perpendicular location of the planes of π^*-orbitals and p_y-orbital occupied by n-electrons. Only due to vibrations of atoms, leading to an increase in the molecule bending and, correspondingly, to an increase in overlapping, the $n \rightarrow \pi^*$-transition can take place. But the main chromophores of the protein are aromatic amino acid residues: tryptophan and, to a less degree, tyrosine and phenylalanine. The tryptophan absorption spectrum, conditioned by its indole ring with a system of conjugated bonds, has two absorption bands with maxima at 220 and 280 nm. Tryptophan has clearly expressed fluorescence with the maximum at 330–350 nm in an aqueous solution and at about 330–350 nm in proteins. The fluorescence quantum yield equal to the ratio of emitted fluorescence quanta to the number of absorbed light quanta is 0.17 in solution and varies from 0.02 to 0.4 with fluorescence lifetime $\tau \sim 2.5 \cdot 10^{-9}$ s in proteins. At low temperatures, when radiationless deactivation of the triplet state is suppressed, tryptophan phosphoresce intensively with $\tau \sim 7$ s displaying three fluorescence maxima at about 410, 440 and 480 nm. Other aromatic amino acids have absorption maxima at 275 and 222 nm (tyrosine) and at 258 nm (phenylalanine). Quantum yields of their fluorescence in proteins are much lower.

The main chromophores in nucleic acids are purine (adenine and guanine) and pyrimidine (cytosine and thymine in DNA, cytosine and uracyl in RNA) nitrogen bases of nucleotides. In addition to $\pi \rightarrow \pi^*$-transitions (the main band at 260 nm), $n \rightarrow \pi^*$-transitions also make a contribution to the net absorption ("shoulders" at 280–320 nm) with the involvement of a lone-electron pair of nitrogen and oxygen

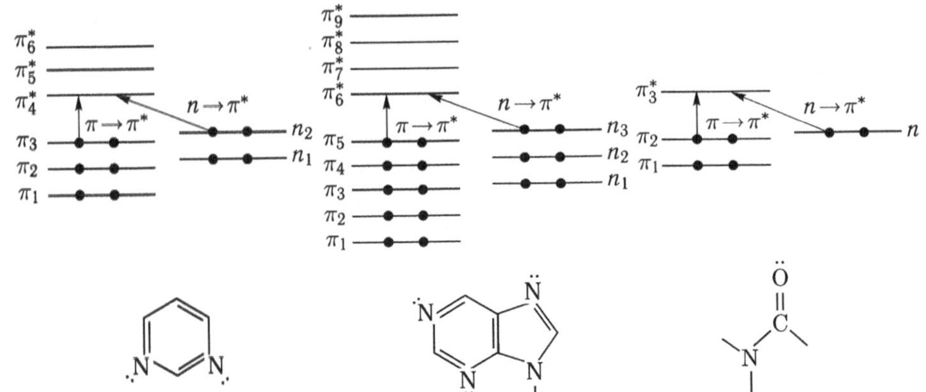

I II III

Figure 24.4. Electron levels and long-wave $\pi \rightarrow \pi^* = n \rightarrow \pi^*$-transitions (shown by arrows) in pyrimidine (*I*) and purine (*II*) bases and the amide group of atoms containing the peptide chain (*III*) (bottom row: forms of corresponding compounds, lone-electron pairs are shown by dots).

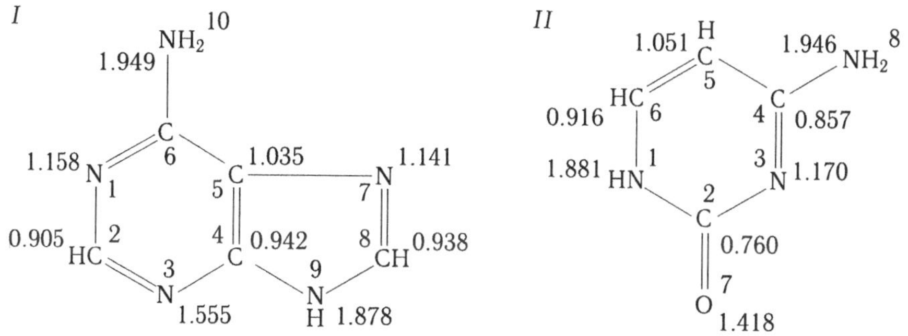

Figure 24.5. Values of charge density for nucleotide bases. *I*, Adenine; *II*, cytosine.

heteroatoms. The electronic structure of nucleotide bases was studied using the method of molecular orbitals, including interactions of π-electrons. As a result, it became possible to obtain values of densities of charges localized on separate atoms. These data allow ascribing the reactivity with separate parts of a molecule (A. Puelman and B. Puelman, 1965). Thus, it was found that electron-acceptor properties of adenine are conditioned mostly by atom C in position 6, while electron-donor properties by atom C in position 8. This regularity is also supported by the example of pyrimidine bases, where the highest π-electron density at atom C in position 5 is the reason for manifestation of electron-donor properties (Fig. 24.5). When in a DNA molecule the base pairs interact with each other, π-electron orbitals overlap causing delocalization of the π-electron system not only in separate bases, but also over the whole base pair. The system of electron levels of a base pair is conditioned by the contribution of each of them. However, calculations show that the probability of transitions between the levels belonging to different bases is much lower than the probability of transitions between levels of the same bases. Therefore the absorption maximum of a gigantic DNA molecule coincides with the maximum that is an average absorption spectrum of separate bases. This is valid also for interactions in dinucleotide of bases located one above the other. However, though the maximum of the first absorption band of DNA at 260 nm coincides with the maximum of the average absorption band of separate bases, here the absorption intensity can drop to 10–30 %. This phenomenon is called hypochromism of a long wave band. At wave lengths of about 260 nm, hypochromism accompanies conformational disordered chain $\rightarrow$ double helix transitions (Fig. 24.6). The explanation of this phenomenon is based on the resonance interactions between nucleotide bases, or more exactly on interactions of their transition dipole moments under different orientations.

According to the laws of electrostatics, the energy of a chain with parallel transition dipoles is higher than their energy at any other orientation. This means that the chain with parallel but oppositely (head-to-tail) or collinearly (head-to-head) located dipoles is characterized by a lower energy as compared to the chain of parallel equally directed dipoles. It is easy to see that the collinear position of dipoles can occur at their orientation along the chain, and the parallel position can take place at their perpendicular position relative to the chain of molecules. Apparently, the presence of a definite and interconsistent orientation of dipoles, provided interactions between all chromosomes take place, should emerge as a displacement of the absorption band of the regular system of dipoles to either side of the spectrum as compared to the isolated molecules.

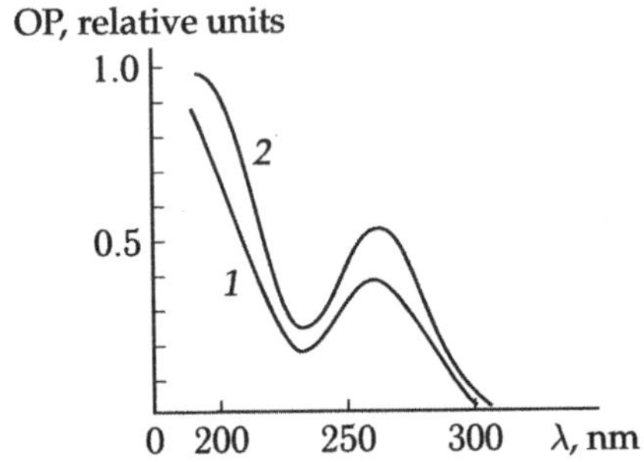

Figure 24.6. Dependence of optical density (OD) on wavelength λ for helical (*1*) and coil-like (*2*) conformation of calf thymus DNA in heavy water.

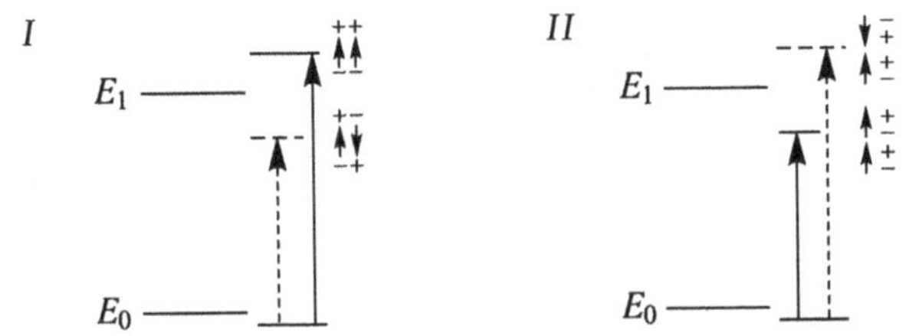

Figure 24.7. Energy levels (E_0 and E_1) for a system of differently oriented transition dipole moments.

I, Parallel position of dipole moments; *II*, collinear position of dipole moments. In all cases the energy level of an isolated molecule is split in two due to interaction of two adjacent dipoles. But in the case of two parallel dipole moments, the intensity of the long-wave band (short dotted arrow) decreases and the intensity of the short-wave band (long solid arrow) increases. As a result, hypochromism of the long-wave band arises. On the contrary, at a collinear position of dipoles, the band with a larger wavelength (short solid arrow) increases its intensity due to the intensity of the short-wave band (long dotted arrow). Hyperchromism in the long-wave length appears.

Figure 24.7 shows a scheme of changes in excitation energy levels and intensities of light absorption bands depending on the mutual orientation of dipoles.

Absorption band splitting is observed at excitation of the chain consisting of dimers; in each of them there may be two opposite orientations of transition dipole moments, included in the dimer (Fig. 24.8).

Thus, the hypochromic effect in DNA emerges because the absorption band at 260 nm is caused by dipole $\pi \to \pi^*$-transitions with their vectors oriented in the planes of nitrogen bases. The latter in the double DNA helix are positioned perpendicular to the helix axis, i.e. dipole moments of $\pi \to \pi^*$-transitions are parallel to each other. On the contrary, the "shoulder" in the absorption spectrum at 282 nm is subjected to a hyperchromic effect, because it is conditioned by the $n \to \pi^*$-transition polarized perpendicular to the base plane. Therefore, in DNA, the dipole moments of the $n \to \pi^*$-transition of two strands are oriented collinearly, and this leads to hyperchromism. In α-helical polypeptides, the hyperchromic effect is observed at 190 nm due to the parallel position of dipole moments of $\pi \to \pi^*$-transitions in α-helical conformations. The disappearance of hyperchromism at the helix–coil transition can yield a quantitative measure of the degree of the protein α-helicity. □

Energy Structure of Polypeptide Chains. A polypeptide group is characterized by a planar position of its atoms (–NH–CO–), when the highest degree of resonance

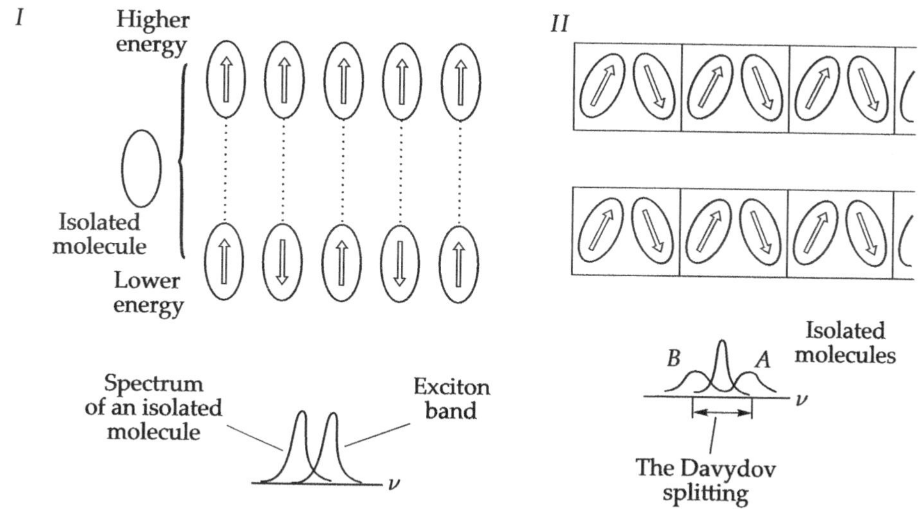

Figure 24.8. Formation of an exciton band (*I*) and Davydov splitting of exciton bands in the presence of two chromophores in the dimer (*II*) (reproduced from P. W. Atkins, 1974).

The magnitude of splitting between the two bands corresponds to the interaction energy of transition dipoles in a dimer cell. A dimer can have different chromophores with different orientation of transition dipole moments. At a parallel position of dipoles, this leads to a decrease in absorption (hyperchromism of the long-wave band) and an increase in absorption (hyperchromism of the short-wave absorption band). At a collinear position, all events will occur inversely. In a system of interacting dipoles, light excitation involves a large area right away, leading to the formation of collective excited states. An excitation wave covering all molecules simultaneously is called an exciton (S. Ya. Frenkel, 1931), and absorption band splitting due to exciton interaction is called the Davydov splitting (A. S. Davydov, 1968). Excitation energy transfer by the exciton mechanism is described in Chapter 25 (Section 5.3).

interaction is achieved due to overlapping of p_α-orbitals, forming a system of π-bonds (see Fig. 9.2). Figure 24.9 shows a scheme of electron levels and orbitals, whose overlapping ensures stability of bonds in the peptide group. Confluence may occur of orbitals of different peptide groups, retained by hydrogen bonds in the protein periodic structure with the formation of common energy bands. In such a system, as a result of interaction of many peptide groups, each molecular level is split not in two ones, as in the case of two molecules (see Section 1), but in a large number of levels. As the number of groups in the chain increases, the difference in the energies between split levels decreases. The distances between them can be neglected, and the area itself formed of confluent levels is called an energy band, i.e. each molecular level in a solid is split in a band. The method of crystalline orbitals, representing generalization of a simple LCAO method, is the basis for calculating the positions of levels in the band. Wave functions of electrons are obtained from orbitals of all atoms of a solid by their summation with corresponding coefficients. The same as in the LCAO method, the values of these coefficients should be such that the total energy of a solid would be minimal. However, since in this case the solid consists of similar repeated units, i.e. has translation symmetry, definite dependence is formed between coefficients of atomic orbitals. Therefore, it becomes possible not to solve the equation with an order equal to the number of atoms in the solid, and do not go beyond the $2n$ power algebraic equations and a system of linear equations with $2n$ unknowns, where n is the number of atoms in a unit cell.

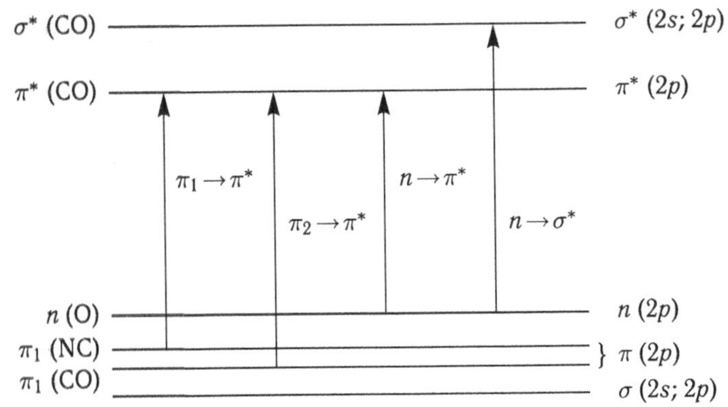

Figure 24.9. Transitions in a peptide group (atoms and their orbitals forming the bonds are given in parentheses).

The energy spectrum of the peptide group depends on the mutual position of binding π-, σ- and antibinding π^*-, σ^*-molecular orbitals and also on energy levels of the lone-electron pair n. The electron energy on σ-orbitals is lower than that on π-orbitals. As a result of strong overlapping of wave functions towards the σ-bonds, the energy of the antibinding σ^*-orbital is higher than that of the π^*-orbital.

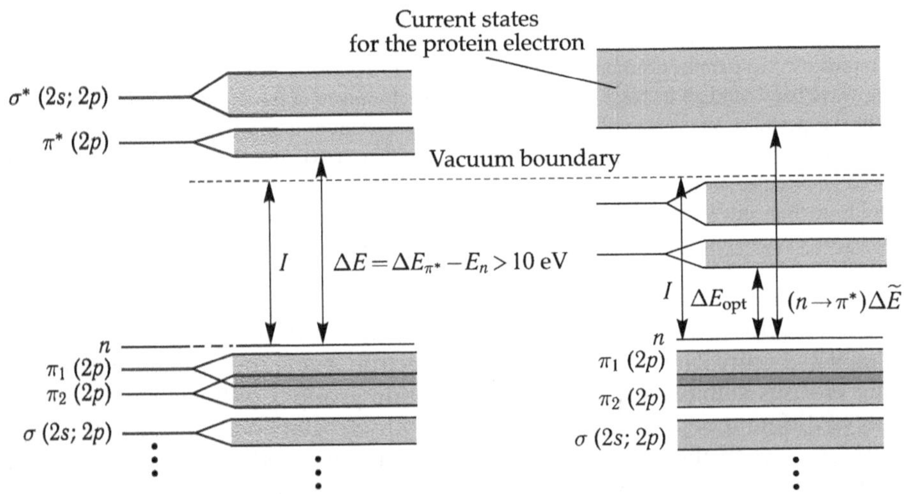

Figure 24.10. Energy scheme of single-electron states in a polypeptide chain (reproduced from E. P. Petrov, 1984).

The scheme is calculated with account for wave functions of atoms C, N and O in states 1s, 2s, 2p and state 1s for atom H. The interaction of the electron and the "hole" in the amide group decreases the energy of π^*- and σ^*-bands to the value of optical excitation (E_{opt} = 5.7 eV). The width of the forbidden band, or the energy gap (ΔE) between the lower filled and the top vacant bands, exceeds 10 eV. This is larger than the optical excitation value (5.7 eV) and ionization potential (9.7 eV). At sufficient energies, the electron can be detached and move independently of the "hole". However, the resultant state has energy higher than the electron energy in vacuum.

Figure 24.10 shows a scheme of the higher filled and lower vacant states of electrons in the polypeptide chain. Because of a large width of the forbidden band, the direct electron transition from the protein valence band to the conductivity band is impossible. This means that under standard conditions, the protein is a dielectric and has no intrinsic spontaneous conductivity.

25

Mechanisms of Charge Transfer and Energy Migration in Biomolecular Structures

Molecular mechanisms of biological processes are in many respects conditioned by electronic properties of macromolecules. A change in the electronic state of one region (active center) of protein gives rise to sequential and directed conformational transformations involving large regions in the macromolecule. Now we will consider physical mechanisms of transformation of electron energy in biostructures. Migration of electron excitation energy and electron transport are of particular significance for biological processes.

Migration of electron excitation energy was discovered in solutions of luminescent dyes. It proceeds according to the general scheme $D^* + A \rightarrow D + A^*$, where D^* and A are molecules of the energy donor and acceptor, respectively. Here the radiationless energy exchange takes place between the electron-excited donor molecule D^* and acceptor molecule A in the ground state. Energy migration in itself is not accompanied by chemical changes in molecules but represents a purely physical process. Excitation transfer goes in parallel with a decrease in the lifetime of the excited state and the quantum yield of donor molecule luminescence. Acceptor molecules play the role of quenchers of the excited state (see Chapter 10).

Among the first reports on migration of electron excitation energy in protein were experiments on photodissociation of carbomyoglobin (the myoglobin-CO complex)

(A. Kaspers and M. Buchner, 1947). Under the influence of light in the presence of oxygen, CO was cleaved from carbomyoglobin with the formation of oxymyoglobin:

$$\underset{\text{Carmyoglobin}}{\text{Protein-heme-CO}} \xrightarrow[O_2]{\text{Light}} \underset{\text{Oxymyoglobin}}{\text{Protein-heme-}O_2} + \text{CO}.$$

The heme-CO bond is disrupted due to photodissociation, or decay, of this compound in the excited state.

The photochemical reaction of CO cleavage from the heme is caused by the light absorbed by the heme and is rather efficient at long waves of about 410 nm. Moreover, at 280 nm where up to 40% energy is absorbed by aromatic groups in proteins, excitation leads to the same effective decay of the heme-CO bond, as the evidence for energy migration from the protein to the heme. Then this energy is spent in photochemical cleavage of carbon oxide from the heme:

$$\text{Protein}^*\text{-heme-CO} \longrightarrow \text{Protein-heme}^*\text{-CO} \xrightarrow{O_2} \text{Protein-heme-}O_2 + \text{CO}.$$

Thus, it was demonstrated that electron excitation energy is transferred between different groups in a protein molecule. Later, excitation energy transfer was observed in protein complexes with the dye — the chromophore (dimethyl aminonaphthol-sulphochloride), where excitation of luminescence of the chromophore group is induced by quanta absorbed by the protein part of the chromoproteins. □

In other experiments, it was found that attachment of an acridine dye to a DNA molecule increases its stability to the action of ultraviolet. In this case, the excitation energy of a DNA molecule is transferred to the dye and emitted as its fluorescence rather than consumed in the photochemical DNA damage.

One of the most important processes of energy migration is realized in photosynthesis (see the details in Chapter 28).

Electron Transfer. Electrons may be transported for relatively large distances and independently of translational motion of electron donor and acceptor molecules that distinguishes these processes from redox reactions in solution.

The semiconductor model of electron migration over large intermolecular distances in the conduction band in the crystal lattice is well known in physics. In this model, D and A molecules can occupy levels of donor and acceptor admixtures, respectively. In spite of its attractiveness, the semiconductor conception has not been corroborated in biological systems. The greatest attention has been paid recently to the concept of electron tunneling proceeding between protein carriers separated by energy barriers. Tunneling transport takes place when the electron energy value is lower than the height of the energy barrier between D and A molecules. This effect is of a purely quantum-mechanical nature and can take place also at low temperatures (liquid nitrogen temperature 77 K and lower). Under such conditions, translational motions of protein molecules are hindered and, consequently, standard physicochemical mechanisms of reactions occurring at the expense of colliding particles with excess kinetic energy cannot be realized at a high rate. Thus, the tunneling mechanism can provide electron transfer under conditions when Arrhenius activation reaction mechanisms "do not work". It can proceed even at normal temperatures successfully competing with usual "activation" reactions.

Low-molecular Redox Reaction of Electron Transfer was for the first time discovered in purple photosynthesizing bacteria. It consists of electron transfer from a cytochrome c molecule to a bacteriochlorophyll molecule excited by the light quantum. This reaction can proceed at the liquid nitrogen temperature (B. Chance, 1960). It was found that the rate of oxidation of cytochrome in *Chromatium* bacteria is independent of temperature at 100–4 K (D. Devault and B. Chance, 1966) and proceeds at characteristic time $\tau_{1/2} \sim 2.3 \cdot 10^{-3}$ s in this temperature range (Fig. 25.1). The curve of temperature dependence has a two-phase character. It is clear that purely radiationless tunneling is dominant at $T < 100$ K. At higher temperatures, the processes requiring activation become competitive.

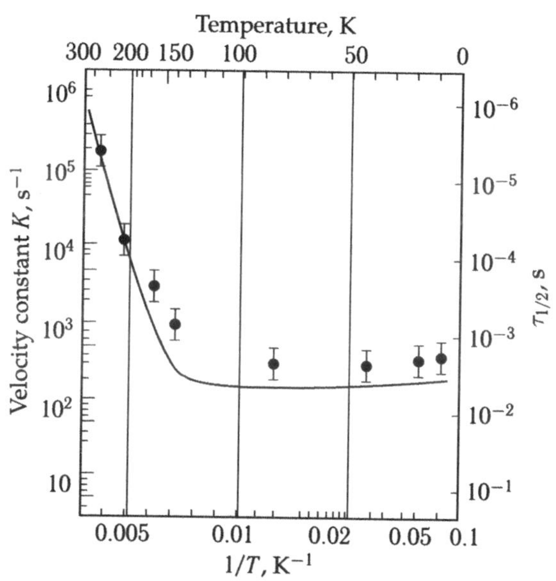

Figure 25.1. Temperature dependence of the cytochrome oxidation rate in photosynthesizing bacteria (reproduced from D. Devault, 1980).

When the room temperature decreases to 140 K, the rate constant decreases by three orders of magnitude and after that remains constant up to the liquid helium temperature.

Large experimental material has been accumulated on other low-temperature stages of electron transfer in photosynthetic reaction centers (Chapters 27 and 28).

No low-temperature reactions of electron transfer occur in mitochondria. At the same time, high transfer rates ($\tau_{1/2} \sim 10^{-3} \div 10^{-6}$ s) at separate regions of the respiratory chain prove that even here the tunneling transfer takes place. In mitochondria, this transfer is also associated with conformational changes in protein components. According to different estimates, distances between active prosthetic groups of carriers, directly transferring the electron, are no less than 0.5–1.0 nm (for cytochromes c and b_5 it is 0.8 nm upon electron transfer between them).

Similar data are obtained also for distances at which the electron penetrates from the surface inside the protein molecule during protein reactions with inorganic reagents.

In preceding chapters, it was shown that the nuclear system of macromolecules is characterized by a large set of different types of intramolecular motions. The problem that will be considered in this chapter is how interaction with vibrational degrees of freedom is realized upon electron transfer or migration of electron excitation energy. The understanding of this problem is necessary for analyzing the mechanism of electron-conformational interactions underlying the functioning of biomacromolecules.

25.1 Tunneling Effect

25.1.1 Quasi-classical Approximation. The Gamow Formula

The directed electron transfer occurring in biological structures as well, proceeds by the tunneling mechanism that will be considered below. As known, the quantum-mechanical description turns into classical under condition when the de Broglie wavelength vanishes. The de Broglie wavelength is determined as follows: $\lambda = h/p$, where h is the Planck constant, $p = mv$ is the impulse, m is the mass, and v is the particle velocity. Consequently, if $p \to \infty$, then $\lambda \to 0$. In this case it is possible to use quasi-classical approximation that is of choice in description of tunneling.

Let us consider two events. (1) The particle is in a potential well (Fig. 25.2, *a*), and (2) the particle meets a potential barrier (Fig. 25.2, *b*). The wave function will be designated $\psi(x)$. Let the particle have potential energy $U(x)$, and its net energy will be designated E. In accord with classical mechanics, we get $E = p^2/2m + U(m)$. Hence the particle impulse will be expressed as

$$p(x) = \pm\sqrt{2m[E - U(x)]}. \tag{25.1}$$

As seen from the above equation, if $E < U(x)$, the impulse value $p(x)$ becomes an imaginary quantity. Thus, the space, where the inequality $E < U(x)$ is valid, is inaccessible for the particle from the classical point of view.

The idea of the quasi-classical approximation is as follows. Consider the stationary one-dimensional Schrodinger equation for a particle moving in potential $U(x)$:

$$\left(-\frac{\hbar^2}{2m}\frac{\partial^2}{\partial x^2} + U(x)\right)\psi(x) = E\psi(x). \tag{25.2}$$

Imagine the desired solution for the wave function as

$$\psi = \exp\left\{\frac{i}{\hbar}\int p(x)\,dx\right\} \tag{25.3}$$

and substitute it in Equation (25.2), then the following equation will be obtained:

$$\frac{p^2}{2m} + U(x) - \frac{i\hbar}{2m}\frac{\partial p}{\partial x} = E. \tag{25.4}$$

Hence it is seen that if

$$\left|\frac{i\hbar}{2m}\frac{\partial p}{\partial x}\right| \ll \frac{p^2}{2m},$$

then this member can be neglected, and we will get a purely classical equation for determining the particle impulse. Consequently, the condition, when Equation (25.2)

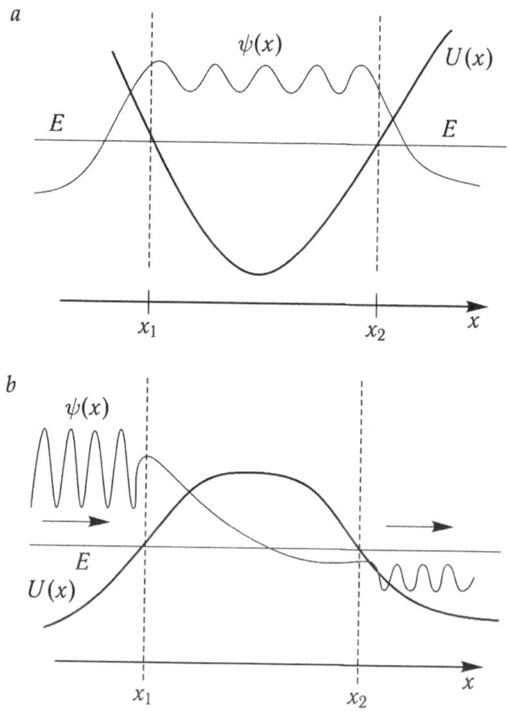

Figure 25.2. Schematic representation of wave function $\psi(x)$ of the particle in potential well $U(x)$ (a) and the particle "bumping" the potential barrier $U(x)$ (b).

Coordinates x_1 and x_2 are called turning points and correspond to condition $U(x_1) = U(x_2) = E$, where E is the total energy of the particle. When approaching these points, the particle impulse $p \to 0$, the length of the de Broglie wave $\lambda \to \infty$. In regions $x < x_1$ and $x > x_2$, where $E < U(x)$, wave function $\psi(x)$ decays.

is approximately valid and turns into a classical equation of motion with substitution of Equation (25.3), is represented as

$$p^2 \gg \hbar \left| \frac{dp}{dx} \right|. \tag{25.5}$$

Expressing the particle impulse as the de Broglie wavelength, $p(x) = h/\lambda(x)$, let us represent this condition as

$$\frac{1}{2\pi} \left| \frac{d\lambda}{dx} \right| \ll 1. \tag{25.6}$$

Thus, we have obtained a quantitative criterion for estimating the quasi-classical approximation validity (25.6) — the de Broglie wavelength of a particle should change little at distances of about the length of the wave itself. If the characteristic dimensions of a classically accessible region are designated as l, then $\frac{d\lambda}{dx} \sim \frac{\lambda}{l}$, and Inequality (25.6) turns into inequality $\lambda \ll l$.

If we return to Expression (25.5) and use Equation (25.1), the condition of applicability of a quasi-classical description can be written as

$$\left| \frac{m\hbar}{p^3} \left(-\frac{dU}{dx} \right) \right| \ll 1.$$

Hence it is seen that classical consideration of quantum-mechanical systems is justified when particles move with large impulses in the potential field with low gradients.

The Gamow Formula. If condition (25.6) is valid, then according to Equation (25.4) the particle impulse is determined by Expression (25.1). Substitution of (25.1) in (25.3) yields the particle wave function $\psi(x)$. As known, the square wave function module is the density of the probability to find a particle in the given region of space. If the particle is in classically accessible regions, i.e. where the impulse $p^2 > 0$, the particle wave function $\psi(x)$ has an oscillatory character as seen from Equation (25.3) (see Fig. 25.2), and its square module $|\psi(x)|^2$ is a constant value independent of the coordinate. This means that the particle can be found with equal probability at any point of the classically accessible region.

An essentially different situation will be in classically inaccessible regions of space, i.e. where $p^2(x) < 0$. In this case, it is possible to separate the imaginary unit, $p(x) = \pm i|p(x)|$, and then it is evident that the wave function will look like a damped exponential,

$$\psi(x) = A \exp\left\{-\frac{1}{\hbar}\int \sqrt{2m(U(x)-E)}\,dx\right\}, \tag{25.7}$$

and correspondingly its square module will be

$$|\psi(x)|^2 = A^2 \exp\left\{-\frac{2}{\hbar}\int \sqrt{2m(U(x)-E)}\,dx\right\}, \tag{25.8}$$

where A is the normalization factor (wave amplitude).

Expression (25.8) underlies the understanding of the tunneling effect nature. If the classically inaccessible region has the final length, say, from x_1 to x_2 (Fig. 25.2, *b*), the probability of finding the particle in point x_2, when it meets the potential barrier in point x_1, generally speaking, is not zero. The non-zero probability of finding the particle in the classically inaccessible region, where $E < U(x)$, means the possibility of the particle to penetrate into this region. This effect has an exclusively quantum (probabilistic) character and cannot be understood from the point of view of classical mechanics.

If the particle turned to be in point x_2, i.e. passed across the potential barrier of width $l = |x_2 - x_1|$, it once again turned to be in the classically accessible region. According to (25.7), the wave function amplitude A_t after passing the barrier will be

$$A_t = A \exp\left\{-\frac{1}{\hbar}\int_{x_1}^{x_2} \sqrt{2m(U(x)-E)}\,dx\right\}. \tag{25.9}$$

The ratio of square amplitudes of the transmitted and incident de Broglie waves $D = A_t^2/A^2$ determines the probability of the particle motion across the potential barrier. This effect was called the tunneling effect. Using Equation (25.9) for calculation of coefficient D, called the barrier transmission factor, we will get the Gamow formula:

$$D = \exp\left\{-\frac{2}{\hbar}\int_{x_1}^{x_2} \sqrt{2m(U(x)-E)}\,dx\right\}. \tag{25.10}$$

This formula is widely used both in physics and chemistry and also in biophysics. However, it should be remembered that this formula is a phenomenological one. This is because for a real multi-particle system, an accurate calculation of the surface profile of potential energy in the given direction $U(x)$ is really an impracticable task. Therefore, the potential function $U(x)$ is usually chosen based on one or another consideration. This choice is sought to be made so that the integral in Equation (25.10) could be calculated analytically.

25.1.2 Calculation of Transmittance of Potential Barriers

Using the Gamow formula, let us estimate the transmittance of barriers of different shapes: rectangular, triangular and parabolic barriers (Fig. 25.3, *a–c*).

Rectangular Barrier. We will consider the simplest case of a rectangular barrier (Fig. 25.3, *a*). The potential energy is assigned as follows:

$$U(x) = \begin{cases} 0, & x < x_1, \\ U_m, & x_1 \leqslant x \leqslant x_2, \\ 0, & x > x_2, \end{cases} \tag{25.11}$$

where U_m is the barrier height, counted out from particle energy value E that for convenience is accepted to be zero (i.e. the energy zero point). The barrier width at the particle energy level will be designated as $d = |x_2 - x_1|$ (in this case, it is the same over the whole barrier height). The integral in the index of the exponential

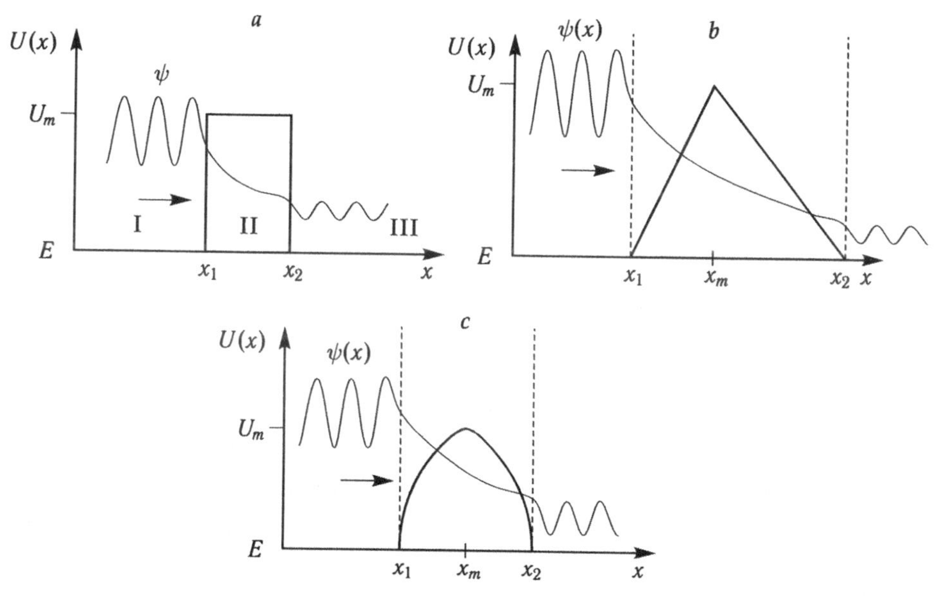

Figure 25.3. Three types of model potential barriers are shown: rectangular (*a*), triangular (*b*), and parabolic (*c*).

Energy is counted out from the total energy value E of the particle bumping the barrier. A schematic representation of the wave function of particles $\psi(x)$ is shown. The arrow indicates the direction of the particle motion. The maximal height of barriers is designated as U_m. Coordinates x_1 and x_2 are turning points, and x_m is the coordinate of the potential apex $U(x)$.

in formula (25.10) is computed elementally and for the transmittance coefficient of a rectangular barrier D_1 we get

$$D_1 = \exp\left\{-\frac{2}{\hbar}\sqrt{2mU_m}\,d\right\}. \tag{25.12}$$

Triangular Barrier. Now let us consider a triangular barrier (Fig. 25.3, *b*). The integral in the exponential index of Equation (25.10) is

$$D_2 = \exp\left\{-\frac{2}{\hbar}\sqrt{2mU_m}\,\frac{2}{3}\,d\right\}, \tag{25.13}$$

where $d = d_1 + d_2 = |x_2 - x_1|$ is the barrier total width at the level of particle energy. From Expression (25.13), it follows that the transmittance of a triangular barrier is independent of the triangle shape at fixed lengths of its base and height.

Parabolic Barrier. Here we consider a parabolic potential barrier (Fig. 25.3, *c*). Calculations of the integral are not difficult, and as a result we get

$$D_3 = \exp\left\{-\frac{2}{\hbar}\sqrt{2mU_m}\,\frac{\pi}{4}\,d\right\}. \tag{25.14}$$

Comparing the transmittance coefficients in all three cases at the same width d and height U_m of barriers, we see that they are connected by inequalities

$$D_2 > D_3 > D_1.$$

The probability of percolation across the barrier can be associated with the characteristic time τ — when it elapses, the particle will necessarily tunnel. Coefficient $D(E)$ is the probability of penetration across the barrier in a single attempt. Let the N-th attempt be successful, then we can write the condition $ND(E) = 1$. If these N attempts were performed for time τ, it can be written that $\tau\frac{N}{\tau}D(E) = 1$. The ratio $\nu = N/\tau$ can be accepted as the frequency of performed attempts to penetrate across the barrier, then

$$\tau = (\nu D(E))^{-1} = \nu^{-1}\exp\left\{\frac{2}{\hbar}\int_{x_1}^{x_2}\sqrt{2m(U(x)-E)}\,dx\right\}. \tag{25.15}$$

Take the barrier height to be $U_m \approx 1$ eV and their width $d \approx 0.7$ E. These parameters are characteristic of the barrier of the double-well potential of a hydrogen bond. Let a proton be the tunneling particle. As the frequency of attempts performed by the proton to overcome the barrier, we may take the frequency of vibrations of the barrier $\nu = 10^{14}\ \text{s}^{-1}$. Then, using Equations (25.15), (25.12), (25.13) and (25.14), we obtain $\tau_1 \approx 8 \cdot 10^{-5}$ s, $\tau_2 \approx 4 \cdot 10^{-8}$ s, $\tau_3 \approx 6 \cdot 10^{-7}$ s for the characteristic times of the proton transfer. These values show that the shape of a potential barrier affects dramatically time characteristics of the process — the transition times differ by the orders of magnitude. Characteristic times of molecular processes are an important factor determining the functioning of biological molecular systems. It should be taken into account upon analyzing molecular processes.

Notice that the tunneling transfer is more efficient for electrons than the over-barrier activation process not only at low but also at usual temperatures. For example, at the barrier height $\epsilon \geqslant 1$ eV and width $d \sim 0.5$ nm this follows from the ratio of the tunneling and over-barrier transfer factors

$$\frac{\exp(-2d\sqrt{2m\epsilon}/\hbar)}{\exp(-\epsilon/k_BT)} > 10,$$

that is valid at $T < 700$ K and at the given above values of parameters d and ϵ.

Particle in One-dimensional Rectangular Potential Well. Let us set the potential where a particle is moving

$$U(x) = \begin{cases} U_m, & x \leqslant x_1, \\ 0, & x_1 \leqslant x \leqslant x_2, \\ U_m, & x \geqslant x_2. \end{cases} \tag{25.16}$$

This potential is as if additional to potential (25.11) (Fig. 25.4, *a*). Analogously to the above, divide the space into three regions: I ($x < x_1$), II ($x_1 < x < x_2$), III ($x > x_2$). Energy values are readily found if the potential well has infinitely high walls. In this case, when $U_m \to \infty$, we get a discrete energy spectrum of the particle in an infinitely deep well:

$$E_n = \frac{\pi^2\hbar^2 n^2}{2mL^2}. \tag{25.17}$$

But if the potential well has the final depth U_m, the following approximate expression can be obtained for energy levels near the well bottom:

$$E_n' \approx E_n\left(1 - \frac{4\hbar}{\sqrt{2mU_m}L}\right). \tag{25.18}$$

In addition, note that for a potential rectangular well of a final depth, the number of energy levels of the particle in this well is a finite number, the maximal one of them can be determined from condition $E = U_m$:

$$n_{\max} = \frac{\sqrt{2mU_m}}{\pi\hbar}L + 1. \tag{25.19}$$

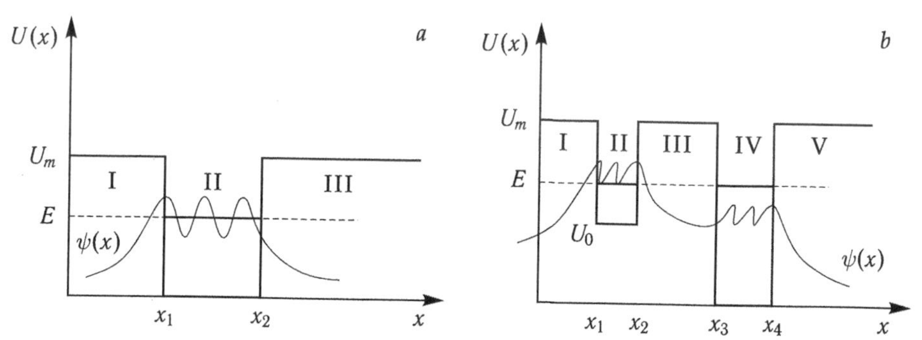

Figure 25.4. Schematic representation of a rectangular potential well (*a*) and a double-well potential (*b*).

The energy of the particle localized in the potential well is E. Outside the well the potential function $\psi(x)$ is damped. In the double-well potential, the particle transition from one well to the other may occur only in the presence of states with the same energy.

Model of Double-well Potential. Let us consider a model system that will be used to discuss some problems of a theoretical description of charge transfer. Set a potential consisting of two rectangular energy wells of different depths and widths (Fig. 25.4, *b*) written as

$$U(x) = \begin{cases} U_m, & x < x_1, & \text{I}, \\ U_0, & x_1 \leqslant x \leqslant x_2, & \text{II}, \\ U_m, & x_2 \leqslant x \leqslant x_3, & \text{III}, \\ 0, & x_3 \leqslant x \leqslant x_4, & \text{IV}, \\ U_m, & x > x_4, & \text{V}. \end{cases} \tag{25.20}$$

Here it is accepted that $U_0 < U_m$. Roman numbers indicate regions in which the space was divided. Designate the width of first (left-hand) and second (right-hand) wells correspondingly as $L_1 = |x_2 - x_1|$ and $L_2 = |x_4 - x_3|$.

Pay attention to an essential circumstance that will be discussed through this chapter. The particle energy should be the same in both potential wells. As has been shown above, the energy spectrum of the particle in the well is discrete. Consequently, the particle transfer from one well to the other can occur only upon realization of condition $E_{n_1} = E_{n_2} = E$, where n_1 and n_2 are numbers of energy levels in the first and second wells, respectively. Provided the wells were the same, these numbers would coincide. However, in a general case of different wells it may occur that there will be no pair of appropriate numbers of levels in these wells. As a result, it is impossible for the particle to transfer from one well to the other, and hence it will remain localized in one of the wells. The transfer will be possible only under the action of some external factor compensating for the difference in energy levels in the two different wells.

Let us underline once again that the particle transfer from one well to the other is possible only upon energy conservation $E_{n_1} = E_{n_2}$. If there are no such numbers, an external effect is required to compensate for the difference in energies upon transition between these energy levels. The role of such an external factor is played, as a rule, by the surrounding with which the particle interacts. As shown below, in real systems the probability of the process, responsible for the preservation of the energy balance upon particle transfer between wells, affects considerably the characteristic rate of charge transfer.

25.1.3 Adiabatic Approximation

In biological systems, electron transport is realized in complex macromolecular constructs. In most cases an elementary act of electron transfer is an electron tunneling transition from one molecule (donor) to the other (acceptor). These carrier molecules are electron binding centers, or localization centers, in the process of its intermolecular transfer. Properties of these molecules as well as parameters of their interaction with the surrounding determine the electron lifetime in the given binding center and, consequently, the efficiency of the whole process of directed charge transfer. Therefore, to understand this process, it is necessary to be able to describe quantum states of molecules. Molecules consist of electrons and nuclei, and the nuclei masses exceed the electron mass by tens of thousand times. This difference in masses underlies the approximate description of quantum states of molecules. At zero approximation, the nuclei are believed to be quiescent, and the following approximations take into account the nuclear motion by methods of the excitation theory. This method is called the adiabatic approximation.

The Zeroth (Rough) Adiabatic Approximation. Let us consider a system consisting of a number of electrons with mass m and atomic nuclei of mass M. The combination of coordinates of all electrons relative to the center of mass of the system will be designated as r, and the combination of nuclear coordinates will be designated as R. The Hamiltonian determining the internal state of the system will be represented as:

$$H = T_R + T_r + V(r, R), \tag{25.21}$$

where operators of the kinetic energy of nuclei (heavy particles) and electrons (light particles) are as follows:

$$T_R = -\frac{\hbar^2}{2M} \sum_i \frac{\partial^2}{\partial R_i^2},$$

$$T_r = -\frac{\hbar^2}{2m} \sum_i \frac{\partial^2}{\partial r_i^2},$$

$V(r, R)$ is the operator of the potential energy of interaction between all particles.

Adiabatic approximation is based on the assumption that the operator of the kinetic energy T_R of heavy particles is a small value as compared to the other operators (25.21)[1)]. Rewrite the Hamiltonian (25.21) as

$$H = H_0 + T_R, \tag{25.22}$$

where

$$H_0 = T_r + V(r, R). \tag{25.23}$$

At the zeroth adiabatic approximation, the nuclei of molecules are considered to be static, i.e. their masses are believed to be infinitely large, and consequently $T_R = 0$. In this case, stationary states of the system are calculated using the Schrodinger equation:

$$H_0 \psi_n(r, R) = \epsilon_n(R) \psi_n(r, R_0) \tag{25.24}$$

for fixed coordinate values (R_0) of heavy particles. Index n determines the totality of quantum numbers characterizing the stationary state of the system. In each of such states, the electron energy $\epsilon_n(R_0)$ and wave functions $\psi_n(r, R_0)$ depend on coordinates of nuclei as on parameters. Therefore, functions $\psi_n(r, R_0)$ characterize electronic states of the molecule at fixed positions of nuclei R_0. It should be underlined that fixed positions of nuclei R_0 do not necessarily coincide with their equilibrium positions — they can be any but fixed positions of nuclei.

The Born–Oppenheimer Approximation. Assume that we know solutions of equation (25.24), i.e. we know the energy spectrum $\epsilon_n(R_0)$ and wave eigenfunctions $\psi_n(r, R_0)$ of the zeroth approximation operator H_0. Wave functions $\psi_n(r, R_0)$ form a complete system of orthonormalized functions. Consequently, the system of these functions makes it possible to expand any function (such expansion is sometimes called the Fourier series expansion).

[1)]The smallness of the operator, as compared to other operators, implies that the mean magnitude of the physical value (24.4), corresponding to this operator in some state of the system, is much smaller than analogous values of other operators. In this case, the physical value is the energy, and therefore the kinetic energy of heavy nuclei is essentially lower than the energy of motion of other parts of the system or the energy of their interaction with each other.

Stationary states of a molecular system with the complete Hamiltonian (25.21) are determined from the solution of the following equation:

$$(T_R + H_0)\Psi(r,R) = E\Psi(r,R). \tag{25.25}$$

One of the possible assumptions facilitating the search for solutions of this equation is an attempt to represent the desired solution as the product

$$\Psi_n(r,R) = \varphi_n(R)\psi_n(r,R_0), \tag{25.26}$$

that corresponds to the assumption on independence of motion of electrons and nuclei. In equation (25.26), functions $\varphi_n(R)$ depend only on coordinates of nuclei, and functions $\psi_n(r,R)$ are eigenfunctions of the zeroth approximation Hamiltonian (25.24) at some fixed values of coordinates of nuclei R_0.

The Hamiltonian in Equation (25.25) differs from the Hamiltonian in Equation (25.23) by additional operator T_R, acting only on nuclear coordinates. Therefore, the desired complete wave function $\Psi(r,R)$ can be represented as expansion in a series of eigenfunctions $\psi_n(r,R_0)$ of operator H_0: $\Psi(r,R) = \sum_n a(R)\psi_n(r,R_0)$, where expansion coefficients $a(R)$ should be dependent on nuclear coordinates R. Generally speaking, *a priori* we do not know the type of these coefficients. However, taking into account Equation (25.26), we can try to use functions $\varphi_n(R)$ as expansion coefficients:

$$\Psi(r,R) = \sum_n \varphi_n(R)\psi_n(r,R_0). \tag{25.27}$$

Thus, $\varphi_n(R)$ play the role of expansion coefficients depending on nuclear coordinates. Consequently, the series expansion (25.27) is equivalent to condition (25.26). The sum symbol in this expression means that the energy spectrum of operator H_0 is assumed to be discrete (in case of a continuous spectrum it is necessary to write an integral).

After that, we perform a standard procedure of rewriting the Schrodinger equation: substitute (25.27) in (25.25), then multiply the obtained left-hand equation by the conjugate function $\psi_m^*(r,R_0)$ and make the term-by-term integration over the electron coordinates.

To understand the below considerations, it is necessary to make an essential remark. Remember that functions $\psi(r,R)$ are parametrically dependent on nuclear coordinates R. It means that we get a solution of Equation (25.24) at some fixed values of nuclear coordinates. Having this in mind, up to the moment we have preserved index "0" in nuclear coordinates R in the independent variable $\psi(r,R)$. However, we can fix different positions of nuclei and, generally speaking, obtain different solutions for Equation (25.24) at a fixed set of nuclear coordinates. Having done this, we get a set of solutions for all possible fixed positions of nuclei that is equivalent to the knowledge, in what way electron wave functions depend on nuclear coordinates. It is just this meaning that is included in the conception of parametric dependence. Hence, as electron wave functions we can use not a single function obtained at the given value $R = R_0$, but rather a whole set of such functions at various nuclear coordinates R. This allows us to consider the parametric dependence as a certain functional dependence and regard electron wave functions as functions of two variables $\psi(r,R)$. Thus as a result of the above procedure, we get the following system of equations:

$$(T_R + \varepsilon_m - E)\varphi_m(R) = \sum_{n \neq m} L_{nm}\varphi_n(R), \tag{25.28}$$

where ε_m is the electron energy in state ψ_m, operator L_{nm} is called the nonadiabacity operator:

$$L_{mn}|_{n\neq m} = \frac{\hbar^2}{M}\sum_j \int_{(r)} \psi_m^*(r,R)\frac{\partial}{\partial R_j}\psi_n(r,R)\,dr\frac{\partial}{\partial R_j} - \\ - \int_{(r)} \psi_m^*(r,R)T_R\psi_n(r,R)\,dr. \tag{25.29}$$

If we had assumed $R \equiv R_0$, the nonadiabacity operator would have been identically equal to zero. However, taking into account the parametric dependence of wave functions on nuclear coordinates, we can "trace" how a change in the position of nuclei alters the electronic state of the molecular system. The integrals included in this operator are matrix elements proportional to the probabilities of electron transfers from states ψ_n to state ψ_m affected by the nuclear impulse operator (the first integral) and the operator of nuclear kinetic energy (the second integral). The operator L_{mn} itself has effect only on nuclear wave functions (see (25.28)). This is indicated by the differential operator $\partial/\partial R_j$ in the first member of operator L_{mn}. The second member of this operator is merely some function of R; its action on the function is reduced to simple multiplication. If electron levels of a molecule are quite remote from each other, the probability that the nuclear motion will cause transitions between electronic states is relatively low, because corresponding operators are small (or, which is the same, the energy of nuclear motion is low). In this case, it is usually stated that the nonadiabacity operator is small. But when some electron levels are close to each other (quasi-degeneration), the energy of nuclear motion can be sufficient for the electron transfer from one state to the other. Then it is necessary to take into consideration this possibility, and the nonadiabacity operator cannot be neglected.

System of equations (25.28) is exact and completely equivalent to the Schrodinger equation (25.25). If the right-hand part of Equation (25.28) is small (i.e. the nonadiabacity operator is small), the equations in this system can be solved by the method of successive approximations (the method of the perturbation theory) using L_{mn} as low perturbation.

If the nonadiabacity operator is neglected at all, i.e. accept $L_{mn} = 0$, system of equations (25.28) disintegrates into a system of independent equations. (It means that we have a situation, when electronic states of a molecule are nondegenerate and the difference between electron energy levels is great as compared to the energy of nuclear motion (no quasi-degeneration takes place). Then

$$\{T_R + \epsilon_m(R)\}\varphi^0_{m\nu}(R) = E^0_{m\nu}\varphi^0_{m\nu}(R), \tag{25.30}$$

for each state of electron motion determined by quantum numbers m. New quantum numbers ν are quantum numbers, determining the states of nuclei in the given electronic state. This approximation is called the Born–Oppenheimer approximation. It is seen from Equation (25.30) that the nuclear motion occurs in the potential field with potential energy $\epsilon_m(R)$, which is the potential electron energy provided that nuclei are fixed in position R (see (25.24)).

Thus, in the Born–Oppenheimer approximation, called simply below the adiabatic approximation, the complete wave function of the system is represented as the product of the nuclear wave function and electron wave function:

$$\Psi_{m\nu} = \varphi^0_{m\nu}(R)\psi_m(r,R). \tag{25.31}$$

This means that states of nuclei $\psi_m(r, R)$, differing in quantum numbers ν, will correspond to each state $\varphi^0_{m\nu}(r, R)$. Below we will discuss conditions for the use of the adiabatic approximation, and now let us analyze Equation (25.30).

Let us consider the ground electronic state of the system, assuming that $m = 0$. The equilibrium configuration of nuclei in this state will be designated as R_0. Energy $\epsilon_0 = \epsilon_0(R)$ will be expanded in degrees of nuclei deviation from equilibrium positions taking only quadratic terms of this expansion:

$$\epsilon_0(R) \approx \epsilon_0 + \frac{1}{2}\sum_{ij}\Gamma_{ij}(R_i - R_{0i})(R_j - R_{0j}), \tag{25.32}$$

where expansion coefficients Γ_{ij} are second derivatives of energy $\epsilon_0(R)$ with displacement of nuclei, $\Gamma_{ij} = \partial^2\epsilon/\partial R_i \partial R_j\big|_{R=R_0}$.

This expansion will have no linear term, because R_0 corresponds to the minimum potential. Limiting ourselves by the quadratic term in the expansion of the potential energy, we turn the task into the system of linked oscillators. System of equation (25.30), describing the motion of coupled oscillators, is split into independent one-dimensional equations upon transition to normal coordinates. Therefore we can represent the complete Hamiltonian of the system in the adiabatic approximation as follows:

$$\widehat{H} = \sum_n \left\{ \widehat{H}_{0n} + \sum_\nu \hbar\Omega_{n\nu}\left(\widehat{a}^+_{n\nu}\widehat{a}_{n\nu} + \frac{1}{2}\right)\right\}, \tag{25.33}$$

where $\widehat{a}^+_{n\nu}$ and $\widehat{a}_{n\nu}$ are the so-called production operator and annihilation operator of vibrational quanta, index n enumerates electronic states, and index ν enumerates vibrational states of nuclei in the n-th electronic state.

Accordingly, the molecule energy is:

$$E_{n\nu} = \sum_n \left\{ \varepsilon_n + \sum_\nu \hbar\Omega_{n\nu}\left(N_{n\nu} + \frac{1}{2}\right)\right\}, \tag{25.34}$$

where $N_{n\nu}$ is the number of vibrational quanta in the ν-th normal mode in the n-th electronic state of the molecule, ε_n is the electron energy in the n-th state at an equilibrium position of nuclei, i.e. $\varepsilon_n = \varepsilon_n(R_0)$.

The wave function of the system in the n-th electronic state is the product from (25.31):

$$\Psi_{n\nu} = \psi_n\Phi_{n\nu} \equiv \psi_n\prod_{\nu'}\varphi_{n\nu'}, \tag{25.35}$$

where the designation is given for the complete wave function describing vibrational states of the molecule nuclear subsystem (i.e. the system of coupled oscillators). Index 0 in function $\varphi_{n\nu}$ is omitted:

$$\Phi_{n\nu} \equiv \prod_{\nu'}\varphi_{n\nu'}. \tag{25.36}$$

Wave functions $\varphi_{n\nu'}$ are functions of normal harmonic oscillators of the molecule, index ν in function $\Phi_{n\nu}$ designates the totality of all quantum numbers ν'.

Thus, we have obtained wave functions (25.35) and energies (25.34) of quantum states of the molecule in the adiabatic approximation. Note here that in this approximation (it is called harmonic approximation) it is possible to obtain an analytical expression for the overlap integral of vibrational wave functions of a molecular system in different electronic states (the so-called Frank–Condon factor). This integral determines the probability of rearrangement of the equilibrium nuclear configuration of molecule upon changing its electronic state. Below we will directly calculate this integral (the Frank–Condon factor) (see Section 25.2.2 "Electron-vibrational Interactions").

Condition for the Applicability of the Adiabatic Approximation. The adiabatic approximation proves to be correct when the solution of exact equation (25.28) differs little from the solution of the adiabatic approximation (25.30). Using the perturbation theory, it is possible to demonstrate that the condition for the applicability of the adiabatic approximation is reduced to the fulfillment of the inequality:

$$|\langle \Phi_{n\nu}|L_{mn}|\Phi_{n\nu'}\rangle| \ll |E_{m\nu} - E_{n\nu'}|, \qquad (25.37)$$

therewith $m \neq n$, and quantum numbers ν and ν' can vary.

Having calculated the left-hand part of this inequality using Equation (25.29), the following statement can be made: a sufficient condition for the applicability of the adiabatic approximation is the smallness of the energy of nuclear vibrations $\hbar\Omega_\nu$ as compared to the energy, corresponding to molecule transitions between electronic states, i.e. as compared to the energy difference between electron levels of the molecule:

$$\hbar\Omega_\nu \ll |\varepsilon_m - \varepsilon_n|. \qquad (25.38)$$

It can be shown that the vibrational energy of nuclei E_v is $\sqrt{\frac{m_e}{M}}$ times lower than the electron energy ε.

Then let us remember that electron wave functions $\psi_n(r, R)$, depending parametrically on nuclear coordinates, are used in the adiabatic approximation at an equilibrium value of parameter $R = R_0$. Such an approximation is possible only if the mean value of the amplitude of zero vibrations of nuclei $\sqrt{\langle x_0^2\rangle}$ is much lower than characteristic dimensions of molecule d. Using the expression for the amplitude of zero vibrations (see Equation (24.16)) $\langle x_0^2\rangle = \frac{\hbar}{M\Omega} = \frac{\hbar^2}{ME_v}$, we get:

$$\eta \equiv \frac{\sqrt{\langle x_0^2\rangle}}{d} = \left(\frac{m_e}{M}\right)^{1/4} \ll 1. \qquad (25.39)$$

Parameter η is a small parameter of the adiabatic approximation introduced by Born and Oppenheimer. Because of the great difference in masses of nuclei and electrons, Inequality (25.39) is well fulfilled. We have already used it when luminescence processes were considered (Chapter 10, Section 2), assuming that due to the great difference in inertia properties of electrons and nuclei, electron photoexcitation of molecules occurs at static nuclei (the Condon principle). Therewith it was supposed that the difference in energies of the ground and first excited (singlet) electronic states of the molecule exceeds significantly the vibrational energy of nuclei. However, as noted above, in the cases of degeneration or quasi-degeneration of electron terms, the adiabatic approximation is yet insufficient and the nonadiabacity operator should be taken into account.

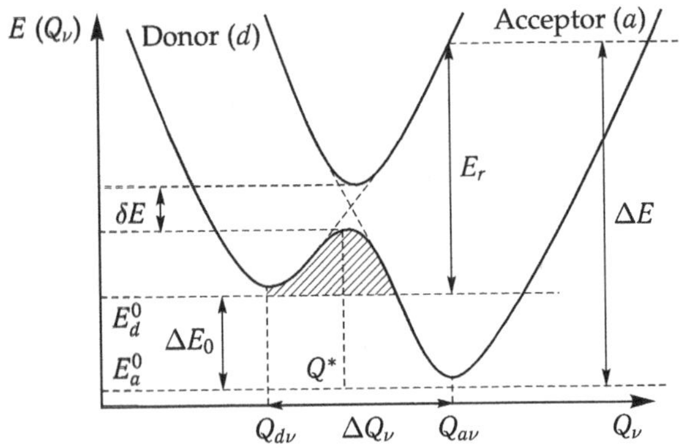

Figure 25.5. Diagram of cross-section of the surface of potential energy of the molecular system along some normal coordinate Q_ν showing the interaction of electron-vibrational terms of the initial (d) and final (a) electronic states of the system.

The energy of the initial state is determined by formula (25.46) and of the final state by formula (25.48). The total difference of energies of these states $\Delta E = \Delta E_0 + E_r$, where $\Delta E_0 = E_d^0 - E_a^0$ is the difference in electron energies of these states, which determines the difference in the free energy of the process $\Delta E_0 = -\Delta G$, E_r is the energy of rearrangement of the environment (see (25.52) and (25.53)) equal to the energy required for the displacement of nuclei from positions $Q_{d\nu}$ in the initial state Ψ_d to positions $Q_{a\nu}$ in the final state Ψ_a provided that the system remains at the initial term Ψ_d. Displacement of equilibrium positions of nuclei is $\Delta Q_\nu = |Q_{a\nu} - Q_{d\nu}|$. Energy δE is the energy of splitting of initial terms, conditioned by their interaction in the intersection point of terms Q^*. If the energy of splitting is comparable to the thermal energy, the description of the system evolution is possible in the framework of the Landau–Zener theory. The shaded region corresponds to the range of nuclei tunneling during rearrangement.

Let us illustrate graphically the condition for application of the adiabatic approximation on the example of electron-vibrational terms (25.34). Expression (25.34) describes the multidimensional potential surface of molecules. Consider the cross-section of this surface along a normal coordinate for two electronic states (Fig. 25.5). The interaction of these terms in the intersection point Q^* leads to their splitting in this point δE, as a result new stationary terms are formed (shown in Fig. 25.5: thick line, lower term; medium line, upper term).

When the splitting energy is considerably higher than the vibrational energy of the normal mode, i.e. $\delta E \gg \hbar\Omega$, the Born–Oppenheimer approximation (i.e. the adiabatic approximation) can be used and, consequently, it is possible not to take into account the nonadiabacity operator L_{mn}. Otherwise, i.e. if the inequality $\delta E \leqslant \hbar\Omega$ is valid, it is required to take into consideration the interaction of nuclear and electron motions (the nonadiabacity operator) in analyzing the evolution of the system (see Fig. 25.5).[2] Thereby let us note one terminological peculiarity. When describing the transition of nuclei from one potential minimum to the other, their own motion is called adiabatic given $\delta E \gg \hbar\Omega$. Otherwise given $\delta E \leqslant \hbar\Omega$, the motion of the nuclei *per se* is called nonadiabatic. This should not be confused with adiabatic and nonadiabatic processes of electron transfer. The latter will be discussed below. The terms "adiabatic" and "nonadiabatic" are widely used both in physics, chemistry and biophysics, therefore

[2]The solution of this task in the quasiclassical approximation upon description of radiationless electron transfers in polyatomic molecules and the phenomenon of predissociation is known as the Landau–Zener approximation.

it is necessary to be careful in using them and in all cases define distinctly to what process this concept is related in a certain situation.

In conclusion let us notice that, as follows from the above estimates, the electron and vibrational energies of molecule differ by η^2 times. This conclusion substantiates inequality (24.2).

25.2 The Charge Transfer Theory

Introduction. The charge transfer is of practical interest only if it is a directed transfer, i.e. electric current. This refers both to technical supplements and biological systems. The practical significance of electric current is determined by its capacity to perform useful work. It is known that in the conducting medium, electric current is generated between points with different electric potentials. Formation of this difference in potentials is associated with the consumption of certain energy — work to separate charges performed by the so-called extraneous forces.

The basis for an elementary act of charge transfer in biosystems is the tunneling transfer. Electron energizing corresponds to the beginning of the process. Then the transfer is performed by consecutive quantum jumps between localization centers — electron donors and acceptors forming an electron transport chain. At each of these centers, the electron "lives" for some period prior to further transition. The electron lifetime at the given donor depends on the probability of its transition to the acceptor determined by many factors, including the interaction with the surrounding. Since inertness is inherent to the medium, the initial state of the system will be inequilibrium when the electron gets to this center. The interaction of the electron with the surrounding will cause relaxation processes transferring the system in an equilibrium state. Because of this, the electron will lose part of its energy. Each stage of the transfer process is accompanied by relaxation of the system from the non-equilibrium state to the equilibrium one. These energy losses make the electron transfer irreversible, i.e. directed.

Relaxation processes can vary both in their nature and in characteristic times. For example, energy dissipation into heat is dependent on energy distribution by vibrational degrees of freedom of the system and proceeds during picoseconds, while conformational changes in the protein molecular structure can be much slower. In many cases namely relaxation processes determine the general rate of electron transport. Thus, at present studies of structural microscopic details of relaxation processes, accompanying charge transfer, have become of great importance for analyzing charge transfer in biosystems. Thereby, at the moment numerical methods of quantum chemistry and computer modeling have become of choice in studying molecular mechanisms of biological processes.

25.2.1 Formulation of the Problem. Localized States

The task of description of electron transport in condensed media with localized states is reduced to solution of the problem of electron transfer from one localization center to the other — from donor D to acceptor A. The complexity of this problem is determined by the fact that in condensed media, the electron transfer between the two localization centers is not a purely electron process. It also includes the motion of the nuclear component manifested in different relaxation processes, affecting the rate of the electron transfer. The electron transfer between localized states can be schematically represented as follows: $D^-A \rightarrow DA^-$. This scheme implies that the multidimensional surface (hypersurface) of potential energy of the whole molecular system has two minima corresponding to states D^-A and DA^- of the system. Transition between these states is the spatial electron transfer for distance R from the donor to the acceptor.

Electron transfer between two localized states D^-A and DA^- can be represented in general by a kinetic scheme as in the case of a chemical reaction:

$$D^-A \leftrightarrows DA^-. \tag{25.40}$$

Direct reaction $D^-A \rightarrow DA^-$ is described by rate constant k_1, and the back reaction $D^-A \leftarrow DA^-$ by rate constant k_2. Let $n_1(t)$ and $n_2(t)$ be populations of the first D^-A and second DA^- states, respectively. Then the rate constant in reaction (25.40) can be compared to the transition rate in a two-level system. This is quite reasonable because in a real system a certain tunneling transition is always realized between two levels.

In a two-level system, populations of the states meet the normalization condition $n_1(t) + n_2(t) = 1$. The system of kinetic equations is as follows:

$$\begin{cases} \dot{n}_1 = -k_1 n_1 + k_2 n_2, \\ \dot{n}_2 = k_1 n_1 - k_2 n_2. \end{cases} \tag{25.41}$$

The dot designates differentiation in time. Choose the initial conditions: $n_1(0) = 1$, $\dot{n}_1(0) = k_1$, $n_2(0) = 0$. Solutions for this system are found easily:

$$\begin{aligned} n_1(t) &= \frac{k_2}{k_1 + k_2}\left(1 + \frac{k_1}{k_2}\exp(-t/\tau)\right), \\ n_2(t) &= \frac{k_1}{k_1 + k_2}(1 - \exp(-t/\tau)), \end{aligned} \tag{25.42}$$

where $\tau = (k_1 + k_2)^{-1}$ is the relaxation time. Reaction (25.40) will be practically irreversible at $k_1 \gg k_2$. Consequently, the problem of the theory is to calculate rate constants k_1 and k_2; for which purpose it is required to estimate the probability of electron tunneling transition from the donor to the acceptor. When considering the model system "hydrogen ion", we determined frequencies of quantum oscillations and, in essence, estimated values of these constants. For this symmetrical model, they are equal to each other, and, as mentioned above, no directed (or irreversible) electron transfer occurs, but rather quantum oscillations take place.

In real molecular systems, realizing electron transport, the electron energy of the initial and final states is, as a rule, different. However the energy of the whole molecular system should remain constant upon an electron transfer. To this end, other molecular processes should occur in the system compensating for the difference in the energies of the initial and final electronic states of the system. The process of vibrational relaxation is one of such processes.

25.2.2 Electron-vibrational Interactions

The theory of vibrational relaxation of the electronic state of molecules, based on the mechanisms of electron-vibrational interactions, has the most developed mathematical apparatus. Results of this theory, formulated as analytical expressions for parameters of electronic states of molecular systems, are in good agreement with experimental data and can be applied to a wide class of electron processes — namely electron-vibrational spectra of molecules, electron transport processes, redox reactions etc.

Our aim is to compute rate constants of direct k_1 and back k_2 electron transfers in the kinetic scheme described by system of equations (25.42) provided the energy balance is realized.

Formulation of the Problem. Let us write the complete Hamiltonian of the system $\widehat{H}_{EN}$ in a general form:

$$\widehat{H}_{EN} = \widehat{T}_r + \widehat{T}_R + V_E(\vec{r}) + V_N(\vec{R}) + W_{EN}(\vec{r}, \vec{R}), \tag{25.43}$$

where $\vec{r}$ and $\vec{R}$ designate the totality of coordinates of all electrons and all nuclei, respectively. Kinetic energy operators are the following:

$$\widehat{T}_r = -\frac{\hbar^2}{2m}\sum_{i=1}^{n}\frac{\partial^2}{\partial r_i^2}, \qquad \widehat{T}_R = -\frac{\hbar^2}{2}\sum_{j=1}^{N}\frac{1}{m_j}\frac{\partial^2}{\partial R_j^2}.$$

Potential energy operators will be represented in general forms, assuming only that they could be written as a sum of double-particle interactions, depending only on the difference of coordinates:

$$V_E(r_{ij}) = \frac{1}{2}\sum_{i\neq j}\frac{e^2}{|\vec{r}_i - \vec{r}_j|},$$

$$V_N(\vec{R}) = \frac{1}{2}\sum_{k\neq l} V_N(\vec{R}_k - \vec{R}_l), \quad W_{EN} = \sum_{i,k} W_{EN}(\vec{r}_i - \vec{R}_k),$$

where $\vec{r}_i$ and $\vec{R}_k$ are the coordinates of electrons and nuclei; and m and m_j are masses of electrons and nuclei, respectively; e is the elementary charge; V_E, V_N and W_{EN} are the potential energy of electron-electron, nuclei-nuclei and electron-nuclei interactions, respectively.

Apparently, generally the Schrodinger equation with such a Hamiltonian cannot be solved. Essential simplifications should be introduced. The first step in such simplification will be the use of single-electron approximation. This can be done if one electron is transferred, while all the other electrons of the system are relatively rigidly linked to their nuclear centers (cores). Then it is possible to describe the interaction of the transferred electron with the other ones by means of some effective potential.

At the next step, we will isolate constant and variable parts from energies V_N and W_{EN}. Therefore, let us represent them as follows:

$$V_N(\vec{R}) = V_N^0(\vec{R}_0) + V_v, \tag{25.44}$$

where $V_N^0(\vec{R}_0)$ is the potential energy of nuclei if they rest in their equilibrium positions $\vec{R}_0$, and V_v is the vibrational energy of the molecular system.

$$W_{EN}(\vec{r}, \vec{R}) = W_{EN}^0(\vec{r}, \vec{R}_0) + W_{el-v}, \tag{25.45}$$

where $W_{EN}^0(\vec{r}, \vec{R}_0)$ is the energy of interaction of electrons and nuclei, resting in their equilibrium positions; W_{el-v} is the energy of electron-vibrational interaction. Namely, this interaction determines transitions between vibrational sublevels of the system in the given electronic state.

Moreover, in the general potential energy of the system we will distinguish energy V_e', determined predominantly by the electrostatic interaction of the transferred electron with the acceptor — a small molecular fragment, where the electron is localized after the quantum jump. Namely, due to this interaction, usually called a resonance interaction, the electron is transferred from the donor to the acceptor. As the electron moves from one potential well to another, the whole system moves from one electronic state to the other. The first of these electronic states of the whole supermolecule is the state of a donor (state d), while the other state is the state of an acceptor (state a).

The electron matrix element referring to the system transition from one electronic state to the other can be calculated (or estimated). Using the apparatus of the nonstationary perturbation theory (the "Fermi's golden rule" (24.31)), we determine the probability of electron transfer from one potential well to another (the probability of electron transfer from the donor to the acceptor). Let us underline that in the Born–Oppenheimer approximation, as well as upon taking into account only the vibrational motion of nuclei (harmonic approximation), the electron tunneling is realized without nuclear motion of the molecular system. In this situation, nuclei take part only in creation of the potential well generation, but their motion does not directly affect the probability of the resonance electron transfer from one well to another. This probability is determined completely by resonance interaction V_e'. The motion (vibration) of nuclei is required for another process — the process of "adjustment" of the system for the electron resonance tunneling. In this approximation, these two events — electron tunneling and creation of conditions for its realization — are independent, because in our approximation we suggest the motion of nuclei and electrons to be independent (see (25.31) and (25.35)). For most bioprocesses, the adiabatic approximation is valid.

If it is suggested that $V_e' = 0$, there will be no electron transfer at all (notwithstanding however intensive the supermolecule nuclei would move). The equality to zero of this potential will correspond to infinitely high walls of the potential well, where the electron is localized, and consequently, the probability of the tunneling or any other electron escape from the well will be zero. Thus, let us once again underline that the electron transfer from the donor to the acceptor (and back) is realized only due to the *electron* interaction with the donor and acceptor. At the same time, the nuclear motions are thought to be frozen (the Condon principle). The states of a supermolecule, corresponding to resonance states of the donor and acceptor, are degenerated, i.e. the system (supermolecule) energy is the same in these two states.

It is natural to propose that an electron is transferred from the upper occupied molecular orbital. Further the energy of this orbital will be designated E_d^0 for the donor state and E_a^0 for the acceptor state. These energies determine electron levels of the system for the given state without account for the vibrational energy of nuclei and energy V_e'. The calculated energy value does not always coincide numerically with the experimental value. Therefore, energy values E_d^0 and E_a^0 should be better taken from experimental data if available.

When describing the tunneling transfer a problematic issue is the question on parameters of the region of electron localization. The wave function of the system contains this information.

The states of the transferred electron, corresponding to term E_j ($j = a,\ d$), are localized, i.e. are realized if there is a spatially restricted potential well with the depth exceeding the electron energy. The width of this potential well at the level of the electron energy determines the spatial region of its localization and is characterized by localization radius a. As a rule, the value of the radius of electron localization does not exceed 1 E. Outside the localization region, the wave function ψ decays. In many cases, this decay is exponential, i.e., has the form of exponential decay $\psi \sim A(r)\exp(-r/a)$, where $A(r)$ is the power function of coordinates. The electron binding centers are frequently remote from each other at distance R, exceeding essentially the radius of localization of the electronic state. Hence the overlapping of wave functions of the electron, localized on the donor and acceptor, depends only on their damped parts.

The initial state D^-A, when the electron is localized on the donor, is characterized by energy

$$E_d(Q) = E_d^0 + \frac{1}{2}\sum_\nu M_\nu \Omega_{d\nu}^2 (Q_\nu - Q_{d\nu})^2, \tag{25.46}$$

where E_d^0 is the purely electron energy of the d-state (the minimal value of the electron term), difference $|Q_\nu - Q_{d\nu}|$ characterizes deviations of normal nuclear coordinates Q_ν from their equilibrium positions $Q_{d\nu}$ in this electronic state, $\Omega_{d\nu}$ is frequencies of normal vibrations in the d-state, M_ν is the reduced masses of normal vibrations, and ν is the number of the normal mode.

The complete wave function of this state of the system (supermolecule) is expressed as the product of the electron wave function and nuclear wave functions (25.35):

$$\Psi_{d\nu} = \psi_d \Phi_d \equiv \psi_d \prod_\nu \varphi_{d\nu}, \tag{25.47}$$

where Φ_d is the product of functions $\varphi_{d\nu}$ that are functions of normal harmonic oscillators.

The energy of the state of the DA^--system when the electron is localized on the acceptor (the final state) is expressed as

$$E_a(Q) = E_a^0 + \frac{1}{2}\sum_\nu M_\nu \Omega_{a\nu}^2 (Q_\nu - Q_{a\nu})^2, \tag{25.48}$$

where E_a^0 is the purely electron energy of the system in the a-state, i.e. the energy of the electron term in the final state minus the vibrational energy, $Q_{a\nu}$ is equilibrium positions of nuclei of the system in this state. Generally speaking, $Q_{a\nu}$ differs from $Q_{d\nu}$. The above expression is written assuming that the reduced masses of normal oscillators M_ν have not changed, though the frequencies of corresponding normal vibrations $\Omega_{a\nu}$ are generally speaking different from $\Omega_{d\nu}$.

The wave function of this state is expressed similar to (25.47):

$$\Psi_{a\nu} = \psi_a \Phi_a \equiv \psi_a \prod_\nu \varphi_{a\nu}, \tag{25.49}$$

where Φ_a is the product of functions $\varphi_{a\nu}$ that are also functions of normal harmonic oscillators, but already in a new electronic state. Note that functions $\varphi_{d\nu}$ and $\varphi_{a\nu}$ are non-orthogonal (i.e. $\int \varphi_{d\nu}\varphi_{a\nu} \neq 0$), because they refer to different electronic states.

Calculation of the Probability of the System Transition from One Electronic State to the Other. Now it is necessary to calculate the probability of the system transition from state $\Psi_{d\nu}$ to state $\Psi_{a\nu}$. This transition is realized as a result of interaction V_e', its energy is assumed to be a small value as compared to energies E_d

and E_a. Generally, electron energy levels $E_d^0 \neq E_a^0$. Conditions for resonance electron tunneling is equivalent to the equality of electron term energy $E_d = E_a$. It is seen from expressions (25.46) and (25.48) that this condition can be satisfied by selection of appropriate vibrational energies of the system. It is here that the role of nuclear motions is displayed. The energy of these motions is consumed to provide the energy balance in a whole system, because the complete energy of the system should be retained upon electron transfer.

To calculate the probability of transition, let us use the "Fermi's golden rule" (24.31). According to this rule, the probability of transition per time unit (i.e. the rate constant in the kinetic scheme considered above (25.42)) is determined as follows:

$$k_{da} = \frac{2\pi}{\hbar} \sum_{\nu,\nu'} P_\nu \left|(V_e')_{\nu\nu'}^{da}\right|^2 \delta(E_{d\nu} - E_{a\nu'}). \tag{25.50}$$

Here the averaging over the initial states of the system and summation over final states (because now vibrational sublevels appear in electron terms) are performed with account of the energy conservation law (as shown by the δ function in Equation (25.50)). Interaction V_e' works as previously — determines the probability of the electron tunneling, corresponding to the system transition from the d-initial electronic state to the a-final state; $(V_e')_{\nu\nu'}^{da}$ is the matrix element of this interaction, and P_ν is the probability of realization of the ν-th initial vibrational state (the Boltzmann factor). The indices ν and ν' contain information on both the number of normal vibrational mode and the number of vibrational quanta excited in it. Note that value k_{da} corresponds to rate constant k_1 of the direct electron transfer in kinetic scheme (25.42).

It is supposed that for a nuclear subsystem, the only consequence of electron transfer is the displacement of equilibrium position of normal coordinates. Therewith possible changes in eigenfrequencies and, generally speaking, in the system of normal coordinates itself are neglected.

The electron transfer from the donor to the acceptor corresponds to the system transition from state $\Psi_{d\nu}$ to state $\Psi_{a\nu'}$. As mentioned above, electron energies in these states are, as a rule, different, i.e. difference $\Delta E^0 = E_d^0 - E_a^0 \neq 0$. Value ΔE^0 can be both positive and negative. However, the complete energy of the system should be conserved upon electron transfer. It means that electron transfer occurs if

$$E_d(Q) = E_a(Q),$$

i.e. the difference in electron energies is compensated by the vibrational energy of the system. This equality does not mean that electron transfer can take place only in the intersection point of terms. It is required only that the energy of electron vibrational states prior to and after the transfer should be the same. Upon electron transfer, the molecular system transits from vibrational state $d\nu$ of the initial electronic state (most frequently from the ground state) to vibrational state $a\nu'$ of the final electronic state. Consequently, to determine the probability of the transition, it is necessary to average the initial electronic state over the vibrational states. After the electron tunneling, the system switches to a new electron term. In this case, it often becomes at a relatively high excited vibrational sublevel ν' of this term, i.e. in non-equilibrium state $a\nu'$. Transition to an equilibrium vibrational state is performed due to the interaction with the thermostat. As a rule, transition to the equilibrium state is associated with the excess energy transferred to the thermostat. In this case, the system will pass to the equilibrium vibrational state, as if coming down along "the steps" (vibrational sublevels) of the electron term.

According to the Condon principle, electron tunneling from the donor to the acceptor (provided all necessary conditions are fulfilled) occurs "instantaneously", i.e. the system "instantaneously" transmits to the new electronic a-state, while the nuclei of the system remain static in their initial positions of the d-state. This in a-electronic state the nuclear configuration of the system is deformed and the whole system is in a non-equilibrium state. The following evolution of the system consists in the transition of the deformation elastic energy into heat, i.e. into vibrational degrees of freedom of the system.

The magnitude of the dimensionless shift between equilibrium positions of nuclei is determined by the following expression

$$\Delta_\nu = \sqrt{M_\nu \Omega_{a\nu}/\hbar}\Delta Q_\nu. \tag{25.51}$$

Difference $\Delta E^0 = E_a^0 - E_d^0 < 0$ is the difference in electron energies of the final and initial states of the system. Implicitly, this is the difference in the free energy of electron transfer ΔG. Note that in the literature ΔE^0 is understood as the reciprocal difference $E_d^0 - E_a^0$. In this case, equality $\Delta E^0 = -\Delta G$ will be fulfilled. The reorganization energy of system is the energy required for the rearrangement of equilibrium configuration of the nuclear subsystem and transition from state Ψ_d to state Ψ_a upon electron tunneling. This energy is necessary to displace normal oscillators from their equilibrium positions in the a-electronic state to positions, corresponding to their equilibrium positions in the d-state ($\Delta Q_\nu = Q_{a\nu} - Q_{d\nu}$). It represents the energy of elastic deformation of the molecular system:

$$E_r = \frac{1}{2}\sum_\nu M_\nu \Omega_{a\nu}^2 (\Delta Q_\nu)^2 \tag{25.52}$$

and at dimensionless variables it is as follows:

$$E_r = \frac{1}{2}\sum_\nu \hbar\Omega_\nu (\Delta_\nu)^2. \tag{25.53}$$

Implicitly, the reorganization energy is equal to the work required to shift nuclei from equilibrium positions in the a-electronic state to equilibrium position in the electronic d-state.

Values ΔE_0 and E_r can be illustrated obviously using the expression for the energy of initial (25.46) and final (25.48) states (Fig. 25.6). The figure shows the cross-section of electronic vibrational terms of the initial and final states of the system along normal coordinate Q_ν.

Franck–Condon Factor. Let us take formula (25.50). The matrix element $(V_e')_{\nu\nu'}^{da}$ with account for wave functions (25.47) and (25.49) is

$$(V_e')_{\nu\nu'}^{da} = \int \Psi_{a\nu'} V_e' \Psi_{d\nu}\, dr\, dQ, \tag{25.54}$$

where r is the electron coordinates and integration is performed over the whole space, Q is the normal coordinates of oscillators and integration is performed over the whole configuration space.

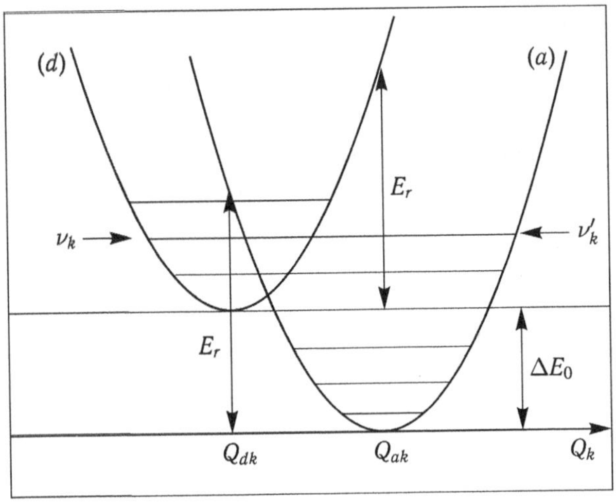

Figure 25.6. Schematic representation of the cross-section of the potential energy surface of molecular system along normal coordinate Q_k.

To explain more clearly the v_k value, the vibrational sublevels of terms in states d and a are shown for the case of energy resonance. $\Delta E_0 = E_d^0 - E_a^0$ is the difference in electron energies of these states, determining the difference in the free energy of the process, and $\Delta E_0 = -\Delta G$; E_r is the reorganization energy of the medium. Transition from term d to term a does not change the normal vibrational mode, however the vibrational sublevel occupied by the electron is changed $\nu_k \to \nu'_k$, consequently, $\nu'_k - \nu_k = v_k$. The activation energy $E_a = (E_2 - \Delta E_0)^2/4E_2$.

By definition, perturbation operator V'_e affects only electron functions ψ_j, therefore the square of the matrix element module (25.54) can be represented as follows:

$$\left|V_{\nu\nu'}^{da}\right|^2 = \left|\int_{(r)} \psi_a^* V'_e \psi_d \, dr\right|^2 \left|\int_Q \Phi_a^* \Phi_d \, dQ\right|^2 \equiv \left|V'_e\right|^2 F_{da}. \tag{25.55}$$

Here designation $|V'_e|^2$ is introduced for the square of the electron matrix element together with the designation:

$$F_{da} = \left|\int_Q \Phi_a^* \Phi_d \, dQ\right|^2.$$

Value F_{da} is the so-called Franck–Condon factor which is the square of the overlap integral of vibrational wave functions of the molecule system.

The Franck–Condon factor averaged over the initial states $W_v = \langle F_{da} \rangle$ represents the probability of the molecular system to transit from the initial electronic state to the final state with the changed totality of vibrational quantum numbers v. Let us write the final expression for the rate constant of electron transfer from the donor to the acceptor (25.50):

$$k_{da} = \frac{2\pi}{\hbar} \left|V'_e\right|^2 \sum_{v=-\infty}^{\infty} W_v \delta(E_d - E_a). \tag{25.56}$$

The calculation of the electron matric element $|V_e'|$ included in this expression will be done after discussion of the temperature dependence of value W_v. Here note that the delta function allows us to choose only such W_v values from the various ones, which make the energies of the system prior to and after electron transfer equal, i.e. when the equation $E_d = E_a$ is fulfilled (the energy conservation law is valid). The procedure of such a choice denotes also summation by the final states.

Summation over Final States. The vibration energy of the system E_{vibr} in the given electronic state minus zero vibrations, i.e. minus the energy of the ground vibrational state (see section "Harmonic Oscillator" in Chapter 24), is expressed as follows:

$$E_{vibr} = \sum_{\nu} \hbar\Omega_{\nu} N_{\nu},$$

where N_ν is the number of vibrational quanta in the ν-th normal mode. Each normal oscillator has an equidistant spectrum with the distance between neighboring levels equal to $\hbar\Omega_\nu$. The number of vibrational quanta N_ν in each normal mode, in principle, can vary, $0 \leqslant N_\nu \leqslant \infty$. Sum E_{vibr} corresponds to the superposition of energy levels of normal modes of molecular systems at variable combinations of occupation numbers N_ν. The minimal distance between energy levels of this superposition is $\Delta\varepsilon_{\min} \leqslant \hbar\Omega_{\min}$, where $\Omega_{\min}$ is the frequency of normal vibration contributing minimally to the electron term energy. Consequently, the estimation of the value of energy density of final states is as follows:

$$\rho \simeq \frac{1}{\hbar\Omega_{\min}}.$$

In its turn, frequency $\Omega_{\min}$ can be determined on the following grounds. Low-frequency vibrational modes correspond to longitudinal acoustic phonons — acoustic waves. Waves that can cause nonhomogeneous deformation in molecular systems and thus change the electronic term energy should have the length of about linear dimensions of the given molecular system. Longer waves do not practically induce such deformation, shifting the molecular system as a whole. Let us designate the characteristic dimension of molecular systems as d_s. Then we obtain the below expression for the vibrational frequency that is of interest for us:

$$\Omega_{\min} \simeq \frac{2\pi c_s}{d_s},$$

where c_s is the velocity of sound. The density of the final states is as follows:

$$\rho \simeq \frac{d_s}{2\pi\hbar c_s},$$

and for the rate constant of electron transfer we get the expression, where summation over final states is given:

$$k_1 \equiv k_{da} = \frac{2\pi}{\hbar}\frac{|V_e'|^2}{\hbar\Omega_{\min}} W_v = \left|\frac{V_e'}{\hbar}\right|^2 \frac{d_s}{c_s} W_v. \tag{25.57}$$

As a result of this reaction, the electron gets to some N-th, probably highly excited, vibrational sublevel of the final electronic term E_a. Two scenarios of the further evolution of this state may take place: either the electron tunnels back to the donor, or as a result of vibrational relaxation it passes to the ground vibrational level of a new term and is localized on the acceptor.

25.2.3 Decay of Excited State of Quantum Systems

Let us consider the evolution of the electronic state of the system, after the electron has attached to the acceptor. If nothing had occurred in the system after the electron jump, i.e. the system had been frozen, the electron would have returned back to the donor at the same rate and quantum oscillations would have emerged in the system. However, occupation by the electron of the high vibrational sublevel of the electronic term signifies emergence of a non-equilibrium situation, because the average nuclear positions still correspond to equilibrium positions characteristic of the initial electronic d-state of the system. Consequently, after the electron quantum jump, the process of relaxation begins — the establishment of equilibrium in the system, realized via the electron vibrational interactions. If the rate of the relaxation process exceeds that of the electron quantum jump, the transfer process can be irreversible, because during its stay on the acceptor, the electron has enough time to scatter part of its energy into vibrations and, accordingly, it will be more difficult for it to return back to the donor. Thus, the relaxation rate will determine the rate constant of the back reaction k_2 (25.41), i.e., the back electron transfer from the acceptor to the donor. Apparently, the efficiency of the back electron transfer is determined by the ratio of the electron lifetime on the acceptor and the characteristic time of electron vibrational relaxation. The electron lifetime on the acceptor is $\tau_e \simeq k_2^{-1}$. If $\tau_e \ll \tau_r$, where τ_r is the characteristic time of disintegration of the excited electron vibrational state, no directed transfer will take place, and quantum oscillations will appear. Otherwise, when $\tau_e \gg \tau_r$, the electron has time during its staying on the acceptor to scatter the rest of its energy, going down to the ground sublevel of the electron term E_a. If $E_a^0 < E_d^0$, the rate constant of the back transfer is proportional to the Boltzmann factor $k_2 \sim \exp(-\Delta E^0/k_{\rm B}T)$, where $\Delta E^0 = E_d^0 - E_a^0$, and, consequently, k_2 can become a rather small value as compared to k_1.

Note that operator W_{el-v} regulates not the external action on the system but reflects the relaxation process going on inside the system. From the point of view of an external observer, our system will spontaneously pass from the excited state to the ground one. In other words, the excited state of our system will not be a stationary state, but will have the final lifetime, i.e. will decay with time. As a result, in some time the system will be in the equilibrium state.

Disintegration of excited states in quantum systems. Quantum systems can spontaneously perform transition from the excited state to a lower energy state. A system cannot be completely isolated from the external world; therefore, excited states are evidently not strictly stationary ones. They are called quasistationary and are characterized by a decay law $L(t)$. Let us introduce function $L(t)$ determining the probability that in time t the system will still be in the given excited state. At relatively high times, as compared to period $T = h/E$ characteristic of oscillations of the complete wave functions in the given excited state, the decay law (or damping) of this state is thought to be exponential:

$$L(t) = \exp(-\Gamma t), \tag{25.58}$$

where value Γ^{-1} is called the lifetime of the excited state.

This problem is closely related to the spectroscopic problem of determining the shape of the emission (absorption) line of quantum objects. Measurements of the energy of the state of a system should be done for a time, not exceeding the lifetime τ of the quantum system in this state (otherwise the state itself will disappear).

Consequently, according to the uncertainty ratio for energies (25.58), the accuracy of determining the energy of a state cannot in principle exceed the error $\Delta E \geqslant \hbar/\tau$. This primary inaccuracy of determining the energy of the state causes the dispersion in the observed transition frequencies $\Delta\omega = \tau^{-1}$. Spontaneous emission (beyond external influence) is determined by the intrinsic lifetime of the excited state, and the observed frequency uncertainty $\Delta\omega$ is called the natural spectral line width. The natural lifetime of the excited state of an isolated atom is about 10^{-8} s. It is clear that if the system (atom) is in a medium, the lifetime of the excited state will be much lower because of the interaction between the system and the surrounding.

In quantum mechanics, the solution of the problem of damped excited states is, as a matter of fact, analogous to the problem of studying the classical damped oscillator.

Thus, it follows from the above-stated that, if the rate constant of the direct reaction of electron transfer from the donor to the acceptor k_1 exceeds essentially the rate of vibrational relaxation τ_n^{-1}, i.e. $k_1 \gg \tau_n^{-1}$, no directed electron transfer during time $T < \tau_n$ will occur, because in this case approximate equilibrium $k_1 = k_2$ will be preserved and, as a result, quantum oscillations of the electron between the donor and acceptor will take place within time T. Note that, if condition $k_1 \gg \tau_n^{-1}$ is fulfilled, the electron transfer occurs at a rate considerably exceeding the relaxation rate of the medium, the latter being as if frozen. If after time $T > \tau_n$ the electron is nevertheless localized on the acceptor, this occurs most likely under the action of other factors. For example, microconformational rearrangements may take place in the structure of the local surrounding of the binding centers, effect of hydrogen bonds and the polarity of the surrounding, etc. If relaxation of the medium were determined only by vibrational relaxation, the directed electron transfer might not have taken place at $k_1 \gg \tau_n^{-1}$.

Otherwise, when $k_1 \ll \tau_n^{-1}$, the electron transfer will become irreversible, because in this case the reverse transfer constant will be $k_2 = k_1 \exp(-\Delta E/k_B T)$, where $\Delta E = v_n\hbar\Omega$. Value ΔE can be compared to the Gibbs free energy of this reaction ΔG. Then we get a standard thermodynamic statement — the reaction will proceed spontaneously towards a decrease in the free energy rather than otherwise. Indeed, constant k_2 can be a small value as compared to k_1, but nonetheless it not zero. Therefore, in principle, the back reaction can take place, but its probability can be so low that the direct reaction can be considered as irreversible.

25.2.4 Analysis of the Temperature Dependence of the Electron Transfer

In previous sections, we obtained basic formulas for the rate constant of electron transfer from the donor to the acceptor (25.56). The formulas contain functions including the medium temperature that makes it possible to represent the dependence of the electron transfer rate constant on temperature $k_{da} = k_{da}(T)$ and verify the theory experimentally.

The vibrational energy of a molecular system can be accumulated as a result of summation of excitation energies of different normal vibrational modes with different occupation numbers. This is clearly seen from formula (25.34): if the vibrational energy has certain value $E_v = \text{const}$, the value of this energy can be fixed in a countless number of ways choosing different Ω_j and N_j. Accordingly, the same vibrational energy can be generated by excitation of a various number of different vibrational quanta (phonons). In practice, mainly for the simplicity, efforts are made to choose one, the so-called normal accept-mode, naming it "soft" or "rigid" depending on its frequency Ω. Such a "single-mode" approach may be justified by the fact that since normal vibrations are independent, the probability of simultaneous excitation of such vibrations is equal to the product of probabilities of excitation of each of them and it can be rather low. Note also that the problem of the number of active normal modes refers mainly to the problem of a "discharge" of the excess vibrational energy upon a change in the electronic state of the system.

When analyzing the temperature dependence of the electron transport rate, let us by tradition separate two extreme cases, determined by the value of dimensionless displacement of the equilibrium position of normal oscillators Δ (25.51): the cases of strong and weak electron vibrational bonds. The classification of these cases is determined by the relation of the amplitude of zero vibrations $x_0 = (\hbar/m\omega)^{1/2}$ (see (24.17)) of normal oscillators and the magnitude of nuclear displacement ΔQ upon relaxation to the new equilibrium state. Amplitudes of zero vibrations of atoms in molecules are determined mainly by their characteristic frequencies (the infrared spectrum). For example, for carbon atoms forming a covalent bond of the $sp^3 - sp^3$ type, the characteristic frequency $\omega = 1.45 \cdot 10^{14}\ \mathrm{s}^{-1}$, and consequently $x_0 \sim 0.06$ E. Thus, if in the process of rearrangement of the nuclear system to a new state, nuclei are displaced by a considerably smaller value than 0.06 E, this corresponds to a weak electron vibrational coupling (equilibrium positions of the nuclei remain practically the same as the positions in the initial state). But if the rearrangement of the nuclear system to a new state requires, on the contrary, a displacement of equilibrium positions of nuclei by a considerably larger value than 0.061 E, this situation corresponds to a strong electron vibrational coupling. Note that for a proton, the lightest nucleus, the amplitude of zero vibrations in a typical covalent bond makes approximately 0.1 E. A graphic illustration of different variants of an electron vibrational bond is given in Fig. 25.7.

Below we will consider a single-mode approximation, i.e. limit ourselves to taking into account one vibrational normal mode of frequency Ω (the accepting-mode). First, from formula (25.53) let us get an expression for the square of a dimensionless displacement of equilibrium positions of nuclei upon electron transfer:

$$\Delta^2 = \frac{\Delta_\Omega}{\sqrt{2}} = \frac{E_r}{\hbar\Omega}. \tag{25.59}$$

If the reorganization energy is known, it is possible to estimate the magnitudes of displacements of nuclei upon the rearrangement. Indeed, using (25.51) we obtain the expression for displacement of normal nuclear coordinates:

$$\Delta Q = \sqrt{\frac{2E_r}{M\Omega^2}}. \tag{25.60}$$

Value $v_k = \nu'_k - \nu_k$ characterizes the change in the number of excited vibrational quanta of the k-th normal mode upon a change in the electronic state of the system. Apparently, the number v_k is associated with the difference in the electron energies $\Delta E^0 = E_d^0 - E_a^0$, corresponding to the difference in the free energy of the process (see Fig. 25.6). Let us use the conventional designation for the difference in free energies of states of the system ΔG, then $-\Delta G = \Delta E^0$, and number v_k is calculated in a simple way:

$$v_k = \frac{\Delta E^0}{\hbar\Omega} = \frac{|\Delta G|}{\hbar\Omega}. \tag{25.61}$$

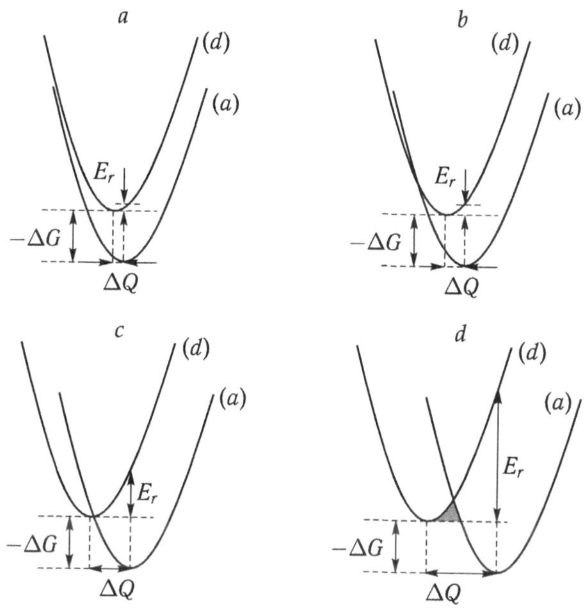

Figure 25.7. Schematic representation of mutual location of electronic terms in the initial (d) and final (a) states: a and b, a weak electron vibrational coupling; c and d, a strong electron vibrational coupling.

The difference of these cases from each other is only in the magnitude of the displacement of the equilibrium position of nuclei upon a change in the electronic state of the system. The free energy value of the process ($\Delta G = \Delta E_0$), the energy of the medium rearrangement (E_r) and the displacement of equilibrium positions of normal coordinates (ΔQ) are shown. Figure c corresponds to the case of a radiationless electron transfer, when $E_r = -\Delta G$. In figure d the region of tunneling of nuclei is shaded.

As concerns designations, it should be noted that in biological literature on the electron transfer, value Δ^2 is frequently designated as S, i.e. $\Delta^2 \equiv S$; reorganization energy E_r is often designated as λ, i.e. $E_r \equiv \lambda$; parameter v_k, simply v for a single-mode approximation, is frequently designated as p, i.e. $v \equiv p$; $n = \exp(-\hbar\Omega/k_B T)$ is the Boltzmann factor, and value $N = (\exp(\hbar\Omega/k_B T) - 1)^{-1}$ is the number of phonons in the given normal mode. The rate constant of the electron transfer is proportional to the relation of probability of phonon emission to the probability of phonon absorption to the extent $v/2 \equiv p/2$. In addition, pay attention to the identical equalities:

$$\frac{1+n}{1-n} \equiv 2N+1,$$
$$\frac{n^{1/2}}{1-n} \equiv (N(N+1))^{1/2}.$$

Now let us derive a general expression for the rate constant of electron transfer (25.56) in the single-mode approximation that will be called tentatively the Jortner formula and indexed by J, i.e. k_J:

$$k_J = \frac{2\pi}{\hbar}\frac{|V_e'|^2}{\hbar\Omega}\exp\left(\frac{v\hbar\Omega}{2k_bT}\right)\exp\left(-\Delta^2\coth\left(\frac{\hbar\Omega}{2k_bT}\right)\right)I_v\left(\Delta^2\sinh^{-1}\left(\frac{\hbar\Omega}{2k_bT}\right)\right). \quad (25.62)$$

Here the property of the δ-function is used:

$$\delta(\Delta E - \hbar\Omega) = \frac{1}{\hbar\Omega}\,\delta\left(\frac{\Delta E}{\hbar\Omega} - 1\right).$$

This expression corresponds to the fact that we substitute the δ-function in expression (25.56) by the density of final states $\rho_E = \frac{1}{\hbar\Omega}$. It is evident for a harmonic oscillator where the energy spectrum is equidistant. As seen from(25.62), the rate constant of electron transfer, excluding the electron matrix element, is the Franck–Condon factor.

In contrast to the electron matrix element, the Franck–Condon factor determines the probability of transition of nuclei of the system to new equilibrium positions. Nuclei can reach new equilibrium positions both due to tunneling (i.e. subbarrier "leakage") and by means of a suprabarrier (activation) transition. Nuclear tunneling should not be confused with electron tunneling, because these are quite different processes. Indeed, the tunneling transition of electrons from the donor to the acceptor is a charge transfer process, whereas the nuclear tunneling characterizes the rearrangement of the molecule nuclear configuration (relaxation). The below limiting cases demonstrate that at low temperatures the Franck–Condon factor is independent of temperature. Therefore, at low temperatures, nuclei are rearranged by means of tunneling, i.e. without activation. In this case, factor value F_{da} is equal to the probability of the process. Another limiting case (high temperatures) shows, on the contrary, that nuclei can overcome the barrier upon activation, because under such conditions their kinetic energy can be high enough for their jump over the barrier. But in this case nuclear tunneling also takes place. However, at high temperatures the probability of their jump over the barrier can be comparable or even higher.

The Case of a Weak Electron Vibrational Bond $\Delta \ll 1$. Electron terms of the initial and target states for this case are given in Fig. 25.7 (*a*, *b*).

1^0. **Low Temperatures.** Using the asymptotic representation of functions, we can derive the following expression in the limiting case of low temperatures (25.62):

$$k_{da}^{0}\Big|_{\Delta \ll 1}^{T\to 0} = \frac{2\pi}{\hbar}\frac{|V_e'|^2}{\hbar\Omega}\,e^{-\Delta^2}\frac{\Delta^{2v}}{v!}. \qquad (25.63)$$

As seen, temperature is not included in this equation. Consequently, within this limit the electron transfer rate does not depend on temperature. In this case, the nuclear system reaches its new equilibrium state by tunneling (without activation).

2^0. **High Temperatures.** Let us consider a situation when inequality $\Delta^2 \ll \hbar\Omega/2k_BT \ll 1$ is fulfilled. This means that displacements of nuclei should be small and the accepting-mode should be of low frequency or of high temperature ($\hbar\Omega \ll k_BT$). Based on inequality $\Delta^2 \ll \hbar\Omega/2k_BT \ll 1$, we will estimate value ΔQ of displacement of equilibrium positions of nuclei. It follows from inequality $\Delta^2 \ll \frac{\hbar\Omega}{k_BT}$ that displacement ΔQ should satisfy condition $\Delta Q \ll \hbar/\sqrt{Mk_BT}$. Then at $T = 300$ K for a carbon atom we obtain $\Delta Q \ll 0.08$ E. This complies with the assumption that the bond is weak. Therefore, the energy of vibrational quanta should obey inequality $\hbar\Omega \ll k_BT$, i.e. at 300 K $\Omega \ll 8 \cdot 10^{13}$ s^{-1}. This condition is quite feasible. Thus, the new equilibrium positions of the nuclei should not differ greatly from the initial equilibrium positions ($\Delta \ll 1$). In this case, strong overlap of nuclear wave functions will be observed. Therefore, the nuclear transition to new equilibrium states will have both a tunneling and an activation character.

Now let us consider another limiting case, when inequality $\hbar\Omega/2k_BT \ll \Delta^2 \ll 1$ is fulfilled. Hence it follows that at 300 K ΔQ should be $0.08 \ll \Delta Q \ll 0.2$ E, i.e. the permissible range of nuclear displacements is rather narrow. Using formula (25.62) for the rate constant, we get the formal approximation:

$$k_{da}^0\Big|_{\Delta\ll 1}^{T\to\infty} \approx \frac{|V_e'|^2}{\hbar}\frac{1}{\Delta}\sqrt{\frac{\pi}{\hbar\Omega k_BT}}\exp\left(\frac{\upsilon\hbar\Omega}{2k_BT}\right). \tag{25.64}$$

At low frequencies of the accepting-mode, inequality $\upsilon\hbar\Omega \ll 2k_BT$ means that the difference in the free energy of electron transfer is rather small, because $-\Delta G = \Delta E^0 = \upsilon\hbar\Omega$. Consequently, under these conditions, the rate constant of electron transfer can be regarded as proportional to the square root of temperature:

$$k_{da}^0\Big|_{\Delta\ll 1}^{T\to\infty} \sim \frac{1}{\sqrt{T}}, \tag{25.65}$$

because the temperature in Expression (25.64) remains only in the pre-exponential multiplier.

Temperature dependence similar to (25.65) should be evidently observed for processes characterized by a quite small difference in the free energy.

But when ΔG of the process is high (i.e. if there is large difference between the electron energies of the initial and target states and, accordingly, number υ is large), the exponential index in (25.64) may also be of a remarkable value so that this exponential cannot be considered equal to unity. Then the rate constant will depend on temperature according to law (25.64) that can be represented as follows:

$$k_{da}^0\Big|_{\Delta\ll 1}^{T\to\infty} \approx \frac{|V_e'|^2}{\hbar}\frac{1}{\Delta}\sqrt{\frac{\pi}{\hbar\Omega k_BT}}\exp\left(-\frac{\Delta G}{2k_BT}\right). \tag{25.66}$$

In this case, the reorganization energy remains of a small value (it can be regarded aspractically equal to zero, see Fig. 25.7, *a*). Consequently, the system transition from state "*d*" to state "*a*" in the scheme of terms corresponds to the situation when, figuratively speaking, remaining static, the system turned out to be over a deep potential well (the well was formed suddenly). Into this well the system is sliding down, and the deeper the well (the higher $-\Delta G$) the faster the sliding.

The Case of a Strong Electron Vibrational Coupling $\Delta \gg 1$.

1^0. **Low temperatures.** If $1 \ll \Delta^2 \ll \delta$, the frequency of the accepting-mode should obey condition $\hbar\Omega \gg 2k_BT\ln(2\Delta^2)$ which I fulfilled, as can be seen, either at high frequencies, or at low temperatures. Using formula (25.62), we get the expression for the rate constant the same as (25.63). Indeed, inequality $1 \ll \Delta^2 \ll \delta$, as can be easily verified, is always valid at relatively low temperatures.

In this case, the rate constant is independent of temperature. In other words, at low temperatures there is no essential difference in whether a strong or a weak electron vibrational coupling is formed upon electron transfer.

2^0. **High temperatures.** At high temperatures and a strong electron vibrational coupling, nuclear transition to new equilibrium positions has both a tunneling and an activation character. The probability of a jump over the barrier at high temperatures can even exceed the probability of tunneling of heavy nuclei for large distances (a strong electron vibrational coupling). In this limiting case of high temperatures and a strong electron vibrational coupling, we get the following expression for the rate constant

$$k_{da}\Big|_{\Delta\gg 1}^{T\to\infty} \approx \frac{2\pi}{\hbar}|V_e'|^2 \frac{1}{\sqrt{4\pi E_r k_B T}} \exp\left(-\frac{(\Delta E_0 - E_r)^2}{4E_r k_B T}\right), \tag{25.67}$$

where $\Delta E_0 = E_d^0 - E_a^0$ is the difference in electron energies prior to and after electron transfer, $E_r = \hbar\Omega\Delta^2$ is the energy of the medium rearrangement. If we introduce designation $\Delta E_0 = -\Delta G$ for the free energy, we get a formula coinciding with the known formula of R. A. Marcus for the rate constant of electron transfer, which will be designated as k_M:

$$k_M \equiv k_{da}\Big|_{\Delta\gg 1}^{T\to\infty} \approx \frac{2\pi}{\hbar}|V_e'|^2 \frac{1}{\sqrt{4\pi E_r k_B T}} \exp\left(-\frac{(E_r + \Delta G)^2}{4E_r k_B T}\right), \tag{25.68}$$

where value

$$U_a = \frac{(\Delta G + E_r)^2}{4E_r} \tag{25.69}$$

means the activation energy of the process. As seen from the given expression, this case has an activation character and is realized with involvement of all possible vibrational modes (both soft and rigid), because inequality $\hbar\Omega/2k_B T \ll 1$ is believed to be fulfilled at any vibrational frequency. Thus, at high temperatures, the curves described by general formula (25.62) and its limiting case (25.68) should be compatible (see Fig. 25.8).

25.2.5 General Analysis of Jortner and Marcus Formulas

In this Section we will discuss in more detail the expressions for rate constants of electron transfer k_J and k_M, most widely used in analysis of experimental data obtained upon studying electron transport in biological systems. Let us write expressions (25.62) and (25.68), explicitly introducing in them the energy of medium rearrangement (E_r) and the difference in the free energy of electron transfer (ΔG). Then the single-mode approximation of the general formula will be as follows (the Jortner formula):

$$k_J = \frac{2\pi}{\hbar}\frac{|V_e'|^2}{\hbar\Omega} \exp\left\{-\frac{E_r}{\hbar\Omega}\cdot\coth\left(\frac{\hbar\Omega}{2k_B T}\right) + \frac{|\Delta G|}{2k_B T}\right\} I_v \left(\frac{E_r}{\hbar\Omega\cdot\sinh\left(\dfrac{\hbar\Omega}{2k_B T}\right)}\right). \tag{25.70}$$

The Marcus formula (25.68) will remain unchanged:

$$k_M = \frac{2\pi}{\hbar}|V_e'|^2 \frac{1}{\sqrt{4\pi E_r k_B T}} \exp\left(-\frac{(\Delta G + E_r)^2}{4E_r k_B T}\right). \tag{25.71}$$

From expression (25.70) it is seen that the value of rate constant k_J depends on five parameters: temperature T, difference in free energy ΔG, reorganization

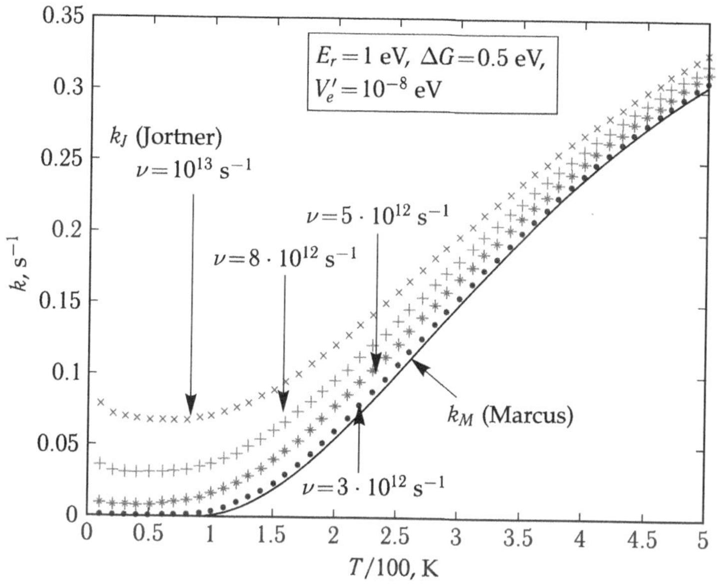

Figure 25.8. Comparison of curves of temperature dependence of rate constant $k_J(T)$ and $k_M(T)$ obtained using formulas (25.70) and (25.71), respectively.

Parameter values used in the calculations are given in the figure. For rate constant k_J, curves are shown for four different frequencies of the normal accepting-mode: 10^{13} (×), $8\cdot10^{12}$ (+), $5\cdot10^{12}$ (∗) and $3\cdot0^{12}$ s^{-1} (•). The bottom solid curve is the curve for $k_M(T)$.

energy E_r, vibrational frequency of the normal accept-mode Ω, and the value of electron matrix element V_e'. Constant k_M is dependent on the same parameters. Only expression (25.71) does not include explicitly the frequency of normal vibrations.

Let us consider these parameters in more detail. Ambient temperature T is a quite definite parameter that can be quite accurately measured experimentally.

The value of the difference in free energy ΔG can be measured, for example, using titration where the midpoint potential of the redox reaction is coordinated with ΔG. The problem discussed in connection with these measurements is that the midpoint potential is determined under equilibrium conditions, wheres upon real electron transport the molecular system may have no equilibrium state. Thus, the experimental magnitude of ΔG is not an exact but rough value.

Reorganization energy E_r cannot be measured in experiments directly and consequently it is an unobservable value and cannot be calculated using this theory. Therefore, the reorganization energy is an adjustable parameter, so that the theory itself is phenomenological.

Information on the frequency of the normal accepting-mode Ω cannot be derived from experimental measurements, though in principle the spectrum of normal vibrations *per se* can be measured (using IR spectroscopy) and even calculated using computer methods. However, it is impossible to isolate from this spectrum the vibration that accompanies the transfer process. Moreover, a real process involves to some extent all normal vibrations of the system.

Finally, the electron matric element, determining the probability of the electron quantum jump from the donor to the acceptor, cannot be measured directly in experiments. When analyzing experimental data, the value of the electron matric element V_e' is "chosen" basing on a quantitative coincidence of the measured and calculated rate constant. Thus, the given theory has at least three adjustable parameters: reorganization energy E_r, frequency of normal vibrations Ω, and the value of electron matrix element V_e'.

The analysis of the experimental temperature dependences of electron transfer rate using formulas (25.70) and (25.71), puts a question: are these parameters themselves dependent on temperature or are their values unchanged?

In the Marcus theory, where the rate constant is described by expression (25.71) derived by Marcus (1985) independently, it is assumed that the reorganization energy of the medium consists of two parts:

$$E_r = E_r^o + E_r^i,$$

where E_r^o is the so-called equilibrium external sphere reorganization energy, determined by the rearrangement of the solvent, and E_r^i is the internal sphere reorganization energy, determined by the change in the lengths of bonds inside the system. The latter is proposed to be independent of temperature and for a certain process is considered constant. The reorganization energy introduced in the general description as the Jortner formula, corresponds namely to this part of the Marcus energy. However, the external sphere of the reorganization energy, corresponding to the "soft" vibrational modes of the solvent, can presumably be dependent on temperature. In the Marcus theory, it is expressed as follows:

$$E_r^o = e^2 \left(\frac{1}{\varepsilon_\infty} - \frac{1}{\varepsilon_s} \right) \left(\frac{1}{2r_1} + \frac{1}{2r_2} - \frac{1}{r_{12}} \right), \tag{25.72}$$

where e is the electron charge, r_1 and r_2 are radii of two reacting molecules, r_{12} is the distance between centers of spherical molecules, ε_∞ and ε_s are permittivity values of the solvent at optical frequencies and at zero (static) frequency, respectively. Generally speaking, static permittivity ε_s can be dependent on temperature and thus makes the reorganization energy E_r^o itself temperature dependent. Consequently, for biological systems a problem arises as to determination of permittivity as such and its dependence on temperature in particular. It becomes principally possible that E_r would depend on temperature. Thus, the reorganization energy is separated in two parts, one of them corresponding to "rigid" modes of normal oscillators and is independent of temperature, while the other corresponds to "soft" modes and may depend on temperature.

The third parameter is the difference in the free energy ΔG of the electron transfer process. Can ΔG be dependent on temperature? In other words, can the difference in energies of electron levels $\Delta E^0 = E_d^0 - E_a^0$ be dependent on temperature (remember that this difference does not include the vibrational energy of molecular system nuclei)? The answer is as follows: ΔG, characterizing the equilibrium initial and final states of the system, cannot be temperature dependent. However, as mentioned above, real experimental conditions can differ from equilibrium conditions, where value ΔG (the midpoint potential) is measured. This is determined by the rate of relaxation processes in the system that can, naturally, depend on temperature. Therefore real value ΔG^* and equilibrium value ΔG can diverge, and the magnitude of this divergence can be temperature dependent. Thus, an apparent dependence of ΔG on temperature can appear in experiments. Responsible for this dependence are relaxation processes that relatively slowly bring the system to an equilibrium state. Here relaxation processes may be any processes "leading" the system to the equilibrium state, excluding processes of vibrational relaxation, because the latter occur at any temperatures rather quickly — during tens, maybe, hundreds of picoseconds. Processes of reorientation of dipole moments of small solvent molecules (for example, water molecules) and polar atomic groups are most frequently referred to as candidates in nonoscillating processes of relaxation, because such a reorientation of dipoles is associated with reorganization of the medium. Among the variety of possible structural changes in molecular systems, special attention is attracted at present to changes involving hydrogen bonds of the system. Hydrogen bonds are a widespread type of bonds in biological molecular systems. They are labile and widely variable in their properties, for example, energy. The hydrogen bond energy in different compounds can vary from 0.065 eV (a very weak bond) to 1.74 eV (a very strong bond, by its energy comparable to a covalent bond). Typical values of the hydrogen bond energy are 0.13–0.35 eV.

Finally, let us consider the fourth parameter of the electron transfer theory — the electron matrix element V_e'. By definition, this parameter cannot be dependent on temperature, because it describes the tunneling electron transition from the donor to the acceptor under conditions of electron resonance. In principle, this transition is independent of temperature. But the magnitude of the electrostatic interaction of the electron with the acceptor depends on permittivity of the medium ε. As seen above, the magnitude of permittivity can be temperature dependent. Consequently, value V_e' can be also temperature dependent, for example, E. W. Knapp (1987) made such assumption.

Based on the possible temperature dependence of parameters included in formulas (25.70) and (25.71), a comparative analysis of the dependence of these constants on each parameter was made with the remaining parameters being constant. The difference between the Marcus and Jortner formulas is that the Jortner formula has an additional variable parameter — the frequency of the normal accepting-mode. In particular, this manifests itself in that at $T \to 0$ the Marcus formula "goes" to zero, while curve k_J $(T \to 0)$ behaves differently depending on the frequency of the so-called normal accepting-mode (Fig. 25.8).

The same situation is shown in Fig. 25.9, where curves are plotted for these constants if $E_r = \Delta G$ that corresponds to the radiationless electron transfer process. In this case, in contrast to Fig. 25.8, we have a decrease in the rate constant with an increase in temperature according to law $k \sim \dfrac{1}{\sqrt{T}}$ mentioned above (25.65). Figure 25.9 shows that by selecting the frequency of a normal mode it is possible to superpose temperature dependence curves of electron transfer rate constants determined by the Jortner–Marcus theory.

In Fig. 25.10, curves are plotted for the dependence of constants k_J and k_M versus the reorganization energy E_r at other constant parameters: $T = 280$ K, $\Delta G = 0.5$ eV; the frequency of the normal mode was chosen to be $6 \cdot 10^{12}$ s^{-1}, the value of the electron matrix element is accepted to be the same ($V_e' = 10^{-8}$ eV). Figure 25.11 shows an analogous dependence of rate constants versus the free energy value (ΔG), where $E_r = 0.5$ eV. The figures demonstrate that these dependencies are practically the same (small differences are at the "tails"). In the literature, such curves are called the "Marcus parabolas": the left-hand branch is called a rising branch, and the right-hand one — the descending branch. It is clear that these curves can be superposed by selecting the frequency of normal vibration. From these curves, it follows that

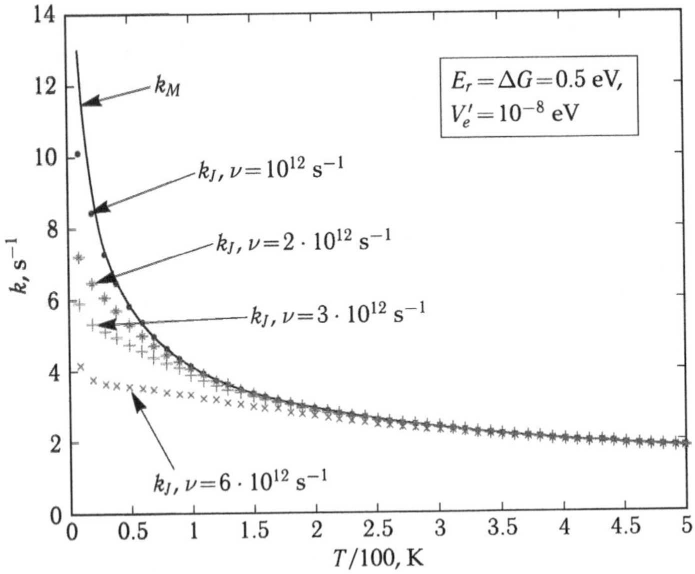

Figure 25.9. Comparison of temperature dependence curves of rate constants $k_J(T)$ and $k_M(T)$ obtained using formulas (25.70) and (25.71), respectively, provided $\Delta G = E_r$ (radiationless process).

Values of parameters used in calculations are given in the figure. For rate constant k_J, curves are given for four different frequencies of the normal accepting-mode: $6 \cdot 10^{12}$ (×), $3 \cdot 10^{12}$ (+), $2 \cdot 10^{12}$ (∗), and 10^{12} s^{-1} (•). The upper solid line is the curve for $k_M(T)$.

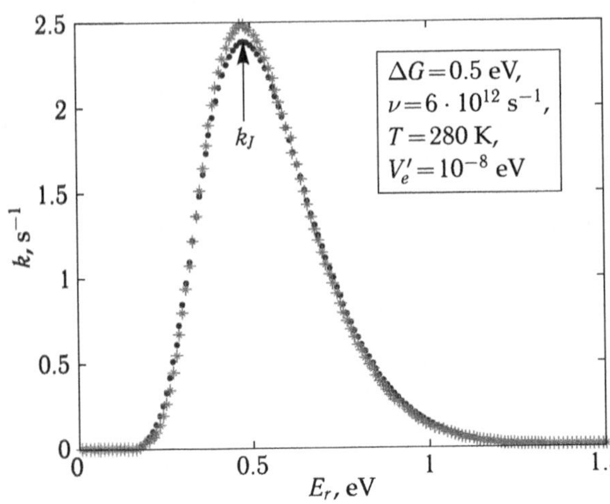

Figure 25.10. Dependence of k_J (•) and k_M (∗) versus the reorganization energy E_r.

the maximum transfer rate is reached, if the reorganization energy and the difference in the free energy are equal ($E_r = \Delta G$). These curves show also that if upon consideration of the temperature dependence of the rate constant it is supposed that E_r or ΔG are also dependent on temperature, then generally speaking, we can obtain different shapes of temperature dependence curves including the non-monotonic one for the

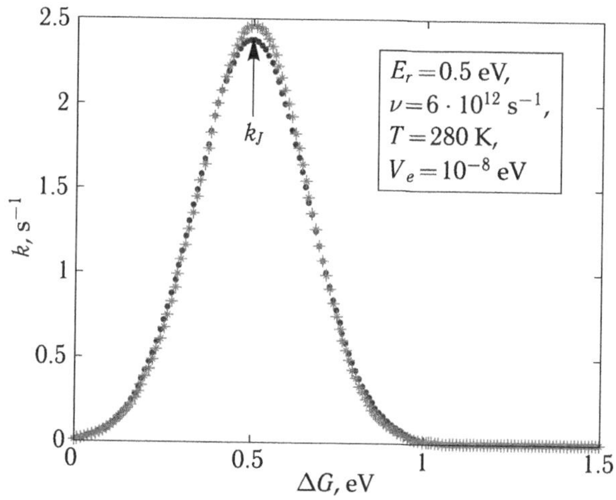

Figure 25.11. Dependence of k_J (•) and k_M (∗) versus the difference in free energy ΔG. This Gaussian distribution Is called the "Marcus parabola".

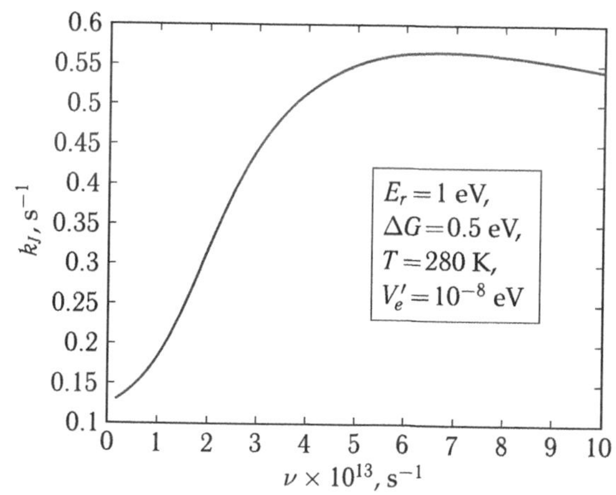

Figure 25.12. Dependence of k_J versus the frequency of the accepting-mode.

rate constants of electron transfer. The problem here concerns the mechanism of E_r or ΔG temperature dependence.

Now let us analyze how rate constant k_J depends on the frequency of the normal accepting-mode. Figure 25.12 shows an example of such dependence for parameter values: $T = 280$ K, $\Delta G = 0.5$ eV and $E_r = 0.5$ eV. This dependency is not monotonic and has a slightly sloping maximum. Consequently, the electron transfer rate, under other equal conditions, is determined by the frequency of the accepting-mode coupled to the electron transfer.

Due to the process of electron vibrational relaxation, the difference in energies of the electronic states of the system before and after donor-to-acceptor electron transfer is compensated. Obviously, some other relaxation processes can play the role of a compensating factor. However, electron vibrational relaxation is of a universal character and to one or another extent is revealed in different electron processes. Electron vibrational relaxation "adjusts" the nuclear component to the new electronic state of the molecular system. This relaxation process, as well as any other process, has a definite *temperature* dependence that affects the rate constant value of electron transfer.

However, it is important to underline that the probability of the intrinsic tunneling electron transition between localization centers is determined utterly by the electron matrix element, which is principally independent of temperature. The electron matrix element determines the probability of electron tunneling, provided the electron energies of the system before and after the tunneling transition are equal, i.e. under conditions of the so-called resonance of energy levels, analogous to that observed in a degenerated quantum system.

The value of the electron matrix element is determined by the electron interaction with the localization centers, and its calculation is a fundamental problem, not solved completely even now. On the one hand, the complexity of the problem is due to the complexity of the electron structure of macromolecular systems, for which it is impossible to derive analytical expressions for electron wave functions. On the other hand, this complexity is explained by the fact that frequently the character of interactions of electron localization centers, i.e. the type of interaction responsible for the electron tunneling transition, is not quite understood.

The localization radii value is very sensitive to the structure of the nearest microsurrounding of the local binding center. Microconformational changes in the structure of this surrounding (for example, changes in the orientation of polar groups, proton displacements in the system of hydrogen bonds, changes in permittivity, etc.) can cause variations in the local potential that will affect the electron localization length and, as a consequence, the rate of electron transport. Note, that the value of the electron localization radii is not sensitive to vibrational state of molecular systems, especially when this state does not differ greatly from the equilibrium state.

At present, it has become clear that many peculiarities of electron transport are caused by changes in the structure of the local microsurrounding of the electron binding centers. Such micrconformational transitions in the local surrounding of the electron binding center can give rise to minor changes in the center-to-center spacing or the localization radii value that can have a dramatic effect on the electron transport rate.

25.3 Role of Hydrogen Bonds in Electron Transport in Biomolecular Systems

As stated above, the rate or efficiency of electron transport in biological molecular structures is influenced greatly by different relaxation processes, caused by a change in the electronic state of molecules, contained in these structures. Hydrogen bonds are a common element of biological molecular structures. Occupying an intermediate position between weak van-der-Waals forces and strong covalent interactions, hydrogen bonds provide, on the one hand, sufficient stability of macromolecular systems, and on the other hand, their relative lability. A peculiarity of hydrogen bonds is not only a very wide energy spectrum, but also, which is most important, the structure of these bonds — not only valence electrons of atoms but also the nucleus of the hydrogen atom (proton) are involved directly in the formation of a hydrogen bond. The state of the proton in the hydrogen bond is determined by the hypersurface of the potential energy of molecular systems. As a rule, the profile (cross-section) of this surface is considered along the line of the hydrogen bond. This profile can have one or, which is most interesting, two potential maxima. The presence of two maxima conditions a possibility of proton migration along the hydrogen bond. If a molecular system has a developed network of hydrogen bonds, proton migration along the bond would lead to the emergence of an electric current under the action of the electric field. Here a striking example is water. But even if there is no branched system of hydrogen bonds, proton migration along its bond can lead to local redistribution of the charge and, as a result, to changes in the local electrostatic potential. This property of hydrogen bonds is combined with their high lability, revealed mainly in that hydrogen bonds can relatively easily bend. In this sense, hydrogen bonds are similar to elements of nanostructures — they possess both high stability and elasticity simultaneously.

A possible relaxation mechanism in molecular systems with hydrogen bonds can be conditioned by redistribution of protons between the two stable positions in the double-well potential following the lines of hydrogen bonds. This redistribution is induced by changes in parameters of the double-well potential of a hydrogen bond due to a change in the electronic state of a molecular system. A special feature of such a relaxation process is that it occurs due to the tunneling transfer of protons along the line of bonds, which means the transfer of the elementary positive charge at a distance of about 0.6–0.8 E. In its turn, such a displacement of the charge (polarization) causes a change in the electrostatic potential in the range of the electron localization — a polaron is formed. This leads to a shift of the energy levels of molecular systems and, as a result, to a change in its redox potential that in its turn may considerably affect the general efficiency of the electron transfer process. The characteristic time of proton polarization is significantly dependent on the molecular system dynamics, in particular, it increases upon deformation (the bending) of hydrogen bonds. Since such deformation enhances with the temperature growth, it drastically decreases the efficiency of redistribution of protons along the hydrogen bond lines. In contrast to the activation process, this results in slowing down of the given relaxation process upon the temperature increase. This can also determine the temperature dependence of energy parameters (the difference of free energy ΔG and/or reorganization energy λ) of the electron transfer reaction (P. M. Krasilnikov et al., 2006–2009).

The most important factor, determining the magnitude of the effect, is a change in the shape of the profile of the hydrogen bond potential upon a change in the electronic state of the system, i.e. when the electron appears or disappears in the localization region. When the electron appears in the localization region upon reduction of the acceptor, the profile of the hydrogen bond potential will change. For example, the minimum of the double-well potential closest to the localization center will become deeper. This is conditioned by the electrostatic interaction of the accepted electron and proton in the hydrogen bond. It enhances the electron stabilization on the binding center.

With a change in the electron state of the system a tunneling transfer of the proton to a new equilibrium state takes place. Thus, such relaxation of the proton populations in the double-well potential of hydrogen bond can be considered as a relaxation process coupled to the reorganization of the nuclear subsystem of a whole molecular system.

This mechanism of proton relaxation of hydrogen bonds matches the interaction which concurrently with a temperature decrease enhances stabilization of the electron on the binding center. If we assume that the given relaxation is associated with reorganization of the surrounding, the reorganization energy should decrease with a temperature decrease. This may be caused by the fact that the analyzed relaxation process includes a tunneling transfer of proton along the line of the hydrogen bond and not only an activation transfer as it is usually implied for relaxation processes.

25.4 Dynamics of Photoconformational Transfer

The elementary act of intermolecular electron transfer is the process of inelastic electron tunneling, and changes in the charge state of a molecular group lead to deformation of the potential energy of nuclei. The probability of electron transfer depends on the character of the electron vibrational coupling. In Section 2 of Chapter 11, it was shown that redescription of microconformational motions of protein with amplitudes $\gtrsim 0.1$ nm requires taking into account the diffusion character of these motions, when the moving molecular protein group collides with the surrounding molecular groups. This means that a displacement of a fragment at a distance ~ 0.1 nm and larger can occur in a condensed medium only coupled to definite fluctuations of density and formation of a microcavity ("hole") near the moving molecular group. In other words, a conformational displacement of a chain fragment suggests overcoming of potential barriers. This is equivalent to the motion in a viscous medium when a system migrates over the variety of conformational substates in the region, limited by some conformational potential (see Fig. 11.4). □

In other words, the case in question is the dependence of the rate of electron transfer on the conformation of certain chain fragments, i.e. on the conformational coordinate.

Let us consider the simplest model (K. V. Shaitan and A. B. Rubin, 1985) — the electron transfer from the external electron source D_∞ to protein group I that later gives the electron to acceptor A_∞:

$$D_\infty \xrightarrow{W_2(x)} I \xrightarrow{W_1(x)} A_\infty,$$

where $W_2(x)$ and $W_1(x)$ are constants of transfer rate, depending on conformational coordinate x. This motion treated as restricted diffusion was analyzed in Section 2 of Chapter 11 and can be described by the Langevin equation:

$$m\ddot{x} + \gamma(x)\dot{x} + \partial U(x)/\partial x = F(t),$$

or, when the inertia member is neglected, by the equivalent Fokker – Planck equation:

$$\frac{\partial p(x,t)}{\partial t} = \frac{\partial}{\partial x} D(x) \left\{ \frac{\partial p(x,t)}{\partial x} + \frac{1}{k_B T} p(xt) \frac{\partial U(x)}{\partial x} \right\}. \tag{25.73}$$

Here $U(x)$ is the conformational potential, and $D(x)$ is the conformational diffusion coefficient:

$$D(x) = k_B T/\gamma(x) = D_0 \exp(-\epsilon(x)/k_B T), \tag{25.74}$$

where $\epsilon(x)$ is the activation energy of conformational mobility, $\gamma(x)$ is the friction coefficient, $p(x,t)$ is the probability density for conformation x at time instant t. Dependence $D(x)$ is determined generally by the dependence of activation energy $\epsilon(x)$ of conformation mobility on coordinate x. In the discussed

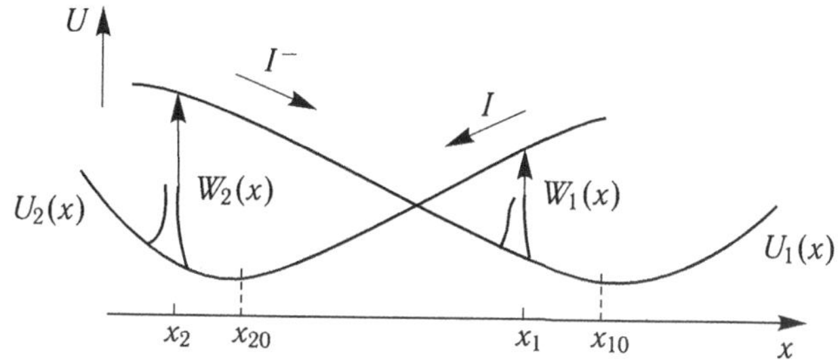

Figure 25.13. Model of a molecular "pump", or electron conformational interaction during electron transport.

Arrows indicate the direction of conformational relaxation. The rate of each stage depends on the conformation of groups I and I^-. With a change in this conformation, or in conformation coordinate x, rate constant $W(x)$ changes also quite drastically. Motion of each group I and I^- proceeds along coordinate x in the region limited by intrinsic conformational potentials $U_2(x)$ and $U_1(x)$.

model (Fig. 25.13) the association of the electron transport rate with the conformational mobility of groups I and I^- is also described by equations with additional members $W_2(x)p_2(x,t)$ and $W_1(x)p_1(x,t)$ reflecting reactions of electron transfer by group I.

Coupling mechanism. The functional activity of group I in electron transfer depends on its mobility in the following way (see Fig. 25.13). When moving along coordinate x in potential field $U_2(x)$, group I gets to point x_2. It is in this point that chemical reaction of reduction of I by D_∞ takes place at rate constant $W_2(x)$. The value of $W_2(x)$ is determined by formula (25.56) of inelastic tunneling, and the stochastic motion along coordinate x is described by the Fokker–Planck equation (25.73). Correlation times τ_c for conformational motions are much higher than the time of vibrational relaxation of high-frequency intramolecular modes, determined by width Γ of vibrational sublevels ($\tau_c \gg 1/\Gamma$). The process of radiationless electron transfer in proteins is characterized by electron vibrational interactions and changes straight inside the contact donor–acceptor complex, as well as by electron vibrational interactions, involving the surrounding of the reacting groups and depending on the protein structure and dynamics. The reduction of group I and formation of group I^- occur during time $1/W_2(x)$ that is much smaller than the time of correlation of conformational motions. It leads to transition of the system from curve $U_2(x)$ for I of the conformational energy to curve $U_1(x)$ for I^- shown in Fig. 25.13 by a vertical arrow. Reverse transition $U_1(x) \rightarrow U_2(x)$ has a similar character upon oxidation of I^- by acceptor A_∞ at rate constant $W_1(x)$ in the point, where conformational coordinate x is equal to x_1. Because of these transitions, the effective thermal effect of each stage $Q_1(x)$ and $Q_2(x)$ depends not only on intramolecular dissipation of electron energy into heat, but also on the change in the conformation energy:

$$Q_1(x) = \Delta E_1 - \Delta U_{12}(x), \quad Q_2(x) = \Delta E_2 - \Delta U_{21}(x),$$

where ΔE_1 and ΔE_2 are thermal effects of intramolecular redox reactions without account for interaction $\Delta U_{12}(x) = -\Delta U_{21}(x) = U_2(x) - U_1(x)$. If $\Delta U_{21}(x) > 0$, as in the case of transition shown in Fig. 25.13, the corresponding portion of internal energy is not immediately dissipated into heat during a rapid vibrational reaction ($\tau \sim 1/\Gamma$), but is transformed into the strain energy of conformation.

Photoconformational transition. Let us consider the regulation mechanism of recombination of charges in the system 'photo-oxidized bacteriochlorophyll dimer — primary quinone acceptor of the reaction center of purple bacteria ($P^+Q_A^- \to PQ_A$) (see in detail Chapter 28). This is one of the most simple and striking examples showing the physical meaning and role of the electron conformational interaction in a biomolecular system.

As known, following the fast charge separation in the RC and formation of an ion-radical pair $P^+Q_A^-$ recombination of charges takes place. Of principal importance here is the high sensitivity of recombination kinetics to conditions of preliminary light adaptation. The effective recombination rate in RC, frozen under saturating light intensities, is close to the recombination rate at room temperature and is almost three-fold lower than the recombination rate in RC, frozen to the same temperature but in the dark. This fact composed with the non-exponential character of recombination kinetics at intermediate temperatures and after active light illumination evidences undoubtedly that relaxation processes go on in RC after rapid charge separation having a direct effect on the recombination kinetics.

After light quantum absorption, the charge state of the donor–acceptor pair is changed as well as the equilibrium conformation of RC adapted to the dark. Because of this change the system moves to a new equilibrium state. In this case, a change in the light intensity will change the effective probability of existence of ion-radical pair $P^+Q_A^-$ generation, and as a result, the conformation of the system will change as well. Therefore, this effect may be called photoconformational transition.

A change in the surrounding gives rise to changes in parameters, determining the intrinsic probability of electron tunneling in recombination $Q_A \to P^+$. This process may be affected by changes in the distances, orientation, parameters of accepting-modes and interactions between π-bridges, virtually involved in charge transfer. These changes are obviously relatively small as compared to their average magnitudes.

Transitions from one conformational state to the other proceed by the mechanism of restricted diffusion and can be described by such equations as the Fokker–Planck one. It is possible to take into account the presence of conformational mobility by introducing conformational coordinate x. This coordinate can describe the mutual arrangement of the donor and acceptor or the group surrounding them. Let us introduce two functions in the consideration: $P_A(x,t)$ and $P_{A^-}(x,t)$. They are probability densities of two random events — neutral and charged states of the acceptor on the conformational coordinate x. Changes in these functions can occur due to charge transfer and motion along x. These processes can be interrelated: in some regions of the conformational space the charge transfer is hindered, while in other, on the contrary, it is facilitated. The time evolution of the introduced functions is described using the system of Fokker–Planck equations

$$\begin{aligned}\frac{\partial P_A}{\partial t} &= D_A \frac{\partial}{\partial x}\left(\frac{\partial P_A}{\partial x} + \frac{1}{k_B T} P_A \frac{\partial U_A}{\partial x}\right) - k_0(x)P_A + k_r(x)P_{A^-},\\ \frac{\partial P_{A^-}}{\partial t} &= D_{A^-} \frac{\partial}{\partial x}\left(\frac{\partial P_{A^-}}{\partial x} + \frac{1}{k_B T} P_{A^-} \frac{\partial U_{A^-}}{\partial x}\right) + k_0(x)P_{A^-} - k_r(x)P_{A^-}.\end{aligned} \tag{25.75}$$

The first members in the right-hand side of the above equation describe the changes in P_A and P_{A^-} due to the diffusion along x in the potential field determined by functions U_A and U_{A^-}, respectively. As mentioned above, potential energies of the acceptor in the neutral and charged states are different.

The second and third members in the right-hand side of equations (25.75), proportional to $k_0(x)$ and $k_r(x)$, describe changes in the probability density function due to charge transfer processes. Function $k_0(x)$ is the rate of charge transfer to the acceptor at the given conformational coordinate x, and $k_r(x)$ is the recombination rate constant with the acceptor under the same conditions. For reaction of photoinduced charge transfer, the value of rate $k_0(x)$ is proportional to the excitation light intensity and can vary in a wide range. Supposing changes in conformational variable are relatively small, let us expand functions $k_0(x)$ and $k_r(x)$ into series in terms of x and restrict ourselves to linear members:

$$k_0(x) = k_0 + \beta x, \quad k_r(x) = k_r + \alpha x. \tag{25.76}$$

Let us discuss briefly stationary solutions of system (25.75). At constant active light, depending on its intensity, equilibrium is established in the system between neutral A and charged A^- forms of the acceptor. Under intensive illumination, equilibrium shifts towards charged form A^- and the region in conformational space of dimension $\sim \sqrt{k_B T/K}$ is occupied. This region corresponds to the charged acceptor, i.e. $P_A \ll 1$, and $P_{A^-}(x) \sim 1$ at $x \sim -\Delta$. In the absence of illumination, a reversed situation is observed: the acceptor is in central state A and the most probable value of the conformational coordinate is Δ. The average value of the acceptor coordinate regardless of its charged state changes gradually with an increase in the incident light intensity from $x \sim \Delta$ to $x = -\Delta$ (see Fig. 25.14). Such behavior of a system represents photoconformational transition.

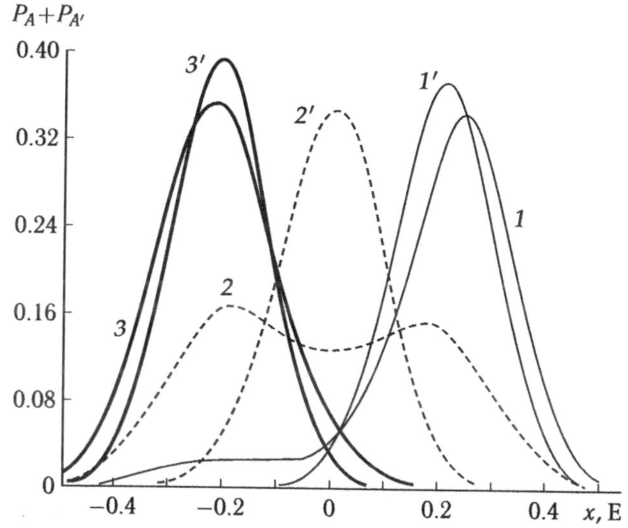

Figure 25.14. Total function of probability density $[P_A(x) + P_{A^-}(x)]$.

Parameter values are as follows: *1* — $k_0/k_r = 0.1, \tau = 20$ ms; *2* — $k_0/k_r = 1, \tau = 20$ ms; *3* — $k_0/k_r = 10$, $\tau = 20$ ms; *1′* — $k_0/k_r = 0.1, \tau = 1.14$ ms; *2′* — $k_0/k_r = 1, \tau = 1.14$ ms; *3′* — $k_0/k_r = 10, \tau = 1.14$ ms. The other parameters are: x_0 (T = 100 K) = 0.1 E; $\Delta = 0.25$ E, $\beta = 0$, $k_r = 0.0207$ ms^{-1}, $\alpha = 0.01$ ms^{-1}E^{-1}.

Upon rapid cooling, the obtained stationary state $(P_A + P_{A^-})$ is fixed. By changing the intensity of illumination of RC samples, we can change parameter k_0/k_r in wide ranges and fix a certain conformational state upon freezing. Kinetic measurements (Fig. 25.15) support the concept of photoconformational transition in the acceptor

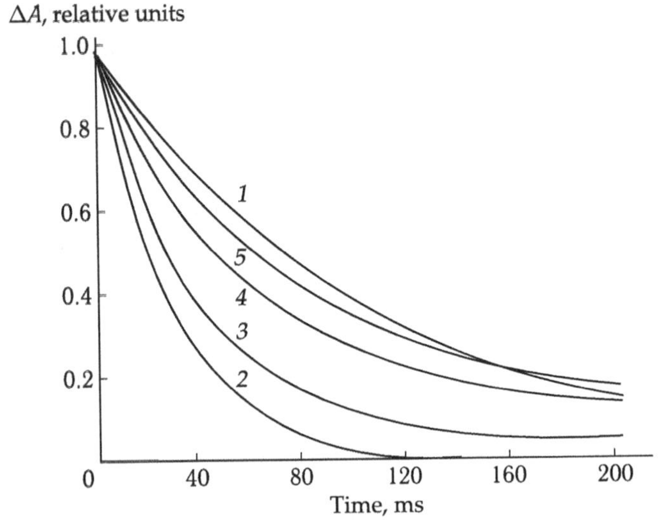

Figure 25.15. Kinetic curves of dark recombination $P^+Q_{A^-}$ under different conditions (A. B. Rubin et al., 1994).

1, At 300 K; *2*, after being frozen in the dark to 80 K; *3–5*, evolution of kinetic curves upon freezing the RC preparation to 80 K in the background light with intensity of $1.1 \cdot 10^2$ W $\cdot$ m^{-2}, $7 \cdot 10^2$ W $\cdot$ m^{-2}, and $2 \cdot 10^3$ W $\cdot$ m^{-2}, respectively. The RC preparation of *Rhodopseudomonas sphaeroides* bacteria was suspended once again in 70 % glycerol and contained 10^{-2} M *o*-phenanthroline.

region of RC. An increase in the intensity of adaptation illumination forms such a state of the system, the recombination kinetics of which even at low temperatures coincides with the kinetics of the process at $T = 300$ K. In other words, at room temperature a conformational transition has enough time to occur in response to a saturated light flash followed by formation of Q_{A^-}. But at low temperatures, the conformational mobility is retarded that determines the difference in the kinetics in Fig. 25.15 (curves *1* and *2*). However, when frozen under illumination and at a gradual increase in the illumination intensity, the system is fixed in the state becoming ever closer to the equilibrium position of Q_{A^-}. It is reflected in the sequential change of kinetic curves in Fig. 25.15 (curves *3–5*).

Figure 25.16 shows the results of processing the recombination kinetics at different temperatures. Four types of points correspond to initial values of the constant of recombination rate at different instants of time. The approach developed above makes it possible to observe directly the process of conformational relaxation. As seen at high temperatures, the recombination constant is independent of time, i.e. conformational relaxation occurs during the time much shorter than the characteristic time of charge recombination in the initial conformational state ($k_r^{-1} \sim 80$ ms). At low temperature ($T < 200$ K), conformational relaxation slows down and the kinetics of photoconformational transition is superimposed on the recombination kinetics. As seen from Fig. 25.16, defrosting of the conformational dynamics takes place at temperatures ~ 200 K. Such a conclusion was made earlier from the Mossbauer spectroscopy analysis. This indicates to a cooperative character of motions of the acceptor and its surrounding.

The analysis of 25.16 allows us to conclude the following. The recombination rate is changed totally in the range from 20 to 100 ms due to both the temperature rise

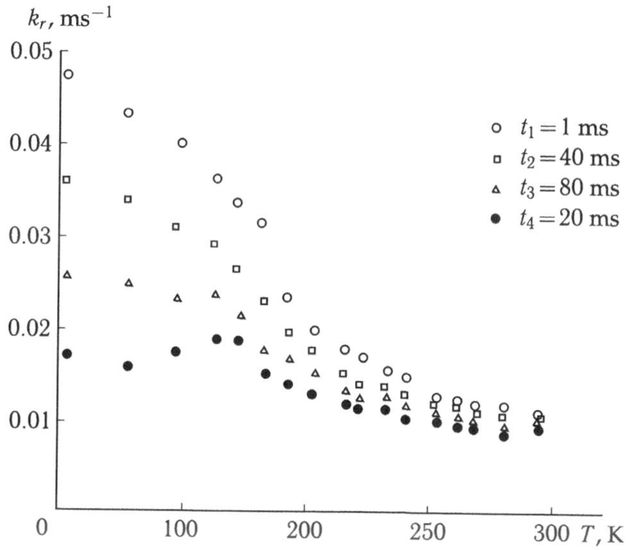

Figure 25.16. Values of instantaneous recombination constant $k_r(t)$ at different intermediate times t_1, t_2, t_3, t_4 (ms) after the flash and at different temperatures (A. B. Rubin et al., 1994).

from 100 K to 300 K and recombinational relaxation for the times exceeding the time resolution of the experiments $t \sim 1$ ms. As seen in time interval $t = 120$ ms practically complete conformational relaxation occurs over the whole temperature range. The amplitude of the temperature dependence of k_r (120 ms) from 100 to 300 K demonstrates that the maximal change in the recombination rate of the relaxed state is observed in the range 0.02–0.01 ms^{-1}. Let us consider now changes in $k_r(t)$ at fixed temperatures as a result of the completeness of the conformational transition at corresponding instants of time. At $T \gtrsim 250$ K practically complete relaxation occurs at $\tau \sim 1$ ms. At $T < 200$ K contributions of the processes with characteristic times ~ 80 ms appear while at $T < 220$ K contributions of the processes with characteristic times $t \sim 40$ ms play their role.

The existence of well pronounced time dependence $k_r(t)$ even at 100 K attracts our attention. It is obvious that processes with times 100 ms take place even at $T \sim 100$ K, thus providing practically complete, though slower relaxation of state Q_{A^-} even at low temperatures. The observed illumination and temperature effects in the kinetics of this process are caused by the conformational transition induced by electron transfer to the acceptor under the action of a light flash. The driving force of the photoconformational transition is electron conformational interactions leading to equilibrium in the system. Such a mechanism of regulation of the kinetics of intraportein processes is of great meaning. The use of donor–acceptor pairs with temperature independent reaction rate makes it possible to control the process only via variations in the macromolecular structure in the nearest surrounding upon conformational motions. Thus, a purposeful change in the kinetic parameters can be conducted to match the rates of biochemical processes. It is achieved not by a substitution of a limited number of reagents, but by varying the structure and microsurrounding related to the amino acid sequence originated in the course of evolution. A possible role of hydrogen bonds in such regulation has been discussed above.

Thus, there are very different mechanisms of association of the conformational mobility and the protein functioning. □

It is clear that definite ways of electron transfer have no strictly fixed character. They depend on the conformational state of the protein globule and can change respectively under different conditions. This corresponds to conceptions of a multiple character of electron transfer pathways in a protein globule, determined by the involvement of different protein groups and bonds between them. The temperature dependencies of electron transfer observed in the experiments reflect the dynamic character of formation of electron transfer pathways in the protein globule.

Therefore, the major factor in electron transfer over protein is the formation of an electron pathway where step-by-step electron tunneling takes place between donor–acceptor pairs included in its structure (H. B. Gray, 1995).

In his latest works, R. Marcus developed a more general approach, where a set of optimal trajectories of electron transfer in protein are considered rather than only one trajectory. When calculating the transfer constant K_e, orbitals of a whole number of electron interacting atoms of amino acid protein residues, making the largest contribution to the super-exchange interaction, were taken into account. It appeared that kinetic characterisitics obtained for some proteins were more correct as compared to modes with only one trajectory. However, in this case different slopes of these dependences are observed for various proteins (in myoglobin and cytochrome the differences in the slope magnitudes were more than twice-fold). This difference indicates to regular changes in the relaxation rates and energy reorganization E_r in proteins with different structures (A. I. Kotelnikov, 1993). In its turn, this factor has an effect on the electron transfer rate if $\tau_r \sim \tau_e$, i.e. upon disturbance of the nonadiabatic approximation.

Consequently, the role of protein in the process of electron transfer is not restricted only to providing rapidly relaxing accepting-modes. Low-frequency intramolecular soft modes, determining rearrangements in protein and formation of donor–acceptor complexes and/or electron pathway in a protein globule, make their contribution to this process. The role of soft modes can manifest also in preventing a back transition by certain conformational reorganization, when after electron transfer the donor and acceptor prosthetic groups separate and terminate interaction with each other (see Chapter 28).

Charge transfer to DNA. The interest to studying propagation of charges in DNA is due to fact that charge transfer contributes to the mechanism of such vital biochemical processes as DNA replication, transcription, destruction and repair. Moreover, radical movement along a DNA molecule plays a considerable role in mutagenesis and carcinogenesis. Studies of the charge transfer process in DNA were initiated at the beginning of the 90s of the last century, when in one of the first experimental investigation (C. J. Murphy et al., 1993) it was demonstrated that charge is transferred in a relatively long DNA fragment containing 15 base pairs. By that time, it had become possible to perform experiments on measuring processes ultra-short in time using nanosecond and femtosecond techniques, and the progress in computer technology allowed making model calculations of such processes in complex biological systems.

In most experiments, electrons or holes (cation radicals) were generated in a special way on DNA fragments with the known sequence of bases. The electron transport rate was calculated either from measurements of fluorescence quenching, or by comparing the number of destructions in nucleotides, caused by the charge transfer, in different fragments of the DNA helix.

As a rule, three types of different mechanisms of charge transfer are considered upon descriptions of charge migration in molecular chains.

The first mechanism suggests incoherent hopping charge transfer, caused by temperature fluctuations. In this case, charges move from the donor along the bridge of sites (nucleotides or nucleotide pairs), hopping from one site to another, staying for some time in each of the sites, thereby moving from the donor to the acceptor. Thus, with a hopping mechanism, each site to which a hop is made is a real chemical intermediate.

The second mechanism is a super-exchange mechanism corresponding to the charge tunneling from the donor to the acceptor in one hop. In this case, the charge is not formed on either of the intermediate sites of the bridge linking the donor to the acceptor — its presence on the bridge sites is barely virtual (see Chapter 24). In accordance with the Marcus theory, the electron or hole move along the DNA fragment at a rate descreasing exponentially with distance (the reaction rate $k \sim \exp\{-\beta R\} \sim \exp\{-\beta N \Delta r\}$, where Δr is the distance between adjacent nucleotide pairs and N is the number of pairs). Experimental results are usually represented by the obtained value of parameter β. The third mechanism is the formation of a self-trapping polaron moving under the influence of temperature fluctuations (P. T. Henderson et al., 1999).

As established, DNA may have different transfer mechanisms and the transfer rate depends not only on the length of the nucleotide sequence along which the transfer occurs, but also on the type of the sequence as well as on characteristics of the donor and acceptor used in the experiments.

In DNA chains of the same length, the charge transfer rate may vary by orders of magnitude depending on the type of sequences. For different experiments, the calculated value β in the super-exchange model changes from 0.1 E^{-1} to 1.6 E^{-1}. To explain small values β different hopping models were proposed, in particular, a model with algebraic dependence of the transfer rate on the number of sites in chain N: $k \sim N^{-\xi}$ (J. Jortner et al., 1998). As a rule, parameter ξ varies from 1 to 2.

In the model (A. S. Davydov and N. I. Kislukha, 1976) of excitation transfer in unidimensional molecular chains, the charge energy in a site depends on the displacement of the site relative to its neighbors. This model was used for calculations of transfer energy upon hydrolysis in protein α-helices and for modeling charge transfer in DNAs. In this model, the classical potential energy depends on the difference in the displacement of sites from equilibrium positions (the sites are "fastened to each other by small springs"). Here a nucleotide pair corresponds to a site. A DNA fragment is considered as a chain of sites — harmonic oscillators with masses M_n, connected to the nearest neighbors by small springs at frequency $\overline{\omega}$. In a simplified DNA model (V. D. Lakhno, 2007), it is proposed that the base planes of nucleotides are parallel to each other, and the distances between the base planes of neighboring sites are constant (Fig. 25.17) that corresponds to the B-form of DNA.

Here the electron transfer to DNA is determined by the overlap of wave functions on the adjacent sites. If the transfer occurs along one of the DNA chains, the overlap integrals do not practically depend on intrasite displacements irrespective of their sizes. Taking into account that the nucleotide mass is several orders of magnitude higher than the electron mass, displacements of Watson–Crick nucleotide pairs can be described by classical equations of motion, and the charge propagation — by the Schrodinger equation.

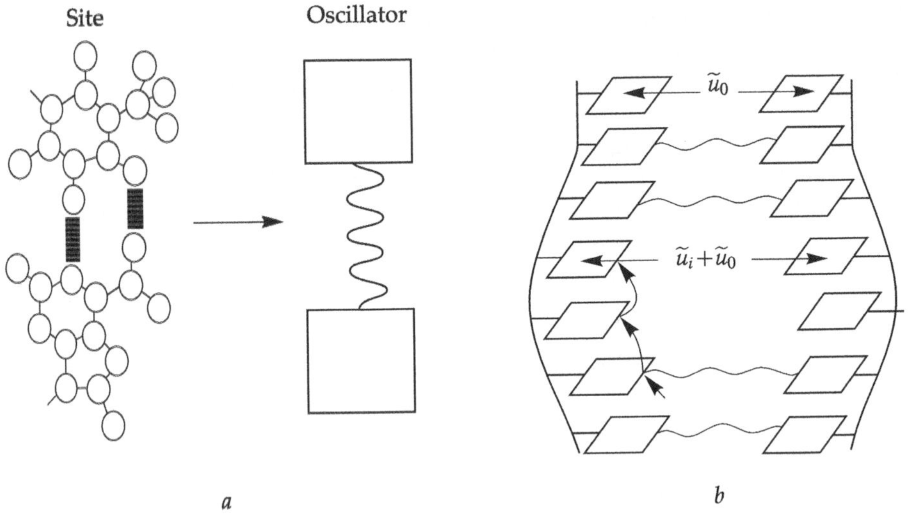

Figure 25.17. *a*, Nucleotide pair and equivalent oscillator. The spring is a hydrogen bond. *b*, Site DNA model. Sites (base pairs) are positioned one above the other.

In most experiments on charge transfer in DNA, the excess charge (hole or electron) is localized initially on one site — the donor. This corresponds to certain initial conditions satisfying the introduction of a charge to the DNA chain and its stay in the donor site.

With equations where members with a random force of the Langevin type are included, it becomes possible to take into account the effect of temperature fluctuations of the surrounding. Calculations of the hole mobility in homogeneous polynucleotides at different temperatures demonstrated that the mobility increases with a temperature decrease. The calculations showed a theoretical possibility of charge transfer in DNA over large distances (thousands of base pairs).

25.5 Mechanisms of Energy Migration

25.5.1 Inductive Resonance Mechanism

Mechanisms of migration of electronic excitation energy. Transformation of electron energy in biostructures is connected not only with electron transfer, but also with migration of electronic excitation energy, not accompanied by electron release from the donor molecule. These processes are of vital biological importance in photobiology, determining the distribution of excitation energy between chromophores.

The main physical ideas, underlying the models of excitation energy migration, are close to the concepts on electron transfer in a two-level system. Depending on the character of interactions of an excited donor molecule and unexcited acceptor molecule, different migration mechanisms are realized.

There are several different mechanisms of electron excitation energy migration. The most important for biological systems are inductive resonance, exchange-resonance and exciton. In Section 2 of this chapter, we have considered general regularities of excitation transfer from an excited donor molecule to an acceptor molecule in a two-level system. The physical reason for excitation transfer and system $D^*A \rightarrow DA^*$ transition from initial state i to final state f is determined by the nature of interactions V_{if}. Excitation transfer over distances much smaller than the emission wavelength occurs usually by means of the Coulomb interaction between electrons in different molecules. During the lifetime of an excited state of molecule D^* due to the interaction of dipole moments of transition of adjacent molecules, the acceptor performs transition to an excited state and the donor returns to the ground state, i.e. radiationless energy transfer takes place. In a two-level system, the irreversible character of energy transfer is determined by relaxation losses of energy due to the connection of the system with the external environment. These processes of energy dissipation from the level of the final state of system DA determine the final width of level Γ and the total time of decay of the initial state upon irreversible energy transfer.

□

In complex luminescent molecules, the width of energy level is determined mainly by the time of the electron vibrational interaction, dissipation of electron energy over the vibrational degrees of freedom and vibrational relaxation. In this case, in the formula for the width of the energy level

$$\Gamma = 1/T_2 \sim 1/T_2^* + 1/(2T_1)$$

value T_1 is determined by time τ^* of electron transfer from the excited state to the ground one, and T_2^*corresponds to the time of vibrational relaxation, $T_2^* \sim \tau_{\text{vib}}$. Since $\tau_{\text{vib}} \ll \tau^*$, consequently, the width of the level depends mainly on intramolecular dissipative processes, i.e at weak interactions

$$1/\Gamma \sim T_2 \sim \tau_{\text{vib}}. \tag{25.77}$$

The inductive resonance mechanism is realized at weak interaction between molecules, when the interaction energy ($E_{\text{int}} < 10\ \text{cm}^{-1} \sim 10^{-3}$ eV) is such that

$$E_{\text{int}} = \sim |V_{if}| \ll \hbar/T_2, \tag{25.78}$$

where V_{if} is determined (24.5) as a matrix element.

In this case, the transfer time, or migration time, of energy τ_{m}

$$\tau_{\text{m}} \sim \hbar/E_{\text{int}} \tag{25.79}$$

exceeds greatly the time of vibrational relaxation ($T_2 \sim \tau_{\text{vib}}$):

$$\tau_{\text{vib}} \sim T_2 \ll \tau_{\text{m}} \sim \hbar/E_{\text{int}}. \tag{25.80}$$

Although the transfer involves excitated vibrational sublevels of electronic states, the vibrational relaxation occurs much faster than the electronic excitation energy transfer can be realized. In other words, here we can neglect the population of the final state level and consider the reverse transfer, if it is possible from the energetic point of view, as independent of the direct transfer. In this case, the processes of excitation energy degradation in the donor and acceptor go on independent of each other as well. The light excited donor molecule relaxes to the lower vibrational sublevel of singlet excited state S_1^*, from where fluorescence light can be emitted. During lifetime τ_D of state S_1^* of the donor, excitation is transferred due to dipole–dipole interaction to the acceptor molecule. It occurs with excitation of one of the upper vibrational sublevels of state S_1 followed by relaxation and fast transfer to lower vibrational levels of state S_1^* of the acceptor in $\tau_{\text{vib}} \sim 10^{-12} \div 10^{-13}$ s.

Thus, the transfer is realized for the time exceeding the thermal relaxation of the excess vibrational energy. This mechanism of energy transfer is called an inductive resonant transfer. Its classical analog is the model of two mechanically connected pendulums, where mechanical excitation of one of them is transferred to the other. The interaction of the two donor and acceptor oscillators is realized via the electromagnetic field generated by the excited donor molecule S^*. T. Ferster was the first to study this mechanism quantitatively in 1948. In Ferster's model, wide spectra are considered as continuous, and the known expression of the perturbation theory for the transfer probability in a continuous spectrum (see (24.31)) is used for the transfer rate:

$$dW = \frac{2\pi}{\hbar}|V_{if}|^2\delta(\Delta E), \tag{25.81}$$

where $\Delta E = (E_D^0 - E_D^1) - (E_A^0 - E_A^1)$, and $|V_{if}|^2$ is the square of the matrix element of the energy of dipole–dipole interaction of the initial (i) and final (f) states. Functions Ψ_i and Ψ_f are constructed in the adiabatic approximation as the product of electron wave function φ and vibrational function χ (25.31). In the single-electron approximation, according to the superposition principle, wave functions of the initial and final states can be written as follows:

$$\begin{aligned}\Psi_i &= \frac{1}{\sqrt{2}}\left[\varphi_D^1(1)\varphi_A^0(2) - \varphi_D^1(2)\varphi_A^0(1)\right]\chi_1(E_D^1)\chi_0(E_A^0),\\ \Psi_f &= \frac{1}{\sqrt{2}}\left[\varphi_D^0(1)\varphi_A^1(2) - \varphi_D^0(2)\varphi_A^1(1)\right]\chi_0(E_D^0)\chi_1(E_A^1).\end{aligned} \tag{25.82}$$

In the Condon approximation at a weak dependence of the electronic state on nuclear coordinates, matrix element of interaction V_{if} is

$$\begin{aligned}V_{if} &= \langle\Psi_i|V|\Psi_f\rangle = \\ &= \left\{\langle\varphi_D^1(1)\varphi_A^0(2)|V|\varphi_D^0(1)\varphi_A^1(2)\rangle - \langle\varphi_D^1(1)\varphi_A^0(2)|V|\varphi_D^0(2)\varphi_A^1(1)\rangle\right\}F_{if},\end{aligned} \tag{25.83}$$

where the first member in the right-hand side takes into account the Coulomb interaction of donor and acceptor electrons, and the second member takes into account the exchange interaction; F_{if} is the overlap integral of vibrational wave functions of states i and f (see (25.55)).

At relatively large distances between D and A, when the overlap of their wave functions can be neglected, the exchange part of the matrix element is much smaller than the Coulomb one:

$$\left|\langle\varphi_D^1(1)\varphi_A^0(2)|V|\varphi_D^0(1)\varphi_A^1(2)\rangle\right| \gg \left|\langle\varphi_D^1(1)\varphi_A^0(2)|V|\varphi_D^0(2)\varphi_A^1(1)\rangle\right|. \tag{25.84}$$

The energy transfer by the inductive resonance mechanism is allowed by the spin selection rules, suggesting the preservation of the electron spin upon excitation energy transfer: $s_D^1 = s_D^0$ and $s_A^1 = s_A^0$, where s is the spin wave function.

The probability of energy transfer for the dipole–dipole intermolecular interaction is as follows:

$$W_k = \frac{kf(\varphi, \Theta)}{\tau_0 R^6} \int\limits_V \frac{1}{V^4} \epsilon(V) f(V)\, dV, \tag{25.85}$$

where k is the coefficient depending on the environment properties (on the refractive index), $f(\varphi, \theta)$ is the factor of relative positions of donor and acceptor transition dipoles; $\tau_0 = 1/p$ is the donor radiative lifetime in state S_1^*, p is the fluorescence probability; R is the distance between interacting molecules; $\epsilon(V)$ and $f(V)$ are the donor emission and acceptor absorption spectra normalized to the unit area, cm^{-1}.

As seen, the probability of transfer is proportional to $1/R^6$ as it is true for dipole–dipole interactions (see Section 1 in Chapter 8) and to the overlap integral of the acceptor absorption and donor emission spectra. The latter circumstance indicates to the presence of wide spectra of interacting molecules, and the approximation of the two-level system with resonance transfer of energy from one level to the other is already not valid. The distances, at which both energy transfer and luminescence emission are equally effective, are about 3–10 nm. Energy transfer rates via the inductive resonance mechanism are 10^6–10^{11} s^{-1}. The probability of energy transfer by this mechanism decreases with a temperature increase due to the narrowing of the spectral bands width and a decrease in their overlap integral. The expression for the temperature dependence of the probability of energy transfer between similar molecules according to the Ferster mechanism is as follows:

$$W = W_0 \exp(-E_a/k_B T), \tag{25.86}$$

where $W_0 = \Phi(R)\kappa_0/\sqrt{2\omega_0^2}$ is the constant dependent on the distance between molecules $\Phi(R)$, absorption intensity k_0 and position of the absorption maximum ω_0; E_a is the activation energy of the transfer process.

The general criterion for applicability of the Ferster theory is weak interaction and rapid relaxation. Consequently, at distances of about the diameter of molecules in solution or about the constant lattice in crystals, this approximation based on the small magnitude of interaction can be not valid. □

Inductive resonance migration is possible not only between singlet donor and acceptor levels. The existence of intercombinative prohibited processes of energy transfer was also demonstrated: triplet-singlet and singlet-triplet processes when the rule for spin preservation in the system is violated. Figure 25.18 shows a diagram of electron levels of energy donor and acceptor molecules at a triplet-singlet energy transfer. Here the spin-prohibited transfer competes well with the intercombinative prohibited transfer from the triplet to the ground state accompanied by phosphorescence emission in the donor. Such a process should be realized at distances of 2.5–5.0 nm characteristic of the inductive resonance mechanism. Intramolecular migration of energy was discovered in protein, pigments, coenzymes and chromoproteins.

Besides the examples given in the introduction to this chapter, there are donor–acceptor pairs, where inductive resonance migration takes place: aromatic amino acids of proteins → ophycobilins (phycoerythrin and phycocyanin that are photosynthetic pigments of red algae and cyanobacteria), tyrosine → tryptophan, tryptophan → tryptophan, tryptophan → ionized tyrosine in different proteins,

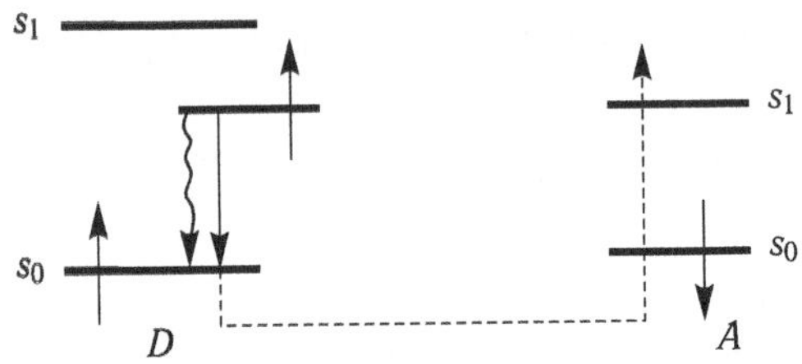

Figure 25.18. Levels at a triplet–singlet energy transfer.

aromatic amino acids → dye in protein–dye complexes, adenyl → nicotinamide ring in NAD, thiazine → pyrimidine ring in thiamine (vitamin B_1) etc. Intermolecular transfer is also widespread in biological systems; an example of this is energy migration between different pigments in photosynthetic membranes.

25.5.2 Exchange Resonance Energy Transfer

Radiationless energy transfer can proceed not only via singlet energy levels of molecules, but also triplet levels. A. N. Terenin and V. L. Ermolayev obtained first evidences for such a transfer for solid solvents of dyes. A formal scheme of such a transfer is as follows:

$$D_T\uparrow\uparrow + A_{S_0}\uparrow\downarrow \rightarrow D_{S_0}\downarrow\uparrow + A_T\uparrow\uparrow.$$

As seen (Fig. 25.19), the transfer occurs with electron spin conversion in molecules D and A. Due to the prohibition rules, transitions $S \rightarrow T$ between levels of different multiplicity in organic molecules are 10^6–10^9 times less probable than transitions between levels of the same multiplicity ($S \rightarrow S_0$, $T \rightarrow T_1$). Therefore, in such cases the Coulomb moiety in V_{if} (see (25.83)) in itself will be extremely small, and an exchange integral determining transitions between states can make the main contribution:

$$\varphi_D^T(1)\varphi_A^0(2) \rightarrow \varphi_D^0(2)\varphi_D^T(1),$$

where φ_D^T and φ_A^T are triplet states of D and A.

Representing electron wave functions of the donor and acceptor as the product of the coordinate wave function $\varphi(r)$ and spin function s, we can transform the exchange integral:

$$I_{\text{exc}} = \langle \varphi_D^1(r_1)\varphi_A^0(r_2)|V|\varphi_D^0(r_2)\varphi_A^1(r_1)\rangle \langle S_D^1(1)|S_A^1(1)\rangle \langle S_A^0(2)|S_D^0(2)\rangle . \tag{25.87}$$

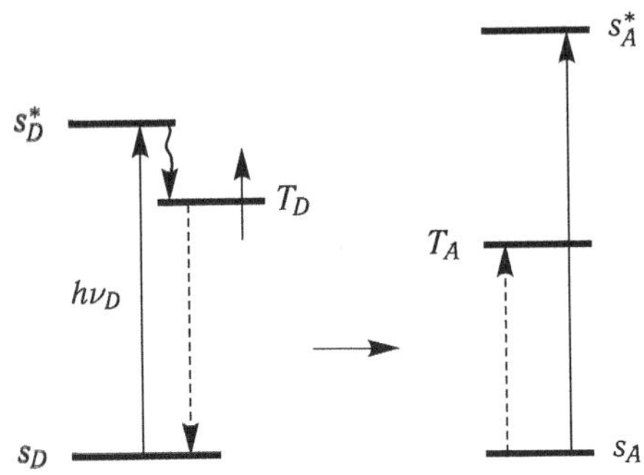

Figure 25.19. Diagram of a triplet–singlet energy transfer.

Hence it follows that $I_{exc} \neq 0$ only when

$$\langle S_D^1(1)|S_A^1(1)\rangle = \langle S_A^0(2)|S_D^0(2)\rangle \neq 0,$$

i.e. when electrons exchange, their spins remain the same. In this case, it can occur that the ground and excited states of both the donor and acceptor have different multiplicity:

$$\langle S_D^1(1)|S_D^0(1)\rangle = \langle S_A^1(2)|S_A^0(2)\rangle = 0, \qquad (25.88)$$

that is observed at a triplet–triplet exchange resonance transfer.

Thus, the electron exchange between D and A in the final state leads to an exchange of their spin states as well. This is taken into account by rule (25.83) of a concurrent change in the donor and acceptor multiplicity and the preservation of the total spin of the system in the process of energy transfer (Fig. 25.20). These interactions are called exchange resonance ones.

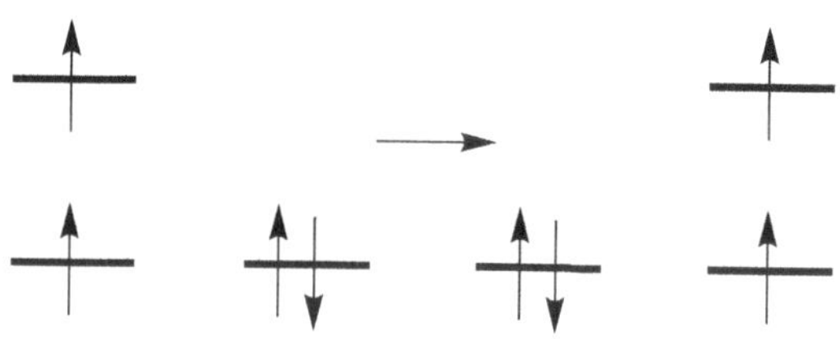

Figure 25.20. Spin conservation at a triplet–triplet transfer.

The physical reason for the exchange resonance interaction is the overlap of electron wave functions of the donor and acceptor. A transfer by this mechanism can occur at distances of about the chemical bond length, i.e. 0.1–0.3 nm. The probability of a transfer by the exchange resonance mechanism is determined by the expression

$$W(R) = \frac{2\pi}{\hbar}\exp(-2R/L)\int f_D(E)\epsilon_A(E)\,dE, \quad \square \qquad (25.89)$$

where L is the average Bohr radius (0.11–0.19 nm for aromatic molecules), $\epsilon_A(E)$ is the normalized absorption spectrum of the acceptor $\left(\int \epsilon_A(E)\,dE = 1\right)$, $f_D(E)$ is the normalized luminescence spectrum of the donor. Exponential dependence $W(R)$ is determined by the corresponding distance-dependent decrease in the wave function of the electron.

The exchange resonance mechanism can work at a singlet–singlet transfer, when molecules form an exchange complex. Of special interest is exchange resonance migration, proceeding with violation of the rule (see Equation (25.91)) of spin preservation in the system including triplet–singlet and singlet–triplet transfers. For example, a singlet–triplet ($D_{S_1} + A_{S_0} \to D_{S_0} + A_T$) transfer occurs when there is no singlet–singlet energy transfer from the donor to the acceptor. Rate constants of this transfer are proportional to the square of the spin-orbital factor in molecules involved in the energy transfer. This means that in this case the reason for energy transfer is spin-prohibited exchange resonance interactions. The latter nevertheless proceeds due to the spin-orbital interaction that removes the prohibition against intercombinative singlet–triplet transitions in a molecule. An example is the exchange resonance energy transfer in photosynthetic membranes from chlorophyll in state S_1^* to the triplet level of carotenoids.

25.5.3 Exciton Mechanism

We have considered the case of energy migration at relatively weak interactions, when the transfer time is high as compared to time T_2 of intramolecular degradation ($\tau_m \gg T_2$). However, if energy E_{int} of interaction between molecules is essential, and transfer time $\tau_m \sim \hbar/E_{int}$ is relatively small as compared to T_2, then

$$\tau_{vib} \sim T_2 \gg \tau_m \sim \hbar/E_{int}. \tag{25.90}$$

In this case, dissipation-free energy transfer becomes possible in crystals, where collective states can be excited due to translational symmetry. Then several hundreds of molecules are involved in the excitation. According to Equation (25.90), the time of vibrational relaxation over the vibrational sublevels of state S_1 is considerably higher than transfer time τ_m. It seems as if excitation "runs" over the upper vibrational sublevels of interacting molecules, having no time to localize in each of them separately. In each of the molecules, excitation stays for a time much smaller than the time of intramolecular vibrational relaxation of an isolated molecule (Fig. 25.21). At energies of molecule interaction of about $E_{int} \sim 100\ \text{cm}^{-1}$ (~ 0.01 eV $\sim 4.6 \cdot 10^{-21}$ J) the time of energy migration τ_m can be calculated as $\tau_m \sim \hbar/E_{int} \sim 10^{-13} \div 10^{-14}$ s that is smaller than the time of vibrational relaxation ($\sim 10^{-12}$ s) of complex fluorescent molecules of dyes.

This type of excitation migration is exciton migration, and the excitation region itself involving a great number of molecules excited simultaneously is called an exciton.

An exciton is formed upon light absorption by molecular crystals, when the electromagnetic field of a light wave with length λ can cause both coherent excitation and displacement of charges in a row of closely positioned molecules ($\lambda \gg a$, where a is the average intermolecular distance). A vigorous interaction of neighboring molecules induces correlation of displacement of their charges, i.e. generation of an excitation wave, or a wave packet, thus involving a large number of molecules. In fact, exciton propagation is equal to spreading of this wave packet.

Let us underline once again that in this case, we have coherent excitation of a great number of molecules, or a coherent exciton, when time of excitation transfer τ_m is much smaller than the time of intramolecular vibrational relaxation ($\tau_m \ll T_2$). Namely the high T_2 value, as compared to τ_m, provides preservation of the excitation phase of the adjacent molecules and generation of a coherent exciton. The resonance interaction between excited molecules, when the whole region of transient dipoles moves in one phase, is manifested in the displacement and splitting of the absorption band (see Section 2 in Chapter 24). □

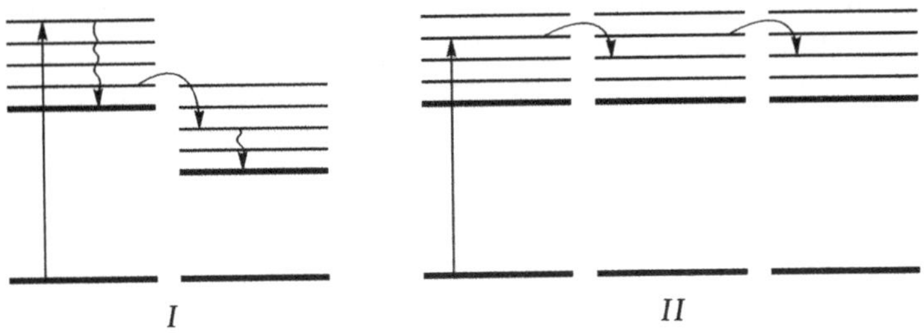

Figure 25.21. Inductive-resonance (*I*) and exciton (*II*) energy transfer.

Up to now, we have not taken into account the exciton–phonon interaction, leading to deformation of molecules, because we assumed that time τ_{def} of displacement of atoms and transfer of molecules to new equilibrium states upon their excitation are much greater than the migration time of the exciton ($\tau_m \ll \tau_{def}$). This is equal to the condition for generation of a coherent exciton ($T_2 \gg \tau_m$). Actually, it is relatively slow relaxation processes of energy redistribution over vibrational sublevels, as compared to the excitation transfer, that provide the invariability of the molecule backbone. However, it may occur that energy E_{def} of the change in local deformation of molecule is such that the characteristic time of atomic displacements ($\tau_{def} \sim \hbar/E_{int}$) becomes smaller than τ_m, i.e.

$$\tau_m \geqslant \tau_{def} \tag{25.91}$$

at $E_{int} \leqslant E_{def}$. In this case, the generated local deformation moves together with excitation along the crystal, and the exciton itself is called "localized", or incoherent. The motion of an incoherent exciton has a character of random jumps, and we cannot describe it in terms of the wave packet spreading. The existence of an exciton–phonon interaction causes spreading of incoherent excitons in the nods of the crystalline lattice, decreasing the length of their free path and can change the spectrum of exciton states as well. At vigorous interactions, migration constant K_m of a coherent exciton depends on the interaction energy,

$$K_{m(coh)} = 4E_{int}/\hbar \quad (E_{int} > 100\ \text{cm}^{-1}), \tag{25.92}$$

and for an incoherent exciton

$$K_{m(incoh)} = (4E_{int}/\hbar)F^S_{\nu\nu'} \quad (E_{int} \sim 10 \div 100\ \text{cm}^{-1}), \tag{25.93}$$

where multiplier $F^S_{\nu\nu'}$ (the square of the matrix element of interaction between vibrational levels of the ground and excited electronic states) reflects the existence of the exciton-phonon interaction. Thus, the probability of migration of an incoherent exciton depends on the Franck–Condon overlap integral of vibrational functions of excited and unexcited states. The time of migration of the incoherent exciton decreases to 10^{-12}–10^{-13} s.

We can represent propagation of coherent excitons as diffusion of exciton excitation over a crystal. It is characterized by the following expression

$$L = \sqrt{D\tau_0}, \tag{25.94}$$

where L is the length of the diffusion displacement, D is the diffusion coefficient, τ_0 is the lifetime of the exciton. For example, for singlet excitons in molecular crystals, value $L \sim 50$ nm at $\tau_0 \sim 10^{-8}$ s, $D \sim 10^{-3}\ \text{cm}^2 \cdot \text{s}^{-1}$.

The transfer of an incoherent exciton has an activation character, because in this case, matrix elements of resonance interactions are relatively small, and the transfer process depends mainly on deformation displacements:

$$D \sim \exp(-E_a/k_B T), \tag{25.95}$$

where E_a is the activation energy of the incoherent exciton hopping. On the contrary, for a coherent exciton, the diffusion coefficient decreases with the temperature growth,

$$D \sim 1/\sqrt{T}, \tag{25.96}$$

because in this case the probability of spreading on phonons increases.

Therefore, the difference of mechanisms of excitation transfer by coherent and incoherent excitons is manifested in the temperature dependence of the diffusion coefficient. A temperature drop alongside with slowing down the incoherent exciton migration can strongly decelerate deformation-induced changes in molecules and thus increase D_{def} so that $\tau_{def} > \tau_m$. In other words, upon a temperature decrease an incoherent exciton can turn into a coherent one. The process of exciton capture in traps goes in parallel with displacement of equilibrium configurations of molecules and phonon emission (absorption) in the lattice and, in general, is dependent on temperature.

As shown further (see Chapter 27), in some cases the character of excitation propagation and quenching in a photosynthetic light-absorbing matrix is essentially dependent on temperature. As a rule, in biological objects temperature affects energy migration mainly by the dependence of the conformational state of the protein carrier on temperature. Changes in temperature cause changes in the distance and mutual orientation of chromophore groups fixed on the protein, which directly transfer the excitation energy (chlorophyll molecules in the photosynthetic membrane, aromatic amino acids in protein). It is accompanied by changes of chromophore interactions. In proteins, the relaxation processes, directly affecting the relation of the rates of vibrational relaxation and energy migration of electron excitation, changes with temperature as well. All these factors can change also the mechanism of migration of electron excitation.

26

Mechanisms of Enzyme Catalysis

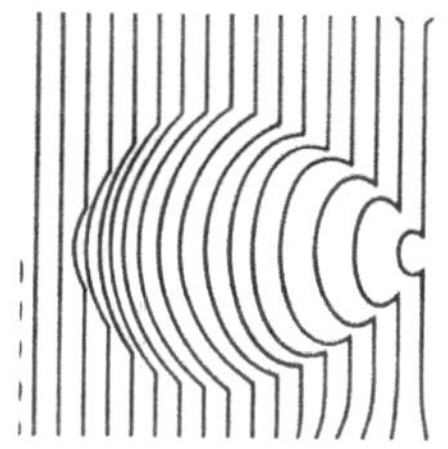

The general scheme of enzyme catalysis proposed by L. Michaelis includes the following stages: formation of enzyme–substrate complex ES, conversion of the substrate into product P, and, finally, desorption of the product from the enzyme and recovery of the latter to initial state E. In Chapters 1–4, problems of kinetics and regulation of enzyme reactions occurring according to the following equation were considred:

$$E + S \underset{k_{-1}}{\overset{k_{+1}}{\rightleftarrows}} ES \xrightarrow{k_{+2}} P + E.$$

The specificity of the enzyme action consists, evidently, in decomposition of complex $(ES) \xrightarrow{k_{+2}} P + E$, i.e. in mechanisms of processes occurring in the enzyme active center, where the substrate is transformed into reaction products. In modern biophysics, the theory of enzyme catalysis has not been formed finally yet. Nevertheless, the concepts of protein dynamics and electron conformational interactions, discussed in previous chapters, are widely used in this field in biophysics.

26.1 Physicochemical Description and Biophysical Models of Enzyme Processes

Thermodynamics of enzyme catalysis. As known, enzymes accelerate reactions by increasing their rate constants. As a rule, this effect is considered within the framework of the transient state or activated complex theory (Fig. 26.1). It is supposed that reagents that are initially in the ground state. They interact to form a complex which

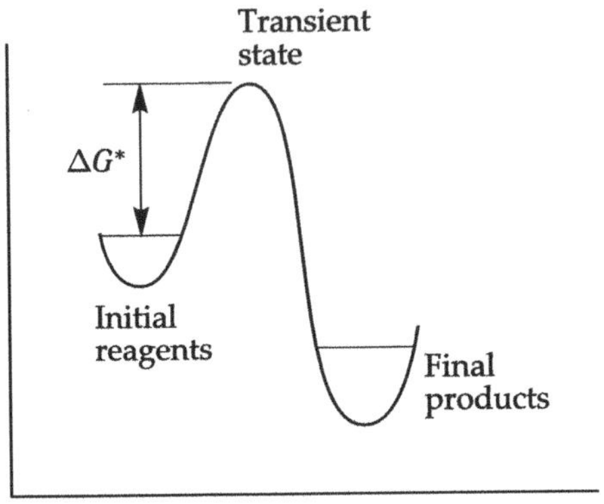

Figure 26.1. Diagram of changes in the reagent energy along the reaction coordinate.

is activated and transfers into a transient state. The maximum on the curve corresponding to changes in the reagent energy along the reaction coordinate refers to this state. Final reaction products are formed following the motion along the reaction coordinate from the transient state. In the activated complex theory, it is proposed that between the transient and ground states thermodynamic equilibrium exists with a corresponding equilibrium constant equal to the raio of concentrations of activated complexes and initial reagents. Then from the difference between free energies of these states it is possible to determine the concentration of the transient complex and after that the reaction rate by multiplying this concentration by the rate constant of its decomposition.

Let the difference in the values of the complete thermodynamic potential G (see Chapter 5) be $\Delta G^{\#}$ for the transient and ground states. Hence the rate constant of the transient complex decomposition, or the reaction rate, will be

$$k_1 = \kappa \frac{k_B T}{\hbar} \exp(-\Delta G^{\#}/RT), \tag{26.1}$$

where κ is the transmission coefficient (~ 1); $k_B T/\hbar = \omega$ determines the frequency of thermal fluctuations. At $k_B T \simeq 0.025$ eW ω is 10^{13} s^{-1}, and correspondingly the lifetime of transient state is $\simeq 10^{-13}$ s.

Representing change $\Delta G^{\#}$ for the activated complex formation as

$$\Delta G^{\#} = \Delta H^{\#} - T\Delta S^{\#}, \tag{26.2}$$

where $\Delta H^{\#}$ is the energy of activation, and $\Delta S^{\#}$ is the entropy of activation, instead of Equation (26.1) we have at $\kappa \sim 1$

$$k_1 = \frac{k_B T}{\hbar} \exp(\Delta S^{\#}/R) \exp(-\Delta H^{\#}/RT). \tag{26.3}$$

Upon formation of the activated complex, the entropy decreases ($\Delta S^{\#} < 0$) due to the formation of a quite definite mutual orientation of molecular groups of reagents as compared to their random arrangement in the initial state. Thus, entropy ($\Delta S^{\#} < 0$) and energy ($\Delta H^{\#} > 0$) factors decrease the reaction rate ($\exp(\Delta S^{\#}/R) \ll 1$, $\exp(-\Delta H^{\#}/R) \ll 1$).

Acceleration of enzyme reactions, as compared to the corresponding non-enzyme ones, according to formal kinetics means that the activation barrier $\Delta G^{\#}$ decreases in enzyme reactions ($\Delta G_e^{\#} < \Delta G_0^{\#}$). This may occur due to the decrease in the absolute value $|S_e^{\#}| < |S_0^{\#}|$ and also due to the decrease in the activation enthalpy $|\Delta H_e^{\#}| < |\Delta H_0^{\#}|$ (indices e and 0 refer correspondingly to enzyme and non-enzyme reactions).

The decrease in the entropy moiety occurs as a result of substrate fixation on the enzyme in the certain configuration of active groups, having lower entropy as compared to a free combination of reagents. Consequently, in such a complex which is initially close to the transient state, the decrease in the entropy upon formation of the transient state itself should not be now as large in its absolute value as in the case of free reagents. So the role of the entropy factor ($e^{\Delta S/R} \ll 1$) in Equations (26.1) and (26.2) that decreases the reaction rate is diminished in the enzymatic process as compared to the usual reaction of free reagents. The excess of energy released upon substrate binding should, at least partially, transform into heat to compensate for the entropy decrease upon the complex formation. Of course, the stabilization and decrease in the free energy of the complex should have slowed down the catalysis. However, in the presented schemes of catalysis it is proposed that not only the fixation of the substrate configuration takes place, but also the generated enzyme-substrate complex becomes strained. It is presumed that the activation energy of a chemical reaction decreases because the strain energy is concentrated on the attacked bond thus facilitating the reaction. In general energy required for such process is derived from the energy of sorption, which, as suggested, is not converted into heat but partially stored in the protein part of the enzyme and concentrated near enzyme–substrate contacts in the activated transient complex.

Models of enzyme catalysis. The Fisher "lock-and-key" and Koshland "hand-and-glove" models used in biochemistry take into account the entropy factors in catalysis. They suggest that the enzyme should fit the substrate structurally in the transient complex. Unlike the "lock-and-key" model, in the "hand-and-glove" (or induced correspondence) model, no requirements for the initial rigid structural complementarity is proposed. It is suggested that the catalytically active conformation in the active center is formed upon interaction of the enzyme and substrate in the enzyme–substrate complex. This strained conformation is less favorable energetically (due to its higher energy). Therefore, part of the free energy of sorption is consumed for thermodynamically disadvantageous approach and orientation of the reaction groups in the active center (Fig. 26.2).

Sorption interaction between reagents that is presumed to accelerate the enzyme reactions can be hydrophobic, electrostatic or dependent on hydrogen bond formation. For example, the "immersion" of the substrate benzyl group from water into the hydrophobic medium of the active center leads to a decrease in the activation energy to −29.4 kJ/mol and acceleration of the reaction to 10^5 times. Side groups of a substrate molecule can interact with the protein, if the reaction particles contain oppositely charged groups that leads to electrostatic stabilization of the transient state (trypsin hydrolysis of methyl groups).

The simplest model summarizing different versions of decreasing the activation energy in the enzyme catalysis is called a "rack" model (H. Eyring, R. Lumry and J. Spikes, 1954). Here sorption forces are used to generate strains (deformations) facilitating the reactions (Fig. 26.3). According to the "rack" model, to form bonds

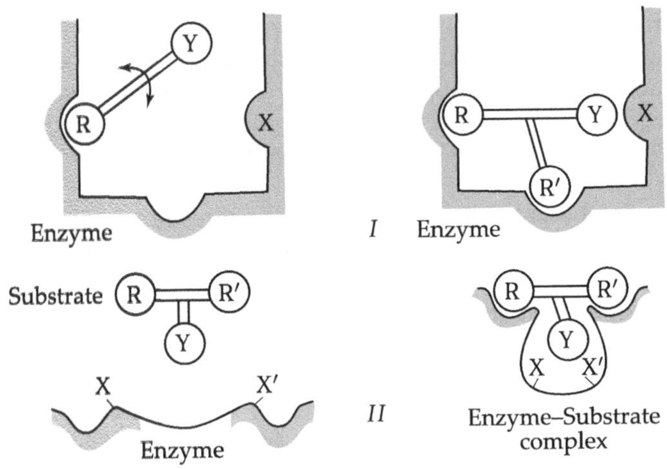

Figure 26.2. Effects of approach and orientation (*I*) and the mechanism of induced correspondence (*II*).

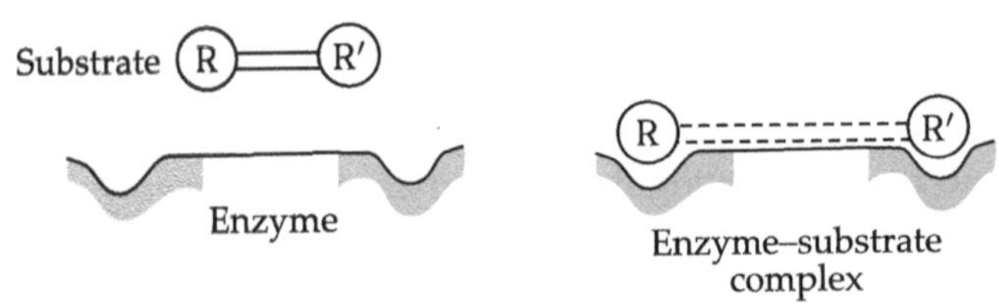

Figure 26.3. The "strain" mechanism.

with the enzyme the substrate undergoes deformation and is activated. In biological systems, both the substrate and enzyme molecules can actually undergo conformational changes that illustrates the general feature of different models of enzyme catalysis.

Thus, the thermodynamic principle of the above considered models is that the free energy of binding (sorption) of the substrate on the enzyme should not dissipate into heat but is consumed for decreasing the activation energy barrier of the chemical reaction. However, in these models it remains unclear in what definite form and by what actual mechanisms the sorption energy can be stored in the protein globule and concentrated on the attacked bond. □

Problem of decreasing the energy barrier. From the thermodynamic point of view, sorption energy of the substrate binding may be sufficient for the substrate conformational deformation. But in order to use this energy for the disruption of strong covalent bonds in the substrate, it is necessary in its turn to form other strong bonds. The question arises: what is the nature of the proposed strained conformational states that should cause such considerable changes in the strain energy, and in general what are direct proves of their existence in the enzyme – substrate complex? The solution of namely this question presents serious difficulties in the traditional physicochemical models of catalysis.

The most extensive information on enzyme-substrate interactions is obtained in studies of crystallized enzymes and their complexes with inhibitors and substrates using the method of X-ray analysis. The atomic coordinates of the enzyme and substrate obtained in this way are then optimized by minimization of energy potentials defined by the sum of interatomic interactions (see Chapters 8 and 9). Based on these data the energy maps (see Section 1 in Chapter 9) of enzyme–substrate complexes make it possible to compare the conformation of separate protein fragments and amino acid residues in the bound and free states.

The method of theoretical conformational analysis was used for studying non-valence interactions of α-chymotrypsin with a number of simplest substrates, lysozyme with triacetyl-glucosamine, ribonuclease with uridine-2,3′-cyclophosphate, carboxypeptidase A with peptide and ether substrates. But due to limited accuracy, this method does not always give an unambiguous answer to the question as to whether such strains exist in the complex. Nevertheless, the general conclusion drawn from the performed theoretical studies is as follows. Though the formation of the Michaelis complex is accompanied by conformational changes, however the fitting of the substrate does not cause either excess conformational strain or formation of a forced conformation in substrate and enzyme molecules. It was shown on α-chymotrypsin that at the precatalytic stage, structural elements of its active center are in an unstrained state.

In particular, the potential curve of side chain *Ser*-195 in the active center of α-chymotrypsin (see Fig. 26.9) contains three shallow minima (8–12 kJ/mol). This shows that rotation around bond $C^{\alpha}-C^{\beta}$ can proceed in the wide range of angle χ without serious steric hindrances. When water is removed (see Fig. 26.9) upon the substrate binding, the serine side chain acquires practically complete freedom of rotation at χ varying from -120 to $120°$. Its new position is determined now by the stabilizing interaction with the target group of the substrate. In addition, the distribution of residues, directly involved in the enzyme–substrate binding in α-chymotrypsin, lysozyme or in the interaction with the heme in myoglobin, is analogous to the energy distribution of other residues.

It was demonstrated that the productive conformation of the substrate is, on the one hand, the most preferable by the energy of non-valence interactions with the active center, and on the other hand, it corresponds to one of the most low-energy forms of a free substrate molecule. This conclusion is confirmed also in the studies of the substrate conformation upon interaction with the active center of lysozyme, as well as upon formation of a complex of carboxypeptidase A with peptide and ether substrates. The mutual orientation necessary for catalysis and the approach of the target bond of the substrate and active groups of the enzyme occur spontaneously and do not require any energetically unfavorable and strained contacts.

In this case, we discuss the conformational strain upon non-valence interactions that should have led to a change in the dihedral angle and distances between atomic groups having no valence bond. Below we consider changes in valence angles of the main chain, accompanying the disruption and formation of valence bonds in the active center, that occur in the chemical act of catalysis.

It means that the strain of the conformation is not a necessary stage, energy source and driving force of catalysis. Apparently, in the Michaelis complex, the mutual approach of functional groups of the enzyme and substrate in the active configuration, required for further electronic and chemical interactions, can be achieved due to usual intramolecular motions of different structural elements with a wide set of characteristic times (see Chapter 10). From the formal kinetic point of view, the formation of such an active configuration, close to the transient state, really decreases the entropy portion of the reaction barrier, but this is just a spontaneous result of intramolecular dynamics of different, including functionally active, molecular groups in the enzyme and substrate. □

According to the models described above, catalysis should be coupled to a decrease in the height of the activation barrier. Let us consider the existing hypotheses on possible mechanisms of a direct decrease in the activation barrier due to energy storage in the protein globule and concentration on the attacked bond in the substrate. In the elastic deformation model, proposed by D. S. Chernavsky, Yu. I. Khurgin, and S. E. Shnol (1967), it is suggested that a protein molecule is a rigid construction. The dimensions of a protein reservoir required to store elastic deformation energy necessary in catalysis were calculated. It was found that if the elasticity limits of a protein body are accepted to be equal to those of most materials, energy storage $\Delta E \sim 0.5$ eV requires propagation of elastic deformation over the whole globule (2.0–5.0 nm).

Deformation energy can be generated also at displacements of electric charges in the protein globule due to a change in the charged state of the active center in the catalysis act. As proposed, it can be stored in the globule up to the beginning of the following act. Calculations show that at dimensions of about $\leqslant 1.0$ nm the electric energy is promptly converted into mechanical and back, while the densities of these energies are equalized. However, even here it is also necessary to state that at present no direct evidences are available for storage of elastic deformation energy.

Discussed is also another possibility of using the conformational energy in the catalysis act due to thermal fluctuations in the protein structure. In these models, the problem is not the static strain, but the transfer of the thermal vibration energy along certain degrees of freedom and its concentration on the target attacked bond in the substrate. This approach is in line with R. Lumry's (1959) concepts of the enzyme protein as a "pool of excess thermal energy", where processes occurring on the protein globule surface are a source of free energy required for the catalysis act.

In the electromechanical model, it is proposed that thermal fluctuations in protein induce vibrations accompanied by a displacement of electric charges in the active center and polarization of the attacked bonds of the substrate. It is postulated in the model that intramolecular interactions in protein stabilize for some time ($\geqslant 10^{-3}$ s) the generated catalytically active conformation with polarized bonds.

In other models, it is underlined also that it is necessary to coordinate in time the thermal conformational fluctuations arising in protein and the concentration of their energy in the protein active center. The probability of generation of thermal fluctuations in protein is proportional to

$$W \sim \exp[-(\Delta G)_F / k_B T], \tag{26.4}$$

where $(\Delta G)_F$ is the change in the free Gibbs energy in protein upon fluctuations. For the catalysis act, value $(\Delta G)_F$ should be at least by an order of magnitude higher than the average energy of thermal vibrations, $k_B T \sim 0.025$ eV. Hence it is possible to determine that the probability of generation of such a fluctuation (see Equation (26.4)) will be 10^{-5} per time unit, and during the rotation time of the enzyme 10^{-3} s such fluctuations can emerge at probability $\sim 10^{-8}$.

Indeed, different types of stochastic motions, occurring near the protein–solvent interface (relaxation of bound water molecules, local changes in conformation, fluctuations in the charge density in the surrounding ion solution, etc.) have correlation times of 10^{-9}–10^{-7} s. To explain the catalysis mechanism, it is proposed that the enzyme is "arranged" in such a way that its structure provides for a coherent character of spreading of fluctuation changes in the conformation from the protein surface to its active center. Thereby the exchange of the free energy between the macromolecule and its surrounding is provided. In the model of B. Somogyi and S. Damjanovich (1975), the enzyme molecule is presented as a number of coupled oscillators, so that a directed transfer of the energy of collisions with solvent molecules on the protein surface to the active center takes place.

The basis of all considered "dynamic" models is an assumption on the generation of a coherent state with one vibrational mode and on the dissipationless energy transfer over the protein globule to the active center along certain degrees of freedom. It is suggested that as a result the energy is concentrated on the attacked bond, leading to a decrease in the activation barrier. Therefore, this seems to be the "secret" of the enzyme. But, at present, no actual proves of this assumption are available. Moreover, the idea essential for all these models on the directed transfer and accumulation of energy of usual thermal fluctuations in the active center without energy losses conflicts with the second law of thermodynamics.

26.2 Electron Conformational Interactions upon Enzyme Catalysis

As has been shown, conformational changes in the protein globule have a relaxation character and are described by a set of different times. As a rule, they proceed much slower than purely electronic transitions. Rapid changes in the electronic state of the protein molecule (for example, reduction of the Fe atom in the active center of cytochrome) distort the initial equilibrium state and lead to a cascade of sequential directed conformational relaxation acts (see Section 1 in Chapter 10). Just this circumstance is the basis of the concept of conformational relaxation in enzyme catalysis (L. A. Blumenfeld, 1972). The reaction product formation is considered as a logical result of electron-conformational interactions in the enzyme–substrate complex. It is proposed that conformational changes in the enzyme–substrate complex, occurring after the change in the electronic state of the substrate in the enzyme active center, have a directed relaxation character and involve processes of conversion of substrate molecules into the product molecules. The elementary act of the enzyme reaction consists in a conformational change of the enzyme–substrate complex, and the rate of the conversion of the substrate into the product is determined by the rate of this conformational change.

The sequence of events is the following. An intial change in the electronic state and local changes in the active center and the substrate occur during the times of vibrational relaxation (10^{-12}–10^{-13} s). They involve only part of the molecule, namely some selected chemical bonds of the substrate and functional groups of the active center but not the remaining main part of the protein globule. However, on the whole at this stage the macromolecular complex is in a conformationally non-equilibrium state. After that slow spontaneous relaxation of the enzyme–substrate complex to a new equilibrium takes place coupled to the substrate conversion into the product:

$$S + E \to SE^* \to \widetilde{E}P,$$

where E^* is the non-equilibrium state, and $\widetilde{E}$ is the new equilibrium conformational states of the protein globule of the enzyme. The chemical change in the substrate, including rearrangement of atoms following the splitting of chemical bonds, is realized here as part of the general conformational change in the macromolecular complex.

At the next stage, the enzyme–product complex decomposes: $E\widetilde{P} \to \widetilde{E} + P$. This stage also goes in parallel with rapid local changes in the active center and changes in solvation of the reaction products, moving into the surrounding volume. The enzyme remains in conformational state $\widetilde{E}$ that now after the product is released becomes again non-equilibrium. Thus, at the final stage, slow conformational relaxation of the free enzyme molecule to the initial equilibrium state $\widetilde{E} \to E$ takes place.

Thus, conformational changes occur under conditions of essential conformational non-equilibrium. The reaction coordinate coincides with the coordinate of conformational relaxation that proceeds on certain degrees of freedom and has a directed character: $E \to E^* \to \widetilde{E} \to E$. □

Namely, the presence of certain "selected" mechanical degrees of freedom makes it possible to consider displacements, occurring in different regions of a macromolecule, as changes performed in one act. The energy, concentrated on these slowly relaxing degrees of freedom, do not dissipate into heat due to the rapid exchange with vibrations providing a directed character of relaxation processes in the enzyme catalysis. From this point of view, the traditional concept of energy and entropy activation, used in the theory of an activated complex, has no direct meaning. The dependence of the enzyme reaction rate on temperature is determined not by the number of active molecules with energy sufficient for overcoming the barrier, but by the effect of temperature on the macromolecule conformation and, consequently, on the pathway and rate of its subsequent relaxation (see Section 2 in Chapter 7). All substrate molecules that form a "correct" complex with the enzyme undergo chemical conversion as a result of spontaneous enzyme relaxation to a new conformational state. Undoubtedly, from the thermodynamic point of view, the general driving force of the process is the difference in chemical potentials of the substrate and the product. However, it determines only the number of collisions of enzyme and substrate molecules, but not the activation mechanism of elementary act of the enzyme–substrate complex transformation.

The formation of the enzyme–substrate complex is connected, at the first stage, with penetration or diffusion of the substrate molecule into the slit, formed by relatively rigid structural elements on the enzyme surface. At the second stage, as the substrate has penetrated into the slit, an active configuration is formed due to conformational motions in the complex. And only at the third stage, after achieving a certain chemically active configuration, a chemical reaction takes place in the system "substrate–active center groups". Within the framework of a single-mode approximation, it is possible to determine the amplitude of α-helices bending due to thermal fluctuations under the action of the stochastic force. It was found that at the length of the α-helical region $L \sim 0.5$ nm, the amplitude of the slit fluctuations is $A \sim 0.5$ nm ($T = 300$ K). At viscosity $\eta \sim 10$ Pa · s, the relaxation time of bending fluctuations is in a microsecond range that is much lower than the enzyme turnover time (10^{-2}–10^{-3} s).

Diffusion of substrate molecules through the fluctuating slit is possible at the first stage only if its opening r^* is adequate (Fig. 26.4). At slit length $L \sim 0.7 \div 2.5$ nm

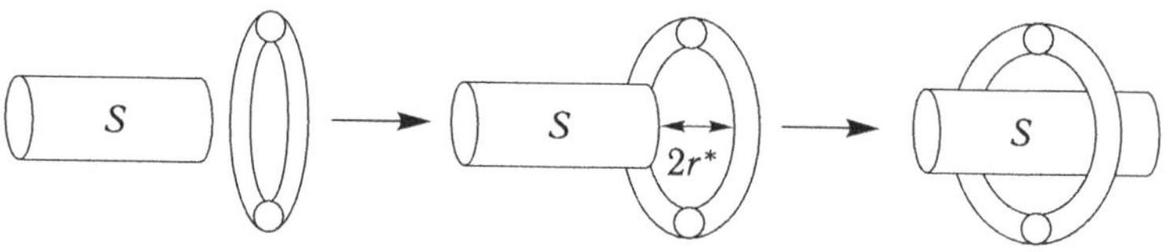

Figure 26.4. Diffusion of substrate S into the fluctuation slit (see explanations in the text).

and Young's modulus values 10^3–10^4 J/cm^3, the time of slit relaxation is $\tau \sim 10^{-6} \div 10^{-8}$ s. The characteristic time of substrate diffusion into such a fluctuating slit is estimated in the model. At microviscosity values $\eta \sim 10$ Pa · s, substrate molecule radius ~ 0.2 nm and slit depth $l \sim 0.5$ nm, this time is about 10^{-6} s.

Penetration of a substrate having an "inappropriate" geometry requires a significantly larger opening of the slit than in the case of geometrical correspondence of the slit and the substrate molecule. As a result, at the stages of the enzyme–substrate complex formation, the structural correspondence can be determined by the size and shape of the fluctuating slit. However, in the kinetic respect, the stages of penetration of the substrate into the enzyme do not restrict the general rate of catalysis.

Conformational dynamics of proteins. Dynamics parameters are directly associated with the mechanism of enzyme catalysis. A catalysis act proceeds only at a strictly determined orientation of several groups, achieved due to conformational motions. Formally, this corresponds to simultaneous encounter of several molecules that is hardly probable in solution. But in protein, the region of conformational motions of separate groups decreases to several cubic nanometers so that reactions, practically unrealized in solution, here become prevailing and highly selective (Fig. 26.5).

Figure 26.6 shows as an example the structure of the active center of alcohol dehydrogenase, a zinc-containing metal enzyme, using NAD$^+$ as a coenzyme. The Zn ion is at the bottom of the hydrophobic "pocket", formed in the site of the catalytic and binding nucleotide domains. Ligands of the Zn atom are atoms S of amino acid residues *cys*-46 and *cys*-174 and atom N of residue *his*-67. The fourth ligand —

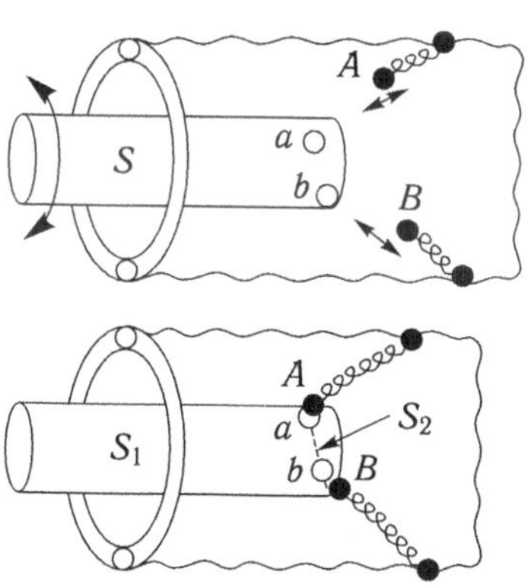

Figure 26.5. Adjustment of active center groups A and B and substrate S in the reactive configuration.

A necessary condition for splitting substrate S to S_1 and S_2 is a simultaneous contact of groups A and a, B and b, respectively.

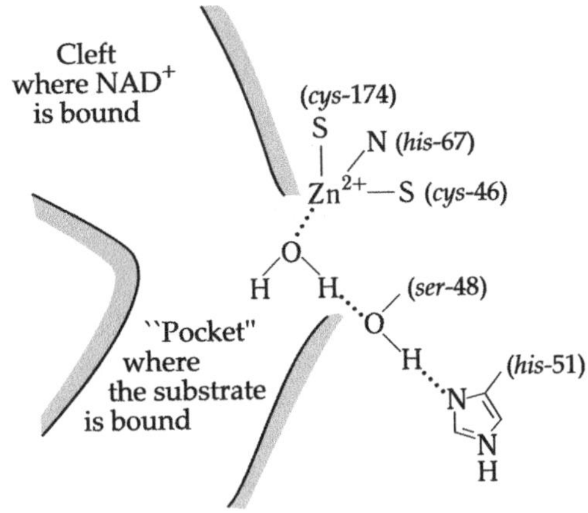

Figure 26.6. Model of the active center of alcohol dehydrogenase (reproduced from F. Fersht, 1980).

a water molecule — is bound to the hydroxyl of residue *ser*-48, which in its turn forms a hydrogen bond with *his*-51.

Oxidation of an alcohol molecule to aldehyde with hydrogen transfer to NAD^+ takes place in the active center. For this reaction to occur, a productive triple complex including molecules of NAD^+, alcohol and a Zn atom should be formed (Fig. 26.7). It is supposed that ionized alcohol substitutes the H_2O molecule bound to Zn, thereby the O_2 atom of alcohol binds directly to the Zn ion, and the hydrophobic side chain binds to the hydrophobic cavity of the "pocket".

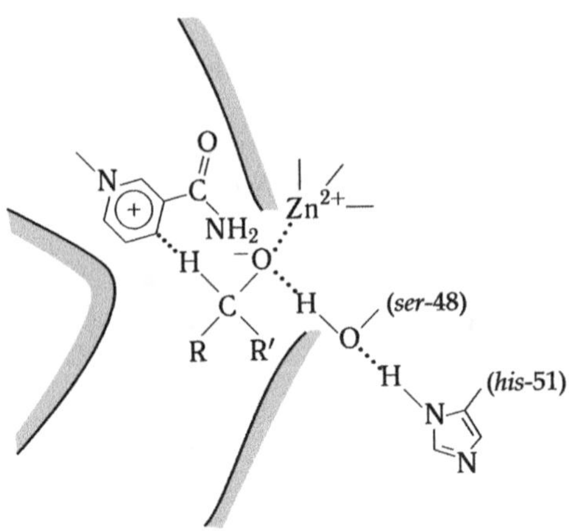

Figure 26.7. Model of a productive triple complex of alcohol dehydrogenase (reproduced from F. Fersht, 1980).

In the case of serine proteases, the functional regions, implementing sorption and nucleophilic catalysis (hydrolysis) of the peptide bond, are spatially separated. According to the X-ray analysis, the active center of α-chymotrypsin contains a rather capacious hydrophobic "pocket", where the aromatic side group of the hydrolyzed peptide is bound and the substrate is fixed (Fig. 26.8). The structure of the active center of α-chymotrypsin prior to its interaction with the substrate is given in

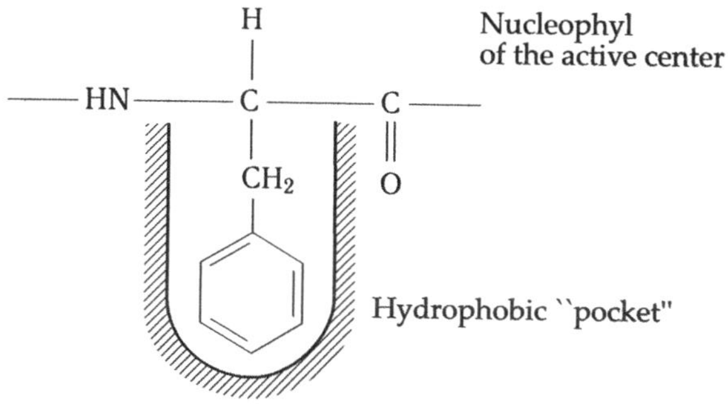

Figure 26.8. Interaction of the peptide substrate with serine protease (α-chymotrypsin), the active center of which consists of spatially separated sorption and catalytic regions.

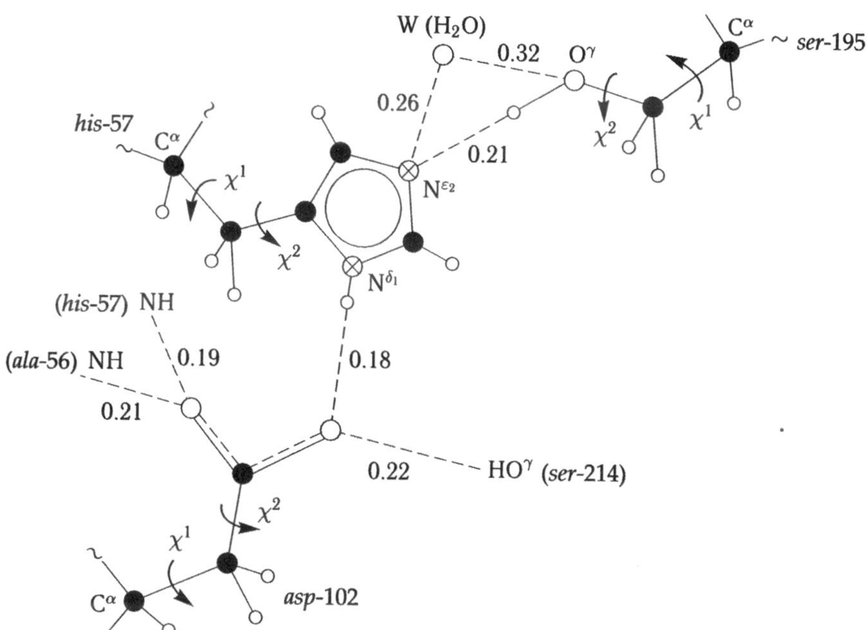

Figure 26.9. Structure of the active center of α-chymotrypsin (numerals show interatomic distances, nm).

Fig. 26.9, where distances and rotation angles for residues *asp*-102, *his*-57, *ser*-195, involved in the formation an active configuration with the substrate, are designated.

Water–protein interactions and enzyme activity. Numerous data indicate that hydrated water has a strong effect on protein properties, and protein has an active effect on the structure-dynamic water properties (see Section 5 in Chapter 9). At low degrees of hydration, water does not practically affect the properties of the enzyme protein. But at a certain level of hydration, this situation changes considerably. At hydration $h = 0.15$ (g water per 1 g protein) bound water appears, when hydrated water molecules are bound to each other via a network of hydrogen bonds. The emergence of bound water enables proton transport over the network of hydrogen bonds. The network of hydrogen bonds involves not only water molecules, but also forms hydrogen bonds with surface protein groups. The results of the Fourier analysis of the intensity of X-ray scattering yield quantitative characteristics of the mechanism of water effect on the formation of protein macromolecules. Upon hydration, water weakens the general water-protein network of hydrogen bonds, loosening the macromolecular structure. As a result, at a slight increase in the hydration degree to $h = 0.20$, the protein thermal stability drops drastically. Approximately at the same degree of hydration, the conformational mobility begins accelerating and the hydrated water mobility increases sharply. Under these conditions, enzymatic activity is generated (Fig. 26.10).

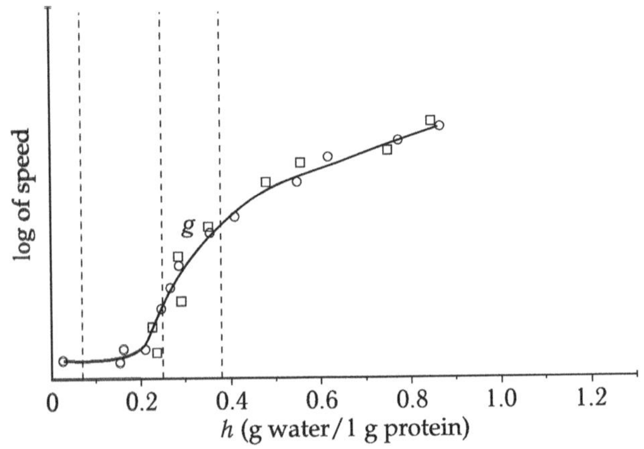

Figure 26.10. Enzymatic activity on a logarithmic scale (squares are $\log \nu_0$, where ν_0 is the reaction rate; circles are the time of rotational relaxation on a logarithmic scale ($\log \tau^{-1}$) for an EPR label TEMPONE) (reproduced from J. A. Rupley et al., 1983).

Thus, at $h = 0.15$–0.20, the general water-protein network of hydrogen bonds weakens, the globule is loosened, the globule stability decreases sharply, enzymatic activity in the protein globule appears and the hydrated water mobility is accelerated greatly. Accordingly, it was found that there is a clear correlation between the weakening of the hydrogen bonds inside the globule, the decrease in the globule stability, the increase in the intramolecular mobility and proton mobility, the appearance of considerable functional activity (Fig. 26.10) in macromolecules and the high acceleration of the hydrated water mobility. At $h = 0.20$–0.38, these processes continue becoming more pronounced up to $h = 0.38$. At $h = 0.38$, the thermodynamic, structural and dynamic properties of protein terminate changing. At this degree of hydration, protein is completely ready to perform its enzymatic function. The protein enzymatic activity makes only 1/10 of the activity in solution at $h = 0.38$. However, this is associated with a low steric accessibility of the enzyme active center for the substrate.

26.3 Electronic Interactions in the Enzyme Active Center

The condition for a successful performance of a catalytic act is formation of an active configuration, where functional groups are properly approached to each other and mutually oriented. Of course, the mechanism and result of their interaction depend on the details of their molecular structure and should be considered particularly in each definite case using quantum chemistry approach. At present, there are not so many of such examples, but even now it is possible to indicate a number of specific peculiarities of fragments, facilitating the substrate conversion in the active center.

Bonds of the active center. Substrate sorption induces transition of its molecule from water medium into the surrounding of amino acid residues in the active center. As a rule, the micromedium of the active center has a lower dielectric permeability as compared to that of water. For example, the dielectric permeability value in the sorption site of the chymotrypsin active center $\varepsilon < 10$ (it is usually accepted that $\varepsilon \sim 3$ for protein and $\varepsilon = 80$ for water). The reason for this is connected with the fixed orientation of protein dipoles relative to the charged groups of the substrate. This makes their difference from "free" dipoles of water, weakening Coulomb interactions of charges upon reorientation. As a result, electrostatic interactions in the hydrophobic medium between the reagents and polar groups of the enzyme are enhanced. In addition, the micromedium of the active center has a strongly decreased polarity as compared to that of water, particularly in those sites of the globule, where hydrophobic interactions of non-polar amino acid residues of the polypeptide chain take place. In the case of reactions accompanied by the charge transfer in the active center, and they are, as a rule, predominant, the low-polarity protein medium is subjected to relatively slight reorganization and partially shields the transferred charge from the action of the polar solvent.

Let us consider in more detail the role of this factor in enzyme reactions with a charge transfer (L. I. Krishtalik, 1988). The activation energy of the reaction, accompanied by a charge transfer and reorganization of the surrounding, is expressed as follows:

$$E_a = (E_r - \Delta E)^2/(4E_r), \tag{26.5}$$

where ΔE is the heat effect of the reaction, E_r is the reorganization energy of the medium (see Chapter 25). In this case, E_r is the energy of reorientation of dipoles from the initial to the final configuration, determined under condition that the system remains in the initial electron term, i.e. the charge is not transferred. The low mobility of protein dipoles shows that they contribute little to the energy of reorganization as compared to the dipole surrounding in solution, thereby decreasing the E_r value in the enzyme. In parallel, the activation energy in Equation (26.5) decreases as well. As a matter of fact, part of the solvent volume with high reorganization energy is excluded from the interaction with the charge of the reacting particles in the protein globule. Apparently, this effect is manifested to a higher degree upon an increase in the globule volume, but only down to a definite limit, depending on the distances between interacting charges. Calculations were done for the case of charge transfer between two points, positioned on the radius of a sphere with dielectric permeability much lower than in the surrounding solvent. It turned out that at a radius of the sphere up to 1.5 nm, the involvement of reagents into it leads to a decrease in the energy of reorganization of the medium by 42–84 kJ and to the corresponding acceleration of the reaction.

Another important factor affecting the E_a value of the activation barrier is generation of local electric fields due to high local concentrations of dipoles in peptide bonds. In such charged regions, the electrostatic potential can reach several tens of millivolts. At standard dimensions of the active center (0.5–1.0 nm), it leads to the field strength of thousands and hundreds of thousands units (V/cm).

Thus, oriented polar groups create an intraglobular electric field, where charge transfer takes place. Along with a great decrease in dielectric permeability, this factor also affects the value of Coulomb interaction of charges or the difference in energies of the initial and final states of charged particles, i.e. the value of thermal effect ΔE of an elementary act of charge transfer. In its turn, this affects the activation energy of the whole process and can lead to its acceleration for a certain class of reactions. This is correct, in particular, for the reactions accompanied by a decrease in the absolute charge value of each of the reactive particles or of the particle localized farther from the globule – water medium interface. □

Electronic transitions. Electronic transitions, induced as a result of interaction of molecular orbitals of reagents, coming together in the active configuration, trigger the elementary act of catalysis. For a quantitation solution of this problem, it is required to apply methods of quantum chemistry, in particular, the method of intermolecular orbitals (IMO). The principle of IMO is that the molecular orbital of a complex can be represented as a linear combination of MO constituting the complex (cf. the LCAO method in Section 6 of Chapter 24). Under the action of perturbation, caused by interaction of electrons and nuclei, MO of the ground and excited states of each molecule will be mixed. The main role will belong to interactions of MO, located close to the boundary orbitals: the upper (antibinding) occupied and the lower (binding) free ones. The energy of interaction of boundary orbitals depends on the degree of overlapping of the upper occupied and the lower free orbitals that is determined by the type of orbitals at mutual orientation of molecules. As a result of such overlapping, the distribution of electron density in molecules changes, and this, in its turn, changes their characteristics and above all the strength of the bonds. If the electron density is added to the antibinding orbitals but decreases in the binding orbitals, the corresponding bonds become weaker.

Such effects are possible if MO of active groups of the enzyme interact with MO of atoms of the attacked bond in the substrate. The additional electron density on the antibinding orbital imparts some propensities of an excited state to this molecule. In many cases, dipole moments in the excited and ground states differ greatly as well as pK values. It is clear that these factors play an additional role, providing specific conditions in the microenvironment of the active center favorable for the catalysis act.

Let us consider the character of interactions in productive configuration in the active center of α-chymotrypsin. These reactions proceed following the mechanism of nucleophilic substitution. Nuclophilic agent (base) *ser*-195 approaches the carbon atom with a deficit of electrons (electrophilic center C′ in the peptide bond) and forms a bond with it, substituting atom N. The substituted atom N with the lone-electron pair moves away together with the proton attached to it. Therefore, the reaction is based on the splitting of the peptide bond in the substrate that results in acylation of the enzyme. The process includes a tetrahedral intermediate state with the formation of valence bond $O^{\gamma}-C'$ between the enzyme and the substrate.

Figure 26.11 shows two limiting cases that can precede the disruption of the N — C bond and acylation of the enzyme. In one case (Fig. 26.11, *I*), the neutral side chain of *ser*-195 occupies the position required for a nucleophilic attack on atom C′ of the hydrolyzed bond. In the other case (Fig. 26.11, *II*), it is proposed that the proton has already moved from the hydroxyl of *ser*-195 to nitrogen $N^{\varepsilon 2}$ of the imidazole ring of *his*-57, so that atoms C′ and N of the peptide bond are attacked by alcoholate

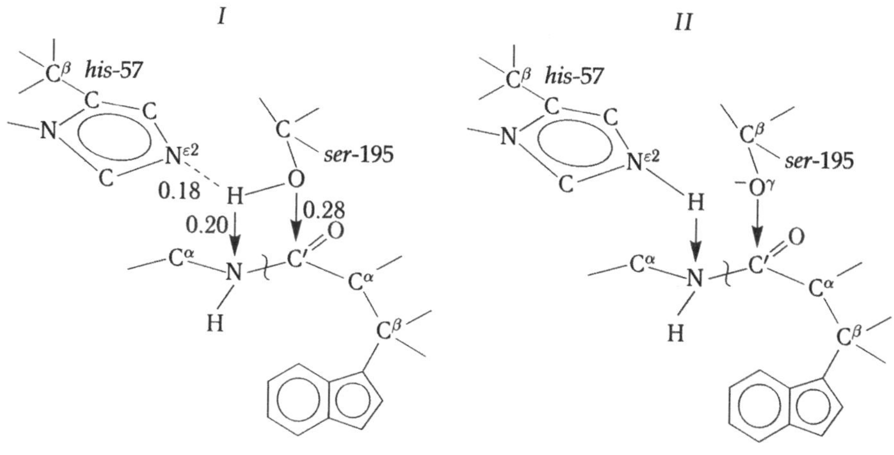

Figure 26.11. Position of the splitted group of the substrate and side chains of *ser*-195 and *his*-57 in a non-valence complex of α-chymotrypsin.

I, Without proton transition over *ser*-195 to *his*-57; *II*, with transition of proton from *ser*-195 to the $N^{\varepsilon 2}$ atom of imidazole group of *his*-57; numerals designate equilibrium interatomic distances (nm) in a non-valence enzyme – substrate complex.

ion $O^{\gamma}_{(195)}$ and group $N^{\varepsilon 2}H$. The deprotonation stage of $O^{\gamma}_{(195)}$ facilitates the disruption of the peptide bond, because the oxygen ion is a stronger nucleophilic agent than the neutral atom.

Let us consider how individual stages of atomic interaction are connected in a tetrahedral complex. The proton transfer between *ser*-195 and *his*-57 is facilitated because bond $O^{\gamma}_{(195)}-H$ is weakened due to interaction of $O^{\gamma}_{(195)}$ and C′. Here the electron density is shifted to the antibinding orbitals in the peptide bond. It is the result of the interaction of the lone-electron pair O^{γ} with π-electrons of atom C′, change in hybridization of atoms C′ and N ($sp^2 \rightarrow sp^3$) leading to the weakening of bond N — C′ (see Section 1 in Chapter 9; Fig. 9.2).

Thus, the weakening of the $O^{\gamma}-H$ bond is accompanied by the attack of $O^{\gamma}_{(195)}$ on carbon C′, expulsion of the lone-pair electrons of nitrogen in the amino group from peptide bond N — C′ and its loosening. The consequence of this is an increase in the alkalinity of the amino group that facilitates its protonation by the enzyme. Therewith the proton moves from $O^{\gamma}_{(195)}$ to *his*-57 that also weakens bond N — C′. This promotes an increase in the efficiency of interaction between $O_{(195)}$ and C′, etc. Therefore, a quite unique situation is established in the tetrahedral complex when several reactions occur concurrently, mutually accelerating each other. The complex represents a charge relay system of charge transfer, where the charge and proton move synchronously between three atomic groups (*ser*-195, *his*-57 and the peptide bond), the enzyme is deacylated and atom $O^{\gamma}_{(195)}$ is protonated, while the water molecule plays the role of the hydroxyl group of *ser*-195.

Hydrolysis of the ester links, CH_3-CO-O, where the trigger mechanism is the draining of negative electron density of the $O^{\gamma}_{(195)}$ lone-pair to the antibinding p_z orbital of the carbon atom in the ester link, proceeds in a similar way. The efficiency of electron interactions is determined from the electronic capacity of antibinding orbitals and, naturally, the degree of overlapping of orbitals of atomic groups in the enzyme – substrate complex.

The principle of formation of a polyfunctional closed system of atomic groups in an active configuration is fulfilled also in the case of other enzyme–substrate complexes.

In enzyme catalysis, the multistage character of substrate conversions is provided due to their synchronous cooperative occurrence in a common polyfunctional system. Namely, this causes a decrease in the activation energy of the whole process that is required from the formal kinetic point of view. In this case, instead of low-efficient consecutive activation stages we have concerted conversion of the substrate in the state equivalent to the transient state of the already prepared activated complex.

Thus, the overall rate of enzyme catalysis is determined by the characteristic time needed to form the required productive conformation in the active center by corresponding atomic groups spontaneously approaching each other. By its order, of magnitude this time is comparable to the times of enzyme turnover. When the distances become close to the chemical bond lengths, the subsequent catalysis proceeds without thermal activation. □

26.4 Molecular Modeling of the Structure of an Enzyme – Substrate Complex

The enzyme structure contains binding sites of small molecules — substrates/products of enzyme reactions, as well as modulators of enzyme activity (cofactors, inhibitors or activators). To study mechanisms of enzyme catalysis, information on the spatial structure of enzyme complexes with these molecules is required. The structure of the enzyme–substrate complex allows assumptions on the role of separate residues of the active center in the catalytic cycle and also interpretation of the experimental data on site mutagenesis. Though in many cases, the structure of the enzyme complex with the studied ligand is unknown, it can be determined for its apoform or for a complex with another ligand (frequently with a substrate-competitive inhibitor).

At present, the three-dimensional structure of protein groups in the active center has been determined experimentally, using mainly the X-ray analysis (XRA) and NMR spectroscopy. The Protein Data Bank contains more than 50,000 protein structures and their complexes, but the number of structures of enzyme complexes with a substrate or inhibitor is far smaller (about 4,000).

If the construction of structural homologs is determined, the target structure can be obtained by molecular modeling, primarily by comparative modeling (Chapter 9). In this case, the quality of structure prediction depends mostly on the degree of homology of the modeled enzyme and the structural pattern. When the structure of a closely-related enzyme (the identity of residues at pairwise alignment > 30–40 %) is used as a pattern, the quality of the three-dimensional model is usually not worse than that of an experimental structure with account for high conservatism of residues forming the active center. However, if the homology is low and the identity of sequences < 25–30 %, the reliability of the model deceases drastically, and it is impossible to predict the structure of the enzyme–substrate complex using it.

Prediction of the structure of an enzyme–substrate complex. During recent years, computer-based methods for prediction of the structure of protein complexes with low molecular compounds (ligands) have been developed actively. Therewith the protein structure should be already known (determined with experimental RSA and NMR techniques or homology modeled). In this case, the modeling of the structure of a protein–ligand complex means the prediction of the ligand conformation or its orientation in the protein active center. Algorithms for realizing this approach are called molecular docking. The development of an effective method of a conformational search is one of the principal directions in upgrading the docking algorithms.

Using molecular dynamics (MD) calculations (see Section 2 in Chapter 9), it is possible to take into account the protein conformational mobility on the level of large-scale changes in the conformation of loop regions or domain motions. In this case, the ligand docking is performed, as a rule, for a representative set of conformers of the same protein. Unfortunately, the modeling of the receptor mobility using MD is an extremely labor- and time-consuming procedure that requires seeking for optimal parameters for MD calculations.

Another important component of a docking algorithm is the search for evaluation functions (EF). The evaluation of the contribution made by individual terms, such as hydrophobic and stacking interactions, for reproducing them adequately in EF makes it possible to increase essentially the effectiveness of EF (R. G. Efremov, 2010).

As an example, let us take the modeling of APT interaction with Ca-ATPase, which uses the energy of ATP hydrolysis for active transport of Ca^{2+} ions across the membrane by the concentration gradient. The three-dimensional structure of Ca-ATPase consists of a transmembrane domain and three extramembrane domains. Extramembrane nucleotide-binding (*N*) and phosphorylation (*P*) domains are involved in the process of ATP binding and hydrolysis. But the distance, separating the binding sites in the *N*- and *P*-domains in the Ca-ATPase structure, was much larger than the size of an ATP molecule (Fig. 26.12, *a*). Therefore, a "direct"

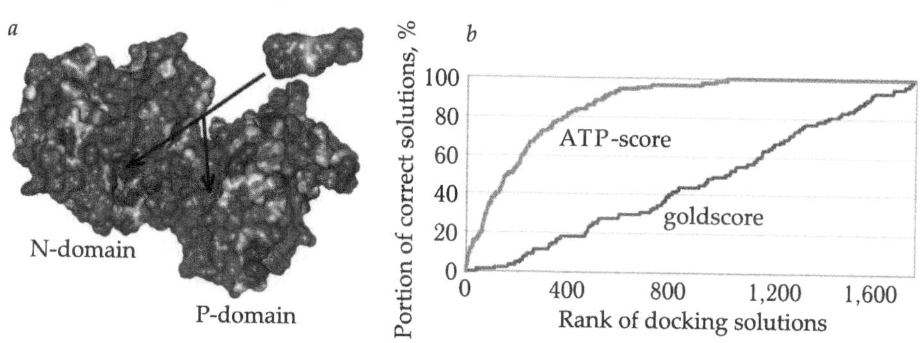

Figure 26.12. *a*, Complementarity of hydrophobic/hydrophilic properties of an ATP molecule and the ATP-binding domain of Ca-ATPase.

The adenine base is responsible for hydrophobic properties (brown) of ATP; hydrophilic properties (blue) are conditioned by phosphate groups and the ribose residue. Arrows indicate the hydrophobic pocket in the *N*-domain site and the hydrophilic active center of the *P*-domain. Calculations of hydrophobic/hydrophilic properties were obtained using web-server http://model.nmr.ru/platinum based on the method of molecular hydrophobic potential (MHP).

b, Curve of accumulation of correct solutions for the task of the ATP docking for MD conformers of Ca-ATPase.

The total number of solutions is ~ 400 (reproduced from Yu. Kosinsky, T. Pyrkov, R. G. Efremov, 2006).

docking without taking into account the molecular mobility of protein would not allow clarifying in what way ATP interacts with two sites of *N*- and *P*-domains. Studies of the conformational mobility of the target-protein using MD calculations revealed the presence of domain motions at times of about several nanoseconds. As a result, the distance between sites of *N*- and *P*-domains changed considerably, and as was later demonstrated using the docking ATP molecules could be bound in the two sites simultaneously. To model an ATP – Ca-ATPase complex, a special ATP-oriented EF was designed that included such terms of intermolecular interactions as stacking and hydrophobic contacts (Fig. 26.12, *a*), not reproduced in the standard EF. This re-ranking of the docking results on the ATP-specific evaluation criterion allowed a significant increase in the accuracy of identification of a correct structure of a protein – ligand complex (Fig. 26.12, *b*). This advantage in the effectiveness, as compared to standard EVs of docking, was achieved due to the creation of special criteria adapted for other ligands (nucleotides and peptides) and also due to a detailed account for hydrophobic, hydrophilic and stacking interactions in complexes (see details in Section 5 in Chapter 9). The results of modeling showed that ATP is bound due to the adenine base and the ribose residue in the *N*-domain and upon "closing" of the interdomain region; phosphate ATP groups oriented towards the P-domain interact with its active center, where hydrolysis of the pyrophosphate bond takes place.

The results of modeling the ATP interaction with Ca-ATPase are on the whole compatible with the later obtained experimentally established structures of this complex.

Estimation of the energy of ligand binding by protein (empirical methods, MD). The understanding of the structure of the protein – ligand complex is required for finding inhibitors of a definite enzyme or high-affinity ligands of the receptor.

Prediction of the free energy of the ligand binding by protein based on the three-dimensional structure of their complex is an interesting and fundamental problem of molecular biophysics. Using the molecular modeling approaches, it is possible to estimate most accurately (from the physical point of view) not the absolute free energy of binding, but its relative change, for example, upon a substitution of one ligand by another similar by the chemical structure. For this purpose, the method of free energy perturbation (FEP) can be used. It was realized in many software packages for calculations of the behavior of molecular systems with MD or Monte Carlo methods. The concept of FEP implies effective scanning of the conformational phase space in protein – ligand complexes in the surrounding of an explicitly definite solvent as well as calculations of the relative free energy of binding based on the averaged (by sampling the conformations) changes in the potential energy of a classical force field. Though the use of FEP requires a significantly larger number of calculations than the computation of the EF docking, in many cases it was shown that the obtained results are well compatible with the experimental data. The limits of the method are conditioned, first of all, by shortcomings of the approach of empirical force fields (molecular mechanics), used in such calculations. In particular, if the protein – ligand complex has a metal ion, the approximation of contact partial charges on atoms in the process of MD will not be correct and the reliability of the modeling results will be dubious.

Quantum-chemical hybrid approach (QM/MM) for studying enzyme catalysis. A reliable pathway of a chemical reaction (including an enzyme reaction) can be modeled only using quantum chemical methods. They make it possible to estimate the energy of molecules, optimize their geometry (nuclear coordinates) and model electron conformational effects. There is also a possibility to search for a transient state of a chemical reaction (the "saddle" point on the energy hypersurface, through which transition from the substrate to the reaction product can be performed) and to estimate the magnitude of the energy barrier, separating the substrate from the product.

The principal problem in modeling an enzyme reaction is a great number of atoms in the enzyme–substrate system that at present makes a direct use of quantum mechanics in studying enzyme catalysis practically impossible. However, the introduction of certain approximations allows us to use the advantages of quantum chemistry in solving complex problems. These assumptions can be formulated as follows.

1) Disregard of the main part of enzyme atoms and taking into account in calculations only the substrate and protein residues, directly involved in the catalytic reaction. A disadvantage of this approach is that in the process of calculations the real geometry of the complex can be distorted in such a "reduced" system.
2) Calculations of hybrid (QM/MM) systems, where the "quantum mechanical" subsystem includes the substrate and the most important residues of the enzyme and the surrounding solvent, while the remaining atoms are considered as a classical system (from the point of view of the method of empirical force fields). These calculations make it possible, at least, to define more exactly interactions in the enzyme–substrate complex and simultaneously to take into account correctly (in contrast to the classical force field) the contribution of metal ions in the complex.

The described techniques of molecular modeling as applied to studying the mechanisms of enzyme reactions, naturally, do not take the place of traditional biochemical experiments, but allow us to understand better the conformation rearrangements and explain the role of residues of the protein active center in these processes. In addition, molecular modeling simplifies essentially the search for more stable forms of enzymes (designing genetic engineering experiments) and identification of promising — high-affinity and selective — ligands (substrates, inhibitors and other modifiers of activity of enzymes). The maximal effectiveness in solving these tasks can be achieved only by combining experimental and theoretical approaches.

PART IV

BIOPHYSICS OF PHOTOBIOLOGICAL PROCESSES

IX
Primary Processes of Photosynthesis

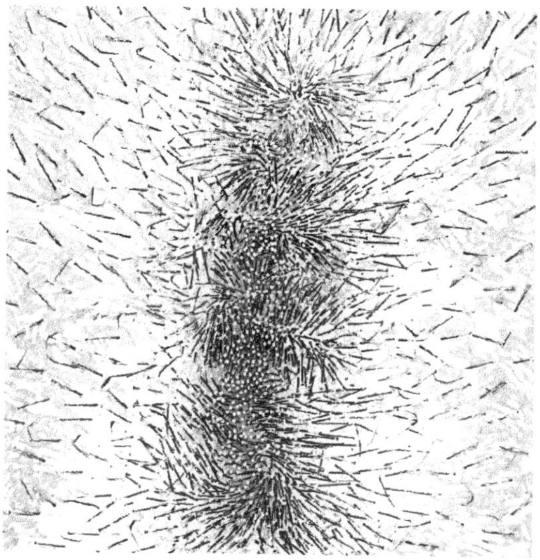

27
Energy Transformation in Primary Processes of Photosynthesis

28
Electron-Conformational Interactions in Primary Processes of Photosynthesis

Photobiology studies the action of light on different biological systems. At present, this field of biology is characterized by profound perception of molecular mechanisms of primary reactions, clarification of internal organization of photoenergetic and photoreceptor systems. Problems of photobiology have appeared to be in the focus of attention of serious biophysical studies, and their results have largely enriched the modern molecular biophysics and, in particular, contributed to discovering mechanisms of electron conformational interactions. The most important in the variety of photobiological processes are photosynthesis, vision, photoreception, the destructive action of ultraviolet light.

In all photobiological processes, light energy is required for overcoming activation barriers in chemical conversions. However, in photosynthesis this is accompanied by direct accumulation of light energy in the form of energy of chemical bonds of final products (glucose), because the latter have a large reserve of free energy as compared to initial substances (CO_2 and H_2O). In the other photobiological processes, light also induces photochemical reactions, but their products have no excess free energy as compared to the initial substances. Nonetheless, even in these cases, at stages following the photochemical stage of the dark reactions, complex physicochemical conversions can be initiated, mobilizing large amounts of free energy previously stored in biostructures. Final results of such conversions (for example, stimulating action of light on morphogenesis, pigment biosynthesis, respiration photostimulation) can be quite high in their energy effect, although in this case no direct storage of light takes place. The sequence of conversions in photobiological processes generally include the following stages: light absorption by the chromophore group and formation of electron excited states → migration of electron excitation energy → primary photophysical act → intermediate stages including charge transfer → generation of primary stable chemical products → physiologo-biochemical processes → final photobiological effect. □

One of the cardinal tasks of biophysics is elucidation of mechanisms and pathways of regulation of initial stages of transformation of electron excitation energy and molecular conversions accompanying it, including changes in chromophore groups, their protein carriers and surrounding membrane. This problem is connected with the role and mechanisms of electron excited state reactions in biological processes. A specific peculiarity of photobiological processes, distinguishing them from other dark biochemical reactions, is that here the source of electron excitation is quanta of light. But in spite of this difference, biophysical investigations in photobiology have led to establishing a number of regularities of a general biological importance.

27

Energy Transformation in Primary Processes of Photosynthesis

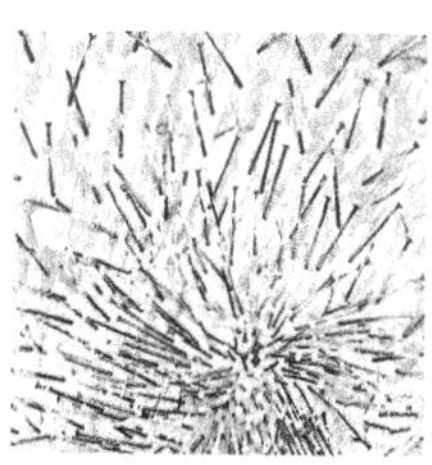

27.1 General Characteristic of Initial Stages of Photobiological Processes

In Chapters 10 and 24, it was demonstrated that following absorption of a light quantum one of the electrons at the lower energy level on the binding orbital moves to the upper energy level, filling the lower antibinding orbital and thereby transferring the molecule to an excited electronic state (singlet S_1 or triplet T, see Fig. 10.4). There are different ways of excitation energy transformation including intramolecular conversion, radiationless transfer of excitation energy to another molecule (energy migration) and use of excitation energy in photochemical reactions:

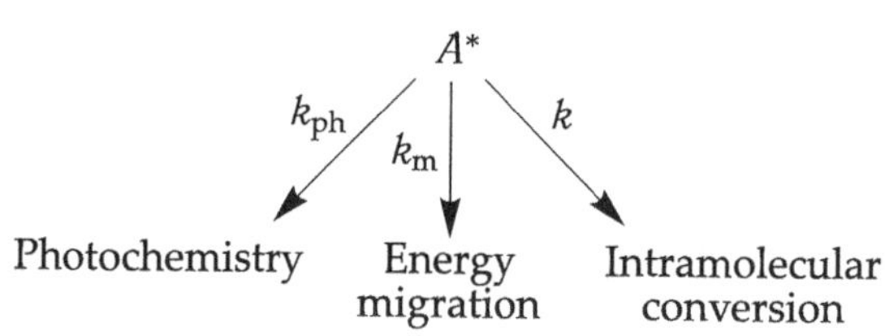

where k_{ph}, k_m and k are rate constants of corresponding processes, k is the total probability of intramolecular deactivation of the excited state (see Equation (10.1)):

$k = p + q + r$, where p, q and r are frequencies of transition per time unit from singlet excited state S_1 to the ground state with emission of fluorescent light (p), thermal deactivation of state $S_1(q)$ and conversion $S_1 \rightarrow T(r)$.

According to Equation (10.2), the fluorescence yield without taking into account energy migration k_m is

$$B = p/(p + q + r) = p/k, \tag{27.1}$$

and with account for the photochemical process

$$B_1 = p/(k + k_{ph}). \tag{27.2}$$

As according to Equation (10.1),

$$\tau = 1/(p + q + r) = 1/k, \tag{27.3}$$

then

$$B = p\tau. \tag{27.4}$$

Upon photochemical deactivation of the excited state

$$\tau_1 = 1/(k + k_{ph}). \tag{27.5}$$

By analogy we get that

$$B_1 = p\tau_1 \tag{27.6}$$

or

$$\tau_1 = (1/p)B_1 = \tau_0 B_1, \tag{27.7}$$

where $p = 1/\tau_0$ is the probability of fluorescence.

Thus, lifetime τ_1 is proportional to the magnitude of fluorescence yield B. Quantum efficiency Φ of a photochemical process is connected with fluorescence yield B. Obviously,

$$\Phi = k_{ph}/(k + k_{ph}). \tag{27.8}$$

Combining Equations (27.1)–(27.7), we get

$$\Phi = 1 - B_1/B \quad \text{or} \quad \Phi = 1 - \tau_1/\tau. \tag{27.9}$$

The reactivity of a light-excited molecule is determined by a number of factors. It is dependent on the position of the excited energy level of electrons that provides overcoming the energy barrier (Fig. 27.1). Of great significance is lifetime τ of the excited state, over which the excess energy is preserved in the molecule. In many chemical processes the high reactivity of the triplet state is realized. It is caused by relatively large lifetimes ($\tau_T \gg \tau_{S1}$) as well as by biradical properties. When in a triplet state, the molecule has two unpaired electrons and thereby it acquires properties of an anion radical for the time of excitation. As known, in reactive anion radical obtained as a result of reduction of neutral molecules, the attached electron occupies the lower vacant antibinding orbital, but cannot decrease its own energy level and move to the system of binding orbitals, because the latter are completely occupied in the neutral molecule.

As a rule, photochemical conversions occur as single-quantum photochemical processes. An exclusion is conditions for powerful laser excitation beams, when still

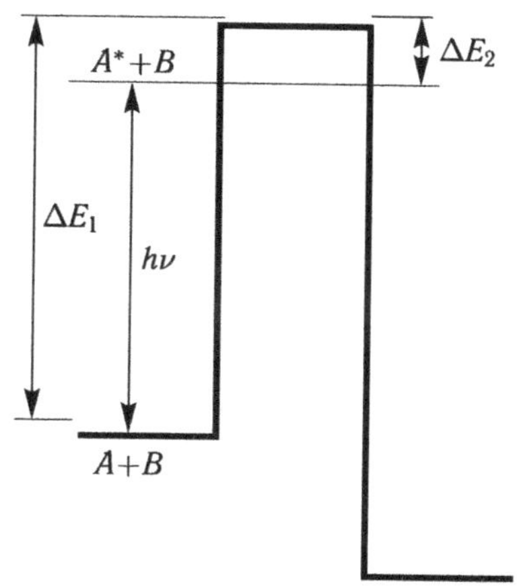

Figure 27.1. Energy diagram of photochemical process $A^* + B \rightarrow P$.
ΔE_1 is the activation energy of dark reaction $A + B \rightarrow P$; ΔE_2 is the activation energy of photochemical reaction $A^* + B \rightarrow P$ ($\Delta E_1 \gg \Delta E_2$).

being in the excited state the molecule has time to absorb the second quantum, so that two-photon excitation of one molecule becomes possible and the molecule moves to the upper excited levels:

$$S_0 + (h\nu_1) \nearrow S_1 + (h\nu_2) \longrightarrow S_2,$$
$$S_0 + (h\nu_1) \searrow T_1 + (h\nu_2) \longrightarrow T_2.$$

Mechanisms of single-quantum photochemical processes are quite different. Apparently, the biomolecular photoreaction will take place, when the second reagent has time to reach the excited molecule prior to its deactivation.

Therewith the relation $1/k_{ph} \lesssim \tau$, where τ is the lifetime of the corresponding state S_1 or T, should be satisfied.

In general, diffusion processes in condensed media such as biological membranes, have no time to provide a high frequency of collisions of reagents during the lifetime of state S_1. However, this difficulty is inessential for the triplet state. It is also facilitated if the molecule forms a complex with the second reagent in the ground or excited state — an exciplex. In this case, an addition reaction can take place, for example, formation of dimers: $A + h\nu \rightarrow A^* + A \rightarrow (A^* \ldots A) \rightarrow A\text{–}A + h\nu_{ph}$.

Excited molecules react usually in a monomolecular way, i.e. dissociate with the formation of ions, radicals:

LHPPC ↗ PPC II ⟶ RC of PS II

LHPPC ↘ PPC I ⟶ RC of PS I,

or isomerize (photo-rearrangement):

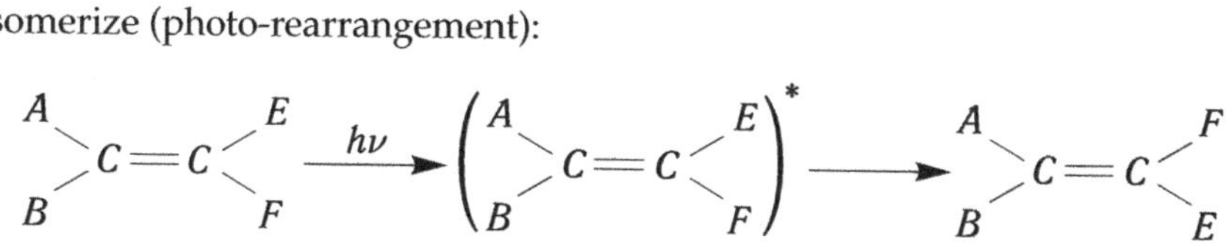

The proton can be transferred in a photochemical way: its attachment, $A^* + BH \rightarrow AH^+ + B^-$, or recoil, $(AH)^* + B \rightarrow A^- + H^+B$, as well as electron phototransfer similar to redox reactions: $A^* + B \rightarrow A^+ + B^-$; $A^* + C \rightarrow A^- + C^+$.

In biological systems, diffusion processes are, as a rule, limited as compared to solutions; rate constants of standard enzymatic processes do not exceed 10^2–10^3 s^{-1}. This makes it possible for initial stages of photobiologicl processes to proceed analogous to intracomplex conversions. In other words, the reactive complex, including the light-absorbing sensitizer molecule and molecules of other reagents, has been formed initially. This major peculiarity of biological systems has an impact on the mechanisms of transformation and use of electron excitation energy in them. □

27.2 General Scheme of Primary Processes of Photosynthesis

The principle of photosynthesis is conversion of light energy into a biochemically accessible form of energy in cells of phototropic organisms; as a result of this organic compounds are finally synthesized in cells.

The unique feature of photosynthesis as a physicochemical process is connected with the light-induced electron transfer against the thermodynamic potential gradient from compounds with a high oxidation potential to compounds with a high reducing ability. Results of these primary processes of photosynthesis, determining mainly its specificity, are photosynthetic phosphorylation and photosynthetic reduction of pyridine nucleotides.

The general photosynthetic equation may be as follows:

$$CO_2 + 2H_2A \rightarrow \langle CH_2O \rangle + H_2O + 2A,$$

where H_2A is the electron donor, and $\langle CH_2O \rangle$ is the elementary unit of carbohydrate.

Different organisms have different nature of the electron donor required for the functioning of the photosynthetic electron transport chain.

A complex combination of all redox reactions of electron transfer between components of the electron transport chain (ETC) underlies the primary processes of photosynthesis (PPPh). Of the utmost interest are mechanisms of three main stages of energy transformation in PPPh: absorption of light by photosynthetic pigments and migration of electronic excitation energy to the reaction center (RC) of the photosystem; primary charge separation and energy transformation in the RC; electron transfer over the ETC and coupled processes leading to the formation of primary stable products (NADPH and ATP), later used in the dark reactions of CO_2 fixation and generation of final products of photosynthesis.

Let us consider general peculiarities of PPPh functioning and organization. In photosynthesis of plants and cyanobacteria, electron transfer from water (the redox potential of half-reduction, $E_{1/2}$, at pH 7 is +0.82 V) to NADP ($E_{1/2} \approx -0.32$ V) takes place at consumption of energy of two light quanta with involvement of correspondingly two photosystems (PS I and PS II), functionally coupled and sequentially operating. The "Z-scheme" of interaction of photosystems and intermediate carries in the ETC of photosynthesis of higher plants is shown in Fig. 27.2. The figure also represents corresponding $E_{1/2}$ values of the ETC components.

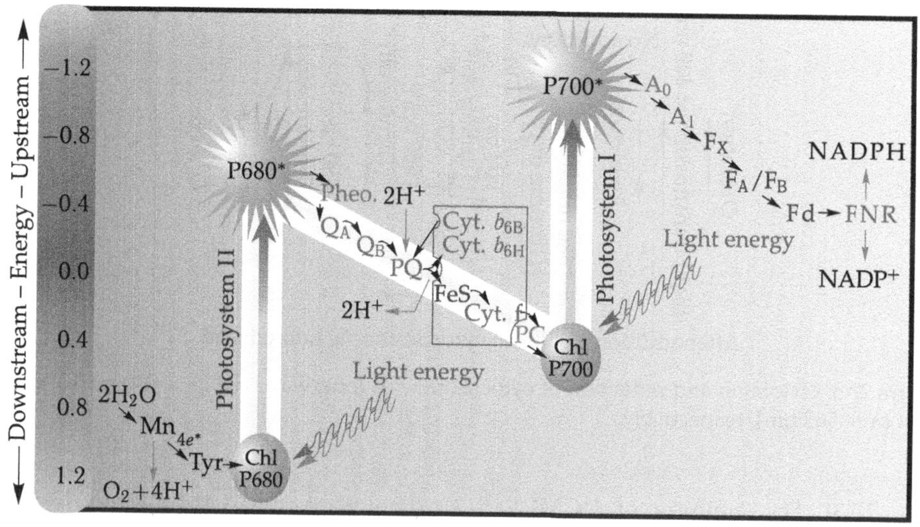

Figure 27.2. Z-scheme of photosynthesis with involvement of the cytochrome b_{6f} complex.

Some general features of the functioning of this system were established prior to deciphering details of its organization.

A remarkable progress in studying PS I and PS II complexes was connected with the X-ray structural analysis (XSA) of crystals of these enzymes from thermophilic cyanobacteria correspondingly at 2.5 and 3.0 E resolution (see below).

In the 1930s, in experiments with intermittent light R. Emerson and V. Arnold measured the amount of oxygen, released by chlorella cells under the action of light flashes of 10^{-5} s. The magnitude of the dark interval (~ 0.02 s) between the flashes, required for the consumption of the absorbed light energy and oxygen release, clearly indicated to the involvement of enzymes, limiting the general rate of PPPh. With a relatively large number of intermediate transfer stages (Fig. 27.2), this evidently shows high electron transport rates at separate stages. It was also demonstrated that the enzymatic system responsible for release of one O_2 molecule, is served concurrently by several hundred chlorophyll molecules. The complex of light-harvesting pigment molecules transducing the energy to one photoactive center (PS I and PS II) was called a photosynthetic unit (PSU).

However, it remained unclear whether the concept of PSU reflects the real molecular organization of photosynthetic membranes, when each light-harvesting pigment complex serves "its own" RC (the unicentral model). Another possibility is that the RCs are "impregnated" in the light-harvesting antenna, and the excitation energy transfer has a statistical character without any topographical restrictions (the multicentral model).

Pigment systems, included in PS I and PS II, differ in their spectral features, and chlorophyll in PS II absorbs light in a more long-range region than that in PS II. This allows us to selectively excite photosystems: separately PS I ($\lambda > 680$ nm) and concurrently PS I + PS II ($\lambda < 680$ nm).

In 1961, L. N. M. Deusense, using methods of absorption differential spectrometry, demonstrated in experiments on algae that long-wave light ($\lambda > 680$ nm) induces oxidation of cytochrome *f* (recorded by the decrease in absorption at 420 nm). The use of short-wave illumination ($\lambda \sim 562$ nm) resulted in partial reduction of cytochrome *f*

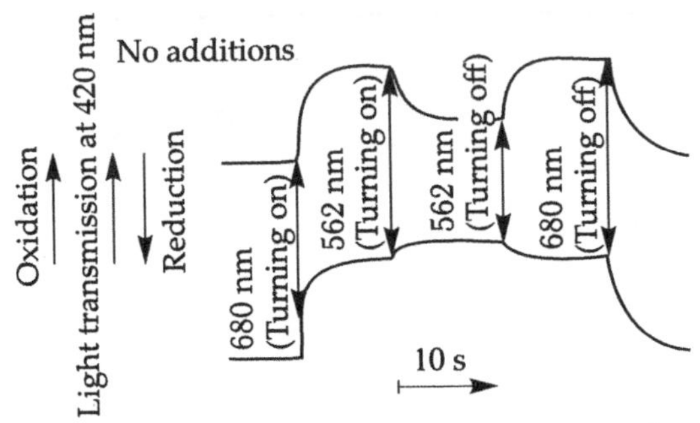

After addition of diuron cytochrome *f* is not reduced.

Figure 27.3. Oxidation and reduction of cytochrome *f* at long-wave red ($\lambda > 680$ nm) and red light ($\lambda \sim 562$ nm), respectively.

(Fig. 27.3). The inhibitor of DCMU (3-(3,4-dichlorophenyl)-1,1-dimethylurea), or diuron, stopped the reducing action of short-wave light. According to the Z-scheme (see Fig. 27.2), long-wave red light oxidizes cytochrome, depriving it of electrons to reduce the primary oxidizer formed in PS I. Short-wave light makes up for the decrease in electrons on the cytochrome due to its reduction by PS II. Diuron inhibits this process leading to inhibition of the oxygen release. The object can be illuminated by two consecutive light flashes: first of red light ($\lambda < 680$ nm) and then of long-wave red light ($\lambda > 680$ nm). In this case, the electrons released in PS II in response to the first flash, will reach over some time cytochrome f which will be oxidized by photosystem I under the action of the second flash. Variations of the time interval between the flashes makes it possible to determine the time necessary for electron transfer from PS II to the cytochrome; it is $\sim 5 \cdot 10^{-3}$ s.

An electron donor for PS I in higher plants is plastocyanine — a copper-containing protein ($E_{1/2} \approx +0.37$ V), while for some algae it is either plastocyanine or cytochrome c_6 (+300 mV) depending on whether copper is present in the surrounding. Many cyanobacteria use only cytochrome c_6 as an electron donor for photooxidized P_{700}, though some of them can also use plastocyanine. Plastoquinones ($E_{1/2} \sim +0.06$ V) take electrons from primary acceptors of PS II; plastoquinones are the only two-electron carriers in the ETC that upon reduction simultaneously attach protons. In addition to cytochrome f, there are also cytochromes b (cytochrome b_6, $E_{1/2} \sim -0.03$ V), having contacts with PS I and involved in the cyclic electron transport. Cytochrome b_{559} is linked to PS II, so that each RC has 1 or 2 molecules of this cytochrome. This is the sequence of donor – acceptor acts accompanying the electron transfer over the ETC in chloroplasts (see Fig. 27.2).

The cardinal importance of the topographical organization of PPPh is that individual electron carriers in the ETC of chloroplasts are united in single macromolecular complexes localized in photosynthetic membranes. □

A scheme in Figure 27.4 shows how these complexes are arranged in the photosynthetic thylakoid membrane. As seen, on its pathway the electron crosses the membrane passing through the b_6–f complex from its inner side, where the donor moiety of PS II is located, to the outer side to the acceptor region of PS I. The interaction between macromolecular complexes of PS I, b_6–f and PS II proceeds by means of mobile carriers — plastocyanine Pc and plastoquinone Pq, therewith the molecules

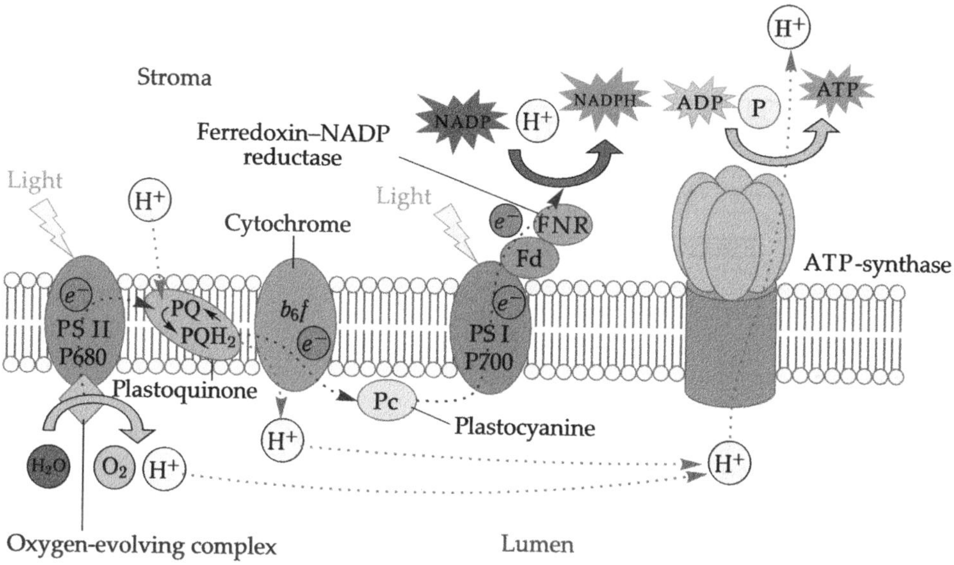

Figure 27.4. The electron transport chain of photosynthesis and the coupling complex in the thylakoid membrane.

Fd, Ferredoxin; electron transport pathways are shown by arrows (see other explanations in the text).

of the latter form a pool, containing no less than 10 molecules of Pq per one Pc. Another mobile carrier — ferredoxin — accepts electron from PS I, associating PPPh with the dark processes. A peculiar combination of usual diffusion and intracomplex reactions in the ETC has a great impact on the kinetics and physical mechanisms of electron transfer.

The scheme given in Fig. 27.4 shows the asymmetric character of transverse position of donor and acceptor components of PS I and PS II in the thylakoid membrane. It has become clear (J. Anderson and J. Barber, 1996) that the lateral position of complexes in the membrane plane also has an explicit heterogeneous character (Fig. 27.5). As seen from the figure, pigment systems PS II and PS I are isolated spatially that impedes migration of electron excitation energy between them.

In bacterial photosynthesis the source of electrons is outer donors of hydrogen (simple organic molecules, hydrogen sulfide).

Key components of the bacterial photosynthetic chain are cytochromes *b* and *c*. In cytochromes *c* the heme is covalently bound to the protein, while in cytochromes *b* it is bound noncovalently. Figure 27.6 shows a simplified scheme of electron transport in bacterial photosynthesis, where components of different types of bacteria are united formally.

In the course of cyclic transport, soluble cytochrome *c* transfers the electron from ubiquinone: cytochrome *c* of oxidoreductase into the photooxidized photosynthetic reaction center. Three stages can be distinguished in this process: (1) diffusion along the intraplasmic membrane or in the periplasmic space (shown by a dashed arrow in 27.6); (2) electrostatic binding to the reaction center; (3) the actual electron transfer. At present, it is believed that this transfer proceeds by the tunneling mechanism with involvement of some aromatic residues in the polypeptide chain of the reaction center protein (for details see Chapter 28).

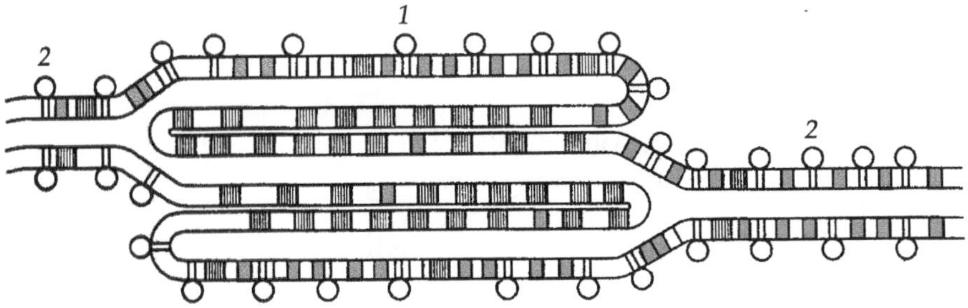

Figure 27.5. Heterogeneous distribution of complexes of photosystems and the coupling factor in the lateral membrane plane: *1*, granal membranes; *2*, stromal membranes.

The granal part including membranes of contacting thylakoids contains complexes of PSP and PS II and a light-harvesting complex (LHC). At the same time, the stromal part of thylakoid (lamellas connecting thylakoid grana) contains mostly PS I complexes and only 10–20% of the total amount of complexes, containing the RC of PSP, and light-harvesting complex LHPPC. Ferredoxin-NADP-reductase and coupling factor CF_0 are also present here.

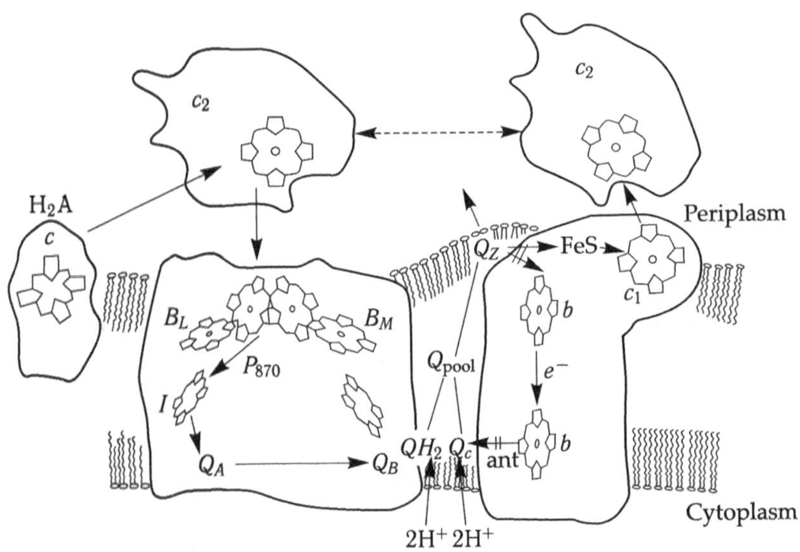

Figure 27.6. Cyclic electron transport and its components in purple bacteria.

The upper part of the figure shows the periplasmic space (cytochrome c_2 able to diffuse into the periplasmic space), and the lower part shows the cytoplasmic space with the photosynthetic membrane between them. In the membrane on the left is shown the reaction center with the primary donor of electrons P_{870} and auxiliary molecules of bacteriochlorophyll B_L, B_M, the primary electron acceptor in bacteriopheophytin I, and also primary and secondary quinones Q_A and Q_B. In the membrane on the right are shown ubiquinone cytochrome c oxireductase with two cytochromes c_1 and b. The exact localization of cytochrome c and Fe–S Rieske protein is not known yet. Cytochrome c is low potential; it provides noncyclic electron transport from outer donor H_2A through c_2. Four ubiquinones are identified: Q_A and Q_B in the reaction center, Q_C and Q_Z on the oxidoreductase. Sites of action of inhibitors of myxothiazol *tuch* and antimycin *ant* are shown by double arrows.

Electron transfer involving the multiheme membrane-bound cytochrome includes only the third stage because the cytochrome subunit exists in a stable fixed state in the membrane.

The photosynthetic chain of electron transfer includes also cytochromes *c* of the third type (c_1) which are the chromophore group of ubiquinone cytochrome *c* oxireductase.

27.3 Structural Organization of Pigment–Protein Antenna Complexes

The structural organization of light-harvesting pigment–protein complexes (LHPPC) determines their functioning in the photosynthetic membrane. These complexes are the main components providing energy transfer to the RC of photosynthesis and its redistribution between photosystems.

Pathways of excitation energy migration. Electronic excitation energy is delivered to the RC of PS I and PS II of higher plants and the RC of bacterial photosynthesis due to energy migration in the light-harvesting antenna. Energy migration in photosynthesis is the most studied type of radiationless transfer of electronic excitation energy in biological systems (see Chapter 25). Its functional biological meaning is to increase the efficiency of consumption of absorbed light quanta. Indeed, the mean time necessary for utilization of the light quantum energy (release of an O_2 molecule) is only 0.01–0.02 s. The time of regeneration of the photoactive pigment of the RC is much lower than this value ($\leqslant 10^{-6}$ s). The intensity of daytime sunlight reaches 10^{15}–10^{16} quantum $\cdot$ cm^2 $\cdot$ s^{-1}. The absorption cross-section of a pigment molecule is 10^{-15}–10^{-16} cm^2. Consequently, each separate chlorophyll molecule will absorb no more than 1–10 quantum $\cdot$ s^{-1} and at standard light intensities even less. In this case, the RC without its own light-harvesting antenna and containing only a photoactive pigment will for the most part "stand idle". Association of many tens of pigment molecules in the so-called photosynthetic unit (PSU) harvest light quanta and maintain the photoactive pigment in "ineffective rest".

Direct redistribution of the absorbed energy occurs between pigment systems PS II and PS I depending on the intensity and spectral composition of illumination. It was shown that illumination by far-red light of PS I provokes such a change in the photosynthetic apparatus when the most quanta are transferred into photosystem II. This induces an increase in fluorescence; its yield in PS II is higher than in PS I. Light of a shorter wavelength causes an inverse effect. The state with a higher yield of fluorescence, induced by illumination with far-red light, is called state 1, and the state with a lower yield of fluorescence is called state 2. Redistribution of quanta between the two photosystems can be caused by both an increase in migration from LHPPC to PPC I and also by redistribution of excitation that is already in PS II by its enhanced migration to PS I. The latter phenomenon was called a "spillover".

Complexes of PS I and PS II are divided laterally in the thylakoid membrane due to different density of their surface charges. The high density of the surface charge in the PS I complex causes their expulsion into the stromal part from the contact region of thylakoids with a low density of surface charges. The stacking of neighboring membranes in grana due to van-der-Waals attractive forces of hydrophobic regions occurs in the presence of less charged complexes of PH II and LHPPC. A change in the charged state of complexes of PS I, PS II, and LHPPC affects their mutual location due to electrostatic interactions and thereby can have a regulatory effect on the energy distribution between them. This is the basis of the "spillover" mechanism.

It has a complex character and is connected mainly with the conformational state of photosynthetic membranes that influences both mutual distances and orientation of pigment complexes of PS I and PS II and their connection with LHPPC.

The main reason for conformational changes in thylakoid membranes and redistribution of excitation energy between PS II and PS I is a change in the charge of LHPPC. Even in the absence of phosphorylation of the light-harvesting complex, but at a low concentration of positive ions, the negative charges on the surface of membrane proteins create a high surface potential on thylakoid membranes and prevent adhesion of thylakoids. Under such conditions, PS I and LHPPC are equally distributed over the membrane surface. Phosphorylation of LHPPC leads to generation of an additional charge on it that causes LHPPC transfer into the stromal region of the membrane and an increase in the energy transfer from it to PS I. Phosphorylation of LHPPC is realized with the help of protein kinase located in granal areas of the membrane.

The reverse reaction of dephosphorylation of LHPPC that turned out to be in the stromal part near PS I occurs due to another enzyme — protein phosphatase in the stromal region of the membrane. By dephosphorylating LHPPC, protein phosphatase causes its return to the granal membrane. Protein kinase is activated upon reduction of plastoquinone, shuttling electrons between PS I and PS II, and passes to an inactive state upon its oxidation. The activity of protein phosphatase remains unchanged. Thus, plastoquinone is a "sensor" for maintaining the excitation energy balance between the photosystems. The excess excitation of PS II induces over-reduction of plastoquinone and activation of protein kinase. Then upon phosphorylation the transfer of part of LHPPC from grana to stromal membranes decreases the excitation flux in PS II and increases it in PS I. As a result, the rates of excitation of PS I and PS II become equal, and the degree of plastoquinone reduction decreases again. On the contrary, excess oxidation of plastoquinone upon predominant excitation of PS I causes inactivation of protein kinase and return of LHPPC to the grana due to its dephosphorylation by protein phosphatase. □

The generation of a negative charge upon phosphorylation of polypeptides in PS II can have also other consequences. Thus, it can weaken the Q_A bond with the protein, leading to the loss of Q_A, and thereby interrupt the linear electron transport.

Redistribution of excitation energy between the photosystems reflects the great role of the relation between noncyclic and cyclic electron flows, predominantly accelerating the latter upon phosphorylation of LHPPC. It was also found that chloroplasts have two types of PS II differing in localization sites and dimensions of their antenna complexes: PS II_α and PS II_β. PS II_α has a larger antenna and therefore is saturated at higher intensity of light, displaying also higher photosynthetic activity. Correspondingly, PS II_β has a smaller antenna and is saturated earlier with an increase in the intensity of light.

Localization of the major part of PS II (PS II_α) in the granal area provides the main noncyclic flow of electrons that reduces plastoqinone molecules, which then diffuse into the stromal area donating electron to PS I. The smaller part of PS II (PS II_β), having also a smaller portion of LHPPC II in the stromal area, induces a weaker flow, interacting with intensive cyclic flow of PS I that is the source of energy for ATP synthesis.

27.4 Mechanisms of Transformation of Excitation Energy in Photosynthetic Membrane

Excitation Energy Trapping in RC. The primary transformation of energy in photosynthesis takes place when the energy of electronic excitation, migrating over the light-harvesting matrix, is trapped by the photoactive pigment of the reaction center. The first measurements of duration and yield of fluorescence of pigments (chlorophyll, bacteriochlorophyll) in solution (τ and B) and in cells of photosynthesizing organisms (τ_l and B_l) demonstrated (S. Brody and A. N. Terenin, 1957) that these values are: $\tau \simeq 5$ ns, $B \simeq 0.3$, $\tau_l \simeq 0.3$ ns, and $B_l \simeq 0.02$.

The quantum yield of primary processes of photosynthesis is relatively high. Therefore the indicated decrease in pigment fluorescence lifetime and yield in living systems should be caused mainly not by thermal losses, but by the process of "photochemical" deactivation of singlet excitation of state S_1 in reaction centers of photosynthesis. Independently of the mechanism of this process, RCs should be considered as natural photochemical quenchers of fluorescence of pigment molecules of the light-harvesting matrix. It is possible to estimate the efficiency of such quenching supposing that in a photosynthetic membrane, constant values p, q and r (see Section 1) remain unchanged, and the process of fluorescence quenching upon energy trapping by reaction centers is equivalent to photochemical deactivation k_{ph} of state S_1 of antenna molecules. Then, substituting values τ, B, τ_l, B_l, in Equation (27.9), we find that the efficiency of using excitation by reaction centers $\Phi \simeq 0.93 \div 0.95$ (A. Yu. Borisov, 1973).

Reaction centers themselves have high photochemical activity and the magnitude of the quantum efficiency of the primary photochemical process in them is close to unity. From this and from Equation (27.9), it follows that the lifetime of the excited state of a photoactive pigment molecule, when the primary act of photosynthesis should occur, does not exceed $\tau_l \simeq 10^{-11}$ s.

Thus, after light absorption in the photosynthetic membrane, at the first stage the excitation energy migrates over the light-harvesting matrix and is trapped by reaction centers for about 0.2–0.3 ns. At the second stage, already directly in reaction centers, the electronic excitation energy is consumed in the primary photochemical act for $\sim 10^{-11}$ s. □

The duration of fluorescence of groups of PPC, directly maintaining reaction centers, should also be determined by the efficiency of excitation energy trapping in the RC, i.e. by the process of "photosynthetic" quenching of fluorescence. The latter circumstance depends not only on the interaction between the RC and the PPC belonging to it, but also on the state of the RC itself. Apparently, active reaction centers efficiently convert electronic excitation energy, giving rise to the electron flow in the ETC of photosynthesis and at the same time are active quenchers of PPC fluorescence. On the contrary, RCs that by some reason cannot transfer electrons into the chain of photosynthesis, cannot utilize the electronic excitation energy, which consequently can be emitted as fluorescence with a high lifetime. □

Photosynthetic reaction centers. The main principle of RC operation is that excitation of a photoactive pigment leads to the release of the electron and reduction of the primary acceptor which in its turn transfers the electron to the secondary acceptor and then farther into the ETC. The general scheme of the RC functioning is as follows:

$$PIQ + h\nu \rightarrow P^*IQ \rightarrow P^+I^-Q \rightarrow P^+IQ^-. \quad \square \qquad (27.10)$$

The subsequent reduction of P^+ occurs from the electron donor in the ETC as shown in Figs. 27.2 and 27.6. In Scheme (27.10), symbol PIQ designates components of the RC: P is the photoactive pigment, I is the primary acceptor, and Q is the acceptor located after the primary acceptor. Porphyrinic molecules of pigments (chlorophyll in the RC of PS I, pheophytin in the RC of PS II, bateriopheophytin in the RC of purple bacteria) play the role of I in the RC. The secondary acceptor is represented by molecules of quinone (the RC of purple bacteria), plastoquinone (the RC of PS II), or iron-sulfur proteins (the RC of PS I). The efficiency of electron release from P^* and its transfer to I and then to Q depend on the state of RC components. Thus, if I is initially reduced in the dark, for example, under rather low redox potential E_h of the surrounding, then under light the electron cannot be released from P. Just in the same way chemical oxidation of a photoactive pigment takes off the electron from P and thereby impedes its ability to give up the electron to primary acceptor I under illumination. Reaction centers in active state PIQ, where electron transfer from P^* occurs in a normal way, are called open centers. The RC cannot give up electron to the photosynthetic ETC, if it is in one of closed centers (P^+I^-Q, P^+IQ^-, P^+IQ, PIQ^- or PI^-Q). In state PIQ^- the generation of a negative charge on Q^- impedes primary separation of charges and appearance of an electron on I ($PIQ^- \rightarrow P^+I^-Q^-$) due to electrostatic repulsion. Therefore, the dark reduction of Q decreases quenching of the electronic excitation by reaction centers and increases the fluorescence yield and lifetime.

In the closed states, where P is initially oxidized in the dark (P^+), the ETC has no primary donor of electrons. When in the dark, the carriers in the acceptor part are initially reduced, vacant sites for electron transfer are occupied and P^* cannot give rise to an electron flow either.

Apparently, when the RC is in a closed state, this increases the yield B and lifetime τ of chlorophyll fluorescence, since the electronic excitation energy cannot be consumed in primary processes of photosynthesis.

However, at closed RCs, the magnitudes of τ and B do not reach the values characteristic of diluted solutions of chlorophyll where no photosynthesis takes place. The matter is that ion radical P^+, formed in oxidized RCs, has its own absorption spectrum overlapping partially with the fluorescence spectrum of PPC pigments. As a result, excitation energy migrates from the PPC molecule to P^+ where it is dissipated into heat. Besides, the electrons that under these conditions have appeared on I cannot move further to Q^- and return back to P^+ giving rise to delayed recombination luminescence of the RC with an additional conversion of excitation energy into heat:

$$PIQ^- \rightarrow P^+I^-Q^- \rightarrow P^*IQ^- \rightarrow PIQ + h\nu.$$

Duration of pigment fluorescence. For the first time, lifetime values of pigment fluorescence in cells were obtained by the method of phase fluorimetry (A. N. Terenin, 1957; L. A. Tumerman and A. Yu. Borisov, 1959). The method is based on measurements of the shift of the phase between a synusoidally modulated exciting light and the fluorescence generated by it in the sample. The phase is shifted as a consequence of the delay of absorbed quanta prior to their emission as luminescence from their fluorescent molecules. The method yield reliable results when the system has only one type of molecules with one overall mean time of exponential fluorescence decay. In the case of several fluorescent components, whose number, times and laws of decay are not known a priori, the possibilities of the method of phase fluorimetry become limited.

The development of laser techniques, in particular generation of picosecond (10^{-12} s) and subpicosecond ($\sim 10^{-13}$ s) light pulses, has led to elaboration of pulse picosecond spectrofluorimetry. In this method, fluorescence of the sample is excited by a light pulse with duration much less than the lifetime of the state of the studied molecules.

As an example, Figure 27.7 shows the recording of the kinetics of the decay of chloroplast fluorescence containing three components with times $\tau_1 \simeq 40$ ps, $\tau_2 \simeq 300$ ps, $\tau_3 \simeq 4{,}500$ ps. In other experiments, τ values of the decay components were 100, 400 and 1100 or 180, 500 and 1400 ps. At present, it is generally accepted that the kinetics of fluorescence decay in intact systems (higher plant chloroplasts, intact cells of green algae) has at least three components: $\tau_1 = 80 \div 300$ ps, $\tau_2 = 400 \div 1{,}500$ ps, and $\tau_3 = 1{,}500 \div 4{,}000$ ps.

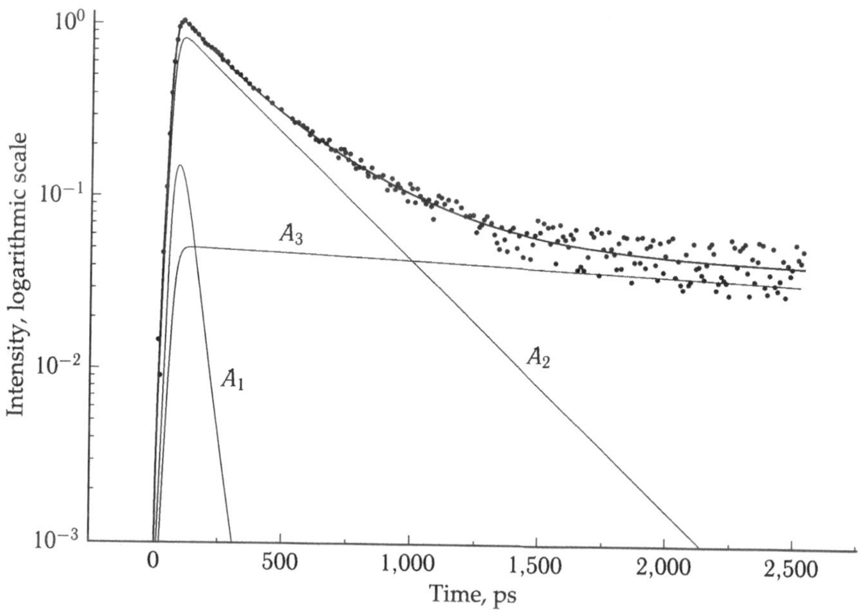

Figure 27.7. Kinetics of fluorescence decay for pea chloroplasts.

•, Experimental points; –, model curve fitting the fluorescence kinetics by the Marquardt method for least nonlinear squares according to expression $\Phi(t) = \int_{-\infty}^{\infty} K(t - t') \times (A_1 \exp(-t'/\tau_1) + A_2 \exp(-t'/\tau_2) + A_3 \exp(-t'/\tau_3))\, dt'$, where $K(t)$ is the instrument function of the method of recording. For parameters $A_1, \tau_1, A_2, \tau_2, A_3, \tau_3$, the following values were obtained: $A_1 = 0.15$, $\tau_1 = 41$ ps; $A_2 = 0.8$, $\tau_2 = 310$ ps; $A_3 = 0.05$, $\tau_3 = 4.6$ ns. Pre-exponential coefficients A_1, A_2, A_3 are proportional to the contribution of the corresponding component to the fluorescence decay kinetics (the approximating curve and components are shown by solid lines).

The nature of long living fluorescence components is rather complex. First, chlorophyll molecules, not linked by some reason to the PSU, possess fluorescence with lifetime τ_3 close to τ of the fluorescence of pigment solutions (5 ns). Second, when the excitation energy hits the closed reaction center PIQ^-, it results in the reduction of the primary acceptor and the RC transfer to state $P^+I^-Q^-$. As further electron transfer from I^- to Q^- is impossible, recombination of charges between P^+ and I^- will take place transmitting P to excited state P^*. The generated delayed fluorescence from the S_1-level is characterized by an increased lifetime corresponding to the time of recombination of separated charges in the RC in state $P^+I^-Q^-$ (V. V. Klimov et al., 1978).

A reverse energy transfer from P^* to the antenna with fluorescence emission can also take place. However, it is still unclear how large the contribution of this recombination luminescence to the total fluorescence yield is when the centers are closed.

The above data allow us to draw a general scheme of energy migration not taking into account all types of LHC of the external antenna, showing times of separate stages of transfer of excitation energy over the pigment omplexes in the photosynthetic membrane of higher plants (Fig. 27.8). Upon light absorption, the excitation energy, hitting the LHPPC, is directed in two fluxes to PPC I and PPC II.

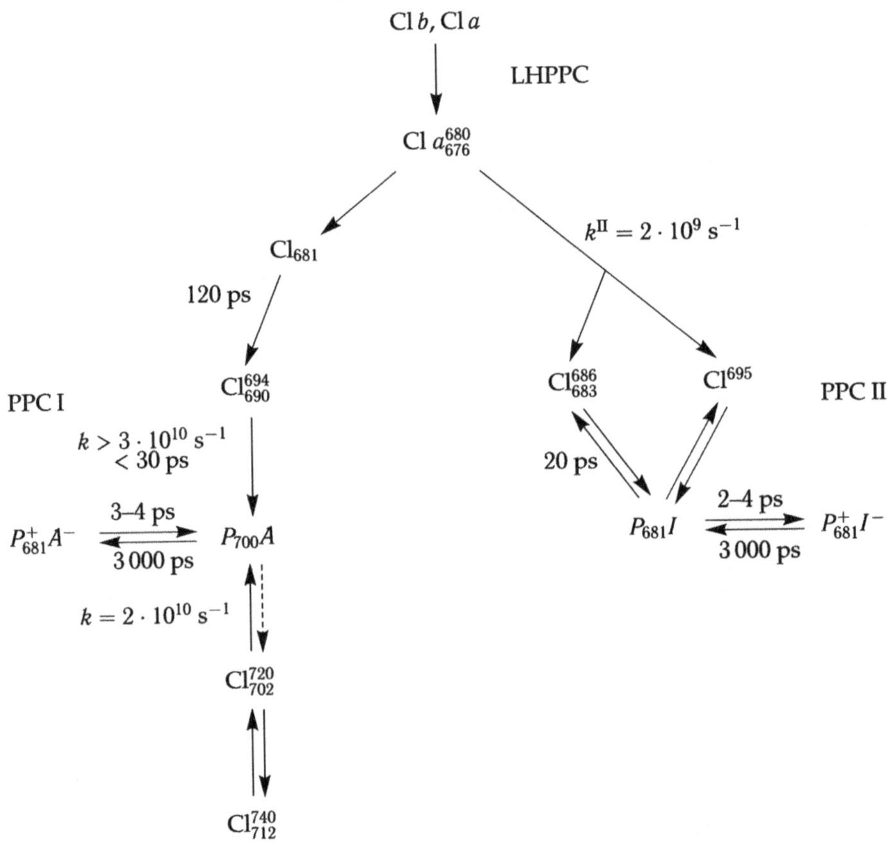

Figure 27.8. Migration of excitation energy in a photosynthetic membrane (times (τ, ps) and constants (k, s^{-1}) of transfer are shown at separate stages); $P_{700}A$ and $P_{680}I$ are components of reactions centers in PS I and PS II.

As seen from the transfer constant values, energy migration between the LHPPC and the core complexes of the two photosystems proceeds by the inductive resonance mechanism (see Chapter 25). Inside the individual clusters of PPCs, energy migration occurs by the exciton type.

Pigment Fluorescence and Trapping of Excitation in RCs. The decrease in the duration and yield of pigment fluorescence in a photosynthetic membrane takes place as a result of trapping of the energy of electronic excitation of the antenna in reaction centers. Measurements of the time of fluorescence decay of separated light-harvesting complexes or in mutants devoid of RCs, gave the value of τ, generally close to the values characteristic of diluted pigment solutions. This is valid both for higher plants ($\tau \sim 2.5$–4 ns) and for photosynthetic bacteria ($\tau \sim 1$ ns). Hence it is clear that a normal photosynthetic membrane has no efficient types of excitation quenching into heat useless for organisms.

The initial models of excitation trapping in RCs implied that the process of excitation trapping in the RC trap is much more efficient than the reverse reaction of excitation return to the antenna (W. L. Butler, 1977). Reaction centers in these models played the role of a "funnel", where excitation energy drains irreversibly. However, lately data have been accumulated that the "bottleneck" in unitization of excitation is the process of its trapping and reversible charge separation.

The analysis of random motions of excitation shows that about 120 excitation hops are required for the exciton to be trapped during its random walk in the RC, surrounded by $N = 80$ antenna chlorophyll molecules. Assuming that interaction energy between chlorophyll molecules in the antenna is $E_{\text{exc}} \sim 100\ \text{cm}^{-1}$ (~ 0.01 eV $\sim 4.6 \cdot 10^{-21}$ J), it is possible to estimate time τ_M of one hop (25.75) as $\tau_M \sim \hbar/E_{\text{exc}} \sim 10^{-13}$–$10^{-14} \sim 50$ fs (~ 0.05 ps). Consequently, the time required for a single hit of the trap is $120 \cdot 50$ fs $= 6$ ps. The total number of hits that the exciton makes during the fluorescence decay of 80 ps prior to the efficient entrapping is 80 ps/6 ps ~ 13. Thus, the "bottleneck" is not excitation migration in the antenna, but the efficiency of its entrapping in the RC trap.

In other words, a reaction center is a shallow trap for excitation, from where it can "get out" back to the antenna, so that excitation energy can return back into the antenna molecules. Namely, due to this for an efficient trapping, several hits (~ 13) to the RC are required during the excitation lifetime (~ 80 ps) in the antenna.

Propagation of excitation over the antenna occurs much faster (by an order of magnitude) than the final charge separation and stabilization in the RC is completed (A. Holzwarth, 1990). In fact, the latter are determined not by the rapid process of energy migration and exciton diffusion in the antenna, but by a reversible excitation trapping and charge separation in the reaction center. A direct experimental proof of this was the fluorescence rise component of a long-wavelength ($\lambda_{\text{fl}} \sim 720$ nm) in PS I with time 10–15 ns, discovered by A. Holzwarth. This corresponds to the propagation time of exciton excitation from the antenna to the pigment molecule, directly belonging to the RC of PS I. Evidently, the excitation that hit P_{700} can return back to the antenna of the RC:

$$\text{Cl}^*P \underset{k_{\text{rev}}}{\overset{k_M}{\rightleftarrows}} \text{Cl} \cdot P^*,$$

where k_M and k_{rev} are constants of excitation trapping and its return to the antenna.

When excitation is localized on P^*, the act of elementary charge separation in the RC takes place with constant k_e:

$$\text{Cl}^*PI \underset{k_{\text{rev}}}{\overset{k_M}{\rightleftarrows}} \text{Cl}P^*I \underset{k_e^-}{\overset{k_e}{\rightleftarrows}} \text{Cl}\,P^+I^- \xrightarrow{k} Z\text{Cl}P^+I \xrightarrow{k_z} Z^+\text{Cl}PI^-,$$

where k_z is the constant of P^+ reduction from donor Z, and k is the constant of further oxidation of I.

This model (G. Schatz and A. Holzwarth, 1988) is based on the assumption that antenna Chl and photoactive pigment P interact intensively, as a result of which very rapid energy transfer (equilibration) occurs between antenna Chl and P (1–2 ps). In this case, it is impossible to measure directly in experiments the equilibration time on the background of long-living Chl* fluorescence (80–100 ps). The kinetics of excited state decay will be determined by the rate of the primary charge separation $P^*I \rightarrow P^+I^-$, since the lifetime of antenna chlorophyll fluorescence is limited by the rate of trapping (the trap-limited model). Another peculiarity of the model is reversible energy exchange between excited state P^* and radical pair P^+I^- ($P^*I \rightleftharpoons P^+I^-$).

State (Chl$^*PI \rightleftharpoons$ ChlP^*I) is a generalized excited state including all molecules of antenna Chl* as well as P^*. Consequently, k_e is the apparent constant of the rate of charge separation. Charge separation rate constants k_e, calculated from the data on the kinetics of fluorescence decay, depend on the rapid reversible exchange of excitation between antenna Chl and P.

In his experiments, Holzwarth obtained the apparent charge separation rate constant, including the process of reversible trapping of excitation (k_M and k_{rev}), which was $k_e \sim 10\ \text{ps}^{-1}$.

Obviously, when the number of the surrounding antenna molecules is minimized, the probability of direct excitation hitting P increases. Under such conditions, elementary charge separation rate constant k_e, calculated by fluorescence decay, approaches k_e values, measured directly by absorption femtosecond spectroscopy (see below) on isolated RC, where it is $\tau_e \sim 1/k_e \sim 2$–4 ps.

Thus, the short-living component of fluorescence rise τ_1 ($\tau_1 \sim$ 10–15 ps) corresponds to the antenna exciton propagation and quenching by pigments of the reaction center, whereas the long-living ($\tau_2 \sim$ 100–200 ps) component of fluorescence decay reflects deactivation of the antenna excitation upon reversible charge separation in the RC. Moreover, a slower stage of charge stabilization and recovery Q_A and P^+ from donor Z,

$$ZP^+I^-Q_A \overset{k_e}{\rightleftharpoons} ZP^+IQ_A^- \overset{k_z}{\rightleftharpoons} Z^+PIQ_A^-,$$

is also in conformity with a separate more prolonged component of fluorescence decay in the antenna $\tau_3 \sim 500$ ps.

Two main components of fluorescence decay with times $\tau_1 \sim$ 50–70 ps and $\tau_2 \sim 250$ ps are found in intact chromophores with active reaction centers. In the absence of a reaction center, the intrinsic time of fluorescence of the antenna is 700–1000 ps. At the same time, the time of one act of excitation transfer between adjacent molecules in the antenna here, like in chloroplasts, is also not high ($\sim 10^{-13}$ s). It means that the time, required for the exciton propagation in the antenna and reaching the RC, should be much lower than the experimentally measured fluorescence lifetime in chromophores. The same as in chloroplasts, it means that excitation hitting the RC can return back to the antenna. The scheme of processes can be represented as

$$A^* \underset{k_{\text{rev}}}{\overset{k_M}{\rightleftharpoons}} P^* \xrightarrow{k_e} P^+\text{Bpp}^- \xrightarrow{k_{PQ}} P^+Q_A^-.$$

Here $k_M = 1/\tau_1$ is the constant of fluorescence quenching of the antenna ($A^* \rightarrow P$), k_{rev} corresponds to the excitation return to the antenna, k_e is the constant of charge separation reactions in the RC and electron transfer from P to Bpp ($k_e^{-1} \sim 3$ ps), k_{PQ} is the constant of electron transfer to the primary quinone.

Consequently, like in chloroplasts, in chromophores excitation rapidly propagates ($\sim$ 10–20 ps) along the antenna in each of the LHPPC II B800–850, LHPPC I B890 complexes, therewith the time of the excitation "jump" between adjacent molecules does not exceed 1–2 ps, and energy migration between the complexes takes 30–40 ps. The bottleneck is the fluorescence quenching in the RC that has a reversible character ($k_1 \sim k_{-1}$).

Attention should be paid to possible effect of protein on fluorescence of pigments. The existence of different forms of pigments with different fluorescence maxima can be associated with the involvement of protein components in relaxation electron vibrational interactions accompanying fluorescence (see Chapter 10). In this case, protein modification in photosynthetic membranes can have a regulatory effect on energy migration both due to a change in the distances between pigments and mutual orientation of dipoles, and due to a more delicate effect on intramolecular relaxation processes and thereby on the relation of fluorescence bands and absorption of pigments in pigment–protein complexes.

Structure of PSU and Trapping of Excitation in RC. Of great importance for efficient trapping of excitation in RC are topographic peculiarities of organization of pigment molecules in the antenna. The efficiency of trapping decreases at large dimensions of the antenna and, correspondingly, an increase in the number of random jumps necessary for excitation hitting the RC.

Among the structural factors that can affect the efficiency of excitation transfer, the most important are the cluster character of the PSU lattice, selective anisotropy of intermolecular distances to support the selected direction of migration and the orientation of vectors of dipole transient moments of molecules in the PSU.

Thus, complexes containing a Bchl 875 antenna with the reaction center have an additional component (with time of 60–80 ps) reflecting transfer of excitations from the antenna to the RC. Therewith the observed rate of transfer to the RC is from 5 to 6 times higher than the rate of transfer between the RC and one Bchl molecule.

To explain these effects, a model of theoretical circular arrangement of N molecules in the antenna was proposed and analyzed (Fig. 27.9). At strong exciton interaction, such an aggregate behaves as a single supermolecule with a set of levels corresponding to collective excitations of the whole aggregate. In this case, the complete dipole force of N molecules is concentrated in two degenerated levels (if dipole moments of transfer of separate molecules are in the plane of a ring). Dipole moments of these two levels are mutually perpendicular in plane XY (Fig. 27.9) and in Y ($N/2$) times larger than monomer dipole moments. The existence of such a pair of levels was shown experimentally by the method of spectroscopy of single complexes (van Oyen), which is a direct confirmation of the exciton model.

Taking into account the energy disorder, the exciton levels of the ring are delocalized over 10–12 bacteriochlorophyll molecules; in this case the superposition of several levels (occupied at room temperature) creates excitation as a wave packet delocalized over 4–6 bacteriochlorophyll molecules.

This explains the three- to four-fold increase in photo-induced absorption changes (band bleaching) in the long-wave range and the proportional increase in positive changes of absorption in the short-wave range. The latter are caused by the additional absorption generated due to transitions from excited exciton levels to double-exciton

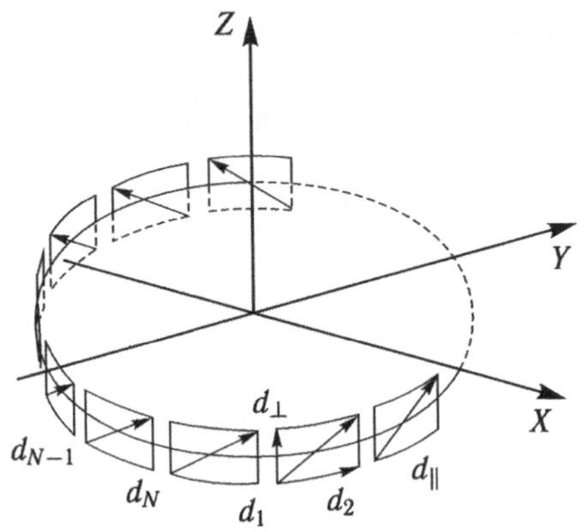

Figure 27.9. Light-harvesting antenna (reproduced from V. I. Novoderezhkin and A. P. Razjivin, 1995).

N is the total number of bacteriochlorophyll molecules; d is the dipole moment of transition of the n-th molecule; X, Y, Z are unit vectors of the coordinate system.

levels (which are also delocalized and have anomalously high dipole moments). The dynamic long-wave shift of the bleaching peak is a result of superfast relaxation between exciton levels as well as the fast (100 fs) component of anisotropy decay. The additional slower anisotropy decay (400 fs) reflects processes of migration of the quasistationary (formed after relaxation) wave packet around the ring.

The efficiency of interactions with the RC and energy transfer to it are proportional to the degree of localization of the exciton wave packet, which explains the five- to six-fold increase in the rate of transfer to the RC as compared to the case of localized excitations in the antenna.

27.5 Reaction Centers in Purple Photosynthesizing Bacteria

The primary act of photosynthesis, leading to charge separation and reduction of the primary acceptor, occurs in reaction centers. At present, due to the development of preparative biochemistry methods it has become possible to isolate a reaction center of photosynthesizing bacteria in the native state that in itself represents a separate pigment – protein complex.

Components of Reaction Centers. A reaction center contains four bacteriochlorophyll (Bchl) molecules, two bacteriopheophytin (Bpp) molecules, one or two ubiquinone (Q) molecules, one Fe atom and a cytochrome complex. Six porphyrin centers (four bacteriochlorophyll and two bacteriopheophytin), two quinones and one non-heme iron atom are included in the reaction center as prosthetic groups.

The nature of the primary photochemical act was first explained in the works of L. N. M. Duysens who discovered small reversible changes in absorption ΔA upon illumination of chromatophores from *Rhodospirillum rubrum*. These changes were characterized by bleaching of the absorption band at 870 nm and a displacement of the band at 800 nm to the blue region. Such small (about 2 %) changes in absorption were obtained also upon chemical oxidation of the RC. This denotes oxidation under the action of light of a small portion of active Bchl *a*, belonging to RC, making only about 2 % of the total content of Bchl *a*, and represents reactive pigment P_{870}. Photo-oxidation of P_{870} proceeds with quantum efficiency, reaching 100 %, and occurs also with high quantum efficiency at 1 K. Figure 27.10, *a* shows the absorption spectrum of preparations of RC from carotenoid-free strain of bacteria *R. sphaeroides*, *R*-26 with designated absorption maxima belonging to Bchl and Bpp pigments. Light absorption induces bleaching of bands at 870, 600 nm and displacement of the band at 800 nm to the blue range (these bands belong to P_{870}) with a concurrent appearance of a new absorption band near 1250 nm.

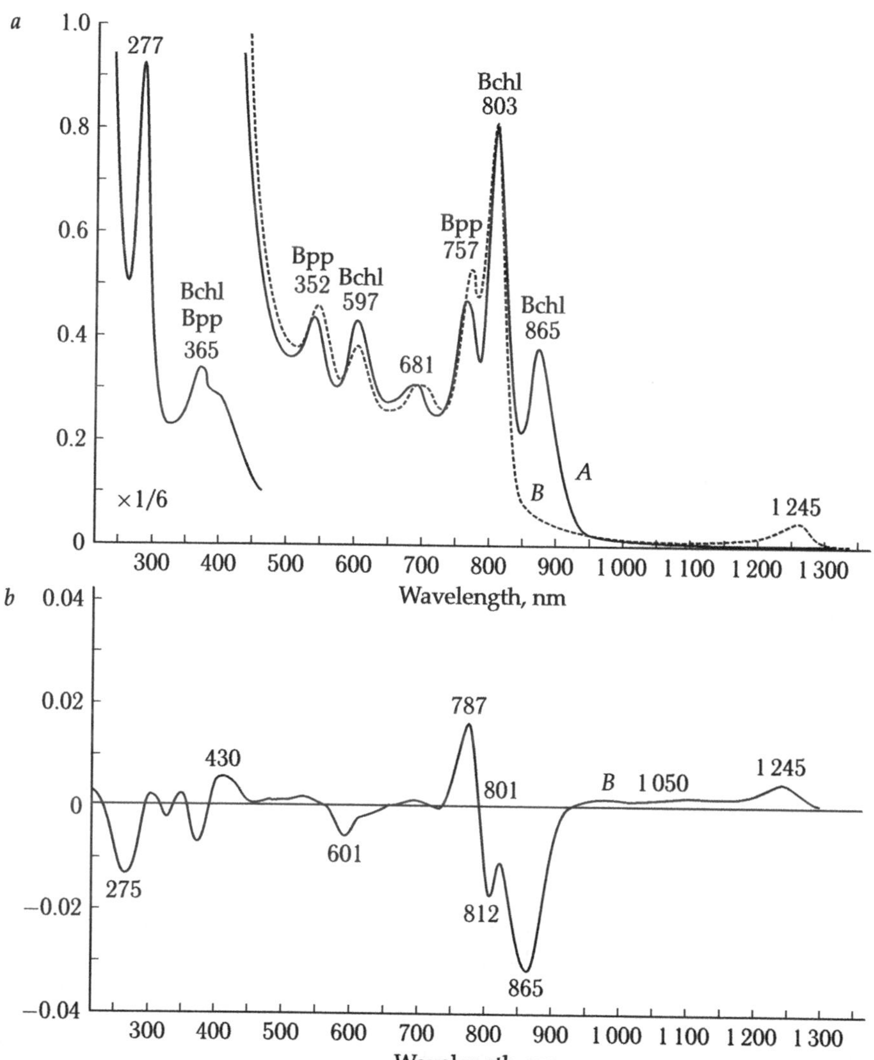

Figure 27.10. Absorption spectrum (*a*) of reaction centers and differential absorption spectrum ("light minus dark") (*b*).

A, Spectrum of RC in the dark; *B*, spectrum of illuminated RC; *C*, the difference between *A* and *B*. Bands of the absorption spectrum (*a*) in the range of wavelengths less than 450 nm have a very high amplitude, therefore they are scaled-down six-fold.

Figure 27.10, *b* shows differential absorption spectra ("light minus dark") upon formation of P^+ with positive maxima at 435, 1250 nm and negative maxima corresponding to bleaching of absorption bands at 600, 870 nm, as well as displacement and bleaching of the band at 800 nm. All these changes are characteristic of the formation of cation radical of a photoactive bacteriochlorophyll dimer ($Bchl^+$). The bleaching of bands at 545 and 760 nm denotes conversion of Bpp, and the positive maximum at 670 nm shows simultaneous emergence of anion radical Bpp (Bpp^-). Besides, an increase in absorption at 670 nm is also characteristic of anion radical Bchl ($Bchl^-$). The bleaching of the band at 275 nm is caused by generation of reduced ubiquinone and generation of ubisemiquinone. After termination of illumination, these spectral changes disappear due to recombination of P^+ and Q^-.

The use of spectral effects made it possible to establish the sequence of direct electron transfer reactions between the above components of RC:

$$PBppQ_A\,Q_B \xrightarrow{h\nu} P^+Bpp^{\bar{\cdot}}\,Q_AQ_B \to P^+BppQ_A^{\bar{\cdot}}\,Q_B \to P^+BppQ_A\,Q_B^{\bar{\cdot}}, \tag{27.11}$$

where P is the bacteriochlorophyll dimer, Q_A and Q_B are the primary and secondary acceptors of the quinone nature (ubiquinones).

Primary quinone Q_A is a single-electron carrier though under certain conditions it accepts two electrons. In contrast to Q_A, secondary quinone Q_B can function as a two-electron carrier under standard conditions. Fe is included in the contents of complex Q_AQ_B that can be represented schematically as $Q_B\text{Fe}Q_B$. The measuring of spectra of the Mossbauer effect demonstrated that Fe does not change its valence and remains in the high-spin Fe^{2+}-state independently of electron transfer in the acceptor quinone complex. However, removal of Fe disturbs electron shuttling between Q_A and Q_B. It is obvious that the iron atom itself is not an electron acceptor but only facilitates electron shuttling between the primary and secondary quinones.

The electron transfer in the quinone complex was studied under short light flashes. The general scheme of the process is as follows:

$$\text{The 1-st flash:}\quad PQ_AQ_B \xrightarrow{h\nu} P^{\dot{+}}Q_A^{\bar{\cdot}}Q_B \longrightarrow {}^{\dot{+}}Q_AQ_B^{\bar{\cdot}},$$

$$\text{The 2-nd flash:}\quad PQ_AQ_B^{\bar{\cdot}} \xrightarrow{h\nu} P^{\dot{+}}Q_A^{\bar{\cdot}}Q_B^{\bar{\cdot}} \xrightarrow{\downarrow 2H^+} P^{\dot{+}}Q_AQ_BH_2 \xrightarrow[\downarrow 2H^+]{} P^{\dot{+}}Q_AQ_B \xrightarrow[+M]{\leqslant 0.1\ s} PQ_AQ_B.$$

Figure 27.11 shows the most probable proton-transport pathways from the cytoplasmic space to the interior of Q_B where amino acid residues serine L233 and glutamine L212 are final links in conducting correspondingly the first and second protons in a relay transfer along chains of hydrogen bonds by the Grotthus mechanism. This scheme was plotted based on the analysis of X-ray data and the results of the effect on the proton transport of the substitution of definite amino acid residues in the surrounding of Q_B by directed point mutagenesis (see Section 8). When the polar residue of aspartate L213 with group COO^- is substituted for nonpolar asparagine L213, the proton transfer is inhibited that indicates to the role of electrostatic interactions in proton stabilization. It is interesting that the functional activity can be recovered again by substituting asparagine M44 by aspartate M44 in another M-subunit. Apparently, the effect of electrostatic interactions can be transferred along the protein globule due to cooperative properties.

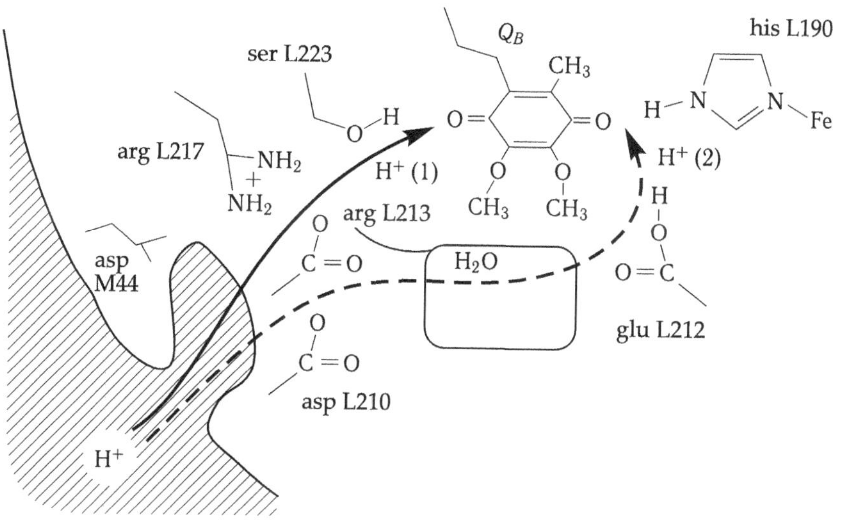

Figure 27.11. Probable pathways of transfer of the first H^+ (1) and second H^+ (2) protons to the reduced secondary quinone in the RC of purple bacteria. The hatching designates the region of localization of the pool of water molecules bound in the RC structure near Q_B. Reproduced from M. Y. Okamura and G. Feher (1992).

As follows from the results of X-ray analysis with high resolution (up to 2.4 E), these interactions may involve water molecules from the RC structure. As seen from Fig. 27.12, at a distance of a hydrogen bond from glutamate L212 in the RC from *Rb. sphaeroides*, there is the beginning of a chain of 12 water molecules identified by the X-ray analysis that form an almost linear structure oriented to the cytoplasmic surface of the RC. It is proposed that under the influence of the difference in the electrostatic potential between Q_B and the RC surface, the protons can move from the cytoplasm to the quinone ring along water molecules, bound with each other and with amino acid residues.

One of the most outstanding achievements in molecular biophysics during recent years is the three-dimensional structure of the protein reaction center resolved by the method of X-ray analysis.

The X-ray analysis of crystalline preparations of complexes of RC from bacterium *Rhosopseudomonas viridis* allowed establishing a three-dimensional pattern of mutual position of 4 hemes of the cytochrome subunit and prosthetic groups of the RC (Fig. 27.13, *a*). As a matter of fact, this complex was one of the first complex hydrophobic membrane-bound proteins with an established detailed three-dimensional structure. In 1988, German scientists H. Michel, J. Deisenhofer and R. Huber were awarded the Nobel Prize for this fundamental research. The structure of a photosynthetic reaction center from bacteria *Rb. sphaeroides*, as found from the X-ray analysis, is shown in Fig. 27.13, *b*.

On the whole, the protein structure of a reaction center forms a rigid framework to which two symmetric chains of pigment molecules are bound. The chains have a common Bchl P dimer: P–Bchl–Bpp–Q. The localization of pigment in Fig. 27.13, *b* is blacked. The circle shows the position of the Fe^{2+} ion. The structure of the RC protein covers rather closely the binding site of Q_A, but in the area of Q_B binding these structures have a cavity through which Q_B can diffuse into the pool of membrane

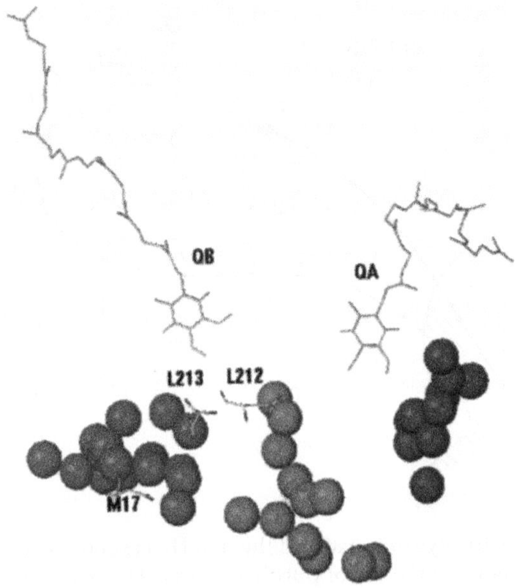

Figure 27.12. Clusters of water molecules identified by the X-ray analysis in the quinone acceptor part of the RC from *Rb. sphaeroides* (reproduced from G. Fritzsch et al., 1998).

quinones, binding the electron-transport chain of the RC to other electron carriers in the membrane. Though the structure of bacterial RCs has two "branches" (*L* and *M*) of carriers forming two electron pathways, the light-induced electron transport proceeds mostly only in one of them — the *L* chain. Apparently, this is connected first of all with the asymmetry both of the protein surrounding and especially of the distribution of aromatic amino acids, playing the role of electron bridges (Chapter 25) in both chains. It is probable that here of certain importance are also the hydrogen bonds formed by electron transfer cofactors with the protein surrounding that play their role in electron stabilization during tunneling (Chapter 25). Hydrogen bonds and aromatic residues for the two branches are shown to be non-equivalent. Thus, the formed branch *L* differs from branch *M* by the presence of hydrogen bonds between the pyrrole ring of Bpp and Glu104 and also by the presence of tyrosine residues in the protein surrounding of the transfer cofactors.

Electron Transfer to the RC. The kinetics of electron transport in purple bacteria cells to the RC was deciphered in detail using methods of absorption and fluorescence spectroscopy of high time resolution up to 10^{-13} s. Pulsed laser excitation of RCs induces characteristic changes in absorption of redox-cofactors (see Fig. 27.10), whose rise and decay times correspond to electron transfer between separate redox components of the RCs.

The use of femtosecond spectroscopy demonstrated that the rise time of the changes in Bpp absorption at 545 and 670 nm, complying with the transfer to Bpp, is about 4 ps.

Recordings of spectra with time resolution < 30 ps reveal also an absorption spectrum at 800 nm different from the above described spectrum. Therefore, an assumption was made on the involvement of an intermediate electron carrier between *P* and Bpp — a Bchl molecule absorbing at 800 nm (V. A. Shuvalov, 1982).

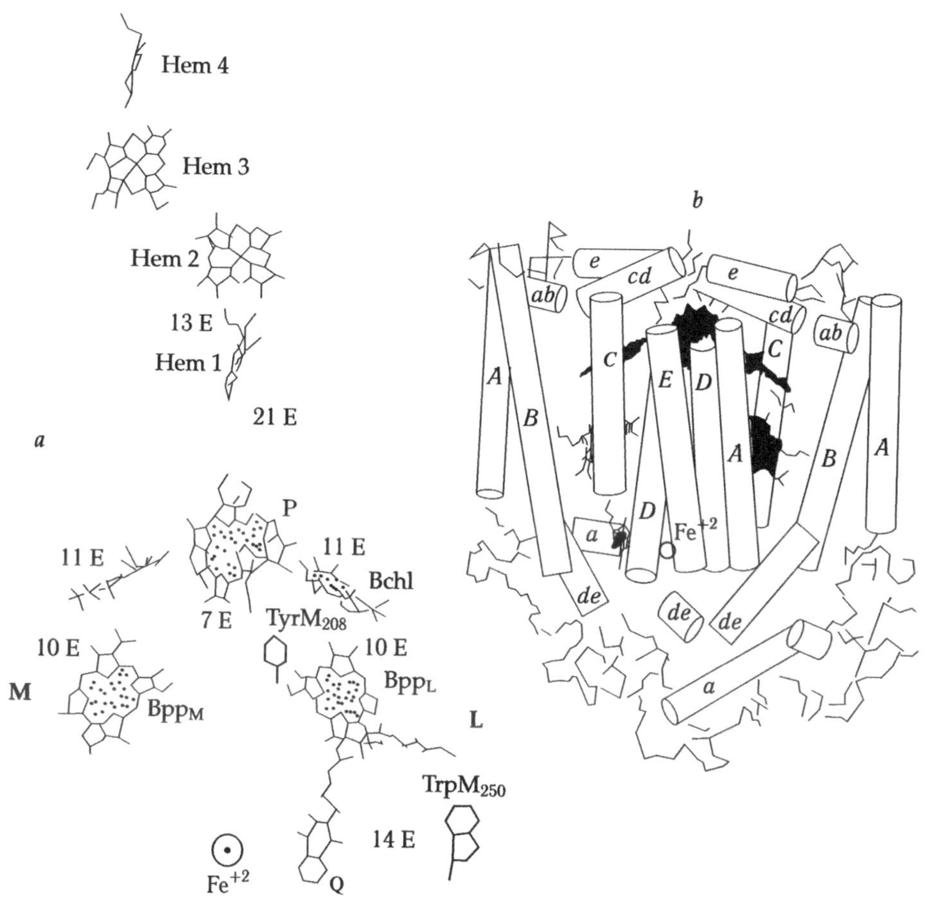

Figure 27.13. *a*, Scheme of arrangement of prosthetic groups and distances between them in the cytochrome-RC complex from *Rps. viridis*; *b*, model of the structure of RC from *Rps. sphaeroides* by the data of X-ray analysis (reproduced from J. Deisenhofer et al., 1987).

In figure *b* cylinders designate helical strands; prosthetic groups are shown in black; the point in a circle shows the localization of non-heme iron; the binding site of Q_A is covered by the protein. Aromatic amino acids $TyrM_{208}$ and $TyrM_{250}$ are bridges for transfer in the $P \rightarrow$ BPP and BPP $\rightarrow Q_A$ regions.

It is believed that a monomer Bchl molecule takes part in electron transfer between P and Bpp. This may be connected with the direct reduction of Bchl and generation of $Bchl^-$.

Another exchange mechanism implies that a Bchl molecule plays the role of an intermediary, enhancing the interaction between P and Bpp without undergoing a real redox conversion.

The most probable scheme of Bchl participation includes two parallel pathways: an exchange transfer (1) and a two-stage redox transfer (2):

$$\begin{array}{ccc} & P^{+}\text{Bchl}^{-}\text{Bpp} & \\ {\scriptstyle (2)}\nearrow & & \searrow{\scriptstyle (2)} \\ P^{*}\text{BchlBpp} & \xrightarrow{(1)} & P^{+}\text{BchlBpp}^{-}. \end{array}$$

In a two-way mechanism, Bchl is reduced (for 3.5 ps) at the first stage, while its oxidation (Bchl→Bpp) occurs much faster (for 0.9 ps) at the second stage thus preventing its accumulation in the RC.

Consequently, at present it is reckoned that the primary state of the RC with separated charges ($P^{\dot{+}}I^{\dot{-}}$) is formed for 4–7 ps. In 150–200 ps the electron is transferred to Q_B now for $\approx 10^{-4}$ s. Thus, direct reaction of electron transfer and their times in the RC can be represented as follows:

$$PBchlBppQ_A\,Q_B \xrightarrow{h\nu} P^*BchlBppQ_A\,Q_B \xrightarrow{<4\text{ ps}} P^{\dot{+}}(BchlBpp)^{\dot{-}}Q_A\,Q_B \longrightarrow$$
$$\xrightarrow{150\text{–}200\text{ ps}} P^{\dot{+}}BchlBppQ_A^{\dot{-}}\,Q_B \xrightarrow{100\ \mu\text{s}} P^{\dot{+}}BchlBppQ_A\,Q_B^{\dot{-}}. \quad (27.12)$$

If prior to excitation Q_A is reduced chemically, i.e. the RC is transferred to state $[P(BchlBpp)Q_A^{\dot{-}}]$, then state $[P^{\dot{+}}(BchlBpp)^{\dot{-}}Q_A^{\dot{-}}]$ formed under light lives much longer (about 10 ns). Apparently, under such conditions $Bpp^{\dot{-}}$ cannot pass the electron to $[Q_A^{\dot{-}}]$ and decay of state $[P^{\dot{+}}BchlBpp^{\dot{-}}Q_A^{\dot{-}}]$ should proceed in another way. Indeed, in due course spectral bands of a new type intrinsic to the triplet state of the photoactive dimer of bacteriochlorophyll ^{T}P (or ^{3}P) appear in the spectrum of photo-induced changes in state $[P^{\dot{+}}BchlBpp^{\dot{-}}Q_A^{\dot{-}}]$. The analysis of spectral and kinetic properties of RC with reduced Q_A shows that in this case triplet state ^{3}P is formed not as a result of a direct (intercombination) transition from the singlet excited state $^{1}P^*$. It appears as a consequence of reorientation of the spins of unpaired electrons P^+ and Bpp^- during the existence of radical pair $[P^{\dot{+}}(BchlBpp)^{\dot{-}}]$ from the antiparallel (singlet) configuration to the triplet one:

$$^{1}[P^{\dot{+}}(BchlBpp)^{\dot{-}}] \rightarrow {}^{3}[P^{\dot{+}}(BchlBpp)^{\dot{-}}], \quad (27.13)$$

with the following recombination of the triplet radical pair.

Figure 27.14 shows a diagram of electron transfer in RCs of purple bacteria. As seen, direct reactions of electron transfer towards $P \rightarrow$ (Bchl, Bpp) $\rightarrow Q_B$ occur for times much shorter than the times of inverse reactions of recombintation and dissipation of corresponding intermediate states. This is the principle of kinetic organization of electron transfer, according to which high efficiency of RC functioning is provided, because direct reactions occur by two or three orders of magnitude faster, than inverse "harmful" reactions that cold have decreased the total quantum yield of primary stages of electron transport. □

In the presence of carotenoids (Car), for 20 ns the energy of the Bchl triplet state migrates to the Car molecule:

$$^{3}Bchl + Car \rightarrow Bchl + {}^{3}Car. \quad (27.14)$$

This process is much faster than the reaction of triplet Bchl molecules with oxygen (10^{-4} s) that can lead to the destruction of the photosynthetic membrane in the light due to the formation of an active oxidizer — singlet oxygen (see Section 4 in Chapter 13). Thereby carotenoids perform a protective function, preventing membrane components from photo-oxidation. The formed ^{3}Car molecules can, in their turn, quench excited states of Bchl, playing the role of quenchers of pigment fluorescence under excess illumination (G. Renger, 1977). Note that the energy of the Car singlet excited state, in contrast to the triplet state, lies higher in Car than in Bchl

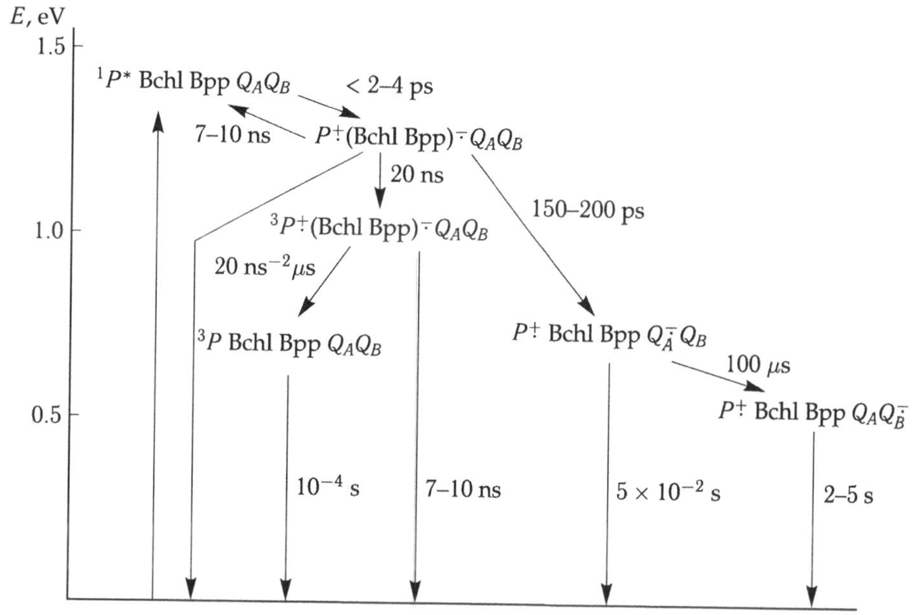

Figure 27.14. Electron transfer in RC of purple bacteria.

and migrates from Car to Bchl molecules of the antenna as from an accessary light harvesting pigment:

$$^{1}\text{Car} + \text{Bchl} \rightarrow \text{Car} + {}^{1}\text{Bchl}. \qquad (27.15)$$

27.6 Pigment – Protein Complex of Photosystem I

Primary electron donor P_{700} in photosystem I was discovered in one of the first experiments with the use of difference optical spectroscopy to study photosynthetic membranes.

B. Cock (1956) gave consideration to small changes in absorption at 700 nm, emerging under action of light on chloroplast suspension (Fig. 27.15). Chemical oxidizers cause a similar spectral effect that allowed an assumption on oxidation of a small fraction of chlorophyll molecules in the membrane.

The difference spectrum of P_{700} is well studied ("light minus dark"); it can be measured also in the presence of other pigments in the surrounding (provided that they do not change their spectral properties in the light). This spectrum includes an intensive "negative" bleaching band at 700 nm (± 3 nm) and at 430 nm, as well as small "positive" bands at 810 nm and 450 nm (Fig. 27.15).

Final electron acceptors in PS I are studied most scrupulously. Such acceptors in all oxygen-releasing photosynthetic organisms (from cyanobacteria and unicellular algae to higher plants) are water-soluble ferredoxins. These relatively small proteins with a molecular mass of about 10 kDa contain iron-sulfur clusters (like 2Fe–2S) and have a rather low redox potential ($E_m \approx -400$ mV). Water-soluble ferredoxin is absorbed by the thylakoid membrane in the locus of PS I localization, and after reduction by electrons coming from PS I it is included in the noncyclic electron flow on $NADP^+$ (the reaction is catalyzed by the enzyme ferredoxin/NADP-reductase) or in the cyclic electron flux in PS I.

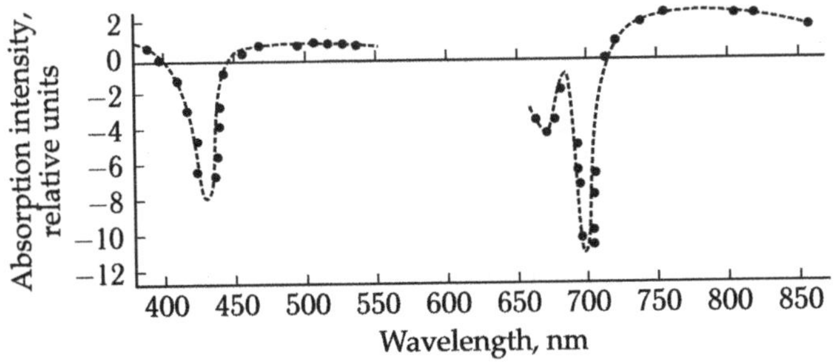

Figure 27.15. Difference spectrum ("light minus dark") for subchloroplast particles with photosystem I.

B. Cock called this fraction P_{700} (P from "pigment", 700 nm is the extremum in a difference spectrum), and this designation has become widely used in the scientific literature. Photochemically active components of reaction centers of other photosystems are also called by analogy with P_{700}, P_{680} in photosystem II, P_{870} or P_{960} in photosynthesizing bacteria, P_{840} in green fluorescent bacteria etc.

This soluble ferredoxin obtains electrons from terminal iron-sulfur clusters F_A/F_B. It was shown that the active center of these proteins is 4Fe, 4S-cluster. The values of redox potentials of F_A and F_B in spinach are −520 mV and −590 mV. Sequential as well as parallel engagement of F_A and F_B centers in the electron transfer can be postulated.

Upon oxidation of P_{700}, an EPR signal characteristic of iron-sulfur proteins with centers 2Fe, 2S is also generated. This signal corresponds to photoreduction of acceptor F_X (or A_2) with the redox potential −705 mV.

At least two more acceptors, called A_0 and A_1, are located between P_{700} and F_X. Evidences for the existence of these acceptors were obtained from the data of picosecond spectroscopy of PS I complexes with chemically reduced or inactivated centers of F_X.

The closest to P_{700} direct acceptor of electrons from the primary donor is the chlorophyll *a* monomer with the absorption band at 693–695 nm (Chl 695). Electron transfer from P_{700} to A_0 occurs faster than for 10 ps and is accompanied by bleaching of the Chl 695 band. After that for 20–50 ps the electron moves to acceptor A_1 that evidently is phylloquinone (vitamin K_1). Two phylloquinone molecules are associated with each reaction center of PS I, and one of them is bound more strongly than the other.

According to different estimates, the time of electron transfer from A_1 to F_X varies from 200 ps to 20–50 ns, and the time of transfer from F_X to F_A and then to F_B is 170 ns.

The kinetics of electron transfer to ferredoxin, as shown by recent data, can be represented by three components with characteristic times from 500 ns to 100 μs, that may be connected with three types of PS I complexes with ferredoxin (Fig. 27.16).

The three-dimensional structure of a pigment–protein complex has been revealed by X-ray analysis of crystals of this complex from thermophilic cyanobacteria at 2.5 E resolution. The existence of trimmers of PS I complex in cyanobacterial thylakoids was proved also by the data of N. V. Karapetyan (1999). He demonstrated on the similarity of fluorescence emission spectra of cells of the *Arthrospira platensis* cyanobacterium and PS I trimmers isolated from them, characterized by the 77 K fluorescence band at 760 nm (F760). The existence of PS I trimers in membranes of other cyanobacteria was established using antibodies and electron microscopy.

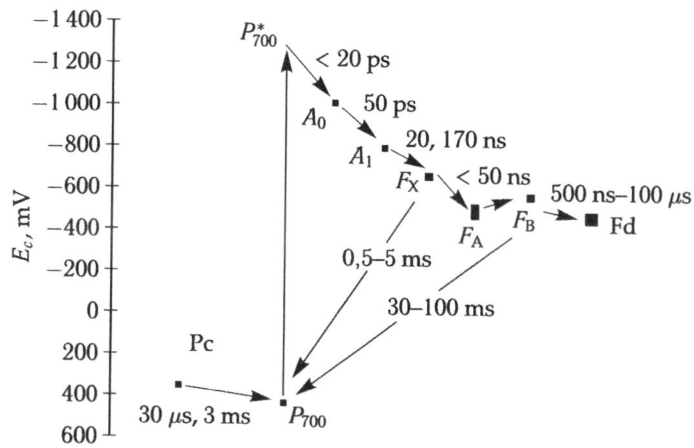

Figure 27.16. Electron transfer to the RC of PS I (reproduced from A. Yu. Semenov et al., 2006). Midpoint potentials and transfer times shown. Variale estimates are caused by the data obtained in different laboratories. Pc is plastocyanin; Fd is ferredoxin.

In addition to the main forms of antenna chlorophyll, PS I in different cyanobacteria cells contains from 3 % to 10 % of long-wave chlorophyll forms, absorbing in the range of 700–750 nm (N. V. Karapetyan et al., 2006). Long-wave chlorophyll forms are thermal traps for energy, and they trap the energy which is efficient in P_{700} photooxidation. Long-wave chlorophyll forms enhance absorption of far-red light upon growing cyanobacteria under weak illumination or protect the PS I complex from excitation energy excess, when the reaction center is inactive.

The times of excitation energy migration from the core antenna complex to the RC in PS I and charge separation with the formation of a radical pair and electron transfer were studied by methods of picosecond spectroscopy. In these experiments, the kinetics of fluorescence decay and the changes in absorption of RC in PS I, excited by picosecond laser pulses, was measured. Let us briefly analyze the results obtained by A. R. Holzwarth (1996–2006) on the excitation transfer times.

At the first stage, equilibrium distribution of excitation energy takes place in the core antenna complex for 500–600 fs. This time grows with an increase in the number of chlorophyll molecules in the antenna (see Section 4). Reverse trapping of excitation occurs for 2 ps, and its equilibrium distribution is established in the RC for 150–200 fs. Then follows a sequence of reversible acts of electron transfer, beginning from primary acceptor A_0 (chlorophyll) with the formation of radical pairs with separated charges at times indicated in Fig. 27.17. Excitation trapping in the RC, the same as electron transfer with formation of radical states (3 states in the scheme), has a reversible character. As seen, the total time of the primary photosynthetic act depends on the initial trapping of excitation and excitation transfer in the antenna. Generation of recombint fluorescence, when accompanying electron transfer from the primary radical pair to the singlet excited level of dimer P_{700}, occurs for $\leqslant 10^{-9}$ s (A. R. Holzwarth, 1996–2006).

Dissipation of Excitation Energy in Cyanobactrial PS I. Quenching of Cation Excitation by Cation Radical P_{700}. Dissipation of excess excitation energy in the PS I complex occurs with involvement of long-wave forms of antenna chlorophyll. In contrast to PS II and photosystem of purple bacteria, fluorescence of PS I of green

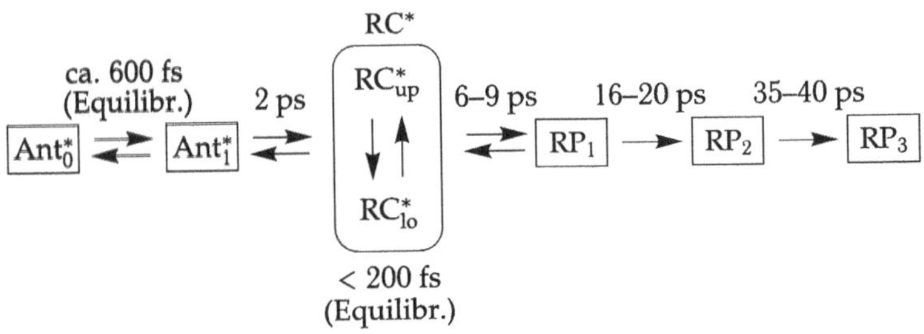

Figure 27.17. Simplified scheme of main components involved in early stages of energy transfer and electron transfer in PS I. Based on data obtained by A. R. Holzwarth (1996–2006).

plants and cyanobacteria is strongly quenched at room temperature. PS I cannot produce intensive variable fluorescence, since at 295 K P_{700} and P_{700}^{+} quench antenna fluorescence with the same efficiency. Thus, photoinduced accumulation of P_{700}^{+} at 77 K leads not to an expected increase, but on the contrary, a decrease in the yield of cyanobacterial PS I fluorescence. The energy absorbed by the main antenna chlorophylls migrates with almost 100 % efficiency to P_{700}. In trimers of cyanobacterial PS I the elementary stages of excitation energy transfer between adjacent molecules of the main antenna chlorophyll occur for 160 fs, the energy equilibrium in this fraction of chlorophyll is established in 360 fs, and for 4–10 ps in the whole spectrally heterogeneous antenna. The fast energy transfer from β-carotene to chlorophyll (105 fs) is determined by the high degree of energy migration (85–90 %) from carotene to chlorophyll. The sequence of excitation energy migration in the PS I antenna (in *Arthrospira platensis*) was estimated by spectra of different components of fluorescence decay in the subpicosecond range. Long-wave forms slow down the energy trapping in PS I by the RC in trimers as compared to monomers: 11 ps in monomers and 15 ps in trimers of PS I.

27.7 Pigment–Protein Complex of Photosystem II

The cardinal function of the RC in PS II is to accept electrons from the oxygen-evolving complex (OEC) and provide their subsequent transfer to the chain of carriers towards PS I. The primary electron donor in PS II is photoactivated pigment P_{680} called so by the analogy with P_{700}. Illumination causes bleaching of the band at 680–690 nm and some increase in absorption at 820 nm, characteristic of the cation radical of chlorophyll. Spectral properties of P_{680} are explained by exciton interaction of photoactive chlorophyll with other six chlorophyll molecules located in the RC at distances larger than in the dimer. Apparently, P_{680} consists of several weakly interacting pigments. Any action disturbing their exciton interactions leads to a sharp decrease in the long-wave shift of the absorption band of P_{680}.

As P_{680} oxidizes a water molecule, its redox potential $E_{1/2}$ should exceed +810 mV. This means that P_{680}^{+} is a strong oxidizer that can oxidize neighboring pigment molecules, if the "normal" donor reducing it has no electrons in the chain.

The general scheme of transfer is as follows:

$$\text{OEC} \xrightarrow[\downarrow]{} \text{Tyr} \rightarrow P_{680} \rightarrow \text{Phph} \rightarrow Q_A \xrightarrow[\downarrow]{\text{Diuron.}} Q_B \rightarrow \text{Quinone pool.} \qquad (27.16)$$

As seen, reduction of P_{680}^+ is provided by the secondary donor tyrosine (Tyr) that accepts electrons from the OEC. Pheophytin plays the role of the primary acceptor.

Figure 27.18 demonstrations a scheme of structural organization of the RC in PS II showing the pathway of electron transfer from H_2O to quinones, the Mn-containing complex responsible for water decomposition and oxygen release (see below) with molecular weights of protein subunits connected with the RC of PS II.

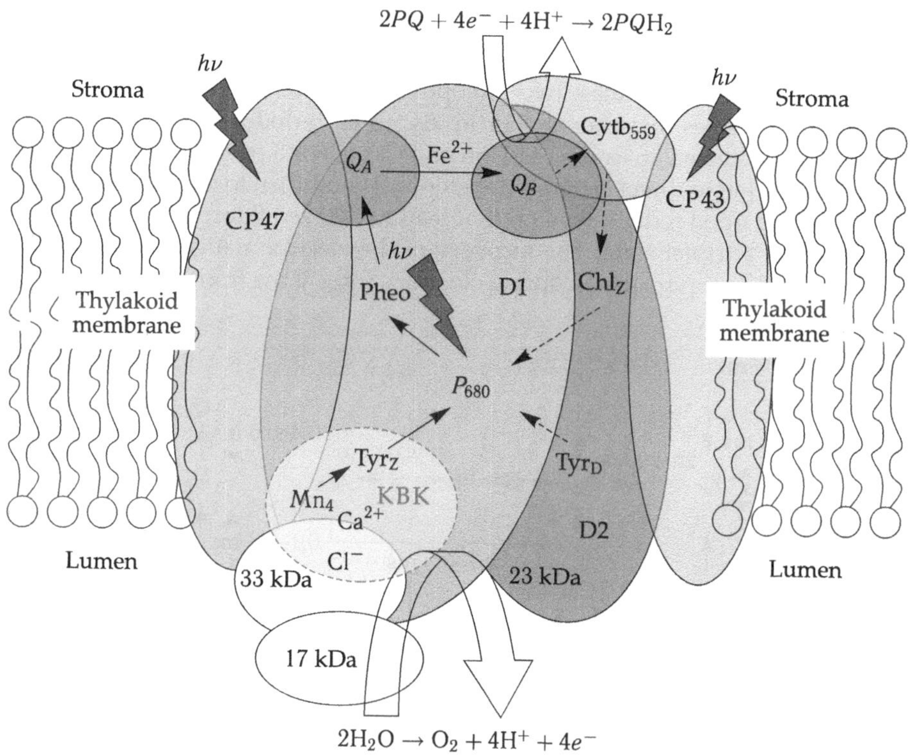

Figure 27.18. Scheme of structural organization of photosystem II (reproduced from G. W. Brudvig, 2007).

In the acceptor part, Q_A and Q_B quinone molecules are plastoquinones. The electron transfer between primary and secondary plastoquinones Q_A and Q_B is inhibited by diuron and low temperature. At low pH values and decreased temperature, reduction of P_{680}^+ is terminated. One of the methods for studying reactions, occurring in PS II, is recording of antenna chlorophyll *a* fluorescence that is dependent on the RC state in PS II. As evident, fluorescence quenching should proceed in state $P_{680}Q_A$, when the center is open (see Section 4). Reduction of Q_A converts the centers to closed state $P_{680}Q_A^-$, where fluorescence quenching is impaired and correspondingly fluorescence intensity enhances. □

However, fluorescence quenching takes place also when the centers are in closed state P_{680}^+, where the photoactive pigment is oxidized. The reason for this is that the absorption spectrum of cation radical Chl a^+ has a wide band in the red and near IR regions (600–850 nm). This provide resonance transfer (see Chapter 25) of excitation energy from antenna molecules to Chl a^+ (P_{680}^+), its transition to the lower excited electronic level near 820 nm with the subsequent energy degradation, in this case not connected with photosynthesis. Thus, an increase in the yield of antenna fluorescence occurs only in state $P_{680}Q_A^-$. If then Q_A^- directs its electron to Q_B and P_{680} remains reduced ($P_{680}Q_A^- Q_B \rightarrow P_{680}Q_A Q_B^-$), the fluorescence decreases.

The general sequence of electron transfer processes in the RC of PS II is shown in Fig. 27.19. The three-dimensional structure of pigment–protein complexes in PS II has been determined by X-ray analysis (XRA) of crystals of this complex isolated from thermophilic cyanobacteria *Thermosynechcoccus elongates* at 3.8 E, 3.2 E and later 3.0 E resolution. Each monomer of the PS II complex contains one copy of 20 different protein subunits and provides the localization site for 35 chlorophyll *a* molecules, 11 β-carotene (CAR) molecules, two pheophytins (Pheo), two plastoquinone (PQ) molecules, two hemes, bicarbonate, 14 lipids, three *n*-dodecyl-β-D-maltose molecules, the Mn_4Cl cluster, an iron (Fe^2+) ion and Ca^2+. For 3 ps radical pair P^+Pheo is formed. In addition to direct electron transfer to Q_A and then to Q_B, inverse reactions are also possible upon reduction of Q_A that lead to delayed fluorescence F_V ($\sim$ 2 nm) or formation of a triplet state. The recovery of the electron deficit occurs for about 30 ns from the donor tyrosine accepting electrons for $\leqslant$ 1 ms from the OEC.

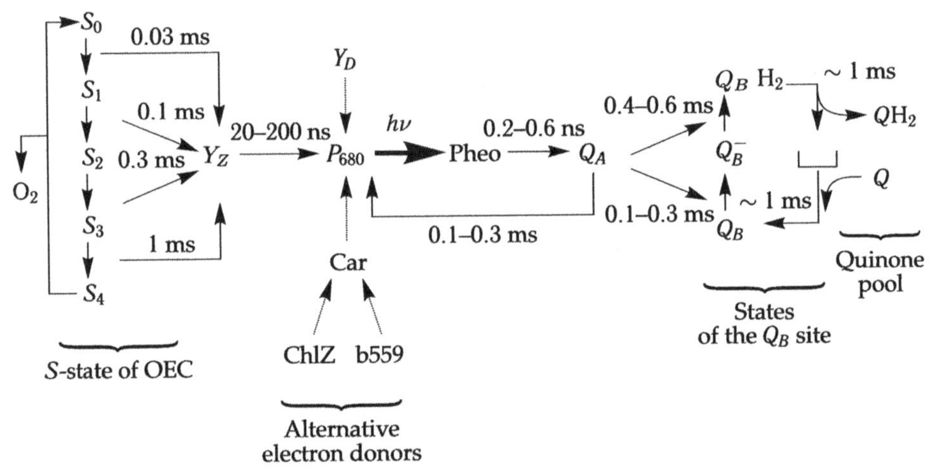

Figure 27.19. Scheme of electron transfer in RC of PS II (reproduced from Shinkarev V. P., 2004). S_0–S_4 are states of the oxygen-evolving complex. Times of transfer and duration of fluorescence are designated F_0.

Details of the kinetics of electron transfer in PS II were reported in the papers by A. R. Holzwarth who used molecular spectroscopy methods in his studies. He found that the kinetics of chlorophyll fluorescence decay in intact particles of PS II with open RCs is characterized by several components. The theoretical analysis analogous to that for studying PS I (Section 6) demonstrated that migration of excitation in a light-harvesting antenna prior to the RC trapping takes $\sim$ 90 ps that is higher than the analogous time of excitation stabilization in the antenna of PS II. Consequent processes of charge separation and electron transfer to RCs take about 90–100 ps. Thus, the processes of migration in a large-size antenna to the RC and charge separation with subsequent electron transfer make an approximately equal contribution to the total time of excitation energy transformation to the RC of PS II. Obviously, the relation between the rates of migration and the overall trapping of excitation depends on the dimensions of the antenna. Holzwarth has also advanced a hypothesis that charge separation in PS II starts with the electron transfer to Pheo from the excited additional chlorophyll molecule Chl^{*}_{add} not included in the composition of P_{680}.

Therefore, according to this scheme the additional chlorophyll molecule plays the role of the primary electron donor. The general sequence and times of separate stages are as follows:

$$\mathrm{ANT} \xrightleftharpoons{90\ \mathrm{ps}} \mathrm{ANT} \xrightleftharpoons{10\ \mathrm{ps}} \mathrm{RC}^* \xrightleftharpoons{18\ \mathrm{ps}} \mathrm{Chl}^{+}_{\mathrm{add}}\mathrm{Pheo}^{-} \rightleftharpoons$$
$$\xrightleftharpoons{35\ \mathrm{ps}} P_{680}\mathrm{Chl}^{+}_{\mathrm{add}}\mathrm{Pheo}^{-} \xrightleftharpoons{200\ \mathrm{ps}} P_{680}Q_A^{-} \rightleftharpoons . \quad (27.17)$$

Oxygen Release. Oxidation of water is catalyzed by the OEC that is one of the main components of PS II. Water is the source of electrons for the synthesis in the photosynthetic apparatus of high-energy compounds (NADP and ATP), involved later in dark reactions. As a result of water photolysis, a by-product (molecular oxygen) is also formed and released into the external environment. Since this reaction is practically the only source of free oxygen on our planet, it is impossible to overestimate its importance.

The catalytic center of the OEC is a tetranuclear manganese cluster. It oxidizes two water molecules and generates one molecule of oxygen, with four protons released into the intrathylakoid space and four electrons moving along the photosynthetic electron-transport chain to be consumed for $NADP^{+}$ reduction. To provide this reaction, successive absorption of four light quanta are required. The absorption of a light quantum by the reaction center is accompanied by charge separation — reduction of phoephytin and oxidation of primary donor P_{680}. Oxidized primary donor P^{+}_{680} has a redox potential exceptionally high for biological compounds (about 1.2 V), sufficient for water oxidation. P^{+}_{680} oxidizes the secondary electron donor (the redox-active residue of tyrosine Y_Z, located in position 161 in the amino acid sequence of integral polypeptide D1 of 33 kDa). Polypeptide D1 together with polypeptide D2 (33 kDa) form the reaction center of PS II. In its turn, oxidized electrons carrier Y_Z (the redox potential of pair $Y_Z/Y_Z\bullet$ near 1 V) oxidizes the manganese cluster. Consecutive absorption of four light quanta is accompanied by successive removal of four electrons from the manganese cluster. Then the oxidized manganese cluster oxidizes two water molecules and is reduced to the initial redox state, thus completing the catalytic cycle. This catalytic cycle is called an *S*-cycle (shown in Fig. 27.20, *a*). This cycle comprises five *S*-states: S_0–S_4. Each *S*-state has a corresponding degree of oxidation of cations in the manganese cluster. The absorption by the reaction center of each of the four light quanta triggers single-electron oxidation of manganese cations and consecutive transition of the cluster from the lowest state S_0 to higher state S_4. Apparently, this process has one exception: upon transition from state S_2 to state S_3 it is not the manganese cation that is oxidized, but the amino acid ligand included in the coordination sphere of the manganese cluster. After successive extraction of four electrons from the manganese cluster, the cluster transits to unstable state S_4, from which it returns to initial state S_0 in1 ms due to its electron-mediated recovery from water. This process is accompanied by the release of one oxygen molecule. Oxidation of water and, evidently, its binding with the manganese cluster occur only in state S_4. The release of protons in the *S*-cycle takes place upon transitions S_0–S_1 (one proton), S_2–S_3 (one proton), and S_4–S_0 (two protons).

The operation of the OEC is successfully studied by measuring the kinetics of oxygen release in response to single light flashes (1–10 μs). The yield of O_2 in response

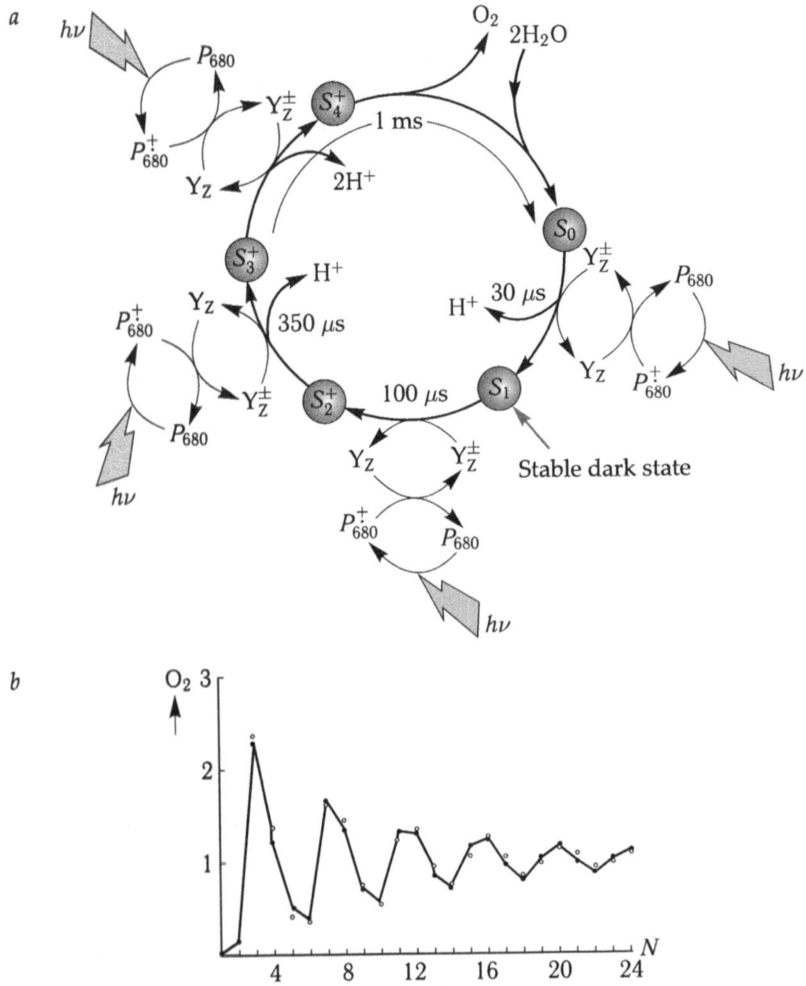

Figure 27.20. *a*, Cycle of *S*-states of the manganese cluster in the OEC; *b*, oxygen evolution depending on the number (*N*) of flashes (reproduced from J. Barber, 2007).

to a separate flash depends on the number of the flash in the series of several consecutive flashes. This dependence is manifested in damped oscillatons of oxygen evolution in response to each flash with a period of four flashes: the maximal yield is observed in response to the 3-rd, 7-th, 11-th etc. flash (Fig. 27.20, *b*).

Dark deactivation of more oxidized to less oxidized states proceeds by a single-step mechanism (S_3–S_2–S_1). After prolonged constant light illumination, uniform distribution is established in states S_0, S_1, S_2, S_3 (25 % in each of the states). As a result of subsequent dark adaption and transitions S_3–S_1 and S_2–S_1, the following distribution is established: 25 % S_0, 75 % S_1, and 0 % S_2 and S_3, because states S_2 and S_3 are unstable in the dark and transit to state S_1 in the time interval from several seconds to several minutes. To activate the OEC, adapted to the dark, and to transit it to state S_3, two flashes are required that will transfer 75 % of state S_1 to state S_3. The third flash will cause transition S_3–S_4 and intensive O_2 evolution that is observed experimentally. Characteristic times of transitions between states S_0–S_1, S_1–S_2, S_2–S_3 are within

200–500 μs that is close to the times of transfer Q_A–Q_B, whereas transition S_3–S_0 takes several ms.

In addition to manganese, necessary components of the OEC are a calcium cation and one or several chloride anions. The role of these cofactors in the mechanism of water oxidation is not clear yet, but in their absence oxygen is not released. Ca^{2+} plays a structural role in early S-transitions of OECs, whereas its functional role is the formation of the O–O-bond in the final S-state (G. W. Brudwig, 2007). In the absence of the chloride ion (Cl^-), transitions $S_2 \rightarrow S_3$ and $S_3 \rightarrow S_0$ are completely inhibited. These data comply with the hypothesis that Cl^- is favorable for the electron transfer between manganese ions (G. M. Ananyev et al., 2001).

The oxygen-evolving complex (OEC) in PS II is located in a large globular domain and localized in the D1 subunit in the form of a "hook" similarly to the Y-shaped structure. To describe the Mn_4Ca-cluster, the model "3 + 1" was proposed (Fig. 27.21), supported in PCA and EXAFS experiments. Manganese ions are numbered (Mn_1 – Mn_2) and are located in the angles of a tetrahedron, and one metal ion is included in the center of the elongated region. The fifth metal is the calcium ion bound to the OEC. The analysis of electron density confirmed the structure of the manganese cluster that implies a cubic construction for Mn_3CaO_4, where each metal ion of this cluster has three μ-oxygen bridges (in the large domain), bound to another manganese ion via a mono-μ-oxygen bridge. The distance between manganese ions was determined using EXAFS. The distance between Mn_1 and Mn_2 and between Mn_1 and Mn_3 is 2.7 E, and probably they are connected with di-μ-oxo bridges, whereas the distance between Mn_1 and Mn_3 and between Mn_3 and Mn_4 is 3.3 E and they are connected with mono-μ-bridges.

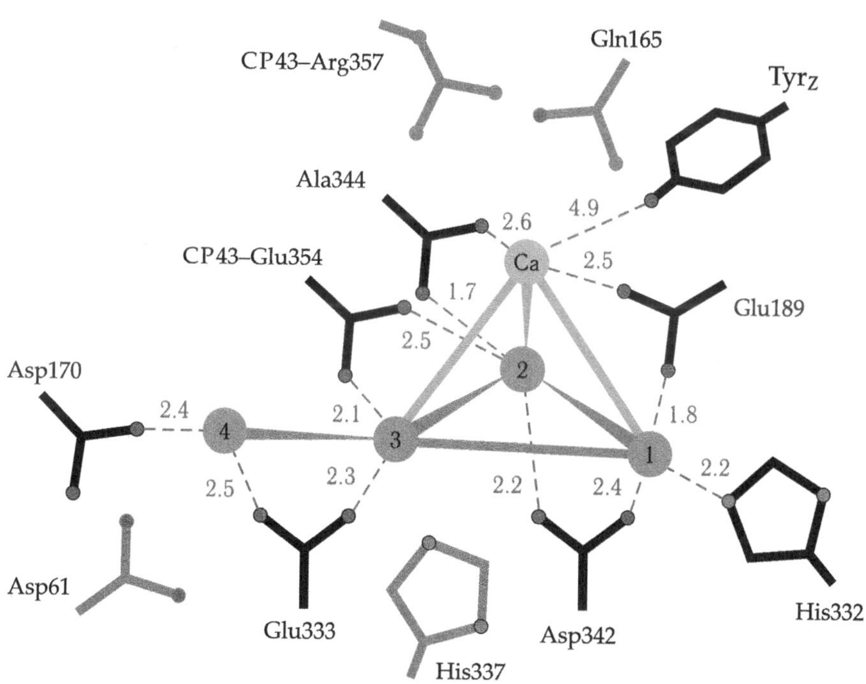

Figure 27.21. Schematic representation of the Mn_4Ca-cluster. See explanations in the text (reproduced from B. Loll et al., 2005).

27.8 Variable and Delayed Fluorescence

The main contribution to chlorophyll fluorescence in chloroplasts is made by PS II. Because of high efficiency of deactivation of excited state P^*_{680} and charge separation in the RC, fluorescence of PS II is characterized by low quantum yields (1–3 %) and decay times (100–200 ps). However, by studying the changes in these values it becomes possible to estimate the efficiency of the "photosynthetic" consumption of electronic excitation energy of P^*_{680} under different conditions. At present, parameters of florescence, as indices of the efficiency of the photosynthetic apparatus, are widely used in fundamental and applied studies.

The main idea is that upon an increase in the efficiency of the light energy consumption in chlorophyll fluorescence intensity should correspondingly enhance. The universe relation should be also true.

Under normal physiological conditions in intact photosynthetic membranes, a change in PS II fluorescence is caused mainly by changes in the electron transfer rate along the chain of carriers that depends on the degree of their reduction and other processes, coupled to the electron transfer.

Variable Fluorescence. Typical changes in PS II fluorescence upon illumination depending on the ETC state in photosynthesis are reflected in the so-called fluorescence induction curves. A typical curve showing the change in chloroplast fluorescence upon illumination of cells preliminarily adapted in the dark is presented in Fig. 27.22. The curve shows that upon illumination the level of chlorophyll fluorescence changes from initial F_0 to maximal F_M passing several intermediate maxima. Let us analyze what information on the state of the ETC can be obtained from the shape of the induction curve.

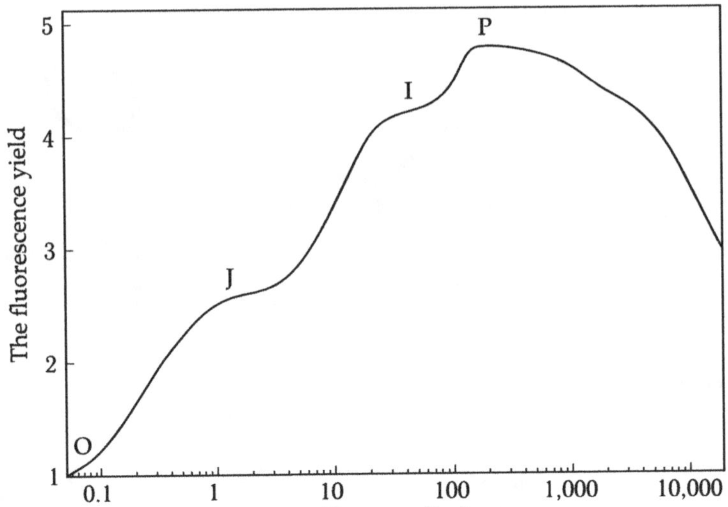

Figure 27.22. Kinetics of light induction of chlorophyll fluorescence (OLR) recorded with high resolution in photosynthetic organisms.

The figure shows the curve measured on a bean leaf using a PEA (Plant Efficiency Analyzer, Hansatech, UK) fluorimeter at exciting light intensity 3,000 μE $m^{-2}s^{-2}$. The fluorescence yield increases from minimal value O (F_0) to maximal P (F_M) passing via intermediate inflexion points J and I.

Essential information is associated with slow changes in fluorescence starting from its initial low level F_0 (point O on the curve). The value of F_0 is determined by chlorophyll florescence, when all RCs are in an "open" active state, all molecules of primary quinone carrier Q_A are oxidized and are ready to accept electrons from P^*_{680}, and the centers quench the antenna fluorescence:

$$P^*_{680}\text{Pheo}Q_A \rightarrow P^+_{680}\text{Pheo}^-Q_A \rightarrow P^+_{680}\text{Pheo}Q^-_A.$$

Here efficient photochemical quenching q_P of antenna fluorescence by open RCs takes place. This is achieved under weak exciting light, when the frequency of light quanta hitting the RC is lower than the frequency of RC "operation" with its return to the initial state, which is completely ready to "accept" a new quantum. Under such conditions, all losses of light energy in the form of fluorescence at level F_0 are determined by the losses in the antenna upon trapping of the excitation energy and do not depend on the state of the RC.

But at high illumination intensities, when the rates of quanta hitting RCs, charge separation and reduction of pheophytin and Q_A in the RC exceed the rates of subsequent electron transfer along the ETC to PS I. As a result, RCs pass to a closed state. In closed RCs, Q_A is reduced, and therefore due to electrostatic repulsion the electron transfer from P_{680} to Pheo^- in the presence of Q^-_A cannot take place. Consequently, the energy of electron excitation of state P^*_{680} for the most part returns to the antenna, and the fluorescence level becomes maximal F_M.

It is seen in the curve (Fig. 27.22) that upon illumination after reaching initial level F_0, at point O the fluorescence increases in the millisecond range, and after passing the intermediate maxima (JI) the intensity reaches the maximum value of F_M at point P. Apparently, region OJ corresponds to the initial reduction of Q_A in photosystem II after switching on the illumination. Then in the process of illumination and "functioning" of the ETC, molecules of the reduced Q_A are accumulated, and point P or maximal level F_M is reached on the curve. The difference between F_M and F_0 ($F_M - F_0$) is called variable fluorescence F_V:

$$F_V = F_M - F_0.$$

The real value of F_M can be determined experimentally by illuminating the samples in the presence of diuron that blocks the electron transfer between photosystems PS II and PS I after the Q_A site. When the peak in point P is reached, a slower change with a gradual decrease in the fluorescence level takes place that is caused by other processes coupled to the ETC.

Measurements of F_M, F_0 and $F_V = F_M - F_0$ values are of special importance in determining the absolute quantum efficiency of charge separation in the RC. It should be reminded (Section 1) that the fate of the excitation energy in the RC depends on the relation of the rate constants of three competing processes of deactivation of excited chlorophyll molecules in the antenna:

$$P^* \xrightarrow{k_f + k_d + k_{ph}} P, \tag{27.18}$$

where P and P^* are the ground and excited states of a chlorophyll a molecule; k_f, k_d and k_{ph} are rate constants of radiative (fluorescence), radiationless (thermal dissipation) and photochemical (in the reaction center) deactivation of excitation.

Quantum yields of the primary charge separation $q\Phi$ in photosynthesis and fluorescence qF_0 are correspondingly

$$q\Phi = \frac{k_{ph}}{k_f + k_d + k_{ph}} \quad \text{or} \quad qF_0 = \frac{k_f}{k_f + k_d + k_{ph}}. \tag{27.19}$$

Constant k_{ph} reflects the rate of excitation trapping by the reaction center in PS II in the process of primary charge separation and electron transfer from the excited pigment molecule P_{680}^{+} in the reaction center to the pheophytin molecule. This is followed by a series of reversible secondary processes of the electron transfer that provide the stabilization of separated charges:

$$\underset{(1)}{P_{680}^{*}} \rightleftharpoons \underset{(2)}{P_{680}^{+}I^{-}} \rightleftharpoons \underset{(3)}{P_{680}^{+}IQ_A^{-}} \rightleftharpoons \underset{(4)}{D^{+}P_{680}IQ_A^{-}} \rightleftharpoons \underset{(5)}{D^{+}P_{680}IQ_A\,Q_B^{-}}. \tag{27.20}$$

At oxidized primary quinone acceptor Q_A (state 1), when the reaction center is in an open state, constant k_{ph} is the largest of the three constants. Therefore, here the excitation energy is consumed in reactions of photosynthesis with quantum yield $q\Phi$, close to unity, and only a small part of excitations (about 0.3 %) is lost as fluorescence during the time of excitation transfer to the reaction center.

The process of energy transformation in the reaction center (transition from state 1 to state 5) takes about 1 ms. During this period (the so-called the time of turnover of the reaction center), the primary products of the photochemical reaction are not used yet in subsequent reactions and the reaction center is in the state, when it cannot use the next portions of energy.

At times higher than 10 μs, state 4 determines changes in the fluorescence yield when illuminated by light. In 10 μs after a short saturating flash of light or under constant light of saturating intensity, all reaction centers of photosystem II are in state 4 and fluorescence reaches maximal value F_M:

$$q\Phi = 0, \quad k_{ph} = 0, \quad qF_M = k_f/(k_f + k_d). \tag{27.21}$$

Thus, the increase in the fluorescence yield upon complete closure of reaction centers ($F_V = F_M - F_0$) is caused by the portion of the light energy that was consumed in primary reactions of photosynthesis at open reaction centers. It is easy to find that the relation of yields of variable and maximal fluorescence is equal to the primary photochemical reaction of charge separation in photosynthetic reaction centers:

$$\frac{qF_M - qF_0}{qF_M} = \frac{k_{ph}}{k_f + k_d + k_{ph}} = q\Phi. \tag{27.22}$$

Thus, measurements of chlorophyll fluorescence intensities F_0 and F_M in relative units allow obtaining the absolute value of quantum efficiency of charge separation in photosynthetic reaction centers.

Now let us consider the shape of the initial region of the induction curve (Fig. 27.22) and the information on the state of the ETC that it provides. The figure shows that in response to illumination with high intensity light, a typical induction curve is characterized by clear transitions O–J–I–P (points on the curve), associated with changes in the redox state of electron carriers. The level designated O in the kinetic curve corresponds to minimal fluorescence F_0, and P is the maximal fluorescence F_M. The reasons for the decrease in the fluorescence level after reaching maximum F_M will be considered lower. Phase O–J is connected with light-induced reduction of Q_A. This is proved by that diuron, inhibiting electron transport between Q_A and Q_B, induces rapid growth of fluorescence to the maximal level for 2 ms, which corresponds to the peak J. It is important to remember that in the absence of DCMU under illumination, phase O–J reflects only in part the accumulation of Q_A, because at the same time electrons flow out to Q_B and further to PS I.

The following region J–I–P of the curve reflects further accumulation of the reduced Q_A^-, but at different states of the quinone pool, i.e. upon successive accumulation of $Q_A^- Q_B$, $Q_A^- Q_B^{2-}$ and PQH_2.

It is also known that PS II system is heterogeneous and contains Q_B-reducing and Q_B-non-reducing centers of PS II. An increase in the fraction of Q_B-non-reducing centers leads to the growth of the amplitude of phase O–J because electron outflow from Q_A to Q_B in these centers is impeded, slowing down the decrease in peak J.

It is also believed that there are additional fractions of the so-called "rapidly" and "slowly" reduced quinones, located correspondingly in grana and stromal thylakoids. It is proposed that phase J–I corresponds to the reduction of the rapid quinone pool, while phase I–P to the reduction of the slow quinone pool. In accord with this, the appearances of peaks J, I and P are separated in time.

In region P–S (Fig. 27.22), the fluorescence level drops from F_M to lower values. Here the so-called nonphotochemical quenching of fluorescence q_N takes place, including the following processes: energy-dependent nonphotochemical quenching or energized quenching q_E; fluorescence quenching by the "spillover" mechanism q_T due to phosphorylation of the LHPPC and its transition in the membrane during several minutes from PS II to PS I (Section 3); fluorescence quenching q_J due to photoinhibition of PS II components under high light (q_E, q_T, q_J).

All of these types of quenching lead to a decrease in F_M under conditions, when Q_A still remains in the reduced state, and namely because of this they are called nonphotochemical q_N.

The drop of F_V occurs due to the decrease in F_M, whereas the level of the initial fluorescence F_0 does not change much. At later stages, relations between the reduction of final products at the light stage and their involvement in the dark cycle of CO_2 fixation play their role.

Nonphotochemical quenching q_N is of significant regulatory meaning, changing the activity of PS II under different physiological conditions, in particular preventing PS II from photo-oxidation under high illumination.

Fluorescence quenching q_E component of q_N depends on the energization of thylakoid membranes related to transmembrane difference in concentrations of hydrogen ions ΔpH and generated by electron transfer in ETC (Section 3, Chapter 21). This energy-dependent quenching occurs because of the increased probability of thermal deactivation of the excited state of chlorophyll in the presence of hydrogen ions. Such effects are known in the photochemistry of dyes, where the quantum yield of dye fluorescence depends on pH of the medium.

It is also proposed that protonation of glutamate and aspartate residues of protein subunits CP29 and CP24 in the light-harvesting complex in PS II induces such conformational rearrangements in the peptide moiety that transfer tightly packed layers of antenna chlorophyll, located in these subunits, to the non-fluorescent state. In this case, mainly radiationless quenching of excitation in the reaction center of PS II is enhanced and only variable fluorescence decreases, while the fluorescence yield in open centers (F_0) does not change. Correspondingly, the ratio F_V/F_M and the efficiency of charge separation in the RC decrease. Upon cessation of illumination, the yield of variable fluorescence returns to the initial level for about tens of seconds, i.e. for the time necessary for the proton gradient to relax. It is believed that this mechanism provides quenching of the excess excitation energy accepted by PS II under saturating light intensity.

A great role in energy-depending quenching belongs to the violoxanthin cycle of conversion of carotenoids. In the thylakoid membrane, violoxanthin de-epoxidase (VDE) is localized near the lumen surface. This protein has the maximum activity at acid pH values. It de-epoxidizes violoxanthin and triggers the operation of the violoxanthin cycle with the formation of zeaxanthin from anteroxanthin. Zeaxanthin epoxidase, localized in the outer surface of the thylakoid, is active at neutral pH values and can convert zeaxnthin back into anteroxanthin and then into violoxanthin, thus closing the cycle.

At active electron transport, protons move into the thylakoid and pH inside the lumen decreases. Such acidification plays a dual role. On the one hand, protonation of VDE activates the enzyme and provides its association with the membrane, inducing conversion of the violoxanthin molecule into anteroxanthin. Anteroxanthin, formed at this first stage of the violoxanthin cycle, is also de-epoxidized already with the formation of zeaxanthin. On the other hand, acidification of the intrathylakoid space facilitates protonation of two carboxyl residues of glutamic acid (*glu*-122 and *glu*-226) into the low-molecular protein Psb (22 kDa), which is involved in nonphotochemical quenching and localized on the surface of the PS II complex. At the same time, due to conformational changes in the core of PS II antenna, symmetric sites for binding zeaxanthin-xantophyll to the de-epoxidized β-ring of the terminal group become activated. The conformation is generated, where excitation transfer from singlet chlorophyll into zeaxanthin is facilitated. Since in a zeaxanthin molecule there are two additional double conjugated bonds conmpared to a violoxanthin molecule, the excited singlet level of zeaxanthin lies lower than that of chlorophyll. Because of this, the excitation energy from the antenna chlorophyll is transferred to the carotenoid, rather than to the reaction center. The time of relaxation of excited singlet states of carotenoids is much shorter than that of chlorophyll. Therefore, as a result of energy transfer from chlorophyll to zeaxanthin, fast thermal dissipation of "excess" energy provided by quanta, absorbed by antenna chlorophyll, takes place. Nonphotochemical quenching associated with acidification of the intrathylakoid space is saturated at high light intensities and relaxes after cessation of illumination for several tens of seconds.

There are also other types of thermal deactivation that can occur not only in the antenna, but also in the reaction center itself. P_{680}^{+} generated upon charge separation is an efficient quencher of delocalized excited states of chlorophyll, but upon active photosynthesis it is rapidly reduced by the electron, coming from the water-splitting complex. Disturbances on the donor side of PS II, decelerating donation of electrons to the reaction center, increase the lifetime of P_{680}^{+} that can quench the excited states of chlorophyll. Acidification of the intrathylakoid space induces the removal of Ca^{2+} ions from the water-splitting complex, which drastically decreases the rate of electron donation to the reaction center. Thus, the probability of quenching the excited states of chlorophyll due to the formation of P_{680}^{+} increases.

The probability of thermal dissipation also increases in the presence of oxidized plastoquinone molecules. In PS I, an oxidized P_{700}^{+} molecule can quench the excited states of chlorophyll if electron donation from PS II is inhibited.

The dissipative cycle around PS II with involvement of cytochrome b_{559} can also serve for "discharge" of excess energy and protection of the complex from photoinhibiting under excess illumination. Cytochrome b_{559} can be present in two redox forms: high-potential (about 350 nm) and low-potential (from 0 to 50 mV). In the low-potential form, the plane of a cytochrome b_{559} molecule is parallel to the membrane plane, whereas in the high-potential form it lies at an angle to the membrane plane. This explains a possibility of redox-transitions of one form to the other, including also the participation of the pheophytin carrier.

When electron transport to Q_A is impeded, reduced pheophytin ($Pheo^-$) can reduce low-potential cytochrome, which then converts to the high-potential form. In its turn, reduced high-potential cytochrome easily donates an electron to P^+_{680}. Thus, passing from one form to the other, cytochrome f can be both a donor and an acceptor of electrons, protecting the PS II complex from the donor and acceptor photoinhibition. However, the "molecular switching" mechanism, inducing transition of one form to the other, is not elucidated yet.

Under prolonged action of light with the intensity much higher than the intensity saturating photosynthesis, photoinhibition of PS II reaction centers (quenching q_J) occurs. In this case, as well as at q_E, mostly variable fluorescence $F_V = F_M - F_0$ is suppressed due to F_M decrease, and the quantum efficiency of PS II decreases. In contrast to energized quenching, the yield of variable fluorescence in the dark after the cessation of the intensive illumination increases for about an hour. It has been established that photoinhibition is connected with irreversible inactivation in the PS II reaction center of polypeptide D1 that is the the plastoquinone docking site. Prolonged reduction of photoinhibited centers is conditioned by the requirement to resynthesize this polypeptide. It is proposed that such protective mechanism of excitation quenching decreases the probability of chlorophyll transition to the triplet state, with the subsequent generation of singlet oxygen and with cell destruction (Section 4, Chapter 13).

Measurements of induction curve parameters are performed not only in laboratories, but also in natural conditions, where it becomes possible to determine the state of the photosynthetic apparatus in leaves and phytoplankton. A method for measuring the F_0 and F_M values as well as quenching parameters q_P and q_N is to apply short flashes of saturating light (Fig. 27.23) on the background of steady illumination after passing maximal level F_M on the induction curve. In this case, each flash converts Q_A to a completely reduced state. However, now these flashes cannot return fluorescence to initial maximum level F_M, because nonphotochemical quenching has already developed in PS II under the background stead illumination. After each saturating flash, lower values F'_M are detected correpoding to photochemical quenching q_P. The difference between values F_M of the maximum initial peak in the induction curve and F'_M indicates the degree of nonphotochemical quenching q_N developed in the system. Hence F_V (initial variable fluorescence) decreases to F'_V:

$$F'_V = F'_M - F_0 = F_V - q_N F_V.$$

By determining initial values $F_V = F_M - F_0$ and q_N at different times as well as value $F'_V = F'_M - F_0$ it is also possible to find quenching coefficients:

$$q_P = \frac{F'_M - F_0}{F_V} \quad \text{or} \quad q_N = \frac{F_V - F'_V}{F_V}. \tag{27.23}$$

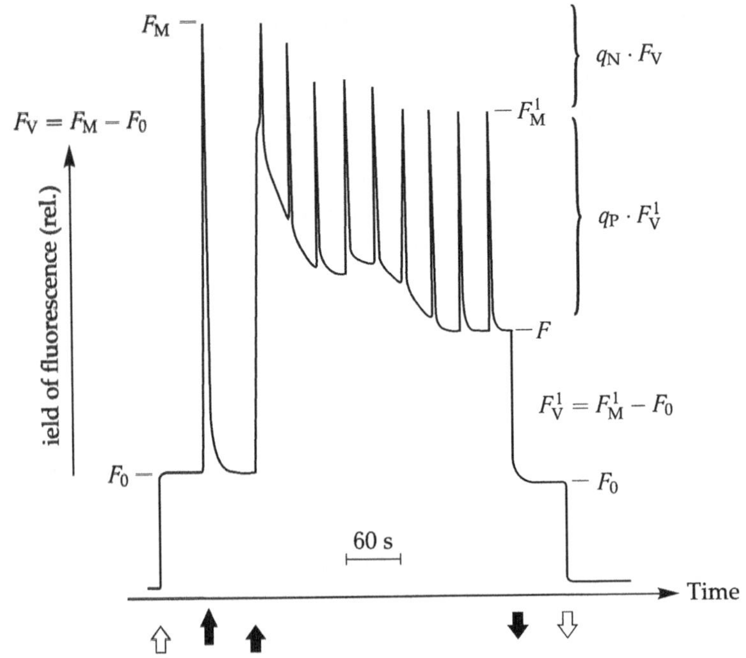

Figure 27.23. Kinetics of fluorescence induction changed upon the action of exciting light flashes on the background of steady illumination (reproduced from C. Büchel and C. Wilhelm, 1993).

Open-headed arrows show switching ON and OFF of the measuring light; short arrows indicate correspondingly the beginning and termination of the steady light; the long arrow corresponds to the first short light flash of saturating intensity.

They determine the efficiency of both the consumption of light energy in electron transport (q_P) and the formation of the electrochemical transmembrane potential (q_N) under real illumination conditions.

Investigations demonstrated that parameters of chlorophyll fluorescence (F_0, F_M, F_V, q_N, q_P) provide important information both on the number of algae cells (F_0), their state, sensitivity to the action of high light intensities, the presence of toxins in the habitat, dependence of the rate of growth on the degree of supply of mineral nutrition elements.

The fluorescence methods and tools developed on this basis provide in real time important information on the abundance of phytoplankton, efficiency of photosynthesis and distribution of these characteristics over the depth in different natural basins. In some cases, this permits to forecast in advance the dynamics of algae populations and changes in their states long before the corresponding morphological manifestation (A. B. Rubin, S. I. Pogosyan and D. N. Matorin, 1995–2007).

Delayed Fluorescence. Prompt fluorescence is conditioned by the excitation energy losses accompanying its migration in photosynthetic reaction centers. However, stabilization of primary separated charges in reaction centers is not absolutely perfect either, i.e. separated charges can recombine. This leads to regeneration of the singlet excited state in the reaction center and consequently to possible radiative energy loss of this secondary excitation in the form of the so-called delayed chlorophyll fluorescence. The decay time of this fluorescence can many orders of magnitude exceed the intrinsic lifetime of the singlet excited state of a chlorophyll molecule.

Notwithstanding such a great difference in the lifetimes, in both cases light emission is generated upon transition from the singlet excited state to the ground chlorophyll state. Thus, both types of fluorescence reflect the dynamics of the excited state generation and decay in the same pool of pigments in photosystem II. Delayed fluorescence (RF) provides important information on the mechanisms of electron transfer in primary processes of photosynthesis. It was discovered by B. L. Streler and V. Arnold (1951), when they studied photophosphorylation, later found practically in all photosynthetic organisms. Strict correlation between DF and the RC activity is demonstrated as suppression of DF upon inhibition of the RC. In mutant strains deprived of RC, there is no delayed fluorescence at all. DF is totally absent in LHPPC preparations and clearly expressed in particles of PS II (LHPPC + PPC II) with active RCs. The coincidence of a DF spectrum with the spectrum of chlorophyll and bacteriochlorophyll fluorescence demonstrates that DF originates from the first singlet excited state of these pigments. Excitation spectra of prompt fluorescence and DF also coincide and both are similar to the action spectrum of photosynthesis. It is obvious that DF occurs as a result of backward recombination of the primary photo-oxidized and photoreduced products of initial stages of photosynthesis in RCs:

$$D_1P^*\text{Pheo} \rightleftarrows D_1P^+\text{Pheo}^-. \tag{27.24}$$

According to the recombination hypothesis, the intensity of afterglow L of DF is proportional to the number of RCs in state $[P^+Q_A^-]$. The L value is dependent on the rate constant of radiative recombination of charges, leading to the formation of P^*, and on the probability of radiation deactivation of the singlet excited state P^*, formed as a result of recombination. □

The intensity of afterglow L is proportional to the probability of the state $P^+Q_A^-$:

$$L = Nk^*\varphi[P^+Q_A^-], \tag{27.25}$$

where N is the concentration of RCs, φ is the quantum yield of radiative deactivation of state P^*, square brackets refers to the probability of the given state, and k^* is the recombination constant. At low intensity of exciting light, all RCs are in an open state, when the electron outflow from Q^- is active. However, in the presence of diuron even at low intensity of exciting light, the centers pass to a closed state and, as the electron outflow from Q^- has already ceased, the intensity of DF becomes proportional to the total number of RCs.

As in the case of variable fluorescence, the process affecting the yield of DF is the energization of the photosynthetic membrane accompanied by generation of the proton gradient. But the mechanism of its influence here is different. The difference between the levels of excited chlorophyll P^* and primary reduced photoproducts is about 0.06 eV. It corresponds to the activation energy of the backward recombination processes and explains why the quantum yield of DF is very low and does not exceed 10^{-5}–10^{-7}. However, the activation energy can be decreased due to the electrostatic field, directed parallel to the generated dipole moment $[P^+Q^-]$. Since recombining charges P^+ and Q^- are located on the opposite sides of the membrane, the difference in the electric potentials Ψ_M across the membrane changes the value of the energy barrier for DF. In this case equation (27.25) reads as follows:

$$L = Nk_p^* \varphi [P^+Q^-] \exp\left(-\frac{\Delta E - F\Psi_M}{RT}\right), \qquad (27.26)$$

where ΔE is the activation energy in the absence of the field, and F is the Faraday number.

According to the above equation, an increase in Ψ_M increases the intensity of DF. This effect is really observed in experiments where diffusion difference in electric potentials on the membrane is obtained by addition of salts or in the presence of, for example, valinomycin that selectively enhances the permeability of the membrane for K^+. It has been noted that an artificially created transmembrane proton gradient ΔpH induces the burst of luminescence. The stimulating effect of ΔpH on DF is determined by a change in the surface electric potential on the outer surface of the thylakoid upon acidification of the medium. Similar to the electric component Ψ_M of the electrochemical potential, gradient pH values can be included in equation (27.26), taking into account that the recombination donor and acceptor are involved in proton equilibrium and change the activation energy of the recombination:

$$L = Nk_p^* \varphi [P^+Q_A^-] \exp\left(-\frac{\Delta E - (F\Psi_M - \lg[RT(\Delta \text{pH})])}{k_B T}\right). \qquad (27.27)$$

The expression in square brackets in the above equation is related to the proton-driving force. Therefore, after calibration we can use the intensity of DF for determining the value of this force, which is about 100 mV. □

A decrease in temperature to $-30 \ldots -40$ °C inhibits the oxygen evolution and direct electron-transport reactions, as a result of which the backward recombination becomes dominating. However, long-living components, resulting from the participation in DF of the reduced products remote from RCs, disappear upon cooling. At rather low temperatures, only two fluorescence components are observed (1–2 and 30 ms), the first one corresponding to the reverse reaction between P_{680}^+ and Q^-. Temperature growth activates RF. This permits estimations of the activation energy for different components of fluorescence (from 0.15 to 0.9 eV). The products formed in the light — precursors of DF — can be stabilized, if the illuminated sample is cooled to a low temperature under illumination. The subsequent heating of the sample in the dark induces the appearance of fluorescence that was called thermoluminescence. Apparently, photoproducts formed in the light and stabilized upon cooling of the sample, are then activated upon heating in the dark and recombine with chlorophyll, leading to generation of P_{680}^* and thus to delayed fluorescence. The curve of the dependence of the thermoluminescence intensity on the temperature of the object following its heating in the dark has several peaks, observed at definite characteristic temperatures. The number and position of the thermoluminescence peaks reflects different stages of energy stabilization at the acceptor side of the ETC after PS II. However, at the same time thermoluminescence depends also on the functioning of the oxygen-evolving complex. The matter is that luminescence is a result of P_{680}^* generation upon recombination of the electron and P_{680}^+ and depends on the concentration of oxidized P_{680}. In the dark, some small oxidation degree of P_{680} takes place due to reversible electron exchange in the chain from OEC to P_{680} (Section 7). These reactions proceed also after illumination and heating of the sample in the dark and are accompanied by P_{680}^+ formation and therefore recombinantn thermoluminescence.

28

Electron-Conformational Interactions in Primary Processes of Photosynthesis

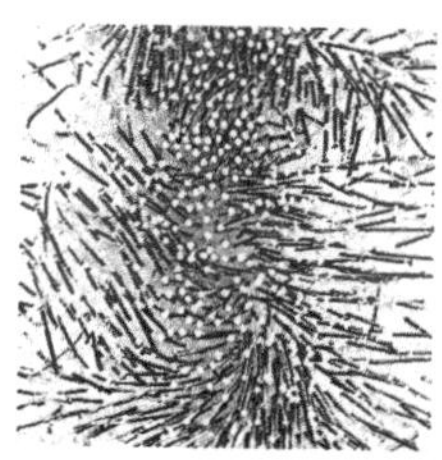

One of the central problems in biophysics is to elucidate mechanisms of primary charge separation in a radical pair (RP) and electron transport between carriers in primary processes of photosynthesis. In its significance, the problem is beyond the scope of photosynthesis, and its solution is of principal importance for understanding intrinsic mechanisms of functioning of biological macromolecules.

A RC of photosynthesis can be compared to a special photoactive enzyme, where prosthetic groups of RC carriers play the role of components of the active center. In place of a typical low-molecular enzyme substrate here are light quanta, the absorption of which "triggers" the functioning of the RC of photosynthesis.

In standard enzymes (Chapter 26), initial changes in the electronic state of the active center take very short periods of time. They give an "impetus" to conformational rearrangements in the enzyme protein globule that vary in scale and time. The experimental data accumulated recently show explicitly that such a principle of close conjugation of electron transfers and conformational rearrangements in the protein is also valid for the RC of photosynthesis.

In this chapter, we will consider mechanisms of electron transfers connected with vibrational and conformational degrees of freedom, responsible for high rates and the efficiency of the initial charge separation and electron transfer in the RC of photosynthesis in higher plants and bacteria. We will rely on the schemes of primary processes described in the previous chapter as well as on experimental data on the kinetics of electron transfer in the RC.

28.1 Studies of Superfast Processes in Reaction Centers of Photosynthesis

In previous chapters, we have considered electron transfer processes in reaction centers of photosynthesis.

Figure 28.1 shows a generalized scheme of electron transport in RCs of bacterial photosynthesis and photosystems I and II. As seen, the three types of RCs are definitely similar in the kinetics of basic stages of the electron transfer. The fast initial charge separation ($\tau \lesssim 3$ ps) and reduction of the primary porphyrin acceptor (Bpheo, Chl, Pheo) by the primary photoactive donor (P_{870}, P_{700}, P_{680}) is accompanied by the subsequent slower ($\tau \sim 150$ ps) charge transfer to secondary quinone acceptors Q_A (ubiquinone, phylloquinone, plastoquinone), and then after $\tau \sim 100$ μs to Q_B (ubiquinone, plastoquinone) or iron-sulfur centers (PS II).

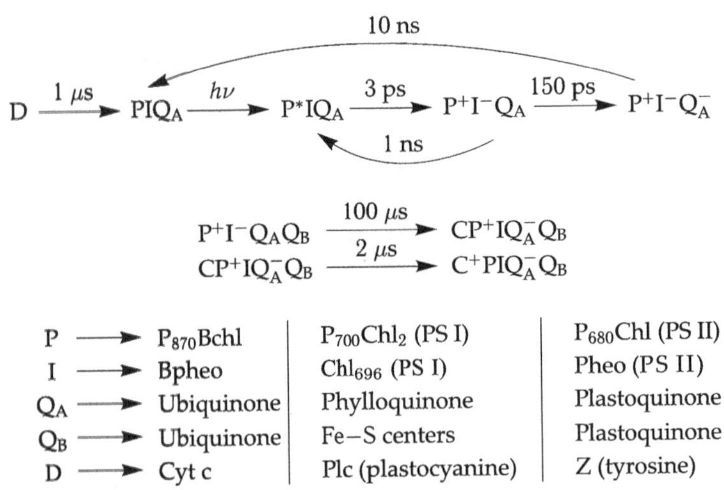

Figure 28.1. Generalized scheme of electron transport in RCs of bacterial photosynthesis and photosystems I and II.

As seen from Fig. 28.1, oxidized photoactive donor P is reduced from secondary donor D that may be cytochrome c, plastoquinone or tyrosine (depending on state S of the water-oxidation complex). Inverse reactions of electron transfer proceed much slower than direct reactions thus providing high efficiency of the total electron flow.

The redox states of intermediate carriers, generated upon reception and donation of electrons, have different absorption spectra with characteristic maxima. Therefore, at definite wavelengths, the kinetics of an absorption change corresponds to the kinetics of redox conversion of a carrier, i.e. reflects reactions of electron "arrival" and "departure".

Figure 28.2, *a* shows the absorption spectrum of purple bacteria (*Rb. sphaeroides*) with designated absorption maxima of separate pigment forms. As seen, Bpheo has maxima at 545 nm and 750 nm, monomer Bchl at 600 nm and 800 nm, and dimer $Bchl_2$ of the photoactive pigment has a maximum at 890 nm (P_{890}). Figure 28.2, *b* shows the difference absorption spectrum obtained as a result of formation of reduced Bpheo under illumination. It represents the difference in the absorption spectra obtained in the light and in the dark ("light minus dark") (cf. Fig. 27.10).

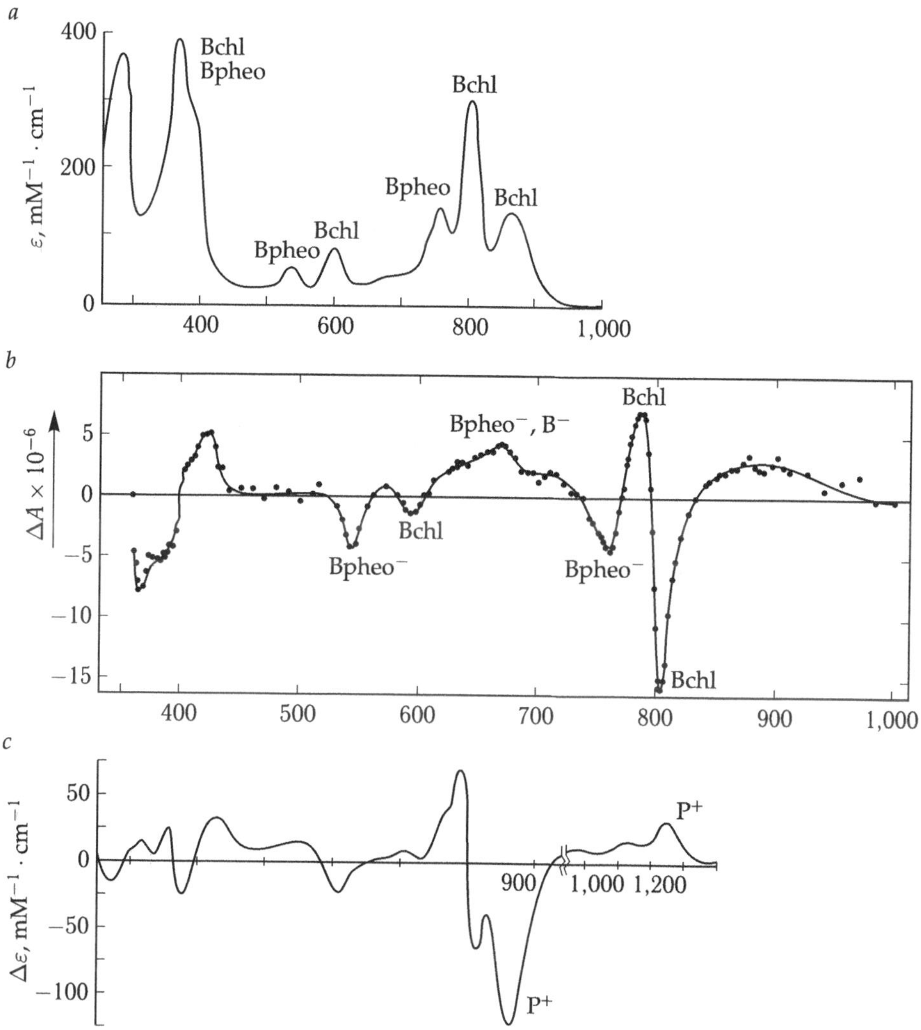

Figure 28.2. Absorption spectrum (*a*) and difference absorption spectra (*b*, *c*) of bacterial reaction centers (see explanations in the text) (*a* and *c*: reproduced from R. Clayton 1980; *b*: reproduced from V. A. Shuvalov and V. V. Klimov, 1976).

The spectrum has negative maxima (545 nm and 750 nm) corresponding to the emergence of anion radical Bpheo as a result of disappearance of neutral Bpheo (Bpheo $\rightarrow$ $Bpheo^{\overline{\cdot}}$). It is also seen that $Bpheo^{\overline{\cdot}}$ is characterized also by the appearance of a positive maximum at 665–675 nm. Moreover, due to an electrochromatic effect, the appearance of a negative charge on Bpheo leads to a displacement of the absorption band of monomer Bchl to the short-wave region, revealed in characteristic bands in the difference spectrum at 600 nm and 800 nm.

The kinetics of absorption bleaching of main maximum *P* will correspond to the formation of cation radical *P* (P^+). At the same time, it is necessary to know also the dynamics of radical Bpheo formation. To this end, absorption at 545 nm and 675 nm wavelengths is measured in the absorption spectrum of anion radical $Bpheo^{\overline{\cdot}}$. After that, the two kinetic curves are compared: bleaching of the long-wave absorption band of P^+ and the change in absorption in the Bpheo spectrum (Bpheo$^{\dot{-}}$). The coincidence of characteristic times of both spectral changes, i.e. the coincidence in time of the emergence of P^+ with the formation of Bpheo$^{\dot{-}}$ will provide grounds for conclusions on direct interaction of carriers P and Bpheo in electron transfer P^*Bpheo $\rightarrow$ P^+Bpheo$^{\dot{-}}$.

Depletion of the ground level of absorption due to transition $S_0 \rightarrow S_1$ and formation of P^*_{890} is coupled to the appearance of new absorption bands characteristic of the excited state and transitions $S_1 \rightarrow S_2$ to higher excited singlet levels. This new photoinduced absorption is generated concurrently with disappearance (photobleaching) of the basic absorption band of P. It exists only during the time of maintaining excited state S_1 in P^*. It disappears with time equal to the time of deactivation of the singlet excited state or quenching of the fluorescence of photoactive dimer P^*.

As the main contribution to quenching the excitation of P^* is made by processes of electron transfer from P to the following acceptor, the time of decay of the fluorescence of P^* should coincide with the time of electron escape from dimer Bchl in the RC. In addition to photobleaching of the main absorption band, observed simultaneously with transition $S_0 \rightarrow S_1$ to P, the subsequent electron transfer from P^* occurs. It induces the formation of cation radical $P^{\dot{+}}$ and, correspondingly, new positive changes in the absorption spectrum at 1,200–1,350 nm. The rise time of these changes is equal to the time of electron transfer from P^* to the adjacent acceptor. Recovery of the absorption of the main band takes place later ($\tau \sim 2\ \mu s$) from the secondary electron donor D.

It is obvious that the highest spectral effects are achieved, when excitation of RCs occurs in the range of the main absorption band of P. However, this complicates experimental recording of the effect of photobleaching of the maximum of P_{890} due to the overlapping of wave lengths of the acting laser flash and the scanning measuring beam.

Therefore, in some cases the method of stimulated, or induced, luminescence is used. If laser excited molecules in state S_1 are illuminated by measuring light, usually absorbed by this molecule in ground state S_0, then due to resonance interaction, transitions from excited level S_1 to ground level S_0 will be induced, accompanied by fluorescence emission. The intensity of this induced fluorescence is evidently dependent on the occupation of excited state S_1 at the moment of application of measuring light. As the number of excited molecules in state S_1 decreases due to their deactivation in RCs, the intensity of induced fluorescence declines. Thus, the recording after a laser flash the time-dependent change in the intensity of fluorescence induced by probing light with wavelength in the main absorption band of P makes it possible to assess the kinetics of deactivation of excitation of P^* in the RC, connected with the electron leaving P.

In real measurements, the emergence of induced fluorescence is related to an apparent decrease in the absorption of RCs of measuring light that is recorded as additional photobleaching of the main absorption band.

Let us consider experimental kinetic curves of the above described spectral changes accompanying the electron transfer in RCs. Figure 28.3, *a* shows the recording of the decay kinetics spontaneous fluorescence of P^* at $\lambda = 940$ nm in the RC upon light excitation $\lambda = 850$ nm. As seen, the decay fluorescence curve, reflecting the kinetics of electron release from P^*, has two components with times $\tau_1 = 2.7$ ps and $\tau_2 = 12.1$ ps. The short component of spontaneous fluorescence decay coincides with the monoexponential curve of induced luminescence decay (Fig. 28.3, *b*).

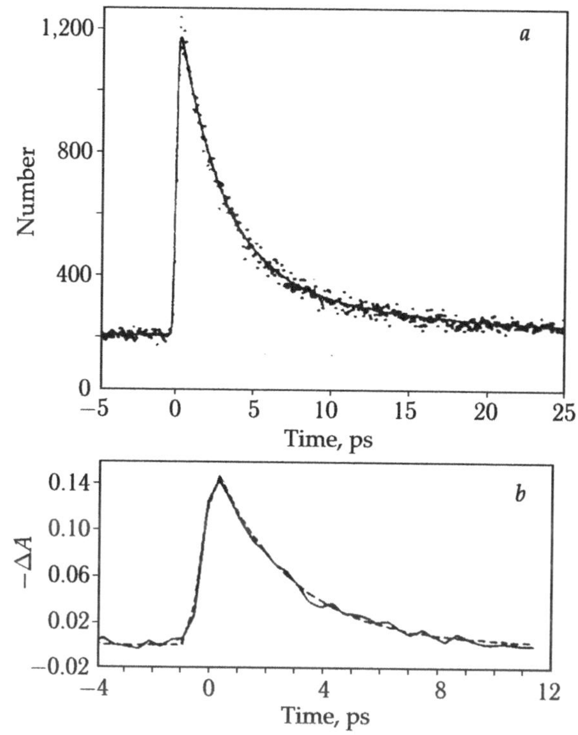

Figure 28.3. *a*, Decay of spontaneous fluorescence from P^* in *Rb. sphaeroides R26* at $\lambda = 940$ nm upon excitation by light at $\lambda = 850$ nm (reproduced from M. Du et al., 1992); *b*, decay of induced fluorescence from P^* in *Rb. sphaeroides R26* was recorded at $\lambda = 926$ nm, excitation at $\lambda = 870$ nm.

In his early studies, J. Breton (1986) obtained data on the kinetics of electron transfer to Bpheo using the methods of femtosecond spectroscopy. In Fig. 28.4, the kinetic curve (*a*) of induced emission of P^* at $\lambda = 1{,}050$ nm is characterized by monoexponential decay with time $\tau = 2.8$ ps. This is compatible with the curve (*b*) of emergence of cation radical P^+ (an increase in absorption at 1,310 nm) and the curve of reduced Bpheo (*c*) (bleaching at $\lambda = 545$ nm). Attention is drawn to the "instantaneous" ($t < 50$ fs) increase in absorption on curves (*b*) and (*c*) immediately after the laser flash action. These changes are ascribed to the absorption in the RC in state P^* that is formed directly upon light absorption. As seen, the times of electron leaving P^* and reducing Bpheo coincide. Hence the authors concluded that monomer Bchl is not a chemical acceptor of the electron on the pathway from P^* to Bpheo, but plays the role of a virtual carrier (a "superexchange" transfer mechanism, Section 5 in Chapter 27).

The possibility to record fast changes in absorption ($\tau < 50$ fs) is used for direct studies of migration of excitation energy between pigments of the light-harvesting antenna. Let us consider, for example, energy transfer in the LHC II of purple bacteria $Bchl_{800} \rightarrow Bchl_{850}$. Upon excitation and formation of $Bchl^*_{800}$, the ground level of this form is depleted and, as a result, absorption (bleaching) of its band at 800 nm weakens. As far as excitation migrates to $Bchl_{800}$, the $Bchl_{800}$ molecules return to the ground level and the absorption at 800 nm enhances. Hence, in particular, the time of migration of $Bchl^*_{800} \rightarrow Bchl_{850}$ was determined ($\tau \sim 1$–2 ps).

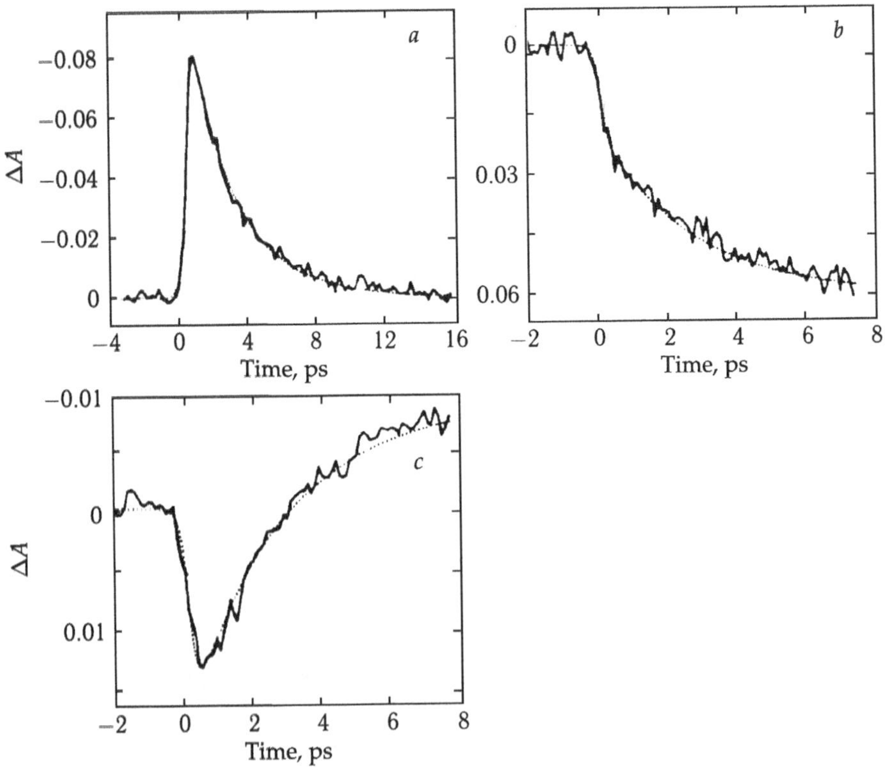

Figure 28.4. *a*, Kinetics of induced fluorescence at 1,050 nm; *b*, kinetics of generation of P^+ at 1,310 nm; *c*, kinetics of generation of $Bpheo^-$ (reproduced from J. Breton et al., 1986).

The use of methods of femtosecond spectroscopy opens principally new perspectives for studying nuclear motions in molecules that are in the electron excited state. In Chapter 10, we have seen that upon absorption and emission of light quantum by a molecule, electron transitions are accompanied by a change in the kinetic energy of nuclei. The excess of vibrational energy formed upon the electron transfer, dissipates swiftly into heat upon nuclear transition to lower vibrational levels (for 1 to 5 ps).

Under stationary conditions, nuclei perform vibrations with the amplitude corresponding to the position of the nuclear vibrational level in the diagram shown in Fig. 10.2. Most of the time, the nuclei are at extreme points of rotation, their coordinates corresponding to the point of the intersection of the vibrational level line with the curve of the molecule electron energy $U^*(R)$. For the time ($\tau \sim 10^{-12}$–$5 \cdot 10^{-12}$ s) when the nuclei stay in these points, fast light emission (10^{-15}–10^{-16} s) occurs as nuclei remain practically immobile in accord with the Frank–Condon principle. The time of retention of electron excitation in molecule S_1 prior to light quantum emission is much higher ($\tau \sim$ 5–10 ns) than the time of nuclear relaxation (1–5 ps).

By exciting a molecule by femtosecond ($\tau < 50$ fs) flashes, it is possible to study the kinetics of the subsequent emission of fluorescence in a subpicosecond range. The kinetics of fluorescence decay should evidently depend on the nuclear oscillations. This can be observed using the phenomenon of induced luminescence at corresponding time resolution. To accomplish it, induced resonance light emission is evoked by additionally illuminating the molecules, excited by a laser flash, with measuring light

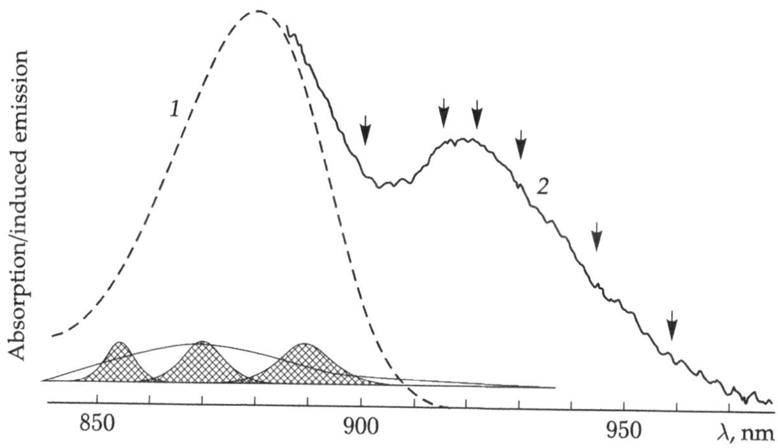

Figure 28.5. Absorption spectrum of the reaction center (*1*) and induced emission spectrum (*2*). Recording is done at wavelengths indicated by arrows (reproduced from M. H. Vos et al., 1993).

in the absorption band. The intensity of this induced emission will be measured in time in accordance with the frequency of nuclear oscillations when nuclei change their position periodically between the extreme turning points. In fact, Martin (1993) discovered oscillations in the intensity of the induced emission recorded at different wavelengths in a fluorescence spectrum (Fig. 28.5).

Figure 28.6 shows the diagram of shuttling of the wave nuclear packet between extreme points on the potential curve of singlet excited state of P^*. At extreme turning points, fluorescence is emitted. This also occurs when nuclei are moving over the potential well, though with a lower intensity and doubled frequency.

Figure 28.7 represents the recording of vibrations of the induced emission intensity at different wavelengths. The Fourier analysis demonstrated that the two main components of induced emission are modulated by 15 and 77 cm^{-1} ($\sim$ 2 ps and $\sim$ 0.4 ps).

Figure 28.6 shows also the intersection of potential curves of states P^* and P^+. The intersection point corresponds to the nuclear coordinate, where the electron is disrupted from P and P^+ is formed. The characteristic time of nuclear oscillations in P of the reaction center coincides with characteristic times of the primary act of electron transfer $P^*I \to P^+I^-$ ($\tau \sim$ 1–3 ps).

Consequently, the nuclear motion is inseparably connected with electron transfers in RCs of photosynthesis that has great impact on superfast mechanisms of electron transport (see Chapter 25 and below in this chapter). As will be seen lower, here the standard approximation used for separation of electronic and relaxation processes is not valid ($\tau_{rel} \ll \tau_{el}$) (Chapter 25).

Sequence of Electron Transfer Acts in Photosynthetic RCs in the Femto-second Time Range. Let us consider the energy scheme describing the reaction of charge separation in RCs between P^* and Bchl, represented as potential surfaces (Fig. 28.8). Here PBchl, P^*Bchl and P^+Bchl$^-$ are potential surfaces of the ground and excited states of the donor (P) and acceptor (Bchl), as well as states with separated charges, respectively. P^+(Bchl$^-$)* is the energy of the excited level with separated charges that is formed as a result of transition (P^+Bchl$^-$) $\xrightarrow{h\nu}$ P^+(Bchl$^-$)*, induced by light excitation in the absorption band of Bchl$^-$ (1,020 nm).

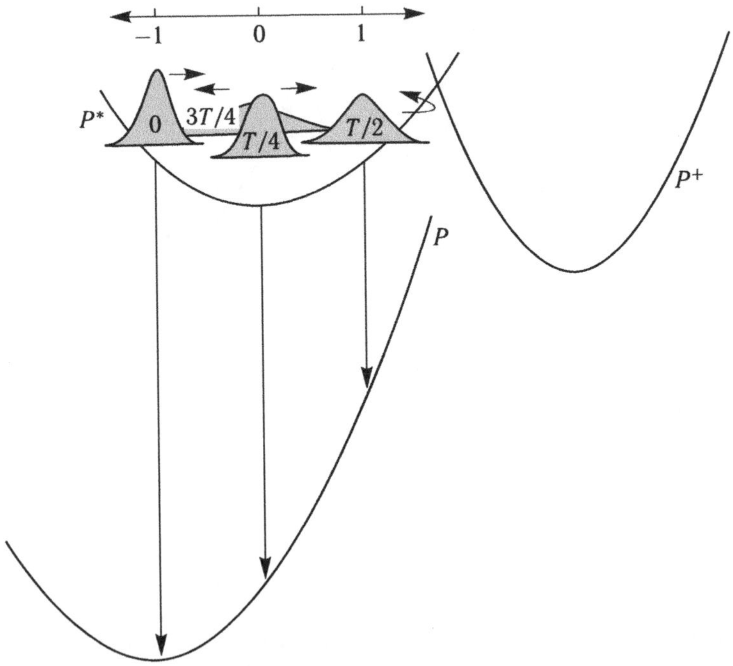

Figure 28.6. One-dimensional diagram of shuttling of nuclei (by the change in the wave packet) over the potential surface of singlet excited state of P^* (reproduced from M. H. Vos et al., 1993).

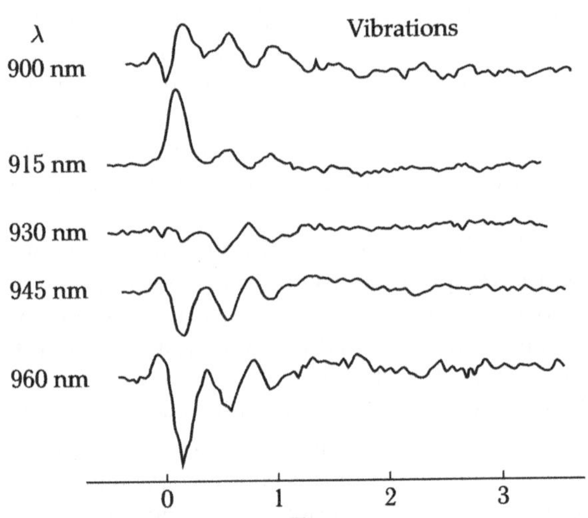

Figure 28.7. Oscillations of intensity of induced emission of P^* recorded at different wavelengths (reproduced from M. H. Vos et al., 1993).

The electronic excitation of the molecule changes the distribution of its electron density, and consequently the force acting from the side of the excited molecule on ions and charged groups of the environment. This effect induces changes in the equilibrium position of ions and charged groups in the environment. As a result, the position of the potential energy of the excited molecule is displaced relative to the

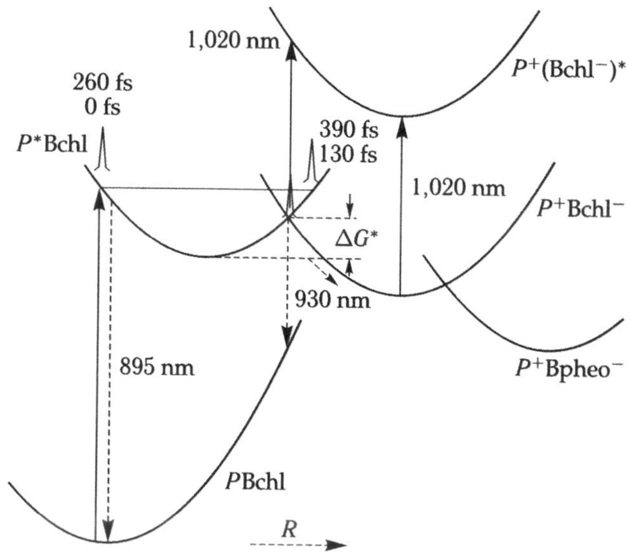

Figure 28.8. Diagram of propagation of excitation over the potential surface of state of P^*Bchl and its emission on opposite sides of the potential surface to the surface of the ground state of PBchl.

The intersection of potential surfaces of P^*Bchl and P^+Bchl$^-$ is also shown. The absorption band of Bchl$^-$ at 1,020 nm as well as the absorption band of surfaces of P^+Bchl$^-$ and P^+Bpheo$^-$ are observed in the intersection (reproduced from A. G. Yakovlev et al., 2002).

surface of the potential energy of an unexcited molecule along the coordinate of reaction R_0. The magnitude of this displacement is a measure of the force of the electron vibrational bond.

Upon excitation of RC preparations by femtosecond light pulses ($\sim$ 20 fs) to the absorption band of P (895 nm), electrons pass to the left-hand branch of the potential surface of P^* (Fig. 28.8). It is important to note that all P molecules in the RC suspension that have absorbed the light quantum, are excited practically instantaneously (for 20 fs), and consequently at the initial moment of reading all excited molecules are in the same position, corresponding to the left-hand branch of the potential energy of P^*. During transition $P^* \to P$ the wavelength of the emitted light is determined by the energy between the "emitting" point on the surface of P^* and the corresponding point on the surface of the ground state of P, connected by vertical lines. Vibrating nuclei stay most time in the two points on the potential surface that are located on the opposite sides of the curve for P^*, where its potential energy is maximal, and the kinetic energy is minimal. Accordingly, the probability of excitation to remain in the extreme points of the potential surface will be maximal, and therefore fluorescence emission will proceed mainly from these points. Thus, induced luminescence will occur at 895 and 935 nm, corresponding to the extreme points on the potential surface of P^*Bchl (Fig. 28.8). V. A. Shuvalov demonstrated that the oscillation period $T = 2\pi/\Delta\omega$ is $\sim$ 260 fs, where $\Delta\omega$ is the cyclic frequency of nuclear vibrations. Let us pass from cyclic frequency $\Delta\omega$ to linear frequency $\Delta\nu$ with dimensionality in [cm^{-1}]. We will use formula $\Delta\omega = 2\pi c\Delta\nu$, where c is the velocity of light. Then, if the oscillation period of excitation on the surface of P^* is 260 fs, the linear part of these oscillations is $\Delta\nu \approx 130$ cm^{-1}. The vibration with this frequency ($\sim$ 125 cm^{-1})

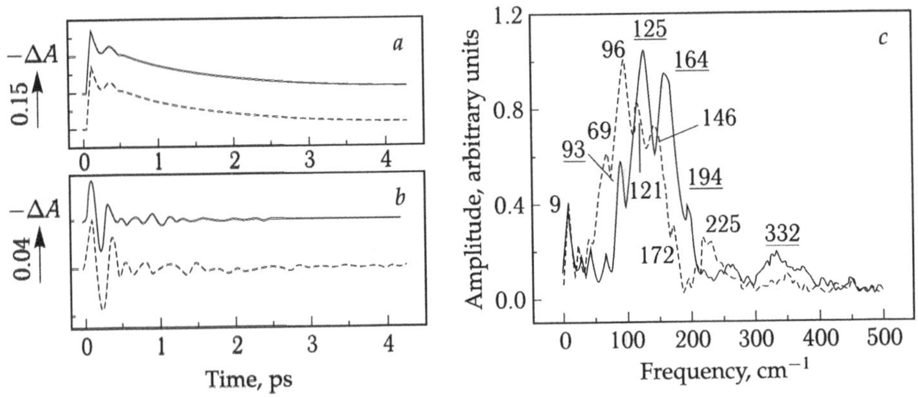

Figure 28.9. Femtosecond kinetics (*a*), their oscillatory portions (*b*) and the spectrum of the Fourier transform of an oscillatory portion (*c*) for the band of induced emission at 935 nm in native RCs from *Rb. spaeroides* (carotenoid-free strain R-26) recorded at 90 K.

Glycerol-H_2O buffer (solid curve) and glycerol-D_2O buffer (dashed curve) were used. Underlined numerals in (*c*) designate frequencies of the Fourier transform spectrum for RCs in glycerol-H_2O buffer; non-underlines numerals designate analogous frequencies for RCs in grycerol-D_2O buffer (reproduced from A. G. Yakovlev et al., 2002).

is the most intensive in the spectrum of the oscillatory moiety of the kinetics of induced emission at 935 nm, reflecting decay of the excited state P^* (Fig. 28.9). It should be noted that changes in fluorescence intensity at 895 and 935 nm will be in opposite phases that was recorded in the kinetic measurements. The complete spectrum of oscillations in the kinetics of induced emission (935 nm) (Fig. 28.9), obtained by the Fourier transform, is concentrated in the range from 10 to 400 cm^{-1} and is represented as oscillations located approximately at 10, 32, 69, 92, 125, 163 and 332 cm^{-1}. The existence of such oscillations is also corroborated in independent hole burning experiments (see below) and resonance combination scattering.

The nature of nuclei, involved in generation of these oscillations, is not utterly clear. However, it is known that the frequency of rotational vibrations of a water molecule is 32 cm^{-1}. It is noteworthy that the set of frequencies in the oscillation spectrum corresponds to the 2-nd (69), 3-rd (92) etc. harmonics of the rotational mode with a water molecule frequency of 32 cm^{-1}. According to structural data (Brookhaven Protein Data Bank, file 1PCR), a special pair is connected to the intermediate acceptor of the active RC branch — a Bchl molecule by the following sequence of atoms: Mg (P)–N–C–N (His)–H–O–H (water)–O = Bchl (Fig. 28.10). It is possible that namely this chain of atoms forms the pathway for the transfer of the electron from P^* to Bchl. A substantial support for the involvement of a water molecule in the electron transfer from P^* to Bchl is that isotopic substitution $H_2O \rightarrow D_2O$ induces a 1.3-fold decrease in all oscillation frequencies.

As shown by femtosecond spectroscopy data, the mechanism of charge separation between P^* and Bchl in RCs of purple bacteria is as follow. Under dark conditions the proposed bridge

$$\text{Mg(P)–N–C–N (His)–H–O–H (water)–O} = \text{Bchl}$$

is in the state optimal for electron transfer. It is believed that the electron density can reversibly shift from P^* via this bridge to Bchl, when excitation of the $\sim$ 260 fs period

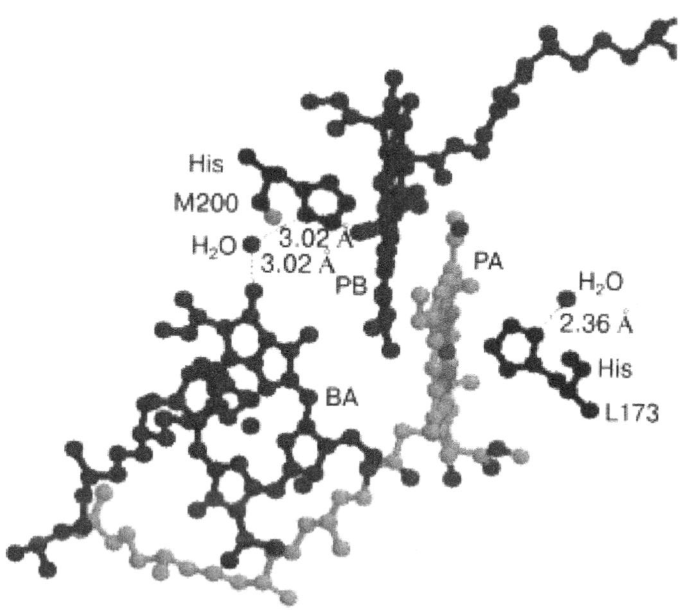

Figure 28.10. View of a special pair of bacteriochlorophyll molecules *PA* and *PB* and monomer bacteriochlorophyll Bchl in RCs of *Rb. spaeroides* (Brookhaven Protein Data Bank, 1 PCR). Molecule *PB* is linked to the Bchl molecule by a chain of polar groups $Mg(P_B)-N-C-N$ (His_{M200})$-H-O-H$ (water)$-O=$ Bchl (A. G. Yakovlev et al., 2002).

is in the site of intersection of the P^* and Bchl surfaces, thus generating the P^+Bchl$^-$ product of charge separation. The electron movement on the surface of P^+ is accompanied by its transfer to Bpheo, when the electron is on the right-hand branch of the potential surface of P^+Bchl$^-$ near its intersection point with the surface of P^+Bpheo$^-$.

The Hole-burning Method has been developed for studying deactivation processes of excited states P^* in RCs. Let us consider briefly this method of high time resolution. At the beginning, the RC samples are irradiated at low temperature (1.5–10 K) by a "burning" laser light. It excites P ($P \rightarrow P^*$) with the subsequent fast deactivation of P^* and charge separation, $P^* \rightarrow P^+$Bpheo$^-$. At low temperatures, radical pair P^+Bpheo$^-$ is stabilized in this state. This makes it possible to study the bleaching spectrum of the main absorption band of P that is a result of depletion of ground level S_0 of P molecules. Evidently P molecules that are in corresponding electron vibrational (vibronic) states and involved in the primary charge separation in RCs, make their contribution to the spectrum of the "burnt out" spectral hole.

During lifetime τ of the excited state, complete thermal equilibrium is achieved due to relaxation processes in the vibrational subsystem, corresponding to the temperature at which the experiment was conducted. In complex molecules, an efficient exchange of vibrational energy takes place, and thermal equilibrium is promptly established at the frequency of the 0–0-transition (ν_{00}).

For typical molecules, placed in a crystalline and glass-like matrix at cryogenic temperatures, at least one narrow hole burning line is observed in the range of the 0–0-transition. Indeed, for porphyrins in glasses or for chlorophyll, introduced in the structure of myoglobin, a narrow hole of 1 cm^{-1} is observed. However, as contrasted to this, in RCs it is not always that narrow holes in the range of the 0–0-transition are observed in the absorption band of P. Instead, there appear wide holes with the width almost equal to the width of the entire absorption band of P. In some cases, weak and narrow lines of holes in the long-wave part of the absorption band P in RCs were also observed against the background of a wide band (Fig. 28.11).

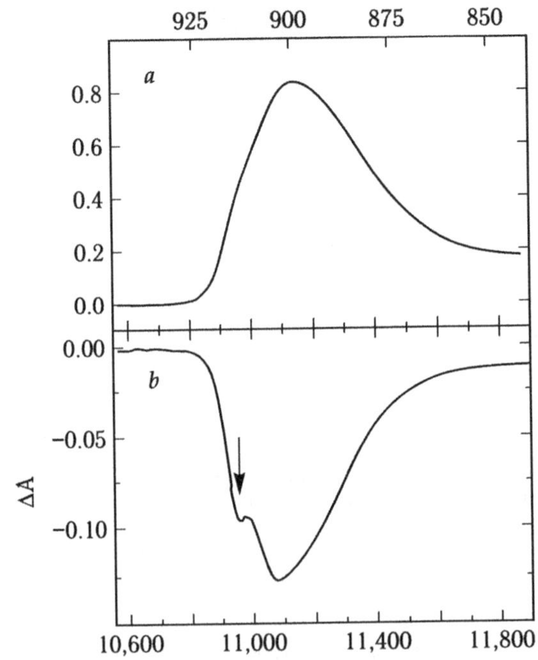

Figure 28.11. Absorption spectrum (*a*) and the spectrum of photochemical hole burning (*b*) for the lower electron transition in RCs of *Rb. sphaeroides* (reproduced from R. Jankowiak and G. J. Small, 1993).

Remember that a homogeneous width of an absorption band, or the width of the energy level Γ of state of P, formed upon light absorption, depends on lifetime τ of the given state by the uncertainty relation

$$\Gamma \cdot \tau \sim \hbar. \tag{28.1}$$

As shown previously (Chapter 10), the level width depends on two processes: deactivation of the electron excitation state P^* (quenching of fluorescence of P^* for time $\tau \sim T_1$) and vibrational relaxation (time T_2) leading to the loss of coherency of the phase of initial vibrational states, generated upon light absorption:

$$\Gamma = 1/2\tau + 1/T_2. \tag{28.2}$$

The value of T_2 is determined by the interaction of the electron excited state of P^* with thermal phonons and is strongly dependent on temperature. For example, at room temperature the width $\Gamma = 1/T_2$ is $\approx kT \approx 200$ cm^{-1}. But at low temperatures, the effect of phase errors on the line width becomes minimal. In the forefront are processes of deactivation of the excited state of P^*. What processes are associated with the narrow and wide lines of holes in the absorption band of RCs?

It was found that the width of narrow lines of holes in RCs corresponds to low-temperature values of the lifetime of the P^* state, i.e. to the time of formation of ion radicals of P^+ and $Bpheo^-$ and of induced emission decay from the level of P^* in the process of electron transfer from P^* to $Bpheo^-$. Hence it seemed that a conclusion could be made on the absence of any processes of fast ($\sim$ 20–200 fs) deactivation of the intrinsic electronic state of P^* prior to the stage of electron detachment. And namely, this preliminary deactivation of P^* could lead to shortening the lifetime of P^* and, consequently, to the observed widening of the general band of hole burning.

At the same time, we have already stated (see above) that during the lifetime of P^*, vibrations of the nuclear wave packet are observed with time $\sim$ 100 fs. This may indicate to the involvement of local modes of the nearest protein surrounding of P in the evolution of the P^* state prior to the electron detachment and transition from P^* to P^+Bpheo^-. In other words, electron vibrational interactions, leading to wide bands of the hole burning process observed experimentally, can be associated with certain fast "structural" changes in RCs upon charge separation (see lower).

Thus, the method of hole burning as well as other methods of studying superfast processes are an important source of information on structural rearrangements accompanying initial stages of electron transfer in RCs.

28.2 Initial Charge Separation in RCs

We have seen that there are evidences for the coincidence of times (1–3 ps) of electron transfer ($P^* \rightarrow Bpheo^-$), fluorescence decay of P^* and value T_1 obtained from the width of the narrow line of the hole. Hence it follows that the initially formed state of P^* should coincide with the $Bchl_2$ dimer state emitting fluorescence. But on the other hand, the simultaneous existence of a wide band of the hole indicates to fast relaxation processes ($\sim$ 20–200 fs) preceding the emission of spontaneous fluorescence from the level of P^*.

This may show that the state of P^* has an admixture of other states, for example, the state with separated charges of P^+Bpheo^-, characterized by large displacements of nuclear configuration relative to the P ground state. It is equivalent to a strong electron vibrational interaction with certain degrees of freedom (see Section 5 in Chapter 25). Thus, the electron transfer involves selective vibrational degrees of freedom in the complicated pigment–protein complex of RC, which interact with electron transfer in the state of P^*.

The relaxation transfer to the state with transfer of the charge of P^+Bpheo^- occurs along the surface of the potential energy of the excited state of P^* (Fig. 28.12) and has a quasiadiabatic character. It is associated with large displacements of the equilibrium nuclear configuration of the ground state. Namely, because of this the narrow holes of 0–0-transitions are poorly seen (cf. Fig. 10.3) against the background of a wide hole. Certainly, a whole number of experimental data, obtained by femtosecond spectroscopy methods in different laboratories, denote rapid relaxation processes in the excited state of P^*. Thus, the maximum bleaching of the main absorption band of P shifts during initial 100–500 fs after formation of P^+. This shift occurs much faster than the subsequent electron transfer.

Absorption changes at 1,000–1,300 cm^{-1}, growing during initial 200 fs period after the flash, were also recorded in the infrared spectral range. Obviously, initial stages of evolution of the excited state of P^* are connected with certain local

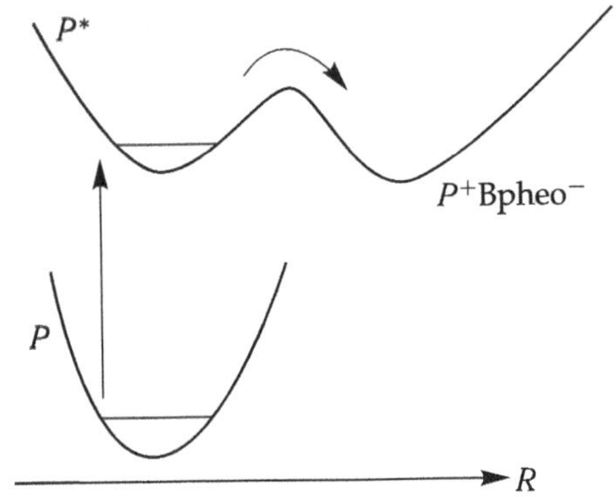

Figure 28.12. Schematic representation of position of potential energy curves of the RC state: P–P^*–P^+Bpheo$^-$, R is the nuclear coordinate.

modes and "microstructural" fast changes in the nuclear configuration of the RC pigment–protein complex. A detailed description of these processes is far from its accomplishment yet. However, we can compare the available data with known principles of molecular spectroscopy.

Remember that in spectroscopy of complex molecules, it is accepted to distinguish relaxation processes with thermal dissipation of energy and electron relaxation connected with certain selective vibrational modes and the change in the excited state during its lifetime (B. S. Neporent, 1970, 1972).

In the first case, thermal relaxation is connected with fast dissipation of the excess vibrational energy, formed on vibrations that took part in transition $P \rightarrow P^*$, and with achieving the equilibrium distribution of all frequencies and amplitudes of normal vibrations. And in the case of electron relaxation, the molecular framework is changed on definite selective degrees of freedom with further evolution of the Franck–Condon electronic state that was initially achieved upon optical transition $P \rightarrow P^*$. The evolution is terminated by generation of a new stationary electronic state self-consistent with the new configuration of the molecule. Electronic relaxation occurs concurrently with the selective vibrational relaxation that includes changes brought about by the break of the twisting of the molecular backbone. As a result, changes may occur in the symmetry, degree of charge transfer, degree of conjugation of π-electron chains of two parts of the molecule, and multiplicity due to enhancement of spin-orbital interactions. Electronic relaxation can lead to smearing of the vibronic structure and emergence of continuous spectra.

It is worth noting that the maximal rate of electronic relaxation may be much higher, than the rate of thermal vibrational relaxation, and take place at times not exceeding 10^{-15}–10^{-13} s.

Effect of Hydrogen Bonds and Low Temperature. In the structure of $Bchl_2$ dimer, of great importance are hydrogen bonds formed between amino acid histidine and carbonyl groups of Bchl molecules in the dimer. One H-bond between His-*L*168 and 2-acetyl group of the first ring in $Bchl_L$ is formed in the wild-type strain of *Rb. sphaeroides*. Methods of genetic engineering make it possible to obtain mutations with a changed number and localization of hydrogen bonds. An increase in the number of hydrogen bonds elevates the E_M value of the midpoint oxidation potential of the P/P^+ pair. The oxidation potential was determined using the method of electrochemical titration following bleaching of the A maximum of the absorption band at 865 nm ($P \rightarrow P^+$).

As a result, this accelerates reduction of P^+ in the course of recombination $P^+Q_A^- \rightarrow PQ_A$. At the same time, the rate of the initial charge separation of $P^* \rightarrow P^+$ slows down as seen from the increase in time $\tau_{ch.\ sep.}$.

The enhancement of the strength and number of hydrogen bonds can be also exhibited by their effect on the rate of relaxation processes proteins and coupled electron transfer in RC.This effect will be caused by a decrease in the intramolecular mobility of the protein matrix on the separated vibrational degrees of freedom and deceleration of electron transport due to the involvement of "unrelaxed" states of carriers.

Figure 28.13 shows changes in absorption at 665 nm, corresponding to the appearance and disappearance of reduced bacteriopheophytin upon isotopic substitution of H_2O for D_2O and the action of dimethyl sulfoxide. As seen, the described modification of the system of hydrogen bonds leads to slowing down the electron transfer at stages $P \rightarrow$ Bpheo (*a*) and Bpheo $\rightarrow Q_A$ (*b*). Concurrently with deceleration of direct transfer $P \rightarrow$ Bpheo there appears a component of reduction of P^+ by recombination $P^+ \leftarrow$ Bpheo$^-$ (Fig. 28.13, *c*, curve *2*) with time ~ 0.7–1.0 ns that is accompanied by generation of delayed luminescence with the same characteristic time.

The determining factor is the relation of characteristic times of electron transfers τ_{el} and relaxation processes τ_{rel}. It should be reminded that relation $\tau_{rel} \ll \tau_{el}$, required for the irreversible electron transfer, is usually fulfilled under standard conditions, because processes of vibrational relaxation occur relatively fast (< 1 ps) as compared to electron transitions and do not affect their character. But at the stages of the initial electron transfer in the PBpheo complex, the times of electron transfers comprise several picoseconds and comparable to the times of vibrational relaxation, $\tau_{rel} \simeq \tau_{el}$. Modification of hydrogen bonds slows down relaxation of the initial state of the complex to a greater extent. As a result, the electron has no time to stabilize adequately on bacteriopheophytin prior to its further transfer from the yet non-relaxed state of [P^+Bpheo$^-$]. From this non-related state of [P^+Bpheo$^-$], the electron has time to return back, recombining with P^+ for time $\sim$ 1 ns that is accompanied by delayed luminescence. Further electron transfer can also occur from the state of [P^+Bpheo$^-$] to Q_A but at a lower rate. Thus, in the relaxation at $\tau_{rel} \simeq \tau_{el}$ in the donor-acceptor complex, microconformational substates are generated that differ in their functional activity (rates and directions of electron transfer). Here the state of hydrogen bonds affects the dynamics of photoinduced relaxation processes in the system.

It was found that at the stage of charge separation in RCs ($P^* \rightarrow P^+$Bpheo$^-$) a drop of temperature accelerates this process, decreasing its time from $\sim$ 3.5 ps at 300 K to $\sim$ 1.2 ps at 4 K.

In Figure 28.14, the diagram shows possible positions of the potential energy curves of initial P^*Bpheo$^-$ and final P^+Bpheo$^-$ states at two temperatures (300 K and 77 K). Deformation and displacement of the curve of the P^*Bpheo$^-$ state upon freezing of the sample leads to the displacement of the intersection point of these curves to the vertex of the parabola of the initial P^*Bpheo$^-$ state (cf. Chapter 25).

In this case, the electron transfer process becomes radiationless. On the contrary, at higher temperatures its rate decreases, because occupation of higher vibrational sublevels increases and, consequently, the system now leaves the intersection point of the curves of initial and final states in the vertex of the parabola, where the probability of transition was maximal.

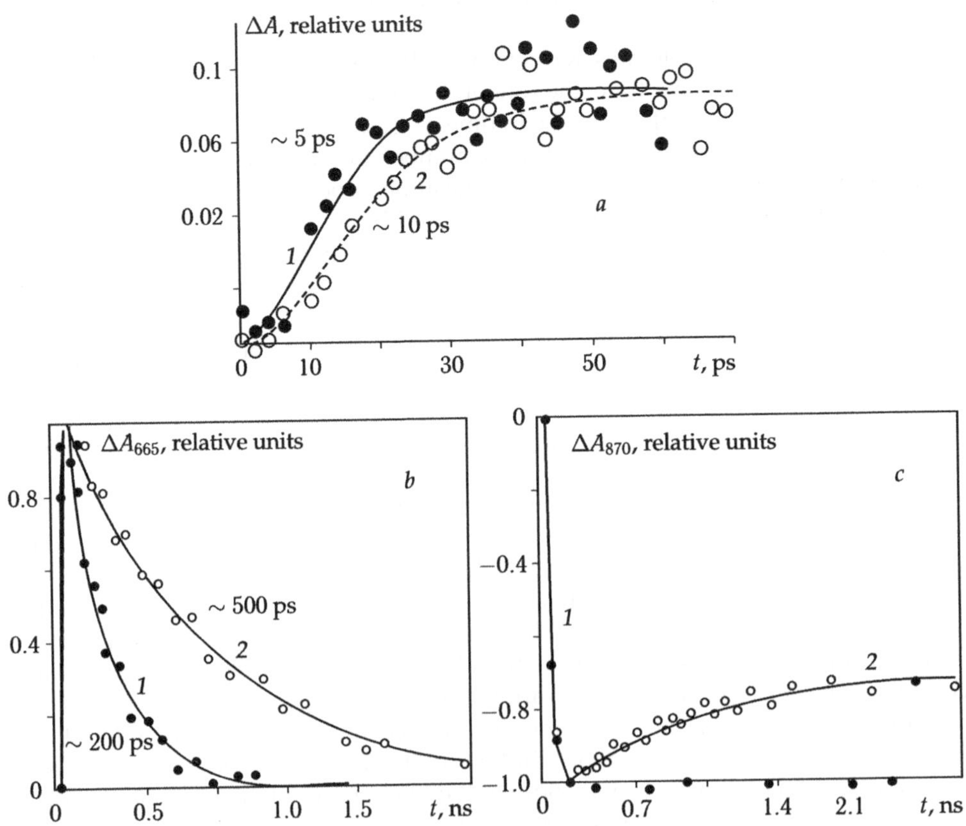

Figure 28.13. Changes in the rate of electron transfer in RCs ($P^* \rightarrow$ Bpheo $\rightarrow Q_A$) in standard samples (curves *1*) upon modification of the system of hydrogen bonds (curves *2*) (A. B. Rubin et al., 1994).

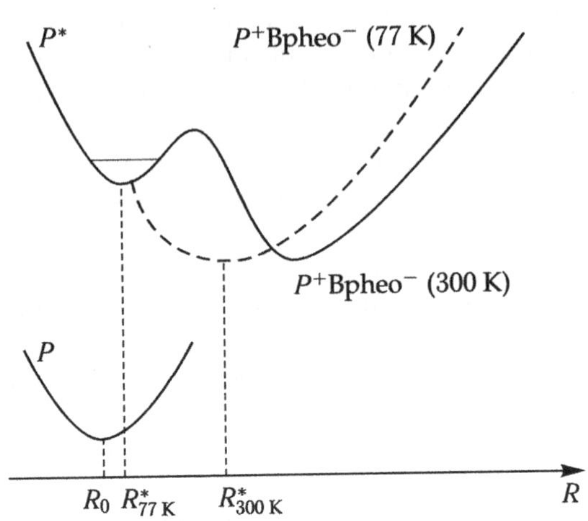

Figure 28.14. Deformation of potential energy curves of RC states at a temperature decrease.

28.3 Mechanisms of Cytochrome Oxidation in Reaction Centers

Devault and Chance measured for the first time the low temperature dependence of cytochrome oxidation by a photo-oxidized bacteriochlorophyll dimer, $C \rightarrow P^+$ (see Fig. 25.1) in purple bacteria (*Chromatium*) cells. It was found that at a decrease in temperature from 300 K to 120 K the characteristic time of cytochrome oxidation increases from 1 μs to 2.5 μs, remaining practically constant upon further temperature drop.

Vast data are available in the literature on low-temperature oxidation of low-potential C_L and high-potential C_H cytochromes by photoactive bacteriochlorophyll P.

Figure 28.15 demonstrates typical two-phase temperature curves of oxidation of cytochrome C_L, diagram of potential surfaces of electronic energy (U_i and U_f) of

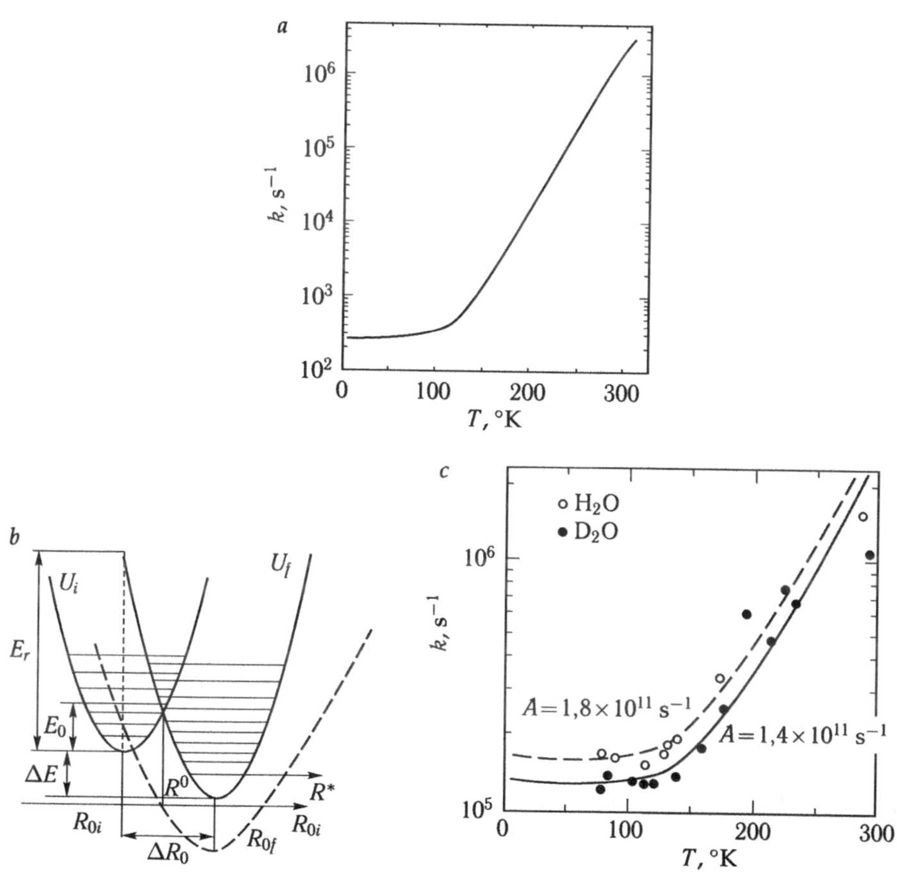

Figure 28.15. *a*, Temperature dependence of the cytochrome oxidation rate in photosynthetic bacteria; *b*, electron vibrational interactions upon electron tunneling; *c*, isotopic effect of substitution of H_2O for D_2O. E_0 is the activation energy; E_r is the reorganization energy of the medium; ΔE is the thermal effect of the electron transfer reaction; ΔR_0 is the shift of the nuclear equilibrium position upon the transition from initial state i to final state f. The dashed line corresponds to radiationless electron transfer E_a (reproduced from R. E. Blankenship and W. W. Parson, 1979).

initial (i) and final (f) states of donor-acceptor pair DA and experimental data on the effect of H_2O for D_2O substitution.

The tunneling of electrons takes place when the energies of states i and f become equal that occurs in point R^0 of the nuclear coordinate, where curves $U_i(R)$ and $U_f(R)$ intersect (Fig. 28.15, b):

$$(D^*A)_i \rightarrow (DA^*)_f.$$

In state $(DA^*)_f$, the system is at one of the highest vibrational sublevels, from where the excess energy is dissipated with the transition to lower vibrational sublevels and excitation of the accepting modes. The approach to point R^0 along the nuclear coordinate occurs with a temperature rise and increased occupation of high vibrational sublevels of the initial state. As it proceeds, the width of the activation barrier of the transition, as seen from Fig. 28.15, b, decreases, consequently, the electron tunneling rate grows with the temperature rise. At low temperatures, only lower vibrational sublevels are occupied in state i, and point R^* can be achieved just due to the tunneling of nuclei themselves (nuclei are rather light particles). However, here the width and height of the barrier are the largest, and therefore the rate of electron tunneling is minimal, but does not depend on temperature. Thus, the two-phase character of the temperature dependence of cytochrome oxidation is explained qualitatively by the effect of temperature on the reorganization of the initial nuclear configuration, which is characterized by the shift of the initial equilibrium nuclear configuration R_{0i} to the new equilibrium value R_{0f} upon electron transfer, $\Delta R_0 = R_{0f} - R_{0i}$. The developed quantitative theory of the electron tunneling in the protein surrounding allows to obtain expressions for the rate constant of tunneling in the case of strong and weak electron vibrational coupling (Chapter 25). Thus, it becomes possible to determine parameters of the electron transfer process, and in particular the frequencies of vibrations of the accepting mode. These frequencies are characterized by values $\sim$ 300–400 s^{-1}. In some cases vibrations of water molecules in hydrate shells of porphyrin molecules can play the role of an accepting mode, in other cases — also vibrations of hydrogen atoms in chemical groups C–H, O–H, N–H_2. The low-temperature limit of the reaction rate is determined by tunneling of H_2O molecules as a whole at distance $\Delta R \approx 0.15$–0.2 E.

Isotopic Effect. At low temperatures, the substitution of H_2O for D_2O in bacterial chromatophores leads to about a 1.4-fold decrease in the oxidation rate constant of C_L by a reaction center. This effect is less pronounced at room temperatures. The frequency 330–350 s^{-1} of vibrations of the accepting mode was determined by the breakpoint position on the temperature dependence curve of the considered reaction. This is very close to the frequency of H_2O vibrations in hydrate shells of Mg and Fe. The substitution of H_2O for D_2O decreases the frequency of vibrations ω to 314 s^{-1} and increases somewhat the constant of the electron vibrational coupling that depends on the relative molecule deformation along the coordinate of the accepting mode ω upon changes in the electronic state.

Characteristic changes in equilibrium distances between Mg^{2+} and H_2O in the coordination sphere upon changes in the ion charge per unit make $\Delta R \sim 0.01 \div 0.12$ nm. Comparing this to the value of the coupling parameter $S_{H_2O} = 6.6$ we can conclude that two or three water molecules are involved in the process (Fig. 28.16).

In this case, the nature of the accepting mode is related to the reorientation of water dipoles. However, other high-frequency vibrations of light-weight atomic groups may also be involved in this process. In particular, the equilibrium position of H in the

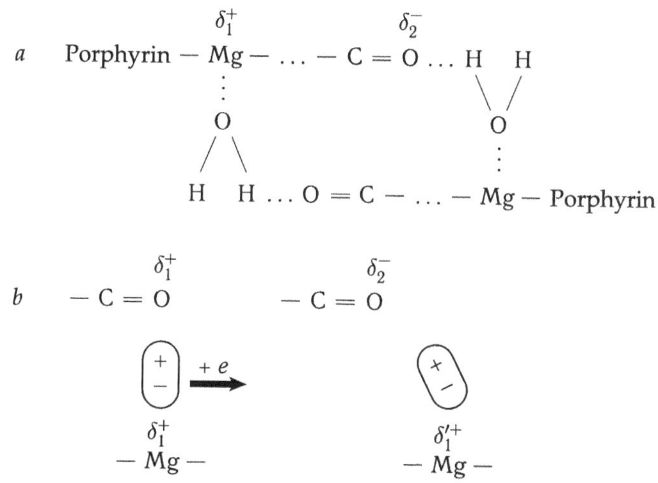

Figure 28.16. Involvement of water molecules in energy dissipation by the accepting mode in cytochrome oxidation in RCs of purple bacteria.

With changes in the charge of the dimer (*a*) the effective charges δ_1 and δ_2 on the Mg and oxygen atoms of the C=O-group are changed. This should induce a change in the equilibrium position of the water dipole (*b*).

hydrogen bond may be displaced along the direction of the asymmetric two-well hydrogen bond that is accompanied by heat dissipation (see Chapters 8 and 25).

Moreover, it was found that the rate of high-potential cytochrome C_H oxidation in RCs (*Ectothiorhodospira shaposhnikovii*) does not at all depend on temperature in the range from 300 K to 120 K ($\tau \sim 2.5$ μs). But at 210–120 K at a constant oxidation rate, a decrease in the signal amplitude was observed, i.e. a decrease in the number of oxidized cytochromes (Fig. 28.17).

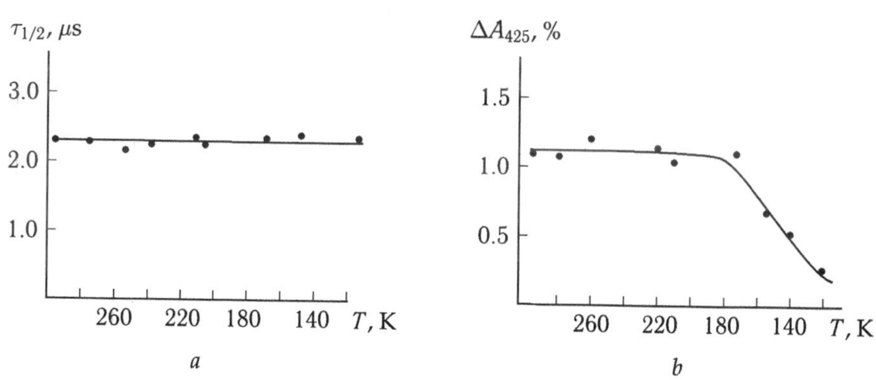

Figure 28.17. Temperature dependence of time $\tau_{1/2}$ and signal amplitude ΔA_{425} of the high-potential cytochrome C_H oxidation (reproduced from S. K. Chamorovsky et al., 1980).

As opposed to low-potential cytochrome C_L, the rate of C_H oxidation does not practically depend on temperature in the range from 300 K to 120 K (*a*). But below some critical temperature, the C_H molecules number of cytochrome liable to photoinduced oxidation decreases (*b*). In the C_H molecules that remained active, the rates of oxidation do not change. Dehydration of the samples has a similar effect on the C_H oxidation.

In this case, soft modes reflect conformational rearrangements in carrier proteins that bring together donor-acceptor groups to generate contact states between them. Namely, the spontaneous motion of donor-acceptor groups in the C_H–P_{870} complex determines the possibility to form such contact state, where electron tunneling with weak electron vibrational coupling proceeds. Thus, the protein chain mediates structural electron tunneling between donor-acceptor groups (Chapter 25).

This conclusion is also substantiated by the results on the effect of dehydration on photoinduced oxidation of high-potential cytochromes. Upon hydration of dehydrated preparations, a sharp enhancement of the mobility of water protons and non-aqueous molecules was observed, as manifested by the appearance of slow components of the spin echo decay $T_2 \simeq 1$ ms (Chapter 10). This enhancement of the mobility in the membrane correlates with the increase in C_H photo-oxidation at the humidity values from 0.5 to 0.8 P/P_s (0.25–0.47 g H_2O/g), when mobile molecules of weakly bound water appear. Upon changes in the water state, the structure of the whole complex also changes due to the modified mutual position of donor-acceptor pairs in the protein matrix. This affects the formation of a contact state between them, where interprotein electron transfer may occur from cytochrome to bacteriochlorophyll.

Thus, in the case of a weak electron vibrational coupling the effect of temperature and humidity is provided via soft modes by influencing the formation of a contact state between functionally active groups. □

28.4 Conformational Dynamics and Electron Transfer in Reaction Centers

Effect of Temperature and Hydration. Figure 28.18 shows temperature dependencies of the photoinduced direct electron transfer efficiency in the system of quinone acceptors, as well as correlation times of rotational diffusion of the spin label, covalently attached to SH-groups of the protein and of the hydrophobic spin probe in chromatophore membranes (Chapter 10). At 140–180 K, the τ' value (effective parameter of correlation time) changes little that shows the absence of fast motions in the system. However, at a temperature rise this value decreases sharply to $\leqslant 10^{-7}$ s, indicating the "defrosting" of motions in the protein–lipid local surrounding of nitroxyl fragments. In the same temperature range, where the intramolecular mobility of the RC protein increases, the functional activity also enhances. At $T < 180$ K, a low (close to zero) electron transport activity corresponds to strongly retarded mobility.

Typical temperature dependence of the probability of recoilless absorption of a γ-quantum or factor $f'(-\ln f')$ for chromatophores is shown in Fig. 28.19.

It is seen that at a temperature rise $T > 170$ K, the f' value drops that evidences for the appearance of the Mossbauer atom ^{57}Fe motions with amplitudes $\langle x_a \rangle \gg 0.03$ nm and the times exceeding 10^{-7} s. Consequently, at $T > 170$ K the motion of the protein macromolecule over the conformational states is accelerated. □

However, in dehydrated preparations, where the water content does not exceed 3 %, the $f'(T)$ curve is monotonic and has no inflection point (solid curve). The amplitude of Mossbauer nuclei displacements increases only slightly upon heating, but the dependence on temperature still remains linear, i.e. typical of crystalline

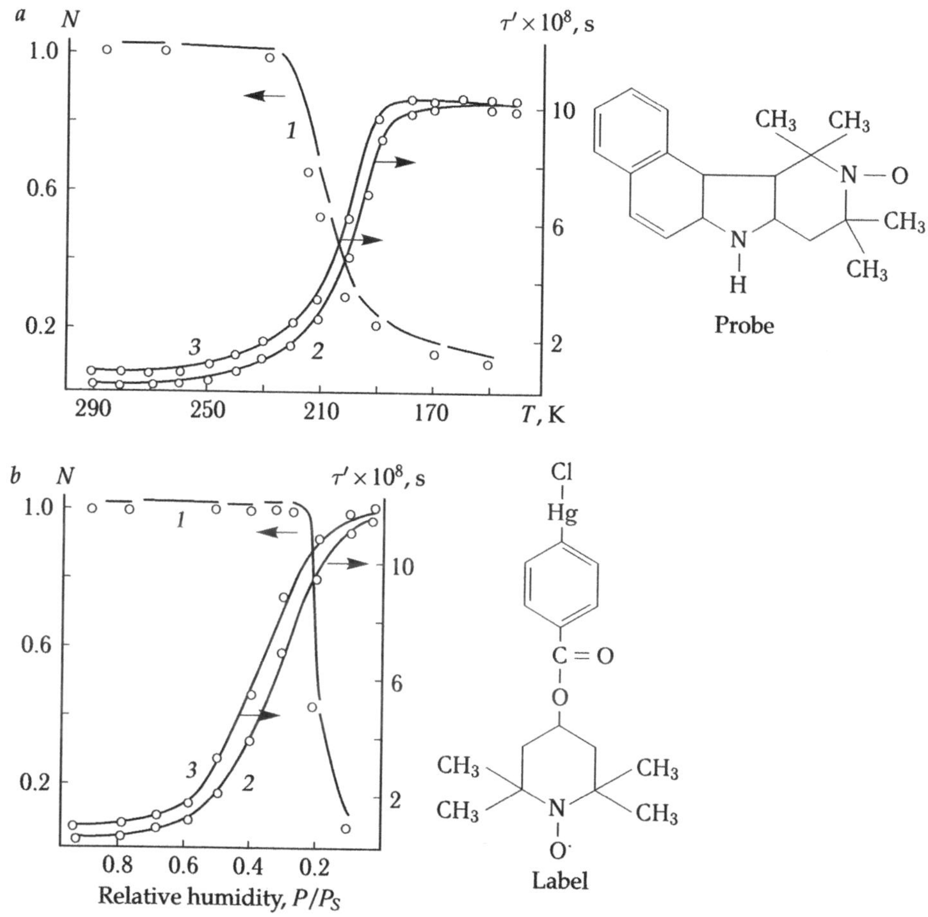

Figure 28.18. Functional activity (*1*) and conformational dynamics (*2, 3*) of reaction centers in spin-labeled chromatophores of *R. rubrum* versus temperature (*a*) and humidity (*b*) (reproduced from A. I. Berg et al., 1979).

N, efficiency of photoinduced electron transfer from Q_A to Q_B (curves *1*); is the effective parameter of the correlation time of rotational diffusion of the hydrophobic probe (2) and of the spin label on SH-groups (3).

solid bodies (see Chapter 10). In moistened preparations, at temperatures below 180–200 K the dependence of $f'(T)$ coincides with the analogous dependence for maximally dried preparations. Upon further heating, anomalous decrease in f' and a corresponding growth of the displacement amplitude are observed.

The observed correlation of the electron transport efficiency and the f' value does not mean a direct connection of the electron transfer namely with those conformational degrees of freedom that contribute to the Mossbauer atom motion. A sharp decrease in factor f' shows only that in the considered temperature range, the correlation times of conformational motions with the amplitude $\geqslant 0.1$ nm are rapidly decreased below 10^{-7} s. Thus, prerequisites are created for accelerating other, large-scale conformational degrees of freedom. As a result, conditions are established for generation of donor-acceptor states, where electron tunneling occurs in the system of quinone acceptors. □

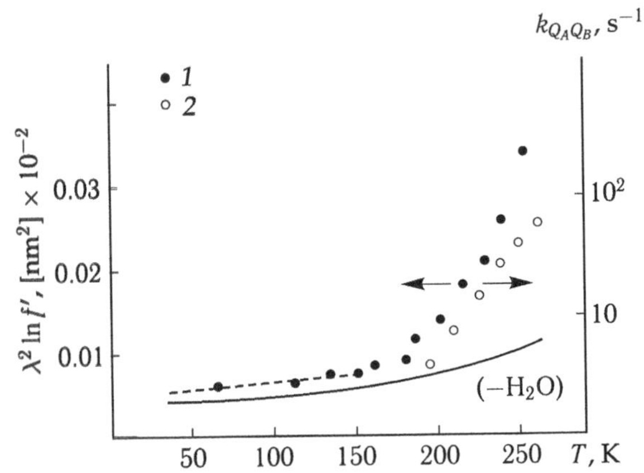

Figure 28.19. Temperature dependence of factor f (*1*) and rate constant $k_{Q_AQ_B}$ (*2*) of photoinduced electron transfer from the primary to the secondary quinone acceptor chromatophores (the value of factor f is the measure of the probability for the Mossbauer effect) (A. B. Rubin et al., 1994).

The temperature dependence of a direct electron transfer in system Q_A–Q_B, localized near the surface of the transmembrane RC protein, correlates better with fast motions in surface layers with characteristic times $\tau_c \sim 10^{-8}$ s. At the same time, changes in the rate of the back reaction proceeding in the internal structures of RCs (recombination of P^+ and Q_A^- ($P^+Q_A^- \rightarrow PQ_A$)), correlate better with the appearance in the samples of low-frequency motions at $\tau_c \gg 10^{-8}$ s manifested upon upon defrosting of the sample. In the temperature range 130–190 K, motions are observed at times $\tau_c \sim 1$ s in the internal parts of membrane proteins, recorded by the shift in phosphorescence spectra maxima of aromatic amino acids.

Conformational mobility is characterized by a wide set of relaxation times (from 1 to 10^{-8} s depending on temperature). This makes the basis for generation of contact states between chain components.

Manifestation at this temperature of one or another type of motion with certain correlation times τ_c does not mean that displacements of corresponding molecular groups were completely absent at other temperatures. The observed temperature effects can be connected with such changes in the already existing motions, when their correlation times fall within the limits of the time resolution of the applied experimental method and thereby can be recorded. □

28.5 Electron Transfer and Formation of Contact States in the System of Quinone Acceptors ($PQ_A\,Q_B$)

In the system of quinone acceptors, the temperature effects on the electron transfer can be realized not only by changing the rate constant (Chapter 25), but also by generation of isolated or contact conformational states between carriers in the $PQ_A\,Q_B$ complex. Note that the rate constants of the electron transfer itself in certain selected states can also be independent of temperature because it is just transitions between these states that are affected by temperature.

Recombination Processes in $PQ_A\,Q_B$. As shown in Chapter 27, under the action of light, the photoactive pigment is oxidized, and the electron moves first to Q_A and then to Q_B. Therefore, firstly the $P^+Q_A^-Q_B$ state is formed and then it disintegrates via two channels:

$$P^+Q_A^-Q_B \to PQ_A\,Q_B, \qquad P^+Q_A^-Q_B \to P^+Q_A\,Q_B^-. \tag{28.3}$$

After cessation of illumination, the observed dark reduction kinetics of photo-oxidized pigment P has two components (Fig. 28.20). The fast component of P^+ reduction corresponds to the electron return from Q_A^- to P^+ from state $P^+Q_A^-Q_B$ ($P^+Q_A^-Q_B \to PQ_A\,Q_B$) and the slow component — to that of Q_B^- from state $P^+Q_A\,Q_B^-$ ($P^+Q_A\,Q_B^- \to PQ_A\,Q_B$).

Thus, the efficiency N of electron transfer from Q_A to Q_B can be estimated by the relation of the slow and fast components of P^+ dark reduction. A decrease in the N value with a decrease in temperature and humidity denotes a drop in the electron transfer efficiency between Q_A and Q_B (see Fig. 28.18).

However, the main reason is not a decrease in the rate constant of electron transfer $Q_A \to Q_B$ itself. This conclusion follows from the difference in the results obtained upon cooling of the studied preparations in the dark or cooling under concurrent activating illumination. It was found that upon cooling in the dark at rather low temperatures ($T < 200$ K), electron transfer from Q_A to Q_B is almost completely inhibited. This is evidenced by the prevalence of the fast component in the kinetics of P^+ reduction after photoexcitation of RCs (Fig. 28.20, *I*). But the situation is quite different if the same preparations are cooled in the light. In this case, in some portion of reaction centers the photomobilized electron is fixed and remains in the acceptor complex for an indefinite period of time. As a result, a large portion of the "irreversible" component of the recorded signal appears in the kinetic curve upon switching off the light after freezing the preparation (see Fig. 28.20, *III*). It corresponds to the non-recoverable portion of P^+. In the remaining portion of the signal, the kinetics of recovery of P^+ does not essentially differ from that at room temperature (Fig. 28.20, *I*). It is characterized only by slow components of P^+ reduction.

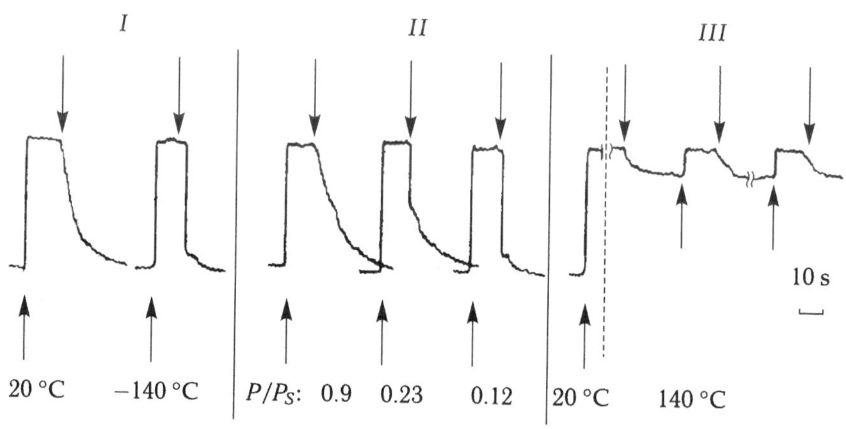

Figure 28.20. Kinetics of dark P^+ reduction in the reaction center of chromatophores depending on temperature (*I*), humidity (*II*) and on cooling the preparations in the light (*III*) (reproduced from P. P. Knox et al., 1979; E. G. Petrov et al., 1983). Upward and downward arrows show correspondingly turning the light on and off.

Consequently, upon cooling in the light, such a structural state of the RC pigment–protein complex is formed, where transfer $Q_A \to Q_B$ occurs fast even at low temperatures and therefore is activationless. As a matter of fact, it corresponds to the formation of a contact state in the Q_AQ_B pair under such conditions.

Crystals of photosynthetic RC preparations were frozen in the dark or in the activating light and subsequently analyzed by the method of X-ray analysis (XRA). As a result, changes were revealed in the position of semi-quinone in Q_B in the structure of light-adapted preparations as compared to the position of neutral quinone in RCs cooled in the dark (Fig. 28.21). In RCs from purple bacteria *Rb. sphaeroides*, cooled to the liquid nitrogen temperature, Q_B was displaced in the protein structure by 5 E closer to the atom of non-heme iron, located between Q_A and Q_B and, apparently, involved in the electron transfer between them. In this case the quinone ring in Q_B is fixed in a position turned by 180° relative to the isoprenoid tail.

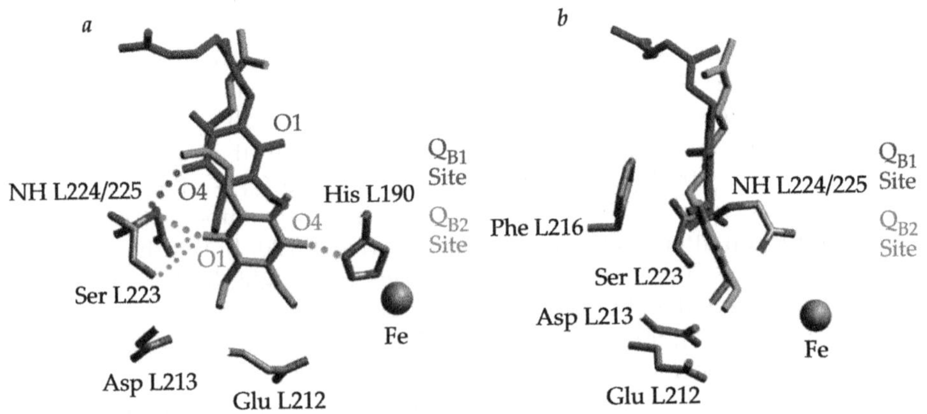

Figure 28.21. Distal (inactive relative to electron transfer from Q_A to Q_B) and proximal (active) position of Q_B in the structure of the reaction center from *Rb. sphaeroides* bacteria according to the X-ray structural analysis (reproduced from M. H. B. Stowell et al., 1997).

01 and 04 are carbonyl oxygen atoms in Q_B. Their positions show the turn of the quinone ring in the proximal position by 180° as compared to the distal position. Dots denote hydrogen bonds in carbonyl oxygen atoms in Q_B with the amino acid surrounding.

At room temperature, a corresponding conformational transition has time to occur in response to illumination and recovery of Q_A. Upon cooling in the dark, the conformational mobility is retarded. But upon cooling in the strong light, the system "restores" its conformation mobility and has time to pass to the position close to the equilibrium at room temperature $(P^+Q_A^-)_{\text{relax}}$. Therefore, by varying the intensity of light illumination and rate of cooling we can fix certain conformational states in RC preparations, achieved successively in the process of relaxation under freezing. The results of numerical modeling of intramolecular dynamics of RCs (K. Schulten et al., 1992) demonstrated that the lengths and angles of the bonds of atomic groups change by values up to 0.4–1.0 E upon electron transfer along the chain $P \to \text{Bchl} \to \text{Bpheo} \to Q_A$.

The readers are referred to a theoretical consideration of the dynamics of photoconformational transition, described in Chapter 25. This pattern coincides qualitatively with the data on interaction between the CO group and the iron atom in myoglobin at different temperatures as presented in Chapter 11.

28.6 Mathematical Models of Primary Electron Transport Processes in Photosynthesis

Let us consider the main results of mathematical modeling of the general system of primary processes of photosynthesis (Fig. 27.4).

As stated in Chapter 27, electron transfer in photosynthetic RCs is performed by both mobile carriers and carriers assembled in molecular complexes. In the first case, the electron transport processes are described by second order kinetic equations, based on the use of the mass action law. In the second case, the mathematical description is based on equations for probabilities of transitions between different states of the molecular complex; these states comprise carriers in corresponding redox states. □

The General Type of Transfer Equations. The mathematical apparatus used in these two types of models was described in detail in Chapters 1–4. An example is the electron transfer reaction between two mobile carriers C_1 and C_2, interacting in an open chain with external donor D and acceptor A,

$$D \xrightarrow{k_1} C_1 \xrightarrow{k_2} C_2 \xrightarrow{k_3} A. \tag{28.4}$$

The kinetic equations look here as follows:

$$dC_1^1/dt = k_1 D^1 C_1^0 - k_2 C_1^1 C_2^0, \qquad dC_2^1/dt = k_2 C_1^1 C_2^0 - k_3 C_2^1 A^0, \tag{28.5}$$

where D^1, C_1^1, C_2^1 are concentrations of reduced forms of chain components and D^0, C_1^0, C_2^0 are concentrations of oxidized forms; k_1, k_2, k_3 are bimolecular rate constants of the corresponding reactions.

As a rule, in chain (28.4) the total concentration of carriers remains unchanged, i.e. $C_1^1 + C_1^0 = C_{10}$, $C_2^1 + C_2^0 = C_{20}$. Under equilibrium conditions in the dark, the initial concentrations of oxidized and reduced forms of carriers are determined by redox potentials E_h of the medium. And according to the Nernst equation,

$$E_h = E_{1/2} + (RT/F)\ln(C_1^1/C_1^0) \tag{28.6}$$

or hence

$$C_1^1 = \frac{C_{10}}{1 + \exp\left(\dfrac{(E_{1/2} - E_h)F}{RT}\right)}, \tag{28.7}$$

where $E_{1/2}$ is the normal potential of half-reduction of the carrier, E_h is the equilibrium redox potential of the medium, and F is the Faraday constant.

The switching on of the light induces electron transfer in accordance with equations (28.5) under initial conditions of equation (28.7). If according to scheme (28.4) electron transfer occurs in a complex of two carriers, the marked graph of states of this complex (C_1, C_2) will look as follows:

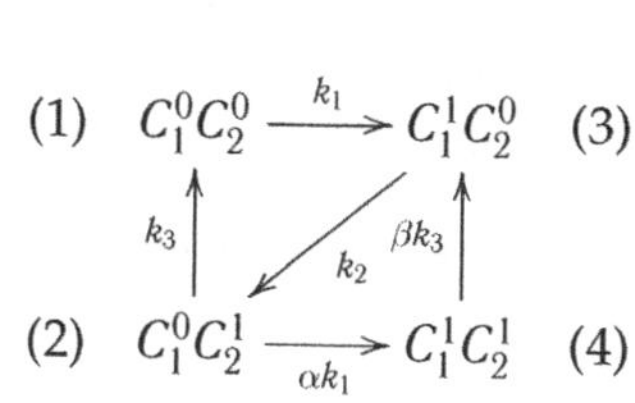

(28.8)

Here the numerals in brackets show the number of the state of complex (C_1C_2), k_i are constants of rates of corresponding transitions designated in scheme (28.4). When cooperativity is absent, the system of differential equations for probabilities of states of complex (C_1C_2) in scheme (28.8) is the following:

$$\frac{dp_1}{dt} = k_3p_2 - k_1p_1, \qquad \frac{dp_2}{dt} = k_2p_3 - (k_1 + k_3)p_2, \qquad \sum_{i=1}^{4} p_i = 1. \tag{28.9}$$
$$\frac{dp_3}{dt} = k_1p_1 + k_3p_4 - k_2p_3, \qquad \frac{dp_4}{dt} = k_1p_2 - k_3p_4,$$

Having solved the above system of equations relative to probabilities p_i for states of complex (C_1C_2), it is possible to find the probability for states of separate carriers. To this end it is necessary to sum up the probabilities of all states of the complex, where the considered carrier is in the state that is of interest for us. Thus, the probability that the first carrier is in an oxidized state C_1^0 is equal to the sum of probabilities of states (1) and (2) of the complex in scheme (28.8):

$$p(C_1^0) = p(C_1^0C_2^0) + p(C_1^0C_2^1) = p_1 + p_2 = 1 - p(C_1^1). \tag{28.10}$$

The total electron transfer rate v between C_1 and C_2 in the complex is proportional to the concentration of complexes in state $C_1C_2^0$ ($v = k_2[C_1^1C_2^0]$). Here both oxidation of C_1 and reduction of C_2^0 occur in a single act. Therefore, if the both carriers included in the complex are either oxidized or reduced, no electron transfer occurs between them. Consequently, components of the same complex interact between each other, and separate components are isolated from each other. In photosynthesis this corresponds to parallel isolated chains of electron transfer. □

In the RC functioning scheme (see Equation (27.11)) for transition

$$P \xrightarrow{k_0} P^*$$

k_0 is equal to the product of intensity I of the incident light to the absorption cross-section σ: $k_0 = I\sigma$ (s^{-1}). In the direct sunlight, I_{max} is $\sim 10^{16}$ quantum $\cdot$ $cm^{-2} \cdot s^{-1}$ and $\sigma \sim 10^{-16}$ cm^2. In this case, the value of k_0 is close to unity. Consequently, under continuous light the value of k_0 is limited by the intensity of the light or the frequency of quanta hitting the RCs. Real times and constants of electron transfer into the ETC can be determined under conditions, when the light intensity is not a limiting factor. This is achieved (see Chapter 27) upon excitation by short laser pulses of 10^{-11} s duration and 10^{22}–10^{25} quantum $\cdot$ s^{-1} power (or 10^{11}–10^{14} quanta per pulse).

The analysis of the models makes it possible to describe how stationary concentrations and transfer curves of redox states of carriers upon switching on and off of the excitation light depend on the transfer constants, in particular on the value of k_0. A comparison of these results with the experimental data allows not only to verify the validity of initial assumptions on the functional organization of the ETC, but also to determine the transfer constants at separate regions of the chain. □

Identification of the models includes both the choice of the structure of the model itself and the determination of parameters of the reaction constants, best compatible with the experimental data.

To identify these parameters, kinetic curves of photoinduced redox conversions of carrier by absorption changes are used as well as fluorescence induction curves (Fig. 28.20). Conditions for uniqueness of parameter identification are formulated only for linear systems, including isolated RC complexes. In the case of more complicated nonlinear systems, additional criteria of the dependence of kinetic characteristics on the illumination regime and action of inhibitors are used. Identification is conducted by minimizing the sum of square weighted deviations of theoretical variables from the experimental ones.

Modeling of Processes in the Photosystem II Complex. One of the approaches to such modeling is to derive equations of transitions between the states in PS II complex under the action of light (Equation (28.8)). Let us consider the scheme of states in the PS II complex shown in Fig. 28.22, where the fluorescence yield is proportional to the concentration of chlorophyll (Chl) molecules in an excited state that emit fluorescence.

In the scheme, each state of PS II includes 4 components: chlorophyll $P680$, pheophythin Pheo, a single-electron carrier quinone Q_A and a binding site of quinone Q_B as a secondary acceptor.

In this model, the excitation energy is rapidly (during picoseconds) distributed over the whole pool of Chl molecules in PS II, including the molecule of $P680$ in the photosynthetic RC. The abbreviation Chl refers to all chlorophyll molecules in PS II. The Chl component may be in neutral (Chl), excited (Chl*) and oxidized (Chl^+) states ($P680^+$). Each of the single-electron acceptors may be in neutral (Pheo, Q_A) and reduced ($Pheo^-$, Q_A^-) states.

The binding site of plastoquinone (PQ) can be empty (the second row in the scheme in Fig. 28.22), contain PQ in a neutral (Q_B) state (the fourth row in the scheme), and also carry one (Q_B^-) or two (Q_B^{2-}) electrons (correspondingly the fifth and first rows in the scheme). Thus, states corresponding to separate rows — x_i, y_i, z_i, g_i ($i = 1, 2, \ldots, 7$) — differ by the state of PQ in the binding site.

Generation of an excited state of Chl in the model is determined by "light" constants of formation of excited states of chlorophyll x_2, x_6; y_2, y_6; z_2, z_6; g_2, g_6. The corresponding constants $k_L = k_i$, $i = 1, 5; 8, 12; 15, 19; 28, 32$, are proportional to the intensity of the current light. Fluorescence in the model is determined by constant $k_F = k_{-i}$. The fluorescence yield in the model is calculated as the ratio k_F/k_L and the sum of "fluorescent" states of chlorophyll in PS II:

$$F = \frac{k_F}{k_L} \cdot (x_2 + y_2 + z_2 + g_2 + x_6 + y_6 + z_6 + g_6). \tag{28.11}$$

Figure 28.22 shows a number of small rectangles that represent states of PS II where the Q_B-site is occupied by diuron molecules (DU_i). In these PS II states, electron transfer to mobile PQ molecules and further electron transport to the cytochrome complex and PS I are inhibited.

By fitting the model and comparing its properties with the experimental curves of fluorescence induction, it became possible to determine the rate constants of radiationless dissipation of energy that could not be calculated directly from experimental data.

Such type of thermal dissipation of energy at the excess illumination and saturation of photosynthesis can play a protective role, preventing generation of active oxygen forms.

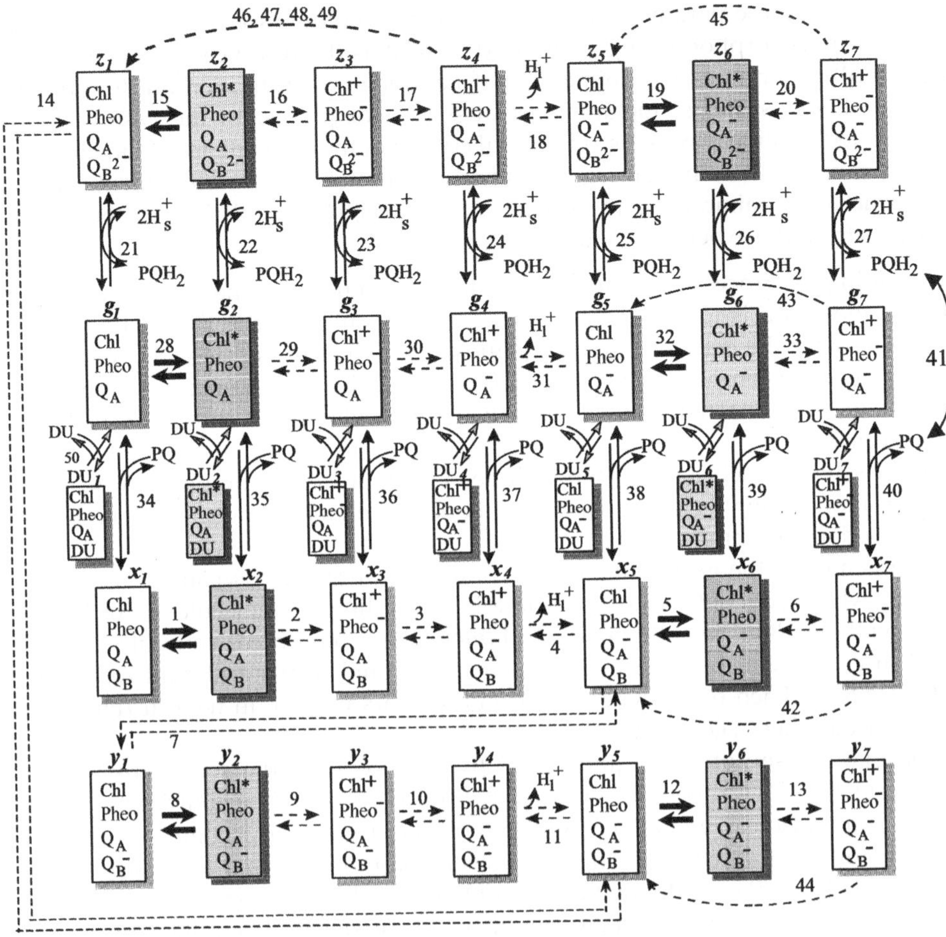

Figure 28.22. Scheme of the catalytic cycle in photosystem II.

Each rectangle represents one of the kinetic states of PS II, determined by the redox-state of electron carriers included in it. Shadowed are the states capable of emitting fluorescence quanta. Chl is all chlorophyll in PS II, including antenna pigments and the photoactive pigment $P680$; Pheo is the primary electron acceptor (pheophythin); Q_A and Q_B are the primary and secondary quinone acceptors. PQ is plastoquinone; PQH_2 is plastoquinol; H_{L^+} is protons released into the lumen; H_{S^+} is protons absorbed from the thylakoid stroma. Dashed arrows designate fast (with characteristic time less than 0.1 ms) stages of the cycle; solid arrows designate slow (with characteristic times no less than 1 ms) stages; thick arrows show light stages. Numerals near the arrows correspond to the number of the reaction, letters over the rectangles ($x_i, y_i, z_i, g_i, du_i, i = 1, \ldots, 7$) correspond to designations of variables in the model. The row with variables du_i (small rectangles) designate states with the Q_B-site occupied by the inhibitor. The inhibitor molecule (DU) occupies the empty Q_B-site with the formation of states du_i, incapable of the electron transfer from the primary quinone to the secondary quinone. Dashed arches show processes of radiationless recombination of Pheo$^-$ with P_{680}^+ (reactions 42–45), Q_A^- with P_{680}^+ (reactions 46–49) (reproduced from N. E. Belyaeva et al., 2006).

Complete Model of Electron Transfer. In such a model, it is necessary to take into account electron transfer between complexes mediated by different mobile carriers, moving in spatial compartments of the system. In the luminal internal space of the thylakoid, these are plastoquinone molecules, in the external stromal space — ferredoxin molecules, and inside the bilayer lipid layer of the membrane — plastoquinone molecules in the neutral and twice protonated forms (plastoquinol). In kinetic models, the interaction of mobile carriers with the complex was described using a traditional set of equations of the mass action law.

As a result, a complete model of processes occurring in the thylakoid membrane is constructed that represents kinetics of changes in the model variables at different values of parameters corresponding to different experimental conditions. Figure 28.23 shows model kinetic curves in real time for three different light constant (I) values, corresponding to the high (1,000 $W \cdot m^{-2}$), middle (100 $W \cdot m^{-2}$) and low (10 $W \cdot m^{-2}$) light intensities. The model makes it possible to adequately describe the time-dependent changes of the values observed experimentally: induction changes in fluorescence, changes in the transmembrane electric potential, kinetics of redox conversions of the photoactive pigment in PS I of $P700$. This enables to represent explicitly the kinetics of the model variables, direct experimental observation of which is impossible: proton flows, potassium ion flows (Fig. 28.23).

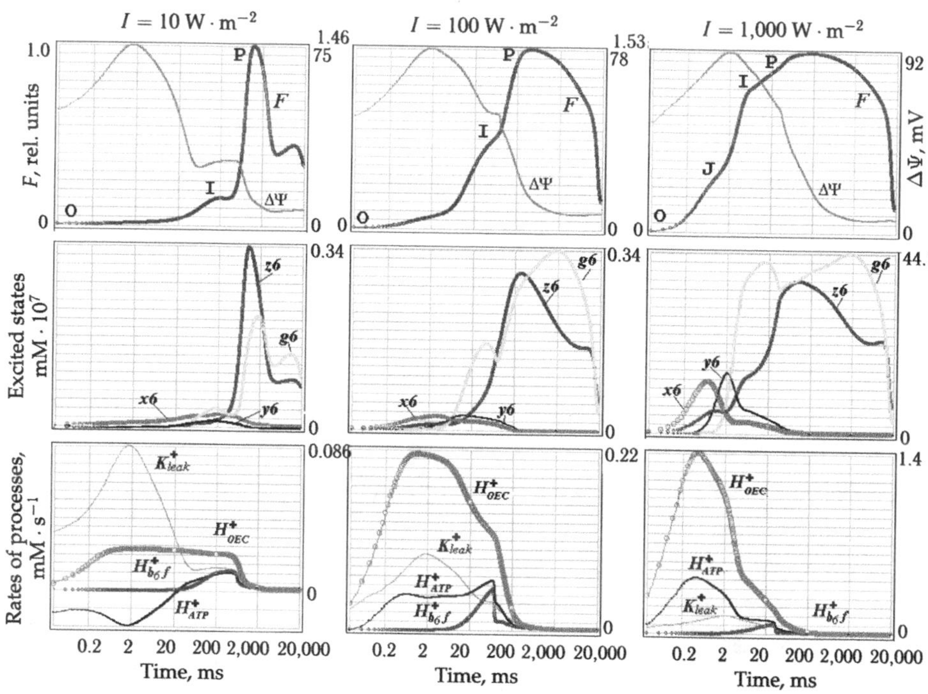

Figure 28.23. Induction curves calculated using the model of primary processes of photosynthesis for three different intensities of light illumination: 1,000, 100 and 10 (1 %) $W \cdot m^{-2}$.

Figures in one column correspond to the same level of illumination. Results are shown on a logarithmic time scale. *a*, Relative yield of fluorescence (F) and transmembrane electric potential ($\Delta\Psi$); *b*, concentrations of different excited states of PS II; *c*, rates of processes generating and consuming the electric charge in the thylakoid lumen; H^{+}_{bf} is the proton flow to the lumen upon oxidation of plastoquinol on the luminal side of the bf-complex; H^{+}_{KBK} is the proton flow to the lumen from the oxygen evolving complex of PS II; H^{+}_{ATP} is the rate of the proton consumption from the lumen in the ATP-synthase reaction; K^{+}_{leak} is the rate of K^{+} ion leakage from the thylakoid lumen (reproduced from G. V. Lebedeva et al., 2002).

Limitations in Kinetic Models. Direct Multiparticle Modeling of ETC. The peculiarities of the photosynthetic apparatus structure and the effect of the surrounding (lumen and stroma) on the motion of mobile carriers cannot be taken into account in models stimulating homogeneous systems, where it is suggested that multienzyme complexes of photosystems I and II and the cytochrome complex are uniformly distributed in the membrane that, generally speaking, is incorrect.

Membrane complexes protrude inside the membrane at a considerable distance. Diffusion cannot be considered either a priori free in the stromal space. Therefore, modeling of processes in a heterogeneous structure and the lack of free diffusion of mobile carriers in it require to develop new approaches.

They can be realized in direct multiparticle computer models. Such an approach has been developed at the Department of Biophysics (Biological Faculty, MSU) in collaboration with the Department of Computational Physics (Physical Faculty, MSU). Alongside with traditional kinetic characteristics, the direct model provides a three-dimensional visual representation of the dynamics of transfer processes on different spatial and time scales and enables simultaneous observation of the behavior of different components. The model is a three-dimensional scene (Fig. 28.24) including the thylakoid membrane, the intrathylakoidal space and the lumen space. There are multienzyme complexes (PS I, PS II, the cytochrome complex, the ATP-synthase complex) and mobile electron carriers (Pc, Fd, PQ).

To model the movement of mobile carriers Pc, Fd, PQ in the space of corresponding compartments, the mathematical apparatus for description of Brownian motion (Chapter 11) is used. It takes into account geometrical restrictions imposed by the generated model scene. It is proposed that a particle moves in a viscous surrounding

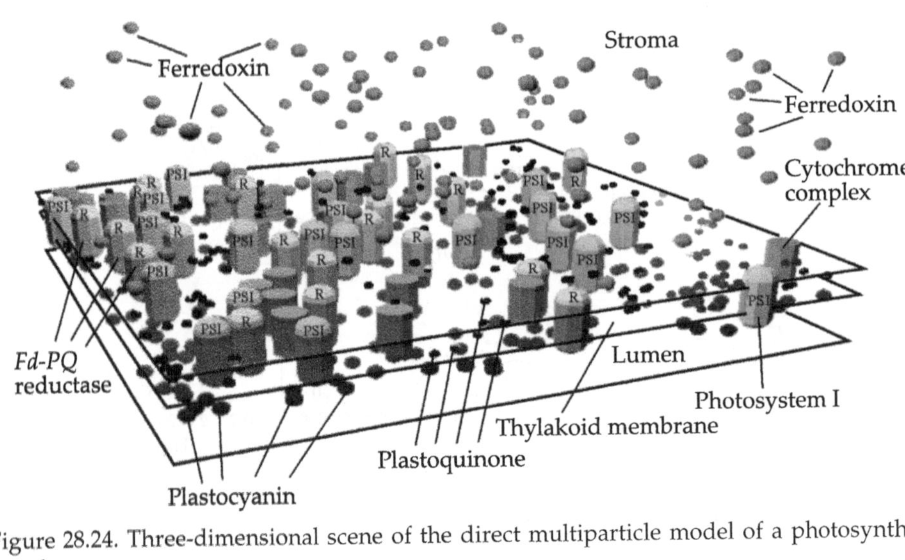

Figure 28.24. Three-dimensional scene of the direct multiparticle model of a photosynthetic membrane.

Part of the thylakoid membrane, the lumen and stromal spaces are shown. The internal part of the thylakoid is limited by the membrane. Inside the thylakoid (in the lumen) particles Pc that can carry an electron on their surfaces are moving. From the outside (in the stroma) particles Fd are moving; they can also carry an electron of their surfaces. The membrane is pierced by complexes of PS I, PS II and cytochrome complexes. The concentrations and dimensions of complexes are chosen following the available literature data.

under the action of a random force, emerging due to collisions with the surrounding molecules. For modeling the movements of a mobile carrier, the Langevin equation was used that describes the change in each coordinate in time under the action of a random force:

$$\xi \frac{dx}{dt} = f(t),$$

where ξ is the friction coefficient, $f(t)$ is the random force distributed normally with the zero mean and dispersion of $2kT\xi$. Here k is the Boltzmann constant, T is the temperature. For a spherical particle the coefficient of friction is determined by formula $\xi = 6\pi\eta a$, where h is the viscosity of the surrounding, a is the particle radius.

The Langevin equation was solved numerically for each mobile particle (Chapter 11). Suppose that as a result of chaotic Brownian motion the molecule of a mobile carrier approaches the protein complex at a distance smaller than some effective radius of interaction between them. In this case with some probability the "docking" takes place, i.e. the carrier binds with the complex. The effective radius of interactions is set as the model parameter characterizing the maximal distance at which docking is possible.

Models of Interaction of Protein Carriers upon Photosynthesis. The process of interaction includes three sequential stages: (1) Brownian diffusion of a mobile carrier in the space of the corresponding compartment; (2) drawing together of the mobile carrier and the multienzyme complex due to electrostatic attractive forces between molecules, mutual orientation of molecules in the space (docking) with the generation of a supercomplex, where electron transfer can take place; (3) electron transfer from one molecule to the other in the supercomplex (tunneling). Figure 28.25 demonstrates the position of a supercomplex: the cytochrome bf-complex — Pc in the thylakoid membrane. In kinetic modeling, the set of these stages is described by a single biomolecular reaction.

The process of "anchoring" of the mobile carrier molecule (Pc) by the multienzyme complex bf is determined not only by the probability of collision with the donor part of the complex, but also by how close the active centers will be to each other. This

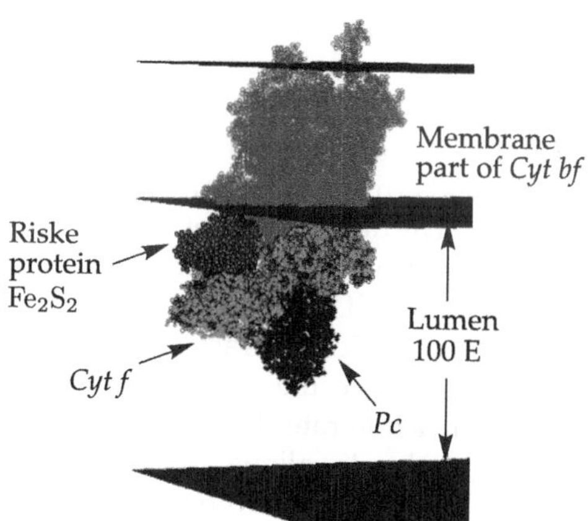

Figure 28.25. Position of supercomplex Pc and cytochrome bf-complex ($Cyt\ bf$) incorporated into the membrane of the thylakoid lumen. The figure is a result of superpositioning of PDB structures $2PCF$ and $1Q90$.

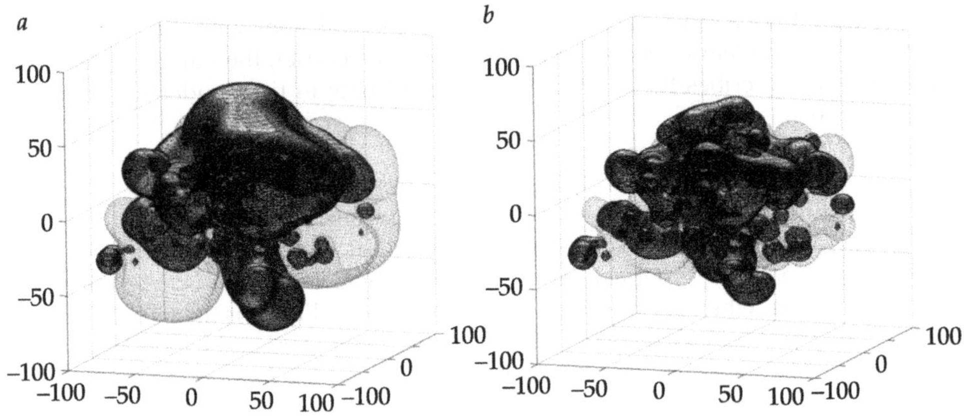

Figure 28.26. Equipotential surfaces (−6.5 mV, gray, and 6.5 mV, black) for PS I. Top, Acceptor side of PS I; bottom, its donor side.

a, The ionic strength of solution is 0 mM; *b*, the ionic strength of solution is 80 mM.

depends on the interaction of those parts of reaction macromolecules that approach each other upon docking. This is greatly assisted by electrostatic interactions of local charges of interacting molecules, included in the consideration when the molecules have already approached each other at closer distances (less than 35 E).

In accord with the available data on local charges on the surface of proteins, equipotential surfaces of interacting proteins are constructed. When approaching other proteins and complexes, the protein is oriented in the electric field, created by these proteins. The process of electrostatic orientation considerably increases the observed kinetic constant of the reaction rate as compared to that upon protein encounter by random sites on their surfaces. The calculated equipotential surfaces for PS I are given in Fig. 28.26.

Rate constants of the reaction depend on the geometrical size of the reaction volume of the thylakoid. The reaction proceeds at the maximal rate, when the distance between the membranes is 8 nm corresponding to experimental estimates under normal conditions. As the distance grows, the rate decreases due to the lowering of the concentration of reactive molecules *Pc*. As the distance becomes smaller, a decrease in the rate is caused by the fact that diffusion of molecules *Pc* in the lumen space becomes hindered and at a small width of the lumen space it may not be possible at all. This result corresponds to the experimentally observed inhibition of the reaction at hyperosmotic stress. The dependence of efficiency of interactions between carriers on geometric parameters can be one of efficient mechanisms of regulation of photosynthetic processes in a plant cell.

By changing the form and geometrical characteristics of the reaction volume, the number and position of interacting molecules and other model parameters, it is possible to study their effect on the reaction rate. The distribution of the electric potential around each of the interacting molecules allows clarification of the role of electrostatic interactions in the process of docking and generation of a supercomplex that is a necessary condition of the effective electron transfer. Direct multiparticle modeling makes it possible to understand in what way physical mechanisms and parameters of elementary molecular interactions (diffusion, electrostatic forces etc.) determine the general dynamics and macroparameters of cell processes.

X
Primary Processes in Biological Systems

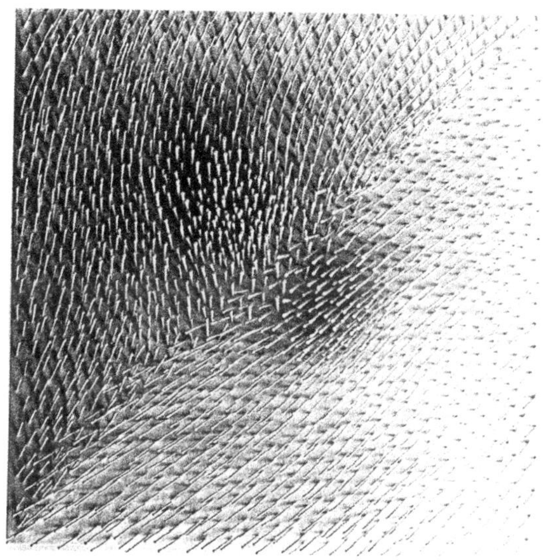

29

Photo-conversions
of Bateriorhodopsin
and Rhodopsin

30

Photoregulatory
and Photodestructive Processes

29

Photo-conversions of Bacteriorhodopsin and Rhodopsin

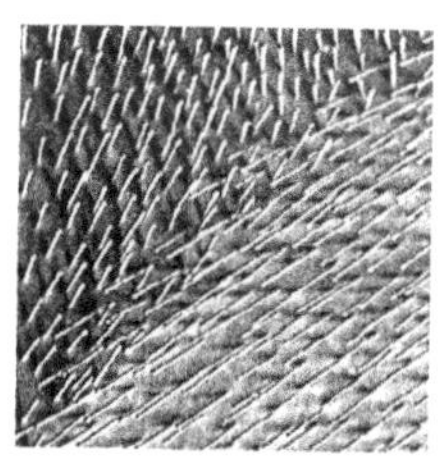

Molecular mechanisms of photo-conversions of rhodopsin and bacteriorhodopsin are an important area in biophysics of photobiological processes that has been markedly developing lately. Bacteriorhodopsin (Brh) was discovered in the purple membrane of halophilic bacteria (V. Stokenius, 1971), which turned out to be a new type of biological membranes capable of converting light energy. This system is evidently the simplest among all studied systems storing light energy in the form of electrochemical potentials of H^+. Each Brh molecule contains one retinal chromophore (polyene aldehyde) in a complex with opsin protein — the only protein contained in the purple membrane. Opsin consumes the light energy for active movement of protons across the membrane, resulting in ATP synthesis and other physiologically important processes. The photochemical cycle of Brh conversions underlies this bioenergy process.

Visual pigment rhodopsin, the same as bacterirhodopsin, is the only protein in the photoreceptor membrane of a visual cell in retina (it comprises up to 80–85 % of all protein in the membrane). Retinal in the complex with opsin serves also as the rhodopsin chromophore. Photo-conversions of rhodopsin are tightly connected with ionic and enzyme processes, underlying visual reception, as well as with generation of transmembrane potential difference across the photoreceptor membrane.

Rhodopsin and bacteriorhodopsin have a number of similar properties: the chemical nature of the chromophore, the topography of the membrane (7 alpha-intramembrane helical strands and hydrophilic loops on both sides of the membrane), and the nature of the bond with the opsin protein in the chromophore center. Common features are inherent also to processes of photochemical conversions, especially at fast initial stages. As will be shown in this chapter, the primary mechanisms of photo-conversions of bacteriorhodopsin and rhodopsin are also in many respects similar to the mechanisms of the processes, occurring in the photosynthetic RC (see Chapter 28). These mechanisms include photoinduced conformational changes in the protein part of photosensitive chromoprotein that occur according to the principle of electron-conformational interactions. □

29.1 Structure and Functions of Purple Membranes

Purple membranes (PM), localized in the cytoplasm of cells of some extremal halophilic bacteria, for example, *Halobacterium salinarium*, contain a single hydrophobic pigment-protein complex (PPC) bacteriorhodopsin (molecular mass ~ 26 kDa), its molecules being located in the PM in a strict order.

Retinal is covalently bound to opsin and forms an aldimine bond with one of lysine residues in the protein chain (the so-called Schiff base (Fig. 29.1) that can be protonated). One Brh molecule contains 7 α-helical regions piercing the PM. Three of these "poles" are positioned almost perpendicular to the membrane plane, the other four form small angles with the normal to the membrane plane. Figure 29.2 demonstrates a three-dimensional model of the Brh structure constructed using the data on electron diffraction in experiments with low-temperature microscopy of samples (R. Henderson et al., 1990). Retinal in Brh is attached to Lys216 that is at the end of the $G\alpha$-helical pole (Fig. 29.2). The site of the retinal attachment to the protein is close to the cytoplasmic side of the PM at a distance of 0.6–1.0 nm from the surface, while the ionone ring of retinal is inside the membrane, almost in its center. The difference in the position of the maximum of retinal absorption with a protonated Schiff base in Brh protein (570 nm) as compared to the position of its maximum in polar solvents (440 nm) is called an "opsin" spectral shift (440 → 570 nm). Let us

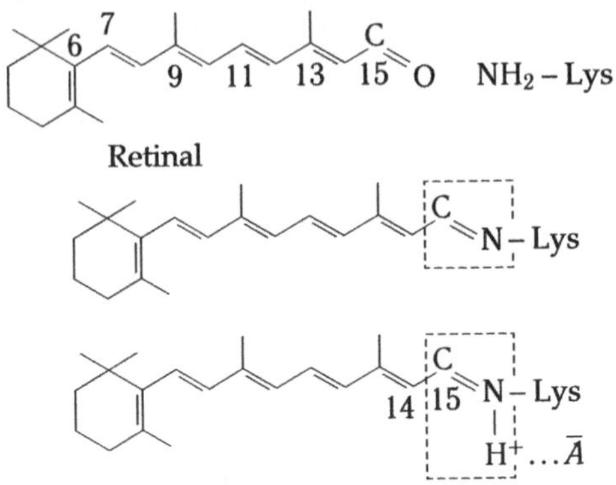

Figure 29.1. Generation of an aldimine bond between retinal and lysine in the protein chain (a Schiff base).

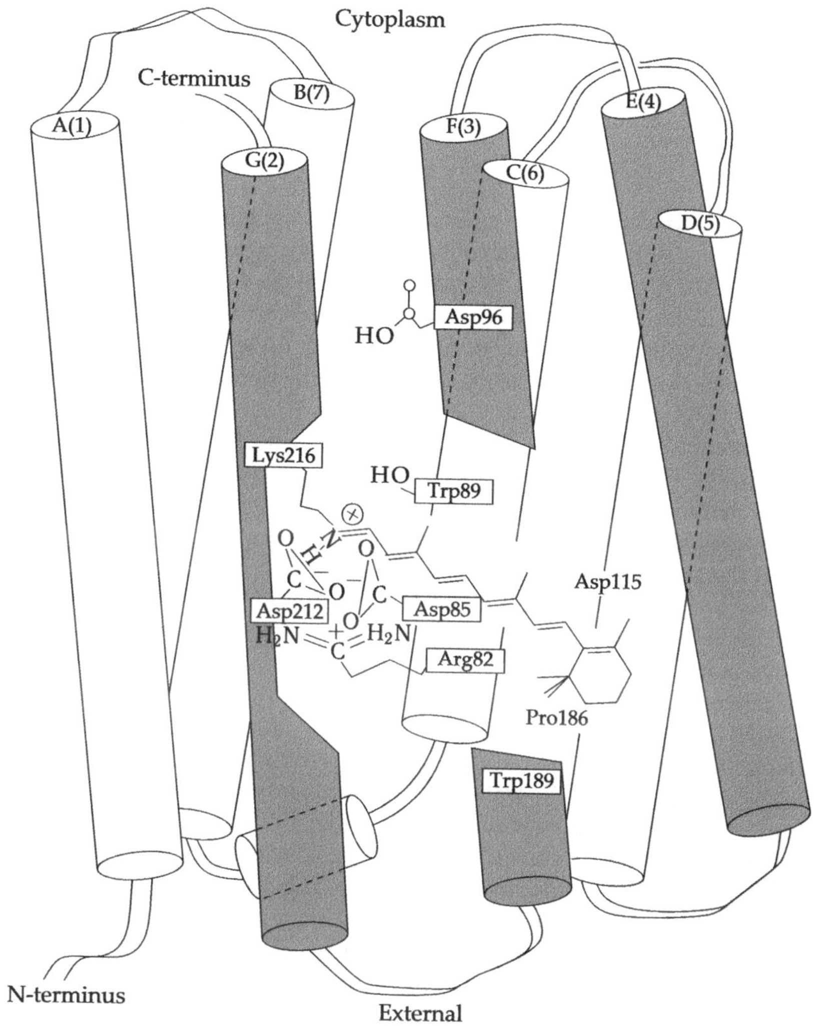

Figure 29.2. Three-dimensional structure of a bacteriorhodopsin molecule.

consider the proton role in this phenomenon. In both cases, the maximum of the spectrum of retinal absorption in the protonated Schiff base is shifted to the long-wave region as compared to its non-protonated form. This is conditioned also by the electrostatic effect of the positive charge of the proton, forming concurrently a hydrogen bond with negatively charged counterion A^- in the nearest surrounding of the Schiff base. The stronger this hydrogen bond, the weaker is the influence of the proton and the smaller the long-wave shift in the position of the maximum of retinal with a protonated Schiff base as compared to the non-protonated form.

Evidently, the effect of the negatively charged counterion A^-, forming a hydrogen bond with the proton in the Schiff base, simultaneously decreases the value of this long-wave shift. In solution, the role of a negative counterion is played by, for example, solvated ions of Cl^-. But its effect is significantly shielded by the solvent, so that the main contribution to the short-wave maximum of absorption, as compared to the absorption of an isolated chromophore, is caused by polar molecules of the solvent.

In the protein, the role of negative counterions (A^-) belongs to ionized amino acid residues.

$$\begin{array}{c} -C = N^+ - Lys \\ | \\ H \\ \vdots \\ A^- \end{array}$$

If the hydrogen bond in the protein becomes weaker, the effect of the positive charge of the proton in the Schiff base in Brh enhances. This contributes to the "opsin" spectral shift (440 → 570 nm) of Brh as compared to model systems. Figure 29.2 shows the structure of bacteriorhodopsin with two water clusters. In addition to basic counterions Asp212 and Asp85, the charges of other residues (Arg82) as well as tryptophan (Trp189) in the hydrophobic pocket of opsin, containing retinal, can also exert their effect.

It was demonstrated that in mutants, where the negatively charged aspartate (Asp85) residue is substituted for neutral asparagine (Asn), the maximum of Brh absorption is displaced farther to the red region to 590 nm. The negative charge on Asp85 disappears also upon protonation of this residue with a corresponding long-wave shift of the position of λ_{max}. This effect can be also induced by a change in the positively charged Arg82 for another neutral residue (Ala, Glu), and as a result of this Asp85 can be protonated. Thus, the positive ion of Arg82 stabilizes Asp85 in its deprotonated state.

The influence of the negative charge of counterion A^- in the protein leads also to stabilization of the ground electronic state of the chromophore molecule due to the Coulomb interaction and to simultaneous destabilization of the excited state, increasing the energy of the S_0S_1-transition. Destabilization of the excited state occurs because upon photo-excitation considerable redistribution of the electron density along the polyene chain proceeds towards the positively charged protonated Schiff base, and the positive charge moves to the ionone ring. Thus, the Coulomb repulsion of the negatively charged counterion itself in the protein and the displaced electron density leads to an additional increase in the energy of the S_0S_1-transition. An important role in the "fine tuning" of the energy value of the S_0S_1-transition is played by the electrostatic field of the protein itself. In this case, its action due to the organized structure leads to the effect that is opposite relative to the action of only one negative counterion.

Therefore, to make a general estimate of the state of the protonated Schiff base in Brh, it is necessary to take into account the effects of both the protein counterion and the amino acid surrounding of retinal with several residues.

Another reason for a long-wave shift in the protein is additional delocalization of π-electrons in the polyene chain of retinal (see Chapter 24). This occurs at 6 *cys* → 6 *trans* (C_6–C_7-bond) isomerization of the free polyene chain of retinal upon its binding to the ionone ring and thereby inclusion of the double bond of the ring in the common system of π-electrons (Fig. 29.1).

Brh in the PM can be in two different states: after adaptation to the dark and to the light. These states differ in spectra of Brh absorption. The maximum of the absorption band of Brh in the dark-adapted samples is positioned at 560 nm. After illumination, the maximum shifts to 570 nm. The data of chemical analyses show that dark-adapted PMs contain equimolar amounts of Brh molecules with retinal in the 13-*cys* and completely *trans*-forms. Light-adapted PMs contain Brh only with completely *trans*-retinal (Fig. 29.3). □

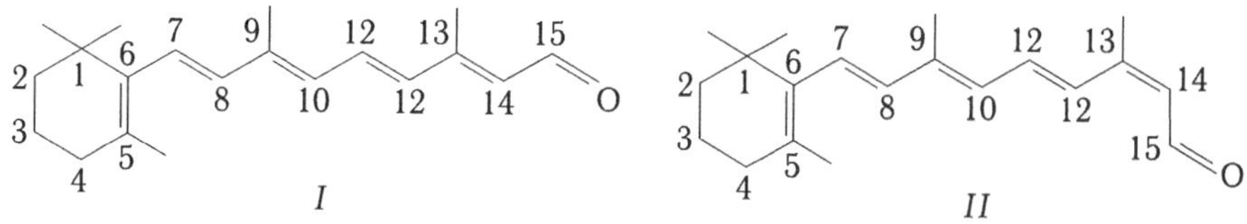

Figure 29.3. Structural formulas for completely *trans*-retinal (*I*) and 13-*cys*-retinal (*II*).

29.2 Photocycle of Bacteriorhodopsin

Common Properties. The functioning of Brh as a proton pump is based on its capacity to transfer protons from the cytoplasmic side to the external side of the cell membrane.

Light absorption by retinal initiates a photochemical cycle of Brh conversions when it undergoes transitions between different forms on its way back to the initial state. Separate stages of the photocycle are accompanied by characteristic spectral changes in Brh that makes it possible to study the kinetics of transitions between the intermediate forms in the photocycle. Each of the stages corresponds to a certain stage in the transmembrane proton transport.

After absorption of a light quantum, retinal is isomerized and transfers from the completely *trans*- to 13-*cys*-form (Fig. 29.3). This is accompanied by the proton transfer from the Schiff base to Asp85. After this the created vacancy is occupied by the proton that came from Asp96. Such an intramolecular proton transfer triggers a relay chain of transport reactions, as a result of which the proton is released on the outer side of the membrane, whereas the proton from the intracellular space is absorbed. This is shown schematically in Fig. 29.4.

During the cycle of photochemical conversions, Brh undergoes conformational transitions that, as a matter of fact, are similar to those in most enzymes (see Chapter 26). These conformational transitions are initiated by Coulomb interactions in donor – acceptor pairs of amino acids: between the positively charged intramolecular donor and the negatively charged acceptor of the proton. Prior to the absorption of a light quantum, the protein structure is stabilized by the binding energy of the complex, formed by the protonated Schiff base, by charged groups in the retinal locus and by bound water (Fig. 29.2). In this conformational state (conformation *E*) the proton in the Schiff base is coordinationally bound to Asp85 and is in contact with the outer side of the membrane. Photoisomerization of retinal displaces the Schiff base relative to Asp85, but the strains, generated in the Brh molecule, promote the proton transfer from the Schiff base to Asp85. The data of structural studies demonstrate that therewith Brh acquires a new conformation (conformation *C*) distinguished by the shift of helix *G* from the cytoplasmic side of the membrane to the molecule center (see Fig. 29.4). In conformation *C* the Schiff base is already in contact with the cytoplasmic side of the membrane, and its contact with Asp85 and, consequently, with the outer side of the membrane is disrupted.

Thus, Brh represents an ion pump with variable access of the active center to the opposite sides of the membrane. In terms of conformational dynamics, Brh is a system with two stable states where the excess conformational energy is stored during a photocycle as the transmembrane gradient of the electrochemical potential of hydrogen ions.

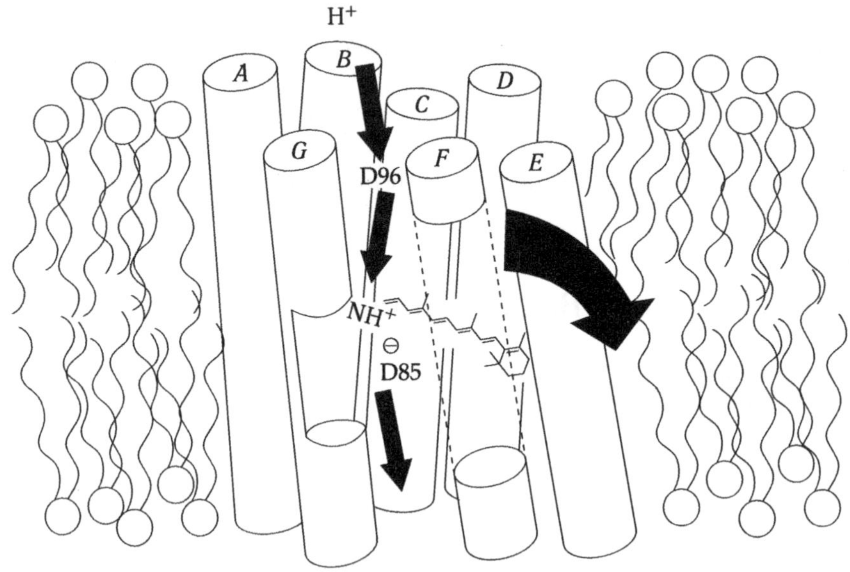

Figure 29.4. Schematic representation of bacteriorhodopsin. The pathway of the proton is shown by straight arrows. The curved arrow shows the direction of helix *F* bending during a photochemical cycle (reproduced from J. K. Lanyi, 1995).

First studies of the cycle of photoinduced conversions of Brh in PM suspension at room temperature, performed using the method of pulse absorption spectroscopy, demonstrated the presence of six intermediate states. Each of these states, designated by letters from *K* to *O* in the order of the increasing lifetimes, is characterized by definite spectral properties. The sequence of transitions between the discovered states of activated Brh is shown in Fig. 29.5. The K_{610} form, generated after light absorption at temperatures below 140 K, is stable (F. F. Litvin and S. P. Balashov, 1975). However, it is photoactive and can be transferred back to the initial state Brh_{570} (*trans-*) by the action of red light with the wavelength > 630 nm. It was found that photoinduced transition $\text{Brh}_{570\ \text{trans-}} \rightarrow K_{610}$ and inverse reaction with quantum yields 0.64 and 0.67, respectively, are independent of temperature (from 77 K to 9 K). It was shown that the generation of the K_{610} form is preceded by the formation of product J_{625} (see below). Initial form Brh_{570} contains retinal only in the *trans*-configuration, and retinal is in the 13-*cys*-state already at the stage of intermediate J_{625}.

According to numerous data, the formation of intermediate form M_{412} in the photocycle of Brh is associated with the proton transfer across the membrane. The process of M_{412} formation is coupled to the proton detachment and its release into the external surrounding. The decay of the M_{412} form is accompanied by the proton binding on the other side of the membrane (V. Stokenius, 1978). □

Mechanisms of Protein Transfer. Up-to-date models of proton transfer in the Brh cycle are based on concepts of two proton channels connecting the retinal with the internal and external sides of the membrane. The retinal is a kind of a molecular switch controlling the general transmembrane transfer of proteins along cytoplasmic and periplasmic channels (Figures 29.5 and 29.6). The primary proton acceptor from the Schiff base is Asp85. Group X is a protonated aqueous cluster $H^+(H_2O)_n$

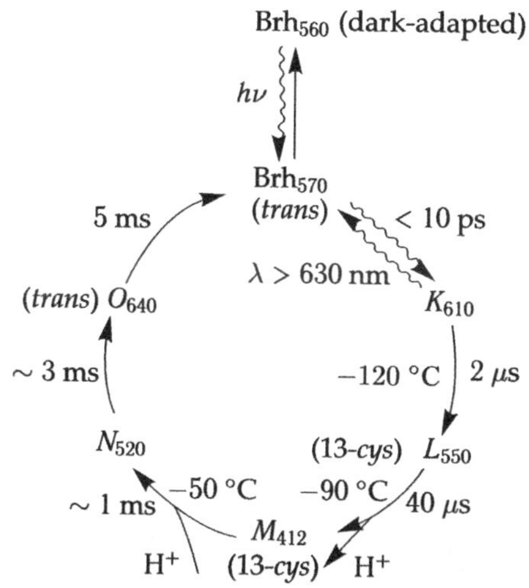

Figure 29.5. Photochemical cycle of bacteriorhodopsin. Numerals show positions of absorption maxima, lifetimes and lower temperature limits of corresponding transitions; the other explanations are given in the text.

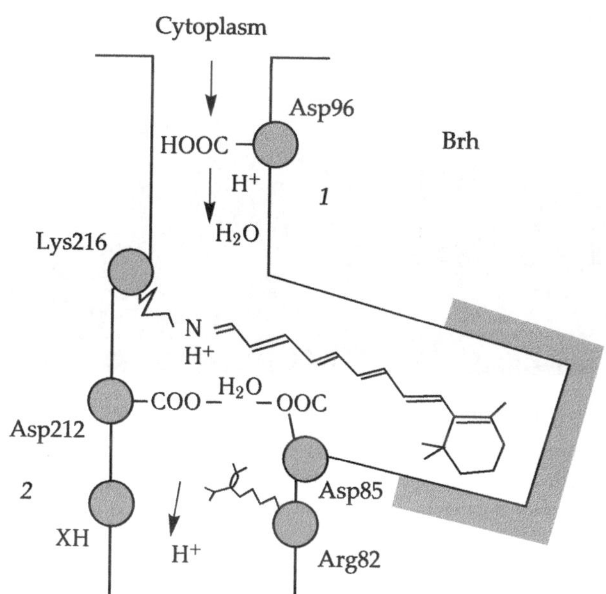

Figure 29.6. Schematic representation of proton transfer in Brh along cytoplasmic (*1*) and periplasmic (*2*) channels. Positions of the main participants of proton transfer are indicated. XH is the terminal proton donor.

containing six H-bonds with side protein chains Arg82, Glu194 and Glu204 (Figs. 29.5 and 29.6). The cluster releases the proton directly in the external surrounding. The Schiff base is reprotonated by residue Asp96, and thus the shortage of protons is filled from the cytoplasmic source. The number of H_2O molecules in $H^+(H_2O)_n$ may vary from 2 to 5 (K. Hervert, 2006).

The principal factor, determining the direction and rate of proton motion between regions of the transport system, is the degree of their affinity to the proton or, as known from biochemistry, the pK_a value. Since amino acid residues are at located short distances from each other, pK_a of each of them is also determined by the effect of electric charges on adjacent groups. The pK_a value of a residue, or its affinity to the proton, increases or decreases if correspondingly negative or positive charges appear in its direct vicinity in the neighboring molecular groups. This factor is determines the driving force and the regulation mechanism of proton transport in Brh.

Figure 29.7 demonstrates the crystalline structure of Brh at 100 K, where amino acid residues, retinal and water molecule are located inside the electron density map. As seen, counterion Asp85 in the external proton semichannel is bound to the positively charged Schiff base by a hydrogen bond via water molecule 402.

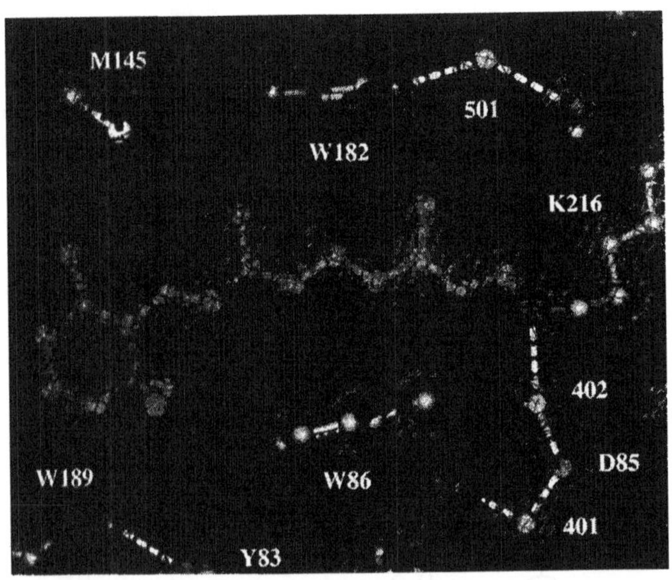

Figure 29.7. Crystalline structure of Brh at 100 K.

Trans-retinal is located between two tryptophan residues (W182 and W856). The hydrogen bond in the Schiff base (SB) is connected with Asp85 (D85) via water molecule 402 interacting with carboxyl Asp85. Resolution 1.55 E (reproduced from H. Luecke et al., 1999).

The system of hydrogen bonds, providing stabilization of the structure and concurrent motion of the proton, involves not only Asp212, but also Arg82, Glu194, Glu204 via water molecules (Fig. 29.8). It was established by methods of molecular dynamics that the total number of water molecules contained inside the Brh structure can be 15–28.

The primary act of photoprocess in Brh (its mechanism will be considered in detail below) includes photoisomerization of *trans*-13-*cys* retinal. As a result of changes in the retinal configuration, the Schiff base is shifted into the hydrophobic region of the protein with a lower dielectric constant. This leads to the enhancement of interaction between the positively charged proton in the base and the negative charge of counterion Asp85 that results in a decrease in the pK_a value of the Schiff base. Now the proton can move to negatively charged residue Asp85 and protonate it. Changes in electrostatic interactions, induced by the act of *trans*-13-*cys*-isomerization of retinal,

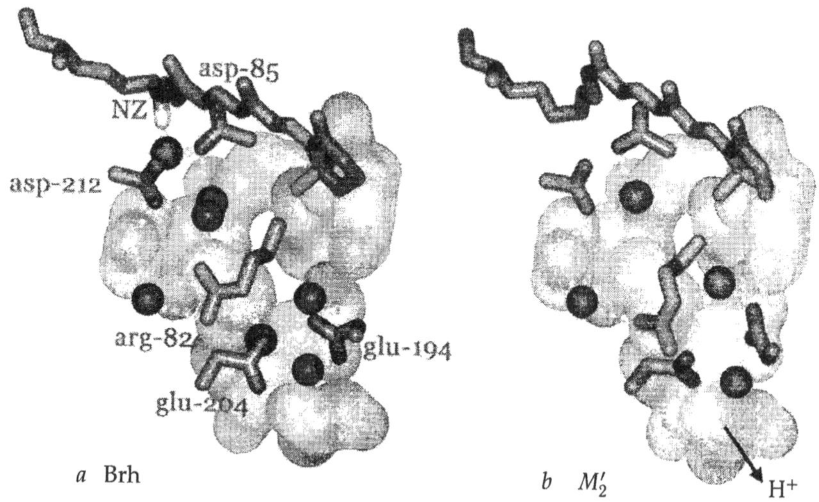

Figure 29.8. Chain of water molecules bound by hydrogen bonds in the extracellular domain of the Brh structure (*a*). Upon proton transfer to the outside, Arg82, Glu204, Glu194 and two interacting water molecules are displaced in the M_{412} intermediate (reproduced from H. Luecke et al., 1999).

are closely connected with polarization and corresponding conformational changes in the Brh protein.

The proton transfer to Asp85 occurs in several stages corresponding to consecutive generation of forms J_{625}, K_{620}, L_{550} and M_{412} in the photocycle (see Fig. 29.5).

In state K_{610}, 13-*cys*-retinal and its nearest protein surrounding are in a strained state, gradually relaxing to a planar position of 13-*cys*-retinal in the *L*-form.

In state L_{550}, the Schiff base is not protonated yet, but the shift of the maximum to the short-wave region (610 nm $\rightarrow$ 550 nm) shows that the effect of acceptor negative counterion Asp85 enhanced. A principal stage in proton transport is connected with the motion of Arg82, decay of form L_{550} caused by deprotonation of the Schiff base, proton transfer to Asp85 resulting in generation of form M_{412}. Therewith the positive charge is displaced to the membrane external side for $\sim$ 25 μs that corresponds to the proton movement to Asp85 and generation of Asp COOH. Form M_{412} has two states, in each of them residue Asp85 is protonated. In the first state of form M_{412} when the proton binds with Asp85, the terminal cluster XH is initially protonated, whereas in the second state XH already releases the proton into the external environment. Thus, proton release on the periplasmic side of the membrane during the initial 1–2 ms occurs while residue Asp still remains in protonated form Asp COOH.

The second part of the Brh photocycle is associated with reprotonation of the Schiff base and decay of M_{412}.

29.3 Primary Act of Bacteriorhodopsin Photoconversions

Then initial stages of Brh photoconversions include generation of an intermediate form J_{625} and the first relatively stable product with spectral properties identical to that of form K_{610} (batoform of Brh):

$$\mathrm{Brh}_{570} \rightarrow \mathrm{Brh}^{*}_{570} \rightarrow J_{625} \rightarrow K_{610}.$$

Namely, at this stage of the photocycle the energy of the absorbed quantum is stored and later is used in proton translocation processes.

The principal experimental approaches and methods of investigation of primary reactions of the Brh photocycle as well as the character of information obtained with their use are in many respects similar to those in the case of photosynthesis. They have been described in detail previously (Chapters 27 and 28).

The use of methods of femtosecond and picosecond laser spectroscopy has allowed thorough examination of the kinetics of fast initial processes in the Brh photocycle. An increase in photoinduced absorption changes at 625 nm occurs for $\sim$ 500 fs that corresponds to the generation of form J_{625}.

Vibrational relaxation of primary photoproduct J_{625} is accompanied by dissipation of the excess of vibrational energy, further structural rearrangement of the molecule backbone during isomerization of retinal and generation of form K_{610} for 3 ps:

$$\mathrm{Brh}^*_{570} \xrightarrow{500\ \mathrm{fs}} J_{625} \xrightarrow{3\ \mathrm{ps}} K_{610}.$$

The dynamics of the initial spectral changes in Brh absorption in the range to 100 fs has a vibrational character, associated with coherent vibrations of the nuclear wave packet upon movement along the potential surface of the excited state during the initial 100–200 fs (cf. Fig. 28.8). The modeling of such motions of nuclei at the frequency of torsion vibrations at 187 cm^{-1} demonstrated that a turn by 90° can be really completed for 100–200 fs.

Apparently, the initial stage of photoisomerization of retinal occurs for 100–200 fs upon the adiabatic transition along the surface of the potential energy of the excited state.

The *trans*-13-*cys*-turn represents electron relaxation of Brh* in the distinguished torsional degree of freedom (Chapter 28). It has a directed microstructural character. The high rate of nuclear motions associated with it leads to widening of absorption bands of Brh, observed in experiments, during the initial 100–200 fs after light excitation. After the *trans*-13-*cys*-turn a radiatonless transition occurs for 500 fs to the potential surface of photoproduct J_{625} with its subsequent vibrational relaxation and transition to K_{610} for 3 ps. This is associated with further structural rearrangements in other parts of the retinal backbone. The time of J_{625} generation is several times ($\sim$ 5-fold) lower than the time of Brh* fluorescence decay or the lifetime of electron excitation state of Brh* that makes 2.5 ps (> 500 fs). Consequently, excitation in Bhr* is retained for some time after J_{625} generation that evidences for a complex character of transition Brh* $\rightarrow J_{625}$.

This is substantiated also by a weak temperature dependence of the rates of initial processes Brh* $\rightarrow J_{625} \rightarrow K_{610}$ in the range from 295 K to 70 K with a small activation energy ($\sim$ 0.5 kJ/mol).

Figure 29.9 shows hypothetical potential energy curves of states Brh* $\rightarrow J_{625} \rightarrow K_{610}$. The curve of state Brh* has a small intermediate barrier of *trans*-*cys*-isomerization, separating the fluorescence state of Brh* and the transition region to J_{625}. It is seen that part of Brh* molecules return to the ground level emitting a fluorescence quantum and are not involved in the J_{625} formation. The other part undergoes photoisomerization being in an excited state.

The rate of retinal isomerization in the compact Brh protein ($\tau \sim$ 0.2 ps) is much higher than the rate of isomerization of isolated retinal in solution ($\sim$ 10 ps). This is determined, evidently, by the peculiarity of the protein surrounding and interaction

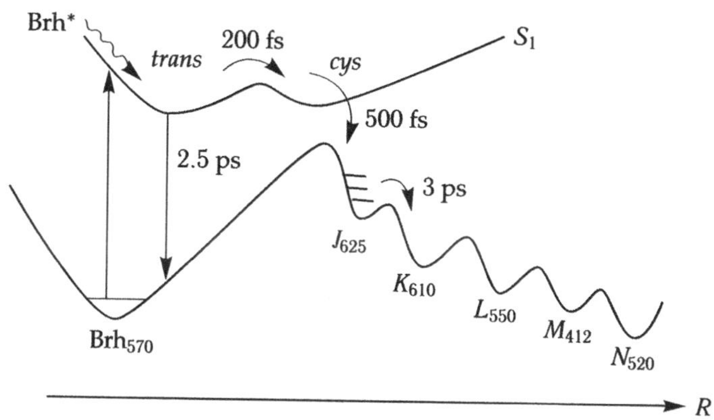

Figure 29.9. Scheme of potential energy curves of states Brh* $\rightarrow J_{625} \rightarrow K_{610}$. R is the nuclear coordinate.

of retinal with amino acid residues, responsible for the actual form of the potential energy curve of state Brh*.

It can be assumed that, as a result of the primary photoact, the system is stabilized at a new ground level of higher energy as compared to the initial Brh_{570}. This is due to intramolecular rearrangements and polarization of the retinal nearest surrounding. Therewith the energy of electron excitation is stored in form K_{610}.

The reverse spontaneous transition to the initial state is hindered now, because it would require rearrangement of molecular groups in Brh that is no longer provided by the external energy. Regeneration of $K_{610} \rightarrow Brh_{570}$ can take place here under the action of light ($\lambda > 630$ nm). As a matter of fact, the whole cascade of Brh conversions proceeds along the curve of the potential energy curve of the Brh_{570} ground state, which is conformationally non-equilibrium.

As a result, not only general stabilization is achieved, but also directed conformational rearrangements are realized in the retinal — opsin macromolecular complex at the expense of the excitation energy as the driving force of the Brh cycle.

Molecular Dynamics of the Photocycle. Voluminous data, partially analyzed above, are available in the literature demonstrating what changes in the location of molecular groups of the protein surrounding of retinal accompany photoconversions of Brh.

The results on numerical simulation of molecular dynamics (Section 3 in Chapter 11) illustrate the cooperative character of conformational motions of the protein in the photocycle of Brh (see papers by K. Schulten et al., 1994–2003). It was found that the spectral shift 570 nm $\rightarrow$ 625 nm, accompanying generation of photoproduct J_{625}, is caused by polarization of the protein matrix for 500 fs under the action of the changed dipole moment of the excited retinal.

The dynamic simulation corroborated also that conversions of intermediate M_{412} are associated with consecutive changes in protein conformations and in the retinal backbone. It turned out that three water molecules binding hydroxyl groups Tyr57 and Tyr89, rather than Asp85, are involved here. This prevents back proton transfer from Asp85 to retinal at the stage of M_{412}. The Schiff base acts as an active regulator switching the proton flow. It disconnects by the *trans-cys*-turn the conduction network of hydrogen bonds between retinal and Asp85 and forms a new network between retinal and Asp96.

29.4 Model Systems Containing Bacteriorhodopsin

Different model systems and especially films of purple membranes (PM) have become of choice in experimental studies of principles of the Brh functioning. These objects are used to examine conversion of light energy to the electrostatic energy of separated charges and the energy of polarization of macromolecular components of Brh. It has been found that the external electric field of 10^7 V $\cdot$ m^{-1}, applied to the PM film in the dark at room temperature, causes reversible changes in the Brh absorption spectrum. The corresponding differential spectrum denotes a long-wave shift of the Brh absorption band under the action of the electric field and is characterized by a negative band at 430–580 nm with a weakly expressed structure and an intensive wide positive band with the maximum at about 630 nm (Fig. 29.10). The obtained electroinduced state of Brh is spectrally similar to the first product (batoform) of the normal cycle of Brh conversions, disclosed at low temperatures.

It is possible that the electric field changes the distribution of electron density in a Brh molecule similarly to that upon light activation that is manifested by the long-wave shift of the absorption band. However, here this effect is not associated with isomerization of retinal itself.

The effect of the emergence of a maximum at 630 nm is caused mainly by polarization rearrangements in the protein as well as, probably, protonation of residue Asp85 by protons in the periplasmic channel not connected with the Schiff base.

The external electric field can affect the photoinduced processes in the PM film not only at initial stages of the photocycle. A significant manifestation of this is electrostimulation of the formation of an intermediate product of the Brh — form M_{412} photocycle.

In the dark, the electric field does not induce changes in the M_{412} absorption band. But when the electric field is switched on the background of blue-green light that causes the $\text{Brh}_{570} \rightarrow M_{412}$ transition, a considerable increase in the stationary concentration of the M_{412} form is brought about (Fig. 29.11).

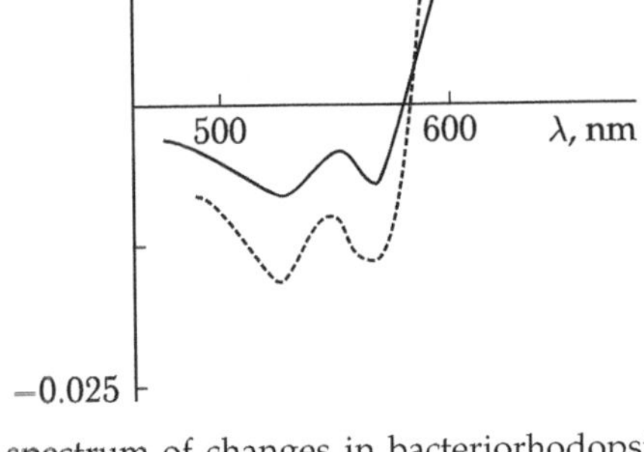

Figure 29.10. Differential spectrum of changes in bacteriorhodopsin absorption, induced by electric fields (*1*) and light (*2*) at −190°C (reproduced from G. Borisevitch et al., 1979).

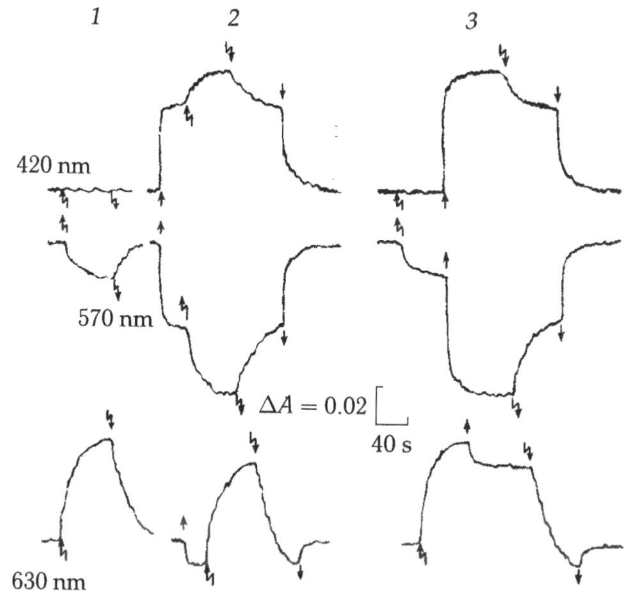

Figure 29.11. Kinetics of electro- and photoinduced changes in Brh absorption at 420, 570 and 630 nm in air-dried PM films.

Zigzag arrows indicate switching on (upward) and off (downward) of the electric field (2×10^7 V · m^{-1}); straight arrows show switching on and off of the constant active light (380 nm $< \lambda <$ 580 nm; 80 Wt·m^{-2}). *1*, Dark changes caused by the field; *2*, electroinduced changes against the background of active light; *3*, photoinduced changes against the background of the applied fields (reproduced from E. Lukashev et al., 1980).

The effect of the electric field is due to the inhibition of the M_{412} decay and regeneration of Brh. It should be, probably, associated with an increase in the activation energy of interaction of the deprotonated Schiff base with the proton-donor groups of amino acid residues (Asp96) to regenerate Brh.

Photopotentials as high as several volts were obtained on dry PM films with a high degree of membrane orientation. The kinetics of slow stages of both generation and discharging of the photopotential and the kinetics of formation and decay of the intermediate M form in the Brh photocycle are rather similar. This result supports the idea that the state M sensitive to the electric field, depends on conformational polarization rearrangements in the Bhr photocycle.

Application in Nanobiotechnology. A peculiar combination of photochrome and photoelectric properties of PM films, containing Brh, makes this protein a long-term material in high-performance biotechnical devices for information recording, storage and reading. The development of this field, called "nanobiotechnology", is based on the use of mechanisms of fast conversion of light energy in protein–chromophore macromolecular photosensitive complexes (protein–chlorophyll, protein–retinal).

Under the action of yellow light, a photopotential of several volts can be generated in dry films of oriented PM as a result of proton displacement from the Schiff base to the primary acceptor of Asp85. Information is recorded under the action of yellow light on Brh and can be read by recording the corresponding electric signal on PM, i.e. in a way not damaging the record. The record is erased upon exposure to blue light and return of Brh to the initial state.

Thin films of oriented PM (less than 50 nm thick) allow resolution of 60 E between memory elements (the distance between the centers of Brh trimers in the PM structure) and also possibly 30 E (the distance between separate Brh molecules in a trimer).

In the long term, nanobiomolecular computers based on elements with extremely small dimensions and high rates of information processing may be developed.

29.5 Molecular Bases of Visual Reception. Visual Cells (Rods)

The ability of receptor cells in the retina to react in response to changes in the light illumination underlies visual reception of vertebrates and invertebrates. The process of transformation of light energy to a photoreceptor signal in vertebrates occurs in light-sensitive cells of the retina — rods and cones. Rods provide twilight vision. Mechanisms of visual reception are studied more thoroughly in rods of vertebrates. A rod can generate a visual signal in response to absorption of one quantum. Figure 29.12 shows a schematic representation of a rod.

In the dark, the retinal chromophore of rhodopsin is in the 11-*cys*-position. Absorption of a light quantum transfers it to the *trans*-state (Fig. 29.13). Photoisomerization of retinal gives rise to a sequence of events leading to generation of a visual signal. The rod membrane has selective permeability for different ions. In the dark state, this leads to polarization of the membrane and generation of the electric potential difference between the protoplasma and the external surrounding (−40 mV) of rod cells. Illumination of the rod induces an increase in the initial negative potential inside the cell, i.e. hyperpolarization of the cell membrane (up to −80 mV). This occurs because the action of light induces a sharp decrease in the permeability of the cell membrane for potassium and calcium ions, carrying a positive charge. Absorption of only one quantum leads to the closure of 100–300 potassium and calcium channels in the external membrane. The generated slow (electrotonic) photopotential — hyperpolarization — propagates along the outer plasmatic membrane of a rod up to its presynaptic ending and decreases the rate of the neuromediator (glutamate) release. In photoreceptor cells of vertebrate retina — rods and cones — a slowly propagating electric (hyperpolarization) potential is generated but not a nerve impulse. Thus, the neurophysiological visual signal is based on the decrease in permeability of the outer membrane of rods for potassium and calcium ions induced by photoisomerization.

Apparently, photoisomerization of retinal in the membrane of discs may lead to closure of ionic channels in the outer membrane of rods only with the involvement of a large number of molecules of a special intramolecular mediator (intermediary) that are in the cytoplasm of the outer segment.

Thus, cyclic nucleotide — cGMP was found to be a low-molecular intracellular mediator. Cyclic nucleotide cGMP ($3', 5'$-cyclic guanosine-monophosphate) can directly interact with potassium-calcium channels of the plasmatic membrane of the external segment in the photoreceptor cell and maintain them in an open state. In the dark-adapted receptor, the concentration of cGMP in the cytoplasm of the external segment is very high, therefore in the dark, cyclic nucleotide cGMP maintains

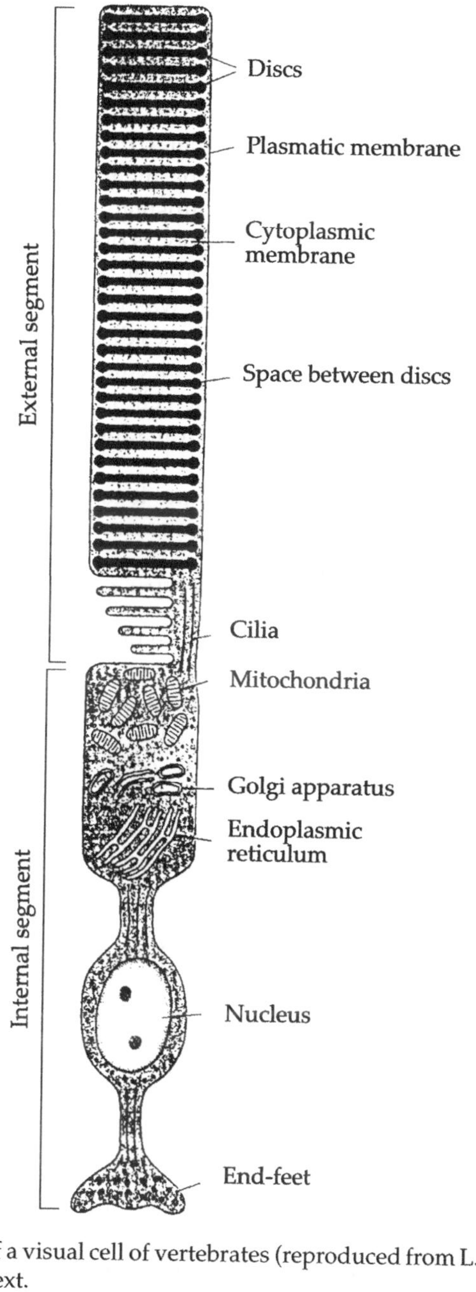

Figure 29.12. Structure of a visual cell of vertebrates (reproduced from L. Stryer, 1987). Detailed explanations are in the text.

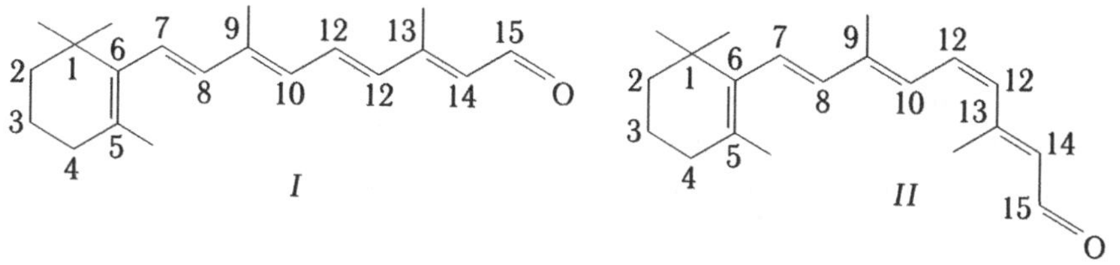

Figure 29.13. Structural formulas of entirely *trans*-retinal (*I*) and 11-*cys*-retinal (*II*).

the Na^+- and Ca^{2+}-channels in an open state. If cGMP is hydrolyzed (by phosphodiestrase — PDE — enzyme), the concentration of cGMP decreases significantly and the Na^+- and Ca^{2+}-channels pass to a closed state.

It is just this that underlies the role of cGMP as a cell mediator. A change in its concentration in the external segment regulates the state of the Na^+-channels in the outer membrane and, as a consequence, the magnitude of the transmembrane electric potential of the rods. A complex multistage functional connection exists between light excitation of rhodopsin and hydrolysis of cGMP by phosphodiesterase, its nature was established at the end of the 90-s of the last century thanks to the works of a number of research laboratories. It was demonstrated that in the light, the level of cGMP in the external segment decreases drastically upon concurrent activation of several hundreds of PDE enzyme molecules. This requires the presence of GTP as well. The factor, binding GTP in the cell, is a special *G*-protein — transducin T, responsible for activation of PDE. A transducin molecule consists of three subunits α, β and γ ($T_{\alpha,\beta,\gamma}$).

Transducin can also bind GDP, but in this state it cannot activate PDE. Transition of transducin from the inactive state to the active one occurs when GDP is substituted for GTP:

$$\text{T–GDP} + \text{GTP} \rightarrow \text{T–GTP} + \text{GDP}.$$

In the formed complex $T_{\alpha,\beta,\gamma}$–GTP, the active α-subunit of transducin T_α and the GTP molecule bound to it are split off. In its turn, the T–GTP complex dissociated from $T_{\alpha,\beta,\gamma}$ activates enzyme PDE. Activation of PDE is also based on different functional roles of subunits α, β and γ, included in the composition of the enzyme molecule, the γ-subunit inhibiting the activity of PDE. It is the splitting of this subunit, inhibiting the activity of the enzyme, that makes the enzyme active.

A complex sequence of reactions is triggered upon photoisomerization of retinal in rhodopsin and transition of rhodopsin R to active state R*. Activated rhodopsin forms a complex with transducin,

$$\text{R}^* + \text{T–GDP} \rightarrow \text{R}^*\text{–T–GDP},$$

where GDP is then substituted for GTP,

$$\text{R}^*\text{–T}_{\alpha,\beta,\gamma}\text{–GDP} + \text{GTP} \rightarrow \text{R}^*\text{–T}_{\alpha,\beta,\gamma}\text{–GTP} + \text{GDP},$$

with its subsequent dissociation

$$\text{R}^*\text{–T}_{\alpha,\beta,\gamma}\text{–GTP} \rightarrow \text{R}^* + \text{T}_{\beta,\gamma} + \text{T}_\alpha\text{–GTP}.$$

Namely, fragment T–GTP activates directly PDE, "depriving" it of the γ-subunit:

$$\text{T}_\alpha\text{–GDP} + \text{PDE}(\alpha,\beta,\gamma) \rightarrow \text{PDE}(\alpha,\beta) + \text{T}_\alpha\text{–GDP–PDE}_\gamma.$$

The sequence of these events is given in Fig. 29.14.

The central place in these cascade reactions belongs to stage IV (Fig. 29.14), where an effect of multiple enhancement of the light quantum action emerges, because one active molecule of PDE(α, β) hydrolyzes about a thousand cGMP molecules.

Deactivation of activated rhodopsin R* occurs during sequential phosphorylation reactions performed by rhodopsin-kinase RK with the formation of R–nP, where n is

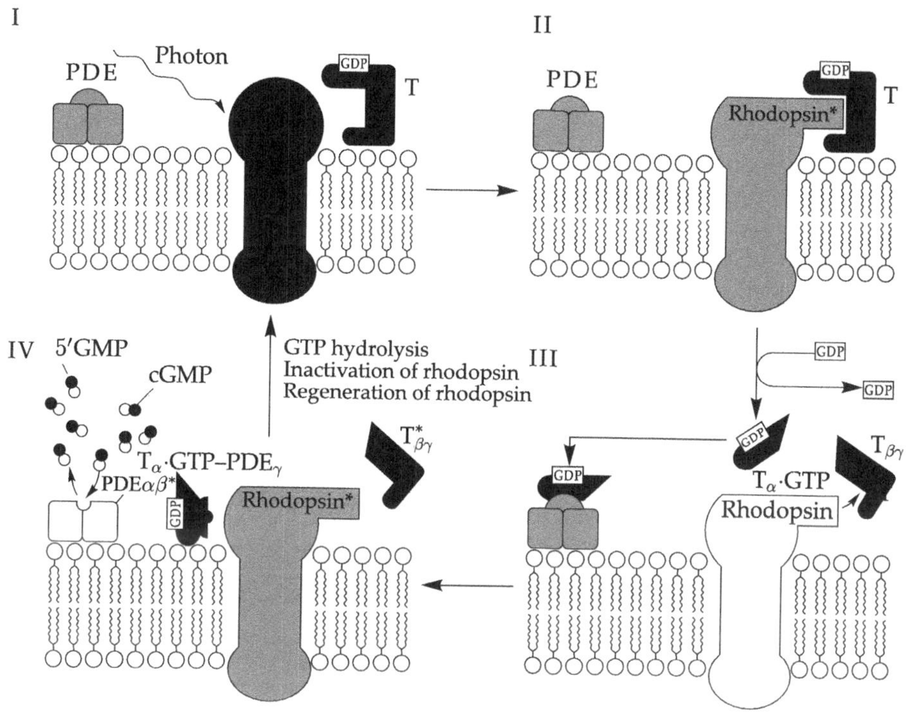

Figure 29.14. Cascade reactions underlying visual reception (reproduced from L. Stryer, 1987). I, Absorption of a light quantum activates rhodopsin R* that later reacts with transducin T; II, GDP is substituted for GTP in the α-subunit of transducin T_α; III, the detached T_α–GTP activates phosphodiesterase PDE, splitting off its γ-subunit; IV, activated PDE hydrolyzes thousands of cGMP molecules. Then GTP in the α-subunit of T_α is split and converted to GDP, and T_α-GDP binds to $T_{\beta\gamma}$. Tranducin and PDE are reduced. Rhodopsin R* is inactivated and regenerated to the initial state.

the number of phosphate groups ($n \simeq 9$). R*–nP binds to protein S, forming the R*–nPS complex. In this state, transducin G-protein is not activated further by rhodopsin that relaxes from the activated state: $R^*–n \rightarrow PS \rightarrow R^0–nPS$. The final regeneration of rhodopsin occurs in $\sim$ 30 min after its dephosphorylation (by phosphatase) and attachment of a new molecule of 11-*cys*-retinal instead of *trans*-retinal released from rhodopsin:

$$R^* \xrightarrow{RK} R^*–nP \xrightarrow{S} R^*–nPS \longrightarrow R^0–nP \longrightarrow \underset{\substack{\downarrow \\ trans}}{R^0} \longrightarrow \overset{\substack{11\text{-}cys \\ \downarrow}}{R}.$$

Attention is attracted to the similarity of the above sequence of events with the general scheme of hormonal reception (Chapter 23) with signal G-proteins, transducing being one of them. Excitation of rhodopsin and transition of transducin to activated state T_α–GTP corresponds to the stage of hormone binding to the membrane receptor and activation of the G-protein and generation of fragment G_α–GTP.

In both cases, active α-subunits of G-proteins (G_α–GTP and T_α–GDP), in their turn, activate corresponding enzymes (adenylate kinase and phosphodiesterase) that by now modify molecules of cell mediators of the reception signal. Cyclic nucleotides play their role: cAMP in the hormonal reception and cGMP in the visual reception.

This is the principal scheme of biochemical conversions underlying visual reception. From the point of view of biophysics, the central role here belongs to phosoisomerization upon light quantum absorption and the mechanisms of rhodopsin activation. This process is based on photochemical conversions in the photochemical cycle of rhodopsin.

In Figure 29.15, the structure of visual protein opsin consisting of several α-helices is presented. Helices II, II and IV are located close to each other. Gly90 (*G*-90) in helix II and Glu113 (*E* 113) in helix III are close to the Schiff base and efficiently interact with it. Helices IV, V and VI are located close to the β-ionic ring of the chromophore. Negatively charged residue Glu113 is a counterion for the positively

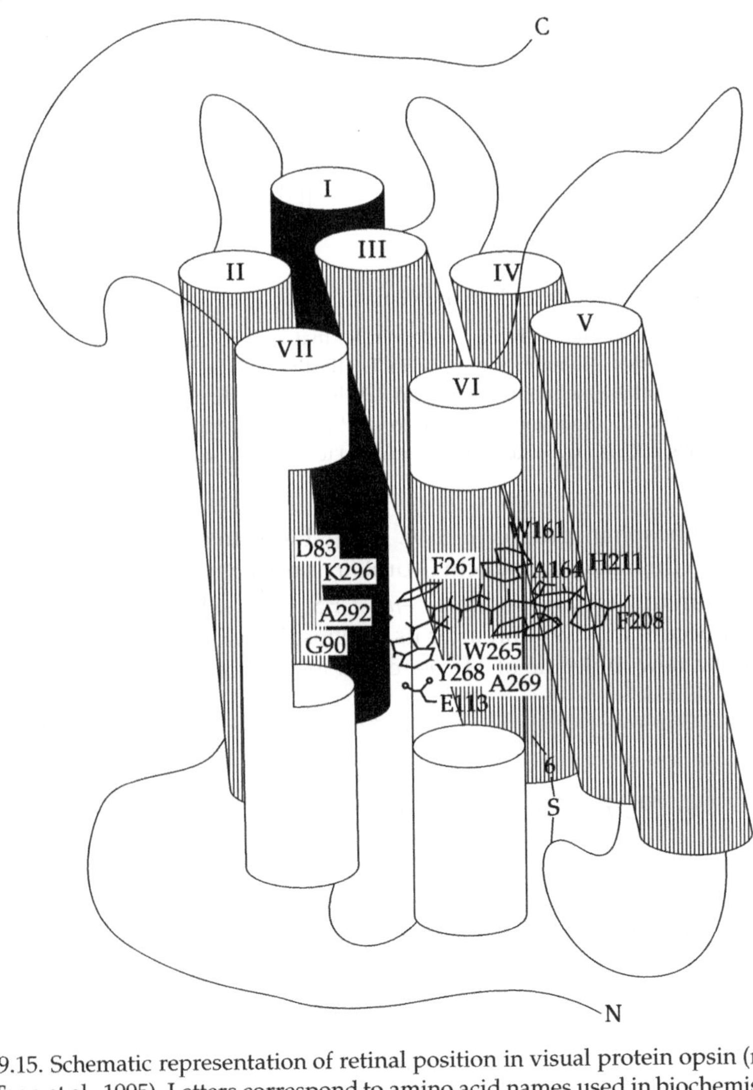

Figure 29.15. Schematic representation of retinal position in visual protein opsin (reproduced from L. Tang et al., 1995). Letters correspond to amino acid names used in biochemistry. Amino acids facing the plane of the figure are designated against the light background.

charged protonated Schiff base in visual pigments of different vertebrate animals. It is located close (~3 E) to the C_{12} atom of retinal. But in contrast to Brh (see Section 1), the hydrogen bond between the protonated Schiff base and the counterion in rhodopsin is not weak. In its properties it is close to hydrogen bonds in model systems, containing Cl^- ions as counterions. Apparently, the magnitude of the batochromic shift in rhodopsin is determined by the interaction with the negative counterion, stabilizing the positive charge near the nitrogen atom, as well as by the effect of the ionic ring connected to the polyene chain at the 6-*s-cys*-bond of retinal.

Rhodopsin Photolysis. Photoconversions of rhodopsin are triggered as a result of visible light absorption. Therewith rhodopsin is bleached and the maximum in the spectrum of chromophore absorption is shifted from 500 nm to a short-wave region. At the same time, the position of the opsin absorption band at 278 nm and its intensity remain practically unchanged.

In the initial state, retinal in rhodopsin is in the 11-*cys*-, 12-*s-cys*-configuration. As a result of light quantum absorption, the *cys-trans*-isomerization of the chromophore followed by deprotonation of the Schiff base takes place. These processes are accompanied by a change in the interaction of retinal with opsin and in the conformation of the latter. At the final stage, the aldimine bond is disrupted and free retinal in the *trans*-configuration is released. Figure 29.16 shows a scheme of rhodopsin photolysis with maxima of absorption spectra of intermediate products, the threshold temperatures above which the next photolytic product is formed, and characteristic times of separate stages at room temperature.

29.6 Primary Act of Rhodopsin Photoconversion

At present, it is completely proved that the initial stage of a photocycle and formation of a bato-product occurs as a result of retinal *cys-trans*-photoisomerization. However, the substantiation of this has required not only wide experimental studies. It was also necessary to understand how can fast photoisomerization of retinal proceed in a compact protein medium at the rates exceeding those in model solutions of low viscosity. Here is a lot in common with corresponding studies in photosynthesis and photoconversons of bacteriorhodopsin.

Y ***Cys-trans*-photoisomerization.** *Cys-trans*-isomerization at the stage of batorhodopsin formation was substantiated by structural studies as well as by the data of circular dichroism and resonance Raman-spectroscopy. □

Light absorption and changes in the electronic state should lead to corresponding molecular rearrangements in the chromophore and its nearest protein surrounding. The use of absorption flash photolysis with picosecond laser excitation made it possible to measure the rate of this process in a wide temperature range. It was found that at room temperature, the time of batorhodopsin formation < 6 ps, and at 4 K it is 36 ps (P. M. Rentzepis et al., 1977). The curve of the temperature dependence of the rate constant has a two-phase shape and, in addition, has a clearly expressed isotope effect on decreasing (by 7-fold) the reaction rate upon deuteration (Fig. 29.17).

Obviously, a change in the proton state occurs at an early stage of photolysis, which thereby cannot be reduced only to a usual *cys-trans*-isomerization of retinal. Indeed, the methods of resonance Raman-spectroscopy allowed demonstration that batorhodopsin formation is accompanied by the appearance of three strong lines

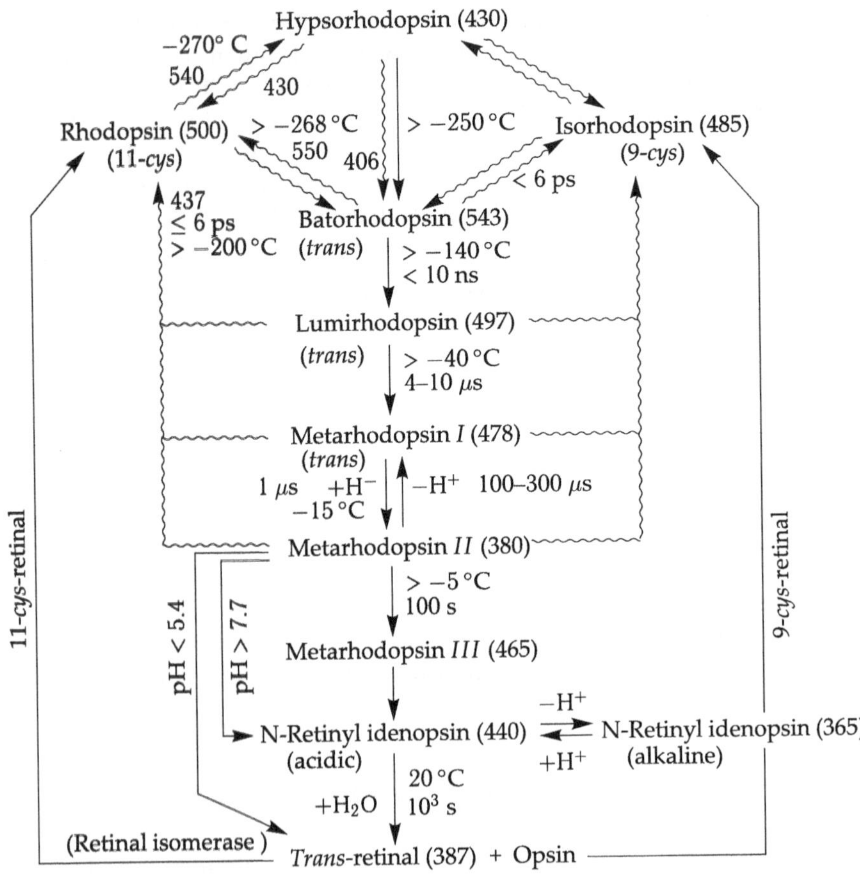

Figure 29.16. Rhodopsin photolysis.

Wavy arrows designate photoreactions, solid arrows — dark transitions; other explanations see in the text.

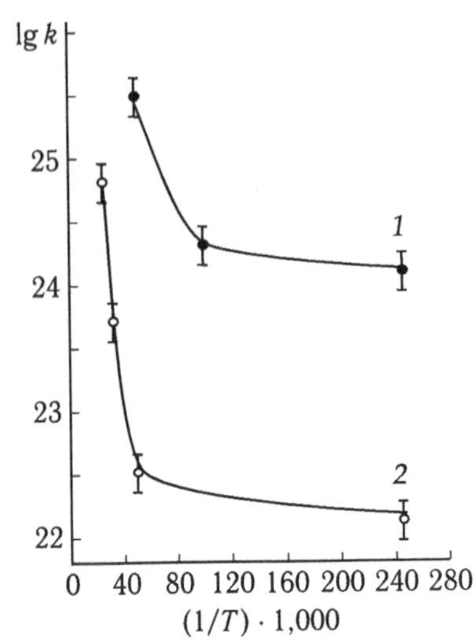

Figure 29.17. Curves of temperature dependence of the rate constant k of the rhodopsin — batorhodopsin transition in Arrhenius coordinates (reproduced from K. S. Peters et al., 1977).

at 853, 875 and 921 cm^{-1} that correspond to high-frequency ($\simeq 10^{-12}$ s) oscillations of protons. The substitution of hydrogen for deuterium in batorhodopsin in carbon atoms C_{10}, C_{11} and C_{12} causes a shift of these lines. These results demonstrate that strong extraplanar vibrations of H atoms, bound to CH_3-groups of retinal, emerge in batorhodopsin.

Not only the formation of the *trans*-configuration (a "rupture" of the retinal chain) by the double bond $C_{11}{=}C_{12}$, but also additional twisting around the single bond $C_{10}{-}C_{11}$ take place in batorhodopsin. As a result, protons in methyl groups of $C_{11}{-}C_{14}$ become exposed to the field of charged groups of this protein that can affect the formation of a batoform. Studies of rhodopsin Raman-spectra revealed that large displacements along several normal coordinates of low-frequency vibrations occur in 11-*cys*-retinal. This is the evidence that photoisomerization caused changes in many parts of the retinal backbone, involving a great number of coordinates of torsional rotation.

Femtosecond Changes. The experiments carried out with the use of the femtosecond technique (R. Mathies and S. Shenk, 1991–1992) made it possible to record changes in rhodopsin absorption in the range of 480–580 nm at times up to 500 fs. It was found that the bleaching of the rhodopsin absorption band at 500 nm and formation of the primary photoproduct are completed during 200–300 fs. It is much shorter than the time of typical electron vibrational relaxation 10^{-11}–10^{-12} s).

Figure 29.18 demonstrates difference spectra of rhodopsin absorption obtained upon excitation at the light wavelength of 500 nm and 35 fs pulse duration. The spectra have two parts corresponding to the action of probe pulses of 10 fs duration at 490 nm and 570–670 nm.

Upon photoexcitation, a system absorbs a light quantum and moves to the surface of the potential energy (PES) of excited singlet electronic state S_1. The transition proceeds in the vertical direction. After that from the Frank–Condon point (see Chapter 10, Section 2) the system moves over the surface of the excited state to the intersection point between the surfaces of ground S_0- and excited S_1-states. The difference spectrum with the lag-time of 33 fs has an intensive band at wavelengths 490–530 nm with the maximum at 510 nm. This band is connected with the absorption in excited state $S_1 \rightleftarrows S_n$ upon the action of the probe pulse. The fast disappearance (bleaching) of this absorption band (already in 100 fs) shows that the system escapes from the excited state and moves to the ground electronic state. It can be noted that after a lag of 200 fs the position of maxima in the spectrum changes only slightly, except for the intensity of the bands and the position of the isobestical point between rhodopsin and the photoproduct. The positive peak formed in 200 fs at 580 nm corresponds to photorhodopsin. At times of about picoseconds, this peak shifts further to a short-wave range.

O. M. Sarkisov and M. A. Ostrovsky (2010) determined the transition times in rhodopsin under the action of femtosecond pulses with wavelengths of 500, 535 and 560 nm using the method of femtosecond laser spectroscopy at 30 fs time resolution.

Transitions were observed from the ground state of rhodopsin R_{498} to excited state R^*_{510} with subsequent formation of photorhodopsin R_{570} and emergence of batoform R_{535}, as well as decay of state R^*_{510} and return to the ground state. Times of transitions are given in the scheme:

$$R_{498} \underset{175\text{–}230\ \text{fs}}{\overset{40\ \text{fs}}{\rightleftarrows}} R^*_{510} \xrightarrow{110\text{–}125\ \text{fs}} R_{570} \xrightarrow{2.4\text{–}2.6\ \text{ps}} R_{535}. \tag{29.1}$$

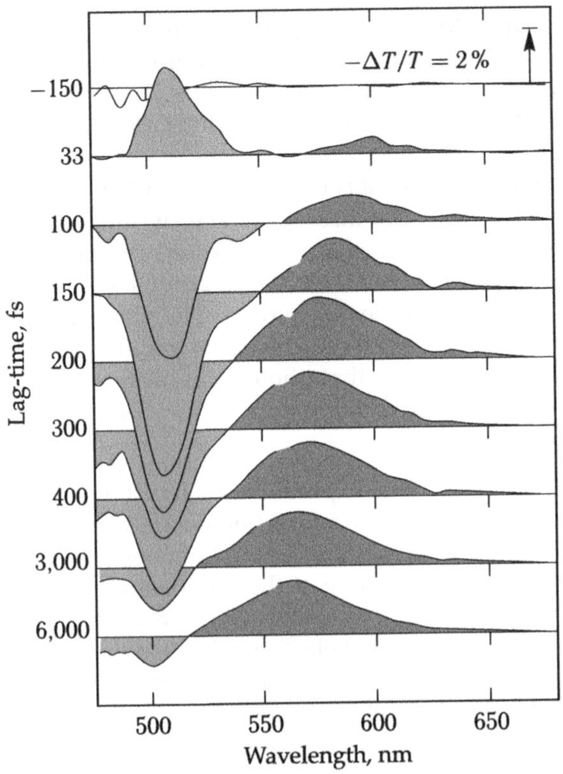

Figure 29.18. Difference spectrum of rhodopsin absorption at different times (lag-times) after light pulse excitation of 35 fs at the wavelength of 500 nm and upon the action of probe pulses of 10 fs duration (reproduced from R. Mathies and S. Shenk, 1991–1992).

In kinetic curves of absorption, changes corresponding to the initial rhodopsin R_{498} molecule and to the formation of photoisomerization products of R_{570} and R_{535}, oscillations of a signal were observed at 2 ps. They are connected with Frank–Condon states generated upon rhodopsin excitation by pulses of different wavelengths that it the system in different points of the potential energy curve following the motion of nuclei (the wave packet) along the reaction coordinate.

The Fourier analysis of these oscillations demonstrated that the power spectrum of the Fourier components contains dominating frequencies of nuclear combinations at 44, 62, 142 and 160 cm^{-1}, corresponding to vibrational modes that are active in the elementary act of isomerization of 11-*cys*-retinal in rhodopsin (62, 160 cm^{-1}) and in the recovery of the initial state of R_{498} (44, 142 cm^{-1}). It turned out also that phases of all oscillation frequencies are independent of the wavelength of probing light in the absorption bands of forms R_{498} and R_{570}. However, upon transition between spectral ranges of these forms the phase of oscillations changes by almost 180°. This means that the transition in the vicinity of the conical intersection S_1/S_0 from the excited state area of R^*_{510} to that of the products R_{570} and R_{535} and state R_{498} proceed with coherent nuclear vibrational modes that pass through the conical section at different times (Fig. 29.19)

In 200 fs after photoexcitaton, the position of bands is observed that is intermediate between 11-*cys*-rhodopsin and completely *trans*-rhodopsin. It is believed

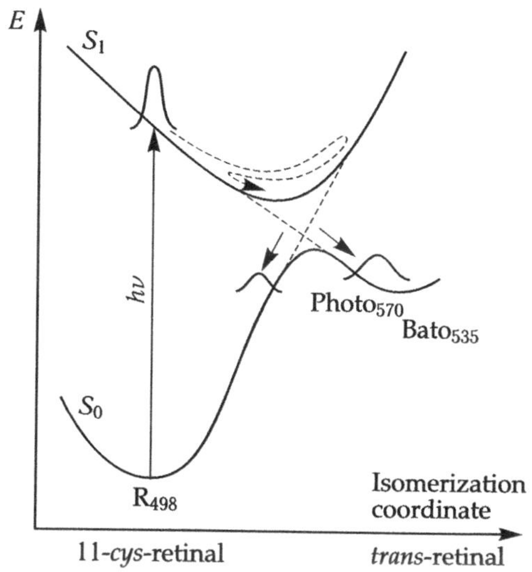

Figure 29.19. Surfaces of potential energy that are involved in the primary reactions of rhodopsin.

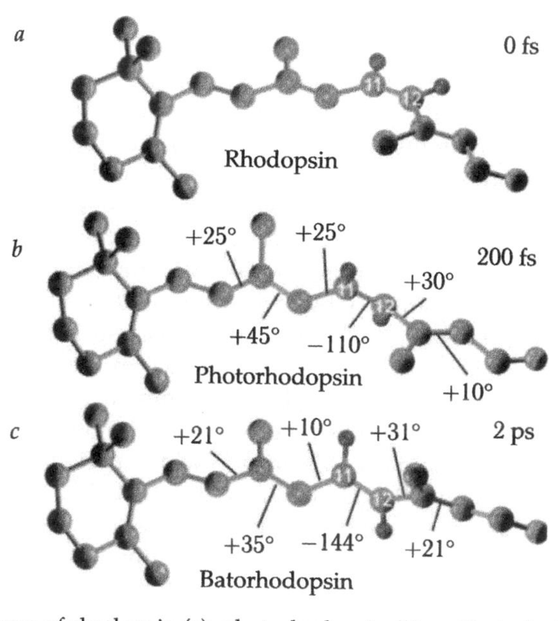

Figure 29.20. Structures of rhodopsin (*a*), photorhodopsin (*b*) and batorhodopsin (*c*) proposed by P. Kukura et al. (2005).

that just this time is required for the primary photoproduct — photorhodopsin not recorded by the XSA — to form. Based on the obtained data, P. Kukura et al. (2005) advanced a hypothesis on the structure of the chromophore group in rhodopsin, photorhodopsin and batorhodopsin (Fig. 29.20).

Figure 29.21 shows a two-dimensional cross-section of the potential energy surface. The reaction starts with fast relaxation (50 fs) on the excited surface and with a transition for 200 fs to the ground state with the formation of photorhodopsin

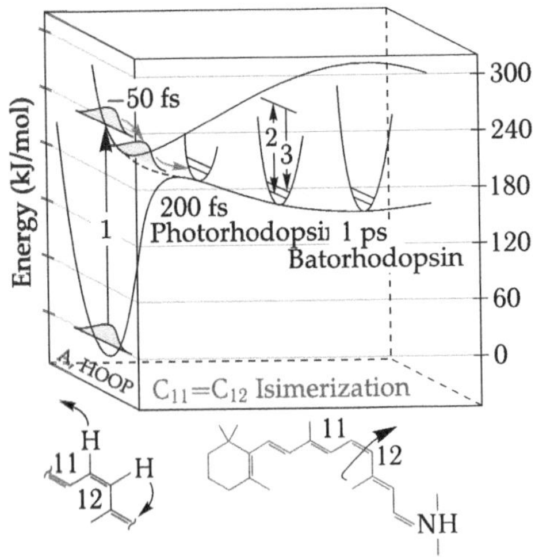

Figure 29.21. Multidimensional representation of the potential energy surface. Torsion angle H−C_{11}=C_{12}−H and the angle of rotation of the carbon backbone around bond C_{11}=C_{12} are chosen as reaction coordinates. Numerals 1, 2 and 3 denote femtosecond excitation pulse, Raman pulse and signal, respectively (reproduced from P. Kukura et al., 2005).

and batorhodopsin. During this time, the dihedral angle H−C_{11}−C_{12}−H can change by no more than 50°. The values of the vibration period (36 fs) and the frequency of vibrations of hydrogen atoms (969 cm^{-1}) are close to the lifetime of the excited state (about 50 fs) around isomerizing bond C_{11}=C_{12}. It can be suggested that namely vibrations of hydrogen atoms around C_{11} and C_{12} correspond to the motion along the reaction coordinate on the excited surface. Moreover, the intensity of the bands in the Raman scattering spectrum, corresponding to extraplanar vibrations of HOOP changes just after photoexcitation. Obviously, the fast transition from the excited surface across the section to the ground state surface is associated mainly with vibrations of H−C_{11}−C_{12}−H. Thus, the first photoproduct on the surface of the ground state is photorhodopsin with the C_{11}=C_{12} bond incompletely isomerized to 110° and with a still strongly distorted general structure at other angles (Fig. 29.20).

In addition, coherent vibrations of absorption with a period of 0.1 ps in the band of bato-product absorption (530–580 nm) were detected. The authors ascribed this vibrational mode to torsion oscillations of the retinal backbone. Figure 29.22 shows a scheme of generation of these vibrations. As seen, upon motion along the torsion coordinate, during 200 fs a transition to the curve of the bato-product ground state S_0 takes place on the surface of potential energy of singlet excited state of rhodopsin S_1. This fast transition in the site of intersection of curves of S_1 and S_0 occurs without the loss of coherence in the vibrations of the wave nuclear packet and is not accompanied by vibrational relaxation in the excited state of rhodopsin (cf. Fig. 29.9). The high rate of reaching the intersection region and the high probability of transition to the surface of S_0 of the bato-product determine a high quantum yield of the bato-product formation. As shown by the experimental data, the ratio of quantum yields of transitions from excited state S_1 to the initial form and to the bato-product is 0.33 : 0.67.

In model systems, the rate of retinal photoisomerization is much lower than in opsin. For example, in methanol the time of formation of a photoproduct of 11-*cys*-retinal in a complex with the protonated Schiff base was ~ 8 ps and the quantum

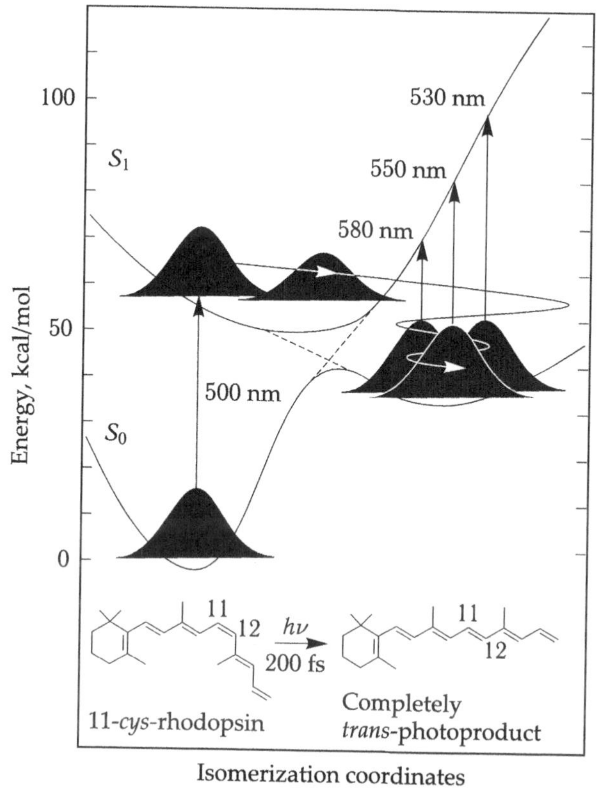

Figure 29.22. Surfaces of potential energies in states S_0 and S_1 upon photoisomerization of retinal in the femtosecond range (reproduced from G. G. Kochenderfer and R. A. Mathies, 1995). Vibrational changes in absorption at 530, 550 and 580 nm are shown.

yield was 0.15. Thus opsin accelerates the retinal photoisomerisation. This is specified by peculiarities of the protein–chromophore interactions. Among the factors, contributing to this, we can indicate electrostatic repulsion between the 13–CH_3-group and 10–H in rhodopsin that accelerates the twisting of the retinal backbone and transition

$$S_1 \rightarrow S_0 \text{ (bato)}.$$

On the whole, the weak temperature dependence (to 4 K) and high rate of photoisomerization in rhodopsin are a consequence of the protein surrounding of retinal.

In solutions, where polar molecules of the solvent create a compact solvation shell, isomerization is impeded due to steric reasons. When the free energy of solvation in solution is taken into account, the potential curve is changed, so that a barrier emerges on it hampering the *cys-trans*-transition in the excited state. This is the reason why the time of photoisomerization in rhodopsin (6 ps) is much less than in solutions ($\sim 1 \div 5$ ns). □

As seen from Fig. 29.23, the free energy ($\Delta F = \Delta U - T\Delta S$) of batorhodopsinis is higher (by $\sim$55 kJ) than the energy of initial rhodopsin. Since at 77 K *cys-trans*-isomerization is not accompanied by great entropy changes, this difference indicates to the direct storage of light energy due to the increase in the internal

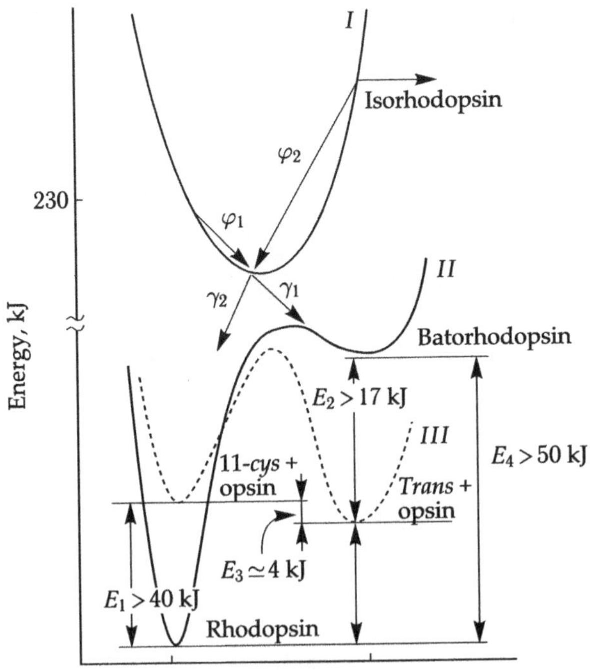

Figure 29.23. Curves of potential energy of the ground and excited states in rhodopsin by the torsion coordinate of C_{11-12}.

I, Excited state of rhodopsin and batorhodopsin. *II*, Ground state of rhodopsin and batorhodopsin. *III*, Ground state of isolated chromophore. φ_1 and φ_2 are probabilities to reach the potential minimum by rhodopsin and batorhodopsin along C_{11-12}. γ_1 and γ_2 are probabilities of transitions from this minimum to the configuration of completely *trans*- and 11-*cys*-ground states.

energy of the batorhodopsin conformation. Free retinal cannot store energy upon *cys-trans*-transition (see Fig. 29.23, *III*). Therefore, this effect is caused by a peculiar microsurrounding of retinal in the protein and, in particular, by its electrostatic interactions with the counterion. For example, the internal energy and the conformation strain will increase, if as a result of the retinal turn upon *cys-trans*-transition, the protonated nitrogen atom ($^+$NH) is fixed in the new position, with the increased distance between it and protein counterion COO^-.

The barrier-free curve of excited state general for rhodopsin and batorhodopsin is a result of the specific protein surrounding of retinal that provides its fast isomerization. □

30

Photoregulatory and Photodestructive Processes

30.1 General Characteristics of Photoregulatory Processes

In addition to photosynthesis, a diversity of photobiological processes exist in nature that are not directly involved in the storage of light quantum energy, but perform a very important function, regulating many aspects of life activity of organisms.

Such photoregulatory processes include photomorphogenesis, phototropism, photoperiodism, phototaxis, photoinduced biosynthesis of pigments and some others. A peculiarity of photoregulatory systems is their high sensitivity to the light action, combined with high biological efficiency. Any photoregulatory process occurs in several sequential stages: light quantum absorption and generation of an electron-excited state of the photoreceptor, photophysical transformation of excitation energy and sensitization of photochemical reaction, formation of intermediate photoproducts and the final manifestation of the photobiological effect. Thus, light is a signal, triggering a complex chain of sequential biochemical changes and as a result leading to a certain photobiological effect.

The first-priority task in studying mechanisms of photoregulatory reactions is determination of the nature of photoreceptors. Phytochrome responsible for the action of red — far red light was identified the first among plant photoreceptors, and after that blue-light receptors — crytpochrome and phototropin — were discovered. Circadian photoreceptors — cryptoreceptors and melanopsin — were found in mammalians. Different pigments (retinal, Flavin, pterin, blefarismin etc.) involved in phototaxis of single-celled organisms were studied.

Notwithstanding the different chemical nature of photoreceptors, primary photophysical and photochemical stages of regulatory processes have a common character. Absorption of a light quantum induces generation of an excited singlet state of the chromophore that can transit to a more long-living triplet excited state. Chromophores staying in one of these states undergo photochemical conversions: isomerization or electron (proton) transfer with a subsequent change in the conformation of the photoreceptor.

As known, consecutive reactions of phosphorylation and dephosphorylation of proteins play a significant role in enhancement and transfer of intracellular signals. In many photoregulatory processes, the generation of a physiologically active signal form includes phosphorylation of the photoreceptor. The important role of phosphorylation is shown by the presence of kinase domains in the structure of photoreceptors — the serin/threonine-kinase domain of phototropin, the C-terminal domain of phytochrome containing a region homologous to histidine kinase. The kinase activity provides the possibility of autophosphorylation of photoreceptors and phosphorylation of other substrates by these photoreceptors. A change in the conformation of the photoreceptor can also induce its phosphorylation by external kinases (for example, in cryptochrome).

The physiologically active form of the photoreceptor can trigger signal cell cascades at the subsequent stages of the light signal transfer or directly modulates the activity of transcriptional and posttranscriptional factors. The domain structure of photoreceptors is shown in Fig. 30.1.

30.2 Photoreceptors and Molecular Mechanisms of Photoregulatory Processes. Phytochromes

Phytochrome is a chromoprotein (Fig. 30.2). Its optical properties and capacity to perform the photoreceptor function are caused by a chromophore — linear tetrapyrrole (or bilitriene). The interaction of the chromophore and the protein moiety of a pigment is of great importance because this determines the characteristic property of phytochrome — its capacity to undergo photoreversible conversion and the presence of correspondingly two interconvertable forms. One form has a long-wave absorption maximum at 660 nm (F_{660}) and the other at 730 nm (F_{730}). Transition of one phytochrome form to the other is induced by red (660 nm) and far red (730 nm) light:

$$F_{660} \underset{730\text{ nm}}{\overset{660\text{ nm}}{\rightleftarrows}} F_{730}.$$

The action spectra of direct and reversible photoconversion of phytochrome are closely compatible with the absorption spectra of F_{660} and F_{730}, respectively. The quantum yield of the $F_{660} \rightarrow F_{730}$ reaction is approximately 1.5-fold higher than the quantum yield of the $F_{730} \rightarrow F_{660}$ reaction.

Iinductive resonance energy migration occurs between aromatic amino acid residues of the protein carrier and the bilitriene group of phytochrome. The efficiency of energy transfer of electron excitation makes 30–35 % with ultraviolet light, absorbed by the protein, being active both for the direct ($F_{660} \rightarrow F_{730}$) and reverse ($F_{730} \rightarrow F_{660}$) reactions. It was established that the energy of electronic excitation

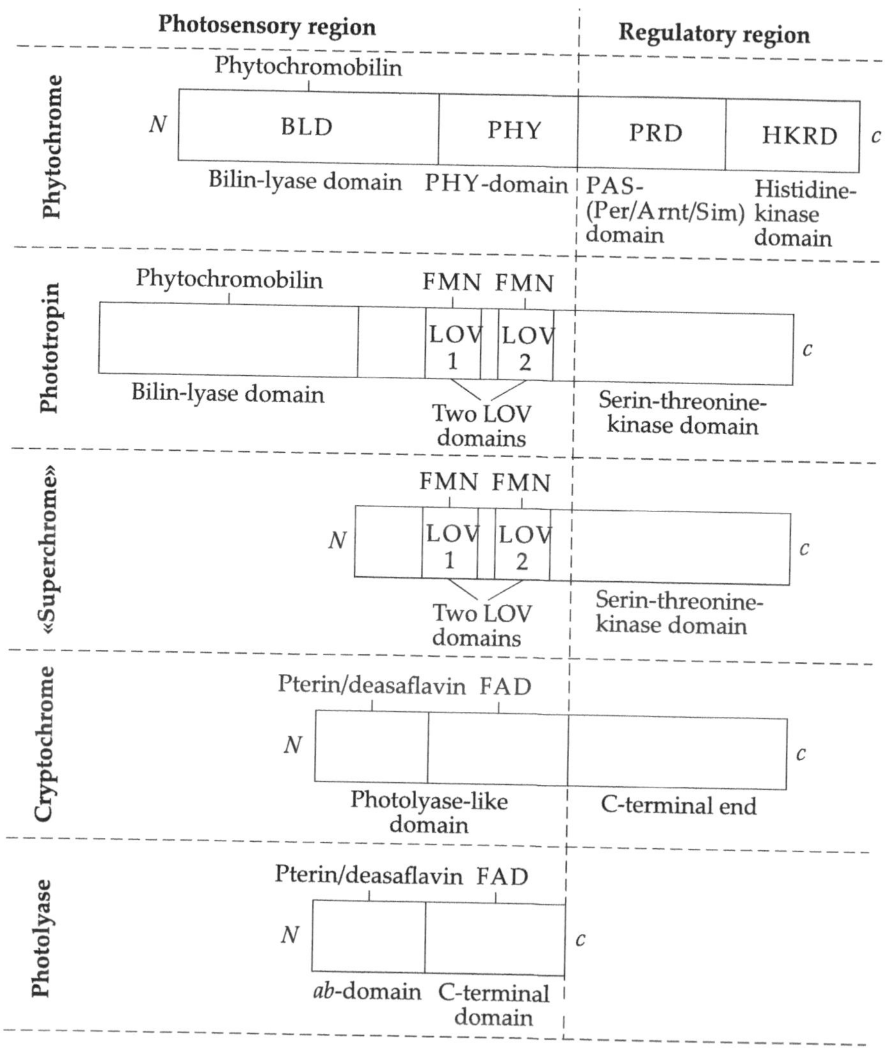

Figure 30.1. Domain structure of photoreceptors.

is transferred from the triplet level of tryptophan residues in the protein to the excited singlet level of bilitriene, which is evidenced by the appearance of delayed fluorescence.

Phototransformation in phytochrome can be separated into two phases. The primary photochemical reaction occurs in the nanosecond time range and consists in photoinduced *Z*–*E*-isomerization of phytochromobilin relative to the double bond at C_{15} between *C* and *D* rings and, probably, a turn of the chromophore around the thioether bond.

Subsequent dark regroupings take milliseconds and make the protein moiety of phytochrome correspond to the new conformation of the chromophore. Due to such photoinduced rearrangement, a far-red form of phytochrome (F_{730}) with a hydrophobic protein moiety that can interact with regulatory and signaling molecules (Fig. 30.2) is generated.

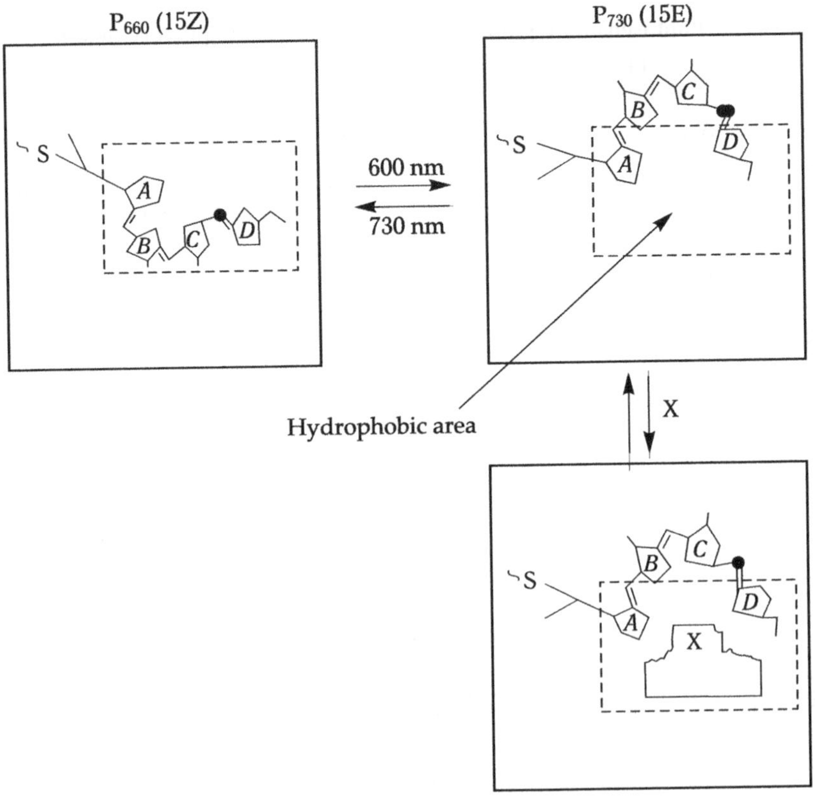

Figure 30.2. Model of photoconversions and regulatory action of phytochrome. Black circles designate C_{15} between *C* and *D* rings of phytochromobilin. X is a regulatory or signal molecule.

Diverse changes in metabolism, enzyme activity, motion, growth and plant differentiation are associated with the photoinduced generation of a physiologically active form of phytochrome. Evidence for the dependence of physiological reaction on phytochrome can be derived from the action spectrum of this reaction with the maximum at 660 nm and the reversibility of the effect upon a sequential alteration of illumination periods by red and far red light. Some of phytochrome-induced processes occur quickly, others have a lag-phase after $F_{660} \rightarrow F_{730}$ photoconversion.

The totality of experimental data available at present shows that "fast" processes are caused by induced F_{730} changes in penetrability of cell membranes for ions, electric potential and adhesive properties of the surface of plant cells. Transduction of a light signal from generation of a physiologically active form of phytochrome to the biological response includes such components of signal cascades as G-proteins, cAMP and Ca^{2+}.

Together with induction of "fast" reactions, the phytochrome system regulates a large number of biochemical and physiological processes, and their realization requires several hours and longer time intervals. As established, activation of genes induced by pigment of F_{730} underlies the mechanism of involvement of the phytochrome system in these so-called "slow" processes (Fig. 30.3). It is supposed that each differentiated cell contains a complete set of genes that with respect to light are subdivided at least in three functional types: active genes, transcribed

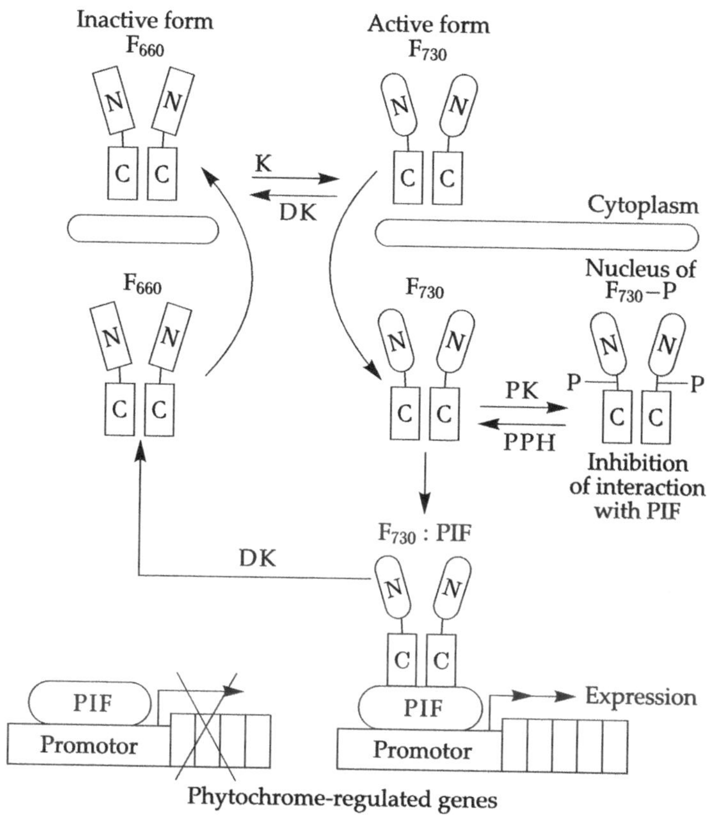

Figure 30.3. Model of regulatory action of phytochrome at the level of expression of "light-dependent" genes; N-, C–N- and C-domains of phytochrome.

PK, Protein kinase; PPH, protein phosphatase. F_{730}–P, Phosphorylated form F_{730}. R, Red light; FR, far-red light. Phytochrome in form F_{660} cannot interact with transcription factors (PIF). After photoactivation, phytochrome form F_{730} interacts with PIF, changing their functional activity by phosphorylation with its kinase domain and regulation of the expression of "light-depending" genes. Form F_{730} phosphorylated by external protein kinase prevents interaction of phytochrome with PIF, dephosphorylation by external protein kinase can reactivate the photoreceptor for further signaling.

similarly in the dark and in the light, non-transcribed either in the dark or in the light; "potentially active" genes activated by phytochrome. Form F_{730} switches on "potentially active" genes and this leads to synthesis of RNA molecules and enzymes that determine the direction of a morphogenetic reaction. The new data, obtained on photoinduced translocation of active forms of phytochrome from the cytoplasm into the nucleus and their specific interaction with a system of phytochrome interacting factors (PIF) make more convincing the conception of phytochrome action as a control regulator of gene expression.

Green fluorescent protein, as well as the group of yellow-orange, red and far-red GFP-like proteins, is at present of choice as molecular fluorescent biosensors.

With the use of chimera proteins, it was demonstrated that in the dark, inactive forms of phytochrome are localized predominantly in the cytoplasm. Red light of definite wavelengths, depending on the type of phytochrome, induces photoreversible translocation of the photoreceptor into the nucleus. When in the nucleus, active forms of phytochromes form complexes with regulatory factors of "potentially active" genes. Under the action of far red light, phytochrome transfers to an inactive form, and the complex dissociates.

Clarification of the role of phosphorylation/dephosphorylation of the phytochrome molecule itself by external protein kinases/phosphatases has advanced greatly during recent years (Fig. 30.4). As demonstrated, in this case the site of phosphorylation, with only active form F_{730} being sensitive to it, is that located between N- and C-terminal domains of apoprotein — phytochrome. In this region, phosphorylation by serine residue (Ser_{598}) goes in parallel with a noticeable conformational change that can evidence for a mediatory role of the process in the interdomain interaction, affecting the functional state of the photoreceptor. Protein phosphatases, including phosphatase PAPP5 localized in the nucleus, can dephosphorylate phosphorylated phytochrome.

Phosphorylation by Ser_{598} weakens the phytochrome-mediated signal transfer due to inhibition of the interaction of the photoreceptor and the chain components of signal transduction, including transcription factors. On the contrary, dephosphorylation reactivates this interaction, because it is a positive mechanism of phytochrome

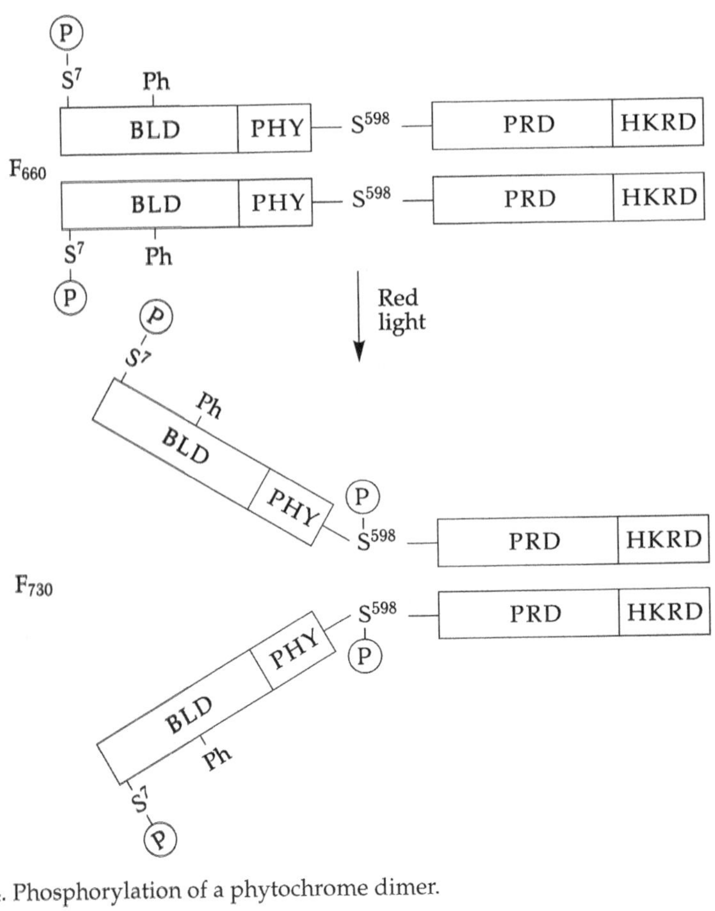

Figure 30.4. Phosphorylation of a phytochrome dimer.

S^7 and S^{598} are serine residues, P is phosphate, Ph is phytochromobilin, F_{660} and F_{730} are inactive and active forms of phytochrome. The domain structure of phytochrome is shown in Fig. 30.1.

signaling. Reversible phosphorylation of phytochrome is one of the key biochemical mechanisms, acting at initial stages of light-depending signaling in plants. At the same time, to understand how the state of phytochrome phosphorylation is regulated in time in different tissues and at different stages of the living cycle in plants, it is necessary to study molecular bases of coordinated actions between protein kinases and phosphatases.

Cryptochromes mediate reactions of blue/longwave ultraviolet light such as de-etiolated seedlings, circadian rhythms.

The protein moiety of cryptochromes consists of two domains: photolyase-like (PHR) and C-terminal (Fig. 30.1). As it follows from its name, the photolyase-like domain is homologous to photolyase that is a photo repair enzyme cleaving dimers and DNA adducts (see below, Section 6). Analogous to photolyase, this domain carries two chromophores — pterin and deaxaflavin with antenna functions and catalytically active reduced flavinadenin dinucleotide (FAD^-). Both proteins are characterized by the presence of conservative tryptophan residues ($Trp_{382} - Trp_{359} - Trp_{306}$).

As demonstrated recently with the differential spectroscopy method of high time resolution, reduction of FAD in an excited state of cryptochrome occurs due to intramolecular electron transfer from the tryptophan residue of apoprotein. It is evident that the reduced flavin chromophore (FAD^-) donates an electron to the C-terminal domain of cryptochrome. The redox transfer changes the conformation of the photoreceptor, as a result phosphorylation of its C-terminal domain (Fig. 30.5) becomes possible. In its turn, this changes the character of interactions of cryptochrome with the regulatory factors and modulation of their activity.

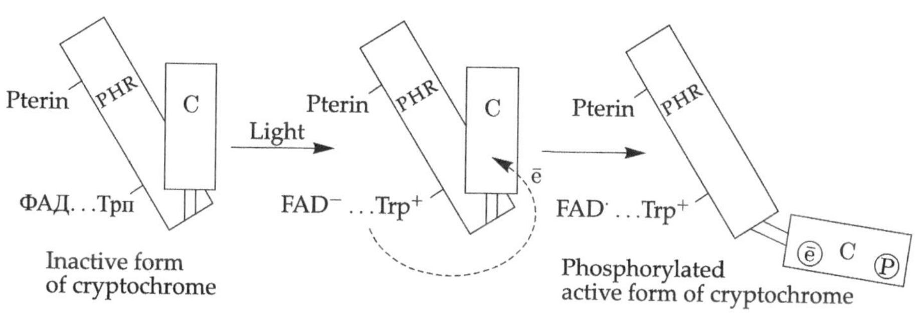

Figure 30.5. Mechanism of cryptochrome activation.

FAD, Flavin adenine dinucleotide; FAD^-, reduced flavin adenine dinucleotide, donating an electron to the C-terminal domain of cryptochrome. The domain structure of cryptochrome is shown in Fig. 30.1.

Regulation of photomorphogenesis by cryptochromes occurs via specific interactions with proteins. These interactions have an effect on the processes of ubiquitin-dependent destruction of proteins in proteasomes. For example, in plants in the absence of illumination, ubiquitin-ligase COP1 (Constitutive Photomorphogenesis Protein 1), localized in the nucleus, induces degradation of positive transcription factors. The active form of cryptochrome induces the release of COP1 into the cytoplasm that results in activation of transcription. Cryptochrome can also directly form complexes with positive regulatory factors, for example, cytoplasmic flowering factors protecting them from proteolytic degradation. As a result, the intracellular concentration of these factors increases and plants go into the flowering period.

Oscillations in transcription of *clock*-genes in animal circadian rhythms are also associated with cryptochrome modulation of the activity of cytoplasmic and nuclear regulatory proteins.

Phototropins. These photoreceptors are associated with the plasma membrane and mediate such reactions of blue/longwave ultraviolet light as phototropism, motion of chloroplasts and opening of stomas.

Phototropins are representatives of the superfamily of proteins with the structure containing the LOV-domain (Light, Oxygen, Voltage-gated potassium channel) carrying a non-covalently bound chromophore — flavin mononucleotide (FMN). Unlike other members of this superfamily, phototropins have two LOV-domains and, correspondingly, two FMNs. In addition to a sensor domain, the molecule contains a serine/threonine kinase domain (Fig. 30.1).

Absorption of a blue light quantum (of about 250 kJ/mol) leads to generation of a number of transient electron excited states of LOV-domains (Fig. 30.6), recorded by the time resolution spectroscopy method. The formation of a singlet excited state and then a triplet state of FMN (LOV_{715}) occurs within 20 ns. The triplet state is stabilized by protonation with involvement of proton transfer from the conservative cysteine residue of the LOV-domain to the excited FMN with the subsequent formation of a transient FMN-cysteinyl-thiol adduct (LOV_{390}). LOV_{390} forms during a microsecond time interval.

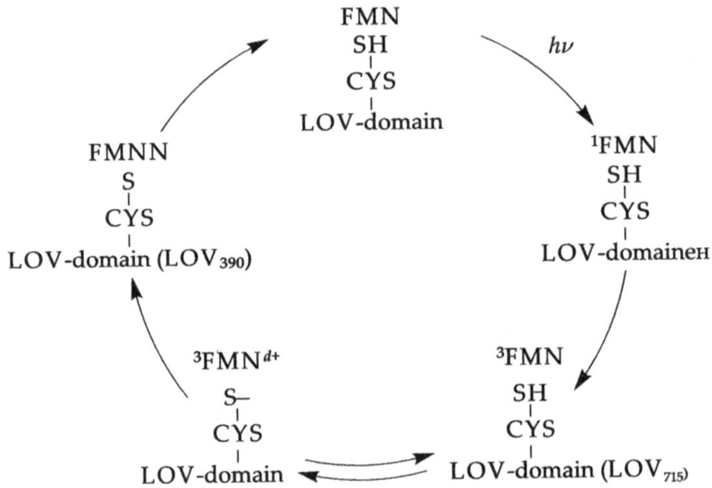

Figure 30.6. Photocycle of the LOV-domain in phototropin: FMN is the ground state of flavin mononucleotide; ^{1}FMN and ^{3}FMN are singlet and excited states of FMN; FMNN$^{\delta+}$ is the protonated triplet excited state of FMN; FMNN-S-CYS is the FMNN-cysteinyl-thiol adduct.

The both states of the LOV-domain (LOV_{715} and LOV_{390}) are high-energy forms storing correspondingly 80% and 70% of the absorbed light quantum energy. Upon formation of LOV_{715}, the conformational changes in the protein moiety of phototropin are insignificant. The transfer to form LOV_{390} is accompanied with a decrease in the entropy and a small conformational shift, while only 5–10% of the energy, related to the change in the enthalpy, is released as heat. The generation of an FMN adduct with a cysteine residue is a signal for autophosphorylation of phototropin. Subsequent stages of phototransduction are associated with activation of Ca^{2+}-channels, and upon opening of the stromas they are supposed to be associated with phosphorylation and activation of the membrane H^+-ATPase. A number of target genes encoding enzymes, were identified. They catalyze reactions of stretching of the cell wall that is a necessary condition for cell stretching at phototropic bending.

Other Photoreceptors. Superchrome is a hybrid photoreceptor of the *Adiantum capillus-veneris* fern, the N-terminal part of which is homologous to the phytochrome domain carrying the chromophore, and the C-terminal part is a phototropin-like protein with two LOV-domains and a serine/threonine kinase domain (Fig. 30.1). This photoreceptor carries three chromophore groups — one phytochromobilin and two FMNs — and performs functions of both phytochrome and phototropin.

Another type of photobiological processes, induced by blue light, is a large group of reactions of photoregulation of microorganism motions called phototaxis.

With the use of a specially developed method of extracellular potentials recording, it became possible to discover and analyze photoelectric reactions connected with reception and transformation of the light stimulus upon phototaxis (O. A. Sineschekov and F. F. Litvin, 1974). These reactions occur on the plasmatic membrane and include the primary response dependent on the electric light intensity and/or the photoreceptor potential of phototaxis, the threshold regeneration reaction and the following slow changes in the emerging difference of potentials. As proposed, the gradual photoreceptor potential directly controls the flagellar motion upon photo-orientation, whereas the electric regeneration reaction is involved in the threshold photophobic response of the cell at a drastic change in illumination. As concerns slow changes in the transmembrane potential, they evidently reflect adaptation of the photoreceptor system to the changed illumination level.

30.3 General Characteristics of Photodestructive Processes

Photodestructive processes include light-induced reactions in biosubstrate molecules leading to such chemical changes in the molecules that are accompanied with distortions or even a complete loss of their functional properties that in the end results in damaging cells or whole organisms.

Photodestructive reactions are induced most efficiently in biological systems by short-wave ultraviolet (UV) emission (< 290 nm) that is connected with direct absorption of this emission by nucleic acids, proteins and some other biologically important intracellular components. At the same time, under certain conditions photodestructive processes can also take place under the more longwave light in the optical spectrum subdivided in three ranges: middle UV range (290–320 nm), longwave UV range (320–400 nm), and visible range (400–700 nm). Unlike short-wave UV emission, absorbed by atmospheric ozone, these types of optical emission reach the Earth surface and, consequently, are ecological components of solar radiation.

Long-wave UV emission and visible light are not practically absorbed by nucleic acids and proteins. Therefore, initiation of destructive reactions in them upon the action of these types of optical radiation occurs predominantly with the involvement of other molecules acting as primary photoreceptors. Such substances are photosensitizers, and the corresponding processes, in which they are involved, are photosensitized by them.

There are several types of primary photochemical conversions of molecules, which are responsible for the destructive action of optical radiation. They include reactions of photo-oxidation, photoionization, photodissociation and photoaddition.

Reactions of photoionization, photodissociation and photoadditiont occur most frequently in proteins and nucleic acids upon their irradiation by shortwave UV light. The main primary photoreaction in the UV photolysis of tryptophan residues (AH) in proteins is its photoionization with the formation of a cation radical ($\dot{A}H^+$) and a solvated electron (e_s^-):

$$AH \xrightarrow{h\nu} \dot{A}H^+ + e_s^-.$$

The primary photochemical reaction in cysteine residues of protein consists in photodissociation of S–S-bonds:

$$R{-}S{-}S{-}R \xrightarrow[+2H]{h\dot{\nu}} 2(R{-}SH),$$

i.e. photolysis of cysteine has a stage of free radicals with localization of the unpaired electron on the serum atom.

In contrast to proteins, photoionization of nitrous bases in nucleic acids takes place only upon absorption of two quanta of UV light. The use of high-intensity pulse laser UV emission allows watching such double-quantum reactions. At relatively low UV light intensities, the principal photochemical reactions, involving bases of nucleic acids, are photoaddition reactions — dimerization and hydration.

In most cases, photosensitized destructive reactions occur with involvement of oxygen. By their mechanism, these photodynamic processes are separated in two types depending on how the energy of light excitation is transferred from the sensitizer to the biological substrate. They can be represented as the following general scheme:

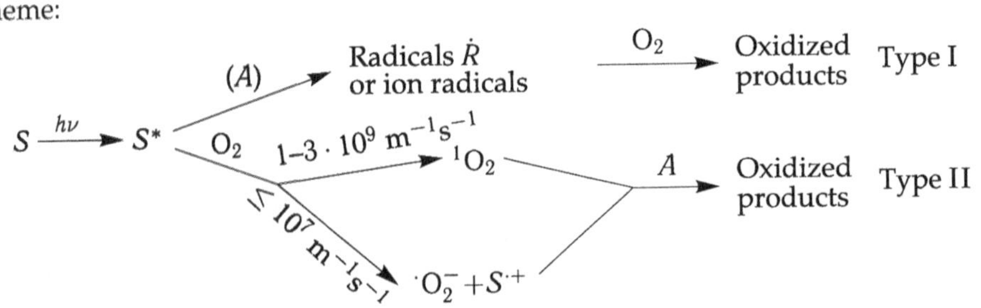

where S^* is the excited photosensitizer, subjected to primary changes, S^+ is the oxidized form of the sensitizer, and A is the biosubstrate molecule involved in the reaction at the next stage.

In type I reactions, the photoexcited sensitizer can undergo redox reactions with different biologically important molecules taking part in hydrogen atom transfer. As a result, reactive radicals (and/or ion radicals) of the sensitizer and biological substrate molecules, entering into further chemical reactions with oxygen, are generated.

Type I photosensitized reactions occur most efficiently at a high concentration of the substrate and a low concentration of oxygen, because oxygen is a strong competitor with the substrate molecule for the interaction with the triplet sensitizer. The substrates that are easily oxidized or reduced are the most efficiently photodegraded by type I mechanism.

In type II photosensitized reactions, energy is transferred from the excited singlet state of to triplet state of the sensitizer molecule and then to oxygen with formation of its electron excited form $^1O_2$; the formed singlet oxygen oxidizes biological substrate molecules. Some triplet sensitizers can realize single-electron reduction of molecular oxygen with the formation of superoxide anion radical $\dot{O}_2^-$.

Singlet oxygen can exist in two excited states: one ($^1\Sigma_g^+$) is with a higher energy (~1.6 eV), and the other ($^1\Delta_g$) with a lower energy (~1.0 eV). The lifetime of state $^1\Sigma_g^+$ in solution is 0.01–0.1 ns, whereas state $^1\Delta_g$ has a much longer lifetime (from 2–4 μs in water to several tens of μs in lipids). Therefore, state $^1\Delta_g$ is the main form of singlet oxygen involved in photodynamic reactions.

The effective transfer of energy from ^3S to $^3O_2$ is possible only if the energy of state ^3S is higher than the energy of the singlet level of oxygen. Since for an oxygen molecule the magnitude of the transfer energy to state $^1\Delta_g$ is rather low, practically any organic molecule in the triplet state can sensitize singlet oxygen formation.

Superoxide radical ($\dot{O}_2^-$) is a relatively long-living form of oxygen (~2 ms at physiological pH) and, in principle, can induce deprotonation of some molecules:

$$\dot{O}_2^- + A\,H \rightarrow H\dot{O}_2^- + A^-,$$

where $H\dot{O}_2^-$ is hydroperoxide.

However, inasmuch as it has a relatively low reactive ability, its damaging action on biostructures is realized mainly due to formation of other reactive particles ($H\dot{O}_2^-$, H_2O_2, $\dot{O}H$) and first of all to hydroxyl radical $\dot{O}H$ that can be formed from $\dot{O}_2^-$ in Haber–Weis reactions catalyzed by iron ions:

$$\dot{O}_2^- + Fe^{3+} \rightarrow O_2 + Fe^{2+},$$

$$H_2O_2 + Fe^{2+} \rightarrow \dot{O}H + OH^- + Fe^{3+},$$

where H_2O_2 (hydrogen peroxide) is a product of either O_2 dismutation:

$$\dot{O}_2^- + H^+ + H\dot{O}_2 \rightarrow H_2O_2 + O_2,$$

or its single-electron reduction:

$$\dot{O}_2^- + e^- \rightarrow O_2^{2-} \xrightarrow{2H^+} H_2O_2.$$

The most reactive radical $\dot{O}H$ promptly (at rate constants close to diffusion-controlled ones) interacts with most biomolecules, by detaching or attaching the hydrogen atom. Since in water, the lifetime of $\dot{O}H$ is less than 1 μs, it reacts with biosubstrate molecules in its nearest microsurrounding similar to singlet oxygen.

30.4 Photochemical Reactions in DNA and Its Components

As generally accepted, DNA is the main intracellular target for lethal and mutagenic short-wave UV emission. In particular, this is supported by the coincidence of the maximum in the action spectra of photobiological effects (260–265 nm) with the maximum in the absorption spectrum of DNA. □

The basic chromophores in DNA are nitrous bases of nucleotides, the quantum yields of photoconversion of pyrimidine components being by an order of magnitude higher than the quantum yields of purine photoconversions. Absorption by nitrous bases of UV light quanta (the absorption maximum at 260 nm) leads to the formation of their electron excited singlet and triplet states, generated predominantly as a result of $\pi \to \pi^*$ transitions.

In S_1- and T_1-states, DNA bases enter different photochemical reactions, among them three reactions of photoattachment process have the highest quantum yield: dimerization, hydration and formation of (6–4) pyrimidine adducts.

Reaction of Photodimerization. For the first time this reaction was discovered upon UV irradiation of frozen thymine solutions. It consists in binding two bases by the 5,6-double carbon bond with the formation of a cyclobutane ring.

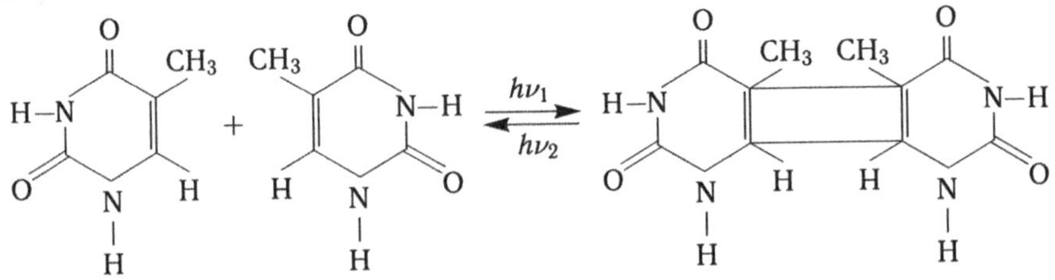

It is accepted that dimers of pyrimidine bases (first of all, thymine), formed in one DNA chain, make 70–80% of all lethal damages, induced by short-wave UV light. A characteristic feature of the dimerization reaction is its photoreversibility. Pyrimidine bases absorb light at 200–300 nm, and their dimers do this in the same range of UV spectra (200–285 nm). Therefore, at UV irradiation of DNA bases for each wavelength of exciting light, dynamic equilibrium is established between dimers and bases, determined by the relation of cross-sections of dimerization of bases and monomerization of dimers.

Because of the biomolecular character of the photodimerization reaction, its quantum yield depends essentially on the level of mutual orientation of monomers upon excitation of one of them. For example, at room temperature the quantum yield of thymine dimerization is about $4.7 \cdot 10^{-4}$ in aqueous solution and $\simeq 1$ in frozen solution. The quantum yield of thymine dimerization in DNA is $\simeq 2 \cdot 10^{-2}$. In accordance with the performed calculations, conditions for thymine dimerization in DNA are optimal, if neighboring monomers are oriented to each other at an angle of 36°. □

Formation of thymine dimers in DNA through a triplet state was demonstrated in experiments on selective occupation of the triplet level with the help of an acetophenone photosensitizer. Positions of singlet excited and triplet levels of the latter relative to the corresponding levels of thymine are shown in the scheme below.

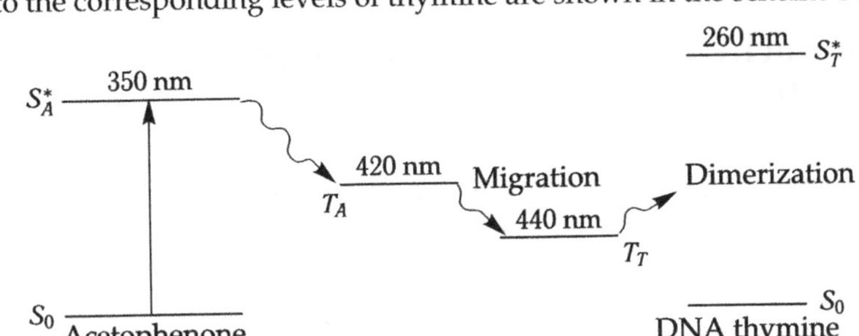

In the scheme the values of energy levels are given in the wavelengths of corresponding quantum. Upon irradiation of the DNA solution in acetophenone by light of the 350 nm wavelength, only the sensitizer absorbs quanta. Then energy is transferred from the triplet level of the sensitizer (acetophenone is characterized by a rather high degree of interconversion) to the triplet (but not to singlet) level of thymine that interacts with the adjacent thymine molecule, forming a cyclobutane dimer.

The results of experiments with acetophenone support the principal possibility of dimerization via the triplet state of thymine. Some acrydine stains can efficiently decrease the yield of UV-induced dimers in DNA. The protective property of acrydines depends on their capacity to intercalate between base pairs in a DNA molecule. A characteristic feature of interaction of acrydines (for example, acrydine orange) with DNA is that upon formation of a complex, the intensity of the dye fluorescence increases sharply in the spectrum maximum at 530 nm. Some researchers believe that this can be caused by singlet-singlet migration of excitation energy from the DNA bases to the dye. Calculations show that the maximal quantum yield of energy migration is 32 % at the ratio of 1 dye molecule per $\sim$ 5 bases.

Generation of (6–4) Pyrimidine Adducts. These photoproducts, characterized by absorption at 315–320 nm and fluorescence at 405–440 nm, were isolated from acidic hydrolysate of UV-irradiated DNA. The structure of adducts was determined using UV-, IR- and NMR spectroscopy and mass-spectroscopy. The quantum yield of (6–4)-adducts is about 10^{-3}, i.e. is about 10 times smaller than the yield of cyclobutane dimers, and, consequently, they contribute insignificantly in the lethal effect of UV-irradiation (254 nm). However, in UV mutagenesis, they can play a considerable role. Unlike lethal damages of DNA, mutation defects emerge much more seldom, and therefore here the requirement of a maximum quantum yield has no principal importance. The reaction of formation of (6–4) pyrimidine adducts is supposed to proceed by the following scheme,

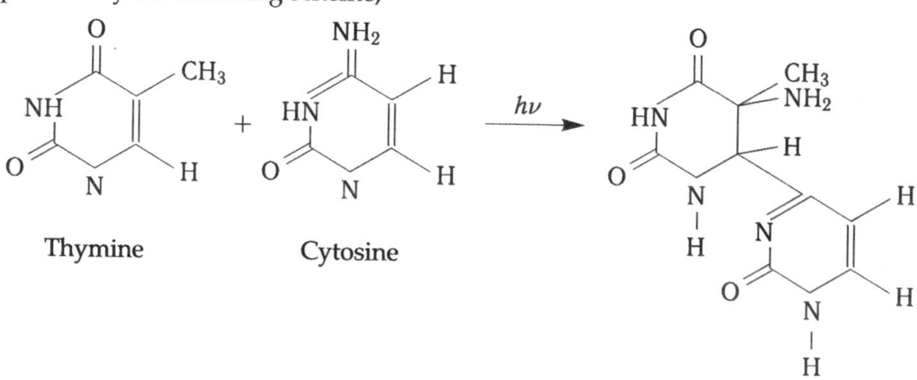

Thymine-cytosine (6-4 adducts)

Reaction of Photohydration. This process is the second important photochemical reaction of pyrimidine bases of DNA consisting in attachment of water to carbon atoms in $C_5(H)$ and $C_6(OH)$ of the pyrimidine ring with a disruption of the double bond between them and formation of 6-oxy-5-hydroderivative bases.

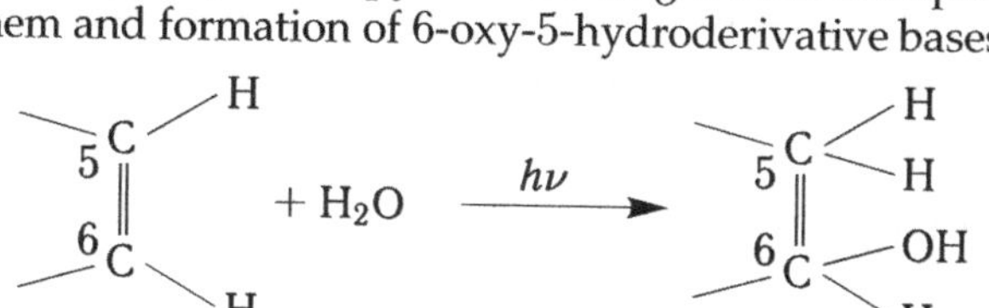

In contrast to dimerization, the hydration reaction is not photoreversible. However, hydrates can disintegrate upon an increase in temperature (> 30 °C) and ionic strength of solution as well changes in pH. The rate of photohydration decreases upon substitution of H_2O for D_2O. Precursors of pyrimidine hydrates seem to be their singlet excited states.

Cross-links with Proteins. This is an example of intermolecular interactions of the third type of photochemical reactions involving DNA bases. The two components are UV light acceptors, because irradiation of both the protein and DNA prior to mixing is accompanied by the formation of cross-links. The mechanism of this process may consist in the attachment of protein amino acid residues via SH- or OH-groups to C_5 or C_6 of cytosine or thymine in DNA. Formation of 5-S-cysteine-6-hydrouracyl by such a mechanism was observed in a model system.

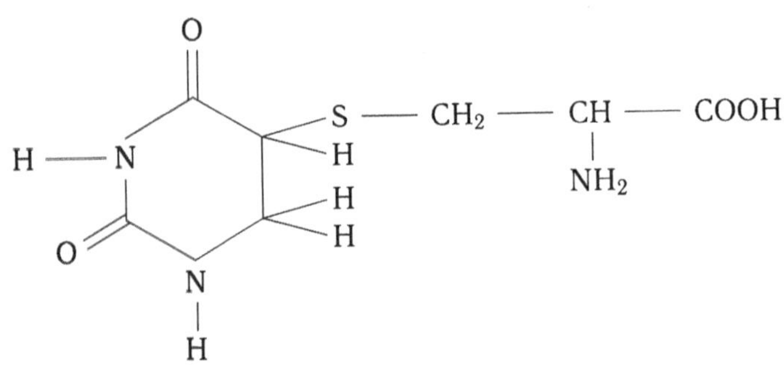

30.5 Effects of High-intensity Laser UV Irradiation on DNA (Two-quantum Reactions)

Prior to the development of laser sources of UV radiation, mainly processes linear with respect to the UV light intensity were analyzed in classical UV photobiology, i.e. only biological effects of single-quantum photochemical reactions were studied.

Two-quantum excitation of nitrous DNA bases in aqueous solution at room temperature is realized by nanosecond ($\tau = 10$ ns) and picosecond ($\tau = 30$ ps) pulses of laser UV radiation with the wavelength of 266 nm, close to the maximum of the first absorption $S_0 \rightarrow S_1$ band of the bases. This is achieved at radiation intensities above 10^8 W/m^2, when the occupation of S_1 and T_1 levels and the rate of their depletion following upward transitions as a result of excitation of high electron levels (S_n and T_n) via intermediate levels S_1 and T_1 occurs. Such double-quantum excitation via a real intermediate state is called two-step excitation.

The level of decomposition of bases is quadratically dependent on the intensity of radiation, evidencing for a double-quantum mechanism of laser-induced photochemical conversions.

The theoretical analysis of processes, occurring upon step-by-step absorption of two light quanta, showed that at picosecond pulses the occupation of high electronic states occurs via the singlet channel ($S_0 \xrightarrow{h\nu} S_1 \xrightarrow{h\nu} S_n$), whereas at nanosecond pulses — via the triplet channel ($S_0 \xrightarrow{h\nu} S_1 \rightarrow T_1 \xrightarrow{h\nu} T_n$) (Fig. 30.7). But regardless of what channel ($S - S$ or $T - T$) is involved in the absorption of the second quantum, it is probable that photoionized states of molecules participate in the reactions. This

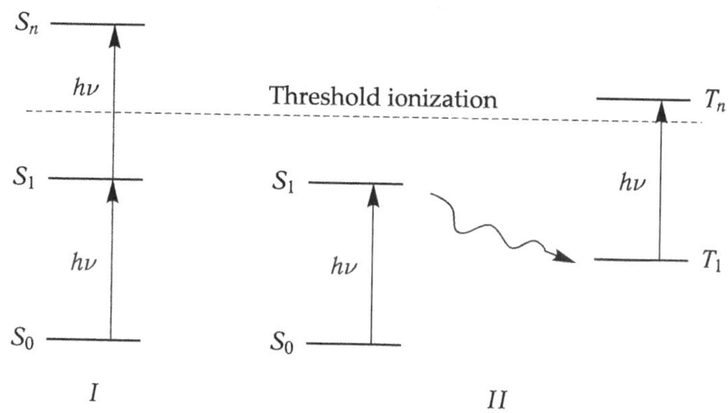

Figure 30.7. Singlet (*I*) and triplet (*II*) channels of two-step excitation.

S_0 is the singlet unexcited level; S_1 is the first singlet excited level; S_n is the high singlet excited level; T_1 is the first triplet excited level; T_n is the high triplet excited level.

conclusion can be verified by experimental data on the similarity of final products of laser photolysis and products of gamma-radiolysis of DNA, the latter are known to proceed via the stage of molecule ionization. Based on these data, a scheme of primary steps of thymine conversion upon laser photolysis was proposed.

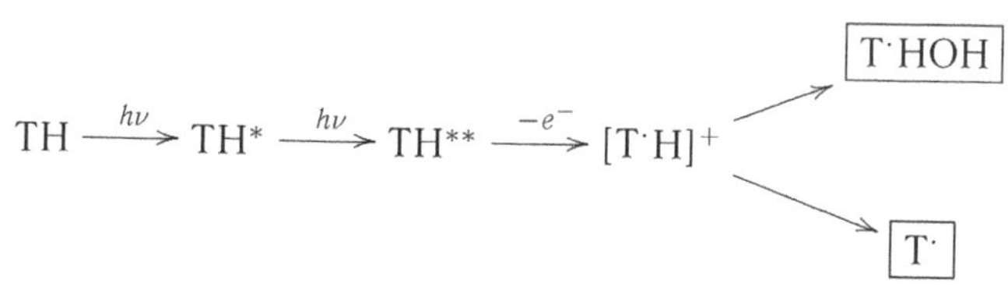

Laser-induced double-quantum excitation of bases included in the DNA structure leads to DNA photochemical conversions that do not take place (or occur with a very low quantum yield) under low-intensity UV light. Along with degradation of DNA bases, disruptions were revealed in the N-glycoside bond with the detachment of thymine from the DNA chain (at low-intensity UV radiation this process does not take place) as well as single-chain disruptions. It was shown that the quantum yield of single-strand disruptions upon transition from low-intensity UV radiation (1 W/m^2) to high-intensity picosecond UV radiation ($4 \cdot 10^{13}$ W/m^2) grows from $(1 \div 2) \cdot 10^{-6}$ to $8 \cdot 10^{-5}$. □

30.6 Photoreactivation and Photoprotection

Photoreactivation. This is a photobiological process aimed at removal of UV-induced lethal photoproducts of DNA — pyrimidine dimers and 6–4-photoadducts. Photoreactivation is performed by light-dependent enzyme photolyase having the maximum efficiency at 350–450 nm of light. Photolyases use the light energy for disruption of the (C_5-C_5)- and (C_6-C_6)-bonds of the cyclobutane ring in the formed pyrimidine dimer. Kinetics of photoenzyme-induced cleavage of dimers corresponds to the kinetics of classical enzyme processes described by the Michaelis–Menten scheme:

$$E + S \underset{k_2}{\overset{k_1}{\rightleftharpoons}} ES \underset{k_3}{\overset{h\nu}{\longrightarrow}} E + P.$$

The formation of the photolyase–pyrimidine dimer complex takes place in the dark at constant $k_1 = 10^6$–$10^7\ M^{-1}s^{-1}$. However, the catalytic act in the enzyme–substrate complex proceeds only under the action of light (320–500 nm): photolyase is active in an electron-excited state.

Molecular masses of apoenzymes of studied photolyases are 54–66 kDa, and all of them have a high content of tryptophan residues. All photolyases contain two chromophores each (the ratio of the chromophore to the apoenzyme is 1 : 1. One chromophore is $FADN^-$ and the other is folate (or deazaflavin)). The folate chromophore (or deazaflavin) plays the role of an antenna, absorbing longwave UV or blue light quanta and transferring the electronic excitation to the $FADN^-$ cofactor. The latter, in its turn, passes to the singlet excited state and donates an electron to the pyrimidine dimer (Fig. 30.8). This dimer form is unstable and through spontaneous decompositon releases pyrimidine and anion radical of pyrimidine. The latter returns

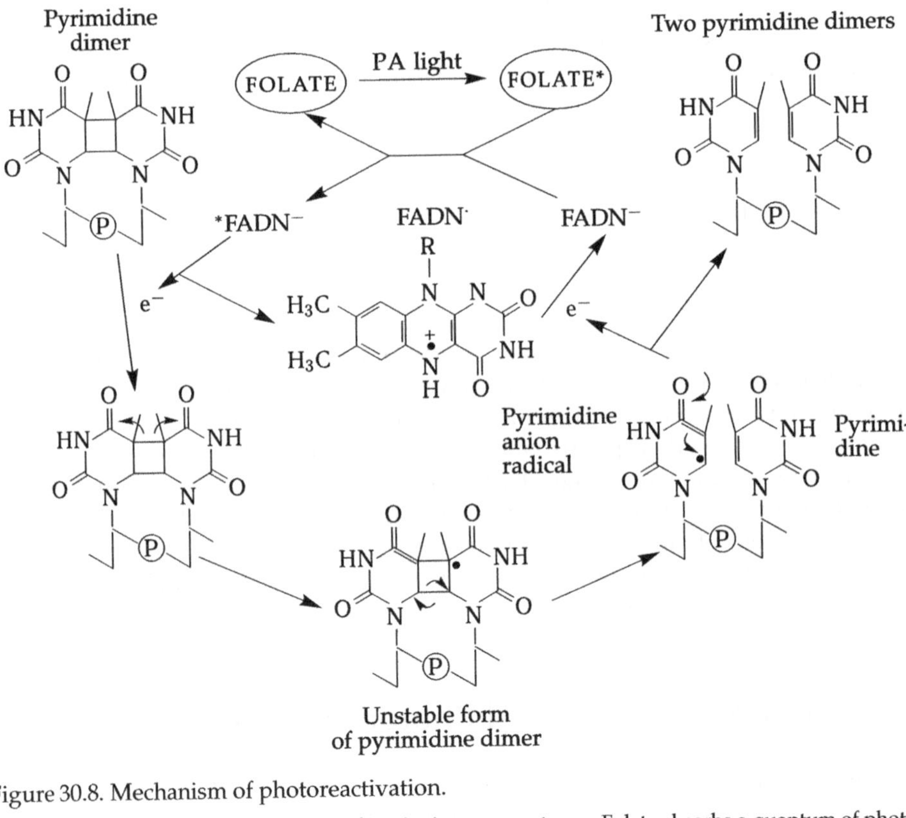

Figure 30.8. Mechanism of photoreactivation.

Folate is the chromophore of photolyase functioning as an antenna. Folate absorbs a quantum of photoreactivating (PA) light and transfers the energy of excitation to the photolyase cofactor — reduced flavin adenine dinucleotide ($FADN^-$). $^*FADN^-$ is the singlet excited state of $FADN^-$. $^*FADN^-$ donates an electron to the pyrimidine dimer, converting it to an unstable state, and turns into a $FADN^{\cdot}$ radical. Regeneration of $FADN^-$ is performed due to the electron transfer from the anion radical of a pyrimidine generated due to the splitting of the cyclobutane ring of the unstable dimer form on $FADN^{\cdot}$.

the electron back to the FADN cofactor with the regeneration of the catalytically active $FADN^-$ form.

Enzymatic photoreactivation of cyclobutane pyrimidine dimers is a widespread photobiological process: it has been found in many organisms, including archaea, prokaryotes and eukaryotes, but is absent in cells of placental mammals and in human cells. In plants, photoreactivation of pyrimidine dimers is considered to be the key defense mechanism protecting DNA from middle UV-induced damages. Photolyase activity in etiolated seedlings is induced by visible light and is controlled by phytochrome.

Photo-protection. In addition to photoenzymatic cleavage of DNA dimers (the photoreactivation process), it is possible to decrease the yield of these photoproducts in cells using another photobiological process — photo-protection. The photo-protective effect is provided by preliminary irradiation of cells with longwave UV light that leads to the significant reduction of their sensitivity to the lethal shortwave UV irradiation. This effect has been found in a number of yeast organisms.

A characteristic peculiarity of the photo-protective effect is that for its manifestation a temperature-dependent time interval should exist between consecutive applications of long- and shortwave UV illumination of cells. During this time interval, photoinduced formation of a certain compound identified as 5-oxytriplamine or serotonin takes place. The role of this compound as the main intermediate in the process of photo-protection is supported by the following results.

1) Photo-protection is not practically observed in cells deficient in the precursors of serotonin synthesis (tryptophan and 5-oxytriptophan), but the effect is again observed upon their exogenous introduction into the cells.
2) A specific inhibitor of the serotonin synthesis completely eliminates the photo-protective effect.
3) Serotonin, introduced exogenously into the cells, induces a photo-protective effect against UV inactivation analogous to photo-protection, i.e. serotonin has a photomimetic effect.

According to the data obtained in experiments *in vivo* and in model experiments (G. Ya. Fraikin et al., 1985), the protective effect of serotonin upon photo-protection is caused as a result of its binding with DNA and a decrease in the yield of UV induced pyrimidine dimers in it. The spectrofluorimetric analysis demonstrated that serotonin binds to DNA in an intercalation mode. Therewith each intercalated serotonin molecule can decrease the yield of dimers in the region of DNA chain with $\sim$4 base pairs.

30.7 Ultraviolet Light Action on Proteins

Proteins are the second intracellular target of UV irradiation of biological systems. UV light inactivates proteins by a single-quantum, single-hit mechanism with a quantum yield for different proteins varying from 10^{-2} to 10^{-3}. To identify the nature of protein chromophores that absorb interacting light, it is necessary to evaluate the degree of photo-damage of different amino acids at different wavelengths and doses of UV irradiation. The use of such an approach makes it possible to reveal the role of certain amino acid residues in maintaining the native structure of proteins and enzymes. □

Using this approach it was demonstrated that, for example, in trypsin the destruction of one tryptophan residue (of four ones) and one cystine residue (of six ones) leads to photo-inactivation of the enzyme. In pepsin, essential for photo-inactivation was destruction of only one tryptophan residue.

Thus, namely photolysis of tryptophan and cystine is of primary importance in the protein photo-inactivation. □

It has been proven that the main primary photoreaction of tryptophan (AH) in protein is its photoionization with the formation of a cation radical and a solvated electron:

$$AH + h\nu \rightarrow AH^* \rightarrow \dot{A}H^+ + e^- . \qquad (30.1)$$

At room temperature, this reaction occurs for 5–20 μs and can be studied only with the use of the flash photolysis method. It was shown that in 5 μs after UV irradiation, characteristic absorption maxima appear in the red and far red ranges of the spectrum, belonging to the solvated electron, i.e. the electron, "knocked out" from the amino acid molecule and caught by the solvent dipole molecules. The solvated electron reacts promptly with other molecules, in particular with solvent and O_2 molecules, as a consequence of which its absorption disappears. However, if frozen tryptophan samples are irradiated at the temperature of liquid nitrogen (under such conditions solvated electrons do not recombine), the absorption of the solvated electron (maximum at 600 nm) can be recorded even using a standard spectrophotometer. Moreover, the existence of the uncompensated magnetic moment (spin) in a solvated electron allows its detection by the EPR method.

It is supposed that the tryptophan cation radical, generated in proteins under the action of UV light, dissociates into a proton and a neutral radical. By interacting with adjacent groups of the polypeptide chain, this neutral radical creates an intermolecular covalent "cross-linking" — a stable photoproduct:

$$\dot{A}H^+ \rightarrow H^+ + \dot{A} \xrightarrow{B} A{-}B. \qquad (30.2)$$

If tryptophan, contained in the active center of the enzyme, undergoes photolysis, this may be sufficient to induce the loss of its enzymatic activity. If tryptophan photodestruction occurs outside the active center, such a cross-linking can change hydrogen, hydrophobic and other weak bonds which are important for maintaining the native protein conformation that will lead to a loss of enzymatic activity.

The considered mechanism of the direct tryptophan photolysis is not the only one responsible for the damaging effect of UV irradiation on proteins. The second mechanism of protein photoinactivation is connected with photosensitized reactions, involving the solvated electron. The latter, as a strong reducing agent, reacts actively with different compounds having high affinity to the electron, and first of all with dissolved O_2, forming $\dot{O}_2$ and hydroperoxide radical $H\dot{O}_2$:

$$\bar{e}_S + O{=}O \rightarrow O{-}O^{\dot{-}} \xrightarrow{+H^+} HO{-}\dot{O}. \qquad (30.3)$$

However, reactions with molecular oxygen make an insignificant contribution to the process of UV inactivation of proteins. Of the greatest importance is the interaction of the solvated electron with amino acid residues. The probability of such interaction depending on the affinity to the solvated electron can be rather different. Inasmuch as cystine has the highest affinity to the solvated electron, it can be expected that namely this amino acid will be destructed as a result of its interaction with the solvated electron.

The use of methods of flash photolysis and EPR made it possible to study the mechanism of interaction of the solvated electron with cystine that includes formation of free radicals by the scheme

$$\underset{\text{Cystine}}{R\text{–}CH_2\text{–}S\text{–}S\text{–}CH_2\text{–}R} + e_S^- \rightarrow \underset{\text{Cystine radical}}{(R\text{–}CH_2\text{–}\dot{S}\text{–}S\text{–}CH_2\text{–}R)^-} \rightarrow$$

$$\rightarrow \underset{\text{Cysteine}}{R\text{–}CH_2\text{–}S^-} + \underset{\text{Cysteine radical}}{\dot{S}\text{–}CH_2\text{–}R}. \quad (30.4)$$

The process of photoinactivation of proteins can be represented as a general scheme (Fig. 30.9).

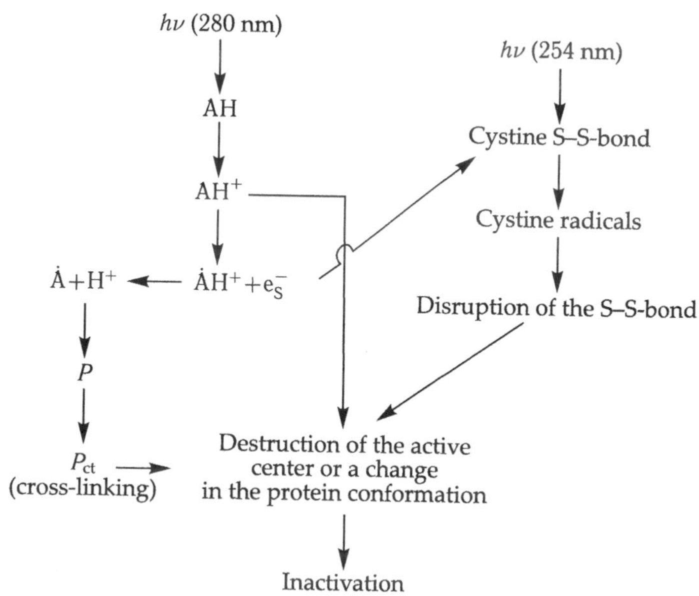

Figure 30.9. UV inactivation of proteins: *A* H-tryptophan.

30.8 Sensitized Damage of Biomolecules in Photodynamic Reactions

Many organic molecules existing in living cells, including biologically important ones, undergo photosensitized changes as a result of photodynamic reactions. The sensitivity of an individual molecule depends on its structure, the photosensitizer properties and conditions of reactions, in particular, on pH of the medium and the solvent nature, as well as on the concentration of reactive substances.

Nucleic Acids. As mentioned above (see Section 4), upon direct absorption of UV light by DNA, predominantly pyrimidine bases enter the photochemical reactions. In contrast to this, it is mainly purine bases that undergo sensitized photo-oxidation, whereas guanine and guanine-containing nucleotides demonstrate the highest sensitivity under physiological conditions.

Photosensitized oxidation of guanine bases in DNA with involvement of singlet oxygen includes the reactions given below.

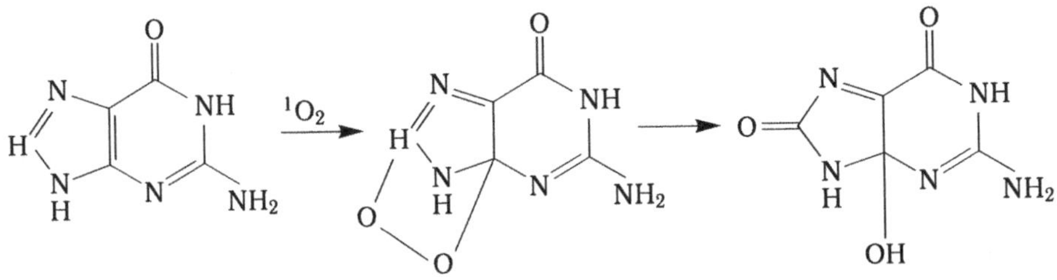

As seen, at the first stage due to the attachment of $^1O_2$, an unstable endoperoxide is formed that turns to the final product. The rate constant of interaction of $^1O_2$ with the guanine base is $5.4 \cdot 10^5\ M^{-1}s^{-1}$.

Sensitized oxidation of guanine bases in DNA induces labilization of glycoside bonds that can lead to single-chain disruptions. Generation of such disruptions due to direct action of $^1O_2$ on the sucrose backbone of DNA is less probable. But some sensitizers and hydroxyl radicals ($OH^{\cdot}$), which are photogenerated by them via the Haber–Weiss reactions, can induce single-chain disruptions by directly interacting with DNA strands.

The principal biological effect, induced by the products of photosensitized oxidation of guanine bases in DNA, is mutagenesis. As follows from the molecular analysis of mutant genes, the main reason of mutations is G–T-transversions, i.e. substitutions of guanine (the purine base) for thymine (the pyrimidine base). G–T-transversions appear upon replication due to the capacity of oxidized guanine to couple not only with cytosine, but also with adenine. At the next replication cycle, on the contrary, thymine is located opposite adenine thus leading to the substitution of guanine for thymine in the chain of nucleotides.

Proteins. Among amino acids in the structure of proteins, histidine, methionine and tryptophan are efficiently photo-oxidized in sensibilization reactions. Histidine is oxidized mostly by type II mechanism (Section 3). Therefore, it is frequently used as a specific acceptor of $^1O_2$ (the bimolecular rate constant of $^1O_2$ quenching by histidine is $7 \cdot 10^6\ M^{-1}s^{-1}$). The primary product of the interaction of histidine with $^1O_2$ is endoperoxide.

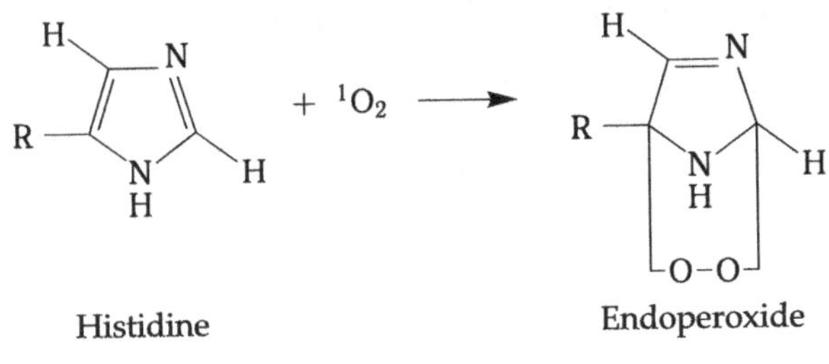

Histidine Endoperoxide

Methionine can be photo-oxidized both in type I and type II reactions. As a result of its type II oxidation by $^1O_2$, mediated, for example, by porphyrin photosensitizers, sulfoxide is formed (the constant of the reaction rate is $5 \cdot 10^6\ M^{-1}s^{-1}$).

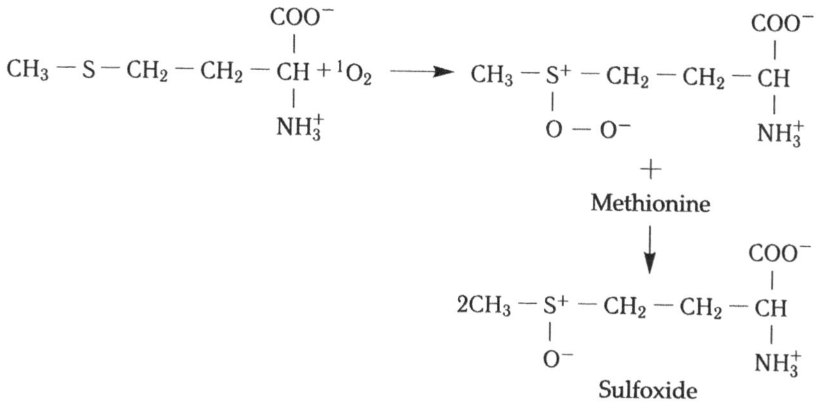

The final product of type I photosensitization reactions is methyonal (β-methylmercaptopropion aldehyde). This product was found in the reaction system when a flavin sensitizer was used. It was demonstrated that in the triplet state flavin disrupts the electron from the sulfur atom in methionine, and the generated methionine radical is deaminated and decarboxylated into methyonal. The flavin radical ($SH^{\cdot}$) regenerates to the ground state (S_0) due to dismutation and subsequent interaction of the product of this reaction (completely reduced flavin–SH_2) with oxygen:

$$SH^{\cdot} + SH^{\cdot} \rightarrow S_0 + SH_2,$$
$$SH_2 + O_2 \rightarrow S_0 + H_2O_2.$$

Mechanisms of sensitized oxidation of tryptophan can include both $^1O_2$ mediated reactions and reactions with involvement of free radicals. Upon photo-oxidation of $^1O_2$ (the rate constant $4 \cdot 10^6\ M^{-1}s^{-1}$), N-formylkynurenine is formed.

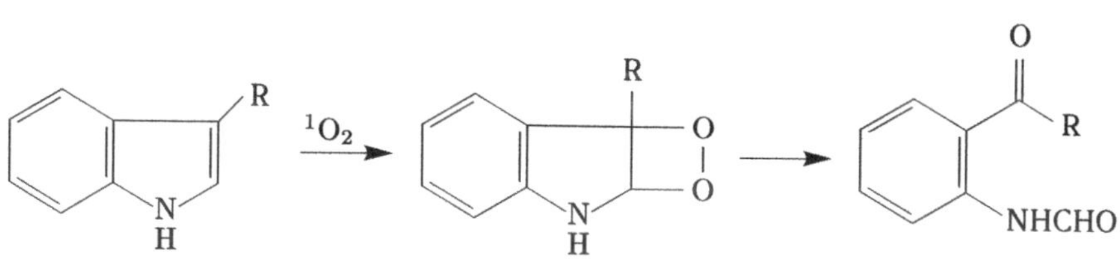

Tryptophan N-formylkynurenine

Photosensitized oxidation of different proteins is caused mainly by photodegradation of the above-considered amino acid residues. These processes are accompanied by noticeable changes in physicochemical properties of proteins, including the capacity to bind cofactors, enhancing sensitivity to proteolytic enzymes etc. Biological functions of proteins are also distorted: enzymatic, hormonal and antigenic activities are inhibited. In some cases, upon irradiation of proteins in the presence of photodynamic sensitizers, protein-protein covalent bonds are formed.

Lipids. In lipids, unsaturated fatty acids (UFA) and cholesterol are most sensitive to photosensitized oxidation. These compounds can be oxidized both by the mechanism of type I and type II photodynamic reactions, in both cases the main intermediate products being hydroperoxides and the final products — aldehydes, ketones and others. The latter can interact with NH_2 and thyol groups of membrane proteins, leading to their inactivation and/or inducing functional changes.

Below is given a general scheme of reactions of photosensitized peroxide oxidation of UFA in lipids (LH), demonstrating the two mechanisms of hydroperoxide formation: type I includes free-radical intermediates and oxygen in the ground state, and type II includes $^1O_2$.

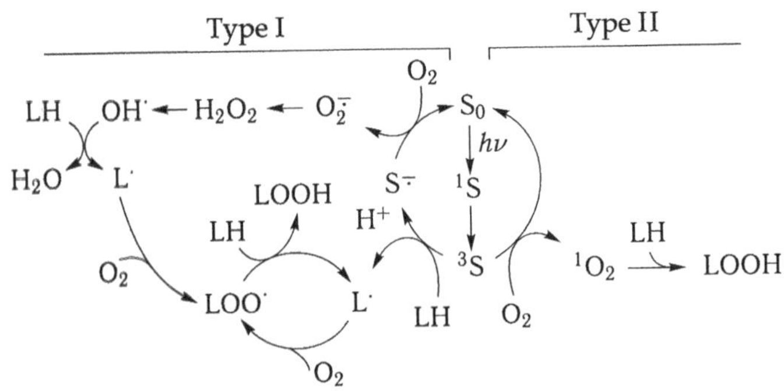

Type I reactions are initiated by hydrogen atom transfer from the substrate (UFA–LH) to photosensitizer S in the triplet state (^3S) with the formation of alkyl radical UFA (L˙) and sensitizer radical (SH˙); deprototation of the latter produces an anion radical ($S^{\dot{-}}$). Interaction of LH with oxygen leads to the formation of peroxide radical (LOO˙); its subsequent reactions with other UFA molecules generate new radical L˙ and hydroperoxides (LOOH). Chain free-radical reactions are limited by the presence of oxygen, UFA and antioxidants of the α-tocopherol type capable of inhibiting these reactions, by competing with LH for the reduction of radical LOO˙.

Upon oxidation of $S^{\dot{-}}$ — the second participant in the primary radical pair — superoxide anion radical of oxygen $O_2^{\dot{-}}$ and the ground state of sensitizer S_0 are formed, the latter can be involved in photoinitiation of radical processes. As a result of dismutation reactions, $O_2^{\dot{-}}$ turns into hydrogen peroxide that in the Haber–Weiss reactions can lead to generation of hydroxyl radical OH˙. The latter initiates free radical peroxide oxidation of lipids by detachment of the hydrogen atom from UFA (LH) (the reaction rate constant is $\sim 10^9\ M^{-1}s^{-1}$).

In type II reactions, at the initial stage the energy is transferred from the triplet sensitizer ^3S to oxygen with generation of $^1O_2$, capable to react directly with UFA (1) and cholesterol (2), by transforming them into hydropeoxides.

(1)

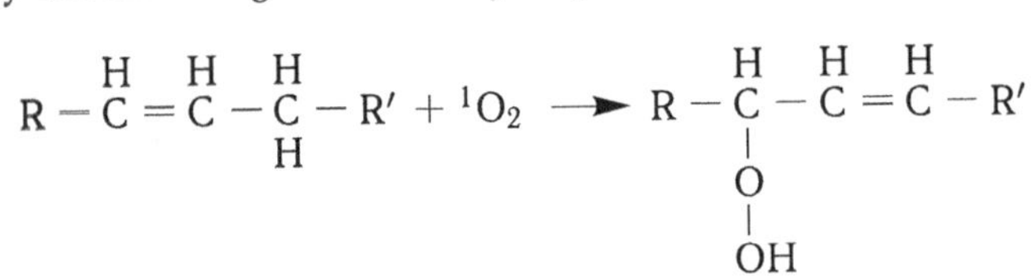

Since 5α–OOH-cholesterol is not formed in reactions of a purely radical nature (type I), this product is used as an evidence for the involvement of $^1O_2$ in photosensitized reactions in biological membranes.

Sensitized photo-oxidation of lipids induces structural changes in membranes, leading to a drastic decrease in barrier properties of lipid bilayers.

(2) 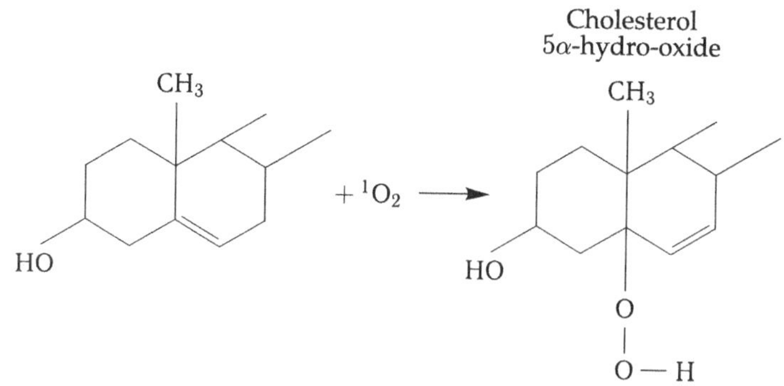

30.9 Photosensitized Effects in Cell Systems

Phototoxic action of long wave length (FUV) and visible light is generally associated with induction of oxidative stress. For example, strong oxygen dependence of the yield of main FUV-induced DNA products, single-chain disruptions, indicates to the involvement of active oxygen species (AOS) in this process.

AOSs are generated in biological systems both in a number of physiological processes and photochemically with involvement of endogenous and exogenous photosensitizers.

Reactions of excited molecules of the sensitizer with the surrounding molecules are divided into two types.

Type I, radical processes (Section 3), includes reactions of direct interaction of an excited sensitizer with substrate and/or solvent molecules accompanied by the formation of radicals or ion radicals during the molecular transfer of a hydrogen atom or electron. The initial form of activated oxygen, generated in type I photo-oxidative processes, is superoxide anion radical ($O_2^{\cdot-}$). In addition to the presence of the inherent self-oxidation potential, the main danger coming from $O_2^{\cdot-}$ generation in living systems is its generation of other AOSs, especially of highly hydroxylic radicals:

$$O_2 + \bar{e} \rightarrow O_2^{\cdot-},$$

$$O_2^{\cdot-} + O_2^{\cdot-} + 2H^+ \rightarrow H_2O_2 + O_2,$$

$$O_2^{\cdot-} + H_2O_2 \rightarrow O_2 + OH^- + OH^{\cdot}.$$

Thus, as a result of reactions of radicals with oxygen, a complex mixture of active intermediates ($O_2^{\cdot-}$, H_2O_2 and $OH^{\cdot}$) is generated that induces oxidative destruction of biomolecules.

Light absorption of NAD(P)N and its generation of a superoxide intermediate in the photoexcited state are of the greatest importance for occurrence of FUV-induced sensitized reactions in cells:

$$\text{UV (340 nm)} \rightarrow NADN^* + O_2 + H_2O \rightarrow NAD^{\cdot} + O_2^{\cdot-} H_3O^+,$$

$$NAD^{\cdot} + O_2 \rightarrow NAD^+ + O_2^{\cdot-}.$$

In type II photodynamic reactions, the triplet-triplet energy transfer from the excited photosensitizer to molecular oxygen in the ground triplet state takes place with the formation of its singlet excited state $^1\Delta_gO_2$ ($^1O_2$). For type II photosensitized reactions, it is required that the energy of the triplet electron-excited state of the photosensitizer would be no less than the energy of the first singlet state of oxygen (7,900 cm^{-1}) with lifetime $\tau_T \geqslant 500$ ns. The second singlet state of oxygen ($^1\Sigma_g{}^+O_2$) with the excitation energy of 13,200 cm^{-1} can also be formed due to energy migration from corresponding triplet levels of photosensitizers, but it has a short lifetime (20 ps) and rapidly passes to the first singlet excited state.

Most natural chromophores, having absorption bands in the visible range of the spectrum, can be ascribed to three large groups depending on their chemical structure: cyclic or linear tetrapyroles, polyenes and aromatic compounds. Representatives of the three classes are characterized by the presence of a system of conjugated double bonds. The primary photochemical reaction of excited molecules of chromophores may consist in photoisomerization (rhodopsin, phytochrome), electron transfer in type I reactions (flavins) or transfer of excitation energy in type II reactions (porphyrins). Porphyrin compounds are the main candidate to play the role of endogenous sensitizers under the visible light. Free photodynamically active endogenous porphyrins are precursors of the synthesis of functionally active metal porphyrins, and their intracellular concentration can increase upon distortions in regulation of metabolism. The energy of the triplet level of porphyrins is sufficient for the formation of the first excitation singlet state of oxygen, and the reactions sensitized by them occur predominantly by the type II mechanism.

In nucleic acids, the main types of photodamages, formed upon the action of FUV light, are chain disruptions and cross-links with proteins.

Singlet oxygen, generated by porphyrin photosensitizers in the visible light, reacts with desoxyguanosine of DNA, leading to formation of several photoproducts, with 8-hydroxy-2-desoxyguanosine as the most frequently generated among them.

Photosensitized Destruction of Proteins with Involvement of AOSs is conditioned by the presence of amino acid residues sensitive to oxidation — tyrosine, tryptophan, methionine, histidine and cysteine. Peripheral amino acid residues are subjected to photo-oxidation most of all. In addition to direct primary oxidation of amino acid residues, secondary AOS-induced reactions also take place — modification of proteins by products of peroxide oxidation of lipids.

Action of Light on Membrane Structure is caused by photo-oxidation not only of protein but also lipid components. Lipids, containing UFAs, undergo peroxide photooxidation (PPOL) in the presence of a photosensitizer, corresponding wavelength light and/or molecular oxygen. PPOL reactions, occurring by the radical mechanism (type I), include the interaction of a sensitizer in the triplet excited state directly with UFA and intermolecular electron transfer. In type II photodynamic reactions, PPOL occurs through oxidation of UFAs by singlet oxygen to produce hydroperoxides.

PPOL processes, occurring in membrane structures, lead to disintegration of mitochondrial, lysosomal and plasmatic membranes, resulting in a decrease in the ATP level, acidification of the cell contents, activation of proteases and cell lysis.

Thus, damages, induced in cells by FUV and visible light, can affect practically all cell structures. This is conditioned by the principal difference between destructive action on biological systems of shortwave and long wave light: the prevalence of direct photochemical reactions upon absorption by biomolecules of shor and middle wavelength UV and the occurrence of photosensitized oxidation reactions in FUV and visible light.

Mechanisms of photosensitization are actively studied not only because of its damaging effect on living organisms, but also for medical use of this phenomenon upon treatment of a number of diseases such as malignant tumors, skin disorders and microbial infections. The nature of the used photosensitizer is determined by the type of cells that should be inactivated, by the selectivity of the sensitizer accumulation, possible ways of its delivery to sensitive cell targets, concentration of oxygen near the localization site and by other parameters. The most studied and actively used are the following groups of sensitizers. Furokumarins, absorbing in the FUV range, are used mainly in treating skin diseases — psoriasis and vitiligo. Phenothiazines, absorbing in the red range of the spectrum, are used for disinfection of blood. Porphyrins and their analogs (chlorines and phthalocyanines) with absorption band shifted to the long wave range, where transmission of light by tissues enhances, are efficient in phototherapy of tumors.

Porphyrins and Their Analogs are widely used in photomedicine. Fluorescent properties of these compounds are used for demarcation of tumors. Inactivation of tumor cells occurs predominantly due to type II photodynamic reactions and is induced by $^1O_2$, generated due to triplet-triplet energy migration from the excited photosensitizer molecule to molecular oxygen.

In order to deliver photosensitizers (producers of active oxygen) to the most sensitive target (the nucleus of a tumor cell), synthetic module nanotransporters were created (A. S. Sobolev, 2009), consisting of the following modules (Fig. 30.10): binding of the damaging agent; binding to tumor cells; penetration into the cytoplasm; penetration into the cell nucleus. As a result, the photosensitizer acquires selective toxicity only relative to tumor cells, and due to penetration into the tumor cell nucleus its selectivity increases 500–3,000 times.

Photosensitization, mediated by porphyrins, underlies not only the method of photodynamic therapy of tumors. Porphyrins are also used as photoactive herbicides, insecticides, and antimicrobial preparations.

The lifetime of $^1O_2$ in the hydrophobic surrounding (20–25 μs in the carbon phase, 7 μs in membranes) is much larger than in aqueous medium (3–4 μs). Therefore, the strongest effect of photodynamic inactivation is observed upon localization of the critical target and sensitizer in the hydrophobic area. The requirement for a tight association of the porphyrin photosensitizer and the target is evident from a small diffusion radius of $^1O_2$ (less than 50 nm) in the intracellular medium, conditioned by high constants of its quenching by cell components. The same reasons can be used to explain inactivation of cells by photosensitizer molecules that are in the plasmatic membrane or in the intracellular space, because generation of singlet oxygen outside the cell leads to its promt quenching by the surrounding components.

Antimicrobial PDT consists in photosensitized selective oxidative destruction of pathogenic microorganisms. The objects of antimicrobial PDT are regarded to include viruses, bacteria, fungi and Protista.

Cell walls of bacteria, especially gram-negative ones, have a strong negative charge. It was demonstrated on synthetic bilayer membranes that the density of the surface negative charges of the lipopolysaccharide layer — the main component of the outer membrane of gram-positive bacteria — is 3.33 elementary negative charges per nm^2. This prevents the binding of lipophilic and hydrophilic anion dyes to bacterial cells, i.e. diminishes the number of photosensitizers with potential antimicrobial activity. Photosensitizers, in which molecules carry positive charged groups, have a wide range and high efficiency of antimicrobial action. The electrostatic interaction of these groups with negatively charged binding centers on the surface of microbial

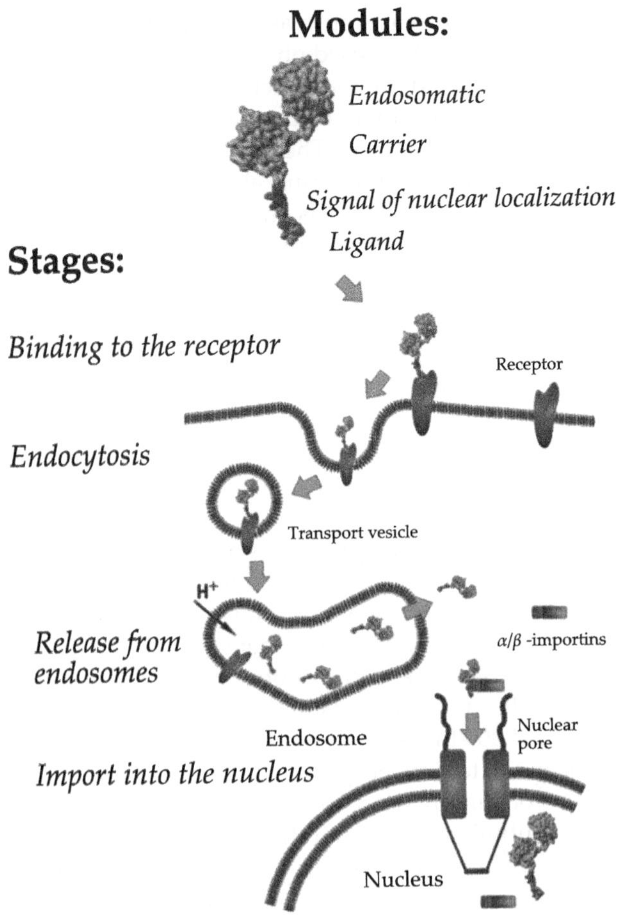

Figure 30.10. Schematic representation of a module nanotransporter (MNT) and stages of its penetration into the target cell.

The first MNT module — ligand — performs a dual function: specific "recognition" of a tumor target cell and penetration into this cell due to receptor-mediated endocytosis. The second (endosomolytic) module allows MNT to "leave" the endocytose pathway before hitting lysosomes, to permit the nanotransporter to interact with importins. To this end, the polypeptide fragment capable of creating defects in membranes at pH of endosomes is used as the second module. The delivery into the cell nucleus is performed by the third module, containing the amino acid sequence of nuclear localization, "recognized" by importins that are in the structureless moiety of the cytoplasm. The forth module of MNT — carrier module — serves for attaching the transported photosensitizer.

cells conditions an increase in the local concentration of photosensitizers near critical target cells. It was shown (M. G. Strakhovskaya et al., 2004) that the efficiency of binding of photosensitizers to cells of gram-negative bacteria and their photobactericide activity depends, on the one hand, on the density of negative charges on the bacteria cell wall, and on the other hand, on the number of positively charged substitutes in dye molecules.

Photodynamic inactivation of pathogenic microorganisms can also be used for sterilization and disinfection. Thus, the field of application of photodynamic impact considerably widens.

Further Reading

Part VIII

1) Fersht, A. *Enzyme Structure and Mechanism*, W.H. Freeman, San Francisco, 1985.
2) Feynman, R.F., Leighton, R.B. and Sands, M. *The Feynman Lectures on Physics*, Addison-Wesley Publishing Company Inc., 1964.
3) Slater, J.C. *Quantum Theory of Molecules and Solids. Volume 1: Electronic Structure of Molecules*, McGraw-Hill, New York, 1963–1974.
4) Devault, D., Quantum Mechanical Tunneling in Biological Systems. *Q. Rev. Biophys.*, 13(4), pp. 387–564, 1980.
5) Marcus, R.A. and Sutin, N., Electron Transfer in Chemistry and Biology. *Biochim. Biophys. Acta*, 811, pp. 265–322, 1985.
6) Jortner, J. Dynamics of Electron Transfer in Bacterial Photosynthesis. *Biochim. Biophys. Acta*, 594, pp. 193–230, 1980.

Part IX

1) Govindjee (Ed.), *Photosynthesis, Volume I: Light Reactions*, Springer International Publishing, 1987
2) Govindjee (Ed.), *Photosynthesis, Volume II: Dark Reactions*, Springer International Publishing, 1987.
3) Clayton, R. *Photosynthesis. Physical Mechanisms and Chemical Models*, Mir, Moscow, 1984.
4) Hall, D.O. and Rao, K.K., *Photosynthesis*, fourth edition, Cambridge University Press (New Studies in Biology Series), Cambridge, 1992.
5) Blankship, R.E., *Molecular Mechanisms of Photosynthesis*, Blackwell Science Ltd, Oxford, UK, 2002.
6) Laisk, A., Nedbal, L. and Govindjee (Ed.), *Photosynthesis in silico: Understandind Complexity from Molecules to Ecosystem*, Springer, Dortrecht, 2009.
7) Renger, R. (Ed.), *Primary Processes of Photosynthesis*. Part 1, RSC, Cambridge, 2008.
8) Renger, R. (Ed.), *Primary Processes of Photosynthesis*. Part 2, RSC, Cambridge, 2008.

Part X

1) Kohen, E., Santus, R. and Hirschberg, J. G., *Photobiology*, Acad. Press, San Diego, N.Y., Boston, London, Sydney, Tokyo, Toronto, 1995.
2) Batschuer, A. (Ed.), *Photoreceptors and Light Signalling*, RSC, Cambridge, 2003.
3) Palumbo, G. and Pratesi, R. (Ed.), *Laser and Current Optical Techniques in Biology*, RSC, Cambridge, 2004.
4) Lanyi, J.K., Bacterirhodopsin. *Annu. Rev. Physiol.*, 66, pp. 665–688, 2004.
5) Briggs, W.R. and Spudich, J.L. (Ed.), *Handbook of Photosensory Receptors*, Wiley VCN, Weinheim, 2005.
6) Rockwell, N. C., Su, Y.-S. and Lagarias, J.C. Phytochrome Structure and Signaling Mechanisms. *Annu. Rev. Plant Biol.*, 57, pp. 837–858, 2006.
7) Renger, R. (Ed.), *Primary Processes of Photosynthesis*. Part 1, RSC, Cambridge, 2008.
8) Renger, R. (Ed.), *Primary Processes of Photosynthesis*. Part 2, RSC, Cambridge, 2008.
9) Britton, G., *The Biochemistry of Natural Pigments*, Cambridge University Press, Cambridge, N.Y., Melbourne, 2009.
10) Muller, M. and Carell, T., Structural Biology of DNA Photolyases and Cryptochromes. *Curr. Opin. Struct. Biol.*, 19, pp. 277–285, 2009.

Index

Also of Interest

By the Same Author:

Fundamentals of Biophysics
By Andrey B. Rubin
Published 2014. ISBN 978-1-118-84245-4

Check out these other published and forthcoming titles from Scrivener Publishing:

Smart Membranes and Sensors
Synthesis, Characterization, and Applications
Edited by Annarosa Gugliuzza
Published 2014. ISBN 978-1-118-42379-0

Advanced Biomaterials and Biodevices
Edited by Ashutosh Tiwari and Anis N. Nordin
Published 2014. ISBN 978-1-118-77363-5

Biosensors Nanotechnology
Edited by Ashutosh Tiwari and Anthony P. F. Turner
Published 2014 ISBN 978-1-118-77351-2

Advanced Sensor and Detection Materials
Edited by Ashutosh Tiwari and Mustafa M. Demir
Published 2014. ISBN 978-1-118-77348-2

Advanced Healthcare Materials
Edited by Ashutosh Tiwari
Published 2014 ISBN 978-1-118-77359-8

Advanced Carbon Materials and Technology
Edited by Ashutosh Tiwari and S.K. Shukla
Published 2014. ISBN 978-1-118-68623-2

Responsive Materials and Methods
State-of-the-Art Stimuli-Responsive Materials and Their Applications
Edited by Ashutosh Tiwari and Hisatoshi Kobayashi
Published 2013. ISBN 978-1-118-68622-5

Nanomaterials in Drug Delivery, Imaging, and Tissue Engineering
Edited by Ashutosh Tiwari and Atul Tiwari
Published 2013. ISBN 978-1-118-29032-3

Biomimetics
Advancing Nanobiomaterials and Tissue Engineering
Edited by Murugan Ramalingam, Xiumei Wang, Guoping Chen, Peter Ma, and Fu-Zhai Cui
Published 2013. ISBN 9781118469620

Encapsulation Nanotechnologies
Edited by Vikas Mittal
Published 2013. ISBN 978-1-118-34455-2

Advances in Contact Angle, Wettability and Adhesion
Volume 1
Edited by K.L. Mittal
Published 2013. ISBN 978-1-118-47292-7

Biomedical Materials and Diagnostic Devices Devices
Edited by Ashutosh Tiwari, Murugan Ramalingam, Hisatoshi Kobayashi and Anthony P.F. Turner
Published 2012. ISBN 978-1-118-03014-1

Intelligent Nanomaterials
Processes, Properties, and Applications
Edited by Ashutosh Tiwari Ajay K. Mishra, Hisatoshi Kobayashi and Anthony P.F. Turner
Published 2012. ISBN 978-0-470-93879-9

Integrated Biomaterials for Biomedical Technology
Edited by Murugan Ramalingam, Ashutosh Tiwari, Seeram Ramakrishna and Hisatoshi Kobayashi
Published 2012. ISBN 978-1-118-42385-1

Integrated Biomaterials in Tissue Engineering
Edited by Murugan Ramalingam, Ziyad Haidar, Seeram Ramakrishna, Hisatoshi Kobayashi, and Youssef Haikel
Published 2012. ISBN 978-1-118-31198-1

The Physics of Micropdroplets
Jean Berthier and Kenneth Brakke
Published 2012. ISBN 978-0-470-93880-0

Introduction to Surface Engineering and Functionally Engineered Materials
By Peter Martin
Published 2011 ISBN 978-0-470-63927-6

Printed and bound by CPI Group (UK) Ltd, Croydon, CR0 4YY
21/09/2026
14965830-0002